FV

WITHDRAWN

 St. Louis Community College

Library

5801 Wilson Avenue
St. Louis, Missouri 63110

NONDESTRUCTIVE TESTING HANDBOOK

Second Edition

VOLUME 5
ACOUSTIC EMISSION TESTING

Ronnie K. Miller
Technical Editor

Paul McIntire
Editor

**AMERICAN SOCIETY FOR
NONDESTRUCTIVE TESTING**

Copyright © 1987
AMERICAN SOCIETY FOR NONDESTRUCTIVE TESTING, INC.
All Rights Reserved.

No part of this book may be reproduced, stored in a retrieval system, or transmitted, in any form or by any means, electronic, mechanical, photocopying, recording, or otherwise, without the prior written permission of the publisher.

Nothing contained in this book is to be construed as a grant of any right of manufacture, sale, or use in connection with any method, process, apparatus, product or composition, whether or not covered by letters patent or registered trademark, nor as a defense against liability for the infringement of letters patent or registered trademark.

The American Society for Nondestructive Testing, its employees, and the contributors to this volume assume no responsibility for the safety of persons using the information in this book.

Library of Congress Cataloging-in-Publication Data

Acoustic emission testing.

  (Nondestructive testing handbook; v. 5)
  Includes bibliographies and index.
  1. Acoustic emission testing. 2. Nondestructive testing.
I. Miller, Ronnie K. II. McIntire, Paul. III. American Society for Nondestructive
Testing. IV. Series: Nondestructive testing handbook (2nd ed.); v. 5.
TA418.84.A264 1987 620.1'127 87-72690
ISBN 0-931403-02-2

Published by the American Society for Nondestructive Testing

PRINTED IN THE UNITED STATES OF AMERICA

PRESIDENT'S FOREWORD

This volume of the *Nondestructive Testing Handbook* is the fifth in ASNT's ten-volume Handbook series. All of the books are intended as reference materials for nondestructive testing practitioners in the field and as documentation of advancements in the various NDT technologies, of their research and applications.

Acoustic Emission Testing was written under the guidance of ASNT's Handbook Development Committee. The members, officers and directors of the Society extend their sincere appreciation to John L. Summers, Handbook Development Director, and to Paul McIntire, editor of the Handbook, for their dedicated efforts on behalf of the Society and the broader NDT community.

The *Nondestructive Testing Handbook* is one element in ASNT's continuing effort to support education and training in NDT and to encourage public awareness of nondestructive testing. It is our belief that *Acoustic Emission Testing* delivers the technical and educational information required for achieving these goals and that the volume stands as an important addition to the literature of the practical sciences.

Don Earl Edwards
ASNT President (1986-1987)

PREFACE

This book is a field practitioner's guide, but it's publication may represent other important concepts. First, the *Nondestructive Testing Handbook* is itself evidence of a successful effort to produce educational materials for the NDT community. That effort is shared by all the books' contributors, *volunteers who provide their expertise and donate their own time for the purpose of compiling this vital information.*

Secondly, this book typifies one of the most important functions of a technical society: to document its related technologies, regardless of apparent or popular interest. When acoustic emission techniques were first being investigated, there were many false starts and the vagaries of a new technology were painfully apparent — there was no expectation of acoustic emission's central role in the quality control of composite materials and none of the present excitement about the possibilities of the technique. Without the continued efforts of the method's early practitioners and their affiliations with technical organizations such as ASNT, communication about the method's applications might have stopped. Presented here, though, are the results of a newly mature and developing technology, including substantiated test data and documentation of effective applications.

This book also represents a new and tangible connection between nondestructive testing and medical diagnosis. When the American Industrial Radium and X-Ray Society was founded in 1941, many of its members were medical radiologists. With time, the society evolved into ASNT and the direct ties with medical applications weakened. An interest in reestablishing those links has recently been initiated and, accordingly, this is the first volume of the *NDT Handbook* to contain a purely medical chapter.

In an attempt to document nondestructive testing applications for an international audience, *Acoustic Emission Testing* was written by authors from many parts of the United States, Japan, Sweden and Canada. Some of the text was provided by the multinational European Working Group for Acoustic Emission. The appeal to an international audience is also apparent in the use of SI units throughout the text.

Special thanks go to Jan van den Andel of Westinghouse Canada for providing the metric conversions in this volume and for serving among our most trustworthy and valued reviewers.

The technical editor of this volume, Ronnie K. Miller, undertook this assignment while working at the United Technologies Research Center and carried the project with him to Physical Acoustics Corporation. Dr. Miller served as the peer contact and final technical authority for all the material in this publication. He is the one person in the acoustic emission testing community who can be credited with this volume's publication and the high quality of its content.

Special recognition also goes to Jack C. Spanner, Sr., author of *Acoustic Emission: Techniques and Applications* (1974), a book that long stood alone in the acoustic emission literature. Spanner willingly brought his publishing experience and his technical expertise to this *NDT Handbook* volume, and coordinated the writing of nearly a third of its chapters.

From among the publishing professionals who contributed to this project, ASNT extends its thanks to Hollis Black, who keyed every character in this volume; Skip Pierce, our drafting source; and Dan Robinson, typographer for Lawhead Press. The quality of the finished volume is representative of their skills and the dedication of all the book's contributors.

Although this volume can legitimately be seen to represent other important things, it serves first and best as a handbook — whose purpose is to explain and encourage the use of acoustic emission nondestructive testing methods. Fulfilling that purpose is the best recognition the makers of this book could receive.

Paul McIntire
Handbook Editor

ACKNOWLEDGMENTS

Acoustic Emission Testing, Volume 5 of the *Nondestructive Testing Handbook*'s second edition, is an effort by many volunteers to present for the first time a comprehensive guide to this technology. As originally outlined and in keeping with the objectives established by ASNT's Acoustic Emission Committee, the volume presents both theory and applications.

Handbook Coordinator

Work on this Volume began at the 1982 ASNT Spring Conference in Boston, Massachusetts. The Handbook coordinator Al Brown of Lawrence Livermore National Laboratories has been with the project from the start. He has without complaint supported its development by supplying text, contributors, reviewers and reviews, and by motivating individuals to meet deadlines when it mattered the most.

Lead Authors

Every Section of this Volume was assigned its own coordinator who in most cases became the lead or primary author. All of these individuals and the contributors they organized (listed below) deserve special recognition. The satisfaction of completing a text that was long overdue was all the reward they received.

Special recognition goes to Jack Spanner of Battelle Pacific Northwest Laboratories. Jack volunteered to be coordinator of the original Section 1, an introduction to acoustic emission technology. He revised my original outline, devised a schedule of deadlines for the contributors and set off on a mission to complete his Section as planned. After many delays, manuscripts began to arrive — in quantity. The contributors' material was of such good technical quality that the original chapter was greatly expanded. The results can be seen in Sections 1 through 6.

Reviewers

Behind each *NDT Handbook* is a large body of individuals who review every single sentence for technical content. Their names are listed below and they deserve much *more* credit, for taking the time and patience to refine these documents for presentation. Thanks to them, many Parts were added to Sections, to complete the book's coverage and to maintain continuity.

Handbook Editor

ASNT's Handbook Editor has no mechanism with which to pat himself on the back when a volume is completed. Paul McIntire has been with this volume from the beginning and deserves as much credit as anyone for its quality and its completion. Although he is not credited elsewhere for doing so, Paul served as coordinator for two of this book's chapters — tasks that were additional to his work as editor and that went beyond our expectations.

It was ASNT that served as the clearinghouse for all the text, the reviews and the preparation of the hundreds of illustrations in this volume. Acoustic emission practitioners are indebted to ASNT for providing this milestone publication.

Ronnie K. Miller
Volume 5 Technical Editor

Handbook Development Committee

William Bailey
Patricia Bouta
Al Brown
Floyd Brown
Lawrence Bryant
Chris Burger
John Cavender
Peter Collins
Gene Curbow
Robert Green, Jr.
Mark Hart
Edmund Henneke
Stephen Lorson
Robert Loveless
Ronnie Miller
Scott Miller
William Mooz
Charles Sherlock
Kermit Skeie
Rod Stanley
Jack Summers
Henry Swain
Frank Vicki

Volume 5 Lead Authors

J.A. Baron, Ontario Hydro
M. Nabil Bassim, University of Manitoba
Thomas Drouillard, Rockwell International
Davis Egle, University of Oklahoma
Don Eitzen, National Bureau of Standards
Allen Green, Acoustic Emission Technology Corporation
George Harman, National Bureau of Standards
Theodore Hopwood, Kentucky Transportation Research Program
Phil Hutton, Battelle Pacific Northwest Laboratory
Stuart McBride, Royal Military College of Canada
Ronnie Miller, Physical Acoustics Corporation
Keith Mobley, Technology for Energy Corporation
Jack C. Spanner, Sr., Battelle Pacific Northwest Laboratory
Sotirios Vahaviolos, Physical Acoustics Corporation
Haydn Wadley, National Bureau of Standards
Timothy Wright, The Hospital for Special Surgery

Volume 5 Contributors

Carol Bailey, Technology for Energy Corporation
Christy Bailey, Dunegan Corporation
J.A. Baron, Ontario Hydro
Harold Berger, Industrial Quality Incorporated
Kevin Bierney, Acoustic Emission Associates
Roberto Binazzi, The Hospital for Special Surgery
P.R. Blackburn, Linde Division of Union Carbide
William Boyce, Sikorsky Aircraft
F.R. Breckenridge, National Bureau of Standards
R.W. Bright, Foundation for Skeletal Research
Al Brown, Lawrence Livermore National Laboratory
James Carr, The Hospital for Special Surgery
Robert Checkaneck, AT&T
C.E. Coleman, Atomic Energy of Canada
Bruce Craig, Metallurgical Consultants
N.O. Cross, NDTech
James Culp, Michigan Department of Transportation
J.R. Dale, Signetics Corporation
Gilbert Dewart, ESD Geophysics
David Dornfeld, University of California
Kjell Eriksson, Karolinska Institutet
Mark Ferdinand, Hartford Steam Boiler
Timothy Fowler, Monsanto Company
S.A. Gee, Signetics Corporation
Mark Goodman, UE Systems
Marvin Hamstad, University of Denver
D. Robert Hay, Tektrend International
Edmund Henneke, Virginia Polytechnic Institute
Phil Hutton, Battelle Pacific Northwest Laboratory
Y. Iwasa, Massachusetts Institute of Technology
Min-Chung Jon, AT&T
John Judd, Vibra•Metrics
Sherwin Kahn, AT&T
J. Lawrence Katz, Rensselaer Polytechnic Institute
Teruo Kishi, University of Tokyo
Hisashi Kishidaka, Kawasaki Steel Corporation
R.F. Klein, Battelle Pacific Northwest Laboratory
Masato Kumagai, Kawasaki Steel Corporation
Charles Lewis, Staveley Technologies
Charles McGogney, Federal Highway Administration
Dennis Miller, AT&T
Ronnie Miller, Physical Acoustics Corporation
James Mitchell, Physical Acoustics Corporation
Vasile Mustafa, Tektrend International
Hiroyasu Nakasa, Central Research Institute of Electric Power Industry
Kenneth Notvest
Emmanuel Papadakis, Ford Motor Company
J. Philip Perschbacher, Sikorsky Aircraft
William Pless, Lockheed Georgia Company
S.J. Poliakoff, Foundation for Skeletal Research
Adrian Pollock, Physical Acoustics Corporation
Dan Robinson, Hartford Steam Boiler
John Rogers, Acoustic Emission Technology Corporation
Mansoor Saifi, Western Electric Company
Thomas Santner, Cornell University

E.B. Schwenk, Battelle Pacific Northwest Laboratory
John Simmons, National Bureau of Standards
Stewart Slykhous, Dunegan Corporation
Michael Soudry, The Hospital for Special Surgery
John Stapleton, AT&T
Richard Stiffler, Virginia Polytechnic Institute
Ryoji Uchimura, Kawasaki Steel Corporation
Jan van den Andel, Westinghouse Canada
C.B.M. van Kessel, Philips Research Laboratories
Governor Ware, Teletype Corporation
David Wypij, Cornell University
Ken Yee, National Bureau of Standards
S.P. Ying, Gilbert/Commonwealth
Hyo Sub Yoon, Rensselaer Polytechnic Institute

Volume 5 Reviewers

Eugene Bailey
Peter Bartle, The Welding Institute
Al Brown, Lawrence Livermore National Laboratory
Rodney Cordeiro, General Dynamics, Electric Boat Division
Tom Crump, American Society for Testing Materials
Gene Curbow, Center for Applied Welding Research
Davis Egle, University of Oklahoma
Sol Goldspiel
Allen Green, Acoustic Emission Technology Corporation
Jim Havens, General Motors Corporation
Robert Kryter, Oak Ridge National Lab
Gary Mancini, American Society for Mechanical Engineers
J.R. Matthews
George Moran, PSE&G Research Corporation
Martin Peacock, ITL Industrial Testing Services
Adrian Pollock, Physical Acoustics Corporation
Jack Spanner, Sr., Battelle Pacific Northwest Laboratory
Rod Stanley, International Pipe Inspectors Association
Terry Swanson, American Society for Testing Materials
Clem Tatro, Lawrence Livermore National Laboratory
Alfred Thiele, Rockwell International Science Center
Patrick Toner, Society for Plastic Industries
Jan van den Andel, Westinghouse Canada
Dennis White, National Board of Boiler and Pressure Vessel Inspectors

Major contributors to individual chapters are listed alphabetically on the Section title page under the name of the primary author.

SYSTEME INTERNATIONALE (SI) UNITS IN ACOUSTIC EMISSION TESTING

Origin and Use of the SI System

The SI system was designed so that all branches of science could use a single set of interrelated measurement units. These units are modified in specified ways to make them adaptable to the needs of individual disciplines. Without the SI system, this *Nondestructive Testing Handbook* volume could have contained a confusing mix of Imperial units, old cgs metric units and the units preferred by certain localities or scientific specialties.

Aside from the consistency provided by the SI system, there is an additional mathematical advantage to its use. Complex equations always balance in the SI system. Base units and their powers, on both sides of the equal sign, are indeed equal. This provides a double-check of accuracy: an equation error will reveal itself not only through an imbalance but through the different units created by the imbalance.

SI Units for Acoustic Emission Testing

Acoustic emission is a shock wave inside a stressed material, where a displacement (length unit) ripples through the material and moves its surface. A transducer on that surface undergoes this displacement as a pressure. The pressure is measured as force per unit area in newtons per square meter ($N \cdot m^{-2}$) or pascals (Pa). The signal from the transducer is sometimes related to velocity ($m \cdot s^{-1}$) and sometimes related to acceleration ($m \cdot s^{-2}$).

Properties of piezoelectric transducers are charge (coulomb) related: a pressure on the element creates a charge on the electrodes. A rapidly changing pressure alters the charge fast enough to allow the use of either voltage or charge amplifiers. After this, signal processing may take the form of displacement (m), velocity ($m \cdot s^{-1}$), acceleration ($m \cdot s^{-2}$), signal strength (V), energy (sometimes V^2 but actually joules) or power (W).

TABLE 1. Base SI Units

Quantity	Unit Name	Unit Symbol
length	meter	m
mass	kilogram	kg
time	second	s
electric current	ampere	a
thermodynamic temperature	kelvin	K
amount of substance	mole	mol
luminous intensity	candela	cd
plane angle	radian	rad
solid angle	steradian	sr

Some applications (see the Section on *Survey of Commercial Sensors and Systems*, use the G for acceleration due to gravity (9.81 $m \cdot s^{-2}$). This phenomenon in itself does nothing to a transducer, but preventing free-fall results in a force:

$$f = m \times a \qquad \text{(Eq. 1)}$$

Where:

m = the mass of the transducer (kilograms); and
a = the acceleration due to gravity.

When the transducer is in use, the force (newtons or N) is spread over the area (m^2) of the element, producing a pressure ($N \cdot m^{-2}$ or Pa). Comparison with G forces is usually a measurement of acceleration or deceleration ($m \cdot s^{-2}$) divided by the acceleration due to gravity (9.81 $m \cdot s^{-2}$). These values could be enormously high if the transducer is dropped on a concrete floor (3,000 G) but very small when used for acoustic emission detection. For this reason, G forces are rarely used in acoustic emission applications.

TABLE 2. Derived SI Units

Quantity	Name	Symbol	Relation to Other SI Units
frequency	hertz	Hz	$1 \cdot s^{-1}$
force	newton	N	$kg \times m \cdot s^{-2}$
pressure (stress)	pascal	Pa	$N \cdot m^{-2}$
energy (work)	joule	J	$N \times m^2$
power	watt	W	$J \cdot s^{-1}$
electric charge	coulomb	C	$A \times s$
electric potential	volt	V	$W \cdot A^{-1}$
capacitance	farad	F	$C \cdot V^{-1}$
electric resistance	ohm	Ω	$V \cdot A^{-1}$
conductance	siemens	S	$A \cdot V^{-1}$
magnetic flux	weber	Wb	$V \times s$
magnetic flux density	tesla	T	$Wb \cdot m^{-2}$
inductance	henry	H	$Wb \cdot A^{-1}$
temperature	degrees Celsius	°C	$K - 273.15$
luminous flux	lumen	lm	$cd \times sr$
illuminance	lux	lx	$lm \cdot m^{-2}$
radioactivity	becquerel	Bq	$1 \cdot s^{-1}$
radiation absorbed dose	gray	Gy	$J \cdot kg^{-1}$
radiation dose equivalent	sievert	Sv	$J \cdot kg^{-1}$

TABLE 3. SI Multipliers

Prefix	Symbol	Multiplier
tera	T	10^{12}
giga	G	10^9
mega	M	10^6
kilo	k	10^3
hecto	h	10^2
deca	da	10
deci	d	10^{-1}
centi	c	10^{-2}
milli	m	10^{-3}
micro	μ	10^{-6}
nano	n	10^{-9}
pico	p	10^{-12}

SI Multipliers

Very large or very small units are expressed using the SI multipliers, prefixes that are usually of 10^3 intervals. The range covered in this text is shown in Table 3. The multiplier becomes a property of the SI unit. For example, a centimeter (cm) is 1/100 of a meter, and the volume unit, cubic centimeter (cm^3), is $(1/100)^3$ or 10^{-6} m^3. Units such as the centimeter, decimeter, decameter and hectometer are avoided in technical uses of the SI system because of their variance from the 10^3 interval.

Every effort has been made to include all necessary SI and conversion specifications in the text of this book and in the tables that follow. If questions remain, the reader is referred to the information available through national standards organizations and to the specialized information compiled by technical societies (see ASTM E-380, *Standard Metric Practice Guide*, for example).

Jan van den Andel
Westinghouse Canada

CONTENTS

INTRODUCTION TO ACOUSTIC EMISSION TECHNOLOGY 1

PART 1: HISTORY OF ACOUSTIC EMISSION RESEARCH 2
    Early Uses of Acoustic Emission 2
    Acoustic Emission in Metalworking 2
PART 2: DETECTING AND RECORDING ACOUSTIC EMISSION 4
    Kaiser's Study of Acoustic Emission Sources . 4
    Acoustic Emission Research in the United States 4
PART 3: ORGANIZING THE ACOUSTIC EMISSION COMMUNITY 7
    The Acoustic Emission Working Group 7
    European and Japanese Working Groups ... 7
    Results of the Acoustic Emission Working Groups 7

SECTION 1: FUNDAMENTALS OF ACOUSTIC EMISSION TESTING 11

PART 1: INTRODUCTION TO ACOUSTIC EMISSION TECHNOLOGY 12
    The Acoustic Emission Phenomenon 12
    Acoustic Emission Nondestructive Testing .. 12
    Application of Acoustic Emission Tests 13
    Successful Applications 14
    Acoustic Emission Testing Equipment 14
    Microcomputers in Acoustic Emission Test Systems 15
    Characteristics of Acoustic Emission Techniques 16
    Acoustic Emission Test Sensitivity 16
    Interpretation of Test Data 17
    The Kaiser Effect 17
    Overview of Acoustic Emission Methodology 18
PART 2: INTRODUCTION TO ACOUSTIC EMISSION INSTRUMENTATION 20
    Choosing an Acoustic Emission Testing System 20
    Expanding the Basic Acoustic Emission Testing System 20
    Computers in Acoustic Emission Systems ... 21
    System Calibration 21
    Pressure Vessel Applications 21
    Spot Weld Monitoring Applications 22
    The Value of Acoustic Emission Instrumentation 22

PART 3: INTRODUCTION TO THE EFFECTS OF BACKGROUND NOISE IN ACOUSTIC EMISSION TESTS 23
    Effects of Hydraulic and Mechanical Noise . 23
    Hydraulic Noise 23
    Mechanical Noise 23
    Noise Associated with Welding 24
    Noise from Solid State Bonding 25
    Electromagnetic Interference 25
    Miscellaneous Noise 26
    Conclusions 26
PART 4: INTRODUCTION TO ACOUSTIC EMISSION SIGNAL CHARACTERIZATION . 27
    The Purpose of Signal Characterization 27
    Interpreting the Data 27
    Characteristics of Discrete Acoustic Emission 27
    Waveform Parameter Descriptions 28
    Acoustic Emission Event Energy 29
    Acoustic Emission Signal Amplitude 29
    Limitations of the Simple Waveform Parameters 29
    Advanced Characterization Methods 29
    Knowledge Representation Method 30
    Feature Analysis and Selection 30
    Pattern Recognition Training and Testing ... 31
    Signal Classification 31
    Signal Treatment and Classification System . 31
    Guidelines for Source Characterization 32
    Comparison of the Classifier Techniques 33
PART 5: INTRODUCTION TO TERMS, STANDARDS AND PROCEDURES FOR ACOUSTIC EMISSION TESTING 34
    Terminology of Acoustic Emission Method .. 34
    Development of Acoustic Emission Terminology 34
    Glossary for Acoustic Emission Testing 34
    Procedures for Acoustic Emission Testing ... 36
    Personnel Qualification 36
    Standards for Acoustic Emission Testing 36
PART 6: ACOUSTIC EMISSION EXAMINATION PROCEDURES 39
    Objectives of the Acoustic Emission Testing Procedure 39
    Technical Characteristics Furnished by the Transducer Manufacturer 39
    Coupling and Mounting of Piezoelectric Transducers 39
    Simplified Characterization and Verification of Piezoelectric Transducers 40
    Verification Precautions 41

SECTION 2: MACROSCOPIC ORIGINS OF ACOUSTIC EMISSION 45

PART 1: FUNDAMENTALS OF ACOUSTIC EMISSION'S MACROSCOPIC ORIGINS 46
Unique Aspects of Acoustic Emission Testing 46
Acoustic Emission Instrumentation 46

PART 2: DEFORMATION SOURCES OF ACOUSTIC EMISSION 47
Plastic Deformation Sources 47
Factors Contributing to Acoustic Emission .. 47
The Kaiser Effect and Deformation 48

PART 3: FRACTURE AND CRACK GROWTH AS ACOUSTIC EMISSION SOURCES 49
Crack Growth Sources 49
Fracture as an Acoustic Emission Source ... 51
Criteria for Fracture 52

PART 4: ACOUSTIC EMISSION FROM FATIGUE 55
Techniques for Monitoring Fatigue Cracking 55
Acoustic Emission Count Rate 55
Relation Between Count Rate and Crack Growth 56
Monitoring Low Cycle Fatigue 56

PART 5: CORROSION AND STRESS CORROSION CRACKING 58
Acoustic Emission from Corrosion 58
Relation Between Count Rate and Corrosion 58
Stress Corrosion Cracking 58
Monitoring Hydrogen Embrittlement and Adverse Test Environments 59
Conclusions 59

SECTION 3: MICROSCOPIC ORIGINS OF ACOUSTIC EMISSION 63

PART 1: FUNDAMENTALS OF ACOUSTIC EMISSION GENERATION 64
Origination of Elastic Waves 64
Detecting Acoustic Emission 64
The Need for Studying Microscopic Origins . 64

PART 2: MICROMECHANICAL MODELING .. 66
Formulation of the Model 66
Sensor Spectral Sensitivity 68
Detectability Criteria 69
Source Directivity 70

PART 3: MICROSCOPIC DISLOCATION SOURCES 71
Detectability of Microscopic Dislocation Sources 71
Single Crystal Deformation 73
Grain Size Effects of Microscopic Dislocation Sources 74
Precipitation Effects 75

PART 4: MICROSCOPIC FRACTURE SOURCES 77
Detectability of Microscopic Fracture Sources 77
Cleavage Microfracture 79
Intergranular Microfracture 79
Particle Microfracture 79
Microvoid Coalescence 81
Environmental Factors 82
Amplification Factors 82

PART 5: MICROSCOPIC PHASE CHANGES ... 84
Solid State Phase Transformations 84
Detectability of Martensitic Transformations . 84
Liquid-Solid Transformations 87

SECTION 4: WAVE PROPAGATION 91

PART 1: WAVES IN INFINITE MEDIA 92
Equations of Motion 92
General Solution of Motion Equations 92
Plane Wave Solutions 93
Simple Harmonic Waves 93

PART 2: SURFACE WAVES 95

PART 3: REFLECTION OF WAVES AT INTERFACES 97
Transverse Waves Incident on a Free Surface 97
Incident Longitudinal Wave 99
Energy Partition at a Free Surface 99
Reflection and Transmission at an Interface Between Two Solids 101

PART 4: ATTENUATION OF WAVES 103
Geometric Attenuation 103
Attenuation from Dispersion 103
Attenuation from Scattering and Diffraction . 104
Attenuation due to Energy Loss Mechanisms 105
Attenuation in Actual Structures 106

PART 5: IDEALIZED WAVE SOURCES 108
Concentrated Force 108
Double Force without Moment 110
Center of Compression 110

PART 6: SOURCES ACTING IN OR ON A HALF-SPACE 112
Concentrated Force Acting Normal To and On the Surface 112
Concentrated Force Below the Surface of the Half-Space 114

PART 7: SOURCES ACTING IN OR ON PLATES 115

PART 8: DIFFUSE WAVE FIELDS 117

Diffuse Wave Field Equations 117
Diffuse Field Solution for Impulsive
    Sources 118

SECTION 5: ACOUSTIC EMISSION SENSORS AND THEIR CALIBRATION ... 121

PART 1: ACOUSTIC EMISSION SENSORS 122
    Definition of Acoustic Emission Sensors 122
    Types of Sensors 122
    Typical Acoustic Emission Sensor Design ... 123
    Characteristic Sensor Dimensions 123
    Couplants and Bonds 124
    Temperature Effects on Acoustic Emission
        Sensors 124
    Special Acoustic Emission Sensors and
        Sensor Mounts 125
PART 2: PRIMARY SENSOR CALIBRATION ... 126
    Terminology of Sensor Calibration 126
    Basic Principles of Sensor Calibration 126
    Step-Force Calibration 128
    Reciprocity Calibration 130
PART 3: SECONDARY CALIBRATION AND
    SENSOR TESTS 131
    Purpose of Secondary Calibration 131
    Development of Secondary Calibration
        Procedures 131

SECTION 6: ACOUSTIC EMISSION SOURCE LOCATION 135

PART 1: FUNDAMENTALS OF ACOUSTIC
    EMISSION SOURCE LOCATION 136
    Comparison of Acoustic Emission to Other
        Nondestructive Tests 136
    Remote Sensors for Source Location 136
PART 2: LOCATING SOURCES OF
    CONTINUOUS ACOUSTIC EMISSION 137
    Characteristics of Continuous Acoustic
        Emission Testing 137
    The Value of Attenuation 137
    Zone Location Method 137
    Signal Amplitude Measurement Method 138
    The Use of Timing Techniques 140
    Cross Correlation Approach 141
    Coherence Techniques 142
    Summary 144
PART 3: LOCATING DISCRETE SOURCES OF
    ACOUSTIC EMISSION 145
    Linear Location of Acoustic Emission
        Sources 145
    Location of Sources in Two Dimensions 146
    Restricting the Active Zone 147

Special Array Configurations 148
Error in Discrete Source Location 148
Source Location in Three Dimensions 152
Conclusions 153

SECTION 7: ACOUSTIC EMISSION APPLICATIONS IN THE PETROLEUM AND CHEMICAL INDUSTRIES 155

PART 1: OVERVIEW OF ACOUSTIC EMISSION
    TESTING IN THE PETROLEUM AND
    CHEMICAL INDUSTRIES 156
    Pressure Vessel Materials and Common
        Causes of Failure 156
    Uses of Acoustic Emission Tests of Pressure
        Vessels 157
PART 2: INDUSTRIAL GAS TRAILER TUBE
    APPLICATIONS 161
    Recertification of Gas Trailer Tubing 161
    Test Procedure for Trailer Tubing Tests 162
    Advantages of Acoustic Emission Testing of
        Trailer Tubes 163
PART 3: ACOUSTIC EMISSION TESTS OF
    METAL PIPING 166
    Acoustic Emission Testing of Gas Pipelines . 166
PART 4: ACOUSTIC EMISSION TESTING OF
    METAL PRESSURE VESSELS 167
    Code for Testing Pressure Vessels 167
    Acoustic Emission Testing of Blowout
        Preventers 168
    Cast Steel Gate Valves 168
    Testing Pressure Vessels to Evaluation
        Standards 170
    Acoustic Emission Testing of Ammonia
        Spheres 170
    Summary of Industrial Acoustic Emission
        Test Procedures of Tanks and Vessels 172
    Testing of Hydroformer Reactor Vessels 173
    Fluid Catalytic Cracking Unit 183
PART 5: ACOUSTIC EMISSION TESTING OF
    METAL STORAGE TANKS 187
    Tests of Flat Bottom Anhydrous Ammonia
        Storage Tanks 187
    Tests of Horizontal Tanks 188
    Tests of Butane Storage Spheres 188
PART 6: ACOUSTIC EMISSION TESTING FOR
    LEAK DETECTION 194
    Metal Pipeline Hydrostatic Testing 194
    Plastic Pipeline Pneumatic Testing 197
    Offshore Crude Oil Pipeline Testing 198
    Tests of Steam Condensate Lines 200
    Acoustic Emission Tests of Flat Bottom
        Tanks 201

PART 7: ACOUSTIC EMISSION TESTS OF
FIBER REINFORCED PLASTIC VESSELS . 203
    Testing Procedures for Pressure, Storage and
        Vacuum Vessels 203
    Applications in the Chemical Industries 203
    Composite Pipe Testing Applications 205
    Zone Location in Fiber Reinforced Plastics . 205
    Felicity Effect in Fiber Reinforced Plastics . . 208
    Acceptance of Acoustic Emission Techniques
        for Testing of Fiber Reinforced Plastics . . 209
PART 8: MACHINERY CONDITION
    MONITORING 213
    Incipient Failure Detection 213
    Applications of Acoustic Emission in
        Machinery Monitoring 214
    Machinery Monitoring Case Histories 215
    Decision-Making as Part of Incipient Failure
        Detection 219
PART 9: ADVANTAGES OF ACOUSTIC
    EMISSION TESTING IN THE
    PETROCHEMICAL INDUSTRIES 220
    Potential Cost Saving from the Use of
        Acoustic Emission Techniques 220
    Outline of Benefits for the Petrochemical
        Industries 220

SECTION 8: ACOUSTIC EMISSION APPLICATIONS IN THE NUCLEAR AND UTILITIES INDUSTRIES 225

PART 1: CONTINUOUS IN-SERVICE
    ACOUSTIC EMISSION MONITORING OF
    NUCLEAR POWER PLANTS 226
    Instrumentation 227
    Detection and Location of Discontinuities . . 228
    Noise Interference 229
    Acoustic Emission Signal Identification 230
    Severity Estimate from Acoustic Emission
        Data 230
    Status and Future of the Technique 231
PART 2: APPLICATION OF ACOUSTIC
    EMISSION TECHNIQUES FOR
    DETECTION OF DISCONTINUITIES IN
    ELECTRIC POWER PLANT STRUCTURES . 233
    Nuclear Power Plant Applications 233
    Fossil Fuel Power Plant Applications 244
    Hydroelectric Power Plant Applications 245
PART 3: ACOUSTIC EMISSION LEAK
    DETECTION ON CANDU REACTORS 253
    Acoustic Emission Identification of Delayed
        Hydride Cracks 253
    Acoustic Emission Identification of Seal
        Leakage 256
    Conclusion 258

PART 4: ACOUSTIC EMISSION FROM
    ZIRCONIUM ALLOYS DURING
    MECHANICAL AND FRACTURE
    TESTING 260
    Zirconium Alloys 260
    Plastic Deformation 260
    Fatigue 261
    Cracking of Surface Oxide 261
    Stress Corrosion Cracking 263
    Delayed Hydride Cracking 263
PART 5: BUCKET TRUCK AND LIFT
    INSPECTION 267
    Acoustic Emission Inspection Development . 267
    Instrumentation for Bucket Truck
        Inspection 268
    Test Procedure for Bucket Truck Inspection . 268
    Typical Test Data 270
    Acceptance Criteria 271

SECTION 9: ACOUSTIC EMISSION APPLICATIONS IN WELDING 275

PART 1: RESISTANCE SPOT WELD TESTING . 276
    Resistance Spot Welding 276
    Principles of Acoustic Emission Weld
        Monitoring 276
    Weld Quality Parameters that Produce
        Acoustic Emission 277
    Acoustic Emission Instrumentation for
        Resistance Spot Welding 279
    Typical Applications of the Acoustic Emission
        Method 280
    Monitoring Coated Steel AC Welds 282
    AC Spot Welding Galvanized Steel 282
    Detecting the Size of Adjacent AC Welds . . . 283
    Control of Spot Weld Nugget Size 283
    Conclusions 284
PART 2: ELECTRON BEAM WELD
    MONITORING 285
    Description of Electron Beam Welding 285
    Discontinuities in Electron Beam Welding . . 285
    Acoustic Emission Method of Testing
        Electron Beam Welds 286
    Advances in Acoustic Emission Testing of
        Welds 286
    Types of Acoustic Emission from Electron
        Beam Welds 287
    Electron Beam Monitoring Instrumentation . 288
    Conclusion 290
PART 3: ACOUSTIC EMISSION FOR
    DISCONTINUITY DETECTION IN
    ELECTROSLAG WELDS 291
    Test Specimens for Acoustic Emission Testing
        of Electroslag Welds 291

The Electroslag Welding Process	291
Acoustic Emission Testing Equipment for Testing Electroslag Welds	292
Acoustic Emission Data Collection for Electroslag Welds	294
Results of Acoustic Emission Tests	294
Acoustic Emission Techniques in Electroslag Welding Process Control	297
Acoustic Emission and Weld Flux	301
Conclusions	301
PART 4: ACOUSTIC EMISSION EVALUATION OF LASER SPOT WELDING	302
Laser Welding Procedures	302
Laser Welding Geometry	302
Laser Welding Mechanics	303
Real-Time Evaluation Theory	305
Experimental Correlation for Laser Spot Welding	307
Conclusions	309

SECTION 10: ACOUSTIC EMISSION APPLICATIONS IN CIVIL ENGINEERING ... 311

PART 1: ACOUSTIC EMISSION GEOTECHNICAL APPLICATIONS	312
Historical Background	312
Geotechnical Applications	312
Laboratory Behavior of Acoustic Emission from Geotechnical Materials	314
Geotechnical Acoustic Emission Equipment	317
Acoustic Emission Geotechnical Field Applications	323
PART 2: ACOUSTIC EMISSION MONITORING OF STRUCTURES	326
Acoustic Emission Activity in Structural Steel Members	326
The Nature of Acoustic Emission in Structures	328
Use of Transducers in Tests of Structures	328
Signal Processing	329
Discontinuity Location	329
Source Activation	331
Acoustic Emission Structural Instrumentation	332
Acoustic Emission Monitoring of Structures	333
PART 3: COMPONENT TESTING	340
Testing of Welded Structures	340
Testing of Steel Wire Rope	340
Testing of Other Components	341

SECTION 11: ACOUSTIC EMISSION APPLICATIONS IN THE ELECTRONICS INDUSTRY ... 347

PART 1: ACOUSTIC EMISSION TESTING OF MULTILAYER CERAMIC CAPACITORS	349
The Ceramic Chip Capacitor and Its Failure Modes	349
Discontinuity Detection Methods	349
Conclusions	350
PART 2: ACOUSTIC EMISSION APPLICATIONS IN UNDERSEA REPEATER MANUFACTURE	352
High Voltage Capacitor in the Repeater Circuitry Unit	352
Instrumentation and Analysis	353
Tubulation Pinchweld on the Repeater Housing	355
PART 3: ACOUSTIC EMISSION DETECTION OF CERAMIC SUBSTRATE CRACKING DURING THERMOCOMPRESSION BONDING	358
The Bonding Process	358
Acoustic Emission System for Detection of Bonding Cracks	359
Conclusions	360
PART 4: ACOUSTIC EMISSION TEST FOR MICROELECTRONIC TAPE AUTOMATED BONDING	361
An Acoustic Emission Nondestructive Bond Quality Tester and Lead Forming System for Tape Bonded Integrated Circuits	362
PART 5: ACOUSTIC EMISSION THERMAL SHOCK TEST FOR HYBRID MICROCIRCUIT PACKAGES	365
Characterization of the Hot Stage Shock Test	365
Acoustic Emission Monitoring of the Shock Test	366
PART 6: EVALUATION OF FRACTURE IN INTEGRATED CIRCUITS WITH ACOUSTIC EMISSION	370
Acoustic Emission Electronics	371
Biaxial Wafer Bending Tests	372
Diamond Indentation Experiment	374
Stable Crack Growth During Die Shear Testing	382
Summary	386
PART 7: ACOUSTIC EMISSION DIAGNOSTIC AND MONITORING TECHNIQUES FOR SUPERCONDUCTING MAGNETS	389
The Phenomenon of Superconductivity	389
Applications for Superconductors	389
Cryostable Versus Adiabatic Magnets: Stability to Quench	390
Sources of Acoustic Emission in Superconducting Magnets	391
Quench Identification	391

Acoustic Emission Signals Induced by Conductor Motion in a Superconducting Dipole 393
Normal Zone Detection in Superconducting Magnets 398
Disturbance Energy Quantification 399
Cryogenic Temperature Acoustic Emission Sensors 399
Conclusion 400

PART 8: PARTICLE IMPACT NOISE DETECTION AS A NONDESTRUCTIVE TEST FOR DETERMINING INTEGRITY OF ELECTRONIC COMPONENTS 402
Definition of PIND 402
PIND Testing History 402
Military Standards Governing PIND Testing . 403
Theory and Instrumentation for PIND Testing 403
Nature of Particles 403
Methods for Elimination of Particles Prior to PIND Testing 406
PIND Test Effectivity 406
Using PIND Testing to Conduct Failure Analysis 407
Alternate Tests for Loose Particle Detection . 407
Conclusion 408

PART 9: DESIGNING MICROELECTRONIC WELDS FOR ACOUSTIC EMISSION TESTABILITY 409
Nature of Welds in Microelectronics 409
Designing Welds for Testability 410

SECTION 12: ACOUSTIC EMISSION APPLICATIONS IN THE AEROSPACE INDUSTRY 417

PART 1: ACOUSTIC EMISSION TESTING OF AEROSPACE METAL ALLOYS 418
Quantitative Observations of Acoustic Emission from Crack Growth in Aircraft Aluminum Alloys 419

PART 2: ACOUSTIC EMISSION MONITORING OF AIRCRAFT STRUCTURES 421
Experimental Observations during Flight ... 421

PART 3: AEROSPACE ORGANIC COMPOSITES 425
Organic Composites..................... 425
Aerospace Composite Applications 425
Applications for Monitoring Organic Composites with Acoustic Emission 426
Composite Testing and Failure Mechanisms . 427
Acoustic Emission Relationships to Composite Mechanisms 428

Correlations of Acoustic Emission and Composite Failure Mechanisms 429
Noise Suppression 429
Standardizing Acoustic Emission Tests 430
Instrumentation for Organic Composite Tests 430

PART 4: METAL MATRIX COMPOSITES 432
Applications of Acoustic Emission Testing to Metal Matrix Composites 432
Proof Load Testing of Metal Matrix Composite Structures 433
Instrumentation for Testing Metal Matrix Composite Materials................. 433

PART 5: CORRELATION OF ACOUSTIC EMISSION AND STRAIN DEFLECTION MEASUREMENT DURING COMPOSITE AIRFRAME STATIC TESTS 434
Detecting Damage During Airframe Structural Tests 434
Effect of Airframe Structure on Testing 434
Analyzing Acoustic Emission Signals 434
Considerations for Acoustic Emission Sensors 435
Acoustic Emission Setup for Tests of Airframes 436
Limit Load Test Techniques 437
Results of Acoustic Emission Tests of Composite Airframes 438
Conclusion 443

SECTION 13: ACOUSTIC EMISSION RESEARCH IN BIOMEDICAL ENGINEERING 447

INTRODUCTION 448

PART 1: ACOUSTIC EMISSION FOR STUDIES OF THE MECHANICAL PROPERTIES OF ENTIRE BONES 449
Methods for Bone Studies 449

PART 2: DIAGNOSIS OF BONE FRACTURES BY A NONINVASIVE AND NONTRAUMATIC ACOUSTIC TECHNIQUE 450
Historical Brief of Medical Applications 450
Experimental Procedure 451
Results and Discussion 451
Conclusions 456

PART 3: ACOUSTIC EMISSION TECHNOLOGY IN EXPERIMENTAL AND CLINICAL PEDIATRIC ORTHOPEDICS 457
In Vitro Model 458
In Vivo Model 459
Summary of Acoustic Emission Application . 459

PART 4: DIAGNOSIS OF LOOSE OR
DAMAGED TOTAL JOINT REPLACEMENT
COMPONENTS 461
   Methods for Diagnosis 461
   Patient Population 461
   Results of the Acoustic Emission Tests 462
   Discussion of the Acoustic Emission
     Technique 463

SECTION 14: PROCESS MONITORING WITH ACOUSTIC EMISSION 467

INTRODUCTION 468
PART 1: ACOUSTIC EMISSION MONITORING
OF AN AUTOMOTIVE COINING
APPLICATION 469
   Sources of Acoustic Emission 469
   Development Monitoring 469
   Benefits of Coining Monitoring with
     Acoustic Emission 470
PART 2: THE USE OF DRILL-UP FOR ON-
LINE DETERMINATION OF DRILL
WEAR 472
   Block Diagram Description of Drill-Up 472
   Use of Drill-Up for Drill Wear Out
     Detection 473
   Experimental Results and Discussion of the
     Use of Drill-Up 476
   Potential Problems and Solutions 477
   Conclusions 478
PART 3: MATERIAL DEPENDENCY OF CHIP
FORM DETECTION USING ACOUSTIC
EMISSION 479
   Acoustic Emission from Chip Formation 479
   Acoustic Emission Cutting Monitor 479
   Experimental Procedure for Cutting
     Monitoring 481
   Results and Discussion of Cutting
     Monitoring 482
   Conclusions 484
PART 4: THE ROLE OF ACOUSTIC EMISSION
IN MANUFACTURING PROCESS
MONITORING 485
   The Function of Automation 485
   Tool Wear and Fracture Monitoring with
     Acoustic Emission 486
   Chip Formation Monitoring 489
   Wheel Contact and Sparkout in Grinding ... 490
   Conclusion 492
PART 5: ACOUSTIC EMISSION FROM
MULTIPLE INSERT MACHINING
OPERATIONS 493
   Acoustic Emission from Milling 493

   Data Analysis of Acoustic Emission from
     Milling 493
   Acoustic Emission from Chip Formation 496
   Conclusion 497
PART 6: DEVELOPMENT OF AN
INTELLIGENT ACOUSTIC EMISSION
SENSOR FOR UNTENDED
MANUFACTURING 499
   Intelligent Acoustic Emission Sensors 499
   Acoustic Emission Intelligent Sensing 500
   Development of Intelligent Acoustic
     Emission Sensors 501
   Intelligent Sensors for Punch Stretching 501
   Intelligent Sensors for Chatter Detection in
     End Milling 502
   Intelligent Sensors for Abrasive Disk
     Machining 502
   Conclusion 503
PART 7: CURE MONITORING OF
COMPOSITE MATERIALS USING
ACOUSTIC TRANSMISSION AND
ACOUSTIC EMISSION 504
   Manufacture of Thermosetting Resin
     Composites 504
   Combination Acoustic Emission and Acoustic
     Transmission Technique 505
   Acoustic Emission Technique 505
   Conclusion 507

SECTION 15: SURVEY OF COMMERCIAL SENSORS AND SYSTEMS FOR ACOUSTIC EMISSION TESTING 513

INTRODUCTION 514
PART 1: ACOUSTIC SENSORS, TRANSDUCERS
AND ACCELEROMETERS 515
   Piezoceramic Crystals Used in Acoustic
     Emission Sensors 515
   Special Geometry Piezoceramic Element ... 516
   Flat Frequency Sensor 516
   Vibration Transducers with Integral
     Electronics 518
   Sensors for Particular Applications 524
   Variety of Sensors from a Single
     Manufacturer 526
PART 2: ACOUSTIC EMISSION AND
ACOUSTIC ANALYSIS SYSTEMS 531
   Acoustic Emission Simulator 531
   Signal Acquisition and Analysis System 532
   Loose Part Monitoring System 533
   Valve Flow Monitoring System 535
   System for Detection of Air-Borne Acoustic
     Signals 537

Assorted Acoustic Testing Systems 542
Variety of Acoustic Emission Systems by a
  Single Manufacturer 545
Particle Impact Noise Detector 546
Acoustic Emission Software 547

SECTION 16: SPECIAL ACOUSTIC EMISSION APPLICATIONS 551

PART 1: ACOUSTIC EMISSION STUDIES OF
  GLACIERS 552
    The Microseismicity of Glaciers 552
    History of Acoustic Emission's Use in Glacier
      Studies 553
    Processes and Problems in the Use of
      Acoustic Emission in Glacier Studies 555
    Suggested Developments 556
    Conclusions 557
PART 2: ACOUSTIC EMISSION STATISTICS
  AND THE CHOICE OF REJECT LIMITS ... 558
    Method for Assignment of Upper Acceptance
      Limit 558
    Batch Analysis Extended to Continuous
      Production 560
    Conclusion 560
PART 3: ACOUSTIC EMISSION IN
  DEVELOPMENT OF COMPOSITE
  MATERIALS 561
    Characteristics of Composite Materials 561
    Acoustic Emission in Composite Tests 561
    Sources of Acoustic Emission in Composite
      Materials 562
    Typical Factors Affecting Acoustic Emission
      Tests of Composites 562
    Measurement of Acoustic Emission Signals
      from Composites 563
    Interpretation of Acoustic Signals from
      Composites 563
    Applications in Research and Development of
      Fiber Reinforced Composites 564
    Applications of Acoustic Emission in the
      Manufacture of Composites 565
    Development of Models for Composite
      Behavior 567
    Determination of Composite Properties 567
    Conclusion 568
PART 4: APPLICATION OF POLYVINYLIDENE
  FLUORIDE AS AN ACOUSTIC EMISSION
  TRANSDUCER FOR FIBROUS COMPOSITE
  MATERIALS 569
    Characteristics of Polyvinylidene Fluoride ... 569
    Transducer Construction 569
    Tests of Sensitivity 570
    Fatigue Tests with the Polyvinylidene
      Fluoride 573
    Conclusions 574
PART 5: STUDIES OF FRACTURE BEHAVIOR
  IN REFRACTORIES USING ACOUSTIC
  EMISSION TECHNIQUES 575
    Experimental Procedure for Study of
      Refractories 575
    Effect of Heating Rate 579
    Effect of Specimen Dimensions 579
    Effect of Restraining Conditions 580
    Conclusions 581
    Testing for Spalling Resistance of
      Refractories 583
PART 6: ENERGY MEASUREMENT OF
  CONTINUOUS ACOUSTIC EMISSION 586
    Definition of Continuous Emission 586
    Theory and Operation of Voltmeters 586
    Advantages of the Techniques 588
    Practical Factors in Energy Measurements .. 589
    Conclusion 591

INTRODUCTION TO ACOUSTIC EMISSION TECHNOLOGY

Thomas F. Drouillard, Rockwell International Corporation, Golden, Colorado

PART 1
HISTORY OF ACOUSTIC EMISSION RESEARCH

Early Uses of Acoustic Emission

Acoustic emission and microseismic activity are naturally occurring phenomena. Although it is not known exactly when the first acoustic emissions were heard, fracture processes such as the snapping of twigs, the cracking of rocks and the breaking of bones were probably among the earliest. The first acoustic emission used by an artisan may well have been in making pottery (the oldest variety of hardfired pottery dates back to 6,500 BC).[1] In order to assess the quality of their products, potters traditionally relied on the audible cracking sounds of clay vessels cooling in the kiln. These acoustic emissions were accurate indications that the ceramics were defective and did indeed structurally fail.

Acoustic Emission in Metalworking

Tin Cry

It is reasonable to assume that the first observation of acoustic emission in metals was *tin cry*, the audible emission produced by mechanical twinning of pure tin during plastic deformation. This phenomenon could occur only after man learned to smelt pure tin, since tin is found in nature only in the oxide form. It has been established that smelting (of copper) began in Asia Minor as early as 3,700 BC.

The deliberate use of arsenic and then tin as alloying additions to copper heralded the beginning of the Bronze Age somewhere between the fourth and third millennium BC. The oldest piece of pure tin found to date is a bangle excavated at Thermi in Lesbos.[2] The tin has been dated between 2,650 and 2,550 BC. It is 41 mm (1.6 in.) in diameter and consists of two strands of pure tin, one wrapped around the other and hammered flat at the end. During the manufacture of this bangle, the craftsman could have heard considerable tin cry.

Early Documented Observations of Tin Cry

The first documented observations of acoustic emission may have been made by the eighth century Arabian alchemist Jabir ibn Hayyan (also known as Geber). His book *Summa Perfectionis Magisterii* (*The Sum of Perfection* or *The Sum of Perfect Magistery*) was published in English translation in 1678; the Latin edition was published in Berne in 1545. In it he writes that Jupiter (tin) gives off a "harsh sound" or "crashing noise." He also describes Mars (iron) as "sounding much" during forging. This sounding of iron was most likely produced by the formation of martensite during cooling.[3]

Since the time of the alchemists, audible emissions have become known and recognized properties of cadmium and zinc as well as tin. Tin cry is commonly found in books on chemistry published in the last half of the nineteenth century. For example, Worthington Hooker in 1882 describes "the cry of tin" as ". . . owing to the friction on minute crystals of the metal against each other."[4] In another text on chemistry published in 1883, Elroy M. Avery states that cadmium ". . . gives a crackling sound when bent, as tin does." He also describes an experiment with tin: "Hold a bar of Sn near the ear and bend the bar. Notice the peculiar crackling sound. Continue the bending and notice that the bar becomes heated. The phenomena noticed seems to be caused by the friction of the crystalline particles."[5]

Around the turn of the twentieth century, metals researchers started to break away from chemistry disciplines and began to develop metallurgy, their own specialized field of science. A great deal of work was done on the study of twinning and martensitic phase transformation. Tin and zinc were historically well documented and two of the best metals for studying these phenomena. During these studies, it was normal to hear the sounds emitted by metals such as tin, zinc, cadmium and some alloys of iron. In 1916, J. Czochralski was the first to report in the literature the association between "Zinn- und Zinkgeschrei" (tin and zinc cry) and twinning.[6] He cited *Gmelin-Kraut's Handbuch der anorganischen Chemie* (1911),[7] which in turn referenced an article by S. Kalischer (1882).[8]

Kalischer in turn cited references to tin cry published prior to 1882 in the *Chemical Handbook* by Gmelin and Berzelius and *Watts' Dictionary of Chemistry*. The following excerpts appear in the 1894 edition of *Watts' Dictionary of Chemistry*. "When a bar of tin is bent, a crackling sound may be heard due to the crystals in the inner parts of the bar breaking against one another." Regarding zinc: "Kahlischer [sic] (B.14, 2747) noticed that rolled zinc ceased to give a ringing sound when struck after it was heated to c. 160°-300°, that it could then be bent easily, and that when bent it emitted a sound like the 'cry' of tin . . ."[9] It may well have been Czochralski who sparked the work of Joseph Kaiser (and his subsequent discovery of the Kaiser effect) since the only reference to acoustic phenomena in Kaiser's thesis is the paper published by Czochralski in 1917.[10]

In 1923, Albert M. Portevin and Francois Le Chatelier reported that "petit bruit sec" (small sharp noises) were clearly audible at a distance of several meters during discontinuous yielding and Lüders band formation in aluminum-copper-manganese alloys.[11] This appears to be an original observation of Portevin and Le Chatelier and they make no reference to anyone else's work.

M. Classen-Nekludowa reported that in 1924 P. Ehrenfest and A. Joffé observed that the process of shear deformation in rock salt crystals and in zinc single crystals was accompanied by very regular "knackenden Lauten" (clicking noises), similar to the ticking of a clock.[12]

Robert J. Anderson published his D.S. thesis in 1925,[13] followed by a paper in 1926[14] in which he stated that during tension testing of an aluminum alloy beyond the yield point and near the failure load, increased loading is accompanied by a series of slips which give rise to a series of serrated lines or Lüders lines. These slips are accompanied by a series of "audible clicks or sounds." For thin sheet, the sounds are high pitched, "not unlike the tinkle of Japanese glass chimes." For thick sheet, the sounds are "low in note, being on the order of a grunt." M.V. Classen-Nekludowa working in 1929 with rock salt, monocrystalline and polycrystalline brass, and polycrystalline aluminum achieved 10^4 optical magnification of deformation steps which occurred at the moment of the "Knackens" (cracking noise).[12]

Erich Scheil reported in 1929 that the formation of martensite in steel is accompanied by a "deutlich hörbaren Geräusch, dessen Klang dem bekannten Zinngeschrei ähnlich war" (clearly audible noise, whose sound was similar to the well-known tin cry).[15] In 1935 and 1937, P.W. Bridgman observed during rotation under compressive loading that few metals and many nonmetals rotate in a noisy manner. Some "chatter," some "squeak," some make a "grinding noise," while others "hiss." He claimed the most important noise, however, was "snapping" which was in most cases superposed on the ordinary phenomena of plastic flow. The metallic elements that exhibited snapping were vanadium, titanium, strontium, cerium, arsenic and osmium.[16,17]

Charles S. Barrett (1947) reported that transformation to the face-centered cubic form for lithium is accompanied by a series of audible clicks, as in the twinning of tin or magnesium and the formation of martensite.[18]

PART 2
DETECTING AND RECORDING ACOUSTIC EMISSION

The transition from the incidental observation of audible tin cry to the deliberate study of acoustic emission phenomena consisted of three separate and unrelated experiments in which instrumentation was used to detect, amplify and record acoustic emission events occurring in the test specimens. The first experiment instrumented specifically to detect acoustic emission was conducted in Germany and the results were published in 1936 by Friedrich Förster and Erich Scheil.[19] They recorded the "Geräusche" (noises) caused by the formation of martensite in 29 percent nickel steel.

In the United States, Warren P. Mason, H.J. McSkimin and W. Shockley performed and published the second instrumented acoustic emission experiment in 1948.[20] At the suggestion of Shockley, experiments were directed toward observation of moving dislocations in pure tin specimens by means of the stress waves they generated. The experiment's instrumentation was capable of measuring displacements of about 10^{-7} mm occurring in times of 10^{-6} seconds.

The third instrumented experiment was performed in England by D.J. Millard in 1950 during research for his Ph.D. thesis at the University of Bristol.[21] He conducted twinning experiments on single crystal wires of cadmium. Twinning was detected using a Rochelle salt transducer.

Kaiser's Study of Acoustic Emission Sources

The early observations of audible sounds and the three instrumented experiments were not directed at a study of the acoustic emission phenomenon itself, nor did the researchers carry on any further investigations in acoustic emission. The genesis of today's technology in acoustic emission was the work of Joseph Kaiser at the Technische Hochschule München in Germany.

In 1950 Kaiser published his Doktor-Ingenieur dissertation where he reported the first comprehensive investigation into the phenomena of acoustic emission.[22] Kaiser used tensile tests of conventional engineering materials to determine: (1) what noises are generated from within the specimen; (2) the acoustic processes involved; (3) the frequency levels found; and (4) the relation between the stress-strain curve and the frequencies noted for the various stresses to which the specimens were subjected.

His most significant discovery was the irreversibility phenomenon which now bears his name, the *Kaiser effect*. He also proposed a distinction between burst and continuous emission. Kaiser concluded that the occurrence of acoustic emission arises from frictional rubbing of grains against each other in the polycrystalline materials he tested and also from intergranular fracture.

Kaiser continued his research at the Institut für Metallurgie und Metallkunde der Technischen Hochschule München until his death in March 1958.[23-26] His work provided the momentum for continued activities at the Institut by several of his coworkers, including Heinz Borchers and Hans Maria Tensi, and also furnished the impetus for further research elsewhere in the world.

Acoustic Emission Research in the United States

Origins of US Research

The first extensive research into acoustic emission phenomena following Kaiser's work was performed in the United States by Bradford H. Schofield at Lessells and Associates. His involvement came about as a result of a literature survey he and coworker A.A. Kyrala had conducted under another contract. They came across Kaiser's article in *Archiv für das Eisenhüttenwesen*.[24] This precipitated a request for a copy of Kaiser's dissertation by John M. Lessells from his friend and the adjudicator of Kaiser's dissertation, Ludwig Föppl. Subsequently, Lessells and Schofield began a correspondence with Kaiser which continued until Kaiser's death.

Acoustic Emission Studies for Materials Engineering

In December 1954, Schofield initiated a research program directed toward the application of acoustic emission to the field of materials engineering. His initial research was to verify the findings of Kaiser and the primary purpose of this early work was to determine the source of acoustic emissions. This research work was published in 1958.[27]

Schofield performed an extensive investigation into how surface and volume effects related to acoustic emission behavior. Experimental data obtained from oriented single crystals of aluminum (both with and without an oxide layer) and from oriented single crystals of gold, helped him conclude that surface condition does have a measurable influence on the acoustic emission spectrum. However, Schofield's most important conclusion was that acoustic emission is mainly a volume effect and not a surface effect.

In the fall of 1956, Lawrence E. Malvern at Michigan State University came across a one-page article entitled "Hörbare Schwingungen beim Verformen" ("Audible Vibrations from Deformation") in *Fliessen und Kriechen der Metalle* by Wilhelm Späth.[28] The article references the observations of "Geräuschen" (noises) by Joffé and Ehrenfest,[29] Klassen-Nekludowa,[12] Becker and Orowan,[30] and Kaiser.[24] Interested in studying the asperity theory of friction, Malvern suggested to a new faculty member, Clement A. Tatro, that this acoustic technique would be interesting to investigate. Consequently, Tatro initiated laboratory studies of acoustic emission phenomena.

Establishment and Expansion of Research Programs

In 1957, Tatro became aware of the work of Schofield and the two began collaborating. Tatro thought that research programs in acoustic emission could follow one of two rather well-defined branches: (1) to pursue studies concerned with the physical mechanisms that give rise to acoustic emission to completely understand the phenomenon; or (2) using acoustic emission as a tool to study some of the vexing problems of behavior of engineering materials.[31] He also foresaw the unique potential of acoustic emission as a nondestructive testing procedure.[32] His enthusiasm for this new technology sparked the interests of a number of graduate students at Michigan State who chose acoustic emission as the subject of their research projects. The students included Paul S. Shoemaker (1961),[33] Robert J. Kroll (1962),[34] Robert G. Liptai (1963)[35] and Robert B. Engle (1966).[36] Tatro left Michigan State in 1962 for Tulane University where he continued research in acoustic emission, fostering two more graduate students, Benny B. McCullough (1965)[37] and Davis M. Egle (1965).[38] In 1966, Tatro joined the staff at Lawrence Radiation Laboratory (now Lawrence Livermore National Laboratory).

Schofield and Tatro encouraged others to become involved in research activity in the field of acoustic emission. That encouragement, combined with the impact of their published work (the first publications in the English language), helped establish acoustic emission research and application around the country.

Julian R. Frederick at the University of Michigan had been engaged in research in ultrasonics since he was a graduate student under Floyd A. Firestone in 1939. His interest in acoustic emission, particularly for studying dislocation mechanisms, was excited in 1948 after reading Mason, McSkimin and Shockley's article "Ultrasonic Observation of Twinning in Tin" in *Physical Review*.[20] Frederick visited both Schofield and Tatro in the late 1950s. But it was not until 1960, when Frederick obtained a National Science Foundation grant for two of his graduate students, that he began his research in acoustic emission. A third student was funded several years later. These students were Jal N. Kerawalla (1965),[39] Larry D. Mitchell (1965)[40] and Anand B.L. Agarwal (1968).[41]

Aerospace Industry

At Aerojet General Corporation, Allen T. Green, Charles S. Lockman and Richard K. Steele were trying to verify the structural integrity of Polaris rocket motor cases fabricated for the US Navy. They noticed that audible sounds were consistently evident during hydrostatic testing and decided to instrument the vessels to detect, record and then analyze the acoustic signals. Their first acoustic emission test was conducted in 1961, using contact microphones, a tape recorder and sound level analysis equipment.

As part of Aerojet's contract with the National Aeronautics and Space Administration, Green, Lockman and Steele instrumented the Thiokol 6.6 m (21.7 ft) diameter SL-1 motor case to monitor the vessel during hydrostatic proof testing in 1965. Analysis of the tape recorded acoustic emission data clearly showed crack initiation and propagation prior to catastrophic failure at about 56 percent of proof pressure. From the recorded test data and the use of triangulation methods, they were able to locate the origin of failure within 0.3 m (1 ft).[42] The fact that the technique had indeed predicted and located failure of the vessel immediately enabled Aerojet to obtain sponsors for the acoustic emission work of Carl E. Hartbower, Phillip P. Crimmins, William W. Gerberich, Walter G. Reuter, George S. Baker and Douglas W. Bainbridge.

Harold L. Dunegan at Lawrence Radiation Laboratory became interested in acoustic emission after hearing a paper by Tatro and Liptai at the Symposium on Physics and Nondestructive Testing held at Southwest Research Institute in 1962.[32] Research began in 1963, as a result of Dunegan's internal report proposing the application of acoustic emission techniques to pressure vessel research.[43] He and coworkers Albert E. Brown and Paul L. Knauss were later joined by David O. Harris, Tatro, Engle and Liptai. Their activities involved a broad spectrum of acoustic emission research and applications.

In 1969, Dunegan and Knauss left the Laboratory and formed Dunegan Research Corporation (now the Dunegan Corporation of Physical Acoustics Group) which was the first company devoted exclusively to the manufacture of acoustic emission equipment.

At the Boeing Company, two groups became involved in

acoustic emission work. In 1965, Harvey L. Balderston initiated research at the Space Division on: (1) incipient failure detection in bearings; (2) signature analysis from other rotating machinery; (3) research on hydraulic systems for leak detection; and (4) cavitation and erosion from fluid flow in valves. This work is summarized in a paper published in 1972.[44]

At Boeing Scientific Research Laboratories, in the summer of 1967, Robert Moss engaged Adrian A. Pollock to add acoustic emission to an existing project on the study of crack growth in titanium.[45] Pollock was a third-year graduate student from Imperial College in England. He had been the recipient of the first grant in acoustic emission placed with Dr. R.W.B. Stephens at Imperial College in October 1964 by Roy S. Sharpe from the NDT Centre, Atomic Energy Research Establishment, Harwell. During his weeks at Boeing, working in several areas of acoustic emission research, Pollock became interested in the practical application of acoustic emission techniques for detecting crack growth. When he returned to England, he directed the remaining research for his Ph.D. thesis (1970) toward the study of acoustic emission technology in nondestructive testing.[46]

Nuclear Industries

Early in 1965, at the National Reactor Testing Station of Phillips Petroleum Company (now EG&G Idaho), Dwight L. Parry was involved in the loss of fluid test (LOFT) program where researchers were looking for a nondestructive testing technique faster than conventional methods used to detect loss of coolant in a nuclear reactor. Based on information in papers by Schofield and Kaiser, Parry applied acoustic emission techniques. In May 1966, he obtained his first definitive acoustic emission data from failure of an embrittled half-scale PM-2A reactor vessel instrumented with accelerometers, charge amplifiers and a tape recorder. Later, discontinuity location was successfully accomplished by triangulation techniques on the full size vessel.[47]

In February 1966, the Division of Reactor Development and Technology of the US Atomic Energy Commission sponsored a two-phase research program between Phillips Petroleum and Battelle Pacific Northwest Laboratories. Phillips Petroleum was assigned the responsibility for developing an incipient failure detection system. Parry and Dan L. Robinson developed an acoustic emission instrumentation system which would detect, locate and characterize incipient failure conditions in the reactor pressure vessel and in the primary pressure system components of a nuclear power plant.

The mission of Battelle Northwest on this program was primarily directed towards basic research of the acoustic emission phenomenon.[48] Philip H. Hutton served as project leader. It was Hutton who had initiated acoustic emission activities at Battelle Northwest in October 1965.[49]

At Argonne National Laboratory, T. Theodore Anderson became involved in acoustic emission in January 1967 after conducting a literature survey of acoustic boiling detection.[50] His research activities, which were directed at boiling detection in sodium cooled reactors, were the first coordinated effort in this area. In 1969 he was joined by A.P. Gavin in developing a high temperature submersible microphone for use in the liquid sodium environment of a reactor.

Robert G. Liptai, who was also at Argonne in 1967, was conducting experiments on martensitic phase transformation in single crystals of gold-cadmium alloy.[51] During this time, he collaborated with Dunegan and Tatro and subsequently joined Tatro's group at Lawrence Radiation Laboratory.

In July 1967, Thomas F. Drouillard went to Rocky Flats Plant, then operated by Dow Chemical Company. His assignment was to investigate acoustic emission as a potential NDT method for production applications. In February 1971, after collaborating with Dunegan, Tatro, Engle and Brown at Lawrence Radiation Laboratory, Drouillard installed the first production applications of acoustic emission at the Rocky Flats Plant. These applications included: (1) monitoring vessels for crack propagation during pressure testing; and (2) monitoring braze joints during a stressing operation to detect fracture of embrittled joints caused by overheating or too long a time at temperature.

Early on, Drouillard began an extensive literature search of the material published on acoustic emission. His first bibliography was published in 1975.[52] His comprehensive bibliography of the world literature on acoustic emission was published in 1979.[53] Since then, Drouillard has published extensive lists of references to acoustic emission literature in the *Journal of Acoustic Emission*. At present the total number of publications on acoustic emission exceeds 5,000.

PART 3
ORGANIZING THE ACOUSTIC EMISSION COMMUNITY

The Acoustic Emission Working Group

Conception of the AEWG

In the spring of 1967, Jack C. Spanner and Allen T. Green observed that a number of researchers were investigating the phenomena of acoustic emission and publishing reports of their work, but there seemed to be a lack of centralized communication. Spanner also observed that there were differences in terminology and experimental techniques, generally reflecting the researcher's educational background and field of expertise.

Jointly Spanner and Green perceived the need to unite and organize these people for the purpose of compiling and exchanging information. Using the constitution and bylaws of the Western Regional Strain Gage Committee as a model (as suggested by Green), Spanner laid the groundwork for the formation of the Acoustic Emission Working Group (AEWG).

The AEWG's First Meeting

Spanner contacted a dozen people with interest or experience in the field of acoustic emission and invited them to participate in the formation and organization of the working group. An informal meeting was subsequently held in the King's Room of the Esquire Inn Motel in Alcoa, Tennessee on November 2, 1967.

Twelve persons attended this formative meeting of the AEWG. In addition to Spanner and Green were Harvey L. Balderston, Harold L. Dunegan, Julian R. Frederick, Philip H. Hutton, Robert W. Moss, Charles W. Musser, Dwight L. Parry, Herb N. Pederson, Bradford H. Schofield and Norman K. Sowards. The first meeting of the AEWG was held in Idaho Falls, Idaho on February 8, 1968. At this meeting, in addition to the founding members, eight persons were elected to form the charter membership. They were T. Theodore Anderson, Thomas F. Drouillard, Robert B. Engle, Robert G. Liptai, D.K. Mitchell, R. Neal Ord, Ronald E. Ringsmith and Richard K. Steele.

European and Japanese Working Groups

Acoustic emission working groups were also organized in Japan and Europe. In Japan, Eiji Isono at the Central Research Laboratories of Nippon Steel Corporation, together with Morio Onoe at the Institute of Industrial Science, University of Tokyo, organized the Japanese Committee on Acoustic Emission and held their first meeting in December 1969.

The first acoustic emission meeting in Europe was the Institute of Physics Conference on Acoustic Emission organized by Adrian A. Pollock and held in March 1972 at Imperial College in London. The success of this conference established both the interest and the need for forming a working group. Consequently, Pollock and Don Birchon organized the European Stress Wave Emission Working Group which held its first meeting in November 1972 at the Admiralty Materials Laboratory in England. During their second meeting at Battelle Institute in Frankfurt am Main, Germany, September 1973, the group formally adopted their present name, European Working Group on Acoustic Emission.

Results of the Acoustic Emission Working Groups

Through the forum provided by these three groups, an international exchange of information and cooperation has developed for addressing specific problem areas. Since the organization of these working groups, there has been a marked increase in all areas of acoustic emission technology, including: (1) research into the phenomenon itself; (2) the application of acoustic emission in general materials research; and, most importantly here, (3) the application of acoustic emission techniques to the broad field of nondestructive testing.

This volume of the *Nondestructive Testing Handbook* contains many chapters that were written and reviewed by the pioneers recognized in this introduction. Because of their participation, the *NDT Handbook* shares with the AE working groups this goal: to compile and present data vital to the development and application of acoustic emission technology.

REFERENCES

1. "Pottery." *The New Encyclopaedia Britannica*, fifteenth edition. Vol. 14. Chicago, IL: Encyclopaedia Britannica, Inc. (1984): pp 897.
2. Aitchison, Leslie. *A History of Metals*. Vol. 1, New York, NY: Interscience Publishers, Inc. (1960): pp 78.
3. *The Works of Geber, Englished by Richard Russell, 1678, A New Edition with Introduction by E.J. Holmyard, M.A., D., Litt.* London, England: J.M. Dent and Sons Ltd. and New York, NY: E.P. Dutton and Company (1928): pp 50, 66, 69, 139 and 149.
4. Hooker, Worthington. "Chemistry." *Science for the School and Family, Part II*, second edition. New York, NY: Harper and Brothers Publishers (1882): pp 252.
5. Avery, Elroy. *Elements of Chemistry: A Text-Book for Schools*. New York, NY: Sheldon and Company (1883): pp 256 and 302.
6. Czochralski, J. "Die Metallographie des Zinns und die Theorie der Formänderung bildsamer Metalle." *Metall und Erz*. Vol. XIII (N.F. IV), No. 18 (September 1916): pp 381-393 (in German).
7. *Gmelin-Kraut's Handbuch der anorganischen Chemie*, seventh edition. C. Friedheim and Franz Peters, eds. Vol. IV, Part 1. Heidelberg, Germany: Carl Winter's Universitätsbuchhandlung (1911): pp 244 (in German).
8. Kalischer, S. "Uber die Molekularstruktur der Metalle." *Berichte der Deutschen Chemischen Gesellschaft*. Vol. 15 (1882): pp 702-712 (in German).
9. Muir, M. and H. Morley. *Watts' Dictionary of Chemistry*. Vol. IV. London, England: Longmans, Green and Company (1894): pp 718, 884.
10. Czochralski, J. "Fortschritte der Metallographie: Einfluss der Formänderung (Progress in Metallography: Effect of Deformation)." *Stahl und Eisen*. Vol. 37, No. 21 (May 1917): pp 502-504 (in German).
11. Portevin, Albert and François LeChatelier. "Sur un Phénomène Observé lors de l'Essai de Traction d'Alliages en Cours de Transformation (On the Phenomenon Observed during Tension Tests on Alloys during Transformation)." *Comptes Rendus Hebdomadaires des Seances de l'Academie des Sciences*. Vol. 176 (February 1923): pp 507-510 (in French).
12. Classen-Nekludowa, M. "Uber die sprungartige Deformation (Sudden Deformation of Crystals under Strain)." *Zeitschrift für Physik*. Vol. 55 (1929): pp 555-568 (in German).
13. Anderson, Robert. *An Investigation on the Constitution, Heat Treatment, and Microstructure of Duralumin*." D.S. thesis. Cambridge, MA: Massachusetts Institute of Technology (1925).
14. Anderson, Robert. "Some Mechanical Properties of Duralumin Sheet as Affected by Heat Treatment." *Proceedings of the Twenty-ninth Annual Meeting of the American Society for Testing Materials*. Vol. 26, Part II. Philadelphia, PA: American Society for Testing Materials (1926): pp 349-377.
15. Scheil, Erich. "Uber die Umwandlung des Austenits in Martensit in gehärtetem Stahl (Transformation of Austenite into Martensite in Hardened Steel)." *Zeitschrift für Anorganische und Allgemeine Chemie*. Vol. 183 (1929): pp 98-120 (in German).
16. Bridgman, P. "Effects of High Shearing Stress Combined with High Hydrostatic Pressure." *Physical Review*. Vol. 48 (November 1935): pp 825-847.
17. Bridgman, P. "Flow Phenomena in Heavily Stressed Metals." *Journal of Applied Physics*. Vol. 8 (May 1937): pp 328-336.
18. Barrett, Charles. "A Low Temperature Transformation in Lithium." *Physical Review*. Vol. 72, No. 3 (August 1947): pp 245.
19. Förster, Friedrich and Erich Scheil. "Akustische Untersuchung der Bildung von Martensitnadeln (Acoustic Study of the Formation of Martensite Needles)." *Zeitschrift für Metallkunde*. Vol. 29, No. 9 (September 1936): pp 245-247 (in German).
20. Mason, W., H.J. McSkimin and W. Shockley. "Ultrasonic Observation of Twinning in Tin." *Physical Review*. Vol. 73, No. 10 (May 1948): pp 1,213-1,214.
21. Millard, D. *Twinning in Single Crystals of Cadmium*. Ph.D. thesis. Bristol, England: University of Bristol (1950).
22. Kaiser, Joseph. "Untersuchungen über das Auftreten Geräushen beim Zugversuch (An Investigation into the Occurrence of Noises in Tensile Tests or a Study of Acoustic Phenomena in Tensile Tests)." Ph.D. thesis (in German). Munich, Germany: Technische Hochschule München (1950).
23. Kaiser, J. *Materialprufverfahren (Material Testing Procedure)*. German Patent No. 852,771 (October 1952).
24. Kaiser, Josef. "Erkenntnisse und Folgerungen aus der Messung von Geräuschen bei Zugbeanspruchung von

metallischen Werkstoffen (Information and Conclusions from the Measurement of Noises in Tensile Stressing of Metallic Materials)." *Archiv für das Eisenhüttenwesen.* Vol. 24, No. 1-2 (January-February 1953): pp. 43-45 (in German).
25. Kaiser, J. "Uber das Auftreten von Geräuschen beim Schmelzen und Erstarren von Metallen (Occurrence of Noises during Melting and Solidification of Metals)." *Forschung auf dem Gebiete des Ingenieurwesens.* Vol. 23, No. 1-2 (1957): pp 38-42 (in German).
26. Borchers, Heinz and Joseph Kaiser. "Akustische Effeckte bei Phasenübergängen in System Blei-Zinn (Acoustic Effects during Phase Transformations in the Lead-Tin System)." *Zeitschrift für Metallkunde.* Vol. 49, No. 2 (Febrary 1958): pp 95-101 (in German).
27. Schofield, B., R.A. Bareiss and A.A. Kyrala. *Acoustic Emission under Applied Stress.* WADC Technical Report 58-194. Boston, MA: Lessells and Associates (April 1958).
28. Späth, Wilhelm. *Fliessen und Kriechen der Metalle* (in German). Berlin-Grunewald, Germany: Metall-Verlag GmbH (1955).
29. Joffé, Abram. *The Physics of Crystals.* Leonard Loeb, ed. New York, NY: McGraw Hill Book Company (1928): pp 50.
30. Becker, R. and E. Orowan. "Uber sprunghafte Dehnung von Zinkkristallen (Discontinuous Expansion of Zinc Crystals)." *Zeitschrift für Physik.* Vol. 79 (1932): pp 566-572 (in German).
31. Tatro, C. *Acoustic Emission from Crystalline Materials Subjected to External Loads.* East Lansing, MI: College of Engineering, Michigan State University (April 1960).
32. Tatro, C. and R.G. Liptai. "Acoustic Emission from Crystalline Substances." *Proceedings of the Third Symposium on Physics and Nondestructive Testing.* San Antonio, TX: Southwest Research Institute (October 1962): pp 145-158.
33. Shoemaker, Paul. *Acoustic Emission, An Experimental Method.* M.S. thesis. East Lansing, MI: Michigan State University (1961).
34. Kroll, Robert. *Stress Waves in Test Specimens due to Simulated Acoustic Emission.* Ph.D. thesis. East Lansing, MI: Michigan State University (1962).
35. Liptai, Robert. *An Investigation of the Acoustic Emission Phenomenon.* Ph.D. thesis. East Lansing, MI: Michigan State University (1963).
36. Engle, Robert. *Acoustic Emission and Related Displacements in Lithium Fluoride Single Crystals.* Ph.D. thesis. East Lansing, MI: Michigan State University (1966).
37. McCullough, Benny. *An Experimental Method for the Investigation of Acoustic Emission.* M.S. thesis. New Orleans, LA: Tulane University (1965).
38. Egle, Davis. *A Comprehensive Analysis of an Acoustic Emission Detection System.* Ph.D. thesis. New Orleans, LA: Tulane University (1965).
39. Kerawalla, Jal. *An Investigation of the Acoustic Emission from Commercial Ferrous Materials Subjected to Cyclic Tensile Loading.* Ph.D. thesis. Ann Arbor, MI: University of Michigan (1965).
40. Mitchell, Larry. *An Investigation of the Correlation of the Acoustic Emission Phenomenon with the Scatter in Fatigue Data.* Ph.D. thesis. Ann Arbor, MI: University of Michigan (1965).
41. Agarwal, Anand. *An Investigation of the Behavior of the Acoustic Emission from Metals and a Proposed Mechanism for its Generation.* Ph.D. thesis. Ann Arbor, MI: University of Michigan (1968).
42. Srawley, John and Jack B. Esgar. *Investigation of Hydrotest Failure of Thiokol Chemical Corporation 260-Inch-Diameter SL-1 Motor Case.* NASA TM X-1194. Cleveland, OH: Lewis Research Center (January 1966).
43. Dunegan, Harold. *Acoustic Emission: A Promising Technique.* UCID-4643. Livermore, CA: Lawrence Radiation Laboratory (December 1963).
44. Balderston, Harvey. "The Broad Range Detection of Incipient Failure Using the Acoustic Emission Phenomena." *Acoustic Emission.* ASTM STP 505. Philadelphia, PA: American Society for Testing Materials (1972): pp 297-317.
45. Pollock, Adrian. *Stress-Wave Emission during Stress Corrosion Cracking of Titanium Alloys.* Document D1-82-0658. Seattle, WA: Boeing Scientific Research Laboratories (October 1967).
46. Pollock, Adrian. *Acoustic Emission from Solids Undergoing Deformation.* Ph.D. thesis. London, England: University of London (April 1970).
47. Parry, D. *Nondestructive Flaw Detection by Use of Acoustic Emissions.* IDO-17230. Idaho Falls, ID: Phillips Petroleum Company (May 1967).
48. Hutton, P., R.N. Ord, H.N. Pedersen and J.C. Spanner. "Crack Detection in Pressure Piping by Acoustic Emission." *Nuclear Safety Quarterly Report* (February, March, April, 1968 for Nuclear Safety Branch of USAEC Division of Reactor Development and Technology). BNWL-885. Richland, WA: Battelle Pacific Northwest Laboratories (October 1968): pp 3.1-3.29.
49. Hutton, P. and J.C. Spanner. *Detection of Metal Overstress by Acoustic Emission.* BNWL-SA-438. Richland, WA: Battelle Pacific Northwest Laboratories (December 1965).
50. Anderson, T., T. Mulcahey and C. Hsu. *Survey and Status Report on Application of Acoustic Boiling Detection Techniques to Liquid Metal Cooled Reactors.* ANL-7469. Argonne, IL: Argonne National Laboratory (April 1970).
51. Liptai, R., H.L. Dunegan and C.A. Tatro. *Acoustic Emission Generated during Phase Transformations in Metals and Alloys.* UCRL-50525. Livermore, CA: Lawrence Radiation Laboratory (September 1968).

52. Drouillard, T. "Acoustic Emission: A Bibliography for 1970-1972." *Monitoring Structural Integrity by Acoustic Emission*. J.C. Spanner and J.W. McElroy, eds. ASTM STP 571. Philadelphia, PA: American Society for Testing Materials (1975): pp 241-284.
53. Drouillard, Thomas. *Acoustic Emission: A Bibliography with Abstracts*. Frances Laner, ed. New York, NY: IFI/Plenum Data Company (1979).

SECTION 1

FUNDAMENTALS OF ACOUSTIC EMISSION TESTING

Jack C. Spanner, Sr., Battelle Pacific Northwest Laboratory, Richland, Washington

Albert Brown, Lawrence Livermore Laboratories, Livermore, California
D. Robert Hay, Tektrend International, Montreal, Quebec, Canada
Vasile Mustafa, Tektrend International, Montreal, Quebec, Canada
Kenneth Notvest, Pensacola, Florida
Adrian Pollock, Physical Acoustics Corporation, Princeton, New Jersey

PART 1
INTRODUCTION TO ACOUSTIC EMISSION TECHNOLOGY

The Acoustic Emission Phenomenon

Acoustic emission is the elastic energy that is spontaneously released by materials when they undergo deformation. In the early 1960s, a new nondestructive testing technology was born when it was recognized that growing cracks and discontinuities in pressure vessels could be detected by monitoring their acoustic emission signals. Although acoustic emission is the most widely used term for this phenomenon, it has also been called *stress wave emission, stress waves, microseism, microseismic activity* and *rock noise*.

Formally defined, acoustic emission is "the class of phenomena where transient elastic waves are generated by the rapid release of energy from localized sources within a material, or the transient elastic waves so generated."[1] This is a definition embracing both the process of wave generation and the wave itself.

Source Mechanisms

Sources of acoustic emission include many different mechanisms of deformation and fracture. Earthquakes and rockbursts in mines are the largest naturally occurring emission sources. Sources that have been identified in metals include crack growth, moving dislocations, slip, twinning, grain boundary sliding and the fracture and decohesion of inclusions. In composite materials, sources include matrix cracking and the debonding and fracture of fibers. These mechanisms typify the classical response of materials to applied load.

Other mechanisms fall within the definition and are detectable with acoustic emission equipment. These include leaks and cavitation; friction (as in rotating bearings); the realignment or growth of magnetic domains (Barkhausen effect); liquefaction and solidification; and solid-solid phase transformations. Sometimes these sources are called *secondary sources* or *pseudo sources* to distinguish them from the classic acoustic emission due to mechanical deformation of stressed materials. A unified explanation of the sources of acoustic emission does not yet exist. Neither does a complete analytical description of the stress wave energy in the vicinity of an acoustic emission source. However, encouraging progress has been made in these two key research areas.

Acoustic Emission Nondestructive Testing

Acoustic emission examination is a rapidly maturing nondestructive testing method with demonstrated capabilities for monitoring structural integrity, detecting leaks and incipient failures in mechanical equipment, and for characterizing materials behavior. The first documented application of acoustic emission to an engineering structure was published in 1964 and all of the available industrial application experience has been accumulated in the comparatively short time since then.

Comparison with Other Techniques

Acoustic emission differs from most other nondestructive methods in two significant respects. First, the energy that is detected is released from within the test object rather than being supplied by the nondestructive method, as in ultrasonics or radiography. Second, the acoustic emission method is capable of detecting the *dynamic* processes associated with the degradation of structural integrity. Crack growth and plastic deformation are major sources of acoustic emission. Latent discontinuities that enlarge under load and are active sources of acoustic emission by virtue of their size, location or orientation are also the most likely to be significant in terms of structural integrity.

Usually, certain areas within a structural system will develop local instabilities long before the structure fails. These instabilities result in minute dynamic movements such as plastic deformation, slip or crack initiation and propagation. Although the stresses in a metal part may be well below the elastic design limit, the region near a crack tip may undergo plastic deformation as a result of high local stresses. In this situation, the propagating discontinuity acts as a source of stress waves and becomes an active acoustic emission source.

Acoustic emission examination is nondirectional. Most acoustic emission sources appear to function as point source emitters that radiate energy in spherical wavefronts. Often, a sensor located anywhere in the vicinity of an acoustic emission source can detect the resulting acoustic emission.

This is in contrast to other methods of nondestructive testing, which depend on prior knowledge of the probable location and orientation of a discontinuity in order to direct a beam of energy through the structure on a path that will properly intersect the area of interest.

Advantages of Acoustic Emission Tests

The acoustic emission method offers the following advantages over other nondestructive testing methods:

1. Acoustic emission is a dynamic inspection method in that it provides a response to discontinuity growth under an imposed structural stress; static discontinuities will not generate acoustic emission signals.
2. Acoustic emission can detect and evaluate the significance of discontinuities throughout an entire structure during a single test.
3. Since only limited access is required, discontinuities may be detected that are inaccessible to the more traditional nondestructive methods.
4. Vessels and other pressure systems can often be requalified during an in-service inspection that requires little or no downtime.
5. The acoustic emission method may be used to prevent catastrophic failure of systems with unknown discontinuities, and to limit the maximum pressure during containment system tests.

Acoustic emission is a wave phenomenon and acoustic emission testing uses the attributes of particular waves to help characterize the material in which the waves are traveling. Frequency and amplitude are examples of the waveform parameters that are regularly monitored in acoustic emission tests. Table 1 gives an overview of the manner by which various material properties and testing conditions influence acoustic emission response amplitudes. The factors should generally be considered as indicative, rather than as absolute.

Application of Acoustic Emission Tests

A classification of the functional categories of acoustic emission applications is given below:

1. mechanical property testing and characterization;
2. preservice proof testing;
3. in-service (requalification) testing;
4. on-line monitoring;
5. in-process weld monitoring;
6. mechanical signature analysis;
7. leak detection and location; and
8. geological applications.

By definition, on-line monitoring may be continuous or intermittent, and may involve the entire structure or a limited zone only. Although leak detection and acoustic signature analysis do not involve acoustic emission in the strictest sense of the term, acoustic emission techniques and equipment are used for these applications.

Structures and Materials

A wide variety of structures and materials (metals, nonmetals and various combinations of these) can be monitored

TABLE 1. Factors that affect the relative amplitude of acoustic emission response

Factors That Tend to Increase Acoustic Emission Response Amplitude	Factors That Tend to Decrease Acoustic Emission Response Amplitude
High strength	Low strength
High strain rate	Low strain rate
Low temperature	High temperature
Anisotropy	Isotropy
Nonhomogeneity	Homogeneity
Thick sections	Thin sections
Brittle failure (cleavage)	Ductile failure (shear)
Material containing discontinuities	Material without discontinuities
Martensitic phase transformations	Diffusion-controlled phase transformations
Crack propagation	Plastic deformation
Cast materials	Wrought materials
Large grain size	Small grain size
Mechanically induced twinning	Thermally induced twinning

FROM SPANNER (ACOUSTIC EMISSION TECHNIQUES AND APPLICATIONS). REPRINTED WITH PERMISSION.

by acoustic emission techniques during the application of an external stress (load). The primary acoustic emission mechanism varies with different materials and should be characterized before applying acoustic emission techniques to a new type of material. Once the characteristic acoustic emission response has been defined, acoustic emission tests can be used to evaluate the structural integrity of a component.

Testing of Composites

Acoustic emission monitoring of fiber reinforced composite materials has proven quite effective when compared with other nondestructive testing methods. However, attenuation of the acoustic emission signals in fiber reinforced materials presents unique problems. Effective acoustic emission monitoring of fiber reinforced components requires much closer sensor spacings than would be the case with a metal component of similar size and configuration. With the proper number and location of sensors, monitoring of composite structures has proven highly effective for detecting and locating areas of fiber breakage, delaminations and other types of structural degradation.

Pressure System Tests

Pressure systems are stressed using hydrostatic or some other pressure test. The level of stress should normally be held below the yield stress. Bending stresses can be introduced to beamed structures. Torsional stresses can be generated in rotary shafts. Thermal stresses may be created locally. Tension and bending stresses should either be unilateral or cyclic to best simulate service induced stresses.

Successful Applications

Examples of proven applications for the acoustic emission method include those listed below.

1. Periodic or continuous monitoring of pressure vessels and other pressure containment systems to detect and locate active discontinuities.
2. Detection of incipient fatigue failures in aerospace and other engineering structures.
3. Monitoring materials behavior tests to characterize various failure mechanisms.
4. Monitoring fusion or resistance weldments during welding or during the cooling period.
5. Monitoring acoustic emission response during stress corrosion cracking and hydrogen embrittlement susceptibility tests.

Acoustic Emission Testing Equipment

Equipment for processing acoustic emission signals is available in a variety of forms ranging from small portable instruments to large multichannel systems. Components common to all systems are sensors, preamplifiers, filters and amplifiers to make the signal measurable. Methods used for measurement, display and storage vary more widely according to the demands of the application. Figure 1 shows a block diagram of a generic four-channel acoustic emission system.

Acoustic Emission Sensors

When an acoustic emission wavefront impinges on the surface of a test object, very minute movements of the surface molecules occur. A sensor's function is to detect this mechanical movement and convert it into a specific, usable electric signal.

The sensors used for acoustic emission testing often resemble an ultrasonic search unit in configuration and generally utilize a piezoelectric transducer as the electromechanical conversion device. The sensors may be resonant or broadband. The main considerations in sensor selection are (1) operating frequency; (2) sensitivity; and (3) environmental and physical characteristics. For high temperature tests, waveguides may be used to isolate the sensor from the environment. This is a convenient alternative to the use of high temperature sensors. Waveguides have also been used to precondition the acoustic emission signals as an interpretation aid.

Issues such as wave type and directionality are difficult to handle in this technology, since the naturally occurring acoustic emission contains a complex mixture of wave modes.

Preamplifiers and Frequency Selection

The preamplifier must be located close to the sensor. Often it is actually incorporated into the sensor housing. The preamplifier provides required filtering, gain (most commonly 40 dB) and cable drive capability. Filtering in the preamplifier (together with sensor selection) is the primary means of defining the monitoring frequency for the acoustic emission test. This may be supplemented by additional filtering at the mainframe.

Choosing the monitoring frequency is an operator function, since the acoustic emission source is essentially wide band. Reported frequencies range from audible clicks and squeaks up to 50 MHz.

Although not always fully appreciated by operators, the observed frequency spectrum of acoustic emission signals is significantly influenced by the resonance and transmission

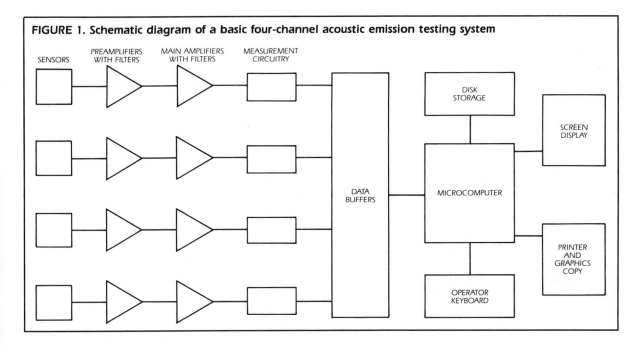

FIGURE 1. Schematic diagram of a basic four-channel acoustic emission testing system

characteristics of both the specimen (geometry as well as acoustic properties) and the sensor. In practice, the lower frequency limit is governed by background noise; it is unusual to go below 10 kHz except in microseismic work. The upper frequency limit is governed by wave attenuation that restricts the useful detection range; it is unusual to go above 1 MHz. The single most common frequency range for acoustic emission testing is 100 to 300 kHz.

System Mainframe

The first elements in the mainframe are the main amplifiers and thresholds, which are adjusted to determine the test sensitivity. Main amplifier gains in the range of 20 to 60 dB are most commonly used. Thereafter, the available processing depends on the size and cost of the system. In a small portable instrument, acoustic emission events or threshold crossings may simply be counted and the count then converted to an analog voltage for plotting on a chart recorder. In more advanced hardware systems, provisions may be made for energy or amplitude measurement, spatial filtering, time gating and automatic alarms.

Acoustic Emission System Accessories

Accessory items often used in acoustic emission work include oscilloscopes, transient recorders and spectrum analyzers, magnetic tape recorders, rms voltmeters, special calibration instruments, and devices for simulating acoustic emission. A widely accepted simulator is the Hsu-Nielsen source, a modified draftsman's pencil that provides a remarkably reproducible simulated acoustic emission signal when the lead is broken against the test structure.

Microcomputers in Acoustic Emission Test Systems

Signal Processing and Displays

Nearly all modern acoustic emission systems use microcomputers in various configurations, as determined by the system size and performance requirements. In typical implementations, each acoustic emission signal is measured by hardware circuits and the measured parameters are passed through the central microcomputer to a disk file of signal descriptions. The customary signal description includes the threshold crossing counts, amplitude, duration, rise time and often the energy of the signal, along with its time of occurrence and the values of slowly changing variables such as load and background noise level.

During or after data recording, the system extracts data for graphic displays and hardcopy reports. Common displays include history plots of acoustic emission versus time or load, distribution functions, crossplots of one signal descriptor against another and source location plots. Installed systems of this type range in size from 4 to 128 channels.

Operator Training and System Uses

Microcomputer based systems are usually very versatile, allowing data filtering (to remove noise) and extensive post-test display capability (to analyze and interpret results). This versatility is a great advantage in new or difficult applications, but it places high demands on the knowledge and technical training of the operator. Other kinds of equipment have been developed for routine industrial application in the hands of less highly trained personnel.

Examples are the systems used for bucket truck testing (providing preprogrammed data reports in accordance with ASTM recommended practices) and systems for resistance weld process control (these are inserted into the current control system and terminate the welding process automatically as soon as expulsion is detected).

Acoustic emission equipment was among the first nondestructive testing equipment to make use of computers in the late 1960s. Performance, in terms of acquisition speed and real-time analysis capability, has been much aided by advances in microcomputer technology. Trends expected in the future include advanced kinds of waveform analysis, more standardized data interpretation procedures and more dedicated industrial products.

Characteristics of Acoustic Emission Techniques

The acoustic emission test is a passive method that monitors the dynamic redistribution of stresses within a material or component. Therefore, acoustic emission monitoring is only effective while the material or structure is subjected to an introduced stress. Examples of these stresses include pressure testing of vessels or piping, and tension loading or bend loading of structural components.

Irreversibility and the Kaiser Effect

An important feature affecting acoustic emission applications is the generally irreversible response from most metals. In practice, it is often found that once a given load has been applied and the acoustic emission from accommodating that stress has ceased, additional acoustic emission will not occur until that stress level is exceeded, even if the load is completely removed and then reapplied. This often useful (and sometimes troublesome) behavior has been named the *Kaiser effect* in honor of the researcher who first reported it.

The degree to which the Kaiser effect is present varies between metals and may even disappear completely after several hours (or days) for alloys that exhibit appreciable room temperature annealing (recovery) characteristics. Some alloys and materials may not exhibit any measurable Kaiser effect at all.

Because of the Kaiser effect, each acoustic signal may only occur once so that inspections have a now-or-never quality. In this respect, acoustic emission is at a disadvantage when compared to techniques that can be applied again and again, by different operators or with different instruments, without affecting the structure or the discontinuity.

The outcome in practical terms is that acoustic emission must be used at carefully planned times: during proof tests, before plant shutdowns or during critical moments of continuous operation. This seeming restriction sometimes becomes the biggest advantage of the acoustic emission techniques. By using acoustic emission during service, production can continue uninterrupted. Expensive and time consuming processes such as the erection of scaffolding and extensive surface preparation can be completely avoided.

Acoustic Emission Test Sensitivity

Although the acoustic emission method is quite sensitive, compared with other nondestructive methods such as ultrasonic testing or radiographic testing, the sensitivity decreases with increasing distances between the acoustic emission source and the sensors. The same factors that affect the propagation of ultrasonic waves also affect the propagation of the acoustic (stress) waves used in acoustic emission techniques.

Wave mode conversions at the surfaces of the test object and other acoustic interfaces, combined with the fact that different wave modes propagate at different velocities, are factors that complicate analysis of acoustic emission response signals and produce uncertainties in calculating acoustic emission source locations with triangulation or other source locating techniques.

Background Noise and Material Properties

In principle, overall acoustic emission system sensitivity depends on the sensors as well as the characteristics of the specific instrumentation system. In practice, however, the sensitivity of the acoustic emission method is often primarily limited by ambient background noise considerations for engineering materials with good acoustic transmission characteristics.

When monitoring structures made of materials that exhibit high acoustic attenuation (due to scattering or absorption), the acoustic properties of the material usually limit the ultimate test sensitivity and will certainly impose limits on the maximum sensor spacings that can be used.

Effects of System Sensors

Sensor coupling and reproducibility of response are important factors that must be considered when applying multiple acoustic emission sensors. Careful calibration checks should be performed before, after and sometimes during the acoustic emission monitoring process to ensure that all channels of the instrumentation are operating properly at the correct sensitivities.

For most engineering structures, sensor selection and placement must be carefully chosen based on a detailed knowledge of the acoustic properties of the material and the geometric conditions that will be encountered. For example, the areas adjacent to attachments, nozzles and penetrations or areas where the section thickness changes usually require additional sensors to achieve adequate coverage. Furthermore, discontinuities in such locations often cause high localized stress and these are the areas where maximum coverage is needed.

Interpretation of Test Data

Proper interpretation of the acoustic emission response obtained during monitoring of pressurized systems and other structures usually requires considerable technical knowledge and experience with the acoustic emission method. Close coordination is required between the acoustic emission system operators, the data interpretation personnel and those controlling the process of stressing the structure.

Since most computerized, multichannel acoustic emission systems handle response data in a pseudo batch procedure, an intrinsic dead time occurs during the data transfer process. This is usually not a problem but can occasionally result in analysis errors when the quantity of acoustic emission signals is sufficient to overload the data handling capabilities of the acoustic emission system.

Compensating for Background Noise

When acoustic emission monitoring is used during hydrostatic testing of a vessel or other pressure system, the acoustic emission system will often provide the first indication of leakage. Pump noise and other vibrations, or leakage in the pressurizing system, can also generate background noise that limits the overall system sensitivity and hampers accurate interpretation.

Special precautions and fixturing may be necessary to reduce such background noise to tolerable levels. Acoustic emission monitoring of production processes in a manufacturing environment involves special problems related to the high ambient noise levels (both electrical and acoustical). Preventive measures may be necessary to provide sufficient electrical or acoustical isolation to achieve effective acoustic emission monitoring.

Various procedures have been used to reduce the effects of background noise sources. Included among these are mechanical and acoustic isolation; electrical isolation; electronic filtering within the acoustic emission system; modifications to the mechanical or hydraulic loading process; special sensor configurations to control electronic gates for noise blocking; and statistically based electronic countermeasures including autocorrelation and cross correlation.

The Kaiser Effect

Josef Kaiser is credited as the founder of modern acoustic emission technology and it was his pioneering work in Germany in the 1950s that triggered a connected, continuous flow of subsequent development. He made two major discoveries. The first was the near universality of the acoustic emission phenomenon. He observed emission in all the materials he studied. The second was the effect that bears his name.

In translation of his own words: "Tests on various materials (metals, woods or mineral materials) have shown that low level emissions begin even at the lowest stress levels (less than 1 MPa or 100 psi). They are detectable all the way through to the failure load, but only if the material has experienced no previous loading. This phenomenon lends a special significance to acoustic emission investigations, because by the measurement of emission during loading a clear conclusion can be drawn about the magnitude of the maximum loading experienced prior to the test by the material under investigation. In this, the magnitude and duration of the earlier loading and the time between the earlier loading and the test loading are of no importance."[2]

This effect has attracted the attention of acoustic emission workers ever since. In fact, all the years of acoustic emission research have yielded no other generalization of comparable power. As time went by, both practical applications and controversial exceptions to the rules were identified.

The Dunegan Corollary

The first major application of the Kaiser effect was a test strategy for diagnosing damage in pressure vessels and other engineering structures.[3] The strategy included a clarification of the behavior expected of a pressure vessel subjected to a series of loadings (to a proof pressure with intervening periods at a lower working pressure).

Should the vessel suffer no damage during a particular working period, the Kaiser effect dictates that no emission will be observed during the subsequent proof loading. In the event of discontinuity growth during a working period, subsequent proof loading would subject the material at the discontinuity to higher stresses than before and the discontinuity would emit. Emission during the proof loading is therefore a measure of damage experienced during the preceding working period.

This so-called *Dunegan corollary* became a standard diagnostic approach in practical field testing. Field operators learned to pay particular attention to emission between the working pressure and the proof pressure, and thereby made many effective diagnoses. A superficial review of the Kaiser effect might lead to the conclusion that practical application of acoustic emission techniques requires a series of ever increasing loadings. However, effective engineering diagnoses can be made by repeated applications of the same proof pressure.

The Felicity Effect

The second major application of the Kaiser effect arose from the study of cases where it did not occur. Specifically in fiber reinforced plastic components, emission is often observed at loads lower than the previous maximum, especially when the material is in poor condition or close to failure. This breakdown of the Kaiser effect was successfully used to predict failure loads in composite pressure vessels[4] and bucket truck booms.[5]

The term *felicity effect* was introduced to describe the breakdown of the Kaiser effect and the *felicity ratio* was devised as the associated quantitative measure. The felicity ratio has proved to be a valuable diagnostic tool in one of the most successful of all acoustic emission applications, the testing of fiberglass vessels and storage tanks.[6] In fact, the Kaiser effect may be regarded as a special case of the felicity effect (a felicity ratio of 1).

The discovery of cases where the Kaiser effect breaks down was at first quite confusing and controversial but eventually some further insights emerged. The Kaiser effect fails most noticeably in situations where time dependent mechanisms control the deformation. The rheological flow or relaxation of the matrix in highly stressed composites is a prime example. Flow of the matrix at loads below the previous maximum can transfer stress to the fibers, causing them to break and emit. Other cases where the Kaiser effect will fail are corrosion processes and hydrogen embrittlement, which are also time dependent.

The Kaiser Principle

Further insight can be gained by considering load on a structure versus stress in the material. In practical situations, test specimens or engineering structures experience loading and most discussions of the Kaiser effect come from that context. But actually, Kaiser's idea applies more fundamentally to stress in a material. *Materials emit only under unprecedented stress* is the root principle to consider. Evaluated point-by-point through the three-dimensional stress field within the structure, this principle has wider truth than the statement that structures emit only under unprecedented load. Provided that the microstructure has not been altered between loadings, the Kaiser principle may even have the universal validity that the Kaiser effect evidently lacks, at least for active deformation and discontinuity growth.

In composite materials, an important acoustic emission mechanism is friction between free surfaces in damaged regions. Frictional acoustic emission is also observed from fatigue cracks in metals. Such source mechanisms contravene both the Kaiser effect and the Kaiser principle, but they can be important for practical detection of damage and discontinuities.

Overview of Acoustic Emission Methodology

This *Nondestructive Testing Handbook* volume contains detailed descriptions of acoustic emission sources, a rich topic that involves the sciences of materials, deformation and fracture. Another topic appearing in this volume is the subject of wave propagation, the process that shapes the signal and brings the information from source to sensor. Attenuation of the wave determines its detectability and must therefore be considered when placing sensors; knowledge of the wave's velocity is also needed for precise source location. These are uncontrolled factors that must be assessed for each structure tested.

Measurement and analysis of the acoustic emission signals is another major component of the technology covered in this book. Acoustic emission signals from individual deformation events may be so rare that a single detected event is enough to warrant rejection of the object under test. Or, they may be so frequent that the received acoustic signal is virtually continuous. Compounding interpretation difficulties are amplitudes of the received signals that range over five orders of magnitude. The time differences used to locate acoustic emission sources range from less than a microsecond to hundreds of milliseconds. In addition to handling all of these variables, an acoustic emission system should allow any of several techniques for reducing background noise and spurious signals that often interfere with acoustic emission measurements.

The acoustic emission technology comprises a range of

powerful techniques for exploiting the natural acoustic emission process and for gaining practical value from the available information. These techniques include methods for characterizing the acoustic emission from particular materials and processes; methods for eliminating noise; for checking wave propagation properties of engineering structures and applying the results to test design; for loading that will optimize the acoustic emission data from a structure without causing appreciable damage; for locating acoustic emission sources, either roughly or precisely; methods of data analysis and presentation; and methods for acceptance, rejection or further inspection of the test structure.

The field of acoustic emission testing is still growing vigorously and presents many challenges. Significant research questions are still unanswered. The mathematical theory of the acoustic emission source has been developed beginning in the mid 1970s and the practical application of this theory is still in the early stages. Advanced techniques for discontinuity characterization by waveform analysis are very promising, but it remains to be determined whether they will significantly affect the way practical acoustic emission testing is performed.

The technology lacks universal frameworks for the description of material emissivities and the interpretation of structural test data. There is a constant need to improve instrumentation performance and noise rejection techniques as acoustic emission is pressed into service in tougher environments and more demanding applications. Code acceptance is continuing but slowly. Perhaps most of all, there is a major need to provide useful information in assimilable form to the many nonspecialists who have a use for acoustic emission testing but find the subject difficult to approach. This volume of the *Nondestructive Testing Handbook* is one way of satisfying that important need.

PART 2
INTRODUCTION TO ACOUSTIC EMISSION INSTRUMENTATION

Choosing an Acoustic Emission Testing System

Financial Considerations

As with any purchase of technical apparatus, choosing an acoustic emission system is a process that includes compromises between cost and performance. The total cost of a system is determined by several factors: (1) the sophistication of the system's software; (2) the value and reliability of the manufactured components; (3) the resources needed to support the nondestructive test; (4) the amount of data required; and (5) the kind of information needed and how it is presented.

Performance Options

Because acoustic emission systems are used in a number of ways, manufacturers provide instrumentation with a variety of options, ranging from simple display options (analog, digital or a combination) to options in size and cost that extend over several orders of magnitude.

Below is a list of generalized questions whose answers will help define the kind of acoustic emission system best suited for a particular application.

1. Is the test primarily to determine incipient failure?
2. Is a potential discontinuity in the part detectable during operating stresses but undetectable otherwise?
3. Is there a metallurgical problem that must be addressed (such as weld or fastener integrity)?
4. Is there damage caused by the manufacturing technique that could be detected during the machining or forming processes?
5. Can the instrumentation be used to materially reduce waste in manufacturing?
6. Is the instrumentation expected to operate in harsh industrial areas or special environments (explosive or high temperature)?
7. Will the instrumentation be used for medical purposes?
8. Will ancillary sources of sound (background noise) obscure the desired acoustic emission sources?
9. Is data analysis necessary for interpretation of test results?

Expanding the Basic Acoustic Emission Testing System

Acoustic emission instrumentation in its most basic form consists of a detector device (sensor) and an oscilloscope. The sensor is acoustically coupled to the test object and electrically coupled to the oscilloscope. The acoustic emission operator watches for an identifiable set of signals emanating from the test part while the part is under an induced stress. These stresses may be thermal, residual, corrosive or mechanical. Occasionally the test part is subjected to dangerously high stresses.

Signal losses in the cable connecting the sensor with the oscilloscope may become excessive, requiring the addition of a preamplifier close to or integral with the sensor. Fixture noise and other spurious electrical or acoustic signals may then become a problem that is overcome through specific filtering techniques.

Attenuation or additional amplification of the signals may be required, depending on the character of the acoustic emission source. Additionally, care must be exercised when selecting the continuous range of signal amplitudes that will be accepted or rejected. Acoustic emission from metal, wood, plastic and biological sources generate signals ranging from one microvolt to one thousand volts. The dynamic range of acoustic emission signal amplitudes from a test object may be 120 dB(V), or a ratio of one million to one. A compromise must be made to choose those amplitudes that fit within the linear range of the instrumentation.

Placement and acoustic coupling of sensors is probably the most exacting challenge in an acoustic emission test and they are aspects of the technique that affect many levels of the test method, including instrumentation. Sensors must be placed on part areas that will not be acoustically shadowed from the source. Equal care must be given to the number of interfaces (acoustic impedances) permitted between the source and sensor. The material under test is intrinsically part of the instrumentation because the material acts as a mechanical filter and a waveguide.

It is important that new test data be related to the recent history of similar signals. Recording devices are needed to ensure that interpretation does not rely on subjective human memory. After conditioning, signals are processed for storage in electronic memory systems.

Computers in Acoustic Emission Systems

Source Location

Before an acoustic emission test can begin, signal calibration is required to establish the sensitivity and selectivity of the acoustic emission instruments and to ensure the repeatability of source location. To narrow an area of acoustic emission activity to a specific location, additional acoustic emission channels may be used on the test part.

Acoustic emission event timing becomes critically important when source location is a requirement. Time of arrival, duration of acoustic emission signals and acoustic emission signal amplitude are important data for interactive process control systems that are dependent on gating or windowing.

Computers have fast response times and are capable of very fast calculations. These characteristics produce high event capture rates, some greater than 100,000 acoustic emission events per second, with the entire waveform digitally recorded. Depending on signal durations and the number of on-line graphics and computational iterations, some computers may be able to capture less than 50 events per second.

Other Computer Uses

Computer algorithms have been written for a number of acoustic emission applications, including: (1) one, two and three-dimensional acoustic emission source location schemes; (2) fast Fourier transform (FFT) calculations; and (3) measurement of event amplitude, duration, rise time, signal energy and power. Computer systems also provide for recording of test parameters such as pressure, strain and temperature.

Few acoustic emission systems are capable of recording all of the test data. A quasicontinuous or interrupted flow of acoustic emission will inhibit most computer systems. Therefore, analog acoustic emission instruments are still essential for the capture of quasicontinuous acoustic emission such as leakage noise. The most common analog instruments consist of true rms voltmeters, frequency counters and event counters.

System Calibration

With all of the features provided by modern instrumentation, calibration of signal input and instrument calibrations can still be a formidable problem. Several standards have been published governing the use of acoustic emission acquisition and data reduction. Among these are documents from the American Society for Testing Materials, including one that specifically treats the calibration of acoustic emission instrumentation (E750).

However, with such a wide selection of instrumentation, no single document can cover all possible contingencies. Sufficiently detailed information on the acoustic emission instrument calibration is often difficult to obtain. The end user has the ultimate responsibility for maintaining proper calibration of the acoustic emission system. Acoustic emission instrument manufacturers strive to produce systems with better quality, higher efficiency and faster speeds, yet more needs to be done to improve the assurance that instruments in the field are properly calibrated.

Greater complexity of instrumentation provides the chance for a greater number of malfunctions. Acoustic emission test personnel need a program to certify the operation of computer hardware and software combinations. Such tests should be designed for field use.

Pressure Vessel Applications

Using acoustic emission systems for monitoring pressurized containers has been met with varying degrees of success. Generally, thin walled vessels are easier to test than thick walled vessels (the locational accuracy for monitoring thick walled vessels is limited to one or two wall thicknesses). The variety of acoustic emission instruments configured to monitor steel and fiber reinforced plastic vessels attests to the complexity of this application.

Special attention is required by (1) the number and location of nozzles, ports and stanchions; and (2) the frequency and spatially dependent attenuation of sound within the vessel. Acoustic emission sensors are often required beyond the number anticipated during preliminary investigations.

The frequency filtering, spatial filtering and amplification required for each individual sensor channel and sensor array may be identical channel to channel or unique to the channel, with consequent effect on the instrumentation configuration or design.

Need for Understanding the System

The acoustic emission testing technician must first be knowledgeable about the test system and must also understand its reaction to acoustic impedance and wave refraction, diffraction and reflection for the specific material and structure under test. In addition, the technician must be able to recognize noise sources, ground loops, instrument malfunction and inappropriate loading control (an acoustic emission test of a pressure vessel may be of such a critical nature that the technician is responsible for terminating the test to eliminate catastrophic consequences).

Especially critical is an understanding of the acoustic emission system's computer. The acoustic emission system operator needs to know a number of performance related characteristics, including:

1. the timing sequences and relationships governing the recording of specific acoustic emission event time;

2. the identification of an acoustic emission event and the derived parameters;

3. the maximum capacity of the buffers in terms of acoustic emission events;

4. the frequency and timing of data transfer to buffers and buffer transfer to the disk;

5. the time-out intervals for internal data transfer;

6. the effects these data transfers have on a continuing incoming data stream; and

7. the relative timing for the recording of additional parametric test data.

Timing becomes critically important when mechanical sequences are monitored with acoustic emission systems and the acoustic emission data must be correlated with the unique sequencing. For example, problems in data interpretation may be caused by some computer systems depending on how the computer stores and reads the buffers and registers. The event parameter registers may continue to update, following the computer's determined end of an event, until the registers are written to the buffer and the registers are finally reset.

The computer may also require additional processing time to complete the transfer of acoustic emission parameters to the buffer and more time to transfer these data to a mass storage device (disk or tape). The time of the event may be delayed until all of the acoustic emission event parameters are transferred to the buffer, causing the time of event to be increased by the event duration time, or more.

Spot Weld Monitoring Applications

Weld monitoring requires correlation between the acoustic emission system data and the metallurgical conditions during the entire weld process. For instance, no metal expulsion is expected until the metal is molten. Acoustic emission examination should be able to isolate whether there is metal cracking during metal chill or coincident with thermal shock. The ability of the system to identify and categorize the acoustic emission signals with the production welding process is vital to the success of a production line.

Windowing and gating the acoustic emission signals relative to the melt, to the chill and for intervals between weld pulses is usually mandatory during pulsed or interrupted welding. Initial contact time and preheat intervals are usually gated out of the acoustic emission records and the system must be designed accordingly.

It should be noted that no reference is made to location detection. If the acoustic emission technique can identify a problem with production control and provide some indication of the nature of the problem, then the location becomes secondary.

The Value of Acoustic Emission Instrumentation

If an acoustic emission system can separate good from faulty parts, the number of parts requiring further investigation will be substantially reduced. When acoustic emission testing is initiated, it is often in response to a lack of production control. Frequently, during the early phase of acoustic emission testing, the system's data will cause a great deal of excitement. Questions will arise regarding the authenticity of the data produced and the manufacturing quality. Faith in the quality of the production technique may have to be subverted in favor of the authenticity of the test results, as supported by hard evidence.

The value of the acoustic emission instrumentation is meaningless if the test is not performed. Manufactured parts under production control generally produce notably uniform acoustic emission data. The continuing absence of dramatic changes in the acoustic emission data may result in neglecting to perform acoustic emission tests.

PART 3
INTRODUCTION TO THE EFFECTS OF BACKGROUND NOISE IN ACOUSTIC EMISSION TESTS

Effects of Hydraulic and Mechanical Noise

Acoustic emission monitoring must always recognize the presence of noise (extraneous or interfering acoustic signals carrying no data of interest). Noise signals may be continuous or intermittent and the source may be either internal or external to the test object. Noise sources should be examined to discriminate between noise and relevant acoustic emission signals. The most effective remedy for noise interference is to identify the noise sources and then remove or inhibit them.

Most acoustic emission monitoring is done using a frequency passband above 100 kHz. Fortunately, most acoustic noise diminishes in amplitude at these higher frequencies. Less noise is detected as the frequency passband becomes narrow and higher, but these frequencies can also inhibit acoustic emission detection. An unacceptable remedy for noise interference is to reduce the instrument gain or raise the detection threshold because this will also exclude valid acoustic emission signals.

As a rule, an indication of the level of interference from electrical or mechanical noise can be obtained by noting the noise level of the acoustic emission system at the test site with the sensors on the test structure and comparing this to the ambient level in a low noise environment. It is important to test for noise in an operating plant, including machinery that will be active during the acoustic emission test. Noise should be measured in terms of peak or peak-to-peak amplitude, because this relates most directly to acoustic emission detection. Other measurement parameters such as root mean square (rms) will generally not recognize noise spikes or very short duration intermittent signals.

Hydraulic Noise

An example of hydraulic noise may occur during acoustic emission monitoring of a hydrostatic pressure vessel test. If the pressure system develops a leak, the leak will result in continuous high amplitude noise that can completely obscure acoustic emission from a discontinuity. In addition to the noise from a leak, there are other noise sources such as boiling of a liquid, cavitation and turbulent fluid flow.

One modern servo hydraulic testing machine can present a serious noise problem during acoustic emission tests.[7] If the valves controlling the hydraulic loading are integral with the yoke of the machine, hydraulic noise will be transmitted to the acoustic emission sensor through the load train. The source of the noise in this machine is the cavitating hydraulic fluid in narrow channels of the load piston and cylinder. Even the high frequency content is often coupled to the specimen. This can be controlled by separating the servo controller from the test frame with high pressure hoses.

Leaking air lines in the test vicinity can also be a source of continuous noise. Turbulent flow or cavitation can also be a serious problem during continuous monitoring of pressurized systems. One cure is to operate the acoustic emission monitoring system in a frequency range above the noise frequency.[8-11]

Mechanical Noise

Any movement of mechanical parts in contact with the structure under test is a potential source of noise. Roller assemblies with spalled bearings or races are examples of noisy parts. Mechanical noise *can* be beneficial when acoustic emission techniques are applied to detect incipient mechanical failure.[12]

Fretting

Rubbing or fretting is a particularly difficult noise source to overcome because it often has a very broad frequency content. A pressure vessel under hydrostatic test will expand elastically in response to the imposed stress and may rub against its supports. Guard sensors can sometimes be employed to identify and obviate this noise source. Riveted, pinned or bolted structures are notoriously noisy. Initially, bearing points on a pinned structure are limited and as these points are loaded to yield stress, the joint will shift to pick up the load and simultaneously produce bursts of noise.

Cyclic Noise

Repetitive noise such as that from reciprocating or rotating machinery can sometimes be rejected by blanking the acoustic emission instrument during the noise bursts. This is especially effective if the repetition rate is low and uniform relative to acoustic emission. Examples of cyclic noise include: positioning of spot welding electrodes; clamping of sheets before shearing; and fatigue testing machines.

Serrated wedge grips used on many physical testing machines are nearly always a source of noise. As the specimen is loaded, two sources of noise can occur: (1) the serrations of the grips bite into the specimen; and (2) the grips slip in the yokes of the testing machine. Pinned load connections will usually alleviate this problem.

Noise Signals and Rise Time

Mechanical noise has characteristics that help distinguish it from the acoustic emission signals of cracks. Noise signals are relatively low frequency with low rise time. The acoustic emission bursts from cracks generally have rise times (from threshold to peak) less than 25 microseconds if the sensor is near the source. Mechanical noise rarely has such a fast rise time.

The rise time of both noise and acoustic emission signals increases with source-to-sensor distance because of the attenuation of the high frequency components. In some cases, a rise time discriminator can be effective in isolating acoustic emission from mechanical noise. Frequency discrimination is usually a more reliable approach.

Control of Noise Sources

When multichannel systems are used for source location, noise signals from outside the zone of coverage can be controlled using signal arrival times (Δt) at the various sensors in the array. To accomplish this, the signal density must be relatively low (less than 10 per second). At higher signal densities, the noise and acoustic emission signals may interact to produce erroneous Δt values.

Another technique for noise rejection is called the *master-slave technique*. Master sensors are mounted near the area of interest and are surrounded by slave or guard sensors relatively remote from the master. If the guard sensors detect a signal before the master sensors, then the event occurred outside the area of interest and is rejected by the instrument circuitry.

Noise Associated with Welding

Effect of Grounding on Welding Noise

The terminals of most welding machines are not grounded to facilitate reversing the polarity. Thus a structure being welded is said to be *floating* with respect to ground and this serves to aggravate the acoustic noise problem. One simple solution for this is to remove the acoustic emission instrument's ground from the main voltage supply and to then connect it along with a blocking capacitor to the structure under test. This brings the welding machine and the structure to the same noise potential and greatly reduces the detected noise. *This procedure may not be legal in some locations.*

Another solution is to use a battery powered acoustic emission system grounded to the structure. An isolation transformer between the instrument and the power supply is also effective. Similarly, care should be taken to ensure that contact between different parts of the acoustic emission system (sensor, preamplifier, main module) and different parts of the structure do not result in different ground potentials or ground loops. Ground potential differences of less than one millivolt can cause noise.

Welding Procedures and Resulting Noise

Metal arc welding gives off spatter that strikes the weldment and may cause appreciable noise. The noise may be enough to make acoustic emission detection impossible with the arc on. If the welding process produces slag on top of the bead, the slag should be removed to enhance acoustic emission detection. Slag fracturing is a true acoustic emission source having no effect on the quality of the weld, but it can obscure acoustic emission originating from within the weld.

Welding requires heat which causes thermal expansion, followed by contraction and warpage with cooling. These movements, especially over a gritty surface, cause random noise bursts until the weldment reaches ambient temperature. Good welding practice is to wipe the weld piece, parts and work table clean before assembly.

Considerable welding is done on hot rolled steel with the mill scale not removed. Such scale (iron oxide) has a different coefficient of thermal expansion than the steel and being brittle tends to fracture with heating and cooling.

Figure 2 shows the acoustic emission detected after heating the edge of a 25 mm (1 in.) plate to 150 °C (300 °F). In this case, the acoustic emission instrument's gain was set at 60 dB. Acoustic emission count rate peaked at 7,500 counts per second when heating was stopped. The count dropped exponentially to zero after four minutes. The best way to

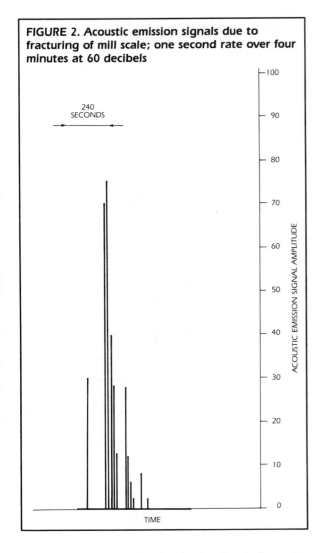

FIGURE 2. Acoustic emission signals due to fracturing of mill scale; one second rate over four minutes at 60 decibels

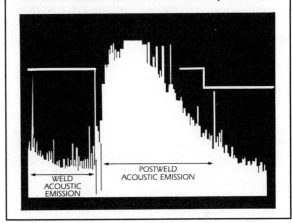

FIGURE 3. Postweld acoustic emission signals from martensitic transformation in a spot weld

eliminate the noise was found to be descaling before welding. However, when this is not possible, recently developed acoustic emission signal pattern recognition methods may be effective in isolating the valid acoustic emission signals in the presence of scale noise.

When welding alloy and high carbon steels, a transformation to martensite is possible. Martensitic transformation is a true and energetic source of acoustic emission and is caused by rapid cooling. Figure 3 is a plot of acoustic emission from spot welding of high carbon steel. The square wave in the upper left portion of the illustration indicates weld power duration. The associated acoustic emission is from nugget formation. When the weld interval ended, the nugget cooled rapidly, causing (1) martensitic transformation and (2) the large envelope of acoustic emission during the immediate postweld interval.[13]

Noise from Solid State Bonding

When clean metallic surfaces are brought into intimate contact, some degree of solid state bonding is possible. This effect increases with temperature and pressure. Such incidental bonding usually does not have much strength and may rupture at the first change of stress, creating a new free surface and a stress wave that is true but undesired acoustic emission.

Fretting is an example of this phenomenon. An underwelded spot weld called a *sticker* causes a similar condition. Even a properly welded spot may be surrounded by solid state bonding and will result in acoustic emission when the structure is first loaded. Such acoustic emission is termed secondary emission[14] and may be an indicator of discontinuity growth.

Electromagnetic Interference

Electromagnetic interference (EMI) consists of noise signals coupled to the acoustic emission instrumentation by electrical conduction or radiation. Sources of EMI include fluorescent lamps, electric motor circuits, welding machines and turning on and off electrical power by relays with inductive loads. Large spikes of electrical noise are thus created and may contain the highest amplitude and frequencies of any signals detected.

The extent to which this is encountered is a function of the environment, shielding and the sensor design. Hostile environments often require better shielding and differential sensors. Optimum standard shielding can be provided by enclosing the sensors, preamplifiers and main amplifiers in high conductivity metal cases, by grounding the cases at a common point, or by using special leads screened with alternate layers of metal and copper.[15]

Commercial acoustic emission instruments usually include features to reject some electrical noise on the supply line and noise radiated as EMI. Use of an isolation transformer on the power supply can also reduce electromagnetic interference. A major problem can arise if the test structure has a ground system different from the acoustic emission instrumentation.

The potential difference between the structure and the acoustic emission system can be substantial and may produce high frequency noise spikes much greater than the valid signals detected by the sensors. No instrument can reject this magnitude of noise. Even a sensor insulated from the structure can pick up capacitively coupled noise. A differential sensor will provide considerable relief from such noise.

Miscellaneous Noise

Metals such as aluminum give off considerable acoustic emission when in contact with acids during etching or pickling. Some solution annealed aluminum alloys will emit acoustic emission when heated to 190 °C (375 °F) because of precipitation hardening (see Fig. 4).

Conclusions

There is a wide range of potential noise sources associated with acoustic emission monitoring. Before performing acoustic emission tests, it is essential to check for background noise and spurious acoustic emission, and their effects on test results. This check should include all electrical equipment and machinery operating during the acoustic emission test. Noise should then be eliminated, either by mechanically removing the identified noise source from the test setup or by selectively analyzing the acoustic emission test data.

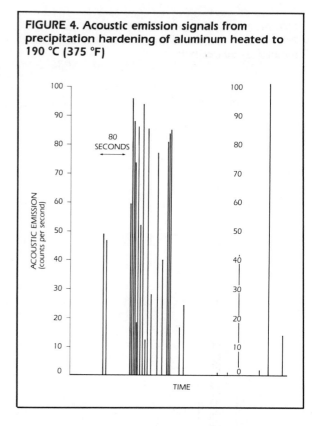

FIGURE 4. Acoustic emission signals from precipitation hardening of aluminum heated to 190 °C (375 °F)

PART 4
INTRODUCTION TO ACOUSTIC EMISSION SIGNAL CHARACTERIZATION

The Purpose of Signal Characterization

The objective of an acoustic emission test is to detect the presence of emission sources and to provide as much information as possible about the source. The technology for detecting and locating sources is well established and acoustic emission signals can provide a large amount of information about the source of the emission and the material and structure under examination.

The purpose of source characterization is to use the sensor output waveform to identify the sources and to evaluate their significance. There is thus a qualitative (source identification) and a quantitative (source intensity or severity) aspect to characterization.

Interpreting the Data

The signal waveform is affected by (1) characteristics of the source; (2) the path taken from the source to the sensor; (3) the sensor's characteristics; and (4) the measuring system. Generally the waveforms are complex and using them to characterize the source can be difficult. Information is extracted by methods ranging from simple waveform parameter measurements to artificial intelligence (pattern recognition) approaches. The former often suffice for simple preservice and in-service tests. The latter may be required for on-line monitoring of complex systems.

In addition to characteristics of the waveforms themselves, there is information available from the cumulative characteristics of the signals (including the amplitude distribution and the total number of signals detected) and from rate statistics (such as rate of signal arrival or energy at the sensor).

There are thus several options available for acoustic emission source characterization. An appropriate approach must be determined for each application.

Characteristics of Discrete Acoustic Emission

Discrete or burst type acoustic emission can be described by relatively simple parameters. The signal amplitude is much higher than the background and is of short duration (a few microseconds to few milliseconds). Occurrences of individual signals are well separated in time. Although the signals are rarely simple waveforms, they usually rise rapidly to maximum amplitude and decay nearly exponentially to the level of background noise. The damped sinusoid in Fig. 5 is often used to represent a burst of acoustic emission.

Acoustic emission monitoring is usually carried out in the presence of continuous background noise. A threshold detection level is set somewhat above the background level (Fig. 6) and serves as a reference for several of the simple waveform properties.

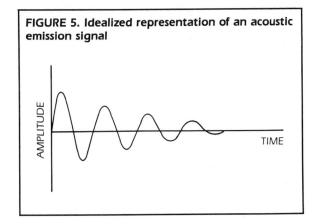

FIGURE 5. Idealized representation of an acoustic emission signal

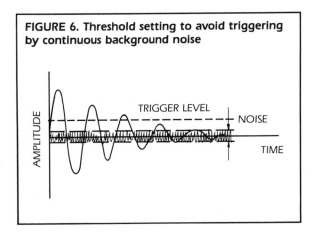

FIGURE 6. Threshold setting to avoid triggering by continuous background noise

Using this model, the waveform parameters in Fig. 7 can be defined:

1. acoustic emission event;
2. acoustic emission count (ringdown count);
3. acoustic emission event energy;
4. acoustic emission signal amplitude;
5. acoustic emission signal duration; and
6. acoustic emission signal rise time.

Cumulative representations of these parameters can be defined as a function of time or test parameter (such as pressure or temperature), including: (1) total events; (2) amplitude distribution; and (3) accumulated energy. Once a specific parameter is selected, rate functions may be defined as a function of time or test parameter: event rate; count rate; and energy rate.

Based on Fig. 7, several methods for characterizing acoustic emission are described below.

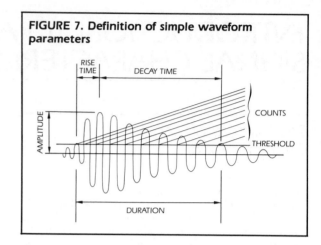

FIGURE 7. Definition of simple waveform parameters

Waveform Parameter Descriptions

Emission Events and Count

Acoustic emission events are individual signal bursts produced by local material changes. The emission count is the number of times a signal crosses a preset threshold. High amplitude events of long duration tend to have many threshold crossings. The number of threshold crossings per unit time depends on (1) the sensor frequency; (2) the damping characteristics of the sensor; (3) the damping characteristics of the structure; and (4) the threshold level.

The idealized signal in Fig. 5 can be represented by Eq. 1 (see reference 16):

$$V = V_o \exp(-Bt) \sin \omega t \quad \text{(Eq. 1)}$$

Where:

V = output voltage of sensor;
V_o = initial signal amplitude;
B = decay constant (greater than 0);
t = time; and
ω = angular frequency.

A common measure of acoustic emission activity is *ringdown counts*: the number of times the sensor signal exceeds a counter threshold. When the time t^* required for the signal to decay to the threshold voltage V_t is long compared to the period of oscillation, then the number of counts N from a given event can be given by:

$$N = \frac{\omega}{2\pi B} \ln \frac{V_o}{V_t} \quad \text{(Eq. 2)}$$

where V_t is the threshold voltage of the counter.

Correlation of Count Rates and Material Properties

As shown in Fig. 7, a single acoustic emission event can produce several counts. A larger event requires more cycles to ring down to the trigger level and will produce more counts than a smaller event. This provides a measure of the intensity of the acoustic emission event.

Correlations have been established between total counts, count rate and various fracture mechanics parameters (such as stress intensity factor)[17-22] as expressed in Eq. 3 or fatigue crack propagation rate[19-22] as expressed in Eq. 4.

$$N \sim K^n \quad \text{(Eq. 3)}$$

Where:

N = total number of counts;
K = stress intensity factor; and
n = constant with a value between 2 and 10.

$$\frac{dN}{dc} \sim \frac{da}{dc} \quad \text{(Eq. 4)}$$

Where:

N = total number of counts;
a = crack size; and
c = number of cycles.

The counts are a complex function of the frequency response of the sensor and the structure (ω), the damping characteristics of the sensor, the event and the propagation medium (B), signal amplitude, coupling efficiency, sensor sensitivity, amplifier gain and the threshold voltage (V_o). Maintaining stability of these parameters throughout a test or from test to test is difficult but it is essential for consistency of interpretation. Nevertheless, counts are widely used as a practical measure of acoustic emission activity.

Acoustic Emission Event Energy

Since acoustic emission activity is attributed to the rapid release of energy in a material, the energy content of the acoustic emission signal can be related to this energy release[23,24] and can be measured in several ways. The true energy is directly proportional to the area under the acoustic emission waveform. The electrical energy U present in a transient event can be defined as:

$$U = \frac{1}{R} \int_0^\infty V^2(t)dt \quad \text{(Eq. 5)}$$

where R is the electrical resistance of the measuring circuit. Direct energy analysis can be performed by digitizing and integrating the waveform signal or by special devices performing the integration electronically.

The advantage of energy measurement over ringdown counting is that energy measurements can be directly related to important physical parameters (such as mechanical energy in the emission event, strain rate or deformation mechanisms) without having to model the acoustic emission signal. Energy measurements also improve the acoustic emission measurement when emission signal amplitudes are low, as in the case of continuous emission.

Squaring the signal for energy measurement produces a simple pulse from a burst signal and leads to a simplification of event counting. In the case of continuous emission, if the signal is of constant amplitude and frequency, the energy rate is the root mean square voltage (rms). The rms voltage measurement is simple and without electronic complications. However, rms meter response is generally slow in comparison with the duration of most acoustic emission signals. Therefore, rms measurements are indicative of average acoustic emission energy rather than the instantaneous energy measurement of the direct approach.

Regardless of the type of energy measurement used, none is an absolute energy quantity. They are relative quantities proportional to the true energy.

Acoustic Emission Signal Amplitude

The peak signal amplitude can be related to the intensity of the source in the material producing an acoustic emission. The measured amplitude of the acoustic emission waveform is affected by the same test parameters as the event counts. Peak amplitude measurements are generally performed using a log amplifier to provide accurate measurement of both large and small signals.

Amplitude distributions have been correlated with deformation mechanisms in specific materials.[25,26] For practical purposes, a simple equation can be used to relate signal amplitudes, events and counts:

$$N = \frac{Pf\tau}{b} \quad \text{(Eq. 6)}$$

Where:

N = cumulative counts;
P = cumulative events;
f = resonant frequency of the sensor;
τ = decay time of the event; and
b = the amplitude distribution slope parameter.

Limitations of the Simple Waveform Parameters

In general, measurements of events, counts and energy provide an indication of source intensity or severity. This is useful information for determining whether the test object is accumulating damage and, in turn, for deciding whether the test should continue or if the structure in question should remain in service.

In many cases, the high sensitivity of acoustic emission testing also detects unwanted background noises that cannot be removed by signal conditioning. In order for acoustic emission to be used effectively in these applications, it is necessary to identify the source of each signal as it is received.

In spite of many attempts to establish fundamental relationships between the simple parameters, correlations between signal and source by analytical methods remain elusive. This dilemma is a result of the complexity of modeling the source, the material, the structure, the sensor and the measurement system. Advanced digital computing methods provide an alternative, permitting successful characterization and source identification.

Advanced Characterization Methods

The objective of source characterization in a specific application is to classify each signal as it arrives at the sensor. In acoustic emission testing, characterization of both noise and discontinuities is desirable for the rejection of noise and the classification of signals by discontinuity type. This permits qualitative interpretations based on information contained in the signal rather than by inference based on filtering, thresholding and interpretation of the simple parameters. It also extends the range of applications for acoustic emission testing to areas where noise interference previously made the method impractical or ineffective.

The area of artificial intelligence known as *pattern recognition* can be used to relate signal characteristics to acoustic emission sources.[27-30] There are several classification methods available in the literature[31-33] which have been used as standard nondestructive testing tools.

Knowledge Representation Method

Signals emitted from an acoustic emission source are highly unreproducible in both the time and frequency domains, and advanced methods are needed to identify common discriminatory features. This can be achieved with *knowledge representation* (also called *adaptive learning*), where information in the acoustic emission pulse is represented in the most complete and effective way.

Statistical methods are used to extract data from a group of signals originating from the same source. It would be virtually impossible to obtain the same information from one signal alone. Pattern recognition techniques using statistical features are particularly suitable for analyzing nonlinear, time varying acoustic emission signals.

The first step in knowledge representation is to extract features from the signal waveform which fully represent the information contained in the signal. In addition to general time domain pulse properties and shape factors, the acoustic emission signal time series is transformed into other domains. Table 2 shows a list of features used in a typical acoustic emission application.

Time domain pulse information is combined with partial power distribution in different frequency bands together with frequency shifts of the cumulative power spectrum. A total of thirty features are extracted in this example. In some cases, more than a hundred features in several domains may be extracted to provide source characterization.

Feature Analysis and Selection

The transformations carried out to generate new features for source characterization do not produce new information. They represent the existing waveform information in new ways that facilitate source characterization. In fact, only a few well-chosen features among the available domains are normally required for characterization.

Selection of the best combination of features often requires special signal classifier development tools to search among the features and select an optimal set. Acoustic emission is sensitive not only to the source but also to sensor, material and structural factors, so that separate signal classifiers must be developed for each application. A combination of computer methods and expert insight is required to select the best feature set.

TABLE 2. Waveform parameters extracted from acoustic emission signals using the knowledge representation technique

- Standard deviation of acoustic emission signal
- Skewness coefficient (third moment) of acoustic emission signal
- Kurtosis coefficient (fourth moment) of acoustic emission signal
- Coefficient of variation of acoustic emission signal
- Rise time of the largest pulse in time domain
- Fall time of the largest pulse in time domain
- Pulse duration of the largest pulse in time domain
- Pulse width of the largest pulse in time domain
- Rise time of second largest pulse in time domain
- Fall time of second largest pulse in time domain
- Pulse duration of second largest pulse in time domain
- Pulse width of second largest pulse in time domain
- Pulse ratio of the second largest pulses
- Distance between the two largest pulses
- Partial power in frequency band 0 to 0.25 MHz
- Partial power in frequency band 0.25 to 0.5 MHz
- Partial power in frequency band 0.5 to 0.75 MHz
- Partial power in frequency band 0.75 to 1.0 MHz
- Partial power in frequency band 1.0 to 1.25 MHz
- Partial power in frequency band 1.25 to 1.5 MHz
- Partial power in frequency band 1.5 to 1.75 MHz
- Partial power in frequency band 1.75 to 2.0 MHz
- Ratio of the two largest partial powers
- Ratio of the smallest and largest partial power
- Frequency of the largest peak in power spectrum
- Amplitude of the largest peak in power spectrum
- Frequency at which twenty-five percent of accumulated power was observed
- Frequency at which fifty percent of accumulated power was observed
- Frequency at which seventy-five percent of accumulated power was observed
- Number of peaks exceeding a present threshold

In a manual approach, the analyst is usually guided by experience when searching for the optimal feature set for a particular application. When several classes of signal are involved, development systems use statistical methods such as interclass and intraclass distances and a biserial correlation estimate of each individual feature as a measure of its discriminating index among the signal classes of interest.[34] Separability arises from differences between class mean feature values.

Pattern Recognition Training and Testing

Generally, a pattern recognition system goes through a learning or training stage, in which a set of decision rules is developed. The system is supplied with a set of representative signals from each source to be classified. The classifier is said to be trained when it can use the decision rules to identify an input signal.

To test the performance of the classifier, sample signals not used during the training process are presented to the system which will identify the class to which each unknown signal belongs. Successful identification of input signals is a good indication of a properly trained system.

Signal Classification

The data used to train a classifier correspond to a group of points in feature space. If two features are used for a system with three source classes, then the feature space might appear as shown in Fig. 8. The classification mechanism must distinguish among the classes.

One simple method of distinguishing among the classes is to place planes between the classes (hyperplanes in a multidimensional space with many features). The classifier is a linear combination of feature elements that defines a hyperplane to separate one class of signals from another in the feature space. This is called a *linear discriminant function classifier*.

When the classes are distributed in a more complex arrangement, more sophisticated classifiers are required. An example would be the *K*-nearest neighbor classifier which considers every member in the training set as a representation point. It determines the distance of an unknown signal from every pattern in the training set and it thus finds the *K*-nearest patterns to the unknown signal. The unknown is then assigned to the class in which the majority of the *K*-nearest neighbors belong. In other cases, the pattern recognition process is modeled statistically and statistical discriminant functions are derived.

Signal Treatment and Classification System

A classification system is shown in Fig. 9. Acoustic emission source characterization and recognition begins with signal treatment. The acoustic emission signals from different sources are first fed into a treatment system that computes the necessary waveform features of each signal. The identity

FIGURE 8. Feature space for two features used for a system with three source classes: (a) linearly separable; (b) nonlinearly separable; and (c) piece-wise linear separation of two regions

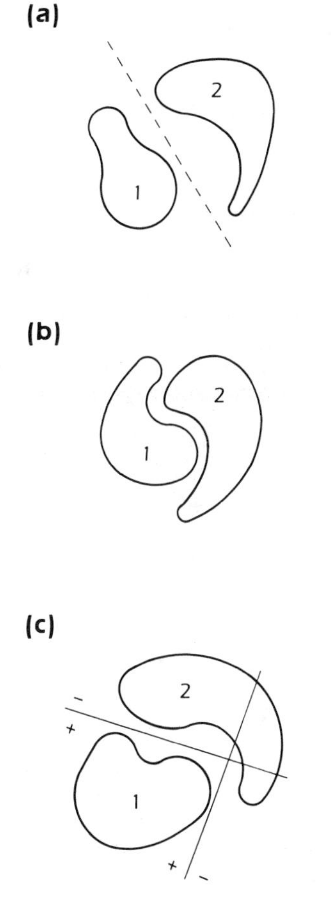

of the signal is tagged to each of the feature vectors. These vectors are then divided into two data sets, one for training the classifier and the other to test its performance.

Generally, a normally distributed random variable is used for assigning the signal to the two data sets and this ensures that each set has the same *a priori* probability. Once the classifier is trained, it is tested by supplying a signal feature vector with a known identity and then comparing the known information to the classification output. The training information and the trained classifier setup can be stored in a decision library that can be used to characterize and classify the acoustic emission sources.

TABLE 3. Performance of a linear discriminant classifier

Source	Size	Class	1	2	3	4	Recognition Rate (percent)
Performance on training							
Stress corrosion	19	1	18	0	1	0	95
Plastic deformation	30	2	0	29	1	0	97
Fracture	27	3	1	1	25	0	93
Machine noise	16	4	0	0	0	16	100
Overall recognition rate							96
Performance on testing							
Stress corrosion	19	1	14	0	4	1	74
Plastic deformation	32	2	0	32	0	0	100
Fracture	27	3	0	1	26	0	96
Machine noise	16	4	0	0	0	16	100
Overall recognition rate							94

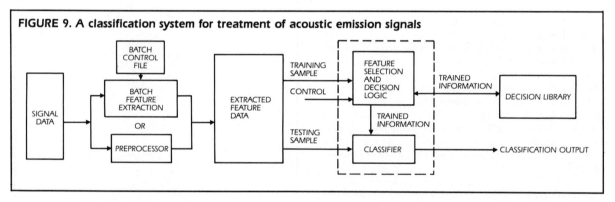

FIGURE 9. A classification system for treatment of acoustic emission signals

Table 3 shows the output of a classifier development system trained to classify acoustic emission from plastic deformation, stress corrosion, fracture and universal testing machine noise in a 7075-T651 aluminum alloy. This method of developing a characterization and classification system shows clearly where the source of error is. In this case, the error lies primarily in identifying the fracture signals.

Guidelines for Source Characterization

The simple waveform parameters and the advanced characterization methods comprise the tools available for acoustic emission source characterization. Simple methods are required to measure the level of acoustic emission activity. More advanced methods are generally required to identify the nature of the emission source.

Of the simple parameters, ringdown counts are sensitive to threshold and gain settings and, for reproducibility from test to test, event measurements are less sensitive to instrument settings.

Several powerful classifiers are available in development tools. However, the three methods outlined above provide an indication of the decisions that have to be made when selecting an advanced characterization method. Factors that must be considered include: (1) available computing power; (2) the sample size available for training; (3) the performance required of the classifier system; and (4) the type of data available for training. It is often difficult to obtain a large enough number of data samples from a particular source.

The linear discriminant function is the simplest method and involves the least computation time. It is therefore readily implemented and rapidly computed. However, the classes of signals must be linearly separable. The K-nearest neighbor and empirical bayesian classifiers are more powerful and address more difficult problems.

The nearest neighbor classifier employs every individual sample in the training set to determine the identity of the unknown signal and each sample bears equal weight in the process of classification. When the training set from a particular source may contain sample signals from other sources, the incorrect sample in the training set carries as much weight as a good sample.

Comparison of the Classifier Techniques

By comparison, the empirical bayesian classifier (using statistical information based on the training set) will produce rather stable recognition performance, so long as the numbers of impure data samples do not dominate the training set. The K-nearest neighbor and the linear discriminant function classifiers are nonparametric classifiers and the selection of features in these two methods depends on the sample size.

Compared to K-nearest neighbor and the linear discriminant function classifiers, the empirical bayesian classifier method is more sensitive to the sample size since the number of samples affects the estimates of the probability distribution. The K-nearest neighbor and the linear discriminant function classifiers do not generally require so large a data set so long as it is representative.

PART 5
INTRODUCTION TO TERMS, STANDARDS AND PROCEDURES FOR ACOUSTIC EMISSION TESTING

Terminology of Acoustic Emission Method

The maturity level of a given technology can never be precisely determined. However, relevant indicators do seem to exist and among these are the availability of industrial standards, professional society activities, the availability of training programs, and international symposia. The following discussions of terminology, standards and procedures indicate that acoustic emission testing has made progress toward acceptance by the industrial community.

Development of Acoustic Emission Terminology

Acoustic Emission Working Group Glossary

The first glossary of terms assembled by the acoustic emission community using a consensus process was developed by the Acoustic Emission Working Group (AEWG) in the United States. This compendium of acoustic emission terminology was drafted by an eight-member subcommittee and was approved during the eighth AEWG meeting held in Bal Harbour, Florida in December 1971. The document contained eight terms and their definitions, and was subsequently published by the American Society for Testing Materials (ASTM) in STP 505.[35]

ASTM Definitions of Terms

The ASTM E07.04 Subcommittee on Acoustic Emission was authorized by the ASTM E7 Committee on Nondestructive Testing in early 1972 and was organized as an operating subcommittee in June of that year. Almost immediately, the Terminology Section (E07.04.01) was charged with the responsibility of developing definitions for acoustic emission terms. This effort culminated in the publication of ASTM E610, "Standard Definitions of Terms Relating to Acoustic Emission" in the *Annual Book of ASTM Standards* in 1977.[36] This document was the second standard generated by the E07.04 subcommittee and consisted of twenty-four definitions unique to acoustic emission technology. The terms developed by the E07.04 Subcommittee on acoustic emission have been widely accepted by the acoustic emission community.

As might be expected, the development of new terms and definitions is a continuing (and often thankless) task. The E07.04.01 Terminology Section continues to meet twice a year to provide new and revised definitions for the terms required by acoustic emission technology.

Glossary for Acoustic Emission Testing

As in other volumes of the *Nondestructive Testing Handbook*, definitions such as those below are provided for reference purposes only and do not carry the acceptance of any consensus group. The European Working Group on Acoustic Emission has independently assembled and published a listing of acoustic emission terminology and many of the definitions below are adapted from that source.

As was stated earlier, the definition of terms is a vitally important task that often goes unrecognized but literally serves as the foundation for developing technologies. One valuable function of such a glossary is that it helps differentiate between terms used in various disciplines.

The words *event* and *burst* illustrate the critical nature of such definitions. One word represents the physical occurrence that causes acoustic emission and the other represents the nature of the corresponding acoustic emission signal. Unfortunately, there have been occasions when the terms were indiscriminately used to represent either phenomenon. Good practice dictates that these two terms always be clearly distinguished, especially when reporting the results of an acoustic emission test.

acoustic emission activity: The number of bursts (or events, if the appropriate conditions are fulfilled) detected during a test or part of a test.

acoustic emission count: The number of times the signal amplitude exceeds the preset reference threshold.

acoustic emission event: A microstructural displacement that produces elastic waves in a material under load or stress.

acoustic emission rate: The number of times the amplitude has exceeded the threshold in a specified unit of time.

acoustic emission signal: The electrical signal obtained through the detection of acoustic emission.

acoustic emission: The transient elastic waves resulting from local internal microdisplacements in a material. By extension, the term also describes the technical discipline and measurement technique relating to this phenomenon.

array: A group of transducers used for source location.

artificial source: A point where elastic waves are created to simulate an acoustic emission event. The term also defines devices used to create the waves.

background noise: The signal in the absence of any acoustic emission events. It has electrical and mechanical origins.

burst counting: A measurement of the number of bursts detected relative to specified equipment settings such as threshold level or dead time.

burst duration: The interval between the first and last time the threshold was exceeded by the burst.

burst emission: A qualitative term applied to acoustic emission when bursts are observed.

burst rate: The number of bursts detected in a specified time.

burst rise time: The time interval between the first threshold crossing and the maximum amplitude of the burst.

burst: A signal, oscillatory in shape, whose oscillations have a rapid increase in amplitude from an initial reference level (generally that of the background noise), followed by a decrease (generally more gradual) to a value close to the initial level.

continuous emission: A qualitative term applied to acoustic emission when the bursts or pulses are not discernible.

count rate: See *acoustic emission rate*.

couplant: A substance providing an acoustic link between the propagation medium and the transducer.

cumulative bursts: The number of bursts detected from the beginning of the test.

cumulative characteristic distribution: A display of the number of times the acoustic emission signal exceeds a preselected characteristic as a function of the characteristic.

cumulative count: The number of times the amplitude of the signal has exceeded the threshold since the start of the test.

cumulative events: The number of events detected from the beginning of a test. Use of this term is restricted in the same way as *event counting*.

delta (t): The time interval between the detected arrival of an acoustic emission wave at two sensors.

event: See *acoustic emission event*.

event counting: A measurement of the number of acoustic emission events. Because an event can produce more than one burst, this term is used in its strictest sense only when conditions allow the number of events to be related to the number of bursts.

event rate: The number of events detected in a specified unit of time. Use of this term is restricted in the same way as *event counting*.

event: A microdisplacement giving rise to transient elastic waves.

felicity effect: The appearance of significant acoustic emission at a stress level below the previous maximum applied.

felicity ratio: The measurement of the felicity effect. Defined as the ratio between (1) the applied load (or pressure) at which acoustic emission reappears during the next application of loading and (2) the previous maximum applied load.

Kaiser effect: The absence of detectable acoustic emission until the previous maximum applied stress level has been exceeded.

location plot: A representation of acoustic emission sources computed using an array of transducers.

maximum burst amplitude: The maximum signal amplitude within the duration of the burst.

parameter distribution: A display of the number of times the acoustic emission parameter falls between the values x and $x + \delta x$ as a function of x. Typical parameters are amplitude, rise time and duration.

pencil source: An artificial source using the fracture of a brittle graphite lead in a suitable fitting to simulate an acoustic emission event.

pulse: Acoustic emission signal that has a rapid increase in amplitude to its maximum value, followed by an immediate return. An example is the signal associated with a wave of a particular mode which has propagated from the source to the transducer as detected using a flat response transducer.

pulser transducer: A transducer used as an artificial source.

reference threshold: A preset voltage level that has to be exceeded before an acoustic emission signal is detected and processed. This threshold may be adjustable, fixed or floating.

ringdown count: See *acoustic emission count*.

source location: The computed origin of acoustic emission signals.

source: The place where an event takes place.

time differential: See *delta (t)*.

transducer, differential: A piezoelectric twin element or dual pole transducer, the output poles of which are isolated from the case and are at a floating potential.

transducer, flat response: A transducer whose frequency response has no resonance with its specified frequency band (the bandwidth to -3 dB being defined) and the ratio between the upper and lower limits of the band being typically not less than 10.

transducer relative sensitivity: The response of the transducer to a given and reproducible artificial source.

transducer, resonant: A transducer that uses the mechanical amplification due to a resonant frequency (or several close resonant frequencies) to give high sensitivity in a narrow band, typically ± 10 percent of the principal resonant frequency at the -3 dB points.

transducer, single ended: A piezoelectric single element transducer, the output pole of which is isolated from the case, the other pole being at the same potential as the case.

transducer, wideband: A transducer that uses the mechanical amplification due to the superposition of multiple resonances to give high sensitivity in several narrow bands within a specified wide band.

transducer: A device that converts the physical parameters of the wave into an electrical signal.

transfer function: Description of changes to the waves arising as they propagate through the medium or, for a transducer, the relationship between the transducer output signal and the physical parameters of the wave.

Procedures for Acoustic Emission Testing

Requirements for Written Procedures

In the United States, most industrial testing applications require that a written procedure be applied during an acoustic emission examination. Whether the testing is done in-house or subcontracted to an acoustic emission field service company, such procedures are usually developed by the organization performing the examination.

A procedure usually specifies: (1) the equipment to be used; (2) the placement of acoustic emission sensors; (3) the process for stressing the component; (4) the data to be recorded and reported; and (5) the qualifications of personnel operating the equipment and interpreting the results. Quite often, the procedure must be approved by the organization that is responsible for the component.

Most written procedures for specific acoustic emission applications are considered proprietary by the organization that prepared them, and many can only be used with the make and model of acoustic emission equipment for which the procedure was written.

Several different data interpretation methods are commonly used throughout the industry, so that it is difficult to directly compare the results obtained from different procedures and different acoustic emission equipment. This difficulty occurs even when considering simultaneous examination of a specific component but the problem may be partially mitigated as ASTM standards and ASME Code rules are incorporated into procedures used for field applications.

Personnel Qualification

One area where standardization is occurring is with the procedures used for qualification and certification of acoustic emission personnel. An Acoustic Emission Personnel Qualification (AE PQ) committee was established in the late 1970s within the Personnel Qualification Division of the American Society for Nondestructive Testing (ASNT). This group has prepared a recommended course outline for training NDT Levels I and II personnel, and is developing certification requirements and examination questions for NDT Levels I, II and III acoustic emission personnel.

These requirements are published in ASNT's *Recommended Practice SNT-TC-1A*.[37] The document also contains a list of recommended references for acoustic emission training courses. A supplement to SNT-TC-1A will contain recommended (typical) certification examination questions for Levels I, II and III acoustic emission personnel. Standardized personnel qualification requirements are an essential prerequisite to responsible application of the acoustic emission method by commercial field organizations.

Standards for Acoustic Emission Testing

Proposed ASME Standard

The American Society for Mechanical Engineers (ASME) Section V Subcommittee on Nondestructive Examination established an Ad Hoc Working Group on Acoustic Emission in 1973. This group issued the first document intended as a US standard for applying acoustic emission technology.

The document is titled *Proposed Standard for Acoustic Emission Examination During Application of Pressure* (ASME E00096) and was issued for trial use and comment in 1975. The issuance of the document in this reviewable form came after a proposal by ASME Section V in 1974 to publish it as a new article in the ASME Code. While this document was widely used as a guideline by many commercial acoustic emission field service companies and was referenced in many procurement specifications (especially within the petrochemical industry), it was not accepted as an ASME Code requirement.

ASTM Acoustic Emission Standards

With the formation of American Society for Testing Materials (ASTM) E07.04 Subcommittee on Acoustic Emission in 1972, a practical forum for the development and distribution of consensus standards was available. The E07.04 subcommittee received final consensus approval for several separate standards covering a variety of acoustic emission applications.[38] More documents are in various stages of development and approval.

Among the ASTM documents available for use by the acoustic emission community are the following.

E569   *Acoustic Emission Monitoring of Structures During Controlled Stimulation*
E610   *Definition of Terms Relating to Acoustic Emission*
E650   *Mounting Piezoelectric Acoustic Emission Contact Sensors*
E749   *Acoustic Emission Monitoring During Continuous Welding*
E750   *Measuring the Operating Characteristics of Acoustic Emission Instrumentation*
E751   *Acoustic Emission Monitoring During Resistance Spot Welding*
E976   *Guide for Determining the Reproducibility of Acoustic Emission Sensor Response*
E1067   *Acoustic Emission Examination of Fiberglass Reinforced Plastic Resin (FRP) Tanks/Vessels*
E1106   *Primary Calibration of Acoustic Emission Sensors*
E1118   *Acoustic Emission Examination of Reinforced Thermosetting Resin Pipe (RTRP)*
E1139   *Practice for Continuous Monitoring of Acoustic Emission from Metal Pressure Boundaries*
F914   *Test Method for Acoustic Emission for Insulated Aerial Personnel Devices*

Special Working Group on Acoustic Emission Testing

The ASME Section V Subcommittee on nondestructive evaluation organized a Special Working Group on Acoustic Emission (SWGAE) and the SWGAE has been more effective than the preceding ad hoc group. For example, acceptance has been achieved for (1) a Code Case on small Section VIII vessels[39] and (2) code rules for acoustic emission examination of fiber reinforced plastic vessels.[40]

Use of Acoustic Emission on Small Pressure Vessels

Code Case 1968 (approved in 1985) is entitled *Use of Acoustic Emission Examination in Lieu of Radiography* and it specifies the conditions and limitations under which acoustic emission examination conducted during the hydrostatic tests may be used in lieu of radiography for examining the circumferential closure weld in pressure vessels. This Code Case applies to small vessels (7 cubic feet maximum) made of P-No. 1 Group 1 or 2 materials with a weld thickness up to 2.5 in.

Use of Acoustic Emission on Fiber Reinforced Plastics

Article 11 of ASME Section V (also issued in 1985) is entitled *Acoustic Emission Examination of Fiber Reinforced Plastic Vessels*. This article describes rules for applying acoustic emission to examine new and in-service fiber reinforced plastic vessels under pressure, vacuum or other applied stress. The examination is conducted using a test pressure not to exceed 1.5 times the maximum allowable working pressure, although vacuum testing can be performed at the full design vacuum. These rules may only be applied to vessels with glass or other reinforcing material contents greater than 15 percent by weight.

Committee on Acoustic Emission from Reinforced Plastics

The Committee on Acoustic Emission from Reinforced Plastics (CARP) was established in the early 1980s by the Society of the Plastics Industry (SPI). The CARP group has issued documents describing guidelines for applying acoustic emission to fiber reinforced plastic components, including: *Recommended Practice for AE Testing of Fiberglass Tanks/Vessels* (1982); and *Recommended Practice for AE Testing of Reinforced Thermosetting Resin Pipe* (1984).

European Working Group on Acoustic Emission

The European Working Group on Acoustic Emission (EWGAE) has been active in the development and issuance of documents to standardize acoustic emission applications. At least seven different codes have been identified for study by this group and several have been finalized and distributed to the international acoustic emission community. Among the documents issued by this group are those below.

Code I   *Acoustic Emission Examination — Location of Sources of Discrete Acoustic Events*
Code II   *Acoustic Emission Leak Detection*
Code III   *Acoustic Emission Examination of Small Parts*
Code IV   *Definition of Terms in Acoustic Emission*
Code V   *Recommended Practice for Specification, Coupling, and Verification of the Piezoelectric Transducers Used in Acoustic Emission*

Japanese Society for Nondestructive Inspection

The Japanese Society for Nondestructive Inspection has published standards pertaining to acoustic emission applications, including those listed below.

NDIS-2106 *Evaluation of Performance Characteristics of Acoustic Emission Testing Equipment* (1979)

NDIS-2109 *Acoustic Emission Testing of Pressure Vessels and Related Facilities During Pressure Testing* (1979)

NDIS-2412 *Acoustic Emission Testing of Spherical Pressure Vessels Made of High Tensile Strength Steel and Classification of Test Results* (1980)

PART 6
ACOUSTIC EMISSION EXAMINATION PROCEDURES

The following text presents useful information about acoustic emission testing procedures. It was liberally adapted from a European Working Group code but should not be considered as a representative code or standard (refer instead to documents such as ASTM E650, E976 or E1139). The text was originally written for pressure vessels examined by acoustic emission techniques and focuses on equipment calibration and data verification.

Objectives of the Acoustic Emission Testing Procedure

In acoustic emission testing, the sensors are generally of the piezoelectric type, to which this text applies. The objectives of this discussion are to:

1. specify the minimum technical characteristics to be furnished by the sensor manufacturer;
2. give the principal rules to ensure correct mounting and coupling of the sensor and verification thereof; and
3. describe some of the methods available to permit a simplified characterization or rapid check of piezoelectric sensors. These methods allow comparison of sensor performance and stability (it is emphasized that this characterization is not a calibration).

These topics are considered below.

Technical Characteristics Furnished by the Transducer Manufacturer

If a transformer, amplifier or acoustic waveguide is combined with the sensor in such a way that the readily accessible terminals include them, then the term *transducer* may apply to the combination. Following is a list of characteristics required from the manufacturer.

1. The type of sensor (resonant, wideband, flat response, single-ended or differential) with reference to applicable code.
2. The type and reference of electrical connectors.
3. Temperature range within which the specified performance characteristics are ensured.
4. Principal resonant frequency and bandwidth at -6 dB or other specified fall-off value.
5. Frequency response curve, derived with respect to a reference source, giving the frequency or frequencies where response is maximum, and the overall bandwidth. The reference source and procedure should be specified.
6. Overall physical characteristics (dimensions, socket).

Optional information from the manufacturer should include the following.

1. Nature and dimensions of the piezoelectric element of the sensor.
2. Material of the coupling shoe and case.
3. Output impedance at resonant frequency, impedance curve.
4. Response to different types of acoustic wave.
5. Directionality (polar response).
6. Resistance to the environment, vibration, shock, in accordance with specified instrumentation standards.

Coupling and Mounting of Piezoelectric Transducers

Coupling ensures transmission of acoustic emission waves from the structure to the sensor by means of an appropriate medium (couplant). The conditions allowing the quality and maintenance of the coupling to be ensured are defined below.

Coupling Methods

Fluid coupling with pressure can be achieved with water, silicone grease or any fluid ensuring continuity in wave transmission with acceptably low attenuation. Pressure can be exerted, for example, by a spring or magnet. Care is needed to avoid introducing errors by excessive pressure or lack of pressure.

Solid coupling can be achieved with glue or cement, ensuring continuity in wave transmission with acceptably low attenuation.

Waveguides are mechanical parts (usually metallic rods of an appropriate shape) secured or welded to the structure and ensuring continuity in wave transmission to the sensor with acceptably low attenuation.

Coupling and Mounting Procedure

The recommended procedure for coupling and mounting piezoelectric sensors is as follows:

1. Select the coupling method, considering conditions such as temperature, access, chemical compatibility with the structure, stability, radiation degradation or acoustical impedance match.
2. Prepare the contact surface to ensure total area of contact between sensor and surface, thus avoiding discontinuity in wave transmission. Cleanliness, roughness, curvature, and so forth, are to be considered.
3. Apply the couplant to ensure minimum transmission loss. Thickness, uniformity of coating, and so on, are considered.
4. Position and fix the sensor.
5. Verify the efficiency of coupling and mounting using appropriate means, such as response to an artificial source (pulser or pencil source, for example) and noting the distance of the source from the center of the sensor.
6. In the case of waveguides, adjust the shape, length, constitutive material and method of attachment to obtain the minimum attenuation in the considered bandwidth. The attenuation introduced should be quoted.

Precautions

The user should pay special attention to the quality of the coupling and the method of mounting in particular, including the following points.

1. Stability of coupling, especially in relation to the environment (risks of desiccation, dilution and so on).
2. Mechanical stability of the attachment (risks of slipping, fall off, protection against vibration or shock effects).
3. Electrical isolation between the sensor and the surface.
4. Susceptibility of the couplant to corrode the structure or the sensor (including consideration of long-term corrosion).
5. Prevention of overstressing the sensor at the signal cable.
6. Appropriate cleaning of the coupling zone before placing the sensor.

Reporting the Coupling and Mounting Procedures

It is recommended that the following items be included in every test report.

1. Couplant type.
2. Attachment of the sensor (mechanical, magnetic).
3. Nature and preparation of the contact surface.
4. Note any eventual cause of transmission loss due to coating, roughness, and so on.
5. In cases where waveguides are used, material geometry.

Simplified Characterization and Verification of Piezoelectric Transducers

Because of the number and complexity of the factors affecting an acoustic emission signal, an absolute measurement is not considered practical for piezoelectric sensors. However, a characterization of the response (limited to its resonant frequency, effective frequency range and sensitivity to a given stimulus) can be achieved with relatively simple methods.

Considering that sensor characteristics are susceptible to variation with time, operation, environment and so forth, it is desirable that the user verify certain important characteristics of the sensor (frequency and sensitivity): (1) at delivery; (2) periodically during service; and (3) after a disruptive incident (for example, shock or abnormal temperature exposure).

The methods below permit a simple characterization or rapid check of piezoelectric sensors.

Simplified Laboratory Characterization and Verification of Piezoelectric Transducers

In the laboratory, it is recommended that a characterization or verification system be set up to include the following elements: (1) a reproducible artificial reference source; (2) a wave propagation medium when appropriate; and (3) a measuring system permitting determination of the sensitivity (response to the stimulus generated by the reference source) and frequency response of the sensors.

Various methods for achieving this are available. They can be tried and adopted according to personal preference and convenience, but the method chosen must be maintained for subsequent characterizations to be meaningful. *Note:* the results obtained will relate only to the particular measurement system and will not constitute an absolute calibration. They will nevertheless be adequate for many practical acoustic emission applications.

Artificial Reference Sources

Three types of reference sources are commonly available and suitable for both laboratory and *in situ* verification.

The *brittle fracture source* or pencil lead (Hsu-Nielsen) source is very intense, using the brittle fracture of a pencil lead of fixed diameter (0.5 mm) and hardness (2H). The optimal length l, angle α and support position are determined experimentally and can then be reproduced using a simple guide ring. The reproducibility of the source-to-sensor distance and orientation of the sensor with respect to the source must be ensured by appropriate means.

The *piezoelectric pulser* is a damped, flat response piezoelectric sensor (such as those used in ultrasonics). It is excited either by an electronic white noise generator or by a pulse generator. The rise time, width and recurrence of the pulses are adapted to the effective bandwidth and the damping of the sensor and are ensured for subsequent tests. The excitation must not lead to saturation of both sensors. The reproducibility of the position and orientation of each sensor must be ensured by appropriate means.

The *gas jet source* is recommended only for the characterization of sensors used in continuous emission measurements. The source of sound is generated by a jet of dry gas, emitted from a nozzle of small output diameter at a convenient pressure. The optimal conditions of pressure, distance and nozzle position must first be determined experimentally and their reproducibility ensured by appropriate means.

Propagation Medium

The sensor can be excited by either of the following methods: (1) direct application of the stimulus on the shoe of the sensor; or (2) through a propagation medium inserted between the source and the sensor. Propagation mediums are generally made of a metallic homogeneous material of an appropriate geometry.

With both of the excitation methods, it is essential that the positioning of the source and sensor can be accurately reproduced. Some possible variants of the propagation medium are described below.

Rectangular or cylindrical blocks are used to simulate the direct incidence of bulk waves. The source and sensor are positioned centrally on opposite faces. It is recommended that the lateral dimensions of the block be greater than five times the wavelength of longitudinal waves associated with the main resonant frequency of the sensor. The height of the block (source-to-sensor distance) should approximately equal the lateral dimension.

A plate medium is used to simulate the incidence of surface or Lamb waves. Usually, the source is placed on the same side as the sensor at a distance greater than ten times the wavelength of surface waves associated with the main resonant frequency of the sensor. The plate thickness should be selected accordingly. The position of the source and lateral dimension of the plate are chosen to minimize reflected waves.

A conical block has been proposed as a propagation medium for characterization procedures. It is designed to simulate the incidence of bulk waves but its shape is intended to restrain the influence of wave reflections.

Measuring System

The system used to characterize the sensor and cable may include conditioning elements. The frequency response and dynamic range of the elements must not modify the sensor response. In particular, the input impedance should be at least 100 times the sensor output impedance at resonance. In addition, the bandwidth should be larger than that of the sensor.

The frequency response curve should be adequate for determining the principal and secondary resonant frequencies as well as the bandwidth. The frequency response (or its envelope) is determined with a spectrum analyzer of adequate dynamic range and sampling rate to capture the full characteristic response of the sensor. The spectrum analyzer gathers data within a time frame chosen to eliminate resonances and reverberation effects introduced by the propagating medium.

Relative transducer sensitivity is conventionally defined as the response of the sensor to a given and reproducible artificial source. It is measured in volts or decibels relative to a specified level. For burst signals, the relative sensitivity is determined using a peak voltmeter or an oscilloscope of adequate frequency response and dynamic range. For a continuous acoustic emission signal (gas jet source), the sensitivity is determined using an rms voltmeter with appropriate frequency response and dynamic range. It is recommended that all measurements be repeated several times to determine the statistical variance in the results.

Verification Precautions

At each verification it is recommended that the following points receive particular attention.

1. Good reproducibility of the sensor positioning and coupling: repeated mounting and demounting of the sensor should not introduce differences in relative sensitivity greater than 3 dB.
2. Calibration and correct working of the measuring system: must be periodically and separately verified.
3. Conformity with the specifications and previous verification procedures.

Quality Control Card

It is also recommended that the user establish a procedure describing the adopted set up and verification method. The procedure should include details on: (1) the acoustic source; (2) the propagating medium; (3) the source-to-sensor distance and orientation; (4) the measuring system; (5) an identification of the apparatus; and (6) the instrument settings.

For each sensor, the user should also establish a delivery card and a control card to include the following elements: (1) identification of sensor; (2) identification of operator; (3) date of verification; (4) method and system used for characterization; (5) source characteristics and position relative to sensor; (6) propagating medium; (7) coupling method; (8) measuring system (identification of the apparatus, settings and so forth); (9) results (relative sensitivity, resonant frequency, bandwidth).

It is recommended that a sensor be rejected when its control card shows a loss in relative sensitivity greater than 6 dB relative to the value at delivery.

REFERENCES

1. *Standard Definition of Terms Relating to Acoustic Emission*. E610. Philadelphia, PA: American Society for Testing and Materials (1982).
2. Kaiser, J. "Erkenntnisse und Folgerungen aus der Messung von Gerauschen bei Zugbeanspruchig von Metallische Werkstoffen (Conclusions and Results from Sound Measurement in the Tensile Stressing of Metals)." *Archiv fur das Eisenhutten Wesen*. Vol. 24, No. 1-2 (January-February 1953): pp 43-45.
3. Dunegan, H.L., D.O. Harris and A.S. Tetelman. "Detection of Fatigue Crack Growth by Acoustic Emission Techniques." *Materials Evaluation*. Vol. 28, No. 10. Columbus, OH: The American Society for Nondestructive Testing (October 1970): pp 221-227.
4. Jessen, E.C., H. Spanheimer and A.J. De Herrera. *Prediction of Composite Pressure Vessel Performance by Application of the Kaiser Effect in Acoustic Emission*. H300-12-2-037. Magna, UT: Hercules Incorporated (June 1975).
5. *Listening to ... Cherry Picker Booms*. Special publication LT-2. Irvine, CA: Dunegan Corporation (1976).
6. Fowler, T.J. and R.S. Scarpellini. "Acoustic Emission Testing of FRP Equipment." *Chemical Engineering* (October 1980 and November 1980).
7. Liptai, R.G., D.O. Harris, R.B. Engle and C.A. Tatro. *Acoustic Emission Techniques In Materials Research*. UCRL Report 72582. Livermore, CA: Lawrence Radiation Laboratory (1970).
8. Hutton, P.H. "Summary Report on Investigation of Background Noise at the San Onofre Nuclear Power Reactor as Related to Structural Flaw Detection by Acoustic Emission." *Southwest Research Institute Project 17-2440*. Biannual progress report No. 6, Vol. I, Projects I-IV. Richland, WA: Battelle Pacific Northwest Laboratory (1972).
9. Smith, J.R., G.V. Rao and J. Craig. *Acoustic Monitoring System Tests at Indian Point Unit 1*. C00/2974-2. Westinghouse Electric Corporation (1979).
10. Ying, S.P., J.E. Knight and C.C. Scott. "Background Noise for Acoustic Emission in a Boiling Water and a Pressurized Water Nuclear Power Reactor." *Journal of the Acoustical Society of America*. Vol. 53, No. 6 (1973): pp 1,627-1,631.
11. Hutton, P.H., et al. *Acoustic Emission Monitoring of Hot Functional Testing, Watts Bar Unit 1 Nuclear Reactor*. NUREG/CR-3693 (PNL-5022). Richland, WA: Battelle Pacific Northwest Laboratory (1984).
12. Balderston, Harvey. "The Detection of Incipient Failure in Bearings." *Materials Evaluation*. Vol. 27, No. 6. Columbus, OH: The American Society for Nondestructive Testing (1969): pp 121-128.
13. Notvest, K.R. "Evaluation and Control of Spot Welding with Acoustic Emission." *Proceedings of the Acoustic Emission Working Group* (1971).
14. Bartle, P.M. and N.R. Stockham. *Acoustic Emission Studies in Offshore Engineering*. Report #3522/7/79. Abington Hall, Cambridge, England: The Welding Institute (1979)
15. Fowler, E.P. *On the Interference Immunity of Coaxial Cables*. MIEE URON (1971).
16. Harris, D.O., A.S. Tetelman and F.A.I. Darwish. *Detection of Fiber Cracking by Acoustic Emission*. ASTM STP 505. Philadelphia, PA: American Society for Testing and Materials (1972): pp 238-249.
17. Dunegan, H.L., D.O. Harris and C.A. Tatro. "Fracture Analysis by Use of Acoustic Emission." *Engineering Fracture Mechanics*. Vol. 1, No. 1 (1968): pp 105-122.
18. Palmer, I.G. and P.T. Heald. "The Application of Acoustic Emission Measurements to Fracture Mechanics." *Materials Science and Engineering*. Vol. 11, No. 4 (1973): pp 181-184.
19. Harris, D.O. and H.L. Dunegan. "Continuous Monitoring of Fatigue Crack Growth by Acoustic Emission Techniques." *Experimental Mechanics*. Vol. 14, No. 2 (1974): pp 71-81.
20. Masounave, J., J. Lanteigne, M.N. Bassim and D.R. Hay. "Acoustic Emission and Fracture of Ductile Materials." *Engineering Fracture Mechanics*. Vol. 8, No. 4 (1976): pp 701-709.
21. Lindley, T.C., I.G. Palmer and C.E. Richards. *Materials Science and Engineering*. Vol. 32 (1970): pp 1.
22. Sinclair, A.C.E., D.C. Connors and C.F. Formby. *Materials Science and Engineering*. Vol. 28 (1977).
23. Harris, D.O. and R. Bell. "The Measurement and Significance of Energy in Acoustic Emission Testing." *Experimental Mechanics*. Vol. 17, No. 9 (1977): pp 347-353.
24. Beattie, A. and R. Jarmaillo. "The Measurement of Energy in Acoustic Emission." *Review of Scientific Instruments*. Vol. 45, No. 3 (1974): pp 352-357.
25. Nakamura, Y., C. Veach and B. McCauley. "Amplitude Distribution of Acoustic Emission Signals." *Acoustic Emission*. ASTM STP 505. Philadelphia, PA: American Society for Testing and Materials (1971): pp 164-186.

26. Pollock, A.A. "Acoustic Emission Amplitude Distributions." *International Advances in Nondestructive Testing*. Vol. 7 (1981): pp 215-239.
27. Harrington, T.P. and P.G. Doctor. "Acoustic Emission Analysis Using Pattern Recognition." *Proceedings of the Fifth Joint Conference on Pattern Recognition* (December 1980): pp 1,204-1,207.
28. Chan, W.Y., D.R. Hay, C.Y. Suen and O. Schwelb. "Application of Pattern Recognition Techniques in the Identification of Acoustic Emission Signals." *Proceedings of the Fifth Joint Conference on Pattern Recognition* (December 1980): pp 108-111.
29. Crostack, H.A. and K.H. Kock. "Application of Pattern Recognition Methods in Acoustic Emission Analysis by Means of Computer Techniques." *Proceedings of the Ninth World Conference on Nondestructive Testing*. Section 4J-11 (1979).
30. Elsley, R.K. and L.J. Graham. "Pattern Recognition Techniques Applied to Sorting Acoustic Emission Signals." *Ultrasonic Symposium Proceedings* (September 1976): pp 147-150.
31. Bae, P., et al. "Pattern Recognition Techniques for Characterization and Classification of Acoustic Emission Signals." *Proceedings of the Fifth Joint Conference on Pattern Recognition* (December 1980): pp 134-136.
32. Ullmann, J.R. *Pattern Recognition Techniques*. London, England: Butterworths Publishing (1973).
33. Tou, J.T. and R.C. Gonzalez. *Pattern Recognition Principles*. Reading, MA: Addison-Welsey Publishing (1974).
34. Hay, D.R. and R.W.Y. Chan. "Identification of Deformation Mechanisms by Classification of Acoustic Emission Signals." *Proceedings of the Fourteenth Symposium on NDE* (April 1983).
35. Liptai, R.G., D.O. Harris and C.A. Tatro. *Acoustic Emission*. STP 505. Philadelphia, PA: American Society for Testing Materials (1972): pp 335-337.
36. "Metallography and Nondestructive Testing." *Annual Book of ASTM Standards*. Part 11. Philadelphia, PA: American Society for Testing Materials (1977).
37. "Personnel Qualification and Certification in Nondestructive Testing." *Recommended Practice No. SNT-TC-1A*. Columbus, OH: The American Society for Nondestructive Testing (August 1984).
38. "Nondestructive Testing." *Annual Book of ASTM Standards*. Section 3, Volume 03.03. Philadelphia, PA: American Society for Testing Materials (1987).
39. "Use of Acoustic Emission in Lieu of Radiography." *Boiler and Pressure Vessels*. Code Case No. 1968, Section VIII, Division 1, Supplement 9. New York, NY: American Society of Mechanical Engineers (1983).
40. *The ASME Boiler and Pressure Vessel Code*, Summer 1985 Addenda. Article 11, Section V. New York, NY: American Society of Mechanical Engineers (1983).

ND
SECTION 2

MACROSCOPIC ORIGINS OF ACOUSTIC EMISSION

M. Nabil Bassim, University of Manitoba, Winnipeg, Manitoba, Canada

PART 1
FUNDAMENTALS OF ACOUSTIC EMISSION'S MACROSCOPIC ORIGINS

The success of acoustic emission as a nondestructive testing technique is intimately related to our understanding the origins of acoustic emission observed during a test. The general principle in setting up an acoustic emission test is to subject a structure to an applied load (this may or may not be higher than the normal operating load) and to observe the resulting acoustic emission activity.

Analysis of the acoustic emission data is carried out either in real-time as the test is proceeding or when the test is completed. In the latter type of analysis, graphs or tables are obtained to illustrate the relationship between conventional acoustic emission parameters and the applied load. These relationships provide a quantitative assessment of damage in the structure as it relates to the presence, location and severity of discontinuities.

Unique Aspects of Acoustic Emission Testing

In acoustic emission testing, the signal originates from the discontinuity itself and this distinguishes it from other techniques. Most other methods of nondestructive testing provide local excitation (energy) to a structure and information about the structure is subsequently gained from the local response to the excitation. Acoustic emission techniques detect the response of the structure to an external loading whose purpose is to create stress throughout the structure, causing significant discontinuities (wherever they are) to emit acoustically.

Because of this difference, acoustic emission is able to monitor large areas of a structure with a relatively small number of sensors. The technique can also provide real-time monitoring under normal operating conditions.

Acoustic Emission Instrumentation

The instrumentation used in acoustic emission has been reviewed at length in the literature (in this volume and in reference 1). A system always contains an acoustic emission sensor which normally operates in the range of 0.1 to 1 MHz. The acoustic signal is subjected to an amplification system and filtering for elimination of extraneous background noise. Maximum amplification is usually up to 120 dB.

The signal is next analyzed and relevant parameters are extracted. These parameters are either in the time domain (total counts, count rate, event counts, event rate, amplitude distribution and energy) or in the frequency domain (where the frequency spectra of the signals is related to sources of acoustic emission in the structure).[2]

The use of computers and microprocessors has contributed significantly to the data analysis of acoustic emission signals. Correlation and coherence techniques, pattern recognition and statistical approaches of signal analysis have all been successfully developed.

This text discusses the macroscopic aspects of acoustic emission testing. It focuses on relating acoustic emission parameters to the phenomena that occur in the structure. The causes of failure in a structure due to loading have been detailed in the literature.[3] They range from excessive deformation (both in the elastic and plastic regions) to cracking. Cracking can take place either during monotonic increased loading (brittle and ductile fracture), by cyclic loading (as in fatigue cracking) or by stress corrosion.

The term *macroscopic* here refers to circumstances where a relatively large (whether in volume or in surface) part of the test material is contributing to the acoustic emission. This is in contrast to acoustic emission related to microscopic mechanisms such as dislocation motion through grains and across grain boundaries. Note, though, that within macroscopic sources (such as cracks, inclusions or voids), microscopic sources of acoustic emission are prevalent. Phenomenological relationships between acoustic emission parameters and macroscopic sources of acoustic emission are developed here and in most cases are sufficient justification for the use of acoustic emission techniques for nondestructive testing.

PART 2
DEFORMATION SOURCES OF ACOUSTIC EMISSION

Plastic Deformation Sources

Plastic deformation is the primary source of acoustic emission in loaded metallic materials. The initiation of plasticity, particularly at or near the yield stress, contributes to the highest level of acoustic emission activity observed on a stress-strain (load-elongation) curve. Plasticity also contributes to the highest levels of activity for count rate or root mean square (rms) parameters of amplitude versus strain. A typical curve for a mild steel is shown in Fig. 1.[4] The observed level of acoustic emission activity depends primarily on the material. To a lesser extent, it also depends on (1) the position of the sensor with respect to the yielded region; (2) the gain level of the system; and (3) the threshold of the system.

Yield Stress

Most acoustic emission occurs at the yield stress of a material. However, it has been reported that in some instances high levels of acoustic emission activity take place before the yield stress. This is attributed to local plastic yielding and demonstrates that acoustic emission provides a good technique for detection of the onset of microyielding in certain materials.

In other instances, the peak of rms or count rate occurs either well into the plastic region or shows spikes of activity similar to those in Fig. 2 for 7075-T651 aluminum alloy.[5] In these cases, micromechanisms such as twinning[6] cause the peak of rms to occur at three percent strain. For the 7075-T651 aluminum alloy, fracture of brittle inclusions causes the spikes of activity that are observed.

Factors Contributing to Acoustic Emission

It is important to examine the factors contributing to the level of acoustic emission activity in materials. As early as 1971, tabulations were developed listing factors that contribute to high or low signal amplitudes (see Table 1 in the Section titled *Fundamentals of Acoustic Emission Testing*).[7] For the practicing acoustic emission technician, it is useful

FIGURE 1. Acoustic emission and stress as a function of strain for a mild steel tension specimen (see reference 4)

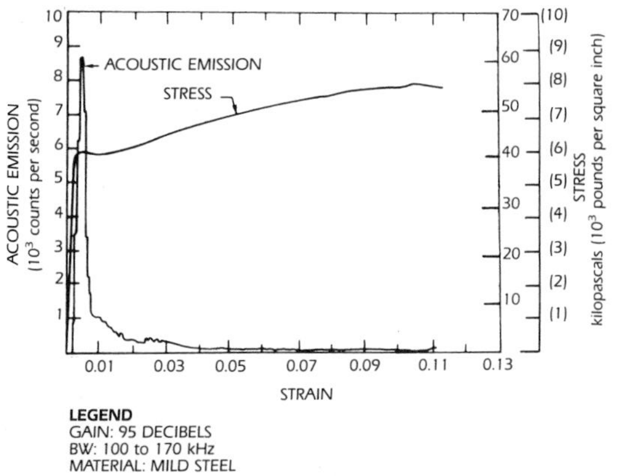

LEGEND
GAIN: 95 DECIBELS
BW: 100 to 170 kHz
MATERIAL: MILD STEEL

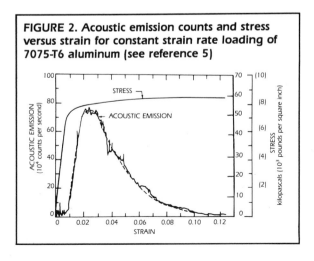

FIGURE 2. Acoustic emission counts and stress versus strain for constant strain rate loading of 7075-T6 aluminum (see reference 5)

Characteristics of Steel

Mild steels or low carbon steels are increasingly used in industrial applications like pipelines, offshore structures and pressure vessels. Such steels generally have low acoustic emission activity. Moreover, in some of these steels, the level of acoustic emission varies depending on the orientation of the specimen (whether it is in the longitudinal or transverse direction).[8]

Characteristics of Aluminum

In aluminum alloys, particularly those containing second-phase particles such as the 2024 and 7075 series, high levels of acoustic emission are observed at the onset of plastic deformation. This factor makes these materials easier to monitor and investigate with acoustic emission techniques.[9]

The Kaiser Effect and Deformation

Another macroscopic manifestation of acoustic emission during deformation is the Kaiser effect (the material shows acoustic emission activity only after the applied load exceeds a previous load). The Kaiser effect is very useful when monitoring steel structures. It has been demonstrated though that this effect is not permanent.[10] The Kaiser effect decreases as a function of (1) the holding time between loads; and (2) the temperature at which the structure is kept between loads.

While most acoustic emission activity is associated with yielding and initiation of the plastic region, some alloys (particularly those containing second-phase particles such as inclusions and precipitates) exhibit significant acoustic emission activity at the onset of plastic instability and up to the fracture strain. Fracturing of inclusions and second-phase particles is responsible for this emission.

(text preceding Characteristics of Steel:)
to realize that the table can be extended to cover other acoustic emission parameters and not just signal amplitude.

PART 3
FRACTURE AND CRACK GROWTH AS ACOUSTIC EMISSION SOURCES

Crack Growth Sources

Early detection of crack growth is critical for the prevention of catastrophic failure, especially for metallic structures subjected to cyclic loads. As early as its inception, acoustic emission technology has been used for such applications.

Acoustic emission phenomena were originally used as research techniques for studying the mechanical behavior of materials. Much effort has since been directed toward characterizing the behavior of notched or discontinuous specimens under load. This type of analysis has in turn lead to correlations between the acoustic emission and parameters characterizing the state of stress at a crack tip. These parameters include crack length; stress intensity factor K; fracture strain at the crack tip and at the plastic zone ahead of the crack.

The development of these correlations grew out of the science of fracture mechanics and detailed analyses of the stress and strain states at the crack tip.[11]

Plastic Deformation Model

The earliest and perhaps the most comprehensive work in this area focused on the relationship between acoustic emission and the stress intensity factor in (1) the crack opening mode of fracture K_1; and (2) the plastic zone.[12] The model proposed in this study is comprehensive and can be applied, as will be shown later, to predict the behavior of cracked materials undergoing both brittle or ductile crack propagation to failure.

The model is based on the fact that acoustic emission is associated with plastic deformation processes. In a cracked specimen, observation of acoustic emission is linked to plastic deformation of the material ahead of the crack tip due to an increase in the stress level beyond the yield stress.

Several assumptions are made in this model. First, it is assumed that a metal or alloy gives the highest rate of acoustic emission when it is loaded to the yield strain. Second, the size and shape of the plastic zone ahead of the crack are determined from linear elastic fracture mechanics concepts, as described in Eq. 1.

$$r_y = \frac{1}{\alpha\pi}\left(\frac{K_I}{\sigma_y}\right)^2 \qquad \text{(Eq. 1)}$$

Where:

r_y = the plastic zone size;
K_1 = the stress intensity factor;
σ_y = the yield stress; and
α = two or six (depending on the plane stress or plane strain conditions at the crack tip respectively).

A third assumption is that strains at the crack tip vary as $r^{-1/2}$ where r is the radial distance from the crack tip. Fourth and finally, the observed acoustic emission count rate N is proportional to the rate of increase for the volume of the material V_p strained between ϵ_y and ϵ_u where ϵ_y is the yield strain and ϵ_u is the uniform strain, or:

$$N \propto V_p \qquad \text{(Eq. 2)}$$

These assumptions lead to development of the following equations for the model, assuming plane stress conditions with $\alpha = 2$.

$$V_p \sim \pi(r_y^2 - r_u^2)B \qquad \text{(Eq. 3)}$$

and

$$V_p = \pi B \left[\frac{1}{2\pi}\left(\frac{K}{E\epsilon_y}\right)^2\right]^2 - \left[\frac{1}{2\pi}\left(\frac{K}{E\epsilon_u}\right)^2\right]^2$$

and

$$V_p = \frac{B}{4\pi}\left[\frac{\epsilon_u^4 - \epsilon_y^4}{4\pi(E\epsilon_u\epsilon_y)}\right]K^4$$

where B is the plate thickness. Thus:

$$V_p \propto K^4 \qquad \text{(Eq. 4)}$$

then

$$\dot{N} = \frac{dN}{dt} \propto \dot{V_p} = \frac{d}{dt}(V_p) = \frac{d}{dt}K^4$$

which leads to:

$$N \propto K^4 \qquad \text{(Eq. 5)}$$

Attempts to verify the validity of the value of the exponent in Eq. 4 were undertaken in single notch specimens loaded in tension. Two materials were considered, beryllium and 7075-T6 aluminum alloy. Representative data of the acoustic emission total count as a function of stress intensity factor for both materials is shown in Fig. 3.

A summary of the experimental values of the exponent in Eq. 5 (which differ from the theoretical value of 4) is shown in Table 1 and indicates that the exponent has constantly higher values than the theoretical value of 4 predicted by the model and generally ranges between 6 and 11. As will be shown later, the importance of this model lies in coupling the observed acoustic emission due to plastic deformation of the crack tip and subsequent crack growth with stress state ahead of the crack. Thus it is assumed that the mechanics at the microscopic level that cause the crack to grow also contribute to the occurrence of acoustic emission in the material.

Crack Growth and Signal Correlation

The types of signals (and their amplitudes) observed during crack growth tests have been categorized.[13,14] In aluminum alloys such as 7075-T6, there is a marked increase in amplitude at the point of unstable crack growth in plane strain (where K_{IC} is measured on the load displacement curve). In more ductile materials such as HY80 steel, the acoustic emission level at the K_{IC} point is barely detectable. Plane strain fracture toughness K_{IC} is measured at the point of unstable crack growth.

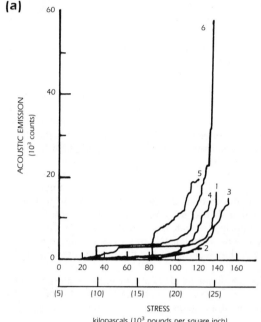

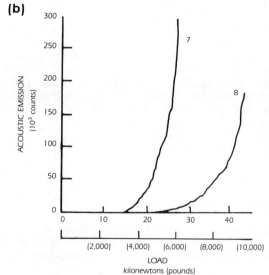

FIGURE 3. Acoustic emission as a function of (a) stress intensity factor for six beryllium fracture specimens; and (b) load for two aluminum specimens (see reference 12)

TABLE 1. Experimentally determined values of the exponent S in the relationship $N \propto K^S$ (see reference 12)

Run Number	S†	S‡	S§
1	11.9	8.7	8.7
2	*		
3	8.6	5.3	4.6
4	10.5	7.6	5.3
5**	7.0	5.9	4.6
6	7.5	6.5	5.7
7	6.0		
8	6.9		

† Calculated from elastic stress intensity factor (no plasticity correction).
‡ Calculated from plastically corrected stress intensity factor using a stress of 275 MPa (40 ksi) in the equation for the plastic zone size.
§ Calculated from plastically corrected stress intensity factor using a yield stress of 207 MPa (30 ksi) in the equation for the plastic zone size.
* Not calculated; did not agree with other runs.
** Calculated after subtracting spurious early jump of 5,000 counts.

Fracture as an Acoustic Emission Source

Linear Elastic Fracture Model

Another model has gained wide acceptance for relating acoustic emission activity observed in loaded notched specimens with the state of stress at the crack tip[15] and it has subsequently been verified experimentally.[16] As with the earlier model, this one uses the concepts of linear elastic fracture mechanics in its analysis. The assumption is that the plastic zone size ahead of the crack is much smaller than the relative dimensions of the specimen under test.

In this model, the total acoustic emission count is proportional to the area of the elastic-plastic boundary ahead of the crack. Thus, acoustic emission is related to either the discontinuity or to the applied stress at fracture by an equation of the form:

$$N = D \cdot S \quad \text{(Eq. 6)}$$

Where:

$N =$ the total acoustic emission count;
$S =$ the size of the plastic zone ahead of the crack; and
$D =$ proportionality constant.

The proportionality constant D depends on strain rate, temperature, thickness of the specimen and on the material microstructure. The plastic zone size is related to the applied stress σ and to the initial crack length by Eq. 7.

$$S = C \left\{ \sec\left(\frac{\pi\sigma}{2\sigma_1}\right) - 1 \right\} \quad \text{(Eq. 7)}$$

Where:

$C =$ half the crack length; and
$\sigma_1 =$ characteristic stress for the material.

The characteristic stress for the material is equal to the yield stress when linear elastic fracture mechanics is applicable (see Eq. 1). From Eqs. 6 and 7 it follows that:

$$N = DC \left\{ \sec\left(\frac{\pi\sigma}{2\sigma_1}\right) - 1 \right\} \quad \text{(Eq. 8)}$$

It has been shown[17] that for appreciable plastic deformation, the critical discontinuity size is related to the fracture stress by:

$$\sigma_f = \frac{2}{\pi} \sigma_1 \sec^{-1} \left\{ \exp\left(\frac{\pi K_{IC}^2}{8\sigma_1^2 C}\right) \right\} \quad \text{(Eq. 9)}$$

Combining Eqs. 8 and 9, the counts for failure N_f is obtained:

$$N_f = D \cdot C \left\{ \exp\left(\frac{\pi K_{IC}^2}{8\sigma_1^2 C}\right) - 1 \right\} \quad \text{(Eq. 10)}$$

For small stresses, Eq. 10 may be reduced to:

$$N_f = D \frac{\pi K_{IC}^2}{8\sigma_1^2} \quad \text{(Eq. 11)}$$

If Eq. 11 is compared with Eq. 5, it can be seen that Eq. 11 is more appropriate for materials with high toughness.

Figure 4 shows a plot comparing the experimental data obtained in the tests with the theoretical model of Eq. 8. It is evident that good agreement for this particular steel exists

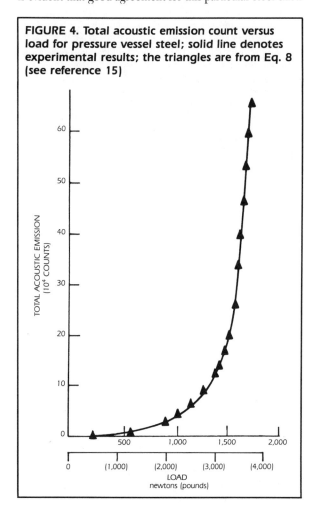

FIGURE 4. Total acoustic emission count versus load for pressure vessel steel; solid line denotes experimental results; the triangles are from Eq. 8 (see reference 15)

between theory and experiment. These data came from experiments on pressure vessel steel alloy. Other conclusions from the same study were (1) that most acoustic emission occurred at yielding, ahead of the crack in a reasonably ductile steel and (2) that the process of ductile crack growth did not produce significant acoustic emission.

Yielding in Ductile Materials

In the 1970s, the design emphasis on the use of ductile materials contributed to the development of theories of general yielding elastic-plastic fracture mechanics. The inadequacy of the critical stress intensity factor K_{IC} to characterize the process of massive yielding and subsequent blunting of the crack tip, brought about other criteria for characterization of the fracture toughness of ductile materials.

Among the proposed criteria were resistance or R-curve analysis for plane stress; the crack opening displacement; and the J-integral. Of these criteria, it appears that the J-integral provides a more unified model for characterization. Both macroscopic and microscopic aspects of the process of blunting are covered by the J-integral as are considerations of initiation of stable crack growth, so that the criterion has gained wide acceptance for ductile fracture.

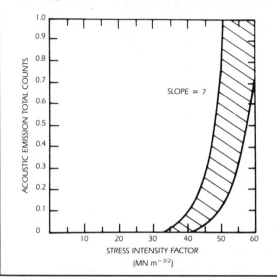

FIGURE 5. Normalized acoustic emission total counts versus the stress intensity factor for four specimens of D6 tool steel with different initial crack lengths (see reference 18)

Criteria for Fracture

Acoustic emission testing of notched specimens kept pace with the development of these criteria for ductile fracture. Correlations between the criteria and acoustic emission parameters were proposed. The most significant of the results are outlined below.

Extension of the Plastic Deformation Model

The initial effort[18] was to extend the analysis of the plastic deformation model to ductile materials that exhibit contraction at the crack tip. It was found that acoustic emission occurs at the point of deviation from linearity on the load displacement curve as well as at the beginning of massive plastic deformation and contraction at the crack tip.

Using this observation, it was proposed to modify the plastic zone size to account for plastic deformation at the crack tip. For the steel tested (namely D6 tool steel in the annealed state), the modification produced a value of 7 for the exponent α' in Eq. 12:

$$N = AK_1^{\alpha'} \qquad \text{(Eq. 12)}$$

It was also reported that the relationship in Eq. 12 was independent of the initial crack length. The plot of Eq. 12 for the D6 steel with specimens of different crack length is shown in Fig. 5.

Detection of general yielding ahead of the crack in A516-70 steel using acoustic emission has also been reported.[19,20] Two sensors (one resonant at 375 kHz and the other a wide band) were positioned on compact tension specimens of this material. Fracture toughness tests were performed to determine J_{IC} according to ASTM Specification E813-81. A plot of the total count versus the J-integral for specimens of different crack lengths is shown in Fig. 6.

It can be seen from this figure that the total count, for all crack lengths (expressed as the ratio of crack length to specimen width) reaches a constant value indicating no further emission at J about 30 kJ•m^{-2}. Yet for this steel J_{IC} is found to be 120 kJ•m^{-2}. It can be concluded that, for this ductile steel, the plastic deformation of the material ahead of the crack seems to be the main source of acoustic emission.

Figure 7 shows the frequency spectra of typical signals obtained at various points of the load-load point displacement curve. It shows that as the load is increased, a shift to lower frequencies is apparent. This observation is in agreement with earlier findings[21] where cracked specimens are characterized by lower frequencies as compared to uniformly deformed uncracked steel specimens.

Calculations Based on Acoustic Emission Data

In another study[22] on 7075-T6 aluminum alloy, attempts were made to use the acoustic emission data to calculate a J_{IAE} for this material. Critical values of K and J determined

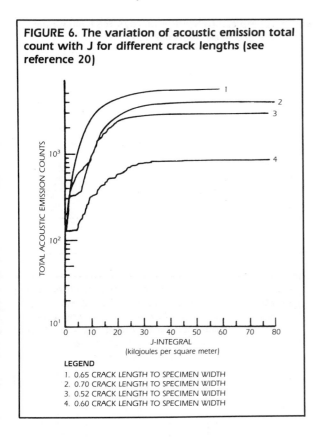

FIGURE 6. The variation of acoustic emission total count with J for different crack lengths (see reference 20)

LEGEND
1. 0.65 CRACK LENGTH TO SPECIMEN WIDTH
2. 0.70 CRACK LENGTH TO SPECIMEN WIDTH
3. 0.52 CRACK LENGTH TO SPECIMEN WIDTH
4. 0.60 CRACK LENGTH TO SPECIMEN WIDTH

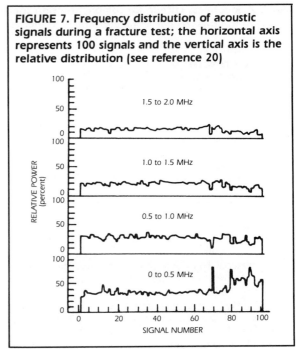

FIGURE 7. Frequency distribution of acoustic signals during a fracture test; the horizontal axis represents 100 signals and the vertical axis is the relative distribution (see reference 20)

by acoustic emission were about 6 and 10 percent lower than values of K_{IC} and J_{IC} measured by fracture mechanics methods. Thus was demonstrated acoustic emission's sensitivity for detecting the start of crack propagation in this alloy.

The agreement of K_{IC} and J_{IC} with K_{IAE} and J_{IAE} for the alloy is shown in Table 2 for both the TL and LT orientations. For all tests, K_{IAE} is less than K_{IC}, and J_{IAE} is less than J_{IC}, indicating that acoustic emission is able to detect the first crack propagation within the material, before any significant changes in the load-load point displacement curve are observed.

The reported results on steels and aluminum alloys indicate that ductile crack propagation mechanisms have low acoustic emission activity. However, emission sufficient for detection is produced by the plastic deformation process (which occurs during the initial blunting of the crack) and general yielding at the crack tip.

Thus, acoustic emission shows great promise for detecting the onset of general yielding in materials that exhibit elastic-plastic behavior at the crack tip, and which can therefore be characterized by J_{IC} or critical crack opening displacement. In fact, this hypothesis was tested[23] on two types of low alloy steel gas pipeline. It was determined that, even for low strength steels with significant ductility, the onset of general yielding can be detected by acoustic emission.

It can be said in summary that mechanical design requires the use of ductile materials with high fracture toughness but that such materials generally exhibit ductile fracture mechanisms (void nucleation and coalescence) and consequently produce low emission activity during testing. This problem has been analyzed at length and one proposal recommends testing these materials in their worst case situations.[24]

Worst Case Acoustic Emission Tests

This procedure is based on the premise that initial cracking in a structure such as a pressure vessel is due to the presence of structural or fabrication anomalies. Such discontinuities include embrittlement from improper heat treatment; segregation of intermetallic compounds; strain aging; environmental effects such as stress corrosion cracking; or hydrogen embrittlement.

The growth of such cracks during the service life of a structure is time dependent because the applied load is nominally constant. Thus the stress intensity factor of the crack increases with time as the crack length increases, until a critical value is finally reached.

In ductile materials, the process of crack growth still relies on brittle mechanisms in the crack region. The ductility

TABLE 2. Toughness criteria for compact tension specimens in TL and LT orientation of 7075-T651 aluminum; J is given in newtons per millimeter and K in megapascals root meter (see reference 22)

RATIO OF CRACK LENGTH TO SPECIMEN WIDTH	J_{IC}	J_{IAE}	K_{IC}*	K_{IAE}*	K_{IC}**	K_{IAE}**
0.430	8.39	7.71	25.8	24.7	26.5	24.7
0.543	7.66	7.06	24.6	23.6	24.4	23.4
0.623	8.27	7.89	25.6	25.3	26.3	25.0
0.040	9.83		27.9		27.8	
0.538	8.73	8.40	26.3	25.5	26.1	25.3
0.545	9.15	8.37	26.9	25.8	26.6	25.5
0.597	8.88	8.28	26.5	25.6	26.0	25.2
0.647	10.36	9.18	28.5	27.0	27.8	26.3

* $K = (1.1)^{1/2} JE$
** K according to ASTM

of the material contributes to crack arrest until further buildup of stress and embrittlement of the crack region forces propagation.

Brittle crack growth mechanisms are known to produce high levels of acoustic emission activity with signals of relatively high amplitude. Crack arrest mechanisms, which are ductile in nature, produce much less acoustic emission activity.

Displacement Control Versus Load Control

Field tests of pressure vessels are performed under load control conditions, where the load is continuously increased as the test proceeds. In contrast, laboratory tests on notched or precracked fatigue specimens are based on displacement control. An example of such a laboratory test is the determination of the J-integral using reference specimens conforming to fracture mechanics specifications for ductile materials.

As shown in Fig. 8 for a typical ductile material, there is a significant difference in acoustic emission activity due to the two types of test control conditions. The acoustic activity exhibits a sudden sharp increase as fracture becomes imminent in the load controlled specimens. This is not observed in the displacement controlled specimens.

These results emphasize the importance of understanding the causal mechanisms of acoustic emission in real structures. Good design and material selection result in new structures that will not fail unless loaded beyond their nominal yield strength. Over time, microstructural and environmentally induced discontinuities result in circumstances that may lead to brittle fracture. Acoustic emission is able to detect the process of brittle crack growth before catastrophic failure takes place and is a powerful technique for determining structural integrity.

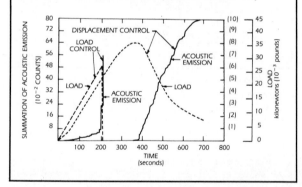

FIGURE 8. Acoustic emission counts and load as a function of time for two identical compact tension specimens of 2219 aluminum alloy, one tested under displacement control and the other tested under load control (see reference 20)

PART 4
ACOUSTIC EMISSION FROM FATIGUE

Acoustic emission is very useful for detecting the initiation and propagation of cracks in fatigued materials. It may also be used to monitor low cycle fatigue.

Acoustic emission techniques have long been used to follow the progress of cracks during fatigue in D6 steel.[25,26] The amplitude and rate of acoustic emission are related to the crack growth rate and the stress intensity factor as shown in Eq. 13:

$$\Delta A \propto (\Sigma g)^2 \frac{E}{K^2} \quad \text{(Eq. 13)}$$

Where:

ΔA = the increase in crack surface; and
Σg = the sum of the acoustic emission occurring during the crack surface increment ΔA.

This equation was obtained from tests of crack growth at rates in excess of 10^{-3} mm per cycle. At lower crack growth rates, where there is high background noise with amplitudes as large as those of the acoustic emission signals, signal characterization is much more difficult.

Techniques for Monitoring Fatigue Cracking

One proposal has proven very useful for solving the problem of noise discrimination in acoustic emission tests at crack growth rates lower than 10^{-4} mm per cycle.[27] An electronic window is used to segregate measurements on valid signals during selected parts of the loading cycle away from the background noise.

Another technique uses two sensors positioned at equal distances from the crack tip. Signals arriving at both sensors at the same time are considered valid.[28] An analog correlator is used to multiply the two coincident signals at the sensors and the results are used for event counts during fatigue.

A third approach uses a fatigue analyzer[29] to show the acoustic emission events as a function of time and of the loading cycles. A photograph obtained on a storage oscilloscope (similar to that in Fig. 9) shows noise due to the testing system as well as the signals from crack propagation occurring at the upper part of the loading cycle.[30]

FIGURE 9. Typical photograph from a fatigue analyzer (see reference 30)

A system for fatigue testing that incorporates the analog correlator and the fatigue analyzer has been used[31] to characterize acoustic emission during crack propagation (see Fig. 10).

Acoustic Emission Count Rate

In acoustic emission monitoring of fatigue crack growth, the most common representation of the data is the acoustic emission count rate per cycle N' against the stress intensity factor ΔK. Other representations have also been reported.[32] On a log-log scale, the relationship between N' and ΔK assumes a straight line and can be described with an equation of the form:

$$N' = A(\Delta K)^n \quad \text{(Eq. 14)}$$

The terms A and n are constants. Typical curves for a variety of steels represented by Eq. 14 are shown in Fig. 11.[33]

Equation 14 bears strong similarity with the Paris law for crack propagation in fatigue[34] namely:

$$\frac{da}{dN} = C(\Delta K)^m \quad \text{(Eq. 15)}$$

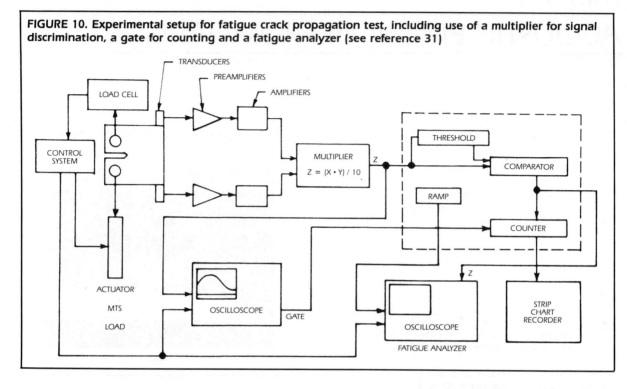

FIGURE 10. Experimental setup for fatigue crack propagation test, including use of a multiplier for signal discrimination, a gate for counting and a fatigue analyzer (see reference 31)

The similarity suggests that the mechanisms for crack growth are also directly responsible for the occurrence of acoustic emission in the material.

Relation Between Count Rate and Crack Growth

Attempts have been made to relate the exponent n Eq. 14 and to the exponent m of Eq. 15. In one case,[35] it was reported that n is equal to 5. In other cases,[36] it was found that $n = m$ or that $n = m + 2$. The mechanisms of acoustic emission in this case are either due to energy released during crack extension or to the deformation and fracture within the plastic zone.

It has also been reported[37] that, depending on the ductility of the material and on the test conditions (the loading ratio $R = \sigma_{min}/\sigma_{max}$), a combination of both plastic deformation and fracture become operational within the plastic zone. In this case, n assumes a higher value than m when R is low. It tends toward m when R is relatively high. The combined contribution of the two mechanisms is as follows for plastic yielding:

$$N_p' = C_p \Delta K^m \frac{\Delta K^2}{(1 - R)^2} \qquad \text{(Eq. 16)}$$

and

$$N_C' = C_s \frac{\Delta K^m}{(1 - R)^m} \qquad \text{(Eq. 17)}$$

for crack tip microdeformation. The overall count rate will therefore be:

$$N' = N_p' + N_C' \qquad \text{(Eq. 18)}$$

The exponent in Eq. 14 will always range between m and $m + 2$. The limits of this range come from the ΔK^m term in Eq. 17 and from Eq. 16, where $\Delta K^m \cdot \Delta K^2 = \Delta K^{m+2}$.

These equations allow the determination of fatigue life curves based on acoustic emission test data. Such curves can be derived for a material or structure and provide a dependable assessment of fatigue damage to the specific material containing a crack.

Monitoring Low Cycle Fatigue

In low cycle fatigue, failure occurs after a relatively low number of cycles at a high level of stress. The curve of acoustic emission total counts versus number of cycles for a given prescribed stress or strain is characterized by the presence of three stages that correspond closely with the stress

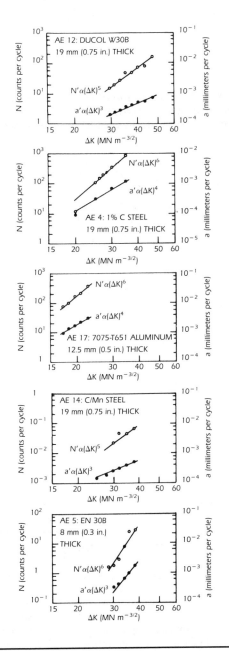

FIGURE 11. Acoustic emission count rate and da/dN as a function of ΔK in fatigue (see reference 33)

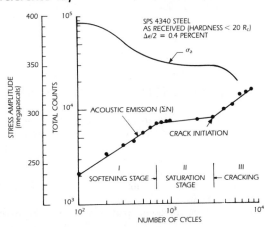

FIGURE 12. Acoustic emission total count and stress amplitude as a function of number of cycles for a 4340 steel in low cycle fatigue (see reference 40)

state in the material.[38] This is shown in Fig. 12 for 4340 steel.

At the beginning of the test, an initial softening (or hardening, depending on the material) results in high acoustic emission activity. Similar results have been reported for other materials, including titanium.[39] The initial increase in total counts is followed by a decrease in activity that accompanies macroscopic crack initiation. Propagation to failure is then observed. A model for analyzing this curve allows prediction of fatigue life from acoustic emission measurements.[40] If the extent of damage is taken as a ratio of the received number of cycles N_b to the number of cycles to failure N_f, then the following equation applies:

$$\Sigma N = \beta \left(\frac{N_b}{N_f} \right)^{m'} \qquad \text{(Eq. 19)}$$

The terms β and m' are constants for the material and depend on the microstructure and the strain amplitude during the test.

For steels, the exponent m' does not exceed a value of one. It is usually higher for quench tempered microstructures than it is for annealed or normalized materials. It also depends strongly on the extent of prescribed total strain during the fatigue test.

Equation 19, with known values of β and m' for a given material, relates acoustic emission test data to the fatigue life of a material and can be used for evaluating the extent of fatigue damage.

PART 5
CORROSION AND STRESS CORROSION CRACKING

Acoustic Emission from Corrosion

Corrosion and fatigue are responsible for the failure of most industrial structures and components. Acoustic emission has potential for monitoring and detecting the initiation and propagation of cracks resulting from the different forms of corrosion. It can perform this function while the structure is in operation and can determine the progress and severity of cracks in real-time.

Many forms of corrosion, particularly stress corrosion cracking, have been studied with acoustic emission. For example, measurements have been made for aluminum alloys that were freely corroding or coupled with copper, steel, cadmium or zinc in sodium chloride solution.[41] It was found that while the acoustic emission from freely corroding aluminum alloys such as 1100 is rather low, coupling with copper or 4340 steel results in an immediate return to the rate observed with freely corroding pure aluminum.

These results are in agreement with separate observations on the corrosion rate of 1100 aluminum alloy; these observations indicate that coupling with copper results in an increase of twelve and that coupling with steels produces an increase of two orders of magnitude.[42] Acoustic emission measurements can be used to monitor the pitting corrosion in these alloys. Hydrogen evolution in acidified pits is reportedly the source of the observed acoustic emission activity.

Relation Between Count Rate and Corrosion

The relation between acoustic emission total count and corrosion rates has been demonstrated.[43] It was found that a linear relationship exists between the total count and the amount of hydrogen collected. This relationship was established using observations of an aluminum wire placed in a hydrochloric acid solution.

In addition, an increase in acoustic emission rate was found to occur when galvanic coupling occurs between aluminum and iron. This is shown in Fig. 13 where the total count increases by several orders of magnitude when iron and aluminum wires are connected in a galvanic cell.

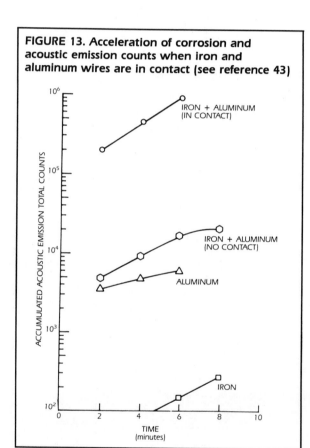

FIGURE 13. Acceleration of corrosion and acoustic emission counts when iron and aluminum wires are in contact (see reference 43)

Stress Corrosion Cracking

Stress corrosion cracking has been widely studied by acoustic emission techniques. One such study[44] used a bolt-loaded 4340 steel specimen to demonstrate that as the crack propagates due to the action of sodium chloride solution acoustic emission burst signals are produced. Another study [45] reported that the emission from a zircaloy-2 specimen of the double cantilever type can be divided into continuous emission due to plastic deformation at the crack tip and high amplitude burst type signals caused by propagation of the crack.

In most cases of stress corrosion cracking, an initiation period occurs and the acoustic emission activity is low. This is followed by a high rate of emission that indicates impending failure of the specimen. A typical acoustic emission curve of stress corrosion cracking is shown in Fig. 14 for a zircaloy specimen in a one percent iodine in methanol solution.

The energy of the signal (obtained by squaring the measured root mean square of the amplitude and integrating over the period of the test) was used as the characteristic parameter of acoustic emission for stress corrosion cracking.[46] Another study[47] reports the tempering of 4340 steel and its effect on the acoustic emission produced during stress corrosion cracking in aqueous solutions. It was found that there exists a correlation between the rate of acoustic emission and the rate of crack propagation, expressed in terms of the stress intensity factor.

Monitoring Hydrogen Embrittlement and Adverse Test Environments

Hydrogen embrittlement is also monitored with acoustic emission techniques.[48] An acoustic emission rate obtained at a value close to that of the critical stress intensity factor can be used to predict the onset of unstable crack growth. It has also been shown that the acoustic emission activity follows a period of initiation where the level is fairly low and that it increases markedly towards the failure of the specimen.

A methane reformer furnace is one example of acoustic emission capabilities for testing structures exposed to adverse environments. In this case, corrosion, hydrogen embrittlement and high temperature conditions are present during thermal cracking of heavy oils. Operating temperatures exceed 870 °C (1,600 °F).[49]

Another example is the detection of hydrogen diffusion and chatter cracking during cooling of rail steel blooms at temperatures exceeding 760 °C (1,400 °F).[50]

The well documented continuous monitoring of nuclear reactor fuel rods in an adverse corrosive environment is another example.[51] Analysis in the frequency domain is performed to detect frequency characteristics associated with the presence of leaks.

Conclusions

This text discusses the macroscopic origins of acoustic emission and indicates the potential of the acoustic emission nondestructive testing technique for monitoring in real-time

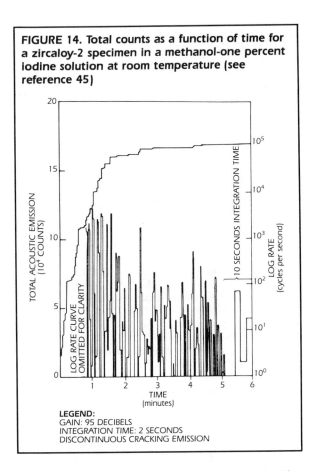

FIGURE 14. Total counts as a function of time for a zircaloy-2 specimen in a methanol-one percent iodine solution at room temperature (see reference 45)

LEGEND:
GAIN: 95 DECIBELS
INTEGRATION TIME: 2 SECONDS
DISCONTINUOUS CRACKING EMISSION

the deterioration and cumulative damage in structures. The technique is also useful for providing a rapid indication of the progress of cracking from the following sources: (1) brittle initiation; (2) following ductile plastic deformation; (3) from cyclic loading; and (4) corrosion.

The macroscopic mechanisms of acoustic emission from fracture and fatigue are directly related to the fracture mechanisms, as analyzed using fracture mechanics concepts. This leads to a predictive approach for determining the expected acoustic emission activity based on the known fracture toughness of the material and the stress distribution in the structure.

The relationships between acoustic emission activity and the fracture parameters K, J and the fatigue parameter ΔK in fatigue are the foundations for quantitative evaluation of the macroscopic aspects of acoustic emission. It is expected that as fracture mechanics develops, parallel extensions of these relationships will be discovered, allowing more accurate ways of predicting the presence, size and severity of existing subcritical discontinuities before they lead to catastrophic failure.

REFERENCES

1. Bassim, M.N. and M. Houssny-Emam. "Time and Frequency Analysis of Acoustic Emission Signals." *Acoustic Emission*, J.R. Mathews, ed. Gordon and Breach Scientific Publishers (1983): pp 139-163.
2. Stephens, R.W.B. and A.A. Pollock. "Waveforms and Frequency Spectra of Acoustic Emission." *Journal of the Acoustical Society of America*. Vol. 50, No. 3 Part 2 (September 1971): pp 904-910.
3. Knott, J.F. *Fundamentals of Fracture Mechanics*. New York, NY: Wiley Publishing (1973).
4. Dunegan, H.L. and A.T. Green. *Acoustic Emission*. ASTM STP 505. Philadelphia, PA: American Society for Testing and Materials (1972): pp 100.
5. Dunegan, H.L. and D.O. Harris. "Acoustic Emission: A New Nondestructive Testing Tool." *Ultrasonics*. Vol. 7, No. 3 (1969): pp 160-166.
6. Handfield, L. Ph.D. thesis. Montreal, Canada: Ecole Polytechnique (1985).
7. Spanner, J.C. *Acoustic Emission: Techniques and Applications*. Evanston, IL: Intex Publishing Company (1974).
8. Bassim, M.N., D.R. Hay and J. Lanteigne. "Effect of Crack Size on Acoustic Emission during Mechanical Testing." *Proceedings of the Second International Conference on Mechanical Behaviour of Materials*. ICM-II (1976): pp 1,636.
9. Blanchette, Y. Masters thesis. Montreal, Canada: Ecole Polytechnique (1981).
10. Bassim, M.N. and M. Veillette. "Acoustic Emission Mechanisms in Armco Iron." *Materials Science and Engineering*. Vol. 50 (1981): pp 285.
11. Rolfe, S.T. and J.M. Barsom. *Fracture and Fatigue Control in Structures*. Prentice-Hall (1977).
12. Dunegan, H.L., D.O. Harris and C.A. Tatro. "Fracture Analysis by Use of Acoustic Emission." *Engineering Fracture Mechanics*. Vol. 1, No. 1 (June 1968): pp 105-122.
13. Schofield, B.H. *Investigation of Applicability of Acoustic Emission*. AFML-TR-65-106. Air Force Materials Laboratory (1965).
14. Hartbower, C.E., W.W. Gerberich and H. Liebowitz. "Investigation of Crack-Growth Stress-Wave Relationships." *Engineering Fracture Mechanics*. Vol. 1, No. 2 (1968): pp 291-308.
15. Palmer, I.G. and P.T. Heald. "The Application of Acoustic Emission Measurements to Fracture Mechanics." *Materials Science and Engineering*. Vol. 11, No. 4 (1973): pp 181-184.
16. Palmer, I.G. "Acoustic Emission Measurements on Reactor Pressure Vessel Steel." *Materials Science and Engineering*. Vol. 11, No. 4 (1973): pp 227-236.
17. Smith, E. *Proceedings of the Royal Society*. Vol. A299. London, England (1967): pp 455.
18. Masounave, J., J. Lanteigne, M.N. Bassim and D.R. Hay. "Acoustic Emission and Fracture of Ductile Materials." *Engineering Fracture Mechanics*. Vol. 8, No. 4 (1976): pp 701-709.
19. Blanchette, Y., M.N. Bassim and J.I. Dickson. "Acoustic Emission Associated with Fracture of A516-70 Steel." *Proceedings of the Fifth Canadian Fracture Conference*, L. Simpson, ed. Pergamon Press (1981): pp 191.
20. Blanchette, Y., J.I. Dickson and M.N. Bassim. "Detection of General Yielding in A516 Steel by Acoustic Emission." *Engineering Fracture Mechanics*. Vol. 17 (1982): pp 227.
21. Bassim, M.N., D.R. Hay and J. Lanteigne. "Waveform Analysis of Acoustic Signals from a Mild Steel." *Materials Evaluation*. Vol. 34, No. 5. Columbus, OH: The American Society for Nondestructive Testing (1976): pp 109-113.
22. Blanchette, Y., M.N. Bassim and J.I. Dickson. "The Use of Acoustic Emission to Evaluate Critical Values of J and K in 7075-T651 Aluminum Alloy." *Engineering Fracture Mechanics*. Vol. 20 (1984): pp 359.
23. Bassim, M.N. and K. Tangri. "Leak Detection in Pipelines with Acoustic Emission." *Proceedings of the International Conference on Pipeline Inspection*. Edmonton, Canada: Canadian Minerals and Energy Technology Center Publication (1983): pp 529-544.
24. Dunegan, H.L. "Progress in Acoustic Emission II." *Proceedings of the Seventh International Acoustic Emission Symposium*. Tokyo, Japan: Japanese Society of Nondestructive Inspection (1984): pp 294-301.
25. Hartbower, C.E., W.W. Gerberich and P.P. Crimmins. *Characterization of Fatigue-Crack Growth by Stress-Wave Emission*. NASA CR-66303. Sacramento, CA: Aerojet General Corporation (1966).
26. Gerberich, W.W. and C.E. Hartbower. "Some Observations on Stress Wave Emission as a Measure of Crack Growth." *International Journal of Fracture Mechanics*. Vol. 3, No. 3 (1967): pp 185-192.
27. Morton, T.M., S. Smith and R.M. Harrington. "Effect

of Loading Variables on the Acoustic Emissions of Fatigue-Crack Growth." *Experimental Mechanics*. Vol. 14, No. 5 (1974): pp 208-213.
28. Smith, S. and T.M. Morton. "Acoustic Emission Detection Techniques for High-Cycle Fatigue Testing." *Experimental Mechanics*. Vol. 13, No. 5 (1973): pp 193-198.
29. Carlyle, J.M. and W.R. Scott. "Acoustic-Emission Fatigue Analyzer." *Experimental Mechanics*. Vol. 16, No. 10 (1976): pp 369-372.
30. Hamel, F. Masters thesis. Montreal, Canada: Ecole Polytechnique (1979).
31. Hamel, F. and M.N. Bassim. "Detection of Crack Propagation in Fatigue with Acoustic Emission." *Proceedings of the Sixth International Conference on the Strength of Metals and Alloys*, R.C. Gifkins, ed. Pergamon Press (1982): pp 839-844.
32. Harris, D.O. and H.L. Dunegan. "Continuous Monitoring of Fatigue Crack Growth by Acoustic Emission Techniques." *Experimental Mechanics*. Vol. 14, No. 2 (1974): pp 71-81.
33. Sinclair, A.C., D.C. Connors and C.L. Formby. *Materials Science and Engineering*. Vol. 28 (1977): pp 263.
34. Paris, P. and F. Erdogan. *Transactions of ASME*. New York, NY: American Society for Mechanical Engineers (1963): pp 528.
35. Morton, H.L., R.M. Harrington and J.G. Bjeletich. "Acoustic Emissions of Fatigue Crack Growth." *Engineering Fracture Mechanics*. Vol. 5, No. 3 (1973): pp 691.
36. Lindley, T.C., I.G. Palmer and C.E. Richards. *Materials Science and Engineering*. Vol. 32 (1978): pp 1.
37. Hamel, F., J.P. Bailon and M.N. Bassim. "Acoustic Emission Mechanisms During High Cycle Fatigue." *Engineering Fracture Mechanics*. Vol. 14 (1981): pp 853.
38. Houssny-Emam, M. and M.N. Bassim. "Study of the Effect of Heat Treatment on Low Cycle Fatigue in 4340 Steel by Acoustic Emission." *Materials Science and Engineering*. Vol. 61 (1983): pp 79.
39. Handfield, L., M.N. Bassim and J.I. Dickson. Unpublished.
40. Bassim, M.N. "Assessment of Fatigue Damage with Acoustic Emission." *Journal of Acoustic Emission*. Vol. 4 (1985): pp 5,224.
41. Mansfield, F. and P.J. Crocker. *Corrosion*. Vol. 35 (1979): pp 541.
42. Mansfield, F., D.H. Hengstenberg and J.V. Kenkel. *Corrosion*. Vol. 30 (1974): pp 343.
43. Rettig, T.W. and M.J. Felsen. "Acoustic Emission Method for Monitoring Corrosion Reactions." *Corrosion*. Vol. 32, No. 4 (1976): pp 121-126.
44. Dunegan, H.L. "Round Robin Testing of the Frequency Spectra of Acoustic Emission Signals." *Proceedings of the Acoustic Emission Working Group* (1976).
45. Cox, B. "A Correlation between Acoustic Emission during SCC and Fractography of Cracking of the Zircaloys." *Corrosion*. Vol. 30, No. 6 (1974): pp 191-202.
46. Harris, D.O. and R.L. Bell. "The Measurement and Significance of Energy in Acoustic Emission Testing." *Experimental Mechanics*. Vol. 17, No. 9 (1977): pp 347.
47. Chaskelis, H.H., W.H. Callen and J.M. Krafft. "Acoustic Emission from Aqueous Stress Corrosion Cracking in Various Tempers of 4340 Steel." *NRL Memorandum Report 2608* (1973).
48. Dunegan, H.L. and A.S. Tetelman. DRC-1-6. Irvine, CA: Dunegan Research Corporation (1972).
49. Hay, D.R., M.N. Bassim and D. Dunbar. "In-Service Acoustic Emission Monitoring of a Steam-Methane Reformer Furnace." *Materials Evaluation*. Vol. 34, No. 1. Columbus, OH: The American Society for Nondestructive Testing (1976): pp 20-24.
50. Lin, P.L., C.W. Bale, A.D. Pelton, D.R. Hay, E.S. Rodriguez, V. Mustapha, S.G. Whiteway and W.F. Caley. *Proceedings of the Nineteenth Annual Meeting of Metallurgists*. Halifax, Nova Scotia: Canadian Institute of Mining and Metallurgy (1979).
51. Kupcis, O.A. and R.T. Hartler. Report 76-478-K. Ontario, Canada: Ontario Hydro Research Division (1976).

SECTION 3

MICROSCOPIC ORIGINS OF ACOUSTIC EMISSION

Haydn N.G. Wadley, National Bureau of Standards, Gaithersburg, Maryland

John A. Simmons, National Bureau of Standards, Gaithersburg, Maryland

PART 1
FUNDAMENTALS OF ACOUSTIC EMISSION GENERATION

Origination of Elastic Waves

Acoustic emission can be thought of as the naturally generated ultrasound created by local mechanical instabilities within a material. Imagine that an object has been placed under load sometime in the past and is now in elastic equilibrium throughout. Suppose a small crack appears within the object at a point distant from that point where the loads were applied. The surfaces of the crack are able to move in such a manner that they become stress free. In doing so, they release some of the stored elastic energy in the object.

This release of energy is in the form of elastic waves that propagate freely throughout the object, experiencing reflections or mode conversions at the object's boundaries. In fact, the propagation of these elastic waves is the mechanism by which the changed elastic state in the immediate vicinity of the crack is transmitted to the rest of the object. The waves enable the entire object to change its shape and accommodate the crack, and each propagating wavefront carries a component of this shape change. In the perfectly elastic object used for mathematical modeling, these waves propagate indefinitely and mechanical equilibrium is never established.

However, in practical materials, absorption (conversion to heat) occurs and the waves eventually dissipate, allowing the object to assume a new (cracked) shape, establishing a new equilibrium with the loads throughout.

Detecting Acoustic Emission

A transducer attached to an object becomes a sensor capable of detecting the motion of the surface with which it is in contact. The transducer's response, following a local crack extension, is the acoustic emission we observe in experiments and tests. Because acoustic emission transducers are usually constructed from piezoelectric slabs and have a resonant behavior, their sensitivity varies with frequency and is usually greatest in the range from 0.1 to 2.0 MHz. Only these frequency components of the emitted waves are sensed. Neither the static surface strains (the difference in shapes of the two equilibrium states discussed above) or the very high frequency components are sensed.

Observed acoustic emission signals, at least close to the source, are dominated by surface displacements associated with wavefront arrivals at the sensor location. If the wavefront contains appreciable frequency components in the transducer bandwidth, then a voltage is created across the faces of the transducer. If this voltage exceeds the background noise, the emission is detected by the transducer.

Time Dependence of Detectability

The detectability of acoustic emission depends on the temporal nature of the source because this determines the amplitude of each spectral component in each wavefront. If the source operates slowly and there is sufficient time for the body to return to quasi equilibrium before the crack has appreciably extended, then it is possible that no signal will be detected (though a static strain gage might register the change of shape if the crack grows sufficiently large).

Conversely, if the crack extends rapidly and then stops so that its growth time is about equal to one divided by the sensor bandwidth, then the emitted wavefronts are dominated by frequency components in the detectable range.

The Need for Studying Microscopic Origins

Understanding the microscopic origins of acoustic emission allows the user to determine the likelihood for detecting various events of potential interest and for distinguishing between them.

The purpose of this discussion is to present the relationship between local mechanical instabilities and the resulting acoustic emission. In particular, expressions will be developed for the surface motion produced by microscopic acoustic emission sources such as dislocations, microcracks and phase transformations, particularly those involving the formation or annihilation of martensite. Using results from tensor wave propagation theory, micromechanical models are used to develop detectability criteria for microscopic sources. Finally, these criteria are used to identify the origin of acoustic emission in materials undergoing deformation, fracture and phase changes.

In this text, a standard assumption of linear isotropic elasticity is used. All metals only approximate these assumptions. In practice, metals exhibiting weak nonlinear behavior due to internal friction mechanisms are anisotropic to a greater or lesser degree, and if preferred grain orientation (texture) is present, the metals will not exhibit spherical wave spreading. In addition, the polycrystalline nature of engineering materials results in grain scattering that can be particularly strong for spectral components whose wavelengths approach or exceed the grain size. None of these effects are considered in the following text because their relative contribution varies from case to case and is very sensitive to detailed aspects of the microstructure.

The elastic waves emitted by an acoustic emission source may also be considered as a superposition of the normal modes of the particular object in which the emission occurs. This alternate way of analyzing acoustic emission is not considered in this chapter.

PART 2
MICROMECHANICAL MODELING

Presented here are the principles for developing detectability criteria for acoustic emission occurring after a general mechanical instability. These principles are later applied to acoustic emission from dislocation motion, crack growth and phase transformations.

Formulation of the Model

The sequence of events giving rise to an acoustic emission signal are summarized in Fig. 1. Elastic waves are generated due to a local change of stress in a region of volume V. These waves propagate (spherically in an isotropic body) as a mechanical disturbance through the structure, causing a time varying surface displacement $u(t)$. The value of $u(t)$ varies with sources and receiver positions for the following reasons: (1) inverse square law decay (note that the wavefront area increases with distance between source and receiver); (2) source directivity; and (3) wavefront reflection or mode conversion at a free surface.

A sensor located on the structure detects the surface disturbance and generates a voltage waveform that is observed as an acoustic emission signal, provided it exceeds the background noise level.

Viewed from the perspective of the surrounding elastic structure, the acoustic emission source appears as an apparent change of stress (stress drop) from within the structure or as a change in tractions at the surface of the structure. Using a Green's function approach[1] the displacement at position r arising from both internal (volume stress) and surface (traction) sources can be written as follows.

FIGURE 1. Schematic diagram of the acoustic emission process

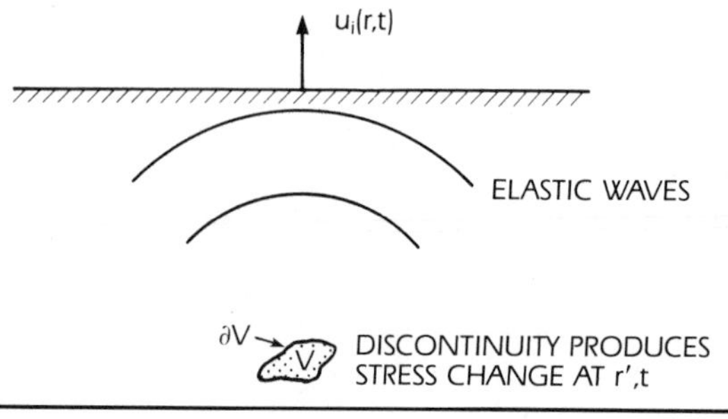

$$u_i(r,t) = \int dr' \int G_{ij,k'}(r,r',t - t')\Delta\sigma_{jk}(r',t')dt' \quad \text{(Eq. 1)}$$
$$- \int dS'_{k'} \int G_{ij}(r,r',t - t')\Delta\tau_{jk}(r',t')dt'$$

Here $G_{ij}(r,r',t)$ are components of the dynamic elastic Green's tensor representing displacement in the x_i direction at r as a function of time t, due to a unit strength force impulse applied at r' and $t = 0$ in the x_j direction. Green's tensor is the solution to the wave equation for a unit force impulse source. The comma notation is used to denote partial differentiation so that $G_{ij,k'}$ is the corresponding wave equation solution for a unit force derivative (dipole) impulse. The terms $\Delta\sigma_{jk}$ and $\Delta\tau_{jk}$ are the volume stress and surface traction changes associated with the source; S' is a vector normal to the surface of the structure.

The time integral in Eq. 1 comprises a convolution. This convolution provides in principle the basis for predicting surface displacement waveforms from stress change sources if the Green's tensor is known (either calculated or measured). Later in this discussion, these stress changes will be deduced for internal microscopic discontinuity sources. Because the source and sensor are of finite size, the representation given by Eq. 1 requires a Green's tensor to be evaluated between every source and every receiver point.

This is a numerically exhausting task beyond normal computing capabilities and it is compounded by the possibility that each stress component might have a different temporal character. To simplify the formulation, the infinitesimal source approximation (wherein the source diameter is considered small compared to $|r - r'|$) and surface sources will be omitted.

For infinitesimal sources, the usual approach[2,3] is to expand Green's tensor as a Taylor's series about the centroid of the source r_o:

$$G_{ij,k'}(r,r',,t) = G_{ij,k'}(r,r_o',t) \quad \text{(Eq. 2)}$$
$$+ G_{ij,k'l'}(r,r_o',t)\Delta r_l' + \ldots$$

where $\Delta r_l' = r_l' - r_{ol}'$. Substituting Eq. 2 into Eq. 1 gives:

$$u_i(r,t) = \int G_{ij,k'}(r,r_o',t - t')\overline{\Delta\sigma_{jk}}(t')dt' \quad \text{(Eq. 3)}$$
$$+ \int G_{ij,k'l'}(r,r_o',t - t')\Delta\Gamma_{jkl}(t')dt' + \ldots$$

where the quantity $\overline{\Delta\sigma_{jk}}(t) = \int \Delta\sigma_{jk}(r',t)dr'$ is the dipole tensor and $\overline{\Delta\Gamma_{jkl}}(t) = \int \Delta\sigma_{jk}(r',t)\Delta\eta_l'dr'$ is the quadrupole tensor. Examples of epicenter displacement waveforms for monopole, dipole and quadrupole sources are shown in Figs. 2, 3 and 4.

At practical frequencies, only small error is introduced for microscopic sources by truncating Taylor's series at the dipole (first) term, provided the source-to-receiver distance is much greater than the source size. The source quantity $\overline{\Delta\sigma_{jk}}(t)$ is the volume integral of the stress change (the average stress drop considered concentrated at the point r_o'). It is also called the *source dipole tensor*[4,5] and is equivalent to the quantity referred to as the *seismic moment* in seismology.[6] Since the stress drop can be thought of as a dipole density, Eq. 1 amounts to a summation of convolutions between the dipole density at each source point r' and Green's tensor between r' and r.

The final step of the formulation is to include the effect of the transduction process on the signal. To date only nondisturbing sensors of finite area S_T have been considered.[1] *Nondisturbing* means that the change in waveform caused by the presence of the sensor can be neglected because the change is small compared to the waveform itself. This approximation is excellent for optical interferometric detection schemes,[7] electromagnetic acoustic and capacitance transducers.[8] Its validity for piezoelectric transducers is yet to be determined, for they undoubtedly load the surface.

The sensor response is here considered in terms of the surface displacement at the location of the sensor S_T, without the transducer attached. The point impulse response is denoted by $T_i(r,t)$, $r\epsilon S_T$ and is defined as the voltage at time t produced from the sensor by a displacement that, without the sensor present, would have been a δ function at point r in the x_i direction at time zero. Under this definition, the voltage at time t due to an infinitesimal source is:

$$V(t) = \iint_{S_T} T_i(r,t-t')G_{ij,k'}(r,r_o',t'-t'')\overline{\Delta\sigma_{jk}}(t'')drdt'' \quad \text{(Eq. 4)}$$

This equation for a point sensor has the form of a convolution between the source function, the impulse response of the body and that of the sensor. In the frequency domain, the convolutions become products and the complex (but scalar) voltage can be written as a function of frequency ω:

$$V(\omega) = T_i(\omega)G_{ij,k'}(\omega)\overline{\Delta\sigma_{jk}}(\omega) \quad \text{(Eq. 5)}$$

Note that under this definition of sensor sensitivity, the impulse response $T_i(t)$ is a vector quantity. The output depends both on the magnitude and direction of the surface displacement. Normally it is assumed that sensitivity exists

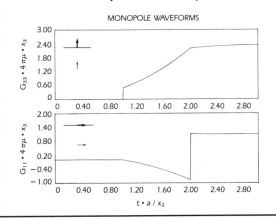

FIGURE 2. Epicenter displacement monopole waveforms for various infinitesimal multiple sources in an isotropic elastic half space

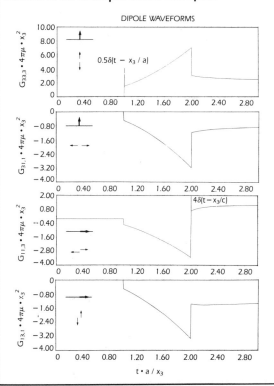

FIGURE 3. Epicenter displacement dipole waveforms for various infinitesimal multiple sources in an isotropic elastic half space

only for one displacement direction, reducing the formulation to a scalar one. Schemes for vector transducer calibration are emerging which can determine the full vector impulse response.[9]

Sensor Spectral Sensitivity

Examination of Eq. 5 indicates that information is transmitted independently, frequency by frequency, from the source to the observed voltage signal. Thus, the detectability of an acoustic emission signal is a frequency dependent concept and is determined by the signal-to-noise ratio at each frequency or band of frequencies. Detectability is naturally very dependent on the frequency dependence of the transducer. Table 1 lists the estimated smallest detectable signals (sensitivities) in a typical laboratory background noise environment.

Common types of transducers used for quantitative acoustic emission measurements are damped lead zirconate titanate (PZT) or air gap capacitors. Both are principally displacement sensitive with sensitivities between 0.1 and 1 MHz of 10^{-13} and 10^{-12} m respectively. More sensitive transducers with narrower bandwidth are frequently used for source detection and location work in the field. Some piezoelectric transducers and all electromagnetic acoustic transducers (EMATs) respond to surface velocity. For an EMAT, the smallest detectable velocity is about 0.1 to 0.2 mm·s^{-1}. A piezoelectric device might be 2 to 3 orders of magnitude more sensitive.

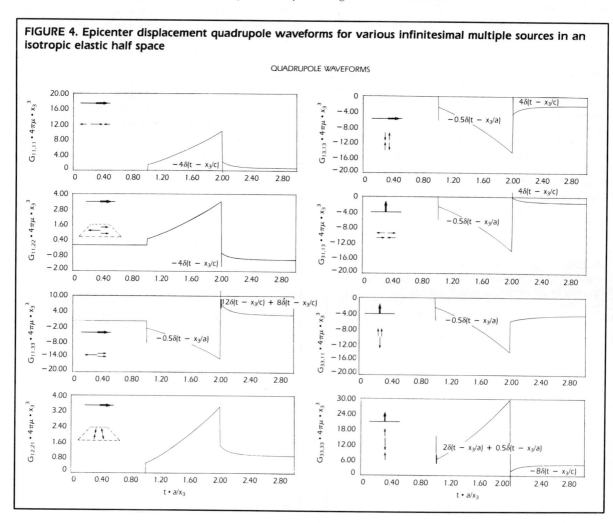

FIGURE 4. Epicenter displacement quadrupole waveforms for various infinitesimal multiple sources in an isotropic elastic half space

TABLE 1. Bandwidths and minimum detectable signals for acoustic emission transducers

Transducer Type	Passband (megahertz)	Sensitivity (meters)
Air gap capacitor	DC to 50	10^{-12}
Electromagnetic/acoustic	0.5 to 15	1 to 2×10^{-4} (per microsecond)
Damped PZT	0.1 to 2	10^{-13}
Resonant PZT	0.1 to 0.3	10^{-14}

Detectability Criteria

With these sensitivities in mind, delimiting the concept of detectable acoustic emission sources can be attempted. For the purpose of this discussion, assume point-like sensors located in the far field of the source on the surface of a half space. Examination of Green's tensor components $G_{jk,l}$ for a linear elastic isotropic half space show that, for a source at a distance x_3 perpendicular to the surface at which the sensor is attached, the waveform is dominated by longitudinal and transverse wavefront arrivals with $\delta(t)$ form ($\bullet$ here implies time derivative). It can be shown that:

$$G_{jk,l}(t) \sim \frac{\delta_{l3}}{2\pi\rho x_3} \left[\frac{(\delta_{j3}\delta_{k3})\dot{\delta}\left(x_3 - \frac{t}{a}\right)}{a^3} + \frac{(\delta_{jk} - \delta_{j3}\delta_{k3})\dot{\delta}\left(x_3 - \frac{t}{c}\right)}{c^3} \right] \quad (Eq.\ 6)$$

where a and c are the longitudinal and transverse wave speeds respectively and δ_{jk} is Kronecker's delta (1 for $j = k$ and 0 otherwise).

Substituting Eq. 6 into Eq. 4 and assuming the source is of infinitesimal size with $\delta(t)$ time dependence allows the deduction of the epicenter displacement waveform. In order to achieve detectability, the maximum value of this signal must exceed the minimum displacement sensitivity d or velocity sensitivity v. That is:

$$d_j \leq \frac{\kappa(\tau)}{2\pi\rho x_3} \max_{t>0} \left[\frac{\delta_{j1}}{c^3} \dot{\Delta\sigma}_{13}\left(x_3 - \frac{t}{c}\right) \right.$$
$$\left. + \frac{\delta_{j2}}{c^3} \dot{\Delta\sigma}_{23}\left(x_3 - \frac{t}{c}\right) + \frac{\delta_{j3}}{a^3} \dot{\Delta\sigma}_{33}\left(x_3 - \frac{t}{a}\right) \right] \quad (Eq.\ 7)$$

or

$$v_j \leq \frac{\kappa'(\tau)}{2\pi\rho x_3} \max_{t>0} \left[\frac{\delta_{j1}}{c^3} \ddot{\Delta\sigma}_{13}\left(x_3 - \frac{t}{c}\right) \right.$$
$$\left. + \frac{\delta_{j2}}{c^3} \ddot{\Delta\sigma}_{23}\left(x_3 - \frac{t}{c}\right) + \frac{\delta_{j3}}{a^3} \ddot{\Delta\sigma}_{33}\left(x_3 - \frac{t}{a}\right) \right]$$

where d_j represents the smallest detectable displacement and v_j the smallest detectable velocity, both in direction x_j.

Since no sensor is of broader band than all microscopic acoustic emission sources, the factors $\kappa(t)$ and $\kappa'(t)$ were introduced to account for sensor filtering. Here, τ is the source duration. If the sensor is of constant sensitivity from DC to $1/\tau$, then κ and κ' are unity. They will be less than unity for sensors with bandwidths smaller than that of the source.

To estimate the value of $\kappa(t)$ for a real sensor, assume the displacement pulse to be detected is of Gaussian form to which is assigned a width τ and magnitude M, so that:

$$|u(t)| = M e^{\frac{-t^2}{\tau^2}} \quad (Eq.\ 8)$$

or with Fourier transform:

$$|u|(\omega) = M\tau\sqrt{\pi}\ e^{\frac{-\omega^2\tau^2}{4}}$$

Assume that the response spectrum of the displacement transducer is Gaussian with bandwidth B by which is meant:

$$R(\omega) = k e^{\frac{-\omega^2}{4B^2}} \quad (Eq.\ 9)$$

where k is a normalization constant and $R(\omega)$ is the voltage produced from an excitation of frequency ω. To determine k, assume that the sensor will produce a signal of magnitude 1, independent of B, if the input is a constant in time of magnitude 1. Thus, if $u(t) = 1$, so that $u(\omega) = 2\pi\delta(\omega)$, then the response is $2\pi k\delta(\omega)$; therefore $k = 1$.

This is equivalent to $R = 1$ at $\omega = 0$ or $\int R(t)dt = 1$. The output from such a sensor has the frequency spectrum $M\tau(\pi)^{1/2} \exp[-\omega^2(1 + \tau^2 B^2)/4B^2]$, whose time response is $[M\tau B/(1 + \tau^2 B^2)^{1/2}] \exp[-t^2 B^2/(1 + \tau^2 B^2)]$. This is a Gaussian whose peak at $t = 0$ has been reduced by the factor of $\tau B/(1 + \tau^2 B^2)^{1/2}$. So that:

$$\kappa(\tau) = \frac{\tau B}{\sqrt{1 + \tau^2 B^2}} \quad (Eq.\ 10)$$

To find $\kappa'(\tau)$ for a velocity sensitive sensor, a differentiating filter is used in the form $k\omega \exp(\omega^2/4B^2)$ where, as above, it can be shown that normalizing with a time function of constant slope 1 gives $k = -i$. Again the time function given in Eq. 8 is used. Its maximum derivative $(2/e)^{1/2}M/\tau$ occurs at $t = \pm\tau/(2)^{1/2}$.

After filtering with the differentiation filter, a frequency spectrum of $-iM\tau\omega(\pi)^{1/2}\exp{-[\omega^2(1 + \tau^2B^2)/4B^2]}$ is calculated. Its inverse transform is $M\tau B/(1 + \tau^2B^2)^{1/2}d/dt\{\exp[-t^2B^2/(1 + \tau^2B^2)]\}$. The maximum derivative for this time response is $(2/e)^{1/2}M\tau^2B^2(1 + B^2\tau^2)$ occurring at $t = \pm[B/2(1 + \tau^2B^2)]^{1/2}$. So that:

$$\kappa'(\tau) = \frac{\tau B^2}{1 + \tau^2 B^2} \quad \text{(Eq. 11)}$$

Source Directivity

The final factor affecting detectability is the directionality or directivity of the source. In linear isotropic elastic materials, the stress change tensor is symmetric ($\Delta\sigma_{ij} = \Delta\sigma_{ji}$). A representation quadric may therefore be used to describe the directivity of the source.

Let the stress drop tensor be represented by its principal values λ_1, λ_2 and λ_3 ($|\lambda_1| < |\lambda_2| < |\lambda_3|$) along principal directions e_1, e_2 and e_3. Then the equation of the directivity surface is:

$$\lambda_1 e_1^2 + \lambda_2 e_2^2 + \lambda_3 e_3^2 = 1 \quad \text{(Eq. 12)}$$

In terms of the regular axes, x_i, $e_i = E_{ij}x_j$ where E is an orthogonal matrix with $E^t = E^{-1}$ whose components E_{ij} are the cosine of the angle between e_i and x_j. Thus, the three stress change components $\overline{\Delta\dot\sigma_{i3}}$ (or $\overline{\Delta\dot\sigma_{i3}}$) are given by:

$$\overline{\Delta\dot\sigma_{i3}} = E_{ji}\lambda_j E_{j3} \quad \text{(Eq. 13)}$$

First consider the stress change component $\overline{\Delta\dot\sigma_{33}}$ that causes out-of-plane motion at epicenter.

$$\overline{\Delta\dot\sigma_{33}} = \lambda_1 E_{13}^2 + \lambda_2 E_{23}^2 + \lambda_3 E_{33}^2 \quad \text{(Eq. 14)}$$
$$= \Lambda + (\lambda_1 - \Lambda)E_{13}^2 + (\lambda_2 - \Lambda)E_{23}^2 + (\lambda_3 - \Lambda)E_{33}^2$$

where $\Lambda = (\lambda_1 + \lambda_2 + \lambda_3)/3$ gives the dilatational component of the stress drop tensor.

The directivity pattern for $\overline{\Delta\dot\sigma_{33}}$ contains an omnidirectional term (Λ) and terms that are proportional to the square of the cosine of the angle between the e_i and x_3 axes. The maximum value in this pattern is λ_3 and occurs when the e_3 axis points in the x_3 direction. Should two of the principal stresses be of opposite sign, there will exist orientations for which there is no vertical motion at the epicenter surface due to $\overline{\Delta\dot\sigma_{33}}$. Provided all three principal stresses are of the same sign, there will then always be a nonzero minimum vertical motion whose magnitude is determined by the smallest principal stress (λ_1).

The analysis of directivity for either of the stress change components $\overline{\Delta\dot\sigma_{13}}$ or $\overline{\Delta\dot\sigma_{23}}$ that produce in-plane epicenter motion is similar. Thus, consider only $\overline{\Delta\dot\sigma_{13}}$

$$\overline{\Delta\dot\sigma_{13}} = \lambda_1 E_{13}E_{11} + \lambda_2 E_{23}E_{21} + \lambda_3 E_{33}E_{31} \quad \text{(Eq. 15)}$$
$$= (\lambda_1 - \lambda_3)E_{13}E_{11} + (\lambda_2 - \lambda_3)E_{23}E_{21}$$
$$= (\lambda_1 - \lambda_2)E_{13}E_{11} + (\lambda_3 - \lambda_2)E_{33}E_{31}$$
$$= (\lambda_2 - \lambda_1)E_{23}E_{21} + (\lambda_3 - \lambda_1)E_{33}E_{31}$$

The latter three equalities follow from the orthogonality of x_1 and x_3. The maximum in-plane motion at epicenter is determined by the maximum of the three quantities $|\lambda_1 - \lambda_2|$, $|\lambda_1 - \lambda_3|$ and $|\lambda_2 - \lambda_3|$. Suppose that $(\lambda_i - \lambda_j)$ is largest and that the direction orthogonal to both e_i and e_j is e_k. Then the maximum epicenter motion will be obtained when $x_2 = e_k$ and when e_i and e_j are in the $x_2 = 0$ plane at ± 45 degrees to the x_1 axis.

The magnitude of the maximum radiation for Eq. 15 is $\pm|\lambda_i - \lambda_j|/2$ and is polarized in the x_1 direction. Should two of the principal stresses be identical, then there will exist an orientation in which there is no horizontally polarized motion anywhere in the x_1,x_2 plane. If all three principle stress values differ, then there is always a horizontally polarized component in the x_1,x_2 plane and this component is greater than the minimum $|\lambda_i - \lambda_j|/2$.

PART 3
MICROSCOPIC DISLOCATION SOURCES

Detectability of Microscopic Dislocation Sources

The stress change ($\overline{\Delta\sigma_{kl}}$) or dipole tensor ($D_{kl}$) created by the formation of a dislocation loop in a linear elastic body has been evaluated and documented in the literature.[10-13] One investigator emphasized the simple relation between the force dipole tensor and the dislocation Burger's vector together with the dislocation loop normal:[13]

$$D_{kl}(t) = C_{ijkl} b_i A_j(t) \qquad \text{(Eq. 16)}$$

Here C_{ijkl} are the elastic stiffness constants, b_i are the components of the dislocation Burger's vector and $A_j(t)$ the projected area swept out on the plane whose normal is in the x_j direction. For the case of isotropic elasticity:

$$C_{ijkl} = \lambda \delta_{ij}\delta_{kl} + \mu(\delta_{ik}\delta_{jl} + \delta_{il}\delta_{jk}) \qquad \text{(Eq. 17)}$$

where λ and μ are Lamé constants. These are related to the longitudinal and transverse wave speeds:

$$\rho a^2 = \lambda + 2\mu \qquad \text{(Eq. 18)}$$

and

$$\rho c^2 = \mu$$

where ρ is the density

To determine detectability criteria for a dislocation source, consider the case of a planar dislocation loop formed on a plane with normal n at a depth r in a half space (see Fig. 5a). Using the first expression from Eq. 7 for the displacement detectability at epicenter:

$$2\pi x_3 d_j \leq \kappa(\tau) \left[\frac{2c^2 b_3 n_3}{a^3} \delta_{j3} \right.$$
$$\left. + \frac{(b_1 n_3 + b_3 n_1)\delta_{j1} + (b_2 n_3 + b_3 n_2)\delta_{j2}}{c} \right] \max_t \ddot{A}(t) \qquad \text{(Eq. 19)}$$

Similarly, the velocity detectability at epicenter is given by:

$$2\pi x_3 v_j \leq \kappa'(\tau) \left[\frac{2c^2 b_3 n_3}{a^3} \delta_{j3} \right.$$
$$\left. + \frac{(b_1 n_3 + b_3 n_1)\delta_{j1} + (b_2 n_3 + b_3 n_2)\delta_{j2}}{c} \right] \max_t \dddot{A}(t) \qquad \text{(Eq. 20)}$$

where $\kappa(t)$ and $\kappa'(t)$ as given by Eq. 10 and Eq. 11 account for filtering by the transducer. If b, n and x_3 are coplanar, these expressions can be more simply expressed in terms of twice the angle between n and the x_1, x_2 plane.

The first term on the right hand side in Eq. 19 and Eq. 20 corresponds to the out-of-plane (normal) motion at epicenter associated with the arrival of a longitudinal wave, while the second term corresponds to in-plane (tangential) motion (parallel to the projection of the Burger's vector) at epicenter associated with the arrival of a shear wave. Note that while acoustic emission due to in-plane motion is sensitive to the Burger's vector and slip plane orientation, it is not (to first order) sensitive to the direction of dislocation motion. That is, it cannot distinguish between edge dislocations slipping in one direction and screw dislocations slipping in the perpendicular direction.

Detectability Criteria

The greatest sensitivity to dislocation motion is obtained from an in-plane sensitive transducer with its poling axis aligned with the Burger's vector direction from dislocations slipping in the horizontal plane, or from dislocations slipping in a vertical plane with vertical Burger's vector and with the poling axis of the transducer aligned normal to the slip plane (see Fig. 5). In either case, if $a = 6.4$ mm•μs^{-1}, $c = 3.2$ mm•μs^{-1}, $b = 2.9 \times 10^{-10}$ m (typical aluminum values) and $r = 40$ mm, then the detectability criteria for horizontally sensitive sensors become (from Eqs. 19 and 20):

$$d_j < 3.6 \times 10^{-10} \kappa(\tau) \max_t [\ddot{A}(t)] \qquad \text{(Eq. 21)}$$

FIGURE 5. The strongest horizontal displacement from an infinitesimal dislocation occurs: (a) when b is coplanar with the surface and n is perpendicular to the surface; or (b) when b is perpendicular to the surface and n is coplanar with the surface

(a) HORIZONTAL DISLOCATION LOOP

(b) VERTICAL DISLOCATION LOOP

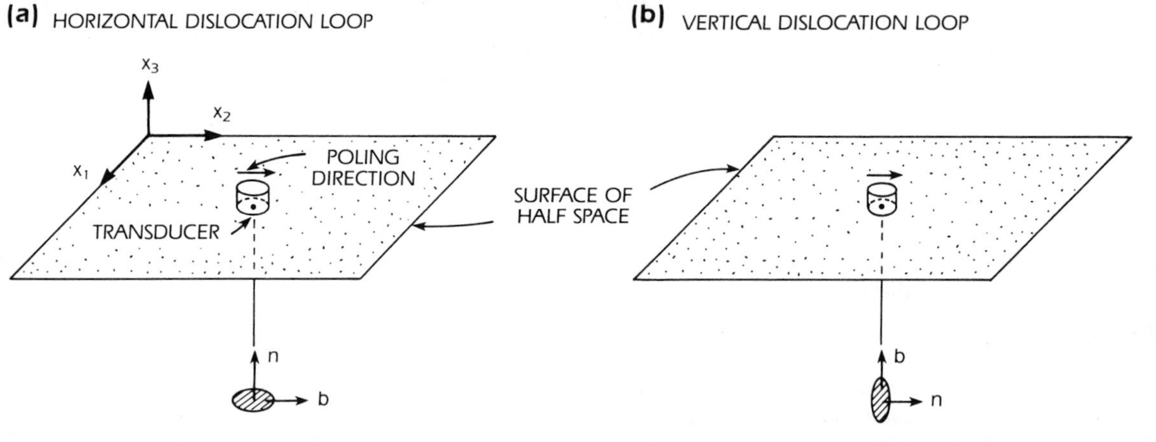

and

$$v_j < 3.6 \times 10^{-10} \kappa'(\tau) \ \max_t[\ddot{A}(t)] \quad \text{(Eq. 22)}$$

where d_j is expressed in millimeters; v_j is in millimeters per microsecond; and $j = 1, 2$.

If the horizontal (d_1, d_2) displacement detection threshold is 10^{-14} m and B is 5 MHz, then $\kappa(\tau) \sim 0.035$ for a dislocation that propagated for a time $\tau = 15$ ns. Then the dislocation loop would have to have a radial velocity of 3 mm·μs^{-1} expanding to a 45 μm radius in order to be detectable.

Such a calculation highlights the often neglected importance of sensor bandwidth (as well as sensitivity) in determining detectability. For instance, it is possible to show that detection could be achieved for a 1.5 μm radius loop if the bandwidth were infinite, provided dislocation speeds were close to the limiting shear wave speed. In fact, if $\tau B \ll 1$, then $\kappa(\tau) \sim \tau B$ and $\tau \sim 1/\dot{r}(18\pi B)^{1/2}$.

While the dislocation configurations shown in Fig. 5 produce the horizontal epicenter displacements, they produce no out-of-plane motion. If however the source is rotated so that the first term in Eq. 19 becomes nonzero for $j = 3$, then finite vertical displacements are produced at epicenter (see Fig. 6). For the situation shown in Fig. 6:

$$b_3 = b \cos \phi \quad \text{(Eq. 23)}$$

$$n_3 = \cos \Theta \quad \text{(Eq. 24)}$$

where b is the magnitude of the Burger's vector and n is the unit normal to the plane of slip. For a glissile dislocation, b is perpendicular to n and the maximum displacement occurs

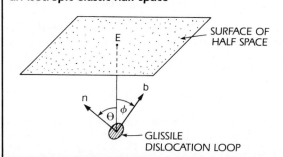

FIGURE 6. An inclined dislocation loop buried in an isotropic elastic half space

at $\Theta = \phi = 45$ degrees. It is one-eighth the maximum horizontal amplitude for a horizontal or vertical loop of the same radius and velocity. Thus, the detectability criterion is:

$$d_3 \leq 0.45 \times 10^{-10} \kappa(\tau) \ \max_t[\dot{A}(t)] \quad \text{(Eq. 25)}$$

in millimeters or:

$$v_3 \leq 0.45 \times 10^{-10} \kappa'(t) \ \max_t[\ddot{A}(t)] \quad \text{(Eq. 26)}$$

in millimeters per microsecond.

If a transducer is used with 10^{-14} m displacement sensitivity and bandwidth such that $\kappa(t)$ is always unity, then the detectability criteria in square meters per second becomes:

$$r\dot{r} \geq 0.03 \quad \text{(Eq. 27)}$$

Detectability criteria for velocity sensitive sensors depend on the sensor's sensitivity (its smallest detectable velocity) and on the acceleration of the source. At present, only very crude estimates exist for the former (even though it appears some conventional acoustic emission transducers are at least partially velocity sensitive). The acceleration history of a dislocation is even less clear, and will be extremely mechanism dependent. Small changes of temperature, composition, heat treatment, cold work and grain size can have a potentially large effect on acceleration.

While this shows promise for those interested in fundamental studies of the dynamics of dislocations (as well as cracks and phase transformations), it raises serious questions about reliability of such sensors for source detection and characterization in nondestructive testing applications.

Restricting further discussion to displacement detectability criteria, it can be seen from Eq. 27 to be $r\dot{r} \geq 0.03$ m^{2}s^{-1} for $x_3 = 40$ mm and aluminum physical property values. Unstable dislocation motion in polycrystalline metals is likely to proceed at velocities $\dot{r}$ of 0.1 to 1 times the shear wave speed ($\sim$ 300 to 3,000 m•s^{-1}). Thus a single dislocation propagating 10 to 100 μm is potentially detectable. This is indicative of the extraordinary sensitivity of acoustic emission instrumentation. The distance of dislocation propagation is, however, strongly influenced by microstructure (discussed below) and not all dislocation motion is detectable.

One further factor can have a very important effect on detectability, and this is the possibility for cooperative slip involving simultaneous motion of large numbers of dislocations. These *microyield phenomena* typically involve thousands of dislocations simultaneously moving on the same slip system. If the dislocations are closely spaced (the shortest wavelength of observation) and if they move synchronously, then their emitted wave fields add. The detectability criterion in square meters per second then becomes:

$$nr\dot{r} \geq 0.03 \quad \text{(Eq. 28)}$$

where n is the number of moving dislocations. This criterion is very easy to satisfy for many metals, especially near the onset of general yield.

Single Crystal Deformation

The best understood case of plastic deformation is that of a single crystal oriented for single slip. This has been the starting point for studies of acoustic emission by a number of groups.[14-16] A typical experimental result is shown in Fig. 7 for high purity aluminum.[14] In the experiment, the sample

FIGURE 7. The variation of acoustic emission with stress and strain for a high purity aluminum single crystal undergoing constant strain rate uniaxial tensile deformation

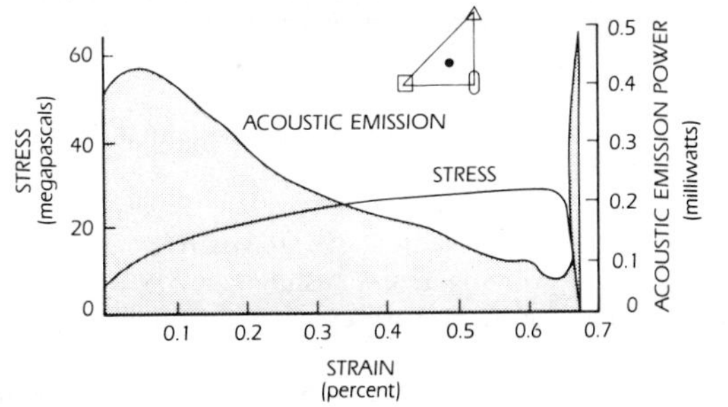

was deformed at a relatively slow constant strain rate of $\sim 10^{-4}$ per second. Acoustic emission in the 0.1 to 1.0 MHz frequency range was recorded with a principally vertical displacement sensitive transducer ($\sim 10^{-14}$ m sensitivity) located on the tensile axis.

The acoustic emission in this experiment was first detected during nominal elastic loading and reached a maximum intensity after straining about two percent beyond general yield. Beyond this point, acoustic emission decreased, even though the slip per unit time (strain rate) remained essentially constant. Other studies[16] indicate that the strain at which the maximum occurs is bandwidth sensitive and occurs at higher strains when larger bandwidths are used.

Microscopic Nature of Slip

For an explanation of this behavior, it is necessary to consider the microscopic nature of slip. In a face-centered cubic single crystal, slip tends to occur in <110> directions on the closed packed {111} planes. There are many possible slip systems that could be activated by the introduction of stress. Typically the system with highest resolved shear stress in the slip direction will be activated.

If the stress is normal to the surface (σ_{33}), then the system with cos Θ cos ϕ closest to 0.5 will be the first activated (that is, the system for which $\Theta \sim \phi \sim 45$ degrees). These dislocations are expected to generate finite vertical displacements at epicenter and the detectability criterion in Eq. 28 will be valid provided $\kappa(t) = 1$.

For example, a typical event might be the motion of a dislocation to form a 150 μm radius loop at an average radial velocity of 300 m•s^{-1}. Then $r\dot{r} = 0.045$ m^{2}s^{-1} and $\tau B \sim 0.5 \times 10^{-6}$ s $\times 10^6$ Hz ~ 1.0, so that the event should create a signal approximately 1.5 times the background noise and therefore be just detectable.

Suppose that a is a large fraction of the crystal radius (a reasonable supposition at the initiation of plastic deformation of a perfect single crystal) or about 1 mm. For a detectable signal from a single loop, $\dot{r}$ must be ~ 35 m·s^{-1}. This is not an unreasonable value to expect. In practice, both r and $\dot{r}$ for slip events are statistically distributed (see Fig. 8). Only those events whose $nr\dot{r}$ products are at the extreme right of the distribution generate detectable events.

At the initiation of deformation, a substantial fraction of slip events may propagate far enough that even moderately slow dislocation groups give detectable signals. However, work hardening reduces the mean free path between sessile dislocations, restricting r and resulting in few events that radiate detectable elastic waves; thus the emission power decreases.

At a strain of about 2 percent and a bandwidth of about 2 MHz, a maximum emission is observed. This is the strain at which the greatest fraction of deformation events satisfy the criteria $nr\dot{r} \geq 0.03$ m^{2}s^{-1} and $\kappa(t) = 1$. However, if B is increased, $\kappa(t)$ can tend to unity for shorter duration events. Shorter duration events tend to become more numerous as work hardening develops (because r decreases and $\dot{r}$ remains constant or increases), and so the maximum in emission shifts to higher strain.[16]

Grain Size Effects of Microscopic Dislocation Sources

It is well known that grain boundaries are effective barriers to moving dislocations. In a polycrystalline metal, the distance that a dislocation loop can expand will be at most that of the radius of the largest grain in the sample. The distribution of slip distances over some range of plastic strains will now be a complicated function of the grain size distribution, grain interior dislocation and dislocation interactions. It seems reasonable to suppose that, at the initiation of slip, a substantial fraction of moving dislocations will propagate over a distance equal to about the mean grain diameter.

As the grain size is reduced, the critical velocity required for a single dislocation to give a detectable signal increases. For example, for a mean grain diameter of 1 mm, $\dot{r} \sim 35$ m·s^{-1} while, for a diameter of about 35 μm, $\dot{r} \sim 1,000$ m·s^{-1}. In pure face-centered cubic crystal metals, dislocation velocities of several hundred meters per second are expected. Emission from small grained material must come from groups of cooperatively moving dislocations. The probability that enough dislocations move together to give a detectable signal decreases with grain size. A disappearance of detectable acoustic emission has been observed (see Fig. 9) by several groups in agreement with the above model.

A note of caution should be given. Some researchers[17] have observed an apparent increase in the acoustic emission at very small grain sizes. This controversial observation has not been fully resolved. If confirmed, the acoustic emission increase may be linked to an increase in the yield stress with a decreasing grain size which increases the average dislocation velocity.

In alloys of aluminum, substitutional impurity segregation to dislocations can occur, even when the alloy element concentration is below the solid solubility limit. This drastically alters the dynamics of dislocations and leads to effects such as yield points, Lüders band propagation and dynamic strain aging during mechanical deformation. The occurrence of these phenomena is almost invariably accompanied by intense acoustic emission signals. Both the incidence of the effects themselves and their associated acoustic emission signals vary greatly with grain size. For example, Fig. 10 shows the acoustic emission energy plotted as a function of grain

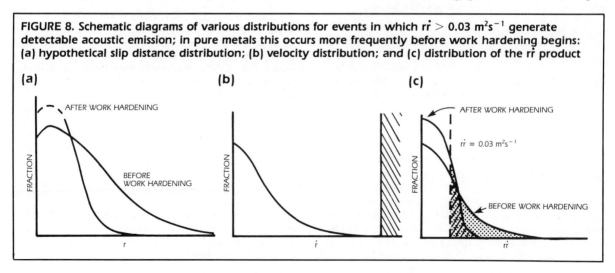

FIGURE 8. Schematic diagrams of various distributions for events in which $r\dot{r} > 0.03$ m^{2}s^{-1} generate detectable acoustic emission; in pure metals this occurs more frequently before work hardening begins: (a) hypothetical slip distance distribution; (b) velocity distribution; and (c) distribution of the $r\dot{r}$ product

size for an Al-1.3Mg alloy solid solution and compared with aluminum.

The peak in the acoustic emission of the solid solution at a grain size of around 80 μm arises from a competition between two effects: as the grain size increases, the duration τ increases but r decreases. The velocity decrease occurs because of the fall in flow stress with increasing grain size and, for grain sizes greater than about 500 μm, the yield stress is insufficient to separate mobile dislocations from the cloud of impurity atoms (Cottrell atmosphere) segregated to their core.[14] This results in the drift-controlled motion of dislocations at velocities too low to give acoustic emission, even though propagation distances may be greater than that required for the radiation of detectable signals in pure aluminum.

Precipitation Effects

There are two extreme classes of dislocation precipitate interaction.

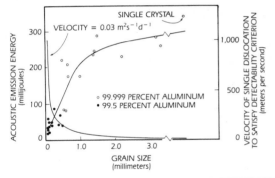

FIGURE 9. Integrated acoustic emission power (energy) for polycrystalline aluminum decreases with decreasing grain size; the necessary dislocation velocity for a single dislocation to generate a detectable acoustic emission signal is also shown

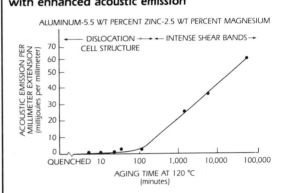

FIGURE 11. Aging supersaturated solid solutions may produce shearable precipitates that in turn cause an intense shear band mechanism of slip with enhanced acoustic emission

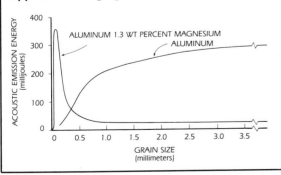

FIGURE 10. Comparison of the grain size dependence of acoustic emission for high purity aluminum and an aluminum-manganese alloy solid solution; the effect of manganese is to enhance the emission at small grain size and suppress it at large grain size

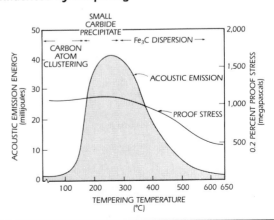

FIGURE 12. Acoustic emission due to plastic deformation in iron based alloys strongly influenced by tempering

1. When the precipitates are strong, widely spaced and incoherent with the matrix, dislocations bow between particles, leaving behind a dislocation loop around the particle.
2. When the precipitates are weak, closely spaced and coherent with the matrix, dislocations may penetrate the precipitates and shear them.

Process-1 encourages uniformly distributed slip while process-2 leads to the formation of intense slip bands. These bands form because the successive passage of dislocations through shearable precipitates reduces their strength, creating a local weak spot in the crystal and a local strain instability then occurs.

Studies have been made in which the precipitate strength, distribution and coherency have been systematically varied in aluminum alloys. The results indicate that the uniform slip process of dispersion hardened alloys (strong, relatively widely spaced incoherent precipitates) gives only moderate levels of detectable acoustic emission. This occurs because r is limited to the interparticle spacing (a few micrometers). In contrast, when precipitate shear occurs, intense acoustic emission signals are observed (see Fig. 11).[18-20] Similar results are also obtained in quenched and lightly tempered ferritic steels (see Fig. 12).[21]

These studies serve to illustrate the complimentary role that acoustic emission plays in the experimental study of plastic flow. In contrast with stress-strain measurements and hardness testing (where the integrated static response of the material to an applied load is measured), acoustic emission measures that part of the slip distribution for which the criterion $nr\dot{r} \geq 0.03$ m^{2}s^{-1} is satisfied (the high speed events). This is very sensitive to metallurgical variables, often much more so than the stress-strain curve.

PART 4
MICROSCOPIC FRACTURE SOURCES

Detectability of Microscopic Fracture Sources

The text below focuses on acoustic emission generated by the formation of isolated microcracks and the extension of macrocracks by the coalescence of microcracks at a crack tip. A crack can be viewed as a material inhomogeneity. The combination of dipole components that are equivalent to the crack depend on the ambient stress in addition to the radius of the crack.

For infinitesimal cracks circular in shape, the dipole tensor is given by:

$$D_{kl}(t) = \frac{8(1-v)}{3\mu(2-v)} C_{klmn} n_n n_q \sigma_{pq} W_{mp} r^3(t) \quad \text{(Eq. 29)}$$

Where:

v = Poissons ratio;
μ = the shear modulus;
C_{klmn} = the elastic stiffness tensor;
n = a unit vector normal to the crack face;
σ_{pq} = the ambient stress;
r = the crack radius; and
W_{mp} = the components of the matrix in Eq. 30.

$$W = \begin{vmatrix} 2 - n_1^2 v & -vn_1n_2 & -vn_1n_3 \\ -vn_1n_2 & 2 - n_2^2 v & -vn_2n_3 \\ -vn_1n_3 & -vn_2n_3 & 2 - n_3^2 v \end{vmatrix} \quad \text{(Eq. 30)}$$

Substituting Eq. 29 into Eq. 7 gives the expression for detectability of a microcrack at a depth x_3 beneath the epicenter of a half space:

$$d_j \leq \frac{4(1-v)\kappa(\tau)}{\pi\rho(2-v)x_3} \max_{t>0} \left\{ [\delta_{j1}(w_1n_3 + w_3n_1) \right. \quad \text{(Eq. 31)}$$
$$+ \frac{\delta_{j2}}{c^3}(w_2n_3 + w_3n_2)] \, r^2\left(x_3 - \frac{t}{c}\right) \dot{r}\left(x_3 - \frac{t}{c}\right)$$
$$\left. + \frac{2\delta_{j3}}{a^3}\left(\frac{v(n \cdot w)}{1-2v} + w_3n_3\right) r^2\left(x_3 - \frac{t}{a}\right) \dot{r}\left(x_3 - \frac{t}{a}\right) \right\}$$

$$v_j \leq \frac{4(1-v)\kappa'(\tau)}{\pi\rho(2-v)x_3} \max_{t>0} \left\{ [\delta_{j1}(w_1n_3 + w_3n_1) \right. \quad \text{(Eq. 32)}$$
$$+ \frac{\delta_{j2}}{c^3}(w_2n_3 + w_3n_2)] \, 2\dot{r}^2\left(x_3 - \frac{t}{c}\right)$$
$$+ r^2\left(x_3 - \frac{t}{c}\right)\ddot{r}\left(x_3 - \frac{t}{c}\right) \delta\left(x_3 - \frac{t}{c}\right)$$
$$+ \frac{2\delta_{j3}}{a^3}\left(\frac{v(n \cdot w)}{1-2v} + w_3n_3\right) 2\dot{r}^2\left(x_3 - \frac{t}{a}\right)$$
$$\left. + r^2\left(x_3 - \frac{t}{a}\right)\ddot{r}\left(x_3 - \frac{t}{a}\right) \delta\left(x_3 - \frac{t}{a}\right) \right\}$$

where w is a vector in the direction of maximum crack face displacement, so that:

$$w_m = 2(2-v) W_{mp}\sigma_{pq}n_q \quad \text{(Eq. 33)}$$

If the force $\sigma_{pq}n_q$ is parallel to n, the crack is termed mode-I. If the force is in the crack plane, it is mode-II/III. For these particular cases, w is parallel to $\sigma_{pq}n_q$. This would not be so for mixed mode cracks since w has differing eigen values of 2 and $2 - v$ in the crack plane and perpendicular to the crack plane, respectively.

The directivity pattern of the acoustic emission from a mode-II/III crack is the same as that for slip (glissile) dislocations with w assuming the role of b. It can be shown that the coefficients of the two horizontal ($j = 1,2$) terms in Eq. 32 correspond to the projection of the vector $w_3n + n_3w$ into the x_1,x_2 plane using one projection factor. The vertical ($j = 3$) term involves the projection of the same vector onto the x_3 axis using a different projection factor plus a vector that is independent of the position of x_3.

The relative epicenter displacement magnitude as a function of angle for both cracks and slip dislocations is summarized in Table 2 for the case where the x_3 axis, w (or b) and n are coplanar.

Microcrack Detectability Criteria

Using methods similar to those shown for dislocation sources, it is possible to determine detectability criteria for microcrack sources. As an example, suppose a sensor sensitive to vertical displacement is used to detect signals from a mode-I horizontal microcrack a depth x_3 below the surface. Then, $w = (0,0,w_3)$, $n = (0,0,n_3)$, and from Eq. 31:

TABLE 2. Directionality factors for σ function wavefront components for mode-I and mode-II/III microcracks

Source Type	Longitudinal Wavefront (Out-of-Plane Displacement)	Shear Wavefront (In-Plane Displacement)
Mode-I microcrack (or prismatic dislocation)	$\alpha(1 - \alpha^2) + \alpha^3 \cos 2\Theta$	$\sin 2\Theta$
Mode-II/III microcrack (or slip location)	$\alpha^3 \sin \phi$	$\cos 2\phi$

$\alpha = c/a =$ ratio of shear to longitudinal wave speeds
$\Theta =$ angle between crack face (dislocation plane) normal and the x_3 axis
$\phi =$ angle between the x_3 axis and w (or b)

$$d_3 \leq \left[\frac{2(2 - \nu)8(1 - \nu)^2}{\pi\rho(1 - 2\nu)a^3} \right] \frac{\kappa(\tau)\sigma_{33}r^2\dot{r}}{x_3} \quad \text{(Eq. 34)}$$

Using the physical properties of steel ($\nu = 0.29$, $\rho = 7.8 \times 10^3$ kg·m^{-3} and $a = 5.9 \times 10^3$ m·s^{-1}), the detectability criteria is: $\sigma_{33}r^2\dot{r} \geq 1.6 \times 10^{14} \kappa(\tau)(x_3d_3)$.

Source factors favoring detectability are high ambient stress, large crack radius and fast crack speed. The criterion is affected by elastic properties so that for the properties of aluminum ($\nu = 0.34$, $\rho = 2.7 \times 10^3$ kg·m^{-3}, $a = 6.4 \times 10^3$ m·s^{-1}), the detectability criteria is: $\sigma_{33}r^2\dot{r} \geq 0.6 \times 10^{14}\kappa(\tau)x_3d_3$

In aluminum, cracks of only a third the area of those in steel are detectable, other conditions being equal.

In metals, the dimensions of microcracks induced by deformation are closely coupled to the dimensions of microstructure constituents such as grain size, inclusion or precipitate diameter and interparticle spacing. For example, in a ferritic steel at or below its ductile-to-brittle transition, cleavage cracks (typically a grain diameter in size) are formed during deformation. Thus, within the statistical variations of the microstructure, it is possible to obtain estimates of r in Eq. 34.

It is known that brittle microcracks such as those associated with the cleavage described above propagate at velocities close to the limiting value of the shear wave speed. At the other extreme, ductile microvoid coalescence occurs under essentially quasistatic conditions. Thus it is possible to estimate crude values for $\dot{r}$ in Eq. 34 for various fracture micromechanisms. If all events are assumed to occur at constant stress (σ_{33}) and constant source-to-receiver distance (x_3), then the fields of each micromechanism can be mapped on a plot of $\dot{r}$ versus πr^2 with the detectability criterion (Eq. 34) superimposed for various d_3 and $\kappa(\tau)$ values. This is shown for steel in Fig. 13 where $\sigma_{33} = 500$ MPa and $x_3 = 0.04$ m.

Only micromechanisms to the right of the detectability criterion are detectable. The intergranular and cleavage brittle fracture are probably very reliably detected, even with insensitive transducers, provided their bandwidth extends up to about 1.0 MHz. There is no possibility of detecting microvoid coalescence with the current generation of sensors. Thus, the reliability of acoustic emission for detecting crack growth depends on the operative fracture micromechanisms.

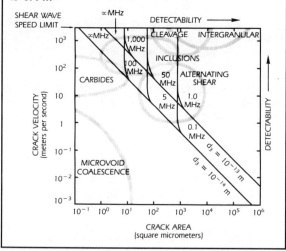

FIGURE 13. Acoustic emission detectability map for the fracture of micromechanisms in steel; solid lines correspond to detectability criteria for transducers of varying sensitivity (d_3) and bandwidth (in MHz); the dotted fields are approximate domains in r and πr^2 space for various fracture micromechanisms; mechanisms above and to the right of a detectability line are detectable; the imposed uniaxial stress is 500 mPa and the distance to the transducer is 0.4 m

Cleavage Microfracture

Cleavage microfracture is a common fracture micromechanism in materials below their ductile-to-brittle transition temperature. Many ceramics, refractory metals and some ferritic steels fail by this mode at ambient temperature. In composite materials where ceramic fibers, whiskers or particles are incorporated in order to reinforce a soft, ductile matrix, reinforcement failure often occurs by cleavage.

Cleavage microcracks occur on specific crystallographic planes, for instance {001} in iron. They are believed to be nucleated either by preexisting discontinuities (such as iron carbide cracks in steels), dislocation pile-up at internal boundaries or at the intersection of slip planes (see Fig. 14). Very limited plasticity may accompany microcrack propagation which may thus proceed at velocities close to that of the shear wave speed (the theoretical upper limit). At low stresses, cracks are often arrested at grain boundaries due to the absence of optimally oriented {100} planes in the adjacent grain at the site of crack impingement.

A horizontal microcrack in steel (with $r = 10$ μm, $\dot{r} = 1{,}000$ m·s^{-1}, $\sigma_{33} = 500$ MPa and $x_3 = 0.04$ m) gives rise to a vertical displacement epicenter signal whose peak amplitude (Eq. 34) is $\mu_3 \sim 2.5 \times 10^{-11}$ m and has a time scale of $\sim 10^{-5}$ m divided by 10^3 m·s^{-1} or 10 ns. Such a displacement signal is readily detectable under laboratory conditions even with narrow band sensors (see Fig. 13).

Intergranular Microfracture

In many engineering alloys, grain boundaries are a site of weakness and may prematurely fail under load. This weakness is often associated with segregation of embrittling chemical species to the interface or with the formation of brittle phases at the boundary. For example, in some low alloy steels, nickel cosegregates to grain boundaries with phosphorous, arsenic, antimony and tin and this causes: (1) a transition from cleavage to intergranular fracture; and (2) appreciable upward shifts in the ductile-to-brittle transition temperature. Glassy grain boundary phases in ceramics such as Al_2O_3 are also linked to the formation of intergranular fracture in ceramics. In aluminum alloys, liquid metals such as indium and gallium promote intergranular fracture.

Once an intergranular crack is nucleated in these systems, it is capable of rapid propagation over considerable distances due to the absence of crack arresting features in the microstructure. Only when the crack reaches a grain boundary triple point and must branch radically from the maximum tensile stress plane is arrest possible in materials with uniformly weak grain boundaries. If the distribution of embrittling agents is not uniform, as occurs in some ceramic systems, then the dimensions of the embrittled region act to control the crack advance distance.

Systems exhibiting brittle intergranular fracture are often copious emitters of detectable acoustic emission. Crack radii are often five to ten times those of cleavage microcracks and crack velocities are at least as high as those during cleavage. Acoustic emission strengths of 20 to 100 times those of cleavage are possible.

In composites, fiber delamination can be likened to intergranular fracture since an interface debond is the basic crack advance mechanism, and the crack advance distance can be many fiber diameters. Since fiber cleavage and debonding are two of the main damage accumulation processes in composite materials, it is clear that their damage evolution has a high acoustic emission detection probability and is the reason for the success of many acoustic emission monitoring activities in these materials.

Particle Microfracture

Some inclusions and precipitates in engineering alloys are brittle at ambient temperature and will undergo brittle (cleavage) fracture under tensile loading. The interface at these discontinuities is often weak and frequently fails at low stress. The generation of acoustic emission from these microstructure constituents depends on the intrinsic properties of the particles, the strength of their interfaces and the particle dimensions.

Inclusions in Steels

In steels, the fracture of manganese sulphide particles appears to be an important emission source, particularly in the absence of cleavage or intergranular failure modes. This source of emission has been examined in detail.[22] The studies found that acoustic emission from manganese sulphide inclusions has a strong orientation dependence. During hot rolling of steel, inclusions are elongated in the rolling direction and flattened in the rolling plane, forming an ellipsoidal shape.

When traction is applied in the rolling plane along the prior rolling direction (L orientation) or perpendicular to the rolling direction (T orientation), very weak signals are observed. These signals apparently originate from cracks or disbonds over the minor axis of the ellipsoid. However, when the steel was tested so that the load was applied normal to the rolling plane (ST orientation), copious acoustic emission was observed, presumably associated with fractures extending over the major axis.

The sulfur content (number of inclusions), size and aspect ratio of the inclusions, together with the stress state, are the factors that control the acoustic emission from inclusions.

FIGURE 14. Schematic diagam of three micromechanisms of cleavage crack nucleation in iron-carbon alloys: (a) nucleation by a grain boundary Fe$_3$C film; (b) grain boundary edge dislocation pile-up nucleation; and (c) nucleation by intersection of edge dislocations on {011} slip planes

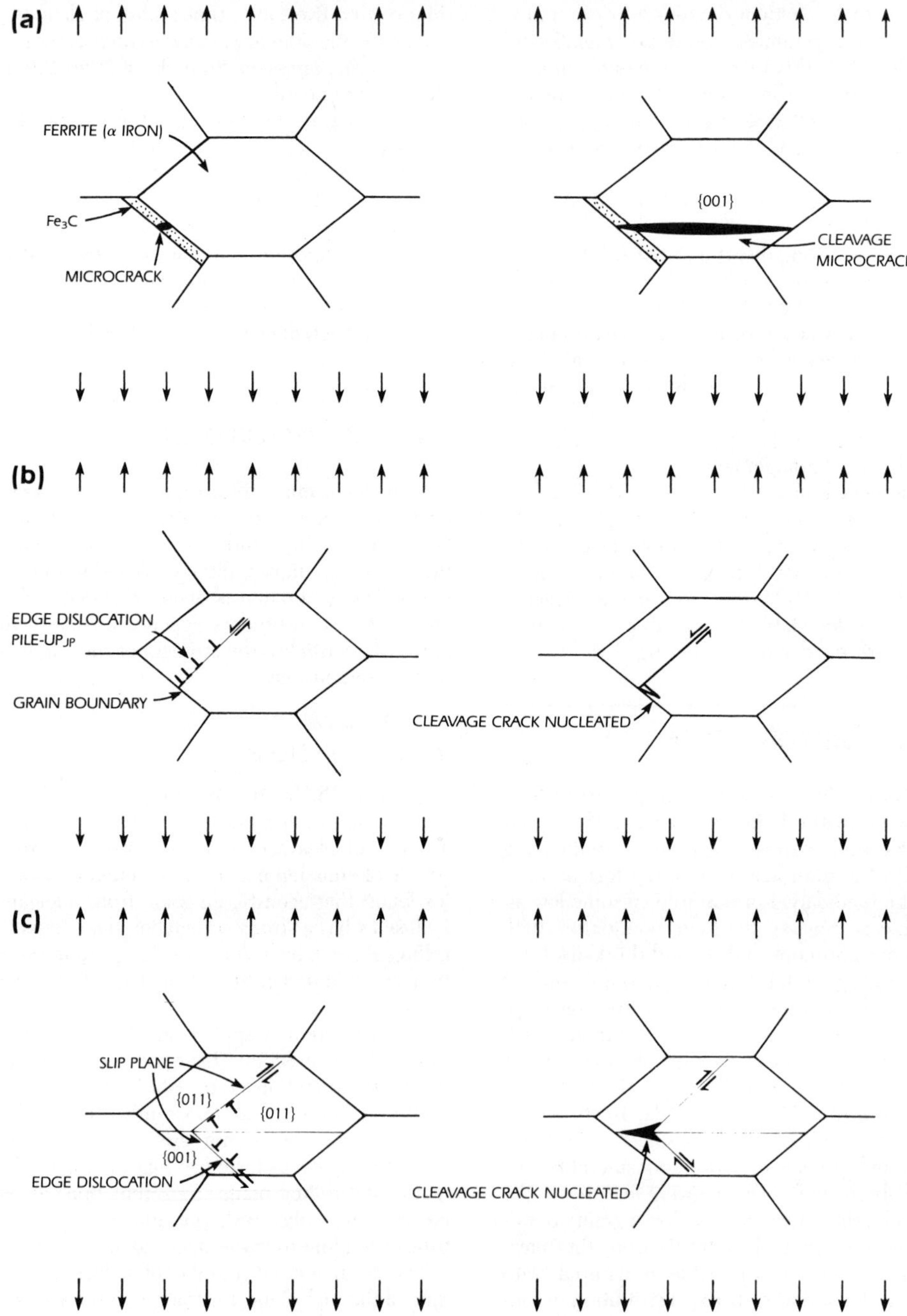

The inclusion size and shape is determined by the solidification pathway and by thermomechanical processing after solidification. In heavily rolled steels, with sulfur contents greater than about 0.06 percent by weight, manganese sulfide fractures associated with short transverse loading can be a significant emission source.

In many steels, crack growth occurs predominantly near welds where manganese sulphide has been melted and then reformed at interdendritic interstices during resolidification. If the sulfur content is high, large elongated inclusions are deposited. These can subsequently become the sites of discontinuities such as lamellar tears. Since discontinuity formation involves crack growth distances greater than 10 μm, they are substantial acoustic emitters.

Inclusions in Aluminum Alloys

Numerous types of inclusions of varying ductility are found in aluminum alloys. Several researchers have found that only those inclusions that are rich in iron ore fracture at ambient temperature.[23] These inclusions are found to give detectable signals and, as with the signals from manganese sulphide in steels, the amplitude distribution of the acoustic emission scales with the size distribution of fracturing particles.

Because of their smaller size and generally reduced aspect ratio, precipitates are less prone to fracture until very high stresses have been attained. Their small size (often less than 1 μm) makes them undetectable acoustic emitters. The exception has been in some specially heat treated steels containing large (5 μm) spheroidal carbides. They are good emitters but are not often used in engineering alloys.

Microvoid Coalescence

Ductile fracture occurs by the sequence of processes depicted in Fig. 15 and outlined below.

1. During loading, cracks or disbonds occur at inclusions, causing a stress concentration in the matrix between inclusions.
2. Plastic deformation of ligaments between inclusions occurs, resulting in the growth of inclusion nucleated voids and an intensification of local stress.
3. Secondary voids are nucleated at finely distributed precipitates within the deforming ligaments.
4. The secondary voids grow and link up, resulting in a crack whose dimension is the interinclusion spacing.

The linking of voids may occur by one of two modes. In materials with low yield strength and high work hardening capacity, the decrease in net load supporting area (due to void growth) is balanced by the increased flow stress (due to work hardening) of the deforming intervoid ligament. Under this stable condition, hole growth continues almost until the voids overlap. This mode of coalescence is likely to be undetectable due to the very low crack growth velocity and small intercarbide ligament thickness.

In materials with high yield strength and limited work hardening capacity, hole growth causes a loss of load supporting area that cannot be compensated by work hardening, unstable strain ensues and results in a premature shear coalescence. This latter process may occur at intermediate velocity over distances determined by the interinclusion

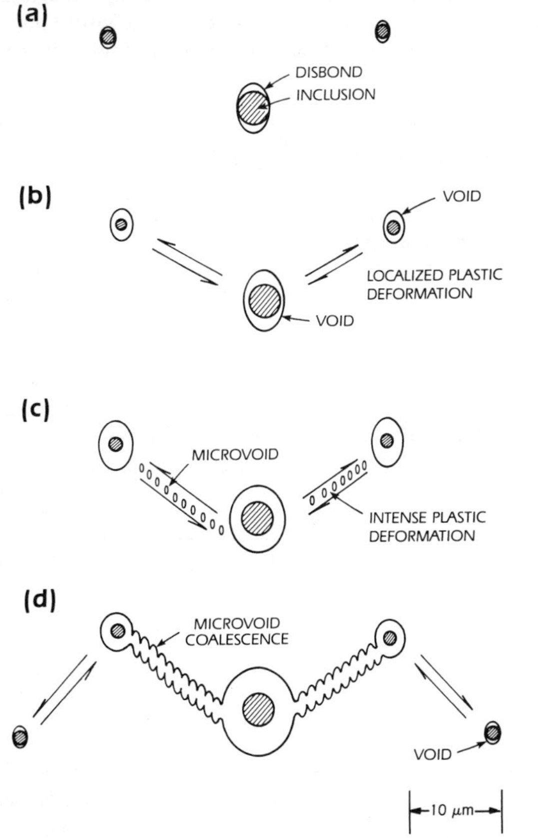

FIGURE 15. Sequence of micromechanisms involved in the growth of a ductile crack: (a) process begins with debonding at inclusion-matrix interfaces; (b) voids grow at the interface; plastic deformation localizes between adjacent voids; (c) intense deformation nucleates further microvoids at precipitate interfaces; and (d) coalescence of the microvoids results in a ductile crack

separation (10 to 100 μm). The alternating shear is a potentially detectable source of acoustic emission.

The design of structures is usually based on the premise that crack growth occurs by stable microvoid coalescence. Efforts are made during materials selection and design to ensure that insufficient stress develops during service for this mode of fracture to exist. Should undetected discontinuities exist in construction of materials, or if a manufactured component is subjected to greater than anticipated stress, ductile crack advance may occur.

In tough low strength steels, only intermittent inclusion fractures indicate this. If these inclusions are absent or if the region ahead of the crack was appreciably prestressed in the past (during proof testing) so that inclusions are already fractured, the possibility exists for silent crack growth. This is a disconcerting phenomenon to those interested in the use of acoustic emission for nondestructive testing.

Given the quality of materials and fracture mechanics analyses, the likelihood of a ductile fracture failure is becoming rare. Rather, the emerging concern is that some kind of embrittling phenomenon occurs due to environmental effects. This reduces the material's resistance to crack growth and changes the mechanism of fracture from a ductile to brittle one that is often detectable. Acoustic emission monitoring for these situations may well be worthy of further investigation.

Environmental Factors

The generation of detectable acoustic emission signals during environmentally assisted fracture (such as hydrogen embrittlement, stress corrosion cracking and corrosion fatigue) will depend on the mechanism of crack extension. As an example of this, the acoustic emission per unit area of crack extension (dE/dA) was measured[24] in three ferritic steels under various environmental conditions. Under vacuum the ambient temperature fracture mode involves microvoid coalescence and is not a major emission source. However, as shown in Table 3, considerable differences in emission per unit area were obtained through variation of grain size and environment.

During intergranular fracture, the acoustic emission activity unit area was approximately proportional to grain size. For a fixed grain size, transgranular cleavage generated an order of magnitude less acoustic emission than the intergranular mode.

A great deal remains to be done to fully understand the role of environmental factors on fracture micromechanisms and the resulting acoustic emission. Nevertheless, it is evident that there exists considerable potential for acoustic emission techniques in these areas, particularly when (1) the favored crack advance mechanism involves intergranular failure and (2) the microstructure contains coarse grains.

Amplification Factors

It was shown earlier that the acoustic emission amplitude is proportional to the crack face displacement w. In determining a detectability criterion, it was assumed that w occurred only by elastic strain and that no plastic deformation occurred. In practice, even brittle fractures have some associated dislocation emission at the crack tip. This plastic deformation is capable of allowing w to increase beyond the value attained purely elastically. Indeed, it is quite common for w to increase one to three orders of magnitude in low yield strength materials by crack tip flow processes.

If this plastic deformation occurs entirely during the period of crack growth, then the acoustic signal could potentially be amplified by the ratio of elastic-to-plastic crack face displacements.

A second source of signal enhancement stems from the

TABLE 3. Effect of metallurgical and environmental variables on acoustic emission per unit area of crack extension (dE/dA) for three ferritic steels

Steel	Grain Size (micrometers)	Fracture Mode	Environment	dE/dA (10^{-2} V^{2}s·mm^{-2})
817 M40	17	intergranular	3.5 percent NaCl	9
	100	intergranular	3.5 percent NaCl	108
	17	intergranular	H$_2$ at 27 kPa (200 torr)	31
897 M39	10	transgranular	3.5 percent NaCl	5
	10	transgranular	H$_2$ at 39 kPa (290 torr)	2
	10	transgranular	H$_2$ at 101 kPa (760 torr)	1.5
AISI 4340	11	intergranular	3.5 percent NaCl	31
	200	intergranular	3.5 percent NaCl	210
	11	intergranular	H$_2$ at 27 kPa (200 torr)	37

relaxation of the faces of a crack when microcrack extension occurs at its tip.[25] Note once again that to first order the acoustic emission amplitude is proportional to the change in crack volume. If a microcrack occurs at the tip of a preexisting crack, the volume in question is now the sum of the microcrack volume and the change in macrocrack volume facilitated by its crack tip extension. This factor may amplify the microcrack signals by one to three orders of magnitude and also seriously affects the spectrum of the emitted wave field.

PART 5
MICROSCOPIC PHASE CHANGES

Solid State Phase Transformations

Solid state phase transformations almost invariably cause the development of an internal stress due to density, modulus and thermal contraction differences between the different phases of an alloy. Consequently, potential sources of acoustic emission include the development of the stress field, microfracture, microplasticity and other mechanisms by which the stress field is relaxed (loss of coherency).

There have been systematic investigations[26] of the effect of cooling rate on the acoustic emission accompanying austenite-ferrite transformation in plain carbon steels. As the cooling rate increases, the reaction products change from pearlite through bainite to martensite. These observations are summarized in Table 4.

From this work it is clear that the diffusion controlled nucleation and growth of ferrite and carbides, while probably causing the development of appreciable stresses, fail to generate detectable elastic waves in the acoustic emission frequency range. The diffusionless transformation of austenite to martensite is easily detectable and provides the basis for a simple nondestructive method of deducing the temperature at which martensite starts to form on cooling (M_s) and the temperature at which martensite formation on cooling is essentially completed (M_f).

Detectability of Martensitic Transformations

Martensitic transformations are diffusionless changes of phase involving local shear and dilatations that cause changes of shape. In some respects they are similar to the formation of deformation twins in that both involve invariant planes of strain. The region that transforms has a characteristic habit plane with normal h. The habit plane is then defined by:

$$h_i x_i = 0 \quad \text{(Eq. 35)}$$

and

$$h_1^2 + h_2^2 + h_3^2 = 1$$

The plane of a twin is usually a low index rational plane. That of martensite however is usually not. In addition to the habit plane, there is also a characteristic deformation direction d defined with $|d| = 1$ and a deformation magnitude m proportional to the distance from the habit plane. The invariant plane strain then takes the matrix form $I + mdh^t$ where d and h are each column vectors and t stands for transpose.

For the case of deformation twinning in steel, the habit plane is a twinning plane ($\{112\}$) and the deformation direction a twinning direction ($<111>$). The subsequent transformed region has the same lattice structure as the parent phase and is only transformed by one of the point group symmetry properties of the parent phase (it may be a mirror image of the surrounding phase).

In a martensitic transformation, the lattice structure of the transformation product is different from that of the parent phase. The habit plane is usually irrational and the deformation direction may not be coplanar with the habit plane. This results in a nonzero volume change (dilatation). Thus, the region that transforms undergoes a change of shape. This change of shape is a mechanism for the generation of acoustic emission and (given md, h and the initial shape) the source function could be evaluated as discussed above.

However, this would fail to incorporate other very important effects. First the change of shape occurs in a constraining medium so that residual stresses form. These interact with the change in elastic constants of the transformed region to generate further emission. If the value of m is small, the residual stress can be accommodated elastically, leading to a thermoelastic transformation. Often the stress is such that considerable plastic deformation and twinning occur. These can be additional emission sources. Also, the change of shape is different from that of the unconstrained case. In attempting to predict acoustic emission from martensite, care must be taken in choosing the appropriate shape change.

The expression for the stress change associated with the martensitic transformation of an ellipsoidal region of volume v is:[27]

$$\overline{\Delta\sigma(t)} = [I + \Delta CD]^{-1}[(C + \Delta C)\beta^* \quad \text{(Eq. 36)}$$
$$- \Delta C\beta°]v(t)$$

TABLE 4. Acoustic emission during continuous cooling of plain carbon steel

Transformation Product	Acoustic Emission Activity	Phase Transformations
Pearlite	None detectable	Diffusion-controlled simultaneous growth of lamellar ferrite and cementite
Bainite	None detectable	Diffusion-controlled growth of small carbides (< 1 μm) and lath ferrite
Martensite	Very energetic signals detected	Diffusionless transformation in which laths or plates typically 20 μm in diameter and several micrometers thick transform from face-centered cubic to body-centered tetragonal structure at about 30 percent of shear wave speed

Where:

C = the stiffness matrix (using Voight notation) of the parent phase;
$C + \Delta C$ = the stiffness matrix of martensite;
β^* = the *unconstrained* shape change;
β° = any preexisting elastic strain (from an imposed stress or nearby source of residual strain); and
D = a shape matrix.

Examination of Eq. 36 reveals that six factors associated with the transformation affect the acoustic emission: (1) the volume of the region transformed; (2) the dilatation strain; (3) the shear or rotational strain; (4) the habit plane; (5) residual stresses (through interaction with ΔC); and (6) the time dependence of the transformation.

Equation 36 can be considerably simplified if ΔC is supposed to be very small (its value is unknown for most transformations):

$$\overline{\Delta \sigma(t)} \sim C\beta^* v(t) \qquad \text{(Eq. 37)}$$

This is an expression very similar to that deduced for a plastic deformation.

Need for Shape Data

To get some idea of the type and magnitude of acoustic emission signals from actual martensitic transformations, data pertaining to the change of shape are needed. Careful measurements of this have been made[28] at the surface of an iron alloy containing 21.89 percent nickel and 0.82 percent carbon.

In one case it was found that:

$$h = \begin{bmatrix} 0.1752 \\ 0.5550 \\ 0.8131 \end{bmatrix} \qquad \text{(Eq. 38)}$$

and

$$d = \begin{bmatrix} -0.2006 \\ -0.6550 \\ 0.7284 \end{bmatrix}$$

with the magnitude of the plane strain vector $m = 0.19$. These values measured at the surface may differ from those in the bulk. Assuming isotropic elasticity, values for λ and μ can be estimated from reference 29:

$$\lambda = 10.5 \times 10^{-2} \qquad \text{(Eq. 39)}$$

and

$$\mu = 7.3 \times 10^{-2}$$

in meganewtons per square millimeter. If $\rho = 8.09 \times 10^{-3}$ grams per cubic millimeter, then $a = 5.57$ millimeters per microsecond and $c = 3$ millimeters per microsecond.

To fully predict the source function, the velocity surface of the transformation must be known. The growth mechanism of a martensitic region is poorly understood, but it is generally thought that the velocity in the habit plane is much greater than that in the perpendicular direction.

Consider the example of an ellipsoidal region with a circular habit plane in which the radial velocity is independent of direction and has a value of 2 mm•μs^{-1}. The rate of semi-axis growth perpendicular to the habit plane is assumed to be 10 percent of the radial velocity, or 0.2 mm•μs^{-1} so that:

$$v = 3.2\pi t^3 \qquad \text{(Eq. 40)}$$

cubic millimeters, and:

$$\dot{v} = 10 t^2$$

cubic millimeters per microsecond. From Eq. 37, the stress change rate is:

$$\dot{\overline{\Delta\sigma}} = t^2 \begin{bmatrix} -0.0287 & -0.0157 & -0.0244 \\ -0.0159 & -0.0623 & -0.0089 \\ -0.0244 & -0.0890 & 0.2032 \end{bmatrix} \qquad \text{(Eq. 41)}$$

Using the method described earlier under *Micromechanical Modeling*, the principle values of the stress change can be identified:

$$\overline{\dot{\Delta\sigma(t)}} = t^2 \begin{bmatrix} 0.176 & 0 & 0 \\ 0 & 0.085 & 0 \\ 0 & 0 & 0.550 \end{bmatrix} \quad \text{(Eq. 42)}$$

where the associated directions are such that the third principle direction is parallel to the x_3 axis (the [001] crystallographic axis), while the others lie in the x_1, x_2 plane, but rotated 80.46 degrees clockwise.

Rotation about the [001] axis enables achievement of maximum vertical and horizontal displacements at epicenter. Both displacement components have a parabolic time dependence, such as that of the microcrack considered earlier.

Assuming the source to be 40 mm beneath the sensor, it can be estimated that (for a vertical displacement transducer with 5 MHz bandwidth, 10^{-11} mm sensitivity and $\tau = 0.3t$) the smallest detectable, if optimally oriented, lath has a 3.25 μm diameter and a growth time of 1.62 ns. Since in this case one principle stress change value is negative, an orientation will exist in which no vertical (out of plane) signal is generated.

The maximum horizontal motion occurs in a plane containing the [001] direction at 80.46 degrees from [100] in a clockwise sense. Using identical bandwidth and sensitivity, the smallest detectable lath of optimal orientation would have a 1.6 μm diameter and a 0.8 ns growth period. Since all the principle stress change values differ, a horizontal displacement occurs for all orientations. It can be shown that a 2.25 μm diameter lath can produce a detectable horizontal displacement even when oriented so that the weakest signal propagates to the receiver.

Dynamics of the Transformations

Acoustic emission techniques are very useful for following the kinetics and measuring the dynamics of martensitic transformations. For example, in Fig. 16 the acoustic emission that occurred during the continuous cooling of a high carbon steel is shown. The M_s temperature for this steel was about 200 °C; only a few acoustic emission counts were generated above this temperature and these might have been associated with a local stress assisted transformation.

As the temperature continued to decrease below M_s, increasing rates of emission were observed, each emission presumably associated with the rapid growth of a martensitic plate across an austenite grain. The maximum transformation rate appeared to be about 60 °C below M_s and emission ceased entirely at about 100 °C below M_s.

The observation that intense acoustic emission signals are emitted at the beginning of martensite transformations has

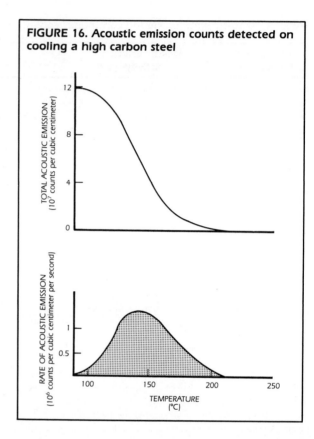

FIGURE 16. Acoustic emission counts detected on cooling a high carbon steel

been utilized to measure M_s values as part of an alloy development program.[30] The M_s values were reported for numerous steel alloys with different compositions and microstructures. The technique was reported to be a simple and accurate method for deducing martensite transformation temperatures. The acoustic emission accompanying the transformation was reported as extremely sensitive to the microscopic processes involved in the transformation.

The carbon concentration of low alloy steels[26] has a very strong effect on both the temperature dependence of the emission during continuous cooling and the number of detectable signals per unit volume (Fig. 17). The temperature dependence of the emission is consistent with the decrease in M_s with increasing carbon concentration. The effect of carbon concentration on the number of detectable signals, however, is more likely to be a manifestation of changes in the dynamics and morphology of the martensite transformation.

Effect of Cold Working

Cold working also influences the acoustic emission during the martensitic transformation of steel. The nature of the

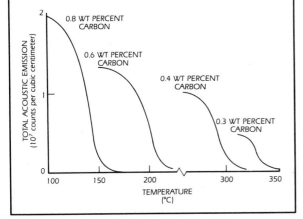

FIGURE 17. Total number of acoustic emission signals per unit volume generated by martensite; total is dependent on carbon content in iron base (4300 series) alloys; M_s temperature decreases with increasing carbon concentration

cold work effect is very sensitive to the carbon and nickel concentrations that control martensite morphology and the transformation kinetics. The cold work effect, like the carbon concentration effect, resides in changes to the morphology and dynamics of individual martensitic transformations.

Further measurements of the individual signals may provide a much improved understanding of this aspect of the martensite transformation. Measurements of the rate at which emission is detected also is a measurement of the rate at which laths form and may help explain the kinetics of the transformations and particularly the autocatalytic phenomena where they occur.

While almost all solid state phase transformations result in the development of internal stresses, only those associated with rapid martensitic transformations appear to generate detectable elastic waves directly. However, plastic deformation and even fracture are sometimes induced by internal stresses, providing indirect detection of their development.

For example, rapidly cooled high carbon steels have a very high internal stress that is normally relieved by a tempering treatment that allows dislocation and impurity atom migration. Using acoustic emission methods, it has been observed that isothermal tempering at too low a temperature will result in microcracking and this is believed to be due to impurity atoms locking of dislocations.[31]

Similarly, while the development of stress in and around a second-phase precipitate has so far not proven directly detectable by acoustic emission methods, signals have been observed[32] from hydride precipitates of niobium, tantalum and vanadium as they relax this localized stress by fracture and adjacent lattice plastic deformation.

These results suggest that there may well be merit in researching the use of acoustic emission techniques to detect the point during aging when the coherency stresses of small precipitates are relaxed by the formation of interfacial dislocation structures around semicoherent precipitates (as when $\Theta'' \rightarrow \Theta'$ in the AlCu system). The prismatic punching of dislocation loops during precipitation and cooling in other alloy systems (such as MoC) might also be worthy of investigation.

Magnetic Effects

Magnetic domain walls in ferromagnetic materials can be induced to move by the application of magnetic fields. This is accompanied by electromagnetic emission (the Barkhausen effect). The motion of these walls under the action of monotonic or alternating magnetic fields has also been found to generate elastic radiation (acoustic emission), the details of which have been found to be dependent on microstructure variables such as dislocation density and carbide distribution.

It has been suggested that measurement of this emission would provide a nondestructive method of microstructure characterization. Because the emission from domain wall motion is sensitive to the stress state, it has also been possible to estimate the residual stress for a given microstructure state.[33] The potential also exists for determining other microstructure information if residual stresses are absent.

Liquid-Solid Transformations

There have been few studies of acoustic emission during solidification. Work on PbSn and SnBi indicates the generation of acoustic emission signals when solidification conditions are such that interdendritic porosity (solidification shrinkage of the final interdendrite liquid) occurs.[34] The emission associated with this process is very intense, although the precise physical mechanism is unclear. This work indicates that the plastic deformation of primary dendrites could also generate low intensity acoustic emission.

Measurements have been made for the acoustic emission during solidification of a titanium alloy containing 4.5 percent aluminum and 0.2 percent copper.[35] It was suggested that the acoustic emission was generated by the formation of porosity. The volume fraction of porosity was varied by adjusting the hydrogen content of the melt. It was found that the solidification acoustic emission was proportional to the volume fraction of porosity (see Table 5).

This study considered the emission to be generated by the unstable formation of hydrogen bubbles in the melt

TABLE 5. Summary of results for solidification emission of aluminum 4.5 percent, copper 0.2 percent titanium alloy

Hydrogen Content ($Pa \cdot m^3$ per kg)	Total Acoustic Emission (10^4 counts)	Pore Fraction (percent)
0.05	1.05	0.19
0.17	2.75	0.46
0.23	6.35	0.63

close to the liquid-solid interface. The lower hydrogen solubility of solid aluminum results in a hydrogen supersaturation in the melt close to the liquid-solid interface. The relief of this supersaturation acts as the driving force for hydrogen bubble formation in a role analogous to that of strain energy reduction in the formation of cracks. This dilatation source then radiates longitudinal elastic waves that are transmitted through the liquid-solid interface and are ultimately detected by a transducer as acoustic emission.

Rapid solidification (occurring at speeds up to 2 $m \cdot s^{-1}$) has become a vigorous area of research because of the advantageous properties that may be achieved through refined microstructure and homogenous distributions of alloy elements. There is a pressing need for process control sensors that could characterize microstructure and measure process variables during solidification.

Electron Beam Melting

Acoustic emission measurements have been made on aluminum alloys during pulsed electron beam melting and resolidification.[36] It was found that acoustic emission signals are emitted during solid state electron beam heating (because of thermoelastic effects similar to those known to be responsible for laser generation of elastic waves), melting and resolidification. The acoustic emission during resolidification, following beam cutoff after attainment of a steady state temperature field, increases in this situation with the electron flux, probably as a result of the increased volume of resolidifying metal. For a given melt depth, the acoustic emission from a 2219 aluminum alloy is up to 100 times more energetic than that from nominally pure aluminum alloy 1100.

Metallographic studies on copper containing 2219 alloy indicate the occurrence of solidification cracking and course slip bands. Both phenomena were absent in the commercially pure aluminum alloy 1100. These results indicate that the large solidification and thermal contraction stresses set up during rapid solidification are responsible for plastic deformation in both materials.

The acoustic emission from dislocations is very weak in the fine grained aluminum alloy 1100 but it is much stronger in the 2219 alloy. Additional work indicates that dislocation motion in aluminum alloys at high temperatures generates more energetic emission than dislocation motion at room temperature.[37] This, together with the additional emission of hot tearing, results in much greater levels of detectable emission.

REFERENCES

1. Simmons, J.A. and R.B. Clough. "Theory of Acoustic Emission." *Proceedings of the International Conference on Dislocation Modelling of Physical Systems*, J. Hirth and M. Ashby, eds. Pergamon Press (1981).
2. Sinclair, J.E. *Journal of Physics D (Applied Physics)*. Vol. 12 (1979): pp 1,309.
3. Simmons, J.A. and H.N.G. Wadley. Unpublished.
4. Scruby, C.B., H.N.G. Wadley and J.E. Sinclair. *Philosophical Magazine*. Vol. A44 (1981): pp 240.
5. Scruby, C.B., C. Jones, J.M. Titchmarsh and H.N.G. Wadley. *Metal Science and Heat Treatment*. Vol. 15, (1981): pp 241.
6. Aki, K. and P.G. Richards. *Quantitative Seismology*. San Francisco, CA: W.H. Freeman Publishing (1980).
7. Drain, L.E., J.H. Speake and B.C. Moss. *SPIE Journal*. Vol. 136. Billingham, WA: Society of Photo-optical Instrumentation Engineers (1977): pp 52.
8. Scruby, C.B. and H.N.G. Wadley. *Journal of Physics D (Applied Physics)*. Vol. 11 (1976): pp 1,487.
9. Simmons, J.A. and H.N.G. Wadley. *Review of Progress in QNDE*, D.O. Thompson and D.E. Chimenti, eds. Vol. 3B. New York, NY: Plenum Press (1984): pp 699.
10. Eshelby, J.D. "The Continuum Theory of Lattice Defects." *Solid State Physics*, F. Seitz and D. Tumbull, eds. Vol. 3. Academic Press (1956): pp 79.
11. Kroner, E. *Erg. Hagew. Math.* Vol. 5 (1958): pp 1.
12. Haskell, N. *Bulletin of the Seismological Society of America*. Vol. 54 (1964): pp 1,811.
13. Burridge, R. and L. Knopoff. *Bulletin of the Seismological Society of America*. Vol. 54 (1964): pp 1,875.
14. Wadley, H.G.N., C.B. Scruby and J.H. Speake. *International Metals Review*. Vol. 25, No. 2 (1980): pp 41.
15. Kiesewetter, N. and P. Schiller. *Physica Status Solidi A (Applied Research)*. Vol. 48 (1978): pp 439.
16. Fleischman, P., F. Lakestani, J.C. Baboux and D. Rouby. *Materials Science and Engineering*. Vol. 29 (1977): pp 205.
17. Bill, R.C., J.R. Frederick and O.K. Felbeck. *Journal of Materials Science*. Vol. 14 (1977): pp 25.
18. Rusbridge, K.L., C.B. Scruby and H.N.G. Wadley. *Materials Science and Engineering*. Vol. 59 (1983): pp 151.
19. Scruby, C.B., H.N.G. Wadley and K.L. Rusbridge. *Materials Science and Engineering*. Vol. 59 (1981): pp 514.
20. Hsu, S. and K. Ono. "Acoustic Emission of Plastic Flow II: Alloys." *Proceedings of the Fifth International Acoustic Emission Symposium* (1980): pp 294.
21. Wadley, H.N.G., S.B. Scruby, P. Lane and J.A. Hudson. *Metals Science and Heat Treatment*. Vol. 15 (1981): pp 514.
22. Ono, K., M. Shibata and M.A. Hamstad. *Metallurgical Transactions*. Vol. 10A (1979): pp 761.
23. Cousland, S.M. and C.M. Scala. *Metal Science and Heat Treatment*. Vol. 15 (1981): pp 609.
24. McIntyre, P. and G. Green. *British Journal of Nondestructive Testing* (1978): pp 135.
25. Wadley, H.N.G. and C.B. Scruby. *International Journal of Fracture Mechanics* (1983): pp 117.
26. Speich, G.R. and A.J. Schwoeble. "Acoustic Emission during Phase Transformation in Steel." *ASTM Special Technical Publication 571*. Philadelphia, PA: American Society for Testing Materials (1975): pp 40.
27. Simmons, J.A. and H.N.G. Wadley. *Theory of Acoustic Emission from Inhomogeneous Inclusions in Wave Propagation in Homogeneous Media and Nondestructive Evaluation*, G.C. Johnson, ed. AMD-62. New York, NY: American Society for Mechanical Engineers (1984): pp 51.
28. Dunne, D.P. and J.S. Bowles. *Acta Metallurgica*. Vol. 17 (1969): pp 201.
29. Ledbetter, H.M. and R.P. Reed. *Materials Science and Engineering*. Vol. 5, No. 6 (1970): pp R-586.
30. Takashima, K., Y. Higo and S. Nunomura. "Acoustic Emission during the Martensitic Transformation of 304 Stainless Steel." *Proceedings of the Fifth Acoustic Emission Symposium* (1980): pp 261.
31. Shea, M.M. and D.J. Harvey. *Scripta Metallurgica*. Vol. 16 (1982): pp 135.
32. Cannelli, G. and R. Cantelli. "Acoustic Emission Stimulated by Hydride Formation in Niobium, Tantalum and Vanadium." *Advances in Acoustic Emission*, H.L. Dunegan and W.F. Hartman, eds. Knoxville, TN: Dunhart Publishing (1981): pp 330.
33. Ono, K. and M. Shibata. *NDT International* (October 1981): pp 227.
34. Tensi, H.M. and W. Radtke. *Metallurgy*. Vol. 32, No. 7 (1978): pp 681.
35. Feurer, J. and R. Wunderlin. *Proceedings of the International Conference on Solidification and Casting*. Vol. 2. London, England: British Metals Society (1977): pp 18.

36. Clough, R.B., H.N.G. Wadley and R. Mehrabian. "Heat Flow, Acoustic Emission, Microstructure Correlations in Rapid Surface Solidification." *Proceedings of the Second International Conference on Use of Electron and Laser Beams for Metals Processing*, E.A. Metzbower, ed. Metals Park, OH: American Society for Metals (1983): pp 37.
37. Hsu, S.Y.S. and K. Ono. "Acoustic Emission of Plastic Flow I: Metals. *Proceedings of the Fifth International Acoustic Emission Symposium* (1980): pp 283.

SECTION 4

WAVE PROPAGATION

Davis Egle, University of Oklahoma, Norman, Oklahoma

PART 1
WAVES IN INFINITE MEDIA

Acoustic emission signals are a sensor's response to sound waves generated in solid media. These waves are similar to the sound waves propagated in air and other fluids but are more complex because solid media are capable of resisting shear forces. This section describes the physical phenomena of sound waves propagating in solid media. Discussions of wave propagation in solids and fluids are available in the literature.[1-6]

The simplest case of wave propagation is that in a medium which has no boundaries, the infinite medium. Two and only two distinct types of waves can exist in infinite media. The waves are called *dilatational* (P) and *distortional* or *equivoluminal* (S). As a dilatational (P) wave propagates past a point, a small imaginary cube within the solid media changes volume, but the angles at the corners of the cube remain at 90 degrees. On the other hand, as a distortional (S) wave passes a point, the angles at the corners of the cube change but the volume remains constant.

Equations of Motion

Given in Eq. 1 are the equations of motion for displacements in a linear, elastic, homogeneous and isotropic body, with an external force applied (this represents three equations because i can be 1, 2 or 3).

$$(\lambda + \mu)u_{k,ki} + \mu u_{i,jj} + \rho f_i = \rho \ddot{u}_i \quad \text{(Eq. 1)}$$

Where:

- i = 1, 2 or 3;
- u_i = the Cartesian components of the particle displacement vector;
- λ, μ = material parameters representing elastic properties of the medium;
- ρ = density of the medium; and
- ρf_i = the applied force.

These expressions are sometimes called *Navier's equations of elasticity*.

Throughout this section, a comma in a subscript denotes partial differentiation, that is: $u_{i,j} = \partial u_i / \partial x_j$. Repeated subscripts denote summation over all values (Einstein convention), that is: $u_{i,kk} = u_{i,11} + u_{i,22} + u_{i,33}$.

The material constants λ and μ are called *Lamb's constants*. Lamb's constants can also be expressed in terms of the more familiar elastic constants E (Young's modulus) and ν (Poisson's ratio) by the following equations.

$$\lambda = \frac{E\nu}{(1 + \nu)(1 - 2\nu)} \quad \text{(Eq. 2)}$$

and

$$\mu = \frac{E}{2(1 + \nu)}$$

These equations can be inverted to yield the following forms.

$$E = \frac{\mu(3\lambda + 2\mu)}{\lambda + \mu} \quad \text{(Eq. 3)}$$

and

$$\nu = \frac{\lambda}{2(\lambda + \mu)}$$

General Solution of Motion Equations

It can be shown that a general solution to Eq. 1 can be written as the sum of two types of displacement functions, as follows:

$$u_i = u_i^d + u_i^e \quad \text{(Eq. 4)}$$

The terms u_i^d and u_i^e are solutions to following equations.

$$u_i^d = \phi_{,i} \quad \text{(Eq. 5)}$$

$$(c_1)^2 \phi_{,ii} = \ddot{\phi}$$

and

$$c_1 = \sqrt{\frac{\lambda + 2\mu}{\rho}}$$

$$u_i^e = \xi_{ijk} \Psi_{j,k} \quad \text{(Eq. 6)}$$

$$(c_2)^2 \Psi_{i,jj} = \ddot{\Psi}_i$$

and

$$c_2 = \sqrt{\frac{\mu}{\rho}}$$

The displacements u_i^d represent motion in which a small cube in the body will change volume (dilate) but will not rotate. The solutions u_i^e in Eq. 6 imply that the cube will not change in volume but will distort, that is, there will be a change in the angles at the corners of the cube.

The solutions to Eq. 5 and Eq. 6 can be shown to represent traveling waves. The dilatational waves propagate with speed c_1 and the distortional waves with speed c_2. Table 1 gives typical values for these wave speeds for a number of common materials.

Plane Wave Solutions

A special case of Eq. 5 and Eq. 6 may be written as follows.

$$u_i^d = \alpha_i \Phi(\alpha_k x_k - c_1 t) \quad \text{(Eq. 7)}$$

$$u_i^e = \xi_{ijk} \alpha_k \Psi_j(\alpha_k x_k - c_2 t) \quad \text{(Eq. 8)}$$

where $\alpha_k \alpha_k = 1$; Φ and Ψ_i are arbitrary scalar and vector functions (they must be twice differentiable).

The arguments of the functions $\Phi(\ )$ and $\Psi_j(\ )$ are constant on any plane where $\alpha_k x_k$ is constant. Hence Eqs. 7 and 8 must represent plane waves traveling in the direction α_k with speed c_1 or c_2. It can be shown that, in the case of Eq. 7, the displacement vector u_i^d acts parallel to the direction of wave travel α_i. This is called a *plane longitudinal wave*. It is a special kind of dilatational wave and travels with speed c_1.

Figure 1 is a diagram illustrating the motion of the particles in a medium in which a plane longitudinal wave is propagating. Because a plane longitudinal wave is a special kind of dilatational wave, the same diagram represents the motion of a dilatational wave as long as the wavefront curvature is small or zero. The circles represent adjacent particles of the medium (atoms in a regular crystalline structure). The wave is propagating from left to right. As the wave passes, the particles come closer together (along an axis parallel to the direction of propagation) and then move apart. After the wave passes, the particle spacing returns to normal.

Because the wavefront is plane and the medium in which the wave is traveling is infinite, the particles do not move in a direction perpendicular to the direction of the propagating wave. However, there will be stresses acting in the perpendicular direction. These stresses result from the Poisson's ratio effect. If the medium were not infinite, there would be a contraction and hence motion in the perpendicular direction.

The displacement vector u_i^e can be shown to act perpendicularly to the direction of wave travel, α_i. Hence this type of wave is called a *plane transverse wave*. Figure 2 is an illustration of the particle motion for a plane transverse wave. Here the particles move in a vertical direction while the wave propagates from left to right. Transverse waves involve shear deformations in the solid and are sometimes called *shear waves*.

Simple Harmonic Waves

A special form of the plane wave solutions are simple harmonic waves that can be generated mathematically by replacing Φ and Ψ_i from Eq. 7 and Eq. 8 with the following expressions.

FIGURE 1. Particle motion for dilatational or longitudinal waves

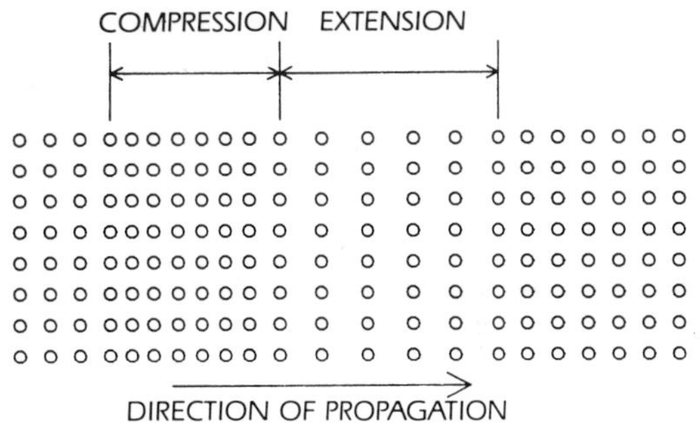

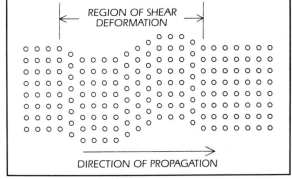

FIGURE 2. Particle motion for distortional or transverse waves

$$\Phi[\ ] = A \cos\left\{\frac{\omega}{c_1}[\ ]\right\} \quad \text{(Eq. 9)}$$

$$\Psi_i[\ ] = B_i \cos\left\{\frac{\omega}{c_2}[\ ]\right\} \quad \text{(Eq. 10)}$$

where A and B_i represent the displacement amplitudes of the longitudinal and transverse plane waves, respectively; ω is the circular frequency and is related to the frequency in cycles per unit time by the following relation.

$$\omega = 2\pi f \quad \text{(Eq. 11)}$$

If the direction of wave propagation is chosen to be coincident with the x_i axis, then the simple harmonic plane wave solutions for a plane longitudinal wave may be reduced as in Eq. 12.

$$u_1 = A \cos[k_1 x_1 - \omega t] \quad \text{(Eq. 12)}$$

where $u_2 = u_3 = 0$. If the particle motion for a transverse wave is assumed to be limited to the x_2 direction, then Eq. 13 is valid.

$$u_1 = 0 \quad \text{(Eq. 13)}$$

and

$$u_2 = B_2 \cos[k_2 x_2 - \omega t]$$

$$u_3 = 0$$

In Eqs. 12 and 13, the parameter k_i is called the *wave number* and is equal to:

$$k_i = \frac{\omega}{c_i} \quad \text{(Eq. 14)}$$

where $i = 1, 2$. A plot of the spatial distribution and the time history of a simple harmonic longitudinal wave is shown in Fig. 3. The spatial distribution is a plot of the displacement as a function of position x with the time fixed. For Fig. 3a, t has been set to 0. The distance between adjacent maxima (or minima) is called the *wavelength* and is denoted by λ. It can be seen from Eq. 12 that $k_1 \lambda = 2\pi$. Combining this expression with Eq. 11 and Eq. 14 results in the frequency-wavelength relationship:

$$f\lambda = c_1 \quad \text{(Eq. 15)}$$

Equation 15 is also valid for transverse waves if the wave speed c_1 is replaced by c_2.

The time history in Fig. 3b shows the displacement as a function of time with x fixed (and equal to 0 for this figure). The time between adjacent maxima is the *period* of the waveform and is usually denoted by T. The period is related to the frequency by the expression in Eq. 16.

$$T = \frac{1}{f} \quad \text{(Eq. 16)}$$

Simple harmonic waveforms are often represented by a complex exponential notation as follows.

$$u_1 = (A)e^{j(k_1 x_1 - \omega t)} \quad \text{(Eq. 17)}$$

where $u_2 = u_3 = 0$. It is understood but rarely stated explicitly that what is meant is:

$$u_1 = A \cdot \text{real part of } \{e^{j(k_1 x_1 - \omega t)}\} \quad \text{(Eq. 18)}$$

where $u_2 = u_3 = 0$. The exponential notation is very convenient and may be used to manipulate the expressions as long as the terms are not subjected to a nonlinear operator.

FIGURE 3. Plots of particle displacement for a simple harmonic longitudinal wave: (a) spatial distribution; and (b) time history

(a)
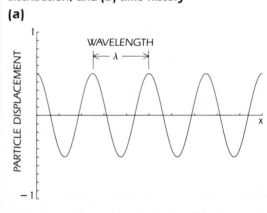

(b)
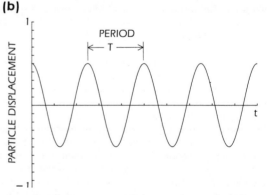

PART 2
SURFACE WAVES

When a free surface is present in a medium, a third type of wave may exist in addition to dilatational and distortional waves. It is called a *surface wave* or *Rayleigh wave*. If the free surface is defined by $x_3 = 0$ and the solid is $x_3 \geq 0$, then the following displacement field satisfies Eq. 1.

$$u_1 = k \left[-\frac{1 - \frac{1}{2}\alpha^2}{\sqrt{1 - \frac{\alpha^2}{\kappa^2}}} e^{-qx_3} \right.$$

$$\left. + \sqrt{1 - \alpha^2}\; e^{-sx_3} \right] e^{j(\omega t - kx_1)} \quad \text{(Eq. 19)}$$

$$u_2 = 0$$

and

$$u_3 = jk\left[\left(1 - \frac{1}{2}\alpha^2\right) e^{-qx_3} - e^{-sx_3} \right] e^{j(\omega t - kx_1)}$$

Where:

$q^2 = k^2(1 - \alpha^2/\kappa^2);$
$s^2 = k^2(1 - \alpha^2);$
$k = \omega / c_r;$
$\alpha = c_r / c_2;$ and
$\kappa = c_1 / c_2.$

Equations 19 represent a simple harmonic wave of frequency ω traveling in the x_1 direction with speed c_r. The speed c_r at which surface waves travel is determined by one of the real solutions to the cubic equation below.

$$\alpha^6 - 8\alpha^4 + (24 - 16\kappa^{-2})\alpha^2 \quad \text{(Eq. 20)}$$
$$+ 16(\kappa^{-2} - 1) = 0$$

Although Eq. 20 has three roots, at least one of them is real and has a value slightly less than unity. Actually this real root varies from about 0.87 for $\nu = 0$ to about 0.95 for $\nu = 0.5$. The two complex roots of Eq. 19 can be shown to be associated with a special case of bulk waves incident on the free surface. Values of c_r computed from Eq. 20 for several materials are shown in Table 1. An approximate value of c_r may be determined from the Bergmann formula below.

$$\alpha = \frac{0.87 + 1.12\,\nu}{1 + \nu} \quad \text{(Eq. 21)}$$

where ν is Poisson's ratio for the medium.

It can be shown that Eq. 19 implies a particle motion that is elliptical. The terms in brackets in the first and third parts of Eq. 19 show that the amplitudes of the waves decrease as the distance from the free surface increases. Figure 4 shows a plot of the amplitudes of u_1 and u_3 as a function of distance from the surface (x_3/λ). Although the computations for the plot were made using $\nu = 0.3$, the form of the displacement functions does not change appreciably for values of Poisson's ratio ranging from 0.2 to 0.4. The out-of-plane displacement u_3 peaks about 0.1λ below the surface and the

TABLE 1. Wave speeds and acoustic impedances of selected materials

Material	c_1	c_2	c_r	c_b	ρc_1
Aluminum	6.3	3.1	2.9	5.1	17
Brass	4.4	2.1	2.0	3.5	36
Cast iron	5.0	3.0	2.7	4.7	36
Copper	4.7	2.3	2.1	3.8	42
Lead	2.2	0.7	0.66	1.2	25
Magnesium	5.8	3.1	2.9	5.0	10
Nickel	5.6	3.0	2.8	4.8	49
Steel	5.9	3.2	3.0	5.1	46

UNITS OF WAVE SPEEDS: kilometer per second
UNITS OF ACOUSTIC IMPEDANCE: 10^6 kilogram per square meter per second

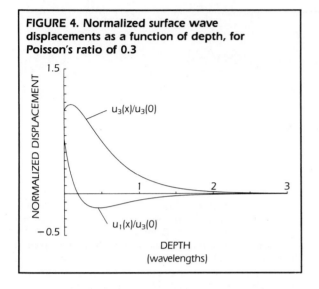

FIGURE 4. Normalized surface wave displacements as a function of depth, for Poisson's ratio of 0.3

surface displacements decrease to negligible levels for depths greater than two wavelengths.

Surface waves may exist on free surfaces as well as on surfaces that are rigidly constrained in both the tangential and the normal directions. Surfaces that have mixed boundary conditions (such as those rigidly constrained from normal motion but with zero in-plane tractions) cannot support surface waves.

The fact that the wave in Eq. 19 is propagating in the x_1 direction is entirely arbitrary. A similar solution exists for a surface wave traveling in any direction parallel to the surface.

Although surface waves and bulk waves are independent of each other in a semi-infinite solid with a free surface, they do couple in solids that have intersecting surfaces (corners) or surfaces with boundary conditions that change with position on the surface. Coupling between surface waves and bulk waves is difficult to avoid in solids of practical geometries.

PART 3
REFLECTION OF WAVES AT INTERFACES

Waves that encounter a change in the media in which they are propagating will change directions or reflect. In addition to reflection, the interface causes the wave to diverge from its original line of flight or refract in the second medium. Also, the mode of the wave may be changed in the reflection process.

A wave incident upon an interface between two media will reflect or refract such that the directions of the incident, reflected and refracted waves all lie in the same plane. This plane is defined by the line along which the incident wave is propagating and the normal to the interface.

Transverse Waves Incident on a Free Surface

For a transverse wave incident upon a free surface, it is convenient to decompose the wave into a component normal to the plane of reflection (SH) and a component that lies in the plane of reflection (SV). The reflection of the SH component is the simplest case because only an SH wave is reflected at an angle equal to the incident angle. The reflected SH wave is equal in magnitude (but of opposite sign) to the amplitude of the incident SH wave. Thus, an SH wave reflects in the same way that a beam of light reflects from a mirror, with only a change in the direction of the wave's travel. The reflection of an SV wave is a more complicated matter.

Consider next a plane transverse SV wave propagating in a solid medium and incident on a free surface as shown in Fig. 5. The direction in which the incident wave is traveling is defined by Θ_i, the angle between the normal to the surface and the direction in which the wave is moving. In order to satisfy the boundary conditions at the free surface, the displacement field must also include two reflected waves A_{rt} and A_{rl} as shown in the figure.

The directions in which the reflected waves travel are functions of the direction of the incident wave Θ_i and wave speeds c_1 and c_2 in the medium. The directions are determined by Snell's law which maintains that the spatial phase of the incident waves and the reflected waves along the surface must be equal. In this case, Snell's law requires that the reflected transverse waves propagate at angle Θ_i from the normal and that the reflected longitudinal waves propagate at the angle Θ_{rl}, which satisfies Eq. 22.

$$\frac{\sin \Theta_{rl}}{c_1} = \frac{\sin \Theta_i}{c_2} \quad \text{(Eq. 22)}$$

Because c_1 is greater than c_2, the angle of the reflected longitudinal wave is greater than the angle of the reflected transverse wave. The angle of the reflected longitudinal wave Θ_{rl} is ninety degrees when Eq. 23 is satisfied.

$$\Theta_i = \sin^{-1} \frac{c_2}{c_1} = \Theta_c \quad \text{(Eq. 23)}$$

This value of the incident angle is called the *critical angle*. At this angle the transverse wave is the only wave reflected back into the medium. The longitudinal wave, which is apparently reflected parallel to the surface, actually has a vanishingly small amplitude away from the free surface. That component can be shown to be equivalent to a disturbance propagating at the same speed as the intersection of the two transverse waves along the surface. For angles greater than Θ_c, the incident transverse wave is totally reflected into the medium.

If the displacement amplitude of the incident transverse wave is unity, the amplitudes of the two reflected waves are as follows.

$$A_{rt} = \frac{\sin(2\Theta_i)\sin(2\Theta_{rl}) - \kappa^2 \cos^2(2\Theta_i)}{\sin(2\Theta_i)\sin(2\Theta_{rl}) + \kappa^2 \cos^2(2\Theta_i)} \quad \text{(Eq. 24)}$$

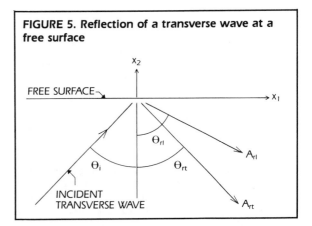

FIGURE 5. Reflection of a transverse wave at a free surface

$$A_{rl} = \frac{\kappa \sin(4\Theta_i)}{\sin(2\Theta_i)\sin(2\Theta_{rl}) + \kappa^2 \cos^2(2\Theta_i)} \quad \text{(Eq. 25)}$$

where $\kappa = c_1/c_2$ is the ratio of the speed of longitudinal waves to the speed of transverse waves.

If Θ_i is less than Θ_c, then Eqs. 24 and 25 are simple and well behaved. If the angle of the incident transverse wave is greater than the critical angle Θ_c, then the two equations change character and become complex functions of the incident angle and the wave speed ratio. If Θ_i is greater than Θ_c, then Eqs. 24 and 25 reduce to the following forms.

$$A_{rt} = e^{j2\alpha} \quad \text{(Eq. 26)}$$

$$A_{rl} = S e^{j\alpha} \quad \text{(Eq. 27)}$$

The values of S and α are determined below.

$$S = \frac{\sin(4\Theta_i)}{\sqrt{\kappa^2 \cos^4(2\Theta_i) + 4[\kappa^2 \sin^2(\Theta_i) - 1]\sin^2(2\Theta_i)\sin^2(\Theta_i)}} \quad \text{(Eq. 28)}$$

$$\tan \alpha = \frac{2\sqrt{\kappa^2 \sin(\Theta_i) - 1}\sin(2\Theta_i)\sin(\Theta_i)}{\kappa \cos^2(2\Theta_i)} \quad \text{(Eq. 29)}$$

In this case, the expressions for the reflected waves are as follows.

and
$$u_1^{rt} = \alpha_2 e^{j[k_2(\alpha_1 x_1 - \alpha_2 x_2) - \omega t - 2\alpha]} \quad \text{(Eq. 30)}$$
$$u_2^{rt} = \alpha_1 e^{j[k_2(\alpha_1 x_1 - \alpha_2 x_2) - \omega t - 2\alpha]}$$

and
$$u_1^{rl} = \beta_1 S e^{\zeta x_2} e^{j(k_2 \beta_1 x_1 - \omega t - \alpha)} \quad \text{(Eq. 31)}$$
$$u_2^{rl} = -\beta_2 S e^{\zeta x_2} e^{j(k_2 \beta_1 x_1 - \omega t - \alpha)}$$

The terms α_1 and α_2 in Eqs. 30 are the direction cosines of the reflected transverse waves and are given by the expressions $\alpha_1 = \sin\Theta_i$ and $\alpha_2 = \cos\Theta_i$. The terms β_1 and ζ in Eqs. 31 are determined from $\beta_1 = \kappa \sin\Theta_i$, $\beta_2 = j\zeta$ and $\zeta = (\kappa^2 \sin^2\Theta_i - 1)$.

The angles of incidence $\Theta_i > \Theta_c$, the amplitude of the reflected transverse wave is equal to the amplitude of the incident transverse wave but the phase of the reflected transverse wave is shifted by 2α. The other term (Eq. 31) was a reflected longitudinal wave for $\Theta_i < \Theta_c$. However, for $\Theta_i > \Theta_c$, it is no longer a plane wave traveling into the medium but rather is a nonhomogeneous disturbance propagating along the surface at the same speed as the projection of the transverse wave along the surface.

Plots of Eqs. 24 and 25 (or for $\Theta_i > \Theta_c$, Eqs. 26 and 27) are shown in Fig. 6. Note that as the incident angle increases from zero, the magnitude of the reflected transverse wave decreases while the amplitude of the reflected longitudinal wave increases. In this region, the incident transverse wave is increasingly converted to a longitudinal wave as the incident angle increases up to about 30 degrees. At an incident angle of about 30 degrees, the amplitude of the reflected transverse wave begins to increase.

The discontinuity in Fig. 6 occurs at the critical angle which is 32.3 degrees for Poisson's ratio of 0.3. For angles

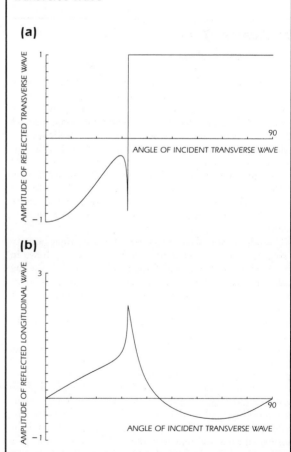

FIGURE 6. Reflection coefficients for an incident transverse wave with Poisson's ratio of 0.3: (a) amplitude of reflected transverse wave for incident transverse wave; and (b) amplitude of reflected longitudinal wave for incident transverse wave

greater than the critical, the amplitude of the reflected transverse wave is equal to the amplitude of the incident transverse wave but the phase of the reflected wave changes as discussed above. The amplitude of the reflected longitudinal wave for $\Theta_i > \Theta_c$ is really the amplitude of the disturbance propagating parallel to the surface and this in turn is dependent on the incident angle.

Incident Longitudinal Wave

As shown in Fig. 7, a plane longitudinal wave incident on a free surface at angle Θ_i produces a reflected longitudinal and a reflected transverse wave. The angle the reflected transverse wave makes with the normal to the surface is determined by Snell's law.

$$\frac{\sin(\Theta_{rt})}{c_2} = \frac{\sin(\Theta_i)}{c_i} \quad \text{(Eq. 32)}$$

The amplitudes of the reflected waves are given by Eqs. 33 and 34.

$$A_{rl} = \frac{\sin(2\Theta_i)\sin(2\Theta_{rt}) - \kappa^2 \cos^2(2\Theta_{rt})}{\sin(2\Theta_i)\sin(2\Theta_{rt}) + \kappa^2 \cos^2(2\Theta_{rt})} \quad \text{(Eq. 33)}$$

$$A_{rt} = \frac{2\kappa \sin(2\Theta_i)\cos(2\Theta_{rt})}{\sin(2\Theta_i)\sin(2\Theta_{rt}) + \kappa^2 \cos^2(2\Theta_{rt})} \quad \text{(Eq. 34)}$$

where $\kappa = c_1/c_2$ as before.

A plot of these amplitude ratios is shown in Fig. 8. The equations are better behaved in this case. The amplitude of the reflected transverse wave gradually increases from 0 to 1.1 at about 45 degrees and then decreases back to zero for an incident angle of 90 degrees. The reflected longitudinal wave is seen to decrease to a minimum at about 65 degrees.

FIGURE 7. Reflection of a longitudinal wave at a free surface

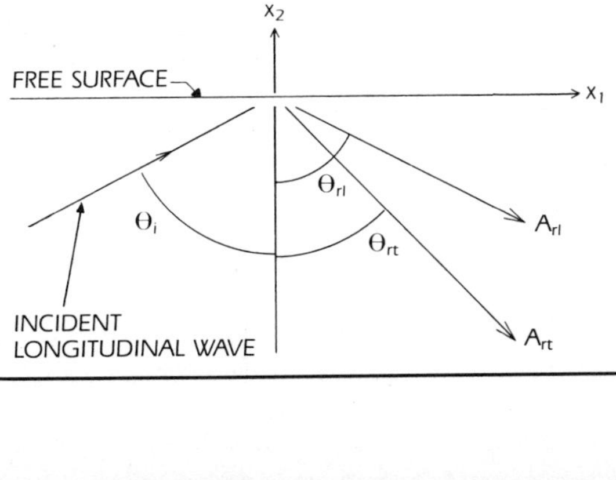

Energy Partition at a Free Surface

Drawing conclusions from the displacement amplitudes about the waves reflected from the free surface can be misleading. Another view from which to examine the reflection problem can be expressed in terms of energy or power in the incident and reflected waves.

The intensity I of a wave is the energy transported by the wave per unit time per unit area, normal to the direction the wave is propagating, as given below.

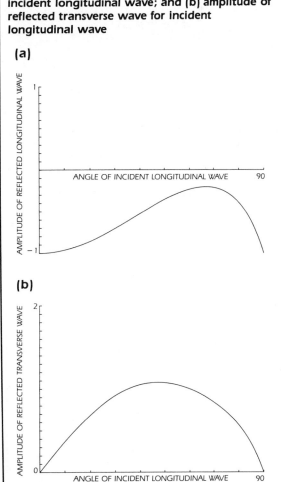

FIGURE 8. Reflection coefficients for an incident longitudinal wave with Poisson's ratio of 0.3: (a) amplitude of reflected longitudinal wave for incident longitudinal wave; and (b) amplitude of reflected transverse wave for incident longitudinal wave

$$I = \frac{1}{2} \rho c (A)^2 \qquad \text{(Eq. 35)}$$

Where:

ρc = the specific impedance (product of density times wave speed) of the medium; and
A = the displacement amplitude of the wave.

This expression is valid for any wave regardless of type.

Using Eq. 35, the intensities of the incident and reflected waves can be computed for the case of an incident longitudinal wave.

$$I_{il} = \frac{1}{2} \rho c_1 (1)^2 \qquad \text{(Eq. 36)}$$

$$I_{rl} = \frac{1}{2} \rho c_1 (A_{rl})^2$$

and

$$I_{rt} = \frac{1}{2} \rho c_2 (A_{rt})^2$$

The amplitudes A_{rl} and A_{rt} are for the reflected longitudinal 2nd transverse waves respectively and are given in Eqs. 33 and 34. The expressions in Eq. 36 give the power per unit area. To calculate the power incident on the surface and the powers in the reflected waves, these expressions must be multiplied by the appropriate area.

A unit area of the incident longitudinal wavefront is shown in Fig. 9. When the two rays of the incident wave reflect from the free surface, the spacing between the rays of the transverse wave changes to w_t. This value can be determined by simple geometry to be:

$$w_t = \frac{\cos(\Theta_{rt})}{\cos(\Theta_i)} \qquad \text{(Eq. 37)}$$

The power incident on or reflected from the free surface is equal to the intensity times the area and thus can be written for the three waves as follows:

$$P_{il} = \frac{1}{2} \rho c_1 \qquad \text{(Eq. 38)}$$

$$P_{rl} = \frac{1}{2} \rho c_1 (A_{rl})^2 \qquad \text{(Eq. 39)}$$

$$P_{rt} = \frac{1}{2} \frac{\rho c_2 (A_{rt})^2 \cos(\Theta_{rt})}{\cos(\Theta_i)} \qquad \text{(Eq. 40)}$$

Proceeding in a similar manner, the intensities and powers associated with the incident and reflected waves for a transverse wave incident on a free surface can be calculated. The powers for this case are:

$$P_{it} = \frac{1}{2} \rho c_2 \qquad \text{(Eq. 41)}$$

$$P'_{rt} = \frac{1}{2} \rho c_2 (A_{rt})^2 \qquad \text{(Eq. 42)}$$

$$P'_{rl} = \frac{1}{2} \frac{\rho c_1 (A_{rl})^2 \cos(\Theta_{rl})}{\cos(\Theta_i)} \qquad \text{(Eq. 43)}$$

The value for Θ_{rl} is computed using Snell's law (Eq. 22) and A_{rt} and A_{rl} are given by Eqs. 24 and 25.

For both of the cases presented above, the principle of the conservation of energy demands the following relationship.

FIGURE 9. Change in area for mode-converted wave

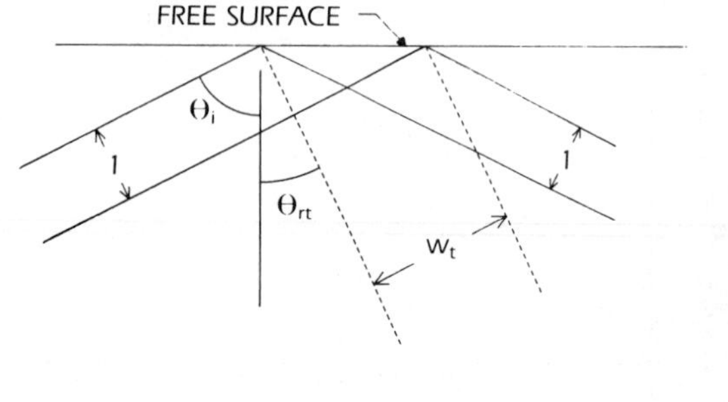

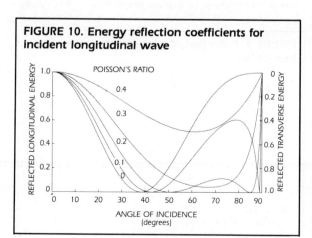

FIGURE 10. Energy reflection coefficients for incident longitudinal wave

$$P_i = P_{rl} + P_{rt} \quad \text{(Eq. 44)}$$

Equation 44 is satisfied by substituting Eqs. 41 through 43 or Eqs. 38 through 40.

Graphs showing the partition of power for an incident longitudinal wave and for several values of Poisson's ratio are given in Fig. 10. The figure shows the ratios P_{rl}/P_{il} on the left hand scale and P_{rt}/P_{il} on the right hand scale. Likewise the ratios P'_{rt}/P_{it} and P'_{rl}/P_{it} are shown in Fig. 11 for the same values of Poisson's ratio.

Reflection and Transmission at an Interface Between Two Solids

Another common interface that appears in many practical solids is an interface formed by joining two different materials as shown in Fig. 12. Considered here is the problem of reflection and transmission of a wave normally incident on the interface (a wave whose direction of propagation is normal to the plane interface). If an incident wave of unit amplitude in medium-1 is longitudinal and if the interface between the solids is the plane $x_2 = 0$, then the solution may be written as below.

$$u_1 = u_3 = 0 \quad \text{(Eq. 45)}$$

and when $x_2 \leq 0$

$$u_2 = e^{j(k_{11}x_2 - \omega t)} + A_r e^{j(-k_{11}x_2 - \omega t)}$$

or when $x_2 \geq 0$

$$u_2 = A_t e^{j(k_{12}x_2 - \omega t)}$$

Where:

A_r = the amplitude of the wave reflected in medium-1;
A_t = the amplitude of the wave transmitted into medium-2;
$k_{11} = \omega/c_1$ for medium-1; and
$k_{12} = \omega/c_1$ for medium-2.

The continuity of stresses and displacements across the interface leads to the following relations.

$$A_t = \frac{2Z_{11}}{Z_{11} + Z_{12}} \quad \text{(Eq. 46)}$$

and

$$A_r = \frac{Z_{11} - Z_{12}}{Z_{11} + Z_{12}}$$

Where:

$Z_{11} = \rho_1 c_{11}$;
$Z_{12} = \rho_2 c_{12}$;
c_{1i} = the longitudinal wave speed in medium-i; and
i = 1 or 2.

The expressions in Eq. 46 also apply to transverse waves incident on a plane interface if the transverse wave speeds are used instead of the longitudinal wave speeds. If the power ratios are computed, the following relations result.

$$\frac{P_{rl}}{P_{il}} = \frac{(Z_{11} - Z_{12})^2}{(Z_{11} + Z_{12})^2} \quad \text{(Eq. 47)}$$

$$\frac{P_{tl}}{P_{il}} = \frac{4Z_{11}Z_{12}}{(Z_{11} + Z_{12})^2} \quad \text{(Eq. 48)}$$

In these equations, P_{il}, P_{rl} and P_{tl} are the powers associated with the incident, reflected and transmitted waves. These same expressions are valid for transverse waves if c_2 is substituted for c_1 in the acoustic impedance terms.

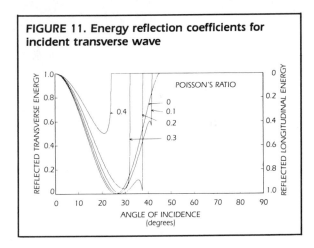

FIGURE 11. Energy reflection coefficients for incident transverse wave

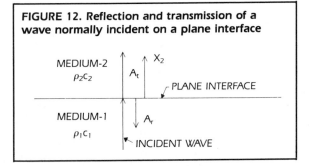

FIGURE 12. Reflection and transmission of a wave normally incident on a plane interface

Equation 47 shows that the only condition necessary for a wave to be transmitted past an interface without reflection is that the acoustic impedance ρc be the same in the two media. Note that the power reflected depends only on the acoustic impedance ratio and is independent of which medium has the larger acoustic impedance. That is, if $Z_{11} = 2Z_{12}$, one ninth of the incident power is reflected back into medium-1. If $Z_{12} = 2Z_{11}$, one ninth of the incident power is reflected back into medium-1. Note also that $P_{il} = P_{rl} + P_{tl}$ as required by the principle of the conservation of energy.

Problems involving the reflections of oblique waves incident upon solid-solid, solid-fluid and fluid-fluid interfaces have been surveyed in the literature.[1]

PART 4
ATTENUATION OF WAVES

Unlike the elementary cases of plane propagating waves considered earlier, waves in real media change in amplitude as they propagate through the solid. The term *attenuation* is used here in a general sense to mean the decrease in amplitude that occurs as a wave travels through a medium.

Attenuation has several causes and may be specifically related to the physical parameter used to specify the amplitude of the wave. The text below discusses several mechanisms that can lead to a decrease in the amplitude of a wave and can therefore be classified as attenuation mechanisms. Not all attenuation mechanisms are associated with energy losses. Some attenuation mechanisms are no more than a redistribution of energy to different modes, with no actual losses occurring.

Geometric Attenuation

When a wave is generated by a localized source, the disturbance propagates outward in all directions from the source. Even in a lossless medium (one in which mechanical energy is conserved), the energy in the wavefront remains constant but is spread over a larger spherical surface. The radius of this sphere is equal to the distance that the wave has traveled from the source. In order for the energy to remain constant, the amplitude of the wave must decrease with increasing distance from the source. For example from the center of a compression source (see *Idealized Sources* later in this Section), the displacement of the generated wave is given as follows.

$$u_r = -\frac{1}{4\pi c_1^2}\left\{\frac{1}{r^2}g\left(t - \frac{r}{c_1}\right) + \frac{1}{rc_1}g'\left(t - \frac{r}{c_1}\right)\right\} \quad \text{(Eq. 49)}$$

Where:

- $g(t)$ = the time history of the applied force;
- r = the distance from the source to the measuring point;
- c_1 = the dilatational wave speed; and
- t = time.

As the wave propagates, the first term decreases inversely as the square of the distance from the source. The second term decreases inversely as the distance from the source. Given that $g'(t)$ is not identically zero, the second term will dominate if the detector is sufficiently far from the source because the amplitude of this term decreases in proportion to $1/r$.

It is generally true for body waves originating from a small source, that if the wave is sufficiently far from the source then the amplitude of the wave will decrease inversely as the distance from the source. Although the example used to illustrate geometric dispersion includes only dilatational waves, the same phenomenon exists for equivoluminal waves. That is, a disturbance consisting of equivoluminal waves will decrease in amplitude as $1/r$ when the detector is far from the source. Thus disturbances that consist of the sum of dilatational and equivoluminal waves will also exhibit geometric attenuation proportional to $1/r$ in the far field.

There is a similar phenomenon for surface waves. However, it can be shown that the amplitude of a surface wave generated by a local source decreases as $1/r^{1/2}$ if the wave is sufficiently far from the source. This can be rationalized intuitively by considering a surface wave as a two dimensional wave expanding in a cylindrical surface. Because the surface area is proportional to the radius of the cylindrical surface (r) and the energy in the wave is proportional to u^2, then the displacement must decrease as $1/r^{1/2}$ in order to maintain constant energy in the wave.

Waves propagating in plates will undergo geometric attenuation if the source is localized to a size small in comparison to the receiver-to-source distance. As the surface or Lamb waves propagate over the surface of the plate, the amplitudes must decrease as $r^{-1/2}$ in the far field.

The only waves that are not subject to geometric attenuation are (1) one-dimensional waves such as plane waves in infinite or semi-infinite media; (2) waves generated by a line source in the plane of a plate; or (3) waves in one-dimensional media such as strings, bars or beams.

Attenuation from Dispersion

Dispersion is a phenomenon caused by the frequency dependence of speed for waves in certain physical systems. For example, the propagation of flexural waves in simple beams is governed by the following equation.

$$a^2 y_{,xxxx} + y_{,tt} = 0 \quad \text{(Eq. 50)}$$

Where:

$a^2 = EI/\rho A$;
y = displacement of the neutral axis of the beam;
x = position along the neutral axis;
t = time;
E = elastic modulus of the beam;
I = second moment of the cross-sectional area of the beam about the neutral axis;
ρ = density of the beam material; and
A = cross sectional area of the beam.

A simple harmonic solution of this equation is given below.

$$y = Be^{i\omega(t - x/c)} \quad \text{(Eq. 51)}$$

This is valid only if:

$$c = \pm \sqrt{a\omega} \quad \text{(Eq. 52)}$$

Because $\omega = kc$, Eq. 52 may also be written as below.

$$c = \pm ak \quad \text{(Eq. 53)}$$

or

$$\omega = ak^2$$

Equations such as Eqs. 53 are sometimes referred to as the *dispersion curve* and the *frequency spectrum* for the type of wave being propagated. The wave speed c is called the *phase velocity*, to distinguish it from the group velocity c_g, which is the speed of a waveform consisting of several frequency components centered about ω. It can also be shown that the energy in the wave propagates with the group velocity c_g.

Although the propagation of flexural waves in beams was chosen to illustrate attenuation due to dispersion, it is not the only physical system for which dispersion exists. In general, waves exhibit dispersion when propagating in any solid medium in which the wavelength is comparable to one or more of the dimensions of the medium. And the dispersion will cause an attenuation as shown in the example of the beam.

Because the wave speed c is a function of frequency, a disturbance made up of various frequency components will change in form as the components propagate along the bar because each component propagates at a different speed. An example of the effect of dispersion on a propagating wave is shown in Fig. 13. The illustration shows the displacement in an aluminum beam caused by an impulsive lateral displacement acting at $x = 0$. The beam has a cross section of 2.5×25 mm (0.1×1.0 in.) and the displacements were monitored at a distance of 50 mm (2 in.) from the source.

The solution shown in Fig. 13 was computed by combining the simple harmonic solutions of Eq. 51 combined with a Fourier transform technique to represent the impulsive source and the resulting wave. The peak amplitude of the waveform clearly decreases as the wave propagates down the beam. In this case, the amplitude has decreased from one unit to approximately 0.2 units in a distance of only 50 mm (2 in.). This decrease is solely due to the effect of dispersion. Notice also that the duration of the wave increases as a result of dispersion, a fact which complicates location of discontinuities using acoustic emission signals. These same effects accompany waves generated by localized sources in thin plates. In addition to the attenuation due to dispersion, the wave in a plate would also be attenuated by the geometric spreading of the beam, scattering and diffraction, and energy loss mechanisms.

Dispersion effects are significant in the waves that comprise the typical acoustic emission events. An example of the effect of dispersion on waves in bars has been documented.[7] Short duration waves were generated in a bar by pulsing a piezoelectric transducer. The resulting waveforms were monitored at several locations downstream from the source and the peak amplitudes showed rapid attenuation with distance from the source.

Attenuation from Scattering and Diffraction

Waves that propagate through media having complex boundaries and discontinuities (such as holes, slots, cavities, cracks and inclusions) interact with these geometric anomalies. In general a wave encountering any type of interface,

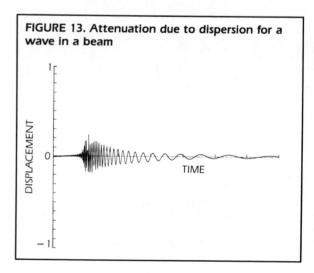

FIGURE 13. Attenuation due to dispersion for a wave in a beam

including a bounded one, will be reflected or at least partially reflected. A plane wave encountering a finite interface such as a cylindrical or spherical void will be partially reflected and partially transmitted past the void. Actually some energy is found propagating in all directions in the general case of a plane wave encountering a finite void or inclusion. Thus the wave is said to be *scattered*.

Likewise a wave encountering a sharp edge such as a crack or a groove in a solid is reflected from the discontinuity. Some of the energy propagating past the sharp edge will be bent or diffracted down into the shadow region of the crack. These diffraction effects are very difficult to compute; introductions to the subject of diffraction effects can be found in the literature.[2,3]

Both scattering and diffraction of a wave can lead to a decrease (or in some cases an increase) in the amplitude of a wave and both phenomena can cause attenuation. A commonly encountered cause of scattering is the inhomogeneities of grain boundaries in some materials. For example, waves in the megahertz frequency range will be significantly scattered by the coarse grain structure in cast iron. Thus the attenuation due to scattering is considerable in this case.

The amplitude decrease (attenuation) of a wave subject to scattering or diffraction is usually so difficult to calculate (if at all possible) that it becomes impractical except in the cases of simplest geometry. This leads to empirical approaches for dealing with attenuation, as discussed below.

Attenuation due to Energy Loss Mechanisms

In all of the mechanisms discussed above, the total mechanical energy of all of the waves (original and reflected, scattered, diffracted, dispersed, etc.) remains constant if the solid is an elastic medium. However, wave propagation in real media is usually not conservative. That is, mechanical energy associated with motion (kinetic energy) and the elastic deformation (potential energy) is not conserved.

Mechanical energy can be converted to thermal energy due to thermoelastic coupling. It can also be lost through (1) plastic deformation if the medium is stressed past its elastic limit; (2) creating new surfaces if a crack is extended; or (3) interactions with dislocation motion.

Losses can be associated with (1) viscoelastic material behavior prevalent in plastics; (2) friction between surfaces that slip; and (3) incompletely bonded inclusions or fibers in composite materials. Losses can be caused by magnetoelastic interactions; interactions with conduction electrons in metals; or paramagnetic electron and nuclear spin systems.[8]

Regardless of the particular mechanism causing the loss of mechanical energy, the amplitude of a wave will decrease as the wave propagates through the medium. If the medium and the loss mechanism are homogeneous, then the losses occur uniformly as the wave travels through the medium. If the wave is a simple harmonic function, the particle displacement can be described as below.

$$y = Ae^{-\alpha x} e^{j(kx - \omega t)} \qquad \text{(Eq. 54)}$$

Where:

x = the coordinate measured in the direction of wave travel;
α = the loss coefficient (must be a positive number);
k = the wave number (equal to ω/c);
t = time; and
A = the amplitude of the wave at $x = 0$.

The coefficient α (which is usually dependent on the frequency ω) has units of length^{-1} but the units are often expressed as nepers per length. A *neper* is a natural logarithmic unit corresponding to a reduction in amplitude of $1/e$ times the initial value. The attenuation coefficient can also be expressed in terms of decibels per length as follows.

$$\alpha_{dB} = 20 \log_{10}(e^\alpha) = 8.7\alpha \qquad \text{(Eq. 55)}$$

Equation 54 may also be written as follows.

$$y(x,t) = Ae^{j(k^*x - \omega t)} \qquad \text{(Eq. 56)}$$

where $k^* = k + j\alpha$.

The complex wave number k^* can be used to represent a wave propagating through a lossy medium. Because the wave number is inversely proportional to the square root of the elastic modulus ($k \propto E^{-1/2}$), then the lossy medium may also be modeled by replacing the elastic modulus E with a complex modulus E^*.

$$E^* = E(1 - j\eta) \qquad \text{(Eq. 57)}$$

The terms α and η are related by the following expression:

$$\alpha = \frac{k\eta}{1 + \eta^2} \qquad \text{(Eq. 58)}$$

The complex wave number approach, as exemplified in Eq. 56, has been used to study the effects of losses encountered in the propagation of waves generated by step forces in plastics.[9] The analytical method compared well to the experimental results.

Attenuation in Actual Structures

The mechanisms discussed here may all contribute to the attenuation of acoustic emission signals in real structures. Under normal circumstances, the attenuation must be measured by tests on the actual structure. Figure 14 shows an example of the attenuation measured in a gas pressure vessel.[10] The vessel was cylindrical with hemispherical caps and had a total length of 12.2 m (40 ft) with an inside diameter of 1.2 m (4 ft). The wall thickness was 125 mm (5 in.) in the cylindrical section and 64 mm (2.5 in.) in the caps. The attenuation was measured by placing a mechanical random noise source at one position and an acoustic emission sensor at a second position on the outer surface of the tank. The plot in Fig. 14 shows the relative amplitude of the received acoustic signals at various frequencies as the separation between the noise generator and the receiving sensor was increased. The relative amplitudes are plotted on a logarithmic scale as attenuation in decibels equal to $20_{10}(A/A_o)$,

where A is the measured amplitude at separation distance r and A_o is the amplitude at $r = 0$.

Each of the vertical scales have been displaced to clearly show the variations with frequency. The measured data are shown by the circles in the figure; the lines are approximations of linear variations of the data.

The plots clearly show an increased attenuation of the signal within the first 1.5 m (5 ft) of separation. Much of this is attributed to the geometrical spreading of the waves as they propagate outward from the source. As noted earlier, the amplitudes of the waves in this platelike structure would vary as $r^{-1/2}$ if geometric attenuation were the only active mechanism. In this case, the amplitude at 1.5 m (5 ft) is 6.9 dB less than the amplitude at 0.3 m (1 ft) and the attenuation at 100 kHz is approximately 7 dB. Once the wave reaches the separation distance of 1.5 m (5 ft) the structure begins to act as a waveguide effectively eliminating the geometric attenuation.

For separations greater than 1.5 m (5 ft), the attenuation per unit length is reduced but still present. Most of this attenuation is probably due to energy loss mechanisms with a smaller component caused by dispersion. The effects of frequency can clearly be seen on the attenuation plots of Fig. 14. These frequency effects are also shown in Fig. 15,

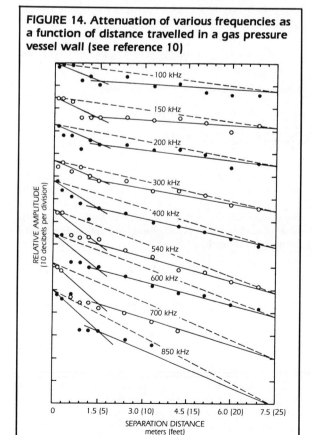

FIGURE 14. Attenuation of various frequencies as a function of distance travelled in a gas pressure vessel wall (see reference 10)

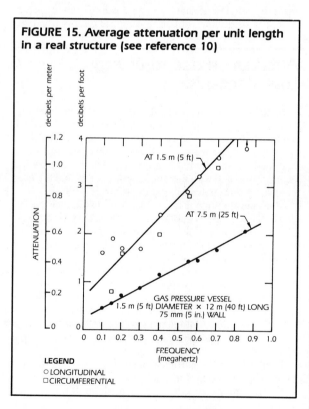

FIGURE 15. Average attenuation per unit length in a real structure (see reference 10)

which is a plot of the average attenuation per unit length obtained from the straight line approximations in Fig. 14 at separations of 1.5 m (5 ft) and 7.5 m (25 ft). This plot also shows attenuation measurements made along the circumference of the vessel very close to the longitudinal attenuation measurements.

Other factors can affect the attenuation of waves in real structures. Figure 16 shows the attenuation of the wave when the source is located on the inside surface of the vessel and the receiver is located on the outside at a distance of 7.5 m (25 ft) from the source. Figure 16 also shows the attenuation of a wave propagated across thick steel compression bands. These large reductions in amplitudes are probably due to reflections caused by the sudden changes in geometry. The large increases in attenuation indicate a potential problem in monitoring specific locations on a structure. Care must be taken when choosing the location of the sensors to avoid such large decreases in the received signals.

An additional factor that can affect attenuation is a surface coating applied to a structure. Figure 17 shows the attenuation effects of foam insulation materials applied to an aluminum panel. The change is quite dramatic and is very sensitive to the technique used to apply the foam.

Attenuation plays a major role in the use of acoustic emission especially on larger structures. It is usually impossible to predict analytically but can often be estimated by experience with similar structures.

Because of attenuation, the monitoring of large structures requires the use of several transducers placed sufficiently close together to ensure that an event emanating from any point on the structure will not be attenuated sufficiently to escape detection. Determining the minimum spacing is usually accomplished empirically, through experience with similar structures or through simulated acoustic emission testing.

FIGURE 16. Additional attenuation due to specific paths: (a) from inner wall to outer wall; and (b) across compression fit joint (see reference 10)

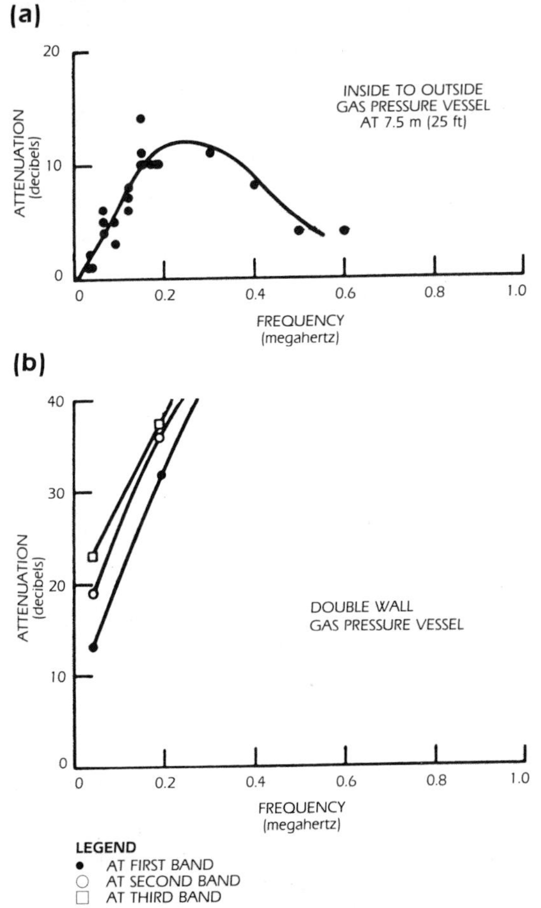

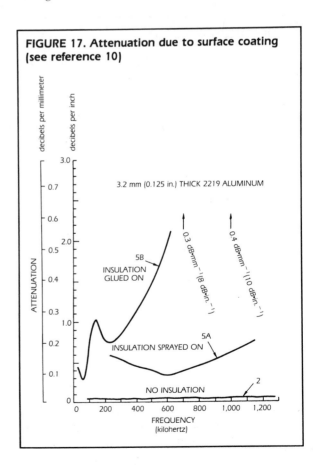

FIGURE 17. Attenuation due to surface coating (see reference 10)

PART 5
IDEALIZED WAVE SOURCES

There are many physical mechanisms that generate the high frequency waves called acoustic emission, yet there are some common characteristics that are helpful in determining a mathematical model representing an acoustic emission source. First, it is generally agreed that acoustic emission sources are small and tend to be localized to relatively small volumes. Secondly, the energy release that occurs when the mechanism activates is short in duration, on the order of microseconds. Thirdly, the real or virtual forces thought of as producing the displacements associated with the acoustic emission event must be self-equilibrating, that is, the forces must satisfy the rigid body equilibrium equations. Some of the sources useful for modeling acoustic emission sources and other sources of wave motion are discussed below.

Concentrated Force

Consider a concentrated force acting in the x_k direction at the origin and having a time history $g(t)$, that is:

$$\rho f_i = \frac{\delta_{ik} \rho g(t) \delta(r)}{4\pi r^2} \quad \text{(Eq. 59)}$$

Where:

$r = (x_1^2 + x_2^2 + x_3^2)^{1/2}$ or the distance from the origin;
$\delta_{ij} = 1$ if i equal to j or 0 if i not equal to j; and
$\delta(r) =$ the Dirac delta function.

The displacements in an infinite medium due to the forces in Eq. 59 are given as follows:[2]

$$4\pi u_i = \left\{ \frac{3x_i x_k}{r^3} - \frac{\delta_{ik}}{r} \right\} \int_{1/c_1}^{1/c_2} s g(t - rs) ds \quad \text{(Eq. 60)}$$

$$+ \frac{x_i x_k}{r^3} \left\{ \frac{1}{c_1^2} g\left(t - \frac{r}{c_1}\right) - \frac{1}{c_2^2} g\left(t - \frac{r}{c_2}\right) \right\}$$

$$+ \frac{\delta_{ik}}{r c_2^2} g\left(t - \frac{r}{c_2}\right)$$

Here $r = (x_1^2 + x_2^2 + x_3^2)^{1/2}$ is the distance from the point of application of the force to the point at which the displacement is being determined.

As a specific example, if the time history of the applied force is impulsive and applied at $t = 0$ so that $\rho g(t) = I_o \delta(t)$ where I_o is the magnitude of the total impulse applied to the body, then the displacements are:

(Eq. 61)

$$\frac{4\pi \rho u_1}{I_o} = \left\{ \frac{3x_1^2 - r^2}{r^5} \right\} t \left[H\left(t - \frac{r}{c_1}\right) - H\left(t - \frac{r}{c_2}\right) \right]$$

$$+ \frac{x_1^2}{r^3} \left\{ \frac{1}{c_1^2} \delta\left(t - \frac{r}{c_1}\right) - \frac{1}{c_2^2} \delta\left(t - \frac{r}{c_2}\right) \right\}$$

$$+ \frac{1}{r c_2^2} \delta\left(t - \frac{r}{c_2}\right)$$

and

$$\frac{4\pi \rho u_2}{I_o} = \left\{ \frac{3 x_1 x_2}{r^5} \right\} t \left[H\left(t - \frac{r}{c_1}\right) - H\left(t - \frac{r}{c_2}\right) \right]$$

$$+ \frac{x_1 x_2}{r^3} \left\{ \frac{1}{c_1^2} \delta\left(t - \frac{r}{c_1}\right) - \frac{1}{c_2^2} \delta\left(t - \frac{r}{c_2}\right) \right\}$$

and

$$\frac{4\pi \rho u_3}{I_o} = \left\{ \frac{3 x_1 x_3}{r^5} \right\} t \left[H\left(t - \frac{r}{c_1}\right) - H\left(t - \frac{r}{c_2}\right) \right]$$

$$+ \frac{x_1 x_3}{r^3} \left\{ \frac{1}{c_1^2} \delta\left(t - \frac{r}{c_1}\right) - \frac{1}{c_2^2} \delta\left(t - \frac{r}{c_2}\right) \right\}$$

Plots of the normalized displacements $4\pi \rho c_1 r^2 u/I_o$ as a function of the normalized time $c_1 t/r$ as given by Eq. 61 are shown in Fig. 18. These plots assume that $c_2 = c_1/2$.

Figure 18a shows the displacement u_1 at a position directly under the applied force ($x_2 = x_3 = 0$). The first disturbance to arrive is an impulse caused by a compressional wave. Following the impulse is a motion that increases linearly with time until the transverse wave arrives at $t = r/c_2$. At that time the displacement u_1 decreases to zero and the point experiences no further motion. The three expressions in Eq. 61 show that the displacements u_2 and u_3 are identically zero at this point.

The displacement u_1 at a point abreast of the force ($x_1 = x_3 = 0$) is shown in Fig. 18b. Note that in this case the displacements u_2 and u_3 are also identically zero. Figure 18b shows no evidence of the compressional wave but the linearly increasing motion starts at $t = r/c_1$. The transverse wave causes an impulsive displacement that arrives at $t = r/c_2$.

The displacements u_1 and u_2 at a point $x_1 = x_2 = r$ and $x_3 = 0$ are shown in Figs. 18c and 18d. These plots show impulsive waves arriving at r/c_1 and r/c_2 as well as the linearly increasing displacement between the two waves.

It should be noted from Eq. 61 that the impulsive terms decrease as r^{-1} while the terms involving the step functions $H(t - \ldots)$ decrease as r^{-2}. While the linearly changing displacement in all of the plots shown in Fig. 18 will be evident when the point is close to the source, this part decreases rapidly as the point moves away from the source. In the far field, only the impulsive terms will be important.

If the displacements in Eq. 61 are replaced by the corresponding velocities ($u_1 \to \dot{u}_1$, $u_2 \to \dot{u}_2$, etc.) and if I_o is replaced by F_o, then the plots shown in Fig. 18 also represent the velocity of a point in the infinite solid subjected to a concentrated force applied at the origin with a step time history [$\rho g = F_o H(t)$].

The typical acoustic emission sensor responds to motion, velocity or acceleration rather than displacement. As a

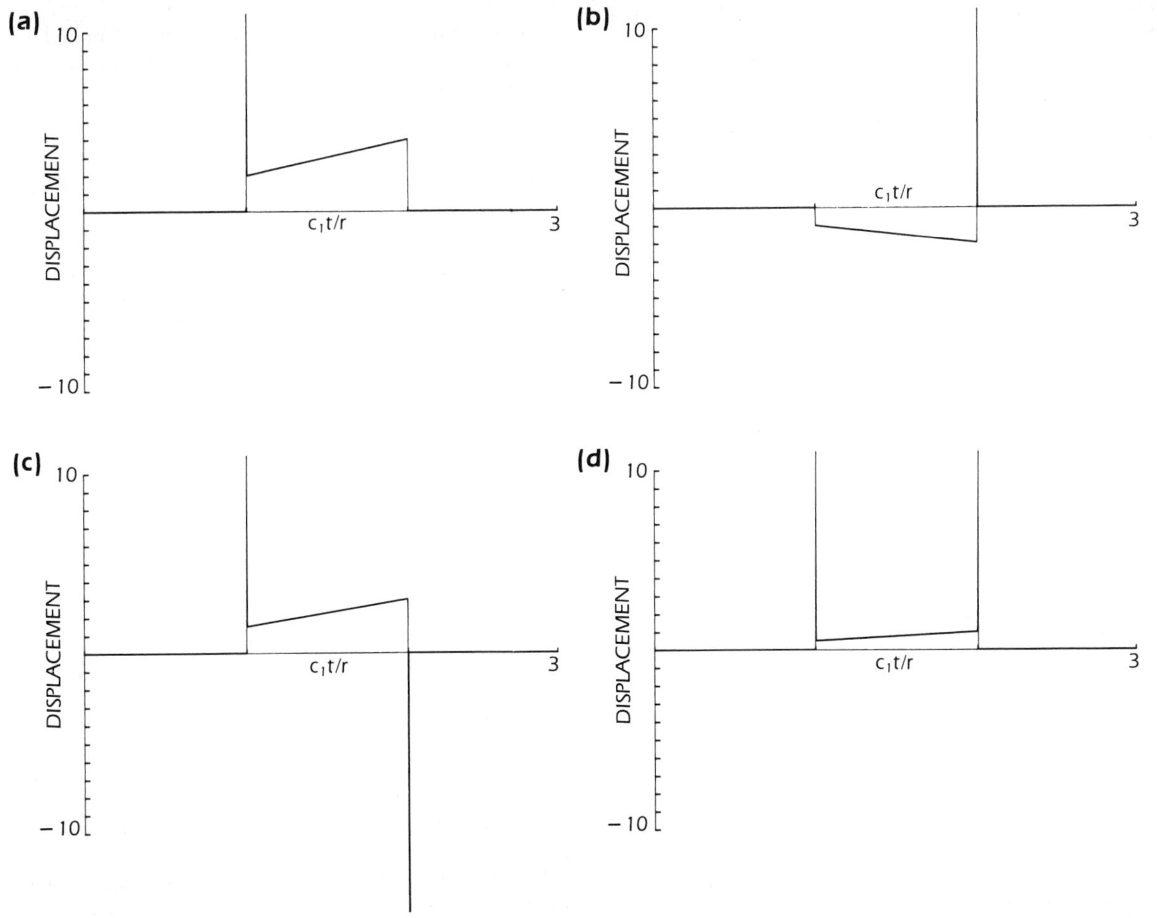

FIGURE 18. Response of an infinite medium to an impulsive force acting in the x_1 direction: (a) displacement u_1 directly under the force; (b) displacement u_2 and u_3 directly abreast of the force; (c) displacement u_1 at position 45 degrees from the vertical; and (d) displacements u_2 and u_3 at position 45 degrees from the vertical

result, the typical sensor will probably see one or two short duration impulses as a result of a step force being applied to the solid.

The concentrated force cannot be used as a model for an acoustic emission event because it is not self equilibrating. However it can be used to model externally applied forces, such as those applied by fracturing a glass capillary or a pencil lead break.

Double Force without Moment

If two forces of equal magnitude but acting in opposite directions are applied simultaneously at a point, the source is said to be a *double force without moment*. The displacements associated with this idealized source may be determined by computing the derivative of Eqs. 60 and 61 with respect to x_1 (double force is assumed acting in the x_1 direction). For the impulsive time history (Eq. 61), the displacements associated with the double force without moment are:

(Eq. 62)
$$\frac{4\pi\rho u_1}{I_o} = \left\{\frac{9x_1}{r^5} - \frac{15x_1^3}{r^7}\right\} t\left[H\left(t - \frac{r}{c_1}\right) - H\left(t - \frac{r}{c_2}\right)\right]$$
$$+ \frac{x_1}{r^4 c_1}\left\{\left(t + \frac{2r}{c_1}\right) - \frac{3x_1^2}{r^2}\left(t + \frac{r}{c_1}\right)\right\}\delta\left(t - \frac{r}{c_1}\right)$$
$$- \frac{x_1}{r^4 c_2}\left\{\left(t + \frac{2r}{c_2}\right) - \frac{3x_1^2}{r^2}\left(t + \frac{r}{c_2}\right)\right\}\delta\left(t - \frac{r}{c_2}\right)$$
$$- \frac{x_1^3}{r^4}\left\{\frac{1}{c_1^3}\delta'\left(t - \frac{r}{c_1}\right) - \frac{1}{c_2^3}\delta'\left(t - \frac{r}{c_2}\right)\right\}$$
$$- \frac{x_1}{r^2 c_2^2}\left\{\frac{1}{r}\delta\left(t - \frac{r}{c_2}\right) + \frac{1}{c_2}\delta'\left(t - \frac{r}{c_2}\right)\right\}$$

and

$$\frac{4\pi\rho u_2}{I_o} = \left\{\frac{3x_2}{r^5} - \frac{15x_1^2 x_2}{r^7}\right\} t\left[H\left(t - \frac{r}{c_1}\right) - H\left(t - \frac{r}{c_2}\right)\right]$$
$$+ \left\{\frac{x_2}{r^3 c_1^2} - \frac{3x_1^2 x_2}{r^6 c_1}\right\}\left(t + \frac{r}{c_1}\right)\delta\left(t - \frac{r}{c_1}\right)$$
$$- \left\{\frac{x_2}{r^3 c_2^2} - \frac{3x_1^2 x_2}{r^6 c_2}\right\}\left(t + \frac{r}{c_1}\right)\delta\left(t - \frac{r}{c_2}\right)$$
$$- \frac{x_1^2 x_2}{r^4}\left\{\frac{1}{c_1^3}\delta'\left(t - \frac{r}{c_1}\right) - \frac{1}{c_2^3}\delta'\left(t - \frac{r}{c_2}\right)\right\}$$

and

$$\frac{4\pi\rho u_3}{I_o} = \left\{\frac{3x_3}{r^5} - \frac{15x_1^2 x_3}{r^7}\right\} t\left[H\left(t - \frac{r}{c_1}\right) - H\left(t - \frac{r}{c_2}\right)\right]$$
$$+ \left\{\frac{x_3}{r^3 c_1^2} - \frac{3x_1^2 x_3}{r^6 c_1}\right\}\left(t + \frac{r}{c_1}\right)\delta\left(t - \frac{r}{c_1}\right)$$
$$- \left\{\frac{x_3}{r^3 c_2^2} - \frac{3x_1^2 x_3}{r^6 c_2}\right\}\left(t + \frac{r}{c_1}\right)\delta\left(t - \frac{r}{c_2}\right)$$
$$- \frac{x_1^2 x_3}{r^4}\left\{\frac{1}{c_1^3}\delta'\left(t - \frac{r}{c_1}\right) - \frac{1}{c_2^3}\delta'\left(t - \frac{r}{c_2}\right)\right\}$$

Center of Compression

If three double forces without moments are applied in directions x_1, x_2 and x_3, then the displacement components are (see reference 2):

$$u_i = \frac{1}{4\pi c_1^2}\frac{\partial}{\partial x_i}\left\{\frac{1}{r}g\left(t - \frac{r}{c_1}\right)\right\} \quad \text{(Eq. 63)}$$

The three double forces without moments combine to give a center of compression. The wave generated by the center of compression is a pure dilatational wave having no transverse components. If Eq. 63 is expanded, the following equation results.

$$u_i = -\frac{x_i}{4\pi r^2 c_1^2}\left\{\frac{1}{r}g\left(t - \frac{r}{c_1}\right)\right. \quad \text{(Eq. 64)}$$
$$\left. + \frac{1}{c_1}g'\left(t - \frac{r}{c_1}\right)\right\}$$

Note that the first term decreases as r^{-3} while the second term decreases as r^{-2}. The first term is only important in the near field (when the point at which the displacement is being measured is near the origin or the point at which the force is applied). In the far field (when $r \gg 1$), the second term dominates. In most acoustic emission tests, the wave travels quite a distance before it is sensed. Thus the second term, the far field term, is expected to provide the dominant part of the solution.

Consider as an example a center of compression in which the time history is a unit step function $g(t) = H(t)$, where $H(t)$ is 0 if $t < 0$ or 1 if $t > 0$. Equation 64 then becomes:

$$u_i = -\frac{x_i}{4\pi r^2 c_1^2}\left\{\frac{1}{r}H\left(t-\frac{r}{c_1}\right) + \frac{1}{c_1}\delta\left(t-\frac{r}{c_1}\right)\right\} \quad \text{(Eq. 65)}$$

In this case the first term represents the static deformation of the solid due to the applied compressive forces. The second term is proportional to the time derivative of the force and is the dynamic portion which tends to dominate in the far field (when r is large). Note that, because $u_i \sim x_i$, the displacement u_i is parallel to a line from the origin to the point r.

PART 6
SOURCES ACTING IN OR ON A HALF-SPACE

While the sources just discussed are useful for understanding waves generated in a solid by various mechanisms, the detection of acoustic emission always involves the presence of a surface. Considered below is the effect of a single surface on waves generated by idealized sources.

Concentrated Force Acting Normal To and On the Surface

A classical solution for the motion of a half-space subjected to a concentrated force acting normal to the surface was formulated and solved for two-dimensional (line load) and three-dimensional (point load) problems.[11] The cases of concentrated forces having simple harmonic time histories were solved and then the solutions for general time histories were expressed using a Fourier integral approach. This class of wave propagation problem (concentrated load acting on a half-space) is now called *Lamb's problem*. A closed form solution to Lamb's problem for a step change in time history (suddenly applied constant force) is available in the literature.[12,13] This solution known as the *Pekeris solution* has been used in several publications on acoustic emission sensors. The solution is presented here with conclusions based on that solution.

Consider a half-space defined by $x_3 \geq 0$. The force is applied through the boundary conditions at $x_3 = 0$ which are given by:

$$\tau_{13} = 0$$
and
$$\tau_{23} = 0$$
and
$$\tau_{33} = \frac{F_o \delta(R) H(t)}{2\pi R}$$

Where:

τ_{ij} = the stress acting in the ij direction;
F_o = the magnitude of the applied force;
$R = [x_1^2 + x_2^2]^{1/2}$; and
$H(t)$ = the unit step function.

An exact solution to this problem[12] (for Poisson's ratio $\nu = 1/4$) is given below.

If $\tau^2 < 1/3$:
$$u_3(R,t) = 0 \qquad \text{(Eq. 66)}$$

If $1/3 < \tau^2 < 1$:
$$u_3(R,t) = \frac{F_o}{32\mu\pi R}\left\{6 - \sqrt{\frac{3}{\tau^2 - 1/4}}\right.$$
$$- \sqrt{\frac{3 \cdot 3^{1/2} + 5}{(1/4)(3 + 3^{1/2}) - \tau^2}}$$
$$\left. + \sqrt{\frac{3 \cdot 3^{1/2} - 5}{\tau^2 - 1/4(3 - 3^{1/2})}}\right\}$$

If $1 < \tau^2 < 1/4[3 + 3^{1/2}]$:
$$u_3(R,t) = \frac{F_o}{16\mu\pi R}\left\{6 - \sqrt{\frac{3 \cdot 3^{1/2} + 5}{(1/4)(3 + 3^{1/2}) - \tau^2}}\right\}$$

If $\tau^2 > 1/4[3 + (3)^{1/2}]$:
$$u_3(R,t) = \frac{3F_o}{8\mu\pi R}$$

If $\tau^2 < 1/3$:
$$u_R(R,t) = 0 \qquad \text{(Eq. 67)}$$

If $1/3 < \tau^2 < 1$:
$$u_R(R,t) = \frac{\sqrt{6}\,F_o\tau}{32\mu\pi^2 R}\left\{6K(k) - 18\Pi[8k^2,k]\right.$$
$$+ (6 - 4\cdot\sqrt{3}\,)\Pi[-(12\cdot\sqrt{3} - 20)k^2,k]$$
$$\left. + (6 + 4\cdot\sqrt{3}\,)\Pi[(12\cdot\sqrt{3} + 20)k^2,k]\right\}$$

If $1 < \tau^2 < 1/4[3 + (3)^{1/2}]$:

$$u_R(R,t) = \frac{\sqrt{6}\, F_o \tau \kappa}{32\mu\pi^2 R} \Big\{ 6K(\kappa) - 18\Pi[8,\kappa] \\ + (6 - 4\cdot\sqrt{3}\,)\Pi[-(12\cdot\sqrt{3} - 20),\kappa] \\ + (6 + 4\cdot\sqrt{3}\,)\Pi[(12\cdot\sqrt{3} + 20),\kappa] \Big\}$$

If $\tau > 1/4[3 + (3)^{1/2}]$:

$$u_R(R,t) = \frac{\sqrt{6}\, F_o \tau \kappa}{32\mu\pi^2 R} \Big\{ 6K(\kappa) - 18\Pi[8,k] \\ + (6 - 4\cdot\sqrt{3}\,)\Pi[-(12\cdot\sqrt{3} - 20),\kappa] \\ + (6 + 4\cdot\sqrt{3}\,)\Pi[(12\cdot\sqrt{3} + 20),\kappa] \Big\} \\ + \frac{F_o \tau}{8\mu\pi R \sqrt{\tau^2 - (1/4)(3 + 3^{1/2})}}$$

Where:

$u_3(R,t)$ = the displacement normal to the surface;
$u_R(R,t)$ = the displacement parallel to the surface;
R = the distance from the point of application of the force to the position at which the displacement is being determined;
$\tau = c_2 t/R$ is the time normalized by the shear wave arrival time;
$k^2 = 1/2(3\tau^2 - 1)$;
$\kappa^2 = 2/(3\tau^2 - 1)$; and
$K(k) = \Pi(0,k)$

$$\Pi[n,k] = \int_0^{1/2\,\pi} \frac{d\Theta}{(1 + n\sin^2\Theta)\sqrt{1 - k^2\sin^2\Theta}} \quad \text{(Eq. 68)}$$

The terms $K(k)$ and $\Pi[n,k]$ are complete elliptic integrals of the first and third kind respectively and tabulations of the values can be found in tabulations of mathematical functions. Employing a simple integration technique, the integral in Eq. 68 may also be evaluated numerically with sufficient accuracy for most purposes.

A plot of the solutions Eq. 66 and Eq. 67 is shown in Fig. 19. The direction of the applied force was taken to be into the surface for these plots. Positive values of u_3 and u_R imply displacements that are away from the solid and away from the point of application of the force, respectively.

In both plots, the first motion is caused by a compressional wave arriving at time R/c_1. Note that even though the load is applied downward (toward the solid) and the steady state displacement for long times is downward, the initial disturbance is upward. The shear wave arrival is marked only by a discontinuity in the slope of u_3. The displacement becomes infinite at the time of arrival of the surface or Rayleigh wave. Not obvious from this solution but a demonstrable fact is that the amplitude of the disturbance associated with the surface wave decreases as $R^{-1/2}$ while the components associated with the P and S waves decrease as R^{-1}. In the far field (as R approaches infinity), the surface wave will dominate the disturbance resulting from the applied force.

Numerical solutions (displacements, velocities and accelerations) have been published[17] for Lamb's problem with arbitrary Poisson's ratio and with a vertical time-dependent point force having time histories of the step and delta functions as well as a finite duration pulse equal to $\cos^2(\pi t/T)$ for $-T/2 \leq t \leq T/2$.

An important conclusion reached in the referenced paper[17] was "... a 10× *reduction* in pulse width at a fixed

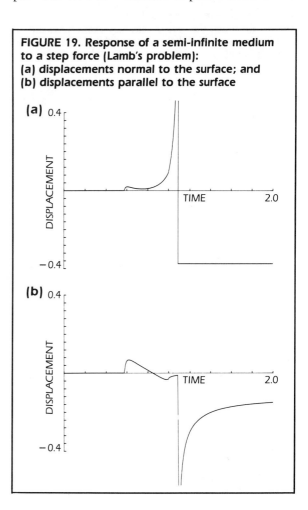

FIGURE 19. Response of a semi-infinite medium to a step force (Lamb's problem):
(a) displacements normal to the surface; and
(b) displacements parallel to the surface

pulse height yielded an *increase* in P and Rayleigh wave amplitudes by factors of 1, 10 and 100 for displacement, velocity and strain, and acceleration, respectively." This points to the fact that the response of the typical acoustic emission sensor (which responds more to velocity or acceleration than to displacement) depends strongly on the rise time or the duration of the source.

Concentrated Force Below the Surface of the Half-Space

The motion at the surface of a half-space excited by a concentrated force acting beneath the surface has also been solved.[18] If a concentrated force of magnitude $F_oH(t)$ acts at the origin (positive F_o implies the force acts away from the free surface) and the free surface is located a distance H above the origin, then the motion of the solid may be expressed as follows.

$$u_3(r,H,t) = \frac{3F_o}{\mu\pi^2 r}(-W_p + W_s) \quad \text{(Eq. 69)}$$

Where:

$u_3(r,H,t)$ = the displacement at the surface; and
r = the distance from the origin to the point on the surface.

The value of the W_p term is determined using Eq. 70.

If $\tau^2 < 1/3$:

$$W_p = 0 \quad \text{(Eq. 70)}$$

If $\tau^2 > 1/3$:

$$W_p = \frac{1}{3} \text{Re} \int_0^{1/2\pi} \frac{\alpha^2\left(2\alpha^2 - \frac{2}{3}\right)d\phi}{\left(2\alpha^2 + \frac{1}{3}\right)^2 - 4\alpha\beta\left(\alpha^2 - \frac{1}{3}\right)}$$

Where:

Re = the real part of the complex expression;
$\tau = c_2 t/r$;
$h = H/r$;
$\nu_0 = [(1 - h^2)(\tau^2 - 1/3)]^{1/2}$;
$\alpha = h\tau + j\nu_0 \sin\phi$; and
$\beta = [\alpha^2 + 2/3]^{1/2}$.

The value of the shear wave component depends on whether the point is within or without the circle $r = H/(2)^{1/2}$. Equation 71 is valid if $r < H/(2)^{1/2}$.

If $\tau^2 < 1$:

$$W_s(r,H,t) = 0 \quad \text{(Eq. 71)}$$

If $\tau^2 > 1$:

$$W_s(r,H,t) = \frac{1}{3} \text{Re} \int_0^\infty \frac{2ab(b^2-1)d\phi}{(2b^2-1)^2 - 4ab(b^2-1)}$$

Where:

$a^2 = b^2 - 2/3$;
$b = h\tau + j\mu_0 \sin\phi$; and
$\mu_0 = [(1 - h^2)(\tau^2 - 1)]^{1/2}$.

If $r > H/(2)^{1/2}$, the expression for W_s is expressed in similar but more involved terms.[18]

Other numerical solutions to this problem have been presented.[19] It has been has shown[20] that the epicenter response to a buried source in a semi-infinite medium can be replaced by an image force approach that superimposes the solution for an infinite space due to the actual force and that due to an image force, as illustrated in Fig. 20. The difference between this approximate approach and the exact solution is small. This technique has been used[9] to study the effects of attenuation and dispersion on acoustic emission waves. Computer programs for Lamb's problem[21] with a surface source and with a buried source are available in the literature and have been used to simulate the acoustic emission waves generated by cracks and slip.

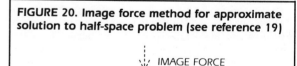

FIGURE 20. Image force method for approximate solution to half-space problem (see reference 19)

PART 7
SOURCES ACTING IN OR ON PLATES

A more practical representation of an actual structure is the infinite plate, a solid bounded by two parallel planes. The analytical problem is much more complicated than the solutions for the half-space or the infinite space because of the multiple reflections that are unavoidable in realistic situations.

Consider the plate and source shown in Fig. 21. Even if the source is a center of compression and radiates only dilatational waves, as each dilatational wave reaches the free surface, it is reflected as both a dilatational (or longitudinal, if curvature is neglected) and an equivoluminal (or transverse) wave. Each of these two components in turn intersects the opposite surface and generates two more waves. It does not take very long to generate a very large number of reflected waves echoing back and forth between the two surfaces.

The lower part of Fig. 21 shows what might be expected at a receiver placed several plate thicknesses from an impulsive center of compression source. The magnitude of the individual components is a function of the path the ray takes to reach the source and can be calculated in ideal cases as noted below. The waveform at the receiver is changed considerably from the original; the task of extracting information about the source wave (given only the received wave) is considerably complicated by the presence of the multiple reflections.

Early research in the effect of structure geometry on the received signal concentrated on simple bar or beam theories[22,23] and attempted to relate spectral content of the received signals to the source time history. Later, plate normal mode analysis was brought into consideration[7,24] and had some success in explaining some of the observed effects.

Perhaps the most notable effect is that the magnitudes of the frequency spectra of signals at various positions along the plate are *relatively* unchanged when compared to the time history of the waveform (displacement, velocity or acceleration). The change in the time history results from the dispersion that occurs in the simple harmonic waves comprising the total signal.

Because the various frequency components travel at different speeds, the phase relationship of the spectral components changes and the components add up to form a different wave. However, the magnitudes of the simple harmonic components do not change, thus the magnitude of the frequency spectrum does not change. In real situations this is modified by energy dissipation mechanisms that reduce the amplitude of the spectral components, especially at the higher frequencies. Thus there is in fact some change that takes place in the magnitude of the spectra as the receiving point moves away from the source.

A second effect is that the apparent wave speed of the disturbance in the plate falls between the speed of longitudinal and transverse waves. It was found that the velocity of the mode carrying the largest portion of the energy corresponded to the first longitudinal plate mode which was about nine percent lower than the plate velocity[24] (c_p) given by the expression below.

$$c_p = c_2 \sqrt{\frac{2}{1 - \nu}} \qquad (Eq.\ 72)$$

While this wave speed represents a good first estimate of the speed of an acoustic emission disturbance propagating along a plate, there are other factors that play a role. First

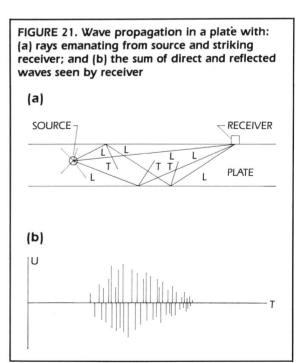

FIGURE 21. Wave propagation in a plate with: (a) rays emanating from source and striking receiver; and (b) the sum of direct and reflected waves seen by receiver

when considering an arbitrary disturbance traveling along a plate, there often is not one particular feature that can be focused on in all situations to determine an unambiguous speed. For example, the first arrival travels with the speed of longitudinal waves c_1. But the amplitude of the first arrival is usually very small compared to the later arrivals. The speed of the waveform peak might be a more realistic feature to focus on but it is not difficult to find waveforms having several distinct peaks and at times the maximum of the waveform will change from one local peak to another local peak. Thus there is no guarantee that the peak of the waveform travels at a constant speed. These difficulties with identifying an unambiguous speed for a disturbance propagating in a plate leads to difficulty in the techniques used to locate the source of acoustic emission events.

Later research on acoustic emission in plates focussed on more exact theories and confirming experiments. The generalized ray theory has been applied to calculate the response of a thick plate due to a variety of ideal sources in the plate.[25-30] The results show the complexity introduced by the multiple reflections even in the near field; however the calculations have been confirmed by experiment. Normal mode theory has also been applied to the simulation of acoustic emission events in plates when the receiver is in the far field.[27] Several aspects of classical wave theory have been applied to the use of acoustic emission tests in the field.[31]

PART 8
DIFFUSE WAVE FIELDS

In most acoustic emission tests, the motion at the detector results from a large number of waves that have been reflected several times before arriving at the detector. The mean length of all the paths that a wave can travel without encountering a surface is given by the expression.[32]

$$L = \frac{4V}{S} \quad \text{(Eq. 73)}$$

Where:

L = the mean travel length;
V = the volume of the medium; and
S = the surface area of the medium.

Consider as an example a compact tension test sample with dimensions 50 × 50 × 25 mm (2 × 2 × 1 in.) and a mean travel length L of 25 mm (1 in.). If the medium is steel or aluminum, the maximum time between reflections is approximately 4 μs for longitudinal waves and 8 μs for shear waves. In a medium this size, the detected acoustic emission signal reaches its peak value in about 100 μs and has a duration on the order of 1 ms. The motion at the detector at the peak (100 μs) consists of waves reflected at least twelve times and as many as twenty-five times on the average.

As a consequence of the large number of reflections that the typical wave undergoes before arriving at the detector and the large number of possible paths from the source to the detector, the typical motion at the detector consists of many waves arriving from many directions. This is one of the factors complicating the analysis of wave motion associated with acoustic emission events.

The complexity of this situation can be reduced considerably if it is assumed that the waves propagating in the medium (1) are large in number; (2) are distributed uniformly in orientation; and (3) are uncorrelated with respect to phase. Such a wave field is called a *diffuse wave field*. For sound waves in fluids, this concept gives rise to a branch of acoustics called *geometric acoustics*.[32]

Diffuse Wave Field Equations

Using the assumptions given above and neglecting contributions from surface waves, it can be shown[33] that the time-average of the acoustic energy per unit volume (energy density) is:

$$\xi = \xi_1 + \xi_2 \quad \text{(Eq. 74)}$$

where ξ_1 and ξ_2 are energy densities of the longitudinal and transverse waves respectively. It can further be shown that each of these energy densities is independent of position in the medium. The equations of motion of the medium (assumed to be homogeneous and isotropic) then reduce to:

$$V\dot{\xi_1} + \frac{1}{4}\gamma_{12}Sc_1\xi_1 - \frac{1}{4}\gamma_{21}Sc\xi_2 = P_1 \quad \text{(Eq. 75)}$$

and

$$V\dot{\xi_2} + \frac{1}{4}\gamma_{21}Sc_2\xi_2 - \frac{1}{4}\gamma_{12}Sc_1\xi_1 = P_2$$

The terms ξ_1 and ξ_2 are the time-averaged energy densities of longitudinal and transverse waves respectively. The terms γ_{12} and γ_{21} are the diffuse field power reflection coefficients (defined as the fraction of incident diffuse field longitudinal or transverse wave power converted to transverse or longitudinal waves due to mode conversion at the boundary of the medium). The terms P_1 and P_2 are the acoustic powers generated by the source in the form of longitudinal and transverse waves respectively.

Equations 75 show that the longitudinal and transverse wave fields generated by the source exist independently, except for the mode conversion that occurs at the boundaries of the medium. The coefficients γ_{12} and γ_{21} determine the extent of the interaction of the two wave types. If the surface of the medium is entirely free, then these coefficients can be calculated by the following integrals:

$$\gamma_{12} = \frac{\int P_{rt}d\Omega}{\int P_{il}d\Omega} \quad \text{(Eq. 76)}$$

and

$$\gamma_{21} = \frac{\int P'_{rl}d\Omega}{2\int P_{it}d\Omega}$$

where P_{il}, P_{rt}, P_{it} and P'_{rl} are given by Eqs. 38, 40, 41 and 42.

The integration in Eq. 74 is over the solid angle of 2π steradians. The factor 2 in the denominator of the second equation results from the fact that SH waves are reflected unchanged from the free surface. The integrals are a function of Poisson's ratio and can be evaluated numerically. Results for selected values of Poisson's ratio are given in Table 2.

TABLE 2. Diffuse field mode conversion power coefficients for selected values of Poisson's ratio

ν	γ_{12}	γ_{21}
0.0	0.601	0.150
0.05	0.650	0.153
0.10	0.708	0.156
0.15	0.761	0.155
0.20	0.789	0.145
0.25	0.769	0.128
0.30	0.686	0.098
0.35	0.536	0.062
0.40	0.337	0.028
0.45	0.131	0.006
0.50	0.0	0.0

Diffuse Field Solution for Impulsive Sources

Consider a source that releases a finite quantity of energy instantaneously at $t = 0$. The subsequent energy densities can be determined by setting P_1 and $P_2 = 0$ and solving Eq. 75 subject to these initial conditions: $\xi_1(0) = E_1/V$ and $\xi_2(0) = E_2/V$. The terms E_1 and E_2 are the longitudinal and transverse energies released by the source. The solutions to Eq. 75 is:

$$\xi_1(t) = K_1 + K_2 e^{-t/t^*} \quad \text{(Eq. 77)}$$

and

$$\xi_2(t) = RK_1 - K_2 e^{-t/t^*}$$

Where:

$K_1 = (E_1 + E_2) / V(1 + R);$
$K_2 = (RE_1 - E_2) / V(1 + R);$
$t^* = 4V / S(\gamma_{12}c_1 + \gamma_{21}c_2);$ and
$R = \gamma_{12}c_1 / \gamma_{21}c_2.$

Equation 77 shows that after a period of time equal to $3t^*$ the ratio of transverse to longitudinal waves in the diffuse field approaches the parameter R. Typical values for a steel medium ($c_1 = 5{,}900$ ms, $\nu = 0.3$) of dimensions $51 \times 51 \times 25$ mm ($2 \times 2 \times 1$ in.) are $R = 13.1$ and $t^* = 5.8$ μs. Thus, for this example, after about 18 μs the diffuse wave field energy density is 93 percent transverse waves and 7 percent longitudinal waves independent of the initial distribution of the source. It has been shown that the ratio R can be determined by counting the number of normal modes associated with dilatational and transverse waves within a small frequency interval.[33] The resulting expression for R is:

$$R = 2 \left\{ \frac{2 - 2\nu}{1 - 2\nu} \right\}^{3/2} \quad \text{(Eq. 78)}$$

The propagation of diffuse waves in plates has also been studied[34] to reveal that a diffuse field in the vicinity of a free surface may be considered an incoherent and homogeneous sum of incident and reflected plane waves.[35] The reciprocity technique was used to estimate diffuse field response of acoustic emission transducers.[36]

REFERENCES

1. Redwood, Martin. *Mechanical Waveguides*. Pergamon Press (1960).
2. Achenback, J.D. *Wave Propagation in Elastic Solids*. North Holland Publishing Company (1975).
3. Graff, Karl F. *Wave Motion in Elastic Solids*. Columbus, OH: The Ohio State University Press (1975).
4. Beranek, Leo L. *Acoustics*. New York, NY: McGraw-Hill Book Company (1954).
5. Kinsler, Lawrence E. and Austin R. Frey. *Fundamentals of Acoustics*, second edition. New York, NY: John Wiley and Sons Publishing (1950).
6. Cagniard, L. *Reflection and Refraction of Progressive Seismic Waves*. New York, NY: McGraw-Hill Book Company (1962).
7. Egle, D.M. and A.E. Brown. "Considerations for the Detection of Acoustic Emission Waves in Thin Plates." *Journal of the Acoustical Society of America*. Vol. 57, No. 3 (1975): pp 591-597.
8. Green, R.E. "Ultrasonic Investigation of Mechanical Properties." *Treatise on Materials Science and Technology*. Vol. Academic Press (1973): pp 145.
9. Kline, R.A. and S.S. Ali. "Methods of Calculating Attenuation and Dispersion Effects on Acoustic Emission Signals." *Journal of Acoustic Emission*. Vol. 4, No. 4 (1985): pp 107-114.
10. Graham, L.J. and G.A. Alers. "Acoustic Emission in the Frequency Domain." *Monitoring Structural Integrity by Acoustic Emission*. ASTM STP 571. Philadelphia, PA: American Society for Testing Materials (1975): pp 11-39.
11. Lamb, H. "On the Propagation of Tremors over the Surface of an Elastic Solid." *Philosophical Transactions of the Royal Society of London*. Series A, Vol. 203 (1904): pp 1-42.
12. Pekeris, C.L. "A Pathological Case in the Numerical Solution of Integral Equations." *Proceedings of the National Academy of Science*. Vol. 26 (1940): pp 433-437.
13. Pekeris, C.L. "The Seismic Surface Pulse." *Proceedings of the National Academy of Science*. Vol. 41 (1955): pp 469-480.
14. Breckenridge, F.R., C.E. Tschiegg and M. Greenspan. "Acoustic Emission: Some Applications of Lamb's Problem." *Journal of the Acoustical Society of America*. Vol. 57, No. 3: pp 626-631 (1975).
15. Hsu, N.N., J.A. Simmons and S.C. Hardy. "An Approach to Acoustic Emission Signal Analysis — Theory and Experiment." *Materials Evaluation*. Vol. 35, No. 10. Columbus, OH: The American Society for Nondestructive Testing (1977): pp 100-106.
16. Breckenridge, F.R. "Acoustic Emission Transducer Calibration by Means of the Seismic Surface Pulse." *Journal of Acoustic Emission*. Vol. 1, No. 2 (1982): pp 87-102.
17. Mooney, H.M. "Some Numerical Solutions for Lamb's Problem." *Bulletin of the Seismological Society of America*. Vol. 64, No. 2 (1974): pp 473-491.
18. Pekeris, C.L. and H. Lifson. "Motion of the Surface of a Uniform Elastic Half-Space Produced by a Buried Pulse." *Journal of the Acoustical Society of America*. Vol. 29, No. 11 (1957): pp 1,233-1,238.
19. Pinney, E. "Surface Motion due to a Point Source in a Semi-Infinite Elastic Medium." *Bulletin of the Seismological Society of America*. Vol. 44 (1954): pp 571-596.
20. Shibata, M. "Theoretical Evaluation of Acoustic Emission Signals — The Rise Time Effect of Dynamic Forces." *Materials Evaluation*. Vol. 42, No. 1. Columbus, OH: The American Society for Nondestructive Testing (1984): pp 107-115.
21. Ohtsu, M. and K. Ono. "A Generalized Theory of Acoustic Emission and Green's Function in a Half Space." *Journal of Acoustic Emission*. Vol. 3, No. 1 (1985): pp 27-40.
22. Egle, D.M. and C.A. Tatro. "Analysis of Acoustic Emission Strain Waves." *Journal of the Acoustical Society of America*. Vol. 41, No. 2 (1967): pp 321-325.
23. Stephens, R.W.B. and A.A. Pollock. "Waveforms and Frequency Spectra of Acoustic Emission." *Journal of the Acoustical Society of America*. Vol. 50, No. 3 (1971): pp 904-910.
24. Fowler, K.A. and E.P. Papadakis. "Observation and Analysis of Simulated Ultrasonic Acoustic Emission Waves in Plates and Complex Structures." *Acoustic Emission*. ASTM STP 505. Philadelphia, PA: American Society for Testing and Materials (1972): pp 222-237.
25. Pao, Y.H. "Acoustic Emission and Transient Waves in an Elastic Plate." *Journal of the Acoustical Society of America*. Vol. 65, No. 1 (1979): pp 96-105.
26. Ceranoglu, A.N. and Y.H. Pao. "Propagation of Elastic Pulses and Acoustic Emission in a Plate." *Journal of Applied Mechanics*. Vol. 48 (1981): pp 125-147.
27. Weaver, R.L. and Y.H. Pao. "Axisymmetric Elastic Waves Excited by a Point Source in a Plate." *Journal of Applied Mechanics*. Vol. 49 (1982): pp 821-836.

28. Pao, Y.H. "Theory of Acoustic Emission." *Elastic Waves and Nondestructive Testing of Materials*. ASME AMD-20. New York, NY: American Society of Mechanical Engineers (1978): pp 107-128.
29. Proctor, T.M., F.R. Breckenridge and Y.H. Pao. "Transient Waves in an Elastic Plate: Theory and Experiment Compared." *Journal of the Acoustical Society of America*. Vol. 74, No. 6 (1983): pp 1,905-1,907.
30. Pao, Y.H. "Transient AE Waves in Elastic Plates." *Progress in Acoustic Emission: Proceedings of the Sixth International Acoustic Emission Symposium* (1982): pp 181-197.
31. Pollock, A.A. *Classical Wave Theory in Practical AE Testing*. Lawrenceville, NJ: Physical Acoustics Corporation (1986).
32. Kuttruff, H. "Room Acoustics." *Applied Science*. London, England (1973).
33. Egle, D.M. "Diffuse Wave Fields in Solid Media." *Journal of the Acoustical Society of America*. Vol. 70, No. 2 (1981): pp 476-480.
34. Weaver, R.L. "On Diffuse Waves in Solid Media." *Journal of the Acoustical Society of America*. Vol. 71, No. 6 (1982): pp 1,608-1,609.
35. Weaver, R.L. "Diffuse Waves in Finite Plates." *Journal of Sound and Vibration*. Vol. 94 (1984): pp 319-335.
36. Weaver, R.L. "Diffuse Elastic Waves at a Free Surface." *Journal of the Acoustical Society of America*. Vol. 78, No. 1 (1985): pp 131-136.
37. Hill, E.V.K. and D.M. Egle. "A Reciprocity Technique for Estimating the Diffuse Field Sensitivity of Piezoelectric Transducers." *Journal of the Acoustical Society of America*. Vol. 67 (1980): pp 666-672.

SECTION 5

ACOUSTIC EMISSION SENSORS AND THEIR CALIBRATION

Donald Eitzen, National Bureau of Standards, Gaithersburg, Maryland

F.R. Breckenridge, National Bureau of Standards, Gaithersburg, Maryland

PART 1
ACOUSTIC EMISSION SENSORS

Definition of Acoustic Emission Sensors

The term *acoustic emission sensor* is accepted by code-writing organizations and several technical societies, including the American Society for Testing Materials. The term is used to describe the type of transducers conventionally used in acoustic emission nondestructive testing techniques. The sensors themselves are used on a test object's surface to detect dynamic motion resulting from acoustic emission events and to convert the detected motion into a voltage-time signal.

This voltage-time signal is used for all subsequent steps in the acoustic emission technique. The electrical signal is strongly influenced by characteristics of the sensor and since the test results obtained from signal processing depend so strongly on the electrical signal, the type of sensor and its characteristics are important to the success and repeatability of acoustic emission testing.

Types of Sensors

A wide range of basic transduction mechanisms can be used to achieve a sensor's functions: the detection of surface motion and the subsequent generation of an electrical signal. Capacitive transducers have been successfully used as acoustic emission sensors for special laboratory tests (see reference 22). Such sensors can have good fidelity, so that the electrical signal very closely follows the actual dynamic surface displacement.

However, the typical minimum displacement measured by a capacitive transducer is on the order of 10^{-10} m. Such sensitivity is insufficient for actual acoustic emission testing.

Laser interferometers have also been used as acoustic emission sensors for laboratory experiments.[1,2] However, if this method is used with a reasonable bandwidth, the method lacks sufficient sensitivity for acoustic emission testing.

Piezoelectric Sensors

Acoustic emission testing is nearly always performed with sensors that use piezoelectric elements for transduction. The element is usually a special ceramic such as lead zirconate titanate (PZT) and is acoustically coupled to the surface of the test item so that the dynamic surface motion propagates into the piezoelectric element. The dynamic strain in the element produces a voltage-time signal as the sensor output. Since virtually all acoustic emission testing is performed with a piezoelectric sensor, the term *acoustic emission sensor* is here taken to mean a sensor with a piezoelectric transduction element.

Direction of Sensitivity to Motion

Surface motion of a point on a test piece may be the result of acoustic emission. Such motion contains a component normal to the surface and two orthogonal components tangential to the surface. Acoustic emission sensors can be designed to respond principally to any component of motion but virtually all commercial acoustic emission sensors are designed to respond to the component normal to the surface of the structure. Since waves traveling at the longitudinal, shear and Rayleigh wave speeds all typically have a component of motion normal to the surface, acoustic emission sensors can often detect the various wave arrivals.

Frequency Range

The majority of acoustic emission testing is based on the processing of signals with frequency content in the range from 30 kHz to about 1 MHz. In special applications, detection of acoustic emission at frequencies below 20 kHz or near audio frequencies can improve testing and conventional microphones or accelerometers are sometimes used.

Attenuation of the wave motion increases rapidly with frequency and for materials with higher attenuation (such as fiber reinforced plastic composites), it is necessary to sense lower frequencies in order to detect acoustic emission events. At higher frequencies, the background noise is lower and for materials with low attenuation, acoustic emission events tend to be easier to detect at higher frequencies.

Acoustic emission sensors can be designed to sense a portion of the whole frequency range of interest by choosing the appropriate dimensions of the piezoelectric element. This, along with its high sensitivity, accounts for the popularity of this transduction mechanism.

In fact, by proper design of the piezoelectric transducer, motion in the frequency range from 30 kHz to 1 MHz (and more) can be transduced by a single sensor.[3-5] This special type of transducer has applications (1) in laboratory experiments; (2) in acoustic emission sensor calibration; and (3) in any tests where the actual displacement is to be measured with precise accuracy (see Fig. 3 and Fig. 8).

Typical Acoustic Emission Sensor Design

Figure 1 is a schematic diagram of a typical acoustic emission sensor mounted on a test object. The sensor is attached to the surface of the test object and a thin intervening layer of couplant is usually used. The couplant facilitates the transmission of acoustic waves from the test object to the sensor. Attachment of the sensor may also be achieved with an adhesive bond designed to act as an acoustic couplant.

An acoustic emission sensor normally consists of several parts. The active element is a piezoelectric ceramic with electrodes on each face. One electrode is connected to electrical ground and the other is connected to a signal lead. A wear plate protects the active element.

Behind the active element is usually a backing material and this is often made by curing epoxy containing high density (tungsten) particles. The backing is usually designed so that acoustic waves easily propagate into it with little reflection back to the active element. In the backing, much of the wave's energy is attenuated by scattering and absorption. The backing also serves to load the active element so that it is less resonant or more broad band (note that in some applications, a resonant transducer is desirable). Less resonance helps the sensor respond more evenly over a somewhat wider range of frequencies.

The typical acoustic emission sensor also has a case with a connector for signal cable attachment. The case provides an integrated mechanical package for the sensor components and may also serve as a shield to minimize electromagnetic interference.

FIGURE 1. Schematic diagram of a typical acoustic emission sensor mounted on a test object

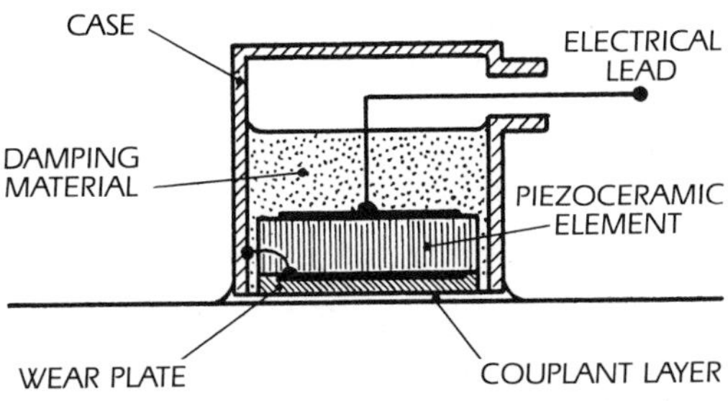

There are many variations of this typical sensor design, including: (1) designs for high temperature applications; (2) sensors with built-in preamplifiers or line-drive transformers; (3) sensors with more than one active element; and (4) sensors with active elements whose geometry or polarization is specifically shaped.

Characteristic Sensor Dimensions

There are two principal characteristic dimensions associated with the typical acoustic emission sensor: the piezoelectric element thickness (t) and the element diameter (D).

Element Thickness and Sensitivity Control

Primarily, element thickness controls the frequencies at which the acoustic emission sensor has the highest electrical output for a given input surface velocity (highest sensitivity). The half-wave resonant frequencies of the sensor define the approximate frequencies where the sensor will have maximum output. These are the frequencies for which: $t = \lambda/2$, $3\lambda/2$, $5\lambda/2$, and so on, where λ is the wavelength of the wave in the element.

The wavelength can be defined as the sound speed in the piezoelectric element divided by the acoustic frequency. Poisson coupling in the element can lead to radial resonances at other frequencies and can also lead to some sensitivity to in-plane motion. For common piezoelectric materials and acoustic emission test frequencies, active elements are typically several millimeters thick. A PZT disk 4 mm thick would normally have a first half-wave resonance of about 0.5 MHz.

Acoustic emission sensors are usually made with backing and with active elements having relatively high internal damping. Because of this design, the variation in sensitivity from resonant to antiresonant (zero output) frequencies is somewhat smoothed out, providing some sensitivity over a significant frequency range.

Active Element Diameter

The other principal characteristic of an acoustic emission sensor is the active element diameter. Sensors have been designed with element diameters as small as 1 mm. Larger diameters are more common.

The element diameter defines the area over which the sensor averages surface motion. For waves resulting in uniform motion under the transducer (as is the case for a longitudinal wave propagating in a direction perpendicular to the surface), the diameter of the sensor element has little or no effect. However, for waves traveling along the surface, the element diameter strongly influences the sensor sensitivity as a function of wave frequency.

If the displaced surface of the test object is a spatial sine wave, then there are occasions when one or more full wavelengths (in the object item) will match the diameter of the sensor element. When this occurs, the sensor averages the positive and negative motions to give zero output. This so-called *aperture effect* has been carefully measured and theoretically modeled.[6]

The consequence of the aperture effect on a transducer can be seen in Fig. 6. For sensors larger than the wavelengths of interest in the test object, the sensitivity will vary with the properties of the test material, depending strongly on frequency and on the direction of wave propagation. Sensor sensitivity is also influenced somewhat by the nonplanar nature of the wavefront.

Because of these complications, it is recommended that the sensor diameter be as small as other constraints allow. For example, when testing steel, a 3 mm sensor works reasonably well below 0.5 MHz.[7]

Couplants and Bonds

For an acoustic emission sensor, the purpose of a couplant is to provide a good acoustic path from the test material to the sensor. Without a couplant or a very large sensor hold-down force, only a few random spots of the material-to-sensor interface will be in good contact and little energy will arrive at the sensor.

Couplants for Sensing Normal or Tangential Motion

For sensing normal motion, virtually any fluid (oil, water, glycerin) will act as a good couplant and the sensor output can often be thirty times higher than without couplant. Note though that in some applications there are stringent chemical compatibility requirements between the couplant and the test object. A sensor hold-down force of several newtons is normally used to ensure good contact and to minimize couplant thickness.

For sensing tangential motion, a suitable couplant is more difficult to find because most liquids will not transmit shear forces. Some high viscosity liquids such as certain epoxy resins are reasonably efficient for sensing tangential motion.

Acoustic Emission Sensor Bonds

An adhesive bond between the sensor and test surface serves to mechanically fix the sensor as well as to provide coupling. Most bonds efficiently transmit both normal and tangential motion.

Depending on the application, bonds are sometimes inappropriate. If for example the test surface deforms significantly because of test loads or if there is differential thermal expansion between the surface, bond or sensor, then the bond or the transducer may break and the coupling is lost. An ASTM standard has been written for sensor mounting.[8]

Temperature Effects on Acoustic Emission Sensors

There can be a strong relation between temperature and the piezoelectric characteristics of the active element in an acoustic emission sensor.[9,10] Some of these effects are important to the use of acoustic emission sensors in testing at elevated or changing temperatures.

Effect of Curie Temperature

Typically, there is a temperature for piezoelectric ceramics at which the properties of the ceramic change permanently and the ceramic element no longer exhibits piezoelectricity. This is known as the *Curie temperature* and is the point at which a material moves from ferroelectric to paraelectric phase.

Piezoelectric ceramic elements have been used successfully within 50 °C of their Curie temperature.[7] The Curie temperature of PZT ceramics is 300 to 400 °C depending on the type of PZT. Other piezoelectric materials have lower (barium titanate at 120 °C) and higher (lithium niobate at 1,210 °C) Curie temperatures.[11]

Testing limitations are therefore encountered in environments where static elevated temperatures cause the loss of piezoelectricity in the sensor's active elements. In addition, failure may occur in other sensor components not designed for high temperature applications.

Effect of Fluctuating Temperature

Special problems are encountered when sensors are placed in environments with widely changing temperatures. Piezoelectric ceramic active elements have small domains in which the electrical polarization is in one direction. Temperature changes can cause some of these domains to flip. This results in a spurious electrical signal that is not easily distinguished from the signal produced by an acoustic emission event in the test object. In a PZT element, a temperature change of 100 °C can cause an appreciable number of these domain flips.[7]

Ceramic elements should be allowed to reach thermal equilibrium before data are taken at differing temperatures. If acoustic emission testing must be done during large temperature changes, then single crystal piezoelectric materials such as quartz are recommended.[7] Acoustic waveguides may also be used to buffer the sensor from large temperature changes.

Special Acoustic Emission Sensors and Sensor Mounts

Acoustic emission sensors are designed for various frequency ranges and are commercially available in a range of sizes with various piezoelectric materials. In addition, sensors or sensor mounts are available for special classes of applications as described below.

Severe Environments

Some acoustic emission sensors are designed for high temperatures and other harsh environments. Sensors are available in which all components are chosen and assembled for elevated temperatures to 550 °C. Sensors for use in harsh environments are fully encapsulated and are available with integral cable.

Sensors with Integral Preamplifiers

Some sensor models combine the sensor and preamplifier functions into one package. These may be miniaturized to the point that they are the same size as conventional sensors. Sensors with integral preamplifiers are reported to have the following advantages: (1) reduced (combined) cost; (2) faster test set up; (3) compatibility with permanent installation for some industrial applications; and (4) lower noise levels (less sensitivity to electromagnetic interference).

Differential Sensors

Differential sensors may be constructed with two or more active elements (or special electrode design) and a positive signal lead, a negative signal lead and a ground lead. The active elements are connected in parallel so that the sensor is less sensitive to electromagnetic interference.

Generally, differential sensors are also relatively insensitive to longitudinal waves arriving at normal incidence to the sensor face. The sensitivity of some models may heavily depend on the direction of propagation in the plane of the surface. Differential sensors are designed for use with differential preamplifiers rather than the single ended preamplifiers normally used with conventional sensors.

Acoustic Waveguides

An acoustic waveguide is a special sensor mount that provides a thermal and mechanical distance between the sensor and the test object. A waveguide is typically a metal rod with one end designed for acoustic coupling with the test object. The other end is constructed to accommodate the mounting of an acoustic emission sensor. Waveguides are used for applications in which an acoustic emission sensor cannot be in direct contact with the test object because of temperature conditions or limited access to the object's surface.

PART 2
PRIMARY SENSOR CALIBRATION

Terminology of Sensor Calibration

Calibration

The *calibration* of a sensor is the measurement of its voltage output into an established electrical load for a given mechanical input. The subject of what should be the mechanical input is discussed under *Basic Principles of Sensor Calibration*. Calibration results may be expressed either as a frequency response or as an impulse response.

Test Block

It is assumed that a sensor is attached to the surface of a solid object either for making measurements of events in the object or for calibration of the sensor. For this discussion, that distinction is not made and the solid object is called the *test block*.

Displacement

Displacement is the dynamic particle motion of a point in or on the test block. Displacement is a function of time and three position variables. Throughout this discussion, *velocity* or *acceleration* could equally well replace *displacement*.

Normal displacement is displacement perpendicular to the face of a sensor or displacement of the surface of a test block perpendicular to that surface. *Tangential displacement* is displacement in any direction perpendicular to the direction of normal displacement.

Basic Principles of Sensor Calibration

If acoustic emission results are to be quantitative, then it is necessary to have a means for measuring the performance of a sensor. Methods of doing this have been the subject of much discussion.

Since there are many types of sensors in use, and since they may be called on to detect waves of different kinds in different materials, it is not possible to have a universal calibration procedure. A sensor calibration, appropriately applied to the signal recorded from a sensor, should provide a record of the displacement of a point on the surface of the object being examined by the sensor. There are several fundamental problems encountered during calibration, as listed below.[12]

1. The displacement of a point on the surface of a test block is a three-dimensional vector but the output of the sensor is a scalar.
2. Displacement is altered by the presence of the sensor.
3. The face of the sensor covers an area on the surface of the test block and displacement is a function not only of time but of the position within this area.

Because of these problems, sensor calibration is not feasible without making some simplifying assumptions. Various approaches have been taken and they all make use of implicit assumptions.

Calibration Assumptions

Regarding the vectoral nature of displacement (problem-1), it is usually assumed that the sensor is sensitive only to normal displacement. Naturally, errors will be introduced if the sensor is sensitive to tangential displacement. Calibrations for other directions of sensitivity are possible and would be useful but are not routine.

The loading effect that the sensor has on the surface motion of a test block (problem-2) is significant but is not subject to any simple analysis. In general, the test block may be considered as having a mechanical impedance (source impedance) at the location of the sensor. The sensor also has a mechanical impedance at its face (load impedance). Interaction between the source and load impedances determines the displacement of the sensor face but both of these impedances are likely to be complex functions of frequency and no method exists for measuring them.

For calibration purposes, the usual solution to this problem is to define the input to the sensor as the unloaded (free) displacement of the test block with no sensor attached. The calibration is then practical because it is the displacement of the test block (and not the interactive effects) that are of interest. The function of the calibration scheme is to determine what the displacement of the surface of the test block would be in the absence of the sensor. It must be noted, however, that when a sensor is attached to different test blocks having different mechanical impedances, it will have different calibrations. Calibrations are transferable only when the test block impedances are the same.

For several calibration procedures, the test block approximates a semi-infinite half space of steel.[13-15] Steel was chosen because it was expected that acoustic emission sensors would be used more on steel than on any other material. The large size of the test block makes the mechanical impedance at its surface a property of the material only, and not of its dimensions within the usual acoustic emission working frequency range. It is demonstrated below that test blocks made of different materials produce significantly different calibrations of the same sensor.

Experiments have been done to determine how much effect the material of the test block has on calibration results. A commercial ultrasonic sensor and a conical transducer[3-5] were calibrated on the NBS steel block and then subjected to surface pulse waveforms in aluminum, glass and methyl methacrylate plastic. The surface pulse waveforms were generated by a pencil break apparatus[16] having the provision for measuring the force. For each material, the surface pulse waveform was calculated at the sensor location, and modified by deconvolution to remove the source characteristics. The results are shown in Figs. 2 and 3. Analysis of the conical transducer has been carried out[17] and the results are shown in Fig. 4.

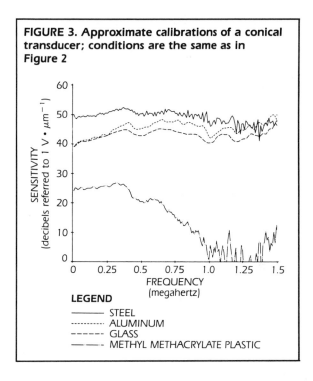

FIGURE 3. Approximate calibrations of a conical transducer; conditions are the same as in Figure 2

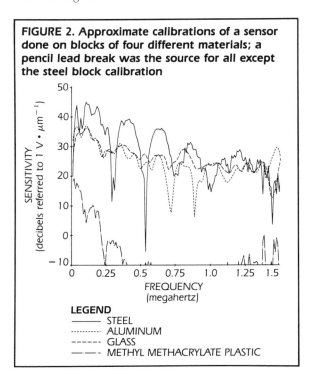

FIGURE 2. Approximate calibrations of a sensor done on blocks of four different materials; a pencil lead break was the source for all except the steel block calibration

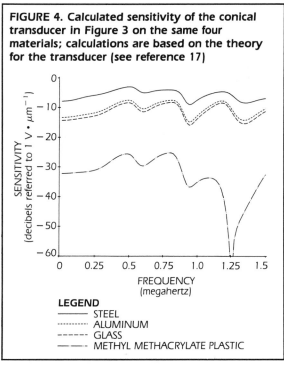

FIGURE 4. Calculated sensitivity of the conical transducer in Figure 3 on the same four materials; calculations are based on the theory for the transducer (see reference 17)

Because the blocks were smaller than optimal, these data are only approximate. The order of magnitude of the effect is clear and in the case of the conical transducer there is reasonable agreement between the theory and the experiment.

The finite size of the sensor's face (problem-3) is often ignored. This is equivalent to (1) assuming that the diameter of the face is small compared to all wavelengths of interest in the test block; or (2) assuming that all motion is in phase over the face. The latter assumption is only true in the case of plane waves impinging on the sensor from a direction perpendicular to its face. In general, a sensor responds to a sort of weighted average of the displacement over its face. This averaging or aperture effect may be considered a property of the sensor and grouped with all sensor properties in calibration. It must be observed though that, as a consequence, the aperture effect and therefore the calibration will differ depending on the type and speed of the wave motion in the test block.

Aperture Effect and Calibration

The aperture effect for an acoustic emission transducer may be described as follows. Neglecting interactive effects between the test block and the sensor, the response of the sensor may be written as follows.

$$U(t) = \frac{1}{A} \iint_S u(x,y,t) r(x,y) dy dx \qquad \text{(Eq. 1)}$$

Where:

S = the region of the surface contacted by the sensor;
A = the area of S;
$u(x,y,t)$ = the displacement of the surface; and
$r(x,y)$ = the local sensitivity of the sensor face.

The XY plane is the surface of the test block. As a special case, assume a straight line wavefront incident on a circular sensor having radius a and uniform sensitivity $r(x,y) = 1$, over its face (see Fig. 5). Assume a wave of the form:

$$U(t) = B \cos(kx - \omega t) \qquad \text{(Eq. 2)}$$

where k is ω/c and c is the Rayleigh speed. Equation 1 then becomes:

$$U(t) = \frac{B}{\pi a^2} \int_{-a}^{a} \cos(kx - \omega t) \sqrt{a^2 - x^2}\, dx \qquad \text{(Eq. 3)}$$

which reduces to:

$$U(t) = \frac{2J_1 ka}{ka} B \cos(\omega t) \qquad \text{(Eq. 4)}$$

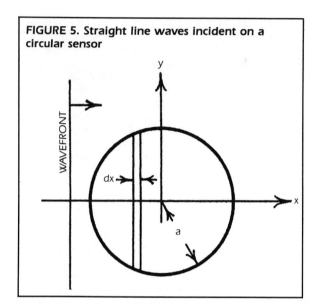

FIGURE 5. Straight line waves incident on a circular sensor

Figure 6 shows this calculated Bessel function response compared to the calibration of an experimental capacitive circular disk transducer.

Surface Calibration

Most calibration systems use a configuration in which the sensor under test and the source are both located on the same plane surface of the test block. The result is a *surface calibration* or *Rayleigh calibration*, so called because most of the propagating energy at the sensor is traveling at the Rayleigh speed. In this case the sensor's calibration is strongly influenced by the aperture effect.

Through Calibration

Other calibration systems use a configuration in which the sensor under test and the source are coaxially located on opposite parallel faces of the test block. All wave motion is in phase across the face of the sensor (except for a negligible curvature of the wavefronts at the sensor) and the calibration is essentially free of any aperture effect. The result is a *through calibration* or *P-wave calibration*. Note that because of the axial symmetry of the through calibration, only normal displacement exists at the location of the sensor under test.

Step-Force Calibration

The basis for the step-force calibration is that known, well characterized displacements can be generated on a plane surface of a test block.[18]

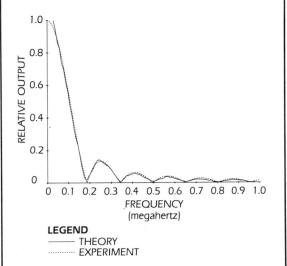

FIGURE 6. Results of the calculation of Equation 4 compared with experimental results from a capacitive disk sensor; source-to-receiver distance (d) is 0.1 m; transducer radius (a) is 10 mm; surface pulse generated by a capillary break on the NBS steel block

FIGURE 7. Schematic diagram of the surface pulse apparatus

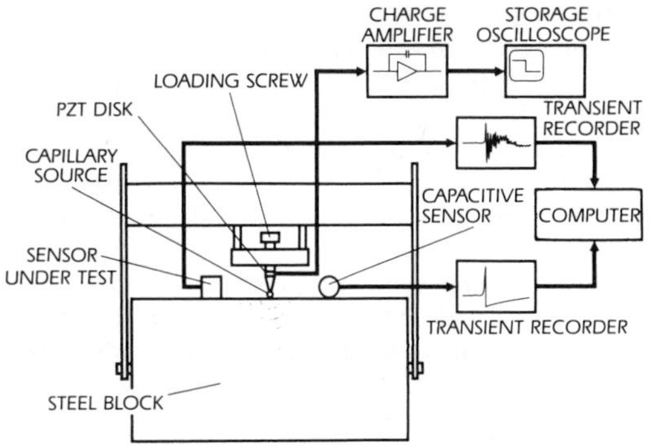

A step-function force applied to a point on one surface of the test block initiates an elastic disturbance that travels through the block. The sensor under test is located either on the same surface (surface calibration) or on the opposite surface at the epicenter of the source (through calibration).

Given the step-function source, the free displacement of the test block at the location of the sensor can be calculated by elasticity theory in both cases.[19-21] The calculated block displacement function is the transfer function (mechanical transfer admittance, when expressed in the frequency domain) or the Green's function for the block. The free displacement of the test block surface can also be measured using a capacitive sensor with a known absolute sensitivity.[22] It is essential to the calibration that the calculated displacement and capacitive sensor measurement agree.

The calibration facility at the National Bureau of Standards (NBS) uses a cylindrical steel test block 0.9 m in diameter by 0.43 m long with optically polished end faces.[13] The step-function force is made by breaking a glass capillary. In the case of surface calibration, free normal displacement of the surface is measured by a capacitive sensor at a location symmetrical to that of the sensor under test with respect to that of the source. The displacement is redundantly determined by elasticity theory from a measurement of the force at which the capillary broke (see Fig. 7). Source and receiver are 0.1 m apart.

For through calibration, the free normal displacement is determined only by the elasticity theory calculation. Both calibrations are absolute: the results are in volts of output per meter of displacement of the (free) block surface.

Following the initiation of the step-function force, an interval of time exists during which the displacement at the location of the sensor under test is as predicted by the elastic theory for the semi-infinite solid (in the case of the surface calibration) or for the infinite plate (in the case of the through calibration). However, as soon as any reflections arrive from the cylindrical surface of the block, the displacement deviates from the theory. The dimensions of the block are large enough to allow 100 μs of working time between the first arrival at the sensor and the arrival of the first reflection. For most sensors, the 100 μs window is long enough to capture most of the information in the output transient waveform in the frequency range of 100 kHz to 1 MHz.

The transient time waveform from the sensor and that from the capacitive sensor are captured by transient recorders and the information is subsequently processed to produce either a frequency response or an impulse response for the sensor under test. The frequency response contains both the magnitude and the phase information.

It is generally assumed that a sensor has only normal sensitivity because of its axial symmetry (an assumption that may not always be justified). Calibration by the surface pulse method for a sensor having significant sensitivity to tangential displacement will be in error because the surface pulse from the step-force contains a tangential component approximately as large as the normal component. It could however be calibrated for the normal component of sensitivity by the through method, because no tangential displacement exists at the location of the sensor under test in the through-pulse configuration. It could also be calibrated (assuming no aperture effect exists) by averaging two surface calibrations with the sensor rotated 180 degrees axially between calibrations. By combining through and surface calibration results judiciously, more information can be gained about the magnitudes of all three components of sensitivity.

The certified frequency range of the NBS calibration is 100 kHz to 1 MHz with less accurate information provided down to 10 kHz. The low end is limited by the fact that the 100 μs time window limits the certainty of information about frequency content below 100 kHz. The expected low frequency errors depend on how well damped the sensor under test is. For sensors whose impulse response function damps to a negligible value within 100 μs, the valid range of the calibration could be extended lower than 100 kHz. The high frequency limitation of 1 MHz is determined by the fact that frequency content of the test pulse becomes weak above 1 MHz and electronic front-end noise becomes predominant at higher frequencies.

Reciprocity Calibration

Reciprocity applies to a category of passive electromechanical transducers that have two important characteristics: (1) they are purely electrostatic or purely electromagnetic in nature; and (2) they are reversible (can be used as either a source or a receiver of mechanical energy). This category includes all known commercial acoustic emission sensors without preamplifiers.

For such a transducer, *reciprocity* relates its source response and its receiver response in a specific way. If two exactly identical transducers are used, one as a source and one as a receiver, both coupled to a common medium, and if the transfer function or Green's function of the medium from the source location to the receiver location is known, then from purely electrical measurements of driving current in the source and output voltage at the receiver, the response functions of the transducers can be determined absolutely. With nonidentical transducers, three such measurements (using each of the three possible pairs of transducers) provides enough information to determine all of the response functions of the transducers absolutely.

The primary advantage of the reciprocity calibration technique is that it avoids the necessity of measuring or producing a known mechanical displacement or force. All of the basic measurements made during the calibration are electrical. It is important to note, however, that the mechanical transfer function or Green's function for the transmission of signals from the source location to the receiver location must be known. This function is equivalent to the reciprocity parameter and is the frequency domain representation of the elasticity theory solution mentioned in the discussion on the step-force calibration.

The application of reciprocity techniques to the calibration of microphones,[23-28] hydrophones[29] and accelerometers[30] has been well established. The use of the reciprocity technique for the calibration of acoustic emission sensors coupled to a solid was proposed in 1976 and was subsequently implemented by Nippon Steel Corporation as a commercial service.[14,31] The Nippon Steel calibration facility uses a cylindrical steel test block 1.1 m in diameter by 0.76 m long to perform Rayleigh calibration (analogous to the NBS surface calibration) and P-wave calibration (analogous to the NBS through calibration).

In the Rayleigh calibration, the sensors are separated by 0.2 m on the same surface of the block and for the P-wave calibration they are located on opposite faces on epicenter. The technique uses essentially continuous wave measurements but the signals are gated to eliminate reflections from the block walls. For a set of three sensors, the three electrical voltage transfer functions and the electrical impedances of all sensors are measured. From these data, receiving response (in volts of output per meter per second of input) and source response (meters per second of output per volt of input) are calculated for the range of 100 kHz to 1 MHz. The *meters per second* of the source response applies to any point on the steel surface located 0.2 m from the source.

The same assumptions about direction of sensitivity that were mentioned under step-force calibration apply to all sensors in a reciprocity calibration. A violation of the assumption by any of the three sensors would contaminate the results. The aperture effect also applies to all sensors in a calibration and the considerations of mechanical loading of the test block are the same as for the step-force calibration.

PART 3
SECONDARY CALIBRATION AND SENSOR TESTS

Purpose of Secondary Calibration

The secondary calibration of an acoustic emission sensor is performed on a system and with a method that has a logical link to the primary calibration system and method. It provides data of the same type as a primary calibration but it may be more limited. For example, there may be no phase information or a narrower range of frequencies or the calibration may be applicable to a different material. Because of the logical link between the methods, the data may be compared if the source-to-sensor geometries are taken into account.

Sensor suppliers usually provide data on the sensitivity of acoustic emission sensors over a range of frequencies. In some cases, these data are developed on a system very similar to a primary calibration facility and can be compared with primary calibration data. More often, the supplied data provide relative response rather than absolute response; such information cannot be logically linked to a primary calibration. It is useful for comparing the response of similar sensors or for checking for changes in sensor response. This information is frequently based on a different physical unit (often pressure) than primary and secondary calibrations (displacement or velocity).

Development of Secondary Calibration Procedures

The development of secondary calibration methods is an area of ongoing research. A secondary method must offer compromises between system complexity and accurate sensor characterization and such a system has not yet been designed, tested and accepted. There are listed below some new tools that may be very useful for developing a secondary calibration procedure.

High Fidelity Transducers

Transducers that accurately measure surface motion with high sensitivity are helpful when developing calibration procedures. Such transducers may be used as transfer standards because of their stability and their uniform sensitivity over the frequency range of interest.

There are transducers having flat frequency response over the range of 10 kHz to 1 MHz or higher. One such transducer (developed at the National Bureau of Standards) has a small conical element backed up by a large brass block.[3-5]

Figure 8 shows the voltage-time output of an NBS conical transducer mounted on a large steel plate and responding to the displacement caused by breaking a glass capillary. The transducer's output is compared with the theoretically predicted displacement of a point on a plate due to a point step-function input. The favorable comparison indicates that the transducer accurately measures transient displacement. Conical transducers are available from NBS with preamplifiers and primary calibration.

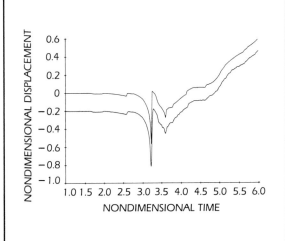

FIGURE 8. Conical transducer's output (lower curve) from a glass capillary breaking on a large steel plate compared to the output of a computer program's calculation (upper curve) of the Green's function of the steel plate

Computer Programming for Secondary Calibration

A second tool useful for secondary calibration is also demonstrated by Fig. 8. A computer program provides a theoretical prediction of surface motion for various (1) plate materials; (2) simulated acoustic emission sources; and (3) source-to-sensor geometries.[32] A mechanical input of known force and time history (such as a breaking glass capillary event[13] or a pencil source[16]) is used and the source is modeled with the computer program. The predicted displacement time history can be used to determine the sensitivity of a sensor as a function of frequency.

Procedures for checking the response of acoustic emission sensors are well developed and extensively documented in the literature. Such procedures are relatively simple and can be used to check sensors for degradation or to identify sensors that have similar performance. These procedures are discussed in detail in ASTM Standard E976.[33] They are not capable of providing sensor calibration or of assuring transferability of data sets between different groups.

Additional Acoustic Emission Sensor Information

A great deal of practical information on sensors is contained in reference 7. Additional detailed information is available in reference 34, particularly for acoustic emission sensors and their characterization. References 11 and 35 contain valuable information about transducers for ultrasonic testing but much of this material is relevant to acoustic emission sensors.

REFERENCES

1. Monchalin, J.P. "Optical Detection of Ultrasound." *IEEE Transactions on Ultrasonics, Ferroelectrics and Frequency Control*. Vol. 33, No. 5. New York, NY: Institute of Electrical and Electronic Engineers (1986): pp 485-499.
2. Palmer, C.H. and R. Green. "Optical Detection of Acoustic Emission Waves." *Applied Optics*. Vol. 16 (1977): pp 2,333.
3. Proctor, T.M., Jr. "An Improved Piezoelectric Acoustic Emission Transducer." *Journal of the Acoustical Society of America*. Vol. 71, No. 5 (May 1982): pp 1,163-1,168.
4. Proctor, T.M. "Some Details on the NBS Conical Transducer." *Journal of Acoustic Emission*. Vol. 1, No. 3 (July 1982): pp 173-178.
5. Proctor, T.M. "More Recent Improvements on the NBS Conical Transducer." *Journal of Acoustic Emission*. Vol. 5, No. 4 (1986).
6. Breckenridge, F.R., T.M. Proctor, N.N. Hsu and D.G. Eitzen. "Some Notions Concerning the Behavior of Transducers." *Progress in Acoustic Emission*. Vol. 3. Tokyo, Japan: Japanese Society for Nondestructive Inspection (1986).
7. Beattie, A.G. "Acoustic Emission, Principles and Instrumentation." *Journal of Acoustic Emission*. Vol. 2, No. 1 (1983): pp 95.
8. "Mounting Piezoelectric Acoustic Emission Sensors." *Annual Book of ASTM Standards*. E650. Vol. 03.03. Philadelphia, PA: American Society for Testing Materials (1986).
9. Cady, W.G. *Piezoelectricity*. New York, NY: Dover Publications (1964).
10. Mason, W.P. *Piezoelectric Crystals and Their Applications to Ultrasonics*. Princeton, NY: Van Nostrand Publications (1950).
11. Krautkramer, J. and H. Krautkramer. *Ultrasonic Testing of Materials*. (1977).
12. Hsu, N.N. and F.R. Breckenridge. "Characterization and Calibration of Acoustic Emission Sensors." *Materials Evaluation*. Vol. 39, No. 1. Columbus, OH: The American Society for Nondestructive Testing (1981): pp 60-68.
13. Breckenridge, F.R. "Acoustic Emission Transducer Calibration by Means of the Seismic Surface Pulse." *Journal of Acoustic Emission*. Vol. 1, No. 2 (April 1982): pp 87-94.
14. Hatano, H. and E. Mori. "Acoustic Emission Transducer and its Absolute Calibration." *Journal of the Acoustical Society of America*. Vol. 59, No. 2 (February 1976): pp 344-349.
15. Hill, R. "Reciprocity and Other Acoustic Emission Transducer Calibration Techniques." *Journal of Acoustic Emission*. Vol. 1, No. 2 (April 1982): pp 73-80.
16. Hsu, N.N. *Acoustic Emission Simulator*. US Patent 4,018,084 (April 19, 1977).
17. Greenspan, M. "The NBS Conical Transducer: Analysis." *Journal of the Acoustical Society of America*. Vol. 81, No. 1 (January 1987): pp 173-183.
18. Breckenridge, F.R., C.E. Tschiegg and M. Greenspan. "Acoustic Emission: Some Applications of Lamb's Problems." *Journal of the Acoustical Society of America*. Vol. 57 No. 3 (1975): pp 626-631.
19. Lamb, H. "On the Propagation of Tremors Over the Surface of an Elastic Solid." *Philosophical Transactions of the Royal Society*. Series A, Vol. 203 (1904): pp 1-42.
20. Pekeris, C.L. "The Seismic Surface Pulse." *Proceedings of the National Academy of Sciences*. Vol. 41 (1955): pp 469-480.
21. Mooney, H.M. "Some Numerical Solutions for Lamb's Problem." *Bulletin of the Seismology Society of America*. Vol. 64, No. 2 (April 1974): pp 473-491.
22. Breckenridge, F.R. and M. Greenspan. "Surface-Wave Displacement: Absolute Measurements Using a Capacitive Transducer." *Journal of the Acoustical Society of America*. Vol. 69, No. 4 (April 1981).
23. Cook, R.K. "Absolute Pressure Calibration of Microphones." *Journal of Research*. Vol. 25. Gaithersburg, MD: National Bureau of Standards (1940): pp 489.
24. W.R. MacLean. "Absolute Measurement of Sound without a Primary Standard." *Journal of the Acoustical Society of America*. Vol. 12 (1940): pp 140-146.
25. Cook, R.K. "Absolute Pressure Calibration of Microphones." *Journal of the Acoustical Society of America*. Vol. 12 (1941): pp 415-420.
26. Foldy, L.L. and H. Primakoff. "A General Theory of Passive Linear Electroacoustic Transducers and the Electroacoustic Reciprocity Theorem (I). *Journal of the Acoustical Society of America*. Vol. 17 (1945): pp 109-120.
27. McMillan, E.M. "Violation of the Reciprocity Theorem in Linear Passive Electromechanical Systems." *Journal of the Acoustical Society of America*. Vol. 18 (1946): pp 344-347.
28. Rudnick, I. "Unconventional Reciprocity Calibration of Transducers." *Journal of the Acoustical Society of America*. Vol. 63 (1978): pp 1,923-1,925.
29. Bobber, R.J. "General Reciprocity Parameter." *Journal of the Acoustical Society of America*. Vol. 39 (1966): pp 680-687.

30. Levy, S. and R.R. Bouche. "Calibration of Vibration Pickups by the Reciprocity Method. *Journal of Research*. Vol. 57, No. 4. Gaithersburg, MD: National Bureau of Standards (October 1956): pp 227-243.
31. Hatano, H. and T. Watanabe. "Calibration on Acoustic Emission Transducers by Reciprocity Technique." Unpublished.
32. Hsu, N. *Dynamic Green's Functions of an Infinite Plate — A Computer Program*. NBSIR 85-3234. Gaithersburg, MD: National Bureau of Standards (August 1985).
33. "Standard Guide for Determining the Reproducibility of Acoustic Emission Sensor Response." *Annual Book of ASTM Standards*. E976. Vol. 03.03. Philadelphia, PA: American Society for Testing Materials.
34. Sachse, W. and N.N. Hsu. "Ultrasonic Transducers for Materials Testing and Their Characterization." *Physical Acoustics*. W.P. Mason and R.N. Thurston, eds. Vol. 14. New York, NY: Academic Press: pp 277-406.
35. Silk, M.G. *Ultrasonic Transducers for Nondestructive Testing*. Bristol, England: Adam Hilger Limited (1984).

SECTION 6

ACOUSTIC EMISSION SOURCE LOCATION

J.A. Baron, Ontario Hydro, Toronto, Ontario, Canada

S.P. Ying, Gilbert/Commonwealth, Reading, Pennsylvania

PART 1
FUNDAMENTALS OF ACOUSTIC EMISSION SOURCE LOCATION

Comparison of Acoustic Emission to Other Nondestructive Tests

One common goal of all nondestructive testing methods is to discover discontinuities. This goal is achieved by interrogating discrete areas of a structure using media such as ultrasound, X-rays, gamma rays, liquid penetrants, magnetic fields or electric fields. Acoustic emission monitoring shares the basic testing goal for many industrial applications but it is a very different kind of nondestructive technique. Instead of generating a probing medium, acoustic emission testing relies on the structure itself to reveal discontinuities through the generation and propagation of stress waves in response to an applied stimulus.

It may not be known where discontinuities are (or indeed if any discontinuities exist) so it also will not be known where the corresponding acoustic emission sources are located. Further, it is not known when acoustic emission will occur or how many signals will be generated. Unlike ultrasonics where the interrogating sound beam may be repeated several hundred times each second, a particular discontinuity may give rise to only one or to many thousand acoustic emissions.

Lacking control over the position, time and quantity of acoustic emission generation during a test is a circumstance that requires a different approach to detecting and locating a source. This approach cannot rely on moving a probe over the surface, as would be done in ultrasonic testing, because of uncertainty in the characteristics of the source. Detection of the source must be achieved using stationary sensors and location must be deduced from the sensor output or from the outputs of a group of sensors.

Advantages of Acoustic Emission Testing

The fundamental premise in locating sources of acoustic emission provides an advantage over other nondestructive testing methods. Because sensors may be left in position, the method becomes amenable to on-line applications, perhaps in unfriendly environments, where large-scale surveillance of a structure is required. It is not necessary to have the close physical proximity between discontinuity and sensor that is required by other methods.

The standard definition of the phenomenon of acoustic emission indicates that it is a materials process.[1] Sensors detect the waves generated by an acoustic emission event and transduce them from a mechanical to an electrical signal. The sensors themselves cannot discriminate between true acoustic emission and mechanical signals that are of an equivalent form.

Susceptibility to signals that are not acoustic emission can be a difficulty in some applications, but it also expands the technique's range of applications, allowing the detection of leakage across a pressure boundary, for example. In this situation, the acoustic signals are not of the transient form associated with acoustic emission events but are instead continuous in nature. Again, the deployment of stationary sensors may be used to detect and locate the source of continuous signals, such as those generated by leaks.

Remote Sensors for Source Location

In general, source characterization using remote sensors is difficult to achieve. There are, for example, certain acoustic emission characteristics that can be used to indicate source severity in some circumstances but this is achieved separately from the techniques for acoustic emission source location.

There are many ways of deducing the location of acoustic emission activity from the electrical output of the sensor-amplifier chain. These techniques may be classified by the type of acoustic source mechanism: (1) continuous emission (such as that generated by a leak) source location; or (2) discrete emission (such as the acoustic emission from a growing crack) source location. Some source locating techniques are common to both categories.

The following text describes the methods of processing both transient and continuous signals to determine the location of acoustic emission sources.

PART 2
LOCATING SOURCES OF CONTINUOUS ACOUSTIC EMISSION

Characteristics of Continuous Acoustic Emission Testing

In a typical acoustic emission test, a load is applied to a structure and discontinuities in the structure's material generate acoustic waves. Sensors mounted on a free surface of the structure transduce the mechanical surface disturbances into electrical signals. The sensor itself has no discrimination function and any surface disturbance of appropriate magnitude and form will be transduced to an electrical signal, regardless of the source mechanism.

Fluid leakage across a pressure boundary is one example of a source mechanism that produces surface disturbances conducive to transduction by acoustic emission sensors. When acoustic emission in its truest sense is being sought, then the surface disturbance caused by leakage is considered noise.[2] Typically, in some applications steps would be taken to eliminate this unwanted source.[3]

The acoustic waves generated by fluid leakage travel throughout a structure and may be detected and perhaps located by an acoustic emission system. If the objective of the test is to maximize system sensitivity to signals originating from a leakage source, then sensors and systems may be optimized for this function. It should be remembered that, when dealing with continuous acoustic sources such as leaks, traditional acoustic emission parameters (count, count rate, amplitude distribution and conventional Δt measurement) become meaningless and are rarely used.

It should also be noted that some materials produce acoustic emission as continuous signals, and this can occur either with or without separate transients (burst emission).[4] Such emission is rarely monitored in industrial applications and the information here does not generally apply to its occurrence.

The Value of Attenuation

Waves propagating away from their source lose energy in a number of ways, including: elongation of the wavefront; heating effects; reflections; refractions; mode conversions; and more. In this discussion, the term *attenuation* refers to the effect of these collective processes, indicating that the signal strength reduces as the distance between source and point of measurement increases. This general principle is modified in situations where resonances of the structure are stimulated, giving rise to standing waves. In this circumstance, attenuation still occurs but the amplitude of the measured signal may not follow the expected distance-amplitude pattern.

Simply recognizing the existence of attenuation permits coarse source location, at least within a defined zone. More substantial knowledge of attenuation characteristics allows finer spatial resolution of the source position. This is one of the approaches for the continuous source location.

Zone Location Method

If a continuous signal source develops on a structure bearing a number of sensors, then the sensor with the highest amplitude output will be closest to the source (assuming equal sensor sensitivities). This occurs because the signal detected at the closest sensor is less attenuated than the others. Thus a map can be developed, as in Fig. 1.

A certain degree of refinement can be achieved if, for example, it is also observed which sensor has the second highest output. This reduces the zone area, as illustrated in Fig. 2.

In many situations, the zone location method, even with the refinement depicted in Fig. 2, is either too coarse or requires more sensors than is practical to reduce the zone size to an acceptable area. If the detailed attenuation characteristics of the structure are known, the location of the source can be computed from observation of the sensor outputs.

It is emphasized that neither the zone location method nor the signal measurement method will work unless the structure exhibits a reasonable degree of attenuation. Similarly, the method will not be practical if the attenuation is too severe.

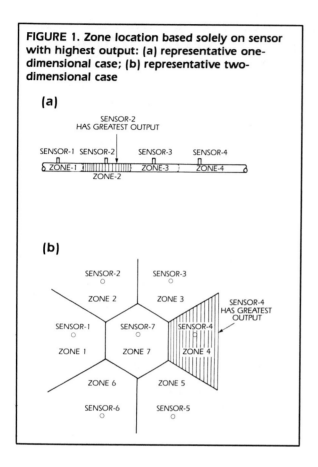

FIGURE 1. Zone location based solely on sensor with highest output: (a) representative one-dimensional case; (b) representative two-dimensional case

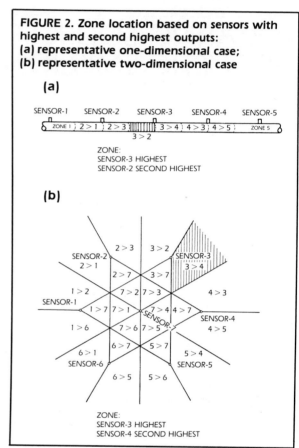

FIGURE 2. Zone location based on sensors with highest and second highest outputs:
(a) representative one-dimensional case;
(b) representative two-dimensional case

Signal Amplitude Measurement Method

With the zone location method, it is only necessary to determine the sensor with the highest signal output and perhaps the next highest. The limitation of the method is that the zone may be unacceptably large. If the sensor outputs are not only rank ordered but also measured, the location of the leak may be determined with a greater degree of precision, provided that the attenuation characteristics of the component are known.

In order to use this amplitude method for continuous source location, it is necessary to measure the attenuation in detail for a particular application (data from similar structures may also be used). It is also important to recognize that sound attenuation in a structure is a complex phenomenon related to wave paths, wave modes and dispersion.

The signal amplitude measurement method of source location comprises the three following stages.

1. Determine the two sensors closest to the leak by noting the highest and second highest outputs from the array. It is necessary that the sensor pair straddle the leak. This method cannot be used with the situation depicted in Fig. 3.
2. Determine the difference (in decibels) between the two sensor outputs and compare that with the attenuation characteristics of the component.
3. On a two-dimensional plane, this process describes a hyperbola passing through the leak site (this is equivalent to using Δt measurements to locate a transient source). A similar process is undertaken using at least one other sensor to generate an intersecting hyperbola. The intersection of the hyperbolae defines the source (equivalent to two-dimensional source location of transient signals using time difference measurements).

The following example illustrates the process for a one-dimensional case. Consider a gas filled steel pipe, 150 mm

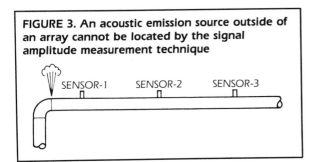

FIGURE 3. An acoustic emission source outside of an array cannot be located by the signal amplitude measurement technique

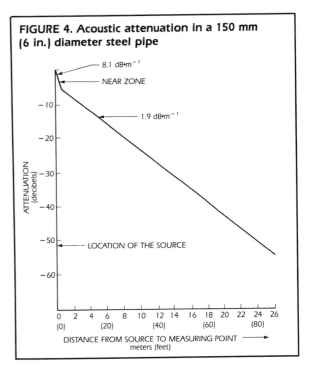

FIGURE 4. Acoustic attenuation in a 150 mm (6 in.) diameter steel pipe

(6 in.) in diameter and 84 m (275 ft) long, . The attenuation characteristics are as illustrated in Fig. 4 (see reference 5). It has been determined that 100 kHz sensors will be used and positioned so that attenuation between sensors will not exceed 25 dB. This level of attenuation occurs over 10.9 m (36 ft). A sensor spacing of 10.5 m then yields 24.3 dB attenuation between sensors. The 84 m (275 ft) long pipe will require nine sensors for complete coverage.

Now consider any sensor pair (sensor-3 and sensor-4 for example). Figure 5 illustrates the respective attenuation characteristics with reference to the source and sensor positions for several arbitrary leak sites. Figure 6 may then be constructed simply by subtracting the attenuation characteristic for one sensor from that of the other for a given position ($A_1 - A_2$ or $B_1 - B_2$). For a source at the midpoint between the two sensors, each sensor will display the same amplitude output; there will be zero decibels difference between the signals generated by the sensors.

For a particular leak, if the output of sensor-3 exceeds that of sensor-4 by 12 dB, then Fig. 6 shows that the leak is located at the 23.1 m (76 ft) position.

This technique relies on the relative amplitude between two participating sensors. The absolute amplitudes of the sensor outputs are irrelevant. However, these assumptions are made: (1) all sensor-amplifier chains are adjusted to identical sensitivities; and (2) ambient noise, either electronic or mechanical, is completely absent. Although it may be quite feasible to acceptably adjust sensitivities, it is rarely possible to eliminate noise.

The Effect of Noise on Signal Amplitude Measurement

Noise pollutes signal measurements and will cause error in calculating source location. Assume that the root mean square (rms) value of the measured signal is the sum of the noise rms and the true signal rms, so that:

$$S_{meas} = \sqrt{S_{true}^2 + n^2} \quad \text{(Eq. 1)}$$

Then if the noise is known, the true signal can be calculated and the described procedure may be followed. Often, a

TABLE 1. Signals at sensors for test with constant 200 mV noise field

Sensor	Position (meters)	Noise (millivolts)	True Signal (millivolts)	Measured Signal (millivolts)
Sensor-1	0	200	4	200
Sensor-2	10.5	200	61	209
Sensor-3	21	200	1,000	1,020
Sensor-4	31.5	200	251	321
Sensor-5	42.0	200	15	201
Sensor-6	52.5	200	1	200
Sensor-7	63.0	200	0	200
Sensor-8	73.5	200	0	200
Sensor-9	84.0	200	0	200

good estimate of noise can be achieved either by prerecording signal levels before leaks occur or observing sensors significantly distant from the leak. The latter approach is not appropriate when only two sensors are used or when the noise is not reasonably constant across the sensor array.

As an example, again consider a leak occurring at the 23.1 m (76 ft) position. The true signal at sensor-3 would be 1,000 mV but the actual signal measured is 1,020 mV. This occurs because the true signal is in a 200 mV noise field (see Table 1).

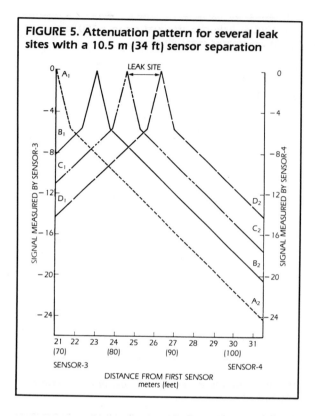

FIGURE 5. Attenuation pattern for several leak sites with a 10.5 m (34 ft) sensor separation

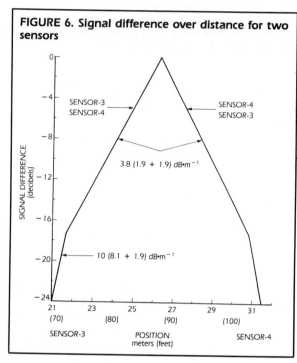

FIGURE 6. Signal difference over distance for two sensors

Observation of outlying sensors (sensor-1 or sensor-6) indicates the noise level that permits the use of Eq. 1 (to determine the true signal) and the subsequent use of Fig. 6. If the noise in this example were ignored, the ratio of the signals measured at sensor-3 and sensor-4 would be 3.17:1 or 10 dB. From Fig. 6, this indicates a source position at 23.6 m (77 ft), an error of 0.5 m (1.5 ft). Such an error may not be acceptable.

If the amplitude of the signal associated with the smallest leak of interest can be determined and the maximum noise is either known or can be measured, then once an allowable error is established, the sensor spacing can be adjusted to suit the circumstances. For leaks larger than the minimum, or noise below the maximum, the actual error in determining the source position will be less than the specified tolerance.

As an example, consider that the minimum leak of interest will generate a 500 mV signal at its source; the maximum noise level is 100 mV. The slope of Fig. 6 is 3.8 decibels per meter and that is twice the measured attenuation (ignore the near field attenuation). For a 1 m (3 ft) uncertainty, the comparison between the measurement (signal plus noise) at a pair of sensors must not differ from the comparison for the true leak signal by more than about 3.8 dB (3.8 dB is the change in attenuation over the specified allowable error of 1 m). Table 2 is then constructed with Fig. 7 following.

This process indicates that a sensor separation of about 6 m (20 ft) is required. This in turn gives rise to Fig. 8 (compare with Fig. 6). If the sensor-1 output exceeds the sensor-2 output by 6 dB, then the source (assuming the above criteria) is between 0.7 and 1.4 m (2 and 4 ft) from sensor-1. The uncertainty is somewhat less than that allowable and this arises because of the near field effect (the higher level of attenuation within 0.4 m (1 ft) of the source).

The Use of Timing Techniques

When acoustic emission signals are transient (have a well defined beginning and end), it is much more apparent that timing techniques can form the basis of a useful source location methodology. If it is possible to identify a particular characteristic of the signal as it is detected by a sensor array, then timing techniques can be used.

The technique develops the time delay of a particular signal between two or more sensors. This procedure may provide the position of the source but it can also be the result of different wave paths, different wave modes or dispersion. This confusion is analogous to the problems encountered in burst emission source location.

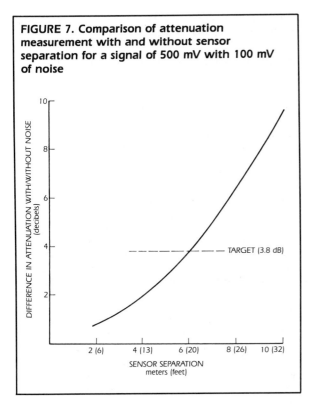

FIGURE 7. Comparison of attenuation measurement with and without sensor separation for a signal of 500 mV with 100 mV of noise

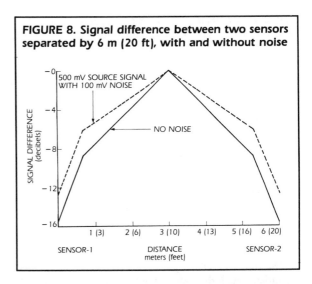

FIGURE 8. Signal difference between two sensors separated by 6 m (20 ft), with and without noise

TABLE 2. Comparison of signal measurement with and without noise at various sensor separations for signal at source of 500 mV and noise of 100 mV

Sensor Separation (meters)	Pure Signal (decibels)	Signal + Noise (decibels)	Difference (decibels)
10	−23.3	−13.7	9.6
8	−19.5	−13.1	6.4
6	−15.7	−11.9	3.8
4	−11.9	−10.0	1.9
2	−8.1	−7.3	0.8

It is necessary to first determine a true sound velocity or the use of this technique on piping, for example, can be particularly difficult. It should be emphasized that the use of timing techniques is not necessarily more problematic than attenuation measurement. Fortunately, both sound attenuation and velocity can be accurately measured even in a complex structure, so that both source location methods may often be used successfully.

Cross Correlation Approach

Conventional techniques for measuring time differences between two burst type acoustic emission waves cannot be used for continuous sources such as leaks. However, cross correlation techniques can be used to measure the time difference or time delay of one wave from another for both discrete waves and continuous waves.[6] The technique has been effectively used in acoustic emission testing.[7] Cross correlation techniques have proved particularly useful in locating leak sites on pipes.[8]

For practical applications, the cross correlation function between an arbitrary wave $A(t)$ and another wave $B(t + \tau)$ with a time delay τ is given by:

$$R_{AB}(\tau) = \frac{1}{T}\int_0^T A(t)B(t + \tau)dt \quad \text{(Eq. 2)}$$

where T is a finite time interval. This definition is different from the general definition, in which T approaches infinity. The reason for using a finite time interval is to allow for practical applications of the technique. Other reasons will become evident later in this discussion.

Equation 2 requires that the product of $A(t)$ and $B(t + \tau)$ be integrated over the time t. Then, if τ is considered as a variable, the cross correlation function $R_{AB}(\tau)$ is a function of τ. In other words, for a given τ_j, $R_{AB}(\tau_j)$ is obtained from the integration of $A(t)B(t)$. The characteristics of $R_{AB}(\tau)$ becomes clear if $R_{AB}(\tau)$ is generated with the following procedure: Let $A(t)$ and $B(t)$ be divided for n small and equal time intervals.

When $t = t_i$, then $A(t) = a_i$, $B(t) = b_i$ and $i = 0, 1, 2 \ldots n$, if $B(t)$ has a time delay τ' with respect to $A(t)$. Then, when $j = 0, 1, 2 \ldots n$:

$$R_{AB}(\tau_j) = \sum_{i=0}^{n} a_{i+j} b_i \quad \text{(Eq. 3)}$$

or when $j = -1, -2, \ldots -n$

$$R_{AB}(\tau_j) = \sum_{i=0}^{n} a_i b_{i-j}$$

and when $j = 0$

$$R_{AB}(0) = \sum_{i=0}^{n} a_i b_i$$

The subscripts of a_{i+j} and b_{i-j} in Eq. 3 shift relatively in accordance with the shift of τ_j for $R_{AB}(\tau_j)$.

The cross correlation function is an integration over a finite time period. Furthermore, in practical applications, a finite portion of each wave is used for data sampling and the wave amplitude beyond the portion used is equal to zero ($a_i = b_i = 0$, if $i > n$).

Then, if $j > 0$ and $i + j > n$, $a_{i+j} = 0$, and if $j < 0$ and $i - j > n$, $b_{i-j} = 0$. Stated otherwise, when $|j|$ increases, then $i + j$ increases, and some of the summation terms in Eq. 3 become zero. Therefore, as $|j|$ increases, the number of summation terms becomes less and less and the amplitude of $R_{AB}(\tau_j)$ decreases. Finally, when $|j| > n$, all a_{i+j} and b_{i-j} are equal to zero so that $R_{AB}(\tau_j) = 0$. When $\tau_j = \tau'$, then $R_{AB}(\tau')$ reaches the maximum value because A and B become in phase. From the maximum peak of $R_{AB}(\tau_j)$, the time difference or time delay τ' of $B(t)$ from $A(t)$ is obtained.

The following example illustrates the calculation procedure and the characteristics of the cross correlation function. Assume $A(t)$ and $B(t)$ are sinusoidal functions so that $A(t) = A_0 \sin \omega t$ and $B(t) = B_0 \sin [\omega t - (\pi/6)]$.

For simplicity, assume $A_0 = B_0 = 1$. As shown in Fig. 9a and 9b, the ωt axes of both $A(t)$ and $B(t)$ are divided into twelve equal intervals, the corresponding $A(t) = a_0, a_1, a_2 \ldots a_{12}$ and $B(t) = b_0, b_1, b_2 \ldots b_{12}$. Then, using Eq. 3, the cross correlation function is calculated at $j = -12, -11 \ldots 0 \ldots 11, 12$. The results are plotted in Fig. 9c.

As shown in Fig. 9c, when $\omega \tau = \omega \tau_{-1} = -\pi/6$, then $R_{AB}(\tau_{-1})$ is maximum. As $\omega \tau$ increases, the absolute values of the $R_{AB}(\tau)$ peaks decrease. When $-2\pi \geq \omega \tau \geq 2\pi$, then $R_{AB}(\tau) = 0$. This example illustrates the characteristics of a cross correlation function in a finite time interval. Note that if the integration time interval approaches infinity, the cross correlation function $R_{AB}(\tau)$ becomes a continuous cosine wave with no distinguishable maximum peak value. This is the reason that a particular finite time interval is necessary to the validity of a cross correlation.

Application to Continuous Source Location

A cross correlation function $R_{AB}(\tau)$ of two functions $A(t)$ and $B(t + \tau')$ in a finite time interval contains a maximum peak value at $\tau = \tau'$ where $A(t)$ is an arbitrary function and τ' is the time delay of $B(t)$ with respect to $A(t)$. This cross correlation methodology can be applied to the location of a continuous source. If a monitoring transducer-A receives a wave $A(t)$ from a continuous source and transducer-B receives a wave $B(t + \tau')$ with a time delay τ', then the time difference for a wave propagating from the source to the transducers is $\Delta t_{AB} = \tau'$ which can be found from the maximum peak value of the $R_{AB}(\tau)$.

Figure 10 shows a typical cross correlation function (CCF) of two wave functions for a continuous source detected by two transducers. The CCF is plotted as function τ in seconds from -0.04 s to $+0.04$ s in a total of 0.08 s time interval. When $\tau = -0.00305$ s (-3.05 ms), the peak CCF value is maximum, as indicated by the cursor position. The time difference is -3.05 ms. The plot in Fig. 10 is actually the average of sixteen CCFs obtained from the same source and the same transducers to reduce possible noise.

Once the time difference for a continuous source is measured by the cross correlation technique, the calculation methods using time differences for source locations are the same as those described earlier. To use correct wave velocity, it is important to pay particular attention to complex wave propagation modes in the structures under test.

The cross correlation function analysis can be found in commercially available dual channel fast Fourier transform (FFT) analyzers. Usually, a cross correlation function in time domain τ is obtained from an inverse Fourier transform of a cross correlation spectrum $G_{AB}(\nu)$ in frequency domain ν. This is:

$$R_{AB}(\tau) = \int_{-\infty}^{+\infty} G_{AB}(\nu) \exp(i2\pi\nu\tau) d\nu \quad \text{(Eq. 4)}$$

where $G_{AB}(\nu)$ is a Fourier transform of $A(t)B(t + \tau)$. However, the frequency range of dual channel FFT analyzers can be limited (often up to 100 kHz). For frequencies above 100 kHz, Eq. 3 or Eq. 4 can be used for hardware design. The equations may also be used in software development for arbitrary continuous waves.

Coherence Techniques

Both the attenuation measurement and cross correlation approaches assume that detection of the leak precedes location of the leak. In some instances it is possible to reverse

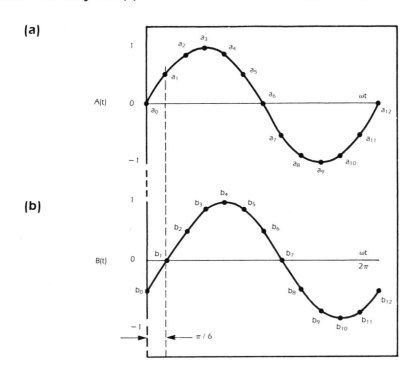
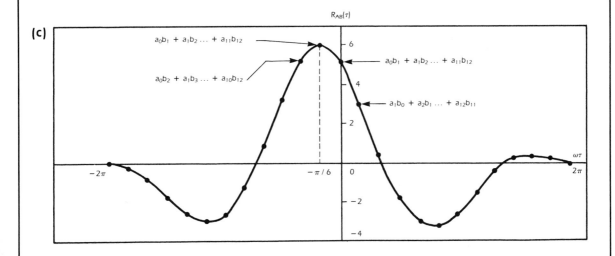

FIGURE 9. Plots of sample cross correlation function: (a) sinusoidal function A(t); (b) sinusoidal function B(t) with $-\pi/6$ delay; and (c) cross correlation function of A(t) and B(t)

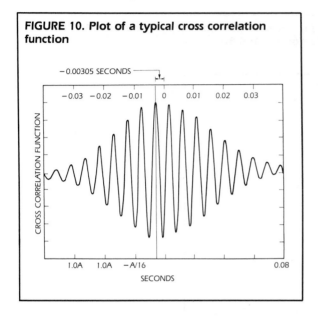

FIGURE 10. Plot of a typical cross correlation function

this sequence. By this reversal, the results of the source location process indicate the presence of the leak. Such an approach has been taken in locating and detecting leaks in a liquid metal heat exchanger.[9]

The method assumes that leakage signals detected by an array of sensors will be coherent. Signals detected in the absence of a leak are noise and tend to demonstrate a low level of coherence. The approach is as follows.

1. Define a position within the space of interest (two or three dimensions).
2. Compute the signal path lengths from the defined position to all sensors. Knowing the velocity of the stress waves, compute the travel times and the delay in reception by all sensors in the array.
3. Capture simultaneously the output from each sensor for a predetermined time. Delay the samples in accordance with the delay times calculated in step 2.
4. Determine the coherence between all delayed sensor outputs. A high level of coherence indicates a leak at the assumed source site.
5. If the coherence is low, assume another site position and repeat procedure from step 2.

This process relies on the predefinition of a source position and then asks whether the signals are consistent with a leak. The sequence may be stated as (1) the site is located by definition; and (2) the source is detected.

Summary

Three methods have been described in detail for determining the source location of continuous signals. A fourth method is described in general terms. These methods progress from the coarse but simple zone location method through the commonly used attenuation measurement method to the significantly more complex cross correlation and coherence methods based on waveform timing. In addition to describing the methods, some indication of potential problems have been presented. A practical application of the attenuation method has been discussed.

Each method has its own valuable attributes depending on the situation and the degree of refinement required.

One topic has not been covered, namely the effect of bends and joints in the structure under test. Generalization of these applications is difficult, given the virtually infinite number of possible configurations (see references 10 and 11). In general, the approach is to make direct measurements on the structure in question and then to follow the procedures described here.

PART 3
LOCATING DISCRETE SOURCES OF ACOUSTIC EMISSION

The ability to locate sources of acoustic emission is one of the most important functions of the multichannel instrumentation systems used in field applications. In some instances, the process may be very simple, such as by defining the sensor closest to the source based either on the first hit or the amplitude of the signal. In other cases, source location can be quite complex, as when defining the full spatial coordinates of an emission source in a large structure. The technique discussed here uses the measurement of time differences for reception of the stress waves at a number of sensors in an array. This method is typically used during proof testing of metallic vessels.

Linear Location of Acoustic Emission Sources

Consider the situation where three sensors are mounted on a linear structure such as a pipe (see Fig. 11). Assume that an acoustic emission event occurs somewhere on the pipe and that the resulting stress waves propagate in both directions at the same constant velocity. The coarsest form of source location would be to note the sensor that received the stress wave first (called the *first hit*). Referring to Fig. 11a, if the first hit occurs at sensor-2, then the source lies in the area from a point half-way between sensor-1 and sensor-2 to a point half-way between sensor-2 and sensor-3. This area can be reduced somewhat by also noting the second hit sensor. If the second hit is sensor-1, then the source lies between sensor-2 and a point half-way between sensor-1 and sensor-2. For evenly spaced sensors, this halves the potential location zone (Fig. 11b).

This procedure is known as *zone source location*. The simple input information only allows the identification of an encompassing zone rather than pinpointing the source site. However, if not only the hit sequence but the time difference between hits is measured (Fig. 11c), more precise source location can be performed. If the time difference between hits at sensor-1 and sensor-2 was zero, it would indicate a site precisely midway between the two sensors.

Or consider this situation, when the hit sequence is sensor-2 then sensor-1. The time difference between hits is equal to the time taken to cross the entire pair separation, or:

$$\Delta t = \frac{D}{V} \quad \text{(Eq. 5)}$$

Where:

D = the distance between sensors;
V = the constant wave velocity; and
Δt = the time difference.

The source would then be located at sensor-2. In general, the source location d is given by:

$$d = \frac{1}{2}(D - \Delta t V) \quad \text{(Eq. 6)}$$

FIGURE 11. Linear source location technique: (a) zone for first hit at sensor-2; (b) zone for first hit at sensor-2 and second hit at sensor-1; (c) hit sequence, time difference measurement $\Delta t = T_2 - T_1$; and (d) source outside of array $\Delta t = T_2 - T_1 = \text{constant} \equiv D$

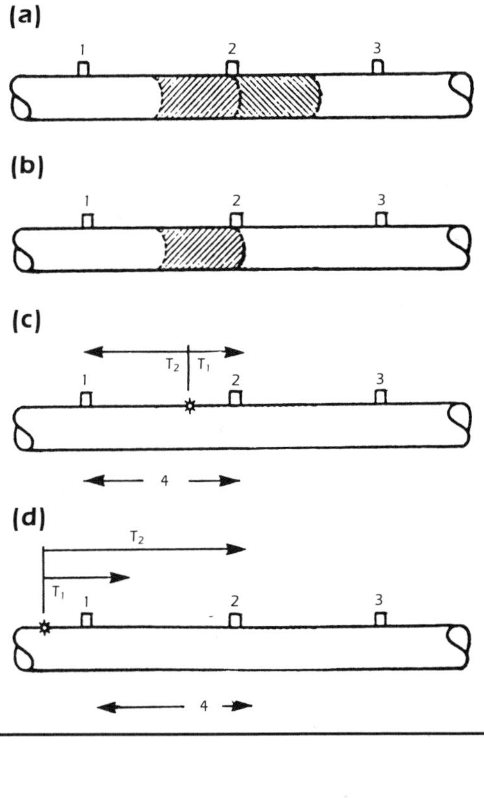

where d is measured from the first hit sensor. Note that if the source is outside of the sensor array (as in Fig. 11d), then the time difference measurement always corresponds to the time of flight between the outer sensor pair; the hit sequence remains constant. In this situation, the most refined location is to state that the source is at or beyond the location of the outer sensor.

The linear case is most appropriate when the sensor separation is large compared to the diameter of the test object. As this ratio reduces, sources close to the sensors can become mislocated if they are away from the direct axial line through the participating sensors.

Location of Sources in Two Dimensions

Consider two sensors mounted on an infinite plane and assume the ideal condition where the stress waves propagating from an acoustic emission source travel at constant velocity in all directions (see Fig. 12).

and

$$\Delta t V = r_1 - R \quad \text{(Eq. 7)}$$

$$Z = R \sin \Theta$$

then

$$Z^2 = r_1^2 - (D - R \cos \Theta)^2 \quad \text{(Eq. 8)}$$

$$R^2 \sin^2 \Theta = r_1^2 - (D - R \cos \Theta)^2 \quad \text{(Eq. 9)}$$

$$R^2 = r_1^2 - D^2 + 2DR \cos \Theta \quad \text{(Eq. 10)}$$

Substituting $r_1 = \Delta t V + R$ from Eq. 7 yields:

$$R = \frac{1}{2} \frac{D^2 - \Delta t^2 V^2}{\Delta t V + D \cos \Theta} \quad \text{(Eq. 11)}$$

Equation 11 is the equation of a hyperbola passing through the source location (X_s, Y_s). Any point on the hyperbola satisfies the input data (the hit sequence and time difference measurement).

The result illustrated in Fig. 12 is insufficient for most two-dimensional source location requirements. However, the addition of a third sensor can improve the situation. The input data now include a sequence of three hits and two time difference measurements (between the first and second hit sensors and the first and third hit sensors). The general situation is illustrated in Fig. 13.

$$\Delta t_1 V = r_1 - R \quad \text{(Eq. 12)}$$

now

$$\Delta t_2 V = r_2 - R \quad \text{(Eq. 13)}$$

which yields

$$R = \frac{1}{2} \frac{D_1^2 - \Delta t_1^2 V^2}{\Delta t_1 V + D_1 \cos (\Theta - \Theta_1)} \quad \text{(Eq. 14)}$$

and

$$R = \frac{1}{2} \frac{D_2^2 - \Delta t_2^2 V^2}{\Delta t_2 V + D_2 \cos (\Theta_3 - \Theta)} \quad \text{(Eq. 15)}$$

FIGURE 12. Result of source location with two sensors on an infinite plane

FIGURE 13. Three sensor array with detection sequence 1, 2, 3

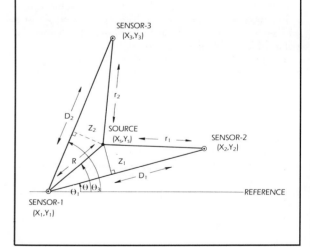

Equations 14 and 15 can be solved simultaneously to provide the location of a source in two dimensions as illustrated in Fig. 14 (this diagram shows an equilateral triangular array but solutions to Eq. 14 and 15 do not impose this requirement).

Restricting the Active Zone

In general, the solutions to the source location equations (Eqs. 14 and 15) are valid over the entire area of the surface on which the three sensors are mounted. There are practical limitations but often the area of interest may represent a small area relative to the whole surface, such as a particular weldment or fixture. Accepting and processing acoustic emission signals from the entire surface may limit the computer processing time available for data from the particular area of interest. There can be significant advantages in rejecting unwanted data as early in the computational process as possible. Also, the geometric arrangement of the structure may tend to cause false location solutions which may be avoided by effectively restricting the operational area of the sensor array. Two such schemes are illustrated in Fig. 15.

In Fig. 15a, each active sensor is paired with a guard sensor. If acoustic emission is detected so that the first sensor hit is a guard, then that signal is rejected and not processed. Only acoustic emission that first hits an active sensor is processed. This limits the operational zone to the shaded area in Fig. 15. Note that the guard sensors are only used to reject data, they do not participate in the measurement function or subsequent calculations.

In Fig. 15b, the active zone is restricted by rejecting all acoustic emission unless the central sensor is hit first. Again, the shaded area indicates the acceptance zone. Unlike the guard sensor approach, all sensors participate (as appropriate) in the source location function. The guard sensors and first hit techniques are effective in practical nondestructive tests. Care must be taken when using guard sensors to ensure that their sensitivities are consistent with the active sensors. When using any technique to restrict the active area, it is recommended that the limits of the area be confirmed using a verification test such as the pencil lead break.[12]

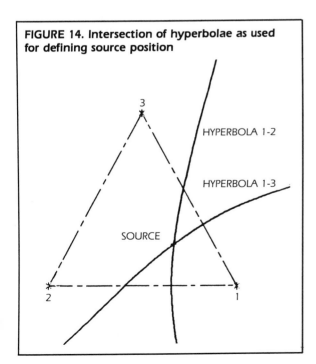

FIGURE 14. Intersection of hyperbolae as used for defining source position

FIGURE 15. Techniques for restricting the area over which source location is effective: (a) eliminate all first hits at guard sensors; and (b) combination of zone and Δt method

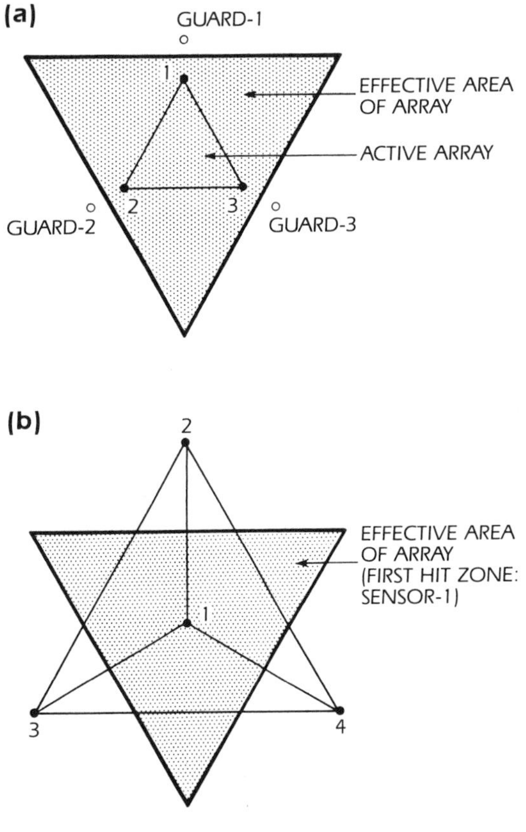

Special Array Configurations

Equations 14 and 15 define the location of an acoustic emission source and were developed without limitation of the sensors' placement. There are some array geometries that are more appropriate than others (for instance, a linear sensor arrangement is not recommended). Also, special configurations can simplify the computation of the hyperbolae intersection.

A four-sensor array has been suggested[13] where the sensors are located at $(X,0)$, $(-X,0)$, $(0,Y)$ and $(0,-Y)$. The two Δt measurements are taken between the pair of X sensors and the pair of Y sensors rather than first hit, second hit and first hit, third hit. The location of the source is then given by:

$$X = aB\sqrt{\frac{b^2 + A^2}{b^2B^2 - a^2A^2}} \quad \text{(Eq. 16)}$$

and

$$Y = Ab\sqrt{\frac{a^2 + B^2}{b^2B^2 - a^2A^2}}$$

Where:

$a = (1/2)\Delta t_x V$;
$A = (1/2)\Delta t_y V$;
$b = (X^2 - a^2)^{1/2}$;
$B = (Y^2 - A^2)^{1/2}$; and
V = acoustic wave velocity.

This solution is amenable to a programmable calculator but because the solution contains the square root form, the quadrant in which the source lies is ambiguous. Observation of the first hit X sensor and first hit Y sensor is sufficient to resolve this ambiguity. Note that the use of special array geometries does not obviate the need for caution regarding sources of error.

Error in Discrete Source Location

There are two types of errors common to locating discrete acoustic emission sources: those due to processing and those due to natural phenomena (attenuation, differing wave modes, reflection, refraction or dispersion). Processing errors are controllable by dictating the number and spacing of sensors, establishing the appropriate clock rate, relying on more than three hits, and so forth. Errors caused by natural phenomena are not controllable and not totally manageable.

By definition, this latter group introduces greater uncertainty of source location computation than does the former. The consequence is that the acoustic emission generated by a single source will not be mapped to a single point but forms a cluster of locations around the true site. The size or concentration of the cluster depends on the source position with respect to the sensor array and all of the other factors discussed earlier.

Ambiguous Solutions

For any given array of three sensors, there exist areas where double intersection of the hyperbolae occur (Fig. 16 illustrates such a case). Both intersections satisfy the input data equally well and the true source cannot be separated from the false source without additional information. The obvious solution is to add a fourth sensor, generate an extra time difference measurement and produce a third hyperbola which should have a single common point of intersection (Fig. 17). In general, this adds quite significantly to computer processing time.

When there is a relatively large array of sensors, another approach is to check both potential sites to see which generates the observed hit sequence. Often the false source can be eliminated in this way. In Fig. 18, if the observed hit sequence is 2, 3, 1, then the false source (s_2) is eliminated. A true source in this location would generate the hit sequence 2, 4, 3 and this was not observed.

If the false source cannot be eliminated based on hit sequences, and the computational cost is unacceptable in

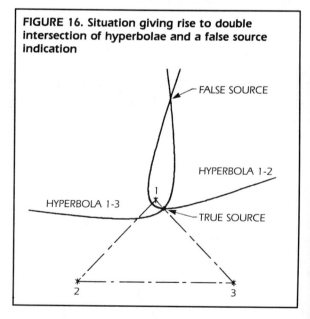

FIGURE 16. Situation giving rise to double intersection of hyperbolae and a false source indication

terms of generating and solving for a third hyperbola, it may be necessary to void the particular input data. Under these circumstances, areas where source location cannot be guaranteed to be unambiguous should be determined and this should be done prior to sensor installation.

Figure 19a illustrates the areas where acceptable source location can be achieved for a particular array, in this case to a precision of 10 mm (0.4 in.) in 10 m (32 ft). The lighter shading in the illustration indicates borderline achievement of the specified precision. Figure 19b shows what happens to the incorrectly located source positions. The lack of systematic error indicates that the particular algorithm used is successfully voiding double intersections. The shading in the lower corners of Fig. 19b indicates marginal achievement of the specified precision, consistent with the lesser degree of shading in the same locations of Fig. 19a.

Mapping Distortions

It is a common procedure to produce a map representing the test object. Unless the structure is linear (such as a flat plane or a pipe with sensor spacing much larger than the diameter), then the representation of the three-dimensional structure on a two-dimensional map will introduce distortion.

It is not initially apparent that unwrapping a cylinder to show the full surface introduces distortion. However, consider the cylindrical vessel in Fig. 20a. If the vessel is unwrapped through a longitudinal cut AA'/BB' to form the map of Fig. 20b, then the adjacent sides of the cut are separated by the full circumference of the cylinder. Also, consider the changed relationship between sensor-1, sensor-8 and sensor-9.

To overcome the distortion of location mapping in two dimensions, one approach is to use *ghost sensors* to maintain the local relationships. Here the true position (two-dimensional) is used for sensor-1 when source location involves sensor-1, 2 and 3. The ghost position for sensor-1 is used when the computation involves sensor-1, 8 and 9.

With little problem, this approach can be extended to vessels such as spheres. The approach is impractical for truly complex structures, such as cylindrical vessels with large nozzles. In these situations, the mapping distortion is accepted or the map is limited to relatively small areas.

In any event, it is prudent to confirm the location of concentrated acoustic emission sites using a direct technique (such as a pencil lead break) and then to confirm the mapped location of this known source.

Weak Acoustic Sources

If the computational process is acceptable (double hyperbolae intersections and mapping distortions are not factors), then the precision of the derived location is dependent on wave velocity, time difference measurement and the accuracy of sensor positions.

However, if a source cannot generate waves of sufficient

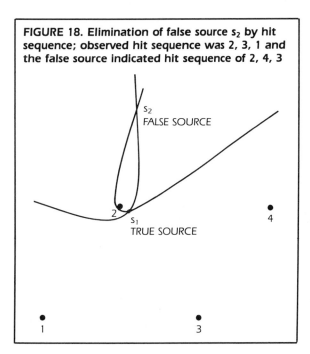

FIGURE 17. Addition of a fourth sensor to define the true source when double intersection of hyperbolae occurs

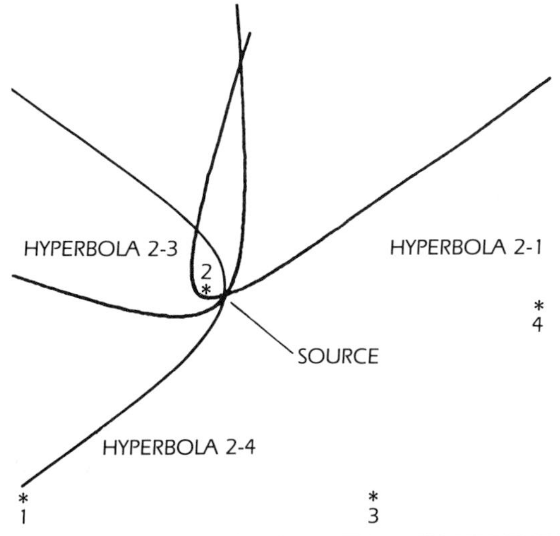

FIGURE 18. Elimination of false source s_2 by hit sequence; observed hit sequence was 2, 3, 1 and the false source indicated hit sequence of 2, 4, 3

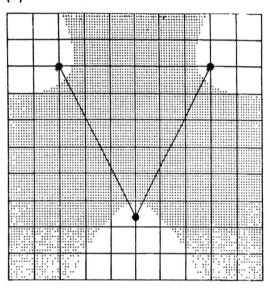
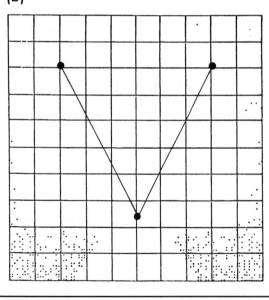

FIGURE 19. Map of acceptable and unacceptable areas of source location: (a) shading indicates areas of unambiguous source location within required precision; (b) shading indicates areas where the required precision is marginal

(a)

(b)

strength for detection by the required number of sensors, then it will not be possible to compute the location of that source. In some applications, this represents the ultimate source location error factor.

Sensor Positions

For purposes of source location, it is rare that the physical size (diameter) of a sensor will play an observable role in the achieved accuracy. Sensor size is a minor factor that will introduce some error but little can be done to resolve the problem without adversely affecting sensitivity.

Where sensor arrays cover large areas, the accuracy of computed locations will be directly dependent on the accuracy of sensor placement. Even in limited areas such as the false source position of Fig. 16, the computed location can be highly sensitive to sensor position.

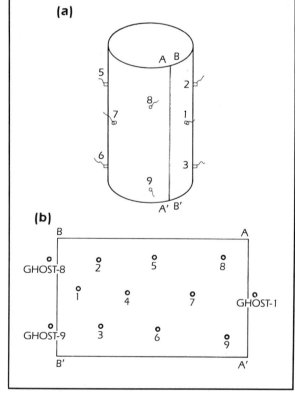

FIGURE 20. Representation of sensors on a two-dimensional map: (a) sensors mounted on cylindrical vessel; and (b) vessel unwrapped at cut through AA'/BB'

Wave Velocity

The basic computational process assumes a known and constant wave velocity. If the true velocity is constant but different from that used in the computation, the resulting positional error will be dependent on the measured time difference values. At the center of an array, where time difference values are small, large discrepancies in velocity will not be noticed. However, the closer the source site is to the sensors (the greater the time difference values), then the greater is the locational error. It is recommended that measured velocities be used rather than published values.

Most multichannel systems cannot accommodate the problem of having different effective velocities in different parts of the structure. In this situation, a compromise has to be made and the test system operator must be aware of the need for compromise.

Nonconstant (or effectively nonconstant) wave velocity can result from situations where one sensor channel is activated by one wave type (compressional waves, for example) while other sensor channels respond only to the Rayleigh wave component. This usually occurs because of the amplitude of the waves at the surface, rather than wave selective characteristics of the sensors.

Another potential problem is dispersion, where the wave velocity becomes a function of frequency.

A problem associated with pipe or tubing installation is spiral waves. Rather than traveling in a purely axial direction, a typical acoustic emission source will generate a wavefront spreading in all directions, including the circumferential direction. The resulting effective wave travel becomes a double (or more) spiral around the pipe. Because of this, the time taken for the wave to activate a sensor channel is variable and is dependent on both axial and circumferential location of the sensor with respect to the source.

Another situation that can give an effectively variable velocity is when the stress waves have to traverse discontinuities such as welds, attached structures, and inset structures such as nozzles and joints. In this case, the true phenomenon may be that associated with change of path rather than change of velocity. However, if dissimilar materials are involved or if there is a situation that causes conversion from one wave propagation mode to another, then the change in path is accompanied by a true change in velocity.

Most commercial sensors simply respond to the motion of the surface beneath them. The sensor output is quite detached from the mechanism or wave type that caused the surface displacement. Given that there is no control of the wave type or the path that a particular acoustic emission source generates, the variability in velocity and the error so produced have to be accepted. Again, the best that can be achieved is the actual measurement of a representative velocity using a known source such as a pencil lead break.

Time Difference Measurement

Time difference measurements are achieved by counting the number of clock pulses between a start trigger (generated by the first hit channel) and a stop trigger (generated by the second or subsequently hit channels). The digital nature of this measurement dictates an expected error up to one clock period at the start and one clock period at the stop. Typical clock rates are in the range 100 kHz to 10 MHz with 1 MHz most prevalent. This gives a potential error of 2 μs at the most common clock rate.

Because the time difference measurements are multiplied by wave velocity to become a difference in wave travel distance rather than simply distance, the resulting error is dependent on sensor spacing and location of the source with respect to the array. Figure 21 indicates the anticipated uncertainty around an array of three sensors assuming a clock rate of 1 MHz.

In this illustration, a wave speed of 3,000 meters per second is assumed and Δt values have been perturbed by ±20 clock cycles. The area of uncertainty in the upper right of Fig. 21 represents a variation of ±0.28 m (11 in.) in X and ±0.33 m (13 in.) in Y. This reduces toward the center of the array where it is ±0.08 m (3 in.) in X and ±0.04 m (1 in.) in Y for the area illustrated.

It is unlikely that normal clock error will be 20 cycles, but error also occurs because the start and stop triggers are initiated when the sensor output achieves a certain value (crosses a threshold). The point of threshold crossing is therefore dependent on the signal rise time. Close to the

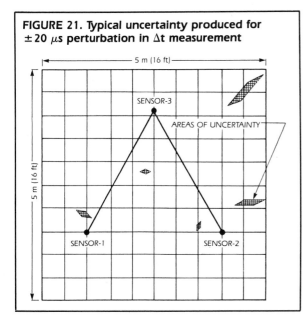

FIGURE 21. Typical uncertainty produced for ±20 μs perturbation in Δt measurement

source, the stress waveform has a fast rise time. As propagation distance increases, separation of the faster wave modes (compressional waves, in particular) plus the higher attenuation rate of the higher frequencies cause the front end of the signal to have a lesser rate of increase.

This creates a delay between the true arrival of the waves at a sensor and the issuing of the stop or start trigger as the output crosses the threshold. The error that results from this process is difficult to quantify because it depends on (1) the originating disturbance and its generation of waves; (2) the dominant wave mode; (3) the material's attenuation characteristics; and (4) the sensitivity of the sensor to the component frequencies of the wave. Again, there is very little that can be done to compensate for this kind of error. Its existence should be recognized especially in relation to sensor spacing.

Source Location in Three Dimensions

Most applications of acoustic emission source location techniques are directed at the problem of locating a source in a practically two-dimensional shell type structure. However, when the wall thickness is substantial or when the area of interest lies internally to the shell, then locating a source in three dimensions becomes important.

One approach is to extrapolate the two-dimensional technique into three dimensions. Each sensor location is defined in full spatial coordinates (X, Y and Z) and the hyperbolae of Eqs. 14 and 15 become surfaces. The solution is more involved than in two dimensions and mapping onto a two-dimensional surface will present its own set of problems. Reference 14 gives some indication of the success that may be expected.

Three-Dimensional Source Location in Cylindrical Test Objects

The following approach is applicable for an intermediate (thick walled) cylindrical vessel. If the outer diameter of the cylinder is not too large, then a distribution of four sensors (as shown in Fig. 22) is sufficient for volumetrically monitoring one half of the vessel. Mathematical calculations for the location of an acoustic emission source in a three-dimensional space become simpler with this arrangement.

As shown in Fig. 22, four sensors are located at $(r_o, \pm \pi/2, 0)$ and $(r_o, 0, \pm h)$ in cylindrical coordinates. When values of $i = 1$ and 2, the distance from an acoustic emission source at an arbitrary position (r, Θ, z) to each sensor is:

$$d_i = \sqrt{z^2 + a_i^2} \quad \text{(Eq. 17)}$$

Where:

$a_1^2 = r_o^2 + r^2 - 2r_o r \sin \Theta$; and
$a_2^2 = r_o^2 + r^2 + 2r_o r \sin \Theta$.

For $i = 3$ and 4, the distance from an acoustic emission source at an arbitrary position (r, Θ, z) to each sensor is:

$$d_i = \sqrt{(h \pm z)^2 + a_i^2} \quad \text{(Eq. 18)}$$

where $a_3^2 = a_4^2$ and $a_3^2 = r_o^2 + r^2 - 2r_o r \cos \Theta$. When $j = 1, 2, 3$ and 4 and the subscripts i and j indicate sensor numbers, then:

$$d_j - d_i = v \Delta t_{ij} \quad \text{(Eq. 19)}$$

The basic relations in Eqs. 17, 18 and 19 lead to a quadratic equation for z in terms of r and Θ for each pair of sensors. The solution of the quadratic equation for sensor-1 and sensor-2 is:

$$z = \sqrt{\frac{4r_o^2 r^2 \sin^2 \Theta}{v^2 \Delta t_{12}^2} + \frac{1}{4} v^2 \Delta t_{12}^2 - r_o^2 - r^2} \quad \text{(Eq. 20)}$$

For sensor-1 and sensor-3:

$$z = \frac{-B \pm \sqrt{B^2 - 4AC}}{2A} \quad \text{(Eq. 21)}$$

Equations 22, 23 and 24 define the terms used in Eq. 21.

$$A = 4(h^2 - v^2 \Delta t_{13}^2) \quad \text{(Eq. 22)}$$

$$B = 4hr_o r(2 \cos \Theta - \sin \Theta) - hA \quad \text{(Eq. 23)}$$

$$\begin{aligned}C = {} & 4r_o^2 r^2 (\cos \Theta - \sin \Theta)^2 \\ & + v^4 \Delta t_{13}^4 - 4v^2 \Delta t_{13}^2 (r_o^2 + r^2) \\ & + 4r_o r(\cos \Theta + \sin \Theta) v^2 \Delta t_{13}^2 - 2h^2 v^2 \Delta t_{13}^2 \\ & + 2h^2 r_o r(\sin \Theta - 2 \cos \Theta) + h^4\end{aligned} \quad \text{(Eq. 24)}$$

To make Eq. 21 valid for sensor-1 and sensor-4, h and Δt_{13} in Eqs. 22 through 24 should be replaced by $-h$ and Δt_{14}, respectively.

The purpose for the development of Eqs. 20 and 21 is to simplify computer programming. Both equations can be used to calculate values of z for each pair of sensors from given values of r and Θ. Therefore, all possible solutions for z can be obtained from two loops in a computer program. The radius r varies from the inner radius to the outer radius and the angle Θ varies from $-\pi/2$ to $\pi/2$. Each loop only needs a limited number of steps to achieve sufficient spatial resolution.

After all solutions for a given Δt_{ij} for each pair of sensors are calculated using either Eq. 20 or 21, then a common solution for at least three pairs of sensors can be found by comparing all possible solutions of one pair of sensors with the others. This is achieved by an if-statement routine. The common solution is the location of the source.

Three-Dimensional Source Location in Containment Structures

Another approach may be taken, particularly when interest is concentrated on the internals of a containment structure rather than the structure itself. This method has been used successfully to locate leaks in a liquid metal heat exchanger.[9] In this situation, the internal medium may be considered homogeneous and nondispersive. Sensors are mounted on the outer surface of the shell (as in Fig. 20a) with outputs fed into a high speed computer. The system works as follows.

1. Assume a potential site of acoustic emission within the vessel.
2. Compute the attenuation and time taken for signals to travel from the potential site to all sensor locations in the array.
3. Correct the sensor outputs for the attenuation and time delay of the signals according to the time of travel for the assumed site.
4. Compute the cross correlation or similar function between all corrected and delayed signals.
5. Based on the results of the cross correlation, decide on the probability of a source of acoustic emission being located at the assumed location.

This technique can be powerful. It is the inverse of the normal procedure, that is, a site is assumed and evidence of signals is evaluated. The common procedure is to assume evidence of a signal and then determine its origin. The technique has been used successfully for determining the presence and location of leaks. It could also be applied to transient source location. The technique would probably be difficult to adapt to situations where inhomogeneity and dispersion are factors.

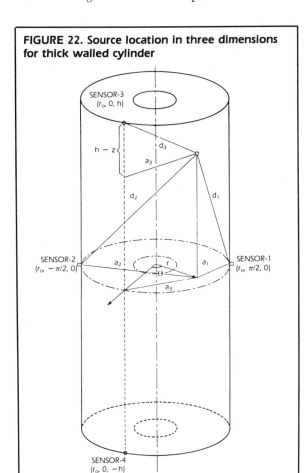

FIGURE 22. Source location in three dimensions for thick walled cylinder

Conclusions

The time difference measurement technique is practical for the location of transient sources of acoustic emission energy. The resulting locations are more accurate than those from the simple zone location technique, but error is still a factor to be considered. The process is straightforward but certain potential problem areas have to be recognized and avoided. Computation of locations using this technique will be slower than the zone location method but may be implemented in a number of ways, including on-line, event by event, off line or through the use of a predetermined look-up table.

REFERENCES

1. *Standard Definition of Terms Relating to Acoustic Emission*. ASTM E610. Philadelphia, PA: American Society for Testing Materials.
2. Bailey, C.D. and W.M. Pless. *Acoustic Emission — Monitoring Fatigue Cracks in Aircraft Structures*." ISA reprint 76239. Pittsburgh, PA: Instrument Society of America (1976).
3. Hutton, P.H. and H.N. Pedersen. "Crack Detection in Pressure Piping by Acoustic Emission." *Nuclear Safety Quarterly Report*. (May - July 1969, for USAEC Division of Reactor Development and Technology) BNWL 1187. Richland, WA: Battelle Pacific Northwest Laboratories (September 1969): pp 3.1-3.18.
4. Fukuzawa, Y., et al. "The Behaviour of AE Second Peak on the 2024 Aluminum Alloy." *Progress in Acoustic Emission II*. M. Onoe, K. Yamaguchi and H. Takahashi, eds. Tokyo, Japan: Japanese Society for Nondestructive Inspection.
5. Kitajima, A., et al. *Acoustic Leak Detection in Piping Systems*. CRIEPI Report 283006. Tokyo, Japan: Central Research Institute of Electric Power Industry (April 1984).
6. Bendat, J.S. and A.G. Piersol. *Engineering Applications of Correlation and Spectral Analysis*. New York, NY: John Wiley and Sons Publishing (1980).
7. Tatro, C.A., A.E. Brown, T.H. Freeman and G. Yanes. *On-Line Safety Monitoring of a Large High Pressure High Temperature Autoclave*. ASTM STP 697. Philadelphia, PA: American Society for Testing Materials (1979).
8. Kupperman, D.S., T.N. Clayton, D. Prine and R. Groenwald. "Acoustic Leak Detection for Light Water Reactors." *Proceedings of Qualtest 1*. Columbus, OH: The American Society for Nondestructive Testing (1982).
9. Greene, D.A., F.F. Ahlgren and D. Meneely. "Acoustic Leak Detection/Location System for Solid Heated Steam Generators." *Proceedings of the Second International Conference on Liquid Metal Technology in Energy Production*. La Grange Park, IL: American Nuclear Society (1980).
10. Wood, B.R.A. and R.W. Harris. "Leak Detection in Pipelines Using Acoustic Emission." *Progress in Acoustic Emission II*. Tokyo, Japan: Japanese Society for Nondestructive Inspection.
11. Wantanabe, K. and K. Shiba. "Propagation of AE Signals at a Bend." *Progress in Acoustic Emission II*. Tokyo, Japan: Japanese Society for Nondestructive Inspection.
12. Hsu, N.M. and S.C. Hardy. *Elastic Waves and Nondestructive Testing of Materials*. AMD29. New York, NY: American Society of Mechanical Engineers (1978).
13. Ying, S.P., D.R. Hamlin and D. Tanneberger. "A Multichannel Acoustic Emission Monitoring System with Simultaneous Multiple Event Data Analyses." *Journal of the Acoustical Society of America*. Vol. 55, No. 2 (February 1974): pp 350-356.
14. Tatro, C.A., A.E. Brown, T.H. Freeman and Yanes. *On-Line Safety Monitoring of a Large High Pressure High Temperature Autoclave*. ASTM STP 697. Philadelphia, PA: American Society for Testing Materials (1979).

SECTION 7

ACOUSTIC EMISSION APPLICATIONS IN THE PETROLEUM AND CHEMICAL INDUSTRIES

Allen Green, Acoustic Emission Technology Corporation, Sacramento, California

P.R. Blackburn, Linde Division of Union Carbide Corporation, Tonawanda, New York
Bruce Craig, Metallurgical Consultants, Denver, Colorado
N.O. Cross, NDTech, Berkeley, New Jersey
Mark Ferdinand, Hartford Steam Boiler, Richland, Washington
Timothy Fowler, Monsanto Company, St. Louis, Missouri
Dan Robinson, Hartford Steam Boiler, Richland, Washington

PART 1
OVERVIEW OF ACOUSTIC EMISSION TESTING IN THE PETROLEUM AND CHEMICAL INDUSTRIES

The first commercial use of acoustic emission testing came shortly after the development of similar tests to monitor aerospace pressure vessels and structures. In 1968, an acoustic emission test was conducted to examine a petroleum facility's cracking vessel which had known welding imperfections after on-site fabrication.[1] Acoustic emission testing has since become an important technique in the nondestructive testing of petroleum and chemical industry pressure vessels.

Acoustic emission testing has a significant advantage over other nondestructive techniques because it has the capacity to locate and evaluate discontinuities in the entire structure at one time rather than selectively testing localized areas. In addition, acoustic emission testing, as part of an overall integrity assessment, provides information about the safe operating pressure of a vessel; other testing methods provide information that must first be evaluated with fracture mechanics or certain simplifying assumptions in order to determine a safe operating pressure.

Acoustic emission testing may also be used to *fingerprint* a vessel by locating discontinuities that can be documented, then evaluated at a later date after a certain service time has elapsed.

This overview covers two important topics. The first is a discussion of common pressure vessel materials and their major causes of failure; the use of acoustic emission testing for predicting these failures is also discussed. The second discussion focuses on the three primary uses of acoustic emission testing for pressure vessels: (1) quality assurance of newly fabricated vessels; (2) short term field evaluation of vessels to determine safe operating pressures; and (3) long term continuous monitoring of pressure vessel performance.

Pressure Vessel Materials and Common Causes of Failure

Most steels used for pressure vessels are medium carbon and low alloy steels. Medium carbon steels such as ASTM A285 grade C, A515, A516, A517, A533, A537 or A612 are frequently used for the construction of pressure vessels. Low alloy steels such as 2.25Cr-1Mo and 3Cr-1Mo are used for hydrogen containment, while 3 to 7 percent nickel steels may be used for cryogenic service. For processes that produce very corrosive environments, titanium, zirconium and high alloy materials such as austenitic stainless steels (304 or 316) and nickel base alloys may be used. Large pressure vessels used in corrosive service are often fabricated from carbon or low alloy steel then overlayed with corrosion resistant materials such as stainless steels, nickel alloys or tantalum.

Consequently, pressure vessels for petroleum and chemical applications may be made from any of a number of materials for operation in a variety of environments. In addition to the many corrosive conditions encountered in service, operating temperatures can range from -40 °C (-40 °F) to over 540 °C (1,000 °F). Pressures can range from atmospheric to greater than 50 MPa (7,300 psi). All of these conditions result in a variety of causes for pressure vessel failure in service.

Temperature Related Failure

Acoustic emission is not only generated when a crack jumps forward. Yielding is a common source of acoustic emission in metals and significant emissions normally begin at stress levels in the range of 60 to 70 percent of the yield point.

As a metal thins down in a high temperature environment, the stress levels increase and an over-pressure test will cause emission if the level of stress approaches local yield of the metal. In practice, areas of geometric discontinuity such as nozzles, knuckle areas and stiffeners have high local flexural stresses. Loss of material in these areas will be detected in an over-pressure test.

Creep and stress rupture are also sources of significant emission. It is common practice during an acoustic emission test to increase the pressure in a series of steps. Monitoring the load hold periods provides valuable information. Often the load hold data are the primary data used in evaluating the condition of the vessel. During the load holds, acoustic emission will gradually decrease with time. This emission is not due to cracks, but is caused by creep of the material as it relaxes in regions of high stress.

A material in a brittle state will give significant emission at very low stress levels. For this reason, acoustic emission has become the standard tool for determining the ductile/brittle transition temperature of operating pressure vessels. The A515, grade 70, steel is a common material for fabrication of vessels that operate at high temperature.

At temperatures approaching freezing, these vessels emit continuously at very low pressures. At higher temperatures, above the ductile/brittle transition point, the vessels will not emit until much higher pressures. A similar pattern of behavior has been documented with a number of steels and with zirconium. Acoustic emission is also used to evaluate the ductility of field welds in low temperature service.

Acoustic emission is probably the most sensitive method for detecting steel degradation due to hydrogen. Frequently, acoustic emission gives an indication of a problem's onset well before the problem can be detected by other nondestructive testing methods. Eventually, blisters and fissures occur and, as would be expected, these can also be detected with acoustic emission. However, the real advantage of acoustic emission techniques is their ability to detect embrittlement at a very early stage. Vessels in hydrogen fluoride service are subject to hydrogen embrittlement and, for this reason, are periodically monitored with acoustic emission methods.

Environmental Cracking

Environmental cracking includes corrosion fatigue cracking, hydrogen embrittlement and stress corrosion cracking. It is probably the failure mode that best lends itself to acoustic emission testing. Most often, this environmental cracking is characterized by slow stable crack growth prior to final failure. Although hydrogen embrittlement frequently occurs by rapid fracture rather than slow crack growth, it has been successfully monitored by acoustic emission in the laboratory. Fatigue cracking and stress corrosion cracking often proceed in a stable fashion with significant acoustic emission activity during crack growth, thus allowing continuous monitoring by acoustic emission.

Thermal Shock

Thermal shock results from a rapid change in pressure vessel temperature. It can produce either (1) catastrophic fractures as a result of reduced impact resistance from various mechanisms or (2) discontinuities that do not cause immediate failure but may be monitored afterward by acoustic emission. Frequently, vessels subject to thermal stress cycles are monitored during temperature cycling, or during start-up or shut-down.

Weld Overlays

Another area of potential acoustic emission monitoring is separation of weld overlays. For many reasons, the corrosion resistant weld overlay applied to pressure vessels may become separated and can continue to separate in service. Acoustic emission testing provides the monitoring of these delaminations that is necessary for continued vessel operation.

Environmental Degradation

Acoustic emission is widely used to detect stress corrosion cracking and other environmentally induced discontinuities. Probably the largest single application for acoustic emission in the petrochemical industry is monitoring for stress corrosion cracking in carbon steel ammonia vessels. The only other nondestructive testing method with comparable sensitivity is wet fluorescent magnetic particle. Acoustic emission is also widely used for detecting stress corrosion cracking in austenitic stainless steels.

Uses of Acoustic Emission Tests of Pressure Vessels

Acoustic emission techniques are most often used for nondestructive testing of pressure vessels and storage tanks in three ways:

1. quality assurance of new pressure vessels;
2. intermittent evaluation of vessels after accumulation of service time; and
3. continuous on-line monitoring of vessels.

Tests of New Pressure Vessels

Most codes require a hydrostatic test of new pressure vessels to 150 percent of the design pressure. In addition to ensuring the structural integrity of the vessels, the test is designed to cause local yielding at geometric discontinuities and to relieve high residual stresses. This local yielding is a desirable feature of the hydrotest and is a source of major acoustic emission. Frequently, such emissions are confused with or mask genuine discontinuities.

To overcome this problem, new vessels are repressurized immediately following the hydrotest and the data from the second cycle are used to evaluate the condition of the vessel. During the second pressure cycle, the level of acoustic emission is greatly reduced and, in most cases, only discontinuities will emit. The problems of noise from initial yielding during the first pressure cycle, and the loss of sensitivity caused by the Kaiser effect during the second cycle, make acoustic emission monitoring of new vessels difficult.

The application of acoustic emission testing to newly fabricated vessels and tanks, while not always performed as a routine test, can be quite valuable both as a means of ensuring the quality of fabrication and as a means for providing a baseline survey of discontinuity locations and sizes. Once a vessel has experienced a certain amount of service, it can be resurveyed on a periodic basis to evaluate the integrity of the vessel and its suitability for continued use, based on the original baseline survey.

Fabrication shop testing of vessels offers nearly ideal conditions for acoustic emission testing: hydrostatic testing facilities are already in place and noise sources that may interfere with acoustic emission signals are minimized. Some of these noise sources are given in Table 1.[2] Acoustic emission testing in the shop also allows easy access for repairs and subsequent retesting or rejection before the vessel has been accepted for service. In addition, shop testing permits

TABLE 1. Noise sources in acoustic emission testing of pressure vessels

Mechanical and Liquid Sources	
Cause	**Possible Corrective Measure**
Movement on supports	Place vessel on wood blocks or other quiet materials. These sources may also be recognized in the data stream and filtered out.
Movement of internal components (trays, rings, baffles or girders)	Remove internal components or place them on wood blocks. Shim with soft plastic if bolted in place. These sources may also be recognized in the data stream and filtered out.
Weather (rain, snow, sleet)	Move vessel under cover or reschedule test.
Leaks in gaskets or screwed fittings	Seal leak.
Flange gaskets and bolting	Tighten bolting.
Hydrostatic test pump and liquid lines to vessel	Increase inside diameter of outlet to reduce noise from flow and cavitation. Reduce rate of pressurization.
	Use high pressure reinforced hose between pump and vessel. Do not pipe pump to vessel.
Mechanical vibration of flow noises transmitted through supporting structure or attached piping	Band-pass filter at frequency outside of offending noise spectrum.
	Disconnect vessel from structure, reduce flow rates or both.
	Conduct test when vessel is out of service.
Nearby sources of mechanical noise (such as overhead cranes, punch presses, railways, rolling mills)	Band-pass filtering.
	Conduct test at a time when noise source is not operative.

Electrical Sources	
Cause	**Possible Corrective Measure**
Nearby welding, especially high frequency start on TIG and MIG	Stop the welding or wait until a time when welding is not scheduled.
Power line noise caused by large motors and controllers	Extensive filtering of instrumental power or use of a portable generator (usually simplest).
Radio station and other detected interference (ground loops)	Shielding of transducer, preamplifier and cable to preamplifier.
	Special care in grounding of all system components. Sometimes use of braided wire or welding cable is necessary between all preamplifiers and then to signal conditioner chassis ground.

detection and repair of discontinuities that, if undetected, could produce catastrophic failure of the vessel at higher stresses.

Acoustic emission testing may also be used as a means for confirming the location and concentration of discontinuities determined by techniques such as radiography and ultrasonic examination. In this way, acoustic emission techniques can provide a measure of the structural significance of discontinuities detected and sized by radiography and ultrasonic methods.

In-Service Tests of Pressure Vessels

Vessels in the field are of concern because of their demanding operating conditions and yet the extent and nature of discontinuities are often unknown. In-service inspection is the single most common application of acoustic emission testing to date. The use of acoustic emission testing with fracture mechanics and ultrasonic examination may allow the complete characterization of a pressure vessel so that life predictions can be made with some degree of certainty.

The basic premise of fracture mechanics is that every structural component (including nonmetals) contains discontinuities and this has been confirmed by failures and nondestructive tests on a variety of structures including pressure vessels. Fracture mechanics presents a mathematical means of defining the critical discontinuity size that will produce fracture under specific loading conditions. Acoustic emission testing provides the location and other NDT methods provide the size of discontinuities that are then treated mathematically to determine (1) under what service conditions failure may occur or (2) if the discontinuities need to be repaired.

Fracture mechanics and acoustic emission testing have limits to their uses, especially for tough, low strength steels. Linear elastic fracture mechanisms, on which much of fracture mechanics analysis is based, assume relatively brittle materials with little plastic deformation during fracture. Application to very ductile, low strength steels may be highly inaccurate. Likewise, acoustic emission studies on steels with different strength ranges and ductilities reveal reduced acoustic emission activity with decreasing yield strength[3] (see Table 2).

Acoustic emission is normally used to detect yielding rather than crack growth. Crack growth during a test is an unusual situation and is only likely to occur in severely deteriorated vessels. Modern pressure vessels are designed to leak before fracture. Fracture mechanics has brought an increased understanding of failure mechanisms and it is now recognized that vessels having the potential of sudden brittle catastrophic failure are unacceptable. Many petrochemical companies are replacing older vessels that fracture before leaking with vessels that leak before fracture, in order to avoid the potentially hazardous situation of a sudden catastrophic failure.

A common means of acoustic emission testing of vessels in the field is to pressurize the vessels with process fluids. Introducing water and oxygen into a vessel, as is required with a hydrotest, often results in damage to the vessel. In addition, a hydrotest may require breaking gaskets, linings and piping. Such actions can lead to additional problems.

During some field tests, the stress level is held until acoustic emission activity stabilizes, then the load is increased incrementally by 0.7 or 1.4 MPa (100 or 200 psi) and held for ten to fifteen minutes to monitor acoustic emission activity. This incremental stressing is continued until a predetermined upper limit is achieved. The vessel stress is then (1) reduced to atmospheric pressure; (2) restressed to the upper limit of the previous test; (3) held to monitor acoustic emission activity; and (4) incrementally increased above this level to a new threshold.

TABLE 2. Relative acoustic emission activities of nine commercial steels during crack propagation at room temperature

Designation (decreasing activity)	Type	Yield Stress ($MN \cdot m^{-2}$)	Ultimate Tensile Stress ($MN \cdot m^{-2}$)
300-M	High strength steel	1,615*	1,960
BS 1501-261	Mo-B steel	405	623
BS 1501-271	Mn-Cr-Mo-V steel	515	638
BS 1501-151	Semi-killed C-Mn steel	232	431
BS 1501-213	Semi-killed, Nb-treated C-Mn steel	410	572
BS 1501-224	Fully killed, Al grain-refined C-Mn steel	284	466
X60	C-Mn pipeline steel	405*	626
BS 1501-821	Austenitic stainless steel	342*	620
BS 1501-622	2.25Cr-1Mo steel	363*	543

*0.2 PERCENT PROOF STRESS

TABLE 3. Typical petroleum and chemical industry uses of acoustic emission testing

Application	Structure
Discontinuity detection[1]	atmospheric vessels pressure vessels piping systems offshore platforms
Leak detection[2]	vessels piping valves
Flow detection[3]	valves piping
Process monitoring[3]	compressors pumps motors fans valves

Certification/ recertification[4]	• ASME Case 1,968
	  metal vessel 0.2 m³ (7 ft³) or less
	  circumferential weld less than 63.5 mm (2.5 in.)
	• SPI 1982
	  fiberglass tanks and vessels
	  62 to 448 kPa (9 to 65 psia)
	• SPI 1983
	  reinforced thermosetting resin pipes
	  less than 34.5 MPa (5,000 psia)
	  0.6 m (24 in.) diameter
	• ASTM E569-78
	• NDIS 2412-1980
	  spherical pressure vessels of 600 MPa steel

1. Tests are applied as shop and field tests; in-service on-line and off-line.
2. Field tests applied in-service on-line.
3. In-service on-line.
4. New or in-service off-line.

Often, in-service pressure vessels are tested by varying the pressure from 0.9 to 1.05 or 1.1 times the design pressure. Higher pressures are undesirable and may require blocking out relief valves and other safety devices.

This procedure provides an assessment of safe operating pressure where no acoustic emission activity is expected to occur based on the pressure cycling results.

Continuous Monitoring of Pressure Vessels

Acoustic emission testing is a nondestructive testing method that may be used as a continuous monitoring technique for pressure vessels. Pressure vessels are frequently in service as long as forty years. They may gradually deteriorate over their entire life or may experience crack initiation and propagation late in their service life as a result of environmental effects or a change in operating conditions. Continuous monitoring of the vessel allows early warning of any acoustic emission activity long before more conventional nondestructive techniques would detect problems.

Plant shutdowns required for other methods of inspection are quite expensive. They are necessary though to inspect the inside of vessels for deterioration or cracking. Continuous on-line acoustic emission testing provides a method of evaluation that safely allows extension of the times between shutdowns to the economic benefit of the plant.

While acoustic emission testing is widely used, other techniques are also used in the petroleum and chemical industry. Typical uses of acoustic emission testing are listed in Table 3.

PART 2
INDUSTRIAL GAS TRAILER TUBE APPLICATIONS

Recertification of Gas Trailer Tubing

Part of the distribution network for industrial gases involves use of steel tubes, mounted on trailers, for transporting gases over highways. Typically, the tubes contain gas at pressures around 18 MPa (2,600 psi). Periodic retesting (every five years) of these tubes is mandated by the US Department of Transportation. Traditionally hydrostatic testing was performed. Since 1983, acoustic emission testing has been used by Union Carbide Corporation on two types of tubes (3AAX and 3T tubes) in place of hydrostatic retesting. This was allowed by exemptions issued by the Department of Transportation.

Characteristics of 3AAX and 3T materials and typical tube dimensions are listed in Table 4. Figure 1 shows a hydrogen service trailer carrying ten 3AAX tubes. Note that the tubes are mounted with free space between them; they do not make contact with one another.

In acoustic emission recertification, the tubes are instrumented and then pressurized to 110 percent of normal filling pressure. Acoustic emission data provide location distributions for each tube. Locations that produce a significant number of events are subject to shear wave ultrasonic inspection. Ultrasonic inspection establishes the presence of a discontinuity, as well as its circumferential location, length and maximum depth. Ultrasonic inspections are performed shortly after the acoustic emission test. Access to discontinuities by the ultrasonic beam has been adequate with the trailer fully assembled. Disassembly of trailers, for the purpose of accomplishing ultrasonic inspection, has not been necessary.

This acoustic emission test is normally performed at a production facility, while the trailer is refilled. Excess gas is then bled into a receiver. Before this test method could be implemented, it was necessary to establish the maximum allowable discontinuity size. In these tubes, the potentially critical discontinuities are longitudinal fatigue cracks. This is because the highest stress is in the circumferential direction. It is first necessary to establish what size discontinuity will grow to critical size *before the next recertification* five years hence.

Analysis[4] indicates that tubes containing discontinuities with a maximum depth of 2.5 mm (0.1 in.) or more should be removed from service. Figure 2 shows plots of years required to reach critical discontinuity size (depth) versus initial discontinuity depth for both 3AAX and 3T tubes. The

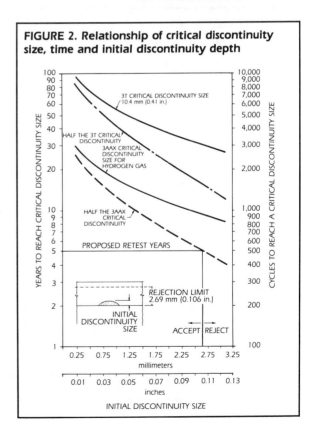

FIGURE 1. Hydrogen service trailer carrying 3AAX tubes

FIGURE 2. Relationship of critical discontinuity size, time and initial discontinuity depth

TABLE 4. Physical properties of 3AAX and 3T industrial gas tubes

	Chemistry (percentage by weight)						
	Carbon	Manganese	Phosphorus (maximum)	Sulfur (maximum)	Silicon	Chromium	Molybdenum
3AAX	0.25/0.35	0.40/0.90	0.04	0.05	0.20/0.35	0.80/1.10	0.15/0.25
3T	0.35/0.50	0.75/1.05	0.035	0.04	0.15/0.35	0.80/1.15	0.15/0.25

	Heat treatment
3AAX	Quenched in oil or other suitable medium and tempered at ≥ 538 °C (1,000 °F)
3T	Quenched in oil or other suitable medium and tempered at ≥ 565 °C (1,050 °F)

	Mechanical properties		
	Yield Strength MPa (ksi)	Ultimate Tensile Strength MPa (ksi)	Elongation in 50 mm (2 in.) Gage (percent)
3AAX	620 to 730 (90 to 106)	725 to 860 (105 to 125)	20
3T	795 to 895 (115 to 130)	930 to 1,070 (135 to 155)	16

	Fabrication
3AAX	forged end seamless pipe
3T	forged end seamless pipe

	Typical Dimensions
3AAX	560 mm (22 in.) diameter 10.4 m (34 ft) length 13.6 mm (0.536 in.) minimum wall thickness
3T	560 mm (22 in.) diameter 10.4 m (34 ft) length 10.5 mm (0.415 in.) minimum wall thickness

worst case is a 3AAX tube in hydrogen service. Hydrogen atmospheres cause accelerated fatigue crack growth.

Test Procedure for Trailer Tubing Tests

Setup for Acoustic Emission Testing

Acoustic emission sensors (140 to 150 kHz resonance) are attached to each end of each tube on the trailer with silicone adhesive; silicone grease is used as a coupling medium. The sensors are connected to a multichannel acoustic emission system that performs detection and linear location.

The sensor outputs are band-passed at 100 to 200 kHz. An acoustic emission system threshold of about 35 dB (one microvolt at the sensing element) is used. Trailer pressure is measured with a transducer connected to the trailer manifold. The pressure transducer output is also recorded by the acoustic emission system.

The sensitivity of each mounted sensor is checked by breaking a pencil lead close to it.[5] The time required for a stress wave to traverse the distance between sensors (typically 10 m or 32 ft) is also measured on each tube by breaking pencil leads or by pulsing outboard of one sensor. The time required for the resulting stress waves to travel from the near sensor to the far sensor is input to the acoustic emission processor for use in location calculations.

The tubes are monitored to test pressure while the trailer is filled. This procedure normally takes four to five hours when using gas plant compression equipment.

Results of Acoustic Emission Tests of Trailer Tubes

Location distributions (plots of acoustic emission event counts versus axial location) for each tube are monitored during the test. Figure 3a shows a four-tube location distribution display. If a sufficient number of events occur at an axial location (for example, ten events in a 200 mm or 8 in.

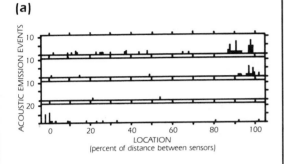

FIGURE 3. Plots of acoustic emission tube tests: (a) event count versus axial location for four tubes; and (b) attenuation versus axial distance for a 3T tube tested with a 150 kHz sensor

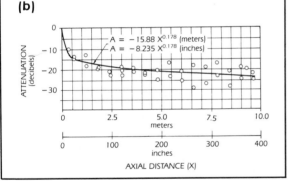

axial distance), a discontinuity probably exists and ultrasonic inspection of that location is performed.

If a sufficient number of events occur at a tube end (say, outside the sensor locations) with sufficient peak amplitude (for example, ten events at 43 dB referenced to 1 μV at the sensing element), then that entire tube end receives ultrasonic inspection. Even though discontinuities in tube ends cannot be located with this acoustic emission setup, their presence is readily detected.

Measured stress wave attenuation data for a 3T tube are plotted in Fig. 3b. A power law curve has been fitted through the data points. Both 3AAX and 3T tubes exhibit about 24 dB of attenuation over the distance between sensors on 10.3 m (34 ft) jumbo tubes.

Figure 4 shows the acoustic emission recertification methodology. Using this approach, discontinuities are detected and located with the acoustic emission system. The discontinuities are then circumferentially located and measured with ultrasonics. Discontinuities that are apparently passive (structurally insignificant) will be measured again during subsequent tests and discontinuity growth will be monitored.

Interpretation of Trailer Tube Test Results

Trailer tubes have a relatively large length-to-diameter ratio and the axial location accuracy is generally very good. Figure 5b shows a location distribution for a long discontinuity on the inside surface of a tube in oxygen service. Figure 5a is a depth profile showing six separate ultrasonic depth measurements.

Five different sources of acoustic emission may occur at a discontinuity. These have been identified and measured during trailer tests and during laboratory tests on compact tension specimens. The five sources are listed in Table 5 in order of decreasing intensity (peak amplitude at the source).

Emission from all or some of the first four sources might be measured during a given acoustic emission test. Mill scale, the source that exhibits the highest peak amplitudes, is always present in natural discontinuities. The sources of mill scale are: (1) oxidation of surface laps during hot forming of seamless pipe at 950 to 1,260 °C (1,750 to 2,300 °F); and (2) residue from lubricants forced into discontinuities during pipe forming operations.

Figure 6 shows a section cut from a 200 mm (8 in.) discontinuity located on the outside surface of a 3T tube. This section shows that the discontinuity (a fold created during fabrication) is filled with mill scale. There is no evidence of fatigue growth at the tip of the discontinuity and the measured acoustic emission was from mill scale.

Because much of the measured emission from discontinuities *is not* from crack propagation, emission can occur at relatively low pressure. Figure 7 shows a plot of events versus pressure from the discontinuity pictured in Fig. 6. In this case, slightly over half of the events occurred at pressures less than normal fill pressure.

Advantages of Acoustic Emission Testing of Trailer Tubes

An acoustic emission recertification test data base has been accumulated since 1983. Other companies in the gas and chemical industries are able to contribute to the data. In time, details regarding various systems, the best signal processor threshold and the relationship between acoustic emission events and discontinuities should become more firmly established.

With the standard hydrostatic test,[6] the trailers are removed from service, transported to a test facility and disassembled. The tubes are then inserted into a water jacket, pressurized to 5/3 of service pressure and the volumetric expansion of the tube is measured. Tubes exceeding allowable volumetric expansion are removed from service.

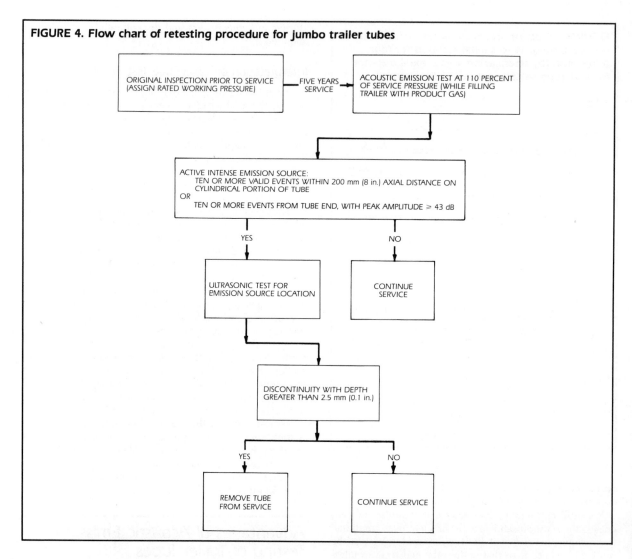

FIGURE 4. Flow chart of retesting procedure for jumbo trailer tubes

The acoustic emission retest, by contrast, can be performed at a production facility. The trailer does not leave service, is not disassembled and the inside surfaces of the tubes do not become wet. The acoustic emission test method outlined in Fig. 4 (when combined with ultrasonic techniques) has these advantages: (1) it produces precise locations and dimensions of discontinuities; (2) it is performed for less cost; and (3) it produces more information about the structural integrity of the tubes.

TABLE 5. Sources of acoustic emission at discontinuities in trailer tubes; listed in order of decreasing intensity

- Mill scale within discontinuities
- Inclusion fracture and delamination at crack tip
- Mechanical interaction at the surfaces of cracks
- Crack growth through parent metal
- Plastic deformation

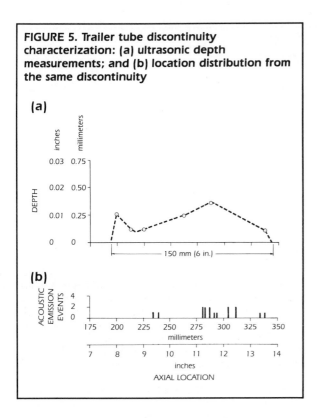

FIGURE 5. Trailer tube discontinuity characterization: (a) ultrasonic depth measurements; and (b) location distribution from the same discontinuity

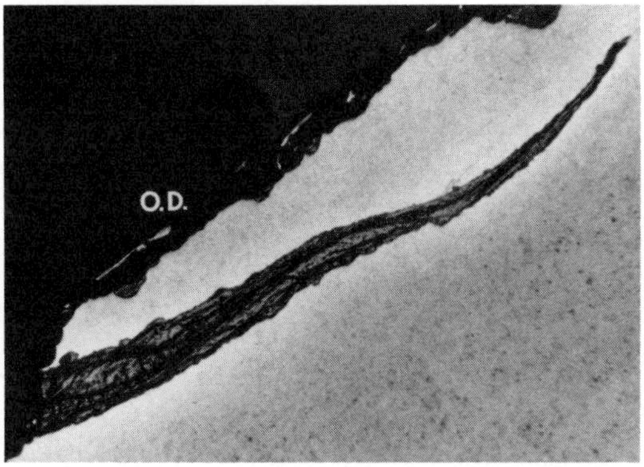

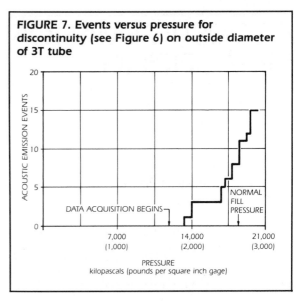

FIGURE 7. Events versus pressure for discontinuity (see Figure 6) on outside diameter of 3T tube

FIGURE 6. Section of discontinuity on outside of 3T tube; maximum depth (radial direction) is 1.4 mm (0.056 in.)

PART 3
ACOUSTIC EMISSION TESTS OF METAL PIPING

Acoustic Emission Testing of Gas Pipelines

Some of the earliest uses of acoustic emission monitoring were accomplished on gas transmission pipelines.[7] The incentive for prevention of failure and leak detection are both extremely high. Not only were the lines themselves tested but pipe mill applications were also explored. While no further implementation was made in the pipe mill area, monitoring pipelines for discontinuities and leak detection continues.

A 30 m (105 ft) section of gas pipeline was evaluated with both 30 kHz and 150 kHz resonant sensors.[8] The pipeline had a 355 mm (14 in.) diameter and a wall thickness of 6 mm (0.25 in.). During cyclic pressurization, the acoustic emission system was able to detect and locate deep corrosion pits at the operating stress level. The deepest pit was 0.5 mm (0.02 in.) and it developed a through-wall leak which was easily detected.

Signal attenuation tests on similar sections of 105 m (350 ft) and 245 m (800 ft) lines were also achieved. A simulated leak signal could be detected at a maximum distance of 105 m (350 ft). Acoustic emission signals of simulated crack growth were detectable and usable for source location at 150 m (500 ft) in buried pipeline. These signals were detectable at 245 m (800 ft) but were so distorted that they were unsuitable for the source location computer.

Acoustic emission source location resolution was within ±100 mm (4 in.) at 32 m (104 ft) sensor spacings, with an estimated ±0.9 m (±3 ft) capability at 300 m (1,000 ft) sensor spacings for the buried pipeline.

PART 4
ACOUSTIC EMISSION TESTING OF METAL PRESSURE VESSELS

Code for Testing Pressure Vessels

The ASME Pressure Vessel and Boiler Code of May 20, 1985, Section VIII, Division 1[9] approves one instance (Case 1,968) for the use of acoustic emission examination in place of radiography.

Case 1,968 is restricted to vessels which "shall not exceed two feet inside diameter length or seven cubic feet capacity with a weld thickness not to exceed 2.5 inches." The material is specified as P-No. 1, Group-1 or Group-2 and the entire case is allowed only for examination of the circumferential closure weld.

In addition, all of the following requirements must be met.

I. The acoustic emission examination shall be performed in accordance with a written procedure that is certified by the manufacturer to be in accordance with *ASME Proposed Standard for Acoustic Emission Examination During Application of Pressure* (1975).[10] The written procedure shall be demonstrated to the satisfaction of the inspector.

II. The manufacturer shall certify that personnel performing and evaluating acoustic emission examinations have been qualified and certified in accordance with their employer's written practice. The American Society for Nondestructive Testing's SNT-TC-1A (1980)[11] shall be used as a guideline for employers to establish a written practice for qualifying and certifying personnel. The qualification records of certified personnel shall be maintained by their employer.

III. The acoustic emission examination shall be conducted throughout the hydrostatic test that is required by UG-99. Two pressurization cycles from atmospheric pressure to the test pressure required by UG-99 shall be used.

IV. Evaluation and acceptance criteria shall be as follows.
  1. During the first pressurization cycle, any rapid increase in acoustic emission events or any rapid increase in acoustic emission count rate shall require a pressure hold. If either of these conditions continues during the pressure hold, the pressure shall be immediately reduced to atmospheric pressure and the cause determined.
  2. During the second pressurization cycle, the requirements of IV.1 above shall apply and, in addition, the following acoustic emission indications shall be unacceptable:
      a. any acoustic emission event during any pressure hold;
      b. any single acoustic emission event that produces more than 500 counts or that produces a signal attribute equivalent to 500 counts;
      c. six or more acoustic emission events detected by a master sensor;
      d. three or more acoustic emission events from any circular area whose diameter is equal to the weld thickness or 1 in., whichever is greater;
      e. two or more acoustic emission events from any circular area (having a diameter equal to the weld thickness or 1 in., whichever is greater) that emitted multiple acoustic emission events during the first pressurization.

V. Welds that produce questionable acoustic emission response signals (such as acoustic emission signals that cannot be interpreted by the acoustic emission examiner) shall be evaluated by radiography in accordance with UW-51. If the construction of the pressure vessel does not permit interpretable radiographs to be taken, ultrasonic examination may be substituted for radiography in accordance with UW-11(a)(7). Final acceptance (or rejection) of such welds shall be based on the radiographic or ultrasonic results, as applicable.

VI. The acoustic emission sensors shall be positioned so that the entire pressure vessel is monitored by the acoustic emission system and all acoustic emission response signals shall be recorded and used in the evaluation. The same acoustic emission acceptance standards shall be applied to the rest of the vessel that are applied to the circumferential closure weld.

VII. The acoustic emission test results and records shall be retained in accordance with the Section VIII requirements for radiographic film.

VIII. Case number 1,968 shall be shown on the data report form.

Acoustic Emission Testing of Blowout Preventers

Blowout preventers are massive valves installed on the wellhead below the drilling rig floor on land rigs. They may also be installed on the sea floor when used with floating offshore rigs. As an oil well is drilled, drilling mud is used to control the pressure exerted by fluids encountered by the drill bit. When more highly pressurized fluids are expected, heavier mud is used. If the drill bit grinds into a geologic formation containing abnormally pressurized fluids, the higher pressures can reverse the flow of fluid and, in the worst cases, can literally blow the drill string out of the well. What protects against this potentially dangerous occurrence is the blowout preventer.

Blowout preventers are always in place in case an emergency arises and when they are needed, their reliability is critical. They are commonly grouped together in stacks of four to six with rated working pressures from 35 to 200 MPa (5,000 to 30,000 psi). These groupings provide a backup or redundant safety system.

The only test required by the industry is specified in the American Petroleum Institute (API) Specification 6-A for wellhead equipment.[12] That standard mandates that the blowout preventer not leak or fail when subjected to a hydrostatic test up to twice the operating pressure or as high as 200 MPa (30,000 psi).

Monitoring Blowout Preventers with Acoustic Emission

Evaluating a blowout preventer with acoustic emission works in the following way. Multiple arrays of acoustic emission sensors are coupled to the blowout preventer at predetermined intervals. Water is pumped into the blowout preventer, gradually increasing the pressure inside. Once the maximum test pressure is reached, it is maintained for three minutes. The acoustic emission activity picked up during this initial pressurization comes from (1) elastic strain and (2) any imperfections in the metal body of the blowout preventer. Those discontinuities in the metal are the focus of the acoustic emission testing.

To take advantage of the Kaiser effect, the pressure is reduced after three minutes and built up again. Holding time for the maximum test pressure is fifteen minutes. Noises registered by the sensors during this phase of the test indicate discontinuities that are propagating.

Readings from the transducers are fed into a computer and analyzed to determine the severity of the imperfection and its location. Nondestructive testing techniques such as ultrasonics play a subsequent role in the exact location and classification of discontinuities for repair.

Cast Steel Gate Valves

Most acoustic emission tests on metallic structures have been performed on relatively large volume pressure vessels, tanks and reactors. The application to structures less than 0.7 m^3 (25 ft^3) has mainly been confined to hydrostatic testing of pipe sections. Source location on piping and similar cylindrical objects is relatively simple but irregular objects such as valves, blowout preventers[13] and other types of pipe fittings do not easily lend themselves to calculated source location. Experimental means are often used to simplify location of acoustic emission sites. Detailed below is an application of this type, involving acoustic emission testing of 355 mm (14 in.) and 405 mm (16 in.) high pressure steel gate valves[14] of the following description.

1. *Material*: carbon steel
2. *Pressure rating*: 4 MPa (600 psi)
3. *Type of valve*: gate, ring joint
4. *Sizes*: 355 mm (14 in.) and 405 mm (16 in.)
5. *Average wall thickness*: 38 mm (1.5 in.)
6. *Operating pressure*: 2 MPa (300 psi)
7. *Operating temperature*: 315 °C (600 °F)
8. *Service*: light hydrocarbon
9. *Hydrostatic test pressure*: 21 MPa (3,000 psi)

The acoustic emission testing system contained twelve channels but only eight channels were used in the valve test. Figure 8 shows the location of sensors on the valve bodies. The complicated geometry of the valves prohibited the use of triangulation for source location. Consequently, a look-up table was used in conjunction with a grid pattern laid out on the valve bodies. An electric pulser was positioned at each grid intersection and time of arrival difference (Δt) coordinates were measured over the entire outside surface of the valve. Figure 9 shows a 405 mm (16 in.) valve with a grid pattern laid out on approximately one-quarter of its surface. This procedure facilitated rapid location of relevant emission sites based on the unique combination of Δt measurements taken during hydrostatic pressurization.

Preparation for the test consisted of locating and cleaning to bare metal the eight sensor locations (Fig. 8). After the magnetically held sensors were attached to the designated locations, individual coaxial cables were strung from sensor preamplifiers to the acoustic emission mainframe components. In the interest of safety, the valve was located in an underground pit about 30 m (100 ft) from the acoustic emission mainframe.

Analysis of Acoustic Emission Responses

During hydrostatic pressurization of the valve, pressure, emission summation, rate and Δt printouts were recorded.

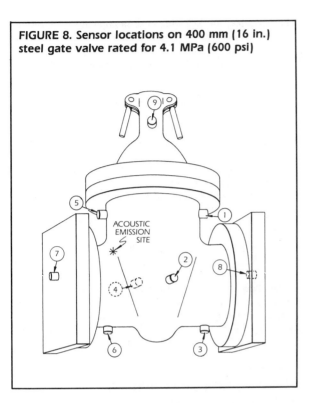

FIGURE 8. Sensor locations on 400 mm (16 in.) steel gate valve rated for 4.1 MPa (600 psi)

FIGURE 9. Valve body with grid pattern laid out on surface

Figure 10 is typical of this display, indicating a pressurization period of approximately eight minutes and maximum hydrostatic pressure of 21 MPa (3,070 psig). There is a slow, gradual rise in acoustic emission count approximately proportional to the pressure increase.

Emission site locations were determined by measuring the Δt coordinates from each combination of three adjacent sensors on the valve. Figure 11 is typical of the emission response from one location on one of the valves. The slow, gradual increase in acoustic emission count with rising pressure is indicative of noncritical discontinuity growth. For comparison, the plot of Fig. 12 shows a rapid emission count increase with rising pressure, indicating that a growing discontinuity is present. In this case, an examination of the suspect area after pressurization revealed a series of shrinkage cracks in the valve body estimated to be about half way through the wall thickness. These cracks were identified on the inside diameter surface by means of magnetic particle examination and their depth was calculated using stereoradiographic methods.

The acoustic emission technique was used successfully on about twenty high pressure valve bodies. The method proved to be highly practical as a production tool used in conjunction with acceptance hydrostatic tests of valve bodies. Acoustic emission tests also proved useful in evaluating

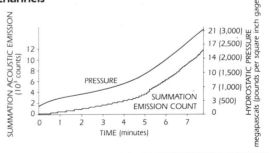

FIGURE 10. Acoustic emission from 406 mm (16 in.) steel gate valve rated for 4.1 MPa (600 psi); summation of emissions from nine channels

fitness for service in valve bodies exposed to adverse operating conditions. Some of these environments were extreme enough to cause a reduction in tensile strength or to induce structural discontinuities.

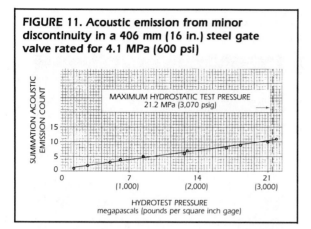

FIGURE 11. Acoustic emission from minor discontinuity in a 406 mm (16 in.) steel gate valve rated for 4.1 MPa (600 psi)

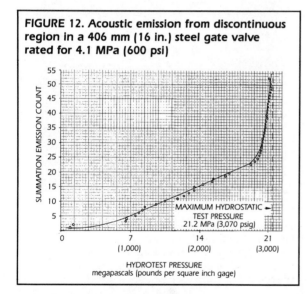

FIGURE 12. Acoustic emission from discontinuous region in a 406 mm (16 in.) steel gate valve rated for 4.1 MPa (600 psi)

Testing Pressure Vessels to Evaluation Standards

Acoustic emission testing for metal tanks and vessels may be achieved by using the numerical evaluation criteria adopted by the American Society for Mechanical Engineers (ASME), the American Society for Testing Materials (ASTM) or the Society for Plastic Industries (SPI). One such application[15] is shown in Table 6, where specific evaluation criteria follow the same format for metal vessels as for fiberglass. The specific numerical values depend on the vessel's material, type, service condition, previous load history and the testing instrument being used.

For example, the following criteria apply for tests with a specific manufacturer's acoustic emission system on an in-service carbon steel tank pressurized with ambient temperature process fluid to 105 percent of its maximum operating level.

1. *Emission during load hold*: not more than two hits per sensor after two minutes.
2. *Count rate*: less than 400 counts per sensor during a load increase of 5 percent of the operating load.
3. *Number of hits*: not more than 60 hits on all sensors up to operating load and not more than 60 hits on all sensors for each subsequent five percent increase in load.
4. *Large amplitude hits*: all hit amplitudes less than or equal to 65 dB.
5. *Relative energy*: does not increase with increasing load.
6. *Activity*: does not increase with increasing load.
7. *Evaluation threshold*: 50 dB.

The above criteria would not hold for a different manufacturer's instrument, or a tank fabricated from another material or operating under different conditions. The criteria shown in Table 6 are based on field experience and were developed from tests of about 1,000 pressure vessels.

Acoustic Emission Testing of Ammonia Spheres

Periodic Tests of Ammonia Spheres

Reference 16 illustrates the value of periodic testing of a vessel to monitor the growth of discontinuities and to provide guidance on the appropriate time for shutdown and repair.

This is also illustrated by the case history of the sphere shown in Fig. 13 which shows a drawing of a carbon steel anhydrous ammonia sphere similar to one tested on a regular cycle with acoustic emission.[18] Stress corrosion cracking had been found in the welds of the sphere during a shutdown. The welds were ground and all detectable cracks were removed before recommissioning the sphere.

Thirty-four permanent sensors were mounted on the vessel and periodic acoustic emission tests were conducted. The first test was conducted during start-up following repair. Minor emissions were detected, suggesting that some fine stress corrosion cracking remained. Figure 14 is a schematic representation of the total acoustic emission for each of the periodic tests. As can be seen, each test gave an increasing amount of emission, with peak emission recorded during the October 1982 test. At this point, it was decided to remove the sphere from service. Stress corrosion cracking was confirmed by magnetic particle inspection.

TABLE 6. Evaluation criteria for acoustic emission tests of pressure vessels

Vessel Type	Emission During Load Hold	Count Rate	Number of Hits	Large Amplitude Hits	Relative Energy or Amplitudes	Activity	Evaluation Threshold (decibels)
Pressure vessels without full post-weld heat treatment on first loading	Not more than E_H hits beyond time T_H	Not applied	Not applied	Not more than E_A hits above a specified amplitude	Do not increase with increasing load	Does not increase with increasing load	V_{TH}
Tanks and pressure vessels other than those covered above	Not more than E_H hits beyond time T_H	Less than N_T counts per sensor for a specified load increase	Not more than E_T hits above a specified amplitude	Not more than E_A hits above a specified amplitude	Do not increase with increasing load	Does not increase with increasing load	V_{TH}

E_H, N_T, E_T and E_A are specified acceptance criteria values.
V_{TH} is the specified evaluation threshold.
T_H is the specified hold time.

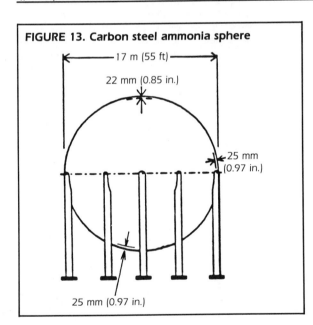

FIGURE 13. Carbon steel ammonia sphere

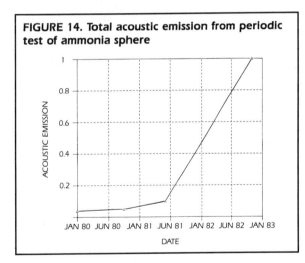

FIGURE 14. Total acoustic emission from periodic test of ammonia sphere

Periodic monitoring provides valuable guidance on the development of discontinuities such as stress corrosion cracking and on the increasing structural severity of such discontinuities. It should be noted that minor discontinuities were present following repair and that more severe discontinuities were present in subsequent tests. However, based on the acoustic emission data, the sphere was deemed safe for continued operation until the emission intensity indicated the need for shutdown, internal inspection and repair.

Stress Cracking in Ammonia Spheres

This application[18] illustrates the use of energy based intensity grading of discontinuities. The intensity evaluation was correlated with subsequent ultrasonic and magnetic particle inspection. It was found that the most severe discontinuities were in the areas with the most intense emission.

Figure 15 shows a developed view of an anhydrous ammonia sphere similar to the one drawn in Fig. 13. The vessel was tested in-service and monitored with acoustic emission. A discontinuity intensity analysis, based on the energy of the most severe events and the event energy trend, was carried out. The results of this analysis are indicated by the letter following the sensor number on the plot. The least severe discontinuities are noted by intensity-*A*, the most severe by intensity-*E*. Sensors without discontinuities do not have a letter following the number. Intensity-*E* indicates that an immediate shutdown of the vessel is recommended. Intensity-*A* indicates a minor discontinuity that should be noted for future reference. Repair of an intensity-*A* discontinuity is not recommended. Intensities *B*, *C* and *D* are intermediate between *A* and *E*.

Following the test, the vessel was shut down, decontaminated and a full internal inspection was performed. The inspection methods included visual, ultrasonic and magnetic particle. The inspection showed a large number of small cracks 3.2 to 9.5 mm (0.125 to 0.375 in.) in length. Most cracks were shallow, to about 1.3 mm (0.05 in.) deep and parallel to the welds. Some of the cracks were perpendicular to the welds in the most severe situations, which corresponded to the *D* and *E* intensity zones. Agreement between the most severe cracking and the highest intensity acoustic emission areas was excellent. In the lower half of the vessel, an isolated area of shallow fine stress corrosion cracking was also detected. The location of this cracking agreed with the data from the sensors noted on Fig. 15.

Summary of Industrial Acoustic Emission Test Procedures of Tanks and Vessels

At least one procedure for acoustic emission testing of tanks and vessels has been developed for use within a large chemical company.[15] Guidelines are provided for acoustic emission testing of metal tanks and vessels under pressure or vacuum to determine structural integrity. The vessels for which this procedure was developed generally conform to American Petroleum Institute (API) standards (atmospheric and up to 105 kPa or 15 psig) or American Society of Mechanical Engineers (ASME) pressure or vacuum code practice.

The procedure is limited to tests not to exceed an internal pressure of 69 MPa (10,000 psig) and/or an external pressure of 3.5 MPa (500 psig) for jacket or vacuum. The procedure is also limited to temperatures ranging from −40 to 540 °C (−40 to 1,000 °F). The technique can be applied to tests of new and in-service equipment fabricated from carbon steel, high strength low alloy steel, zirconium, titanium, austenitic and martensitic stainless steel or aluminum.

The procedure document contains thirteen sections and five appendices. Following sections on *Scope*, *Applicable*

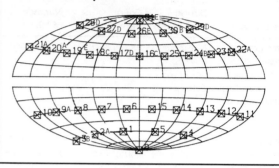

FIGURE 15. Density of discontinuities in carbon steel ammonia sphere

Documents, *Definitions*, *Summary of Method* and *Personnel Qualification*, Section 6 on *Significance* makes the following points.

The acoustic emission test method detects discontinuities and damage in metal equipment. Among the failure modes that generate acoustic emission in both parent metal and weld associated regions are: (1) crack growth; (2) the effect of corrosion, including cracking of corrosion products or local yielding; (3) stress corrosion cracking; (4) certain metallurgical changes, including yielding and crystal dislocations; and (5) embrittlement.

In typical weld associated regions, the failure modes that generate acoustic emission include: (1) incomplete fusion; (2) lack of penetration; (3) voids and porosity; (4) inclusions; and (5) contamination. In parent metal, delamination is the important acoustic emitting discontinuity. Discontinuities in unstressed areas and passive (structurally insignificant) discontinuities will not generate acoustic emission.

This procedure is suitable for on-line use to determine the structural integrity of in-service equipment, usually with minimal process disruption. The acoustic emission test method will detect major discontinuities that have grown or are newly developed since previous tests.

Discontinuities detected by acoustic emission should be examined by other nondestructive testing techniques and may be repaired and retested as appropriate.

A summary of the method indicates that the procedure provides guidelines for determining the nature, severity and location of structural discontinuities. It also lists evaluation criteria for assessing the structural integrity of the metal equipment.

The procedure provides guidelines for acoustic emission testing of metal equipment within the limits previously stated. Maximum test pressure (or vacuum) for a metal vessel will be determined by agreement between owner, manufacturer or test agency.

For new pressure vessels operating at internal or external pressures greater than 0.1 MPa (15 psig) or designed to the provisions of the ASME Code, the normal test pressure is as specified in Section VIII of the ASME Code. For new atmospheric tanks, the test pressure will normally be over the range of 0 to 105 percent of the design pressure. New and in-service vacuum vessels will normally be tested over the range from 90 percent to 105 percent of the maximum operating value. In-service pressure vessels will normally be tested over the range from 90 percent or less to 110 percent of the maximum operating value.

The technical sections of the procedure cover the instrumentation, test preparations, test setup, instrumentation system performance check, testing procedure, interpretation of results and the report. Stressing sequences are illustrated for new pressure vessels, new tanks, new and in-service vacuum vessels, in-service pressure vessels and in-service tank tests.

The evaluation criteria shown in Table 6 are based on extensive experimental data. The criterion for increasing activity is an indication of a growing defective area. The criteria based on increasing hit amplitudes and activity are an indication that the discontinuity is a significant structural discontinuity responding to increases in stress. Departure from a linear count per load or relative energy should signal caution. If the acoustic emission count or relative energy rate increases rapidly with load, the vessel should be unloaded and either the test terminated or the source of the emission determined and the safety of continued testing evaluated. A rapidly (exponentially) increasing count or relative energy rate may indicate uncontrolled, continuing damage indicative of impending failure.

The criterion based on emission during load hold is particularly significant. Continuing emission indicates continuing yield or damage due to creep or a growing discontinuity. Fill and other background noise will generally be at a minimum during a load hold period.

The criterion based on count rate is a measure of the total activity. This is particularly important for detecting widespread damage such as general corrosion or stress corrosion cracking.

The criterion based on number of hits gives an indication of the size and severity of the defective region. This criterion is particularly important for vessels in-service, where hits below operating stress indicate the presence of significant discontinuities. The degree to which hits below operating stress are present is an indication of the severity of a discontinuity.

High amplitude hits often indicate the presence of a growing crack. Increasing hit amplitudes are an indication of a significant structural discontinuity and are often associated with growth of fatigue cracks from an original weld discontinuity.

Threshold of detectability for an instrument is determined by breaking a 0.3 mm pencil lead (2H) on a 99 percent pure lead sheet. The dimensions of the lead sheet are 1,200 × 1,800 × 13 mm (48 × 72 × 0.5 in.). A decibel calibration curve is then developed for pencil lead breaks a certain distance from the sensor for various acoustic emission systems. Figures 16 through 25 illustrate typical acoustic emission sensor locations for various applications.

The remaining sections of the procedure are appendices that discuss instrumentation performance requirements, instrument calibration, sensor placement guidelines, attenuation and background information. The test procedure is based on a data bank of results from previously tested vessels and tanks.

Testing of Hydroformer Reactor Vessels

A hydroformer reactor was acoustic emission tested during periods of temperature and pressure fluctuations associated with reactor regeneration and shutdown.[17] The purposes of the test were (1) to observe the overall behavior of the vessel as normal process thermal and pressure loads were applied; (2) to identify areas of high stress concentration; and (3) to identify high acoustic event activity in specific locations such as weld lines, vessel penetrations or reinforcements.

A total of twelve channels were used to monitor the vessel. As shown in Fig. 26, the sensors were placed on the vessel in a pattern containing a dome sensor (channel-1), two sensor bands (channel-2 to channel-6 for band-1 and channel-7 to channel-11 for band-2) and a base sensor (channel-12). By offsetting the sensor positions of band-2 with respect to band-1, an interlocking arrangement of equilateral triangles was created with a sensor located at each point of a triangle. In this manner, total coverage of the vessel is possible while ensuring that acoustic emission events always occur within a triangle section. Figure 27 depicts this sensor array pattern as presented by computer. The sensors were held in place using magnets. A high temperature grease was used as acoustic couplant except on the dome and base channels where stand-off waveguides were used because of extreme surface temperatures.

Once the sensors were placed on the vessel, system calibration was performed. Some averaging of the received time of arrival difference (Δt) was utilized. The obtained velocity of sound in the vessel was determined to be 5.3×10^6 mm•s^{-1} (2.1×10^5 in.•s^{-1}).

Hydroformer Reactor Test Data and Interpretation

Data from the test are divided into two parts: (1) Fig. 28 through Fig. 33 present data from the start of regeneration through the coke burn; and (2) Fig. 34 through Fig. 40 are from the end of coke burn to the vessel cool-down at about 150 °C (300 °F).

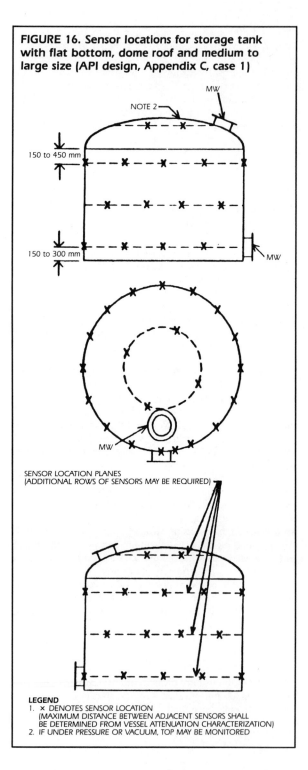

FIGURE 16. Sensor locations for storage tank with flat bottom, dome roof and medium to large size (API design, Appendix C, case 1)

FIGURE 17. Sensor locations for storage tank with flat bottom, dome roof and small to medium size (API design, Appendix C, case 2)

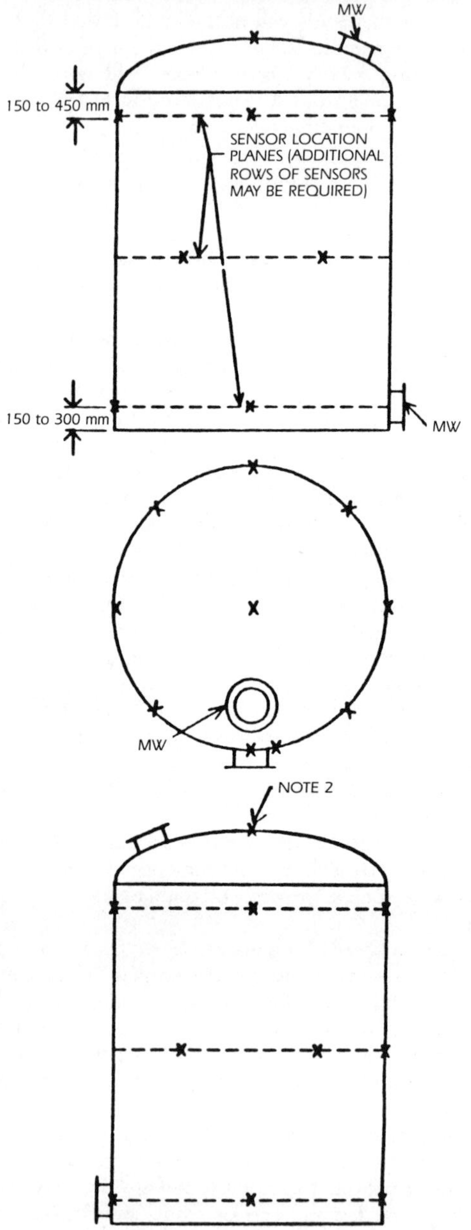

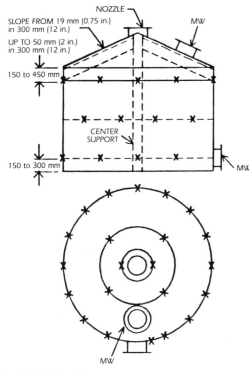

FIGURE 18. Sensor locations for storage tank with flat bottom, cone roof and medium to large size (API design, Appendix C, case 3)

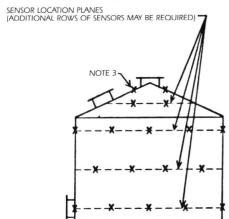

LEGEND
1. × DENOTES SENSOR LOCATION
   (MAXIMUM DISTANCE BETWEEN ADJACENT SENSORS SHALL BE DETERMINED FROM VESSEL ATTENUATION CHARACTERIZATION)
2. ADDITIONAL SENSORS MAY BE PLACED AT POINTS OF SPECIAL INTEREST
3. IF PRESSURE OR VACUUM INVOLVED, TOP MAY BE MONITORED

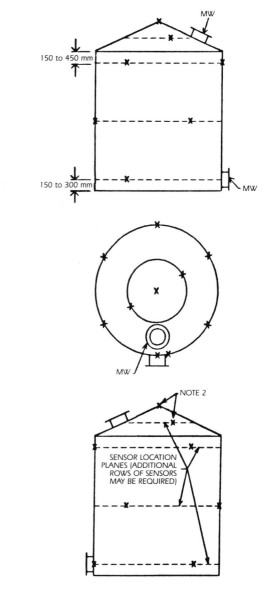

FIGURE 19. Sensor locations for storage tank with flat bottom, dome roof and small to medium size (API design, Appendix C, case 4)

LEGEND
1. × DENOTES SENSOR LOCATION
   (MAXIMUM DISTANCE BETWEEN ADJACENT SENSORS SHALL BE DETERMINED FROM VESSEL ATTENUATION CHARACTERIZATION)
2. IF PRESSURE OR VACUUM INVOLVED, TOP MAY BE MONITORED

FIGURE 20. Sensor locations for storage tank with flat bottom, flat top (API design, Appendix C, case 5)

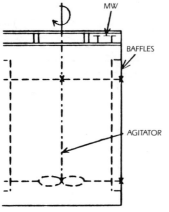
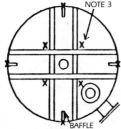
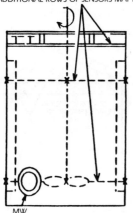

LEGEND
1. x DENOTES SENSOR LOCATION (MAXIMUM DISTANCE BETWEEN ADJACENT SENSORS SHALL BE DETERMINED FROM VESSEL ATTENUATION CHARACTERIZATION)
2. AGITATOR SHOULD BE RUN FOR SHORT PERIOD WHEN FULL
3. TOP MAY BE MONITORED IF PRESSURE OR VACUUM INVOLVED OR TO CHECK TOP UNDER AGITATOR LOADS

FIGURE 22. Sensor locations for vertical pressure vessel with dished heads, agitator, baffled lug or leg supports (ASME Section VIII design, Appendix C, case 7)

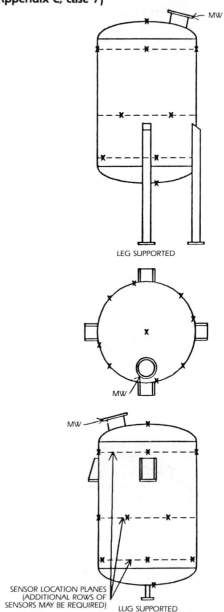

LEGEND
x DENOTES SENSOR LOCATIONS (MAXIMUM DISTANCE BETWEEN ADJACENT SENSORS SHALL BE DETERMINED FROM VESSEL ATTENUATION CHARACTERIZATION)

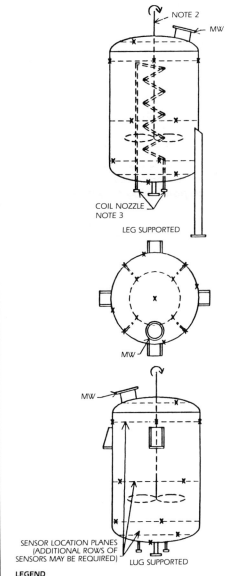

FIGURE 22. Sensor locations for vertical pressure vessel with dished heads, agitator, baffled lug or leg supports (ASME Section VIII design, Appendix C, case 7)

LEGEND
1. x DENOTES SENSOR LOCATIONS
   (MAXIMUM DISTANCE BETWEEN ADJACENT SENSORS SHALL BE DETERMINED FROM VESSEL ATTENUATION CHARACTERIZATION)
2. AGITATOR SHOULD BE RUN FOR A SHORT PERIOD WHEN FULL
3. SENSORS MAY BE LOCATED ON OUTLET TO DETECT DISCONTINUITIES IN COIL

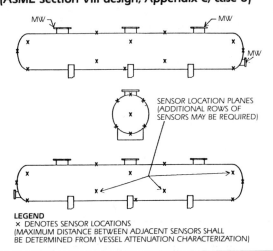

FIGURE 23. Sensor locations for vertical pressure vessel with dished heads and saddle supports (ASME Section VIII design, Appendix C, case 8)

LEGEND
x DENOTES SENSOR LOCATIONS
(MAXIMUM DISTANCE BETWEEN ADJACENT SENSORS SHALL BE DETERMINED FROM VESSEL ATTENUATION CHARACTERIZATION)

As can be seen in Figs. 28 and 29, most of the event activity concentrated in the reinforcement pad area from −18 to 32 °C (0 to 90 °F) and occurred early in the monitoring period before the coke burn (note locations marked A and B).

The statistic graph of Fig. 30 shows this clearly with section-16 and section-17 reporting a majority of the activity. Pulse height (relative amplitude) of these received events shows a large number above a scale of 12 (greater than 5.0 volts peak). Figures 31 and 32 are displays of section-16 and section-17, showing a large amount of scatter associated not only with the reinforcement pad but also with the vessel welds in this area. A total of 720 events were received during a monitoring period of 27.8 hours. Of those events, 520 were reported as valid (Fig. 33).

The monitoring period during cooling produced the highest event activity for the test. In Fig. 34, the activity can be seen to continue along the reinforcement pad area (locations marked B and C) with new areas reporting events in section-17 and section-8 (marked as location A) and along the shroud weld from 32 to 82 °C (90 to 180 °F). The graph of total events in Fig. 35 shows a higher event activity in the earlier portion of the monitoring period. This activity tapered off as time progressed.

During the later period, control operators reported a slowing in the temperature reduction rate at about 205 °C (400 °F). Figure 36 shows that the vast majority of events for this monitoring period was located in section-8 and that the average pulse height dropped significantly when compared to the previous test run. An enlargement of the most active sections is presented in Figs. 37 through 39. A total of 1,884 events were reported in just over eight hours of monitoring with 1,590 events recorded as valid (Fig. 40).

Table 7 is a list of areas on the vessel which were marked for inspection with other nondestructive testing methods. A diagram of these locations is presented in Fig. 41.

Specifications of the test were as follows.

1. *Structure tested*: hydroformer reactor
2. *Test dates*: July 16 to 19
3. *System channels*: twelve in equilateral triangles
4. *Intersensor distance*: 4 m (13.5 ft)
5. *Sensor frequency*: 375 kHz with 250 to 500 kHz filter band-pass
6. *System gain*: 80 dB reference to 1.0 volt floating threshold
7. *Velocity of sound*: 5.3×10^6 mm·s^{-1} (2.1×10^5 in.·s^{-1})

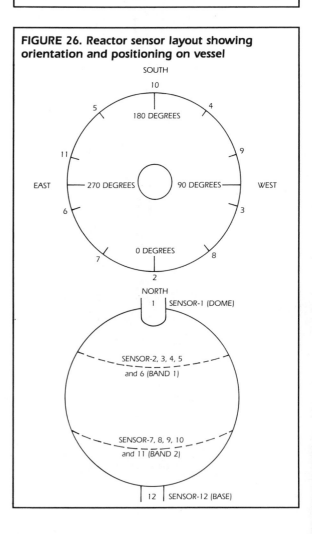

FIGURE 24. Sensor locations for vertical pressure vessel with packed or trayed column, dished heads and lug or skirt supports (ASME Section VIII design, Appendix C, case 9)

FIGURE 25. Sensor locations for spherical pressure vessel with leg supports (ASME Section VIII design, Appendix C, case 10)

FIGURE 26. Reactor sensor layout showing orientation and positioning on vessel

TABLE 7. Hydroformer locations to be inspected with other nondestructive methods

Array Section	Horizontal Displacement[1]	Vertical Displacement[2]
8	70 to 80 degrees	8.2 to 11 m (27 to 36 ft)
16, 17, 18	0 to 90 degrees	10.7 m (35 ft) along reinforcement weld
2,3	90 to 180 degrees	3 m (10 ft) along shroud weld

1. Measured in degrees from 0 degrees (north)
2. Measured from vessel inlet (dome)

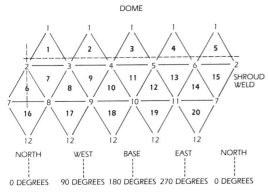

FIGURE 27. Sensor layout (flat plate representation) as displayed by acoustic emission computer system; intersensor distance equals about 4 m (13 ft) in equilateral triangles; numbers within triangles are section numbers

FIGURE 29. Events versus time (between start of regeneration through coke burn) for acoustic emission tests of hydroformer reactor

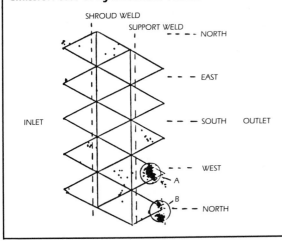

FIGURE 28. Event locations (between start of regeneration through coke burn) for acoustic emission test of hydroformer reactor

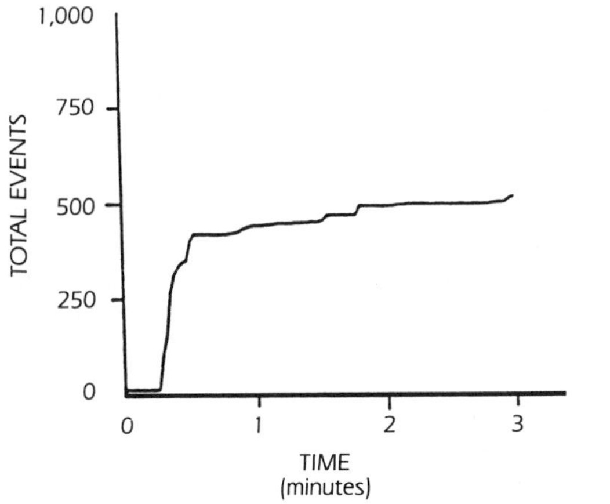

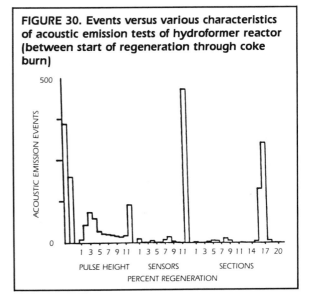

FIGURE 30. Events versus various characteristics of acoustic emission tests of hydroformer reactor (between start of regeneration through coke burn)

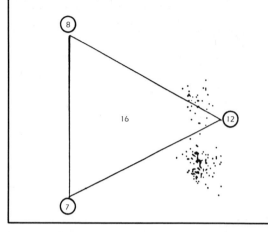

FIGURE 31. Acoustic emission event locations for hydroformer section-16 (sensor 7, 8 and 12) between start of regeneration through coke burn

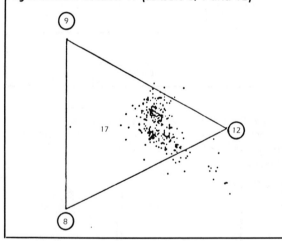

FIGURE 32. Acoustic emission event locations for hydroformer section-17 (sensors 8, 9 and 12)

FIGURE 33. Printout of data from test of hydroformer between start of regeneration through coke burn

```
LIST STATUS
MAX DELTA T=700 INTERSENSOR DISTANCE=135 VELOCITY OF SOUND=192
REJECT OUTPUT FREQUENCY=1000
TIMER RATES IN SECONDS: EVENT=1 PLOT=1000 EVENT RATE=1
EVENTS: TOTAL=720 GOOD=520 REJECT=200 LOST=0 PROCESS=0
EVENT TIMER=96532 PLOT TIMER=96
THRESHOLD =4995
```

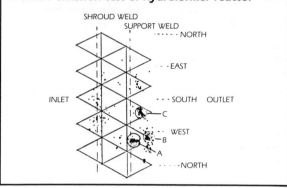

FIGURE 34. Event locations (between end of coke burn through part of cool down) for acoustic emission test of hydroformer reactor

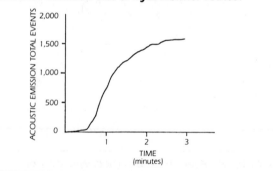

FIGURE 35. Time versus events (between end of coke burn through part of cool down) for acoustic emission test of hydroformer reactor

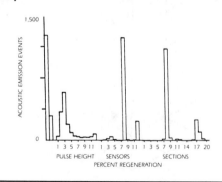

FIGURE 36. Events versus various characteristics of acoustic emission tests of hydroformer reactor (between end of coke burn through part of cool down)

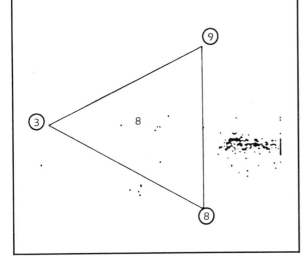

FIGURE 37. Event locations in hydroformer section-8 during acoustic emission tests (sensors 3, 8 and 9) between end of coke burn through part of cool down

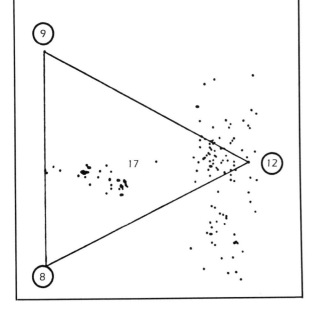

FIGURE 38. Event locations in hydroformer section-17 during acoustic emission tests (sensors 8, 9 and 12) between end of coke burn through part of cool down

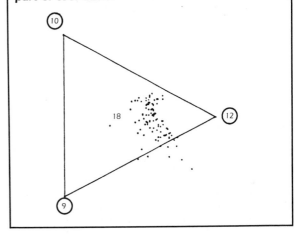

FIGURE 39. Event locations in hydroformer section-18 during acoustic emission tests (sensors 9, 10 and 12) between end of coke burn through part of cool down

FIGURE 40. Printout of data from test of hydroformer between end of coke burn through part of cool down

```
LIST STATUS
MAX DELTA T=700 INTERSENSOR DISTANCE=135 VELOCITY OF SOUND=192
REJECT OUTPUT FREQUENCY=1000
TIMER RATES IN SECONDS: EVENT=1 PLOT=300 EVENT RATE=1
EVENTS: TOTAL=1884 GOOD=1590 REJECT=294 LOST=0 PROCESS=0
EVENT TIMER=29064 PLOT TIMER=96
THRESHOLD =4995
```

FIGURE 41. Locations of acoustic emission event activity on hydroformer reactor vessel

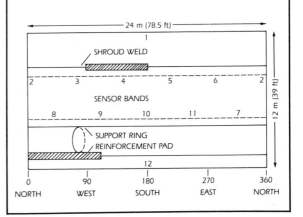

Fluid Catalytic Cracking Unit

Upper Regenerator Section and Plenum Weld Area Start-Up Monitoring

The purpose of this test was to develop condition data for the upper shell and plenum weld area of a regenerator in a fluid catalytic cracking unit.[19] These data were collected using acoustic emission techniques capable of detecting and analyzing active cracking and other activity originated by discontinuities.

The plenum weld area was suspected because cracking had recently been found and repaired in that weld. In addition, experience with similar mechanical designs (hydroformers) had shown that such welds are subject to high stresses and premature failure.

The test was run over a three-day period during which the unit went from shutdown status to an on-line condition. Details of the start-up sequence are given in Table 8. Testing instrumentation and setup are described in Table 9 and Fig. 42 respectively.

Test results and interpretation are shown in Figs. 43 and 44. Several graphs of selected process variables are included mainly because this type of testing relies on the stresses that develop during normal operation as variables are changed. These stresses are usually sufficient to promote discontinuity activity so that a good idea of condition can be developed with acoustic emission techniques. These data can be quite useful because the discontinuity signature is so closely related to realistic operating conditions.

Fluid Catalytic Cracking Unit Test Data and Interpretation

No major (concentrated) discontinuity areas were found. Some minor acoustic emission activity was found on the south section of the regenerator dome. Since this correlated with recent ultrasonic inspection, further testing was recommended to determine changes in condition.

Overall acoustic emission activity was relatively low, indicating a smooth start-up involving few mechanical shocks.

Both temperature changes and pressure changes were sufficient to promote discontinuity activity. This means that if discontinuities were present, there was a high probability of finding them with this type of acoustic emission testing. Delta pressure changes across the plenum were significant in stimulating acoustic emission activity, indicating that fatigue failure of the plenum weld will eventually occur though it is not an immediate concern.

High frequency acoustic attenuation in the shell indicates that further acoustic emission testing should be done at lower frequencies and probably with more sensors at shorter spacings.

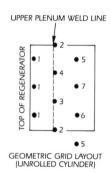
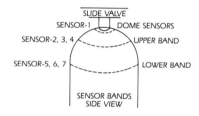
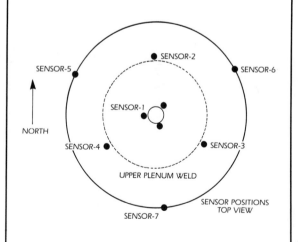

FIGURE 42. Sensor layout for acoustic emission test of hydroformer reactor vessel; sensor-2, 3 and 4 are on 4 m (13 ft) rods attached to the regenerator shell at full penetration welds

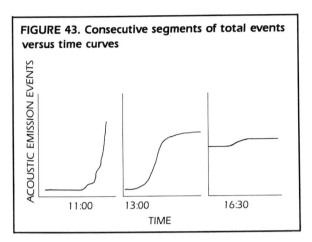

FIGURE 43. Consecutive segments of total events versus time curves

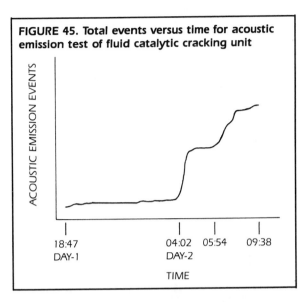

FIGURE 45. Total events versus time for acoustic emission test of fluid catalytic cracking unit

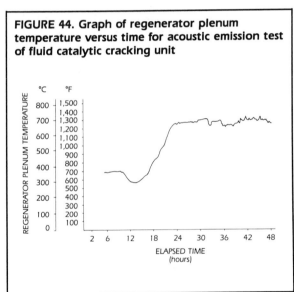

FIGURE 44. Graph of regenerator plenum temperature versus time for acoustic emission test of fluid catalytic cracking unit

The pictures in Fig. 43 are consecutive segments of the total events versus time curves. The initial knee in the first curve occurs at about 11:00, the point at which the plenum temperature began to drop from a previously stable level of 370 °C (700 °F). The total events curve then leveled off, as shown at the start of the second picture, around 13:00 when the plenum temperature bottomed out at about 290 °C (550 °F). The curve then increased as the temperature began to rise and then leveled off again around 15:00. Another increase in acoustic emission activity occurred around 16:30 when the torch was lit.

The fact that acoustic emission activity correlates with temperature changes indicates that thermal stressing was sufficient to promote discontinuity activity. The graph of regenerator plenum temperature is shown in Fig. 44.

Figure 45 shows total events versus time over the period from 18:47 on August 26 to 09:38 on August 27. The step changes in activity at about 03:00 and 06:00 could not be correlated to process activity because of lack of process data. However, there is some indication that this was the time in which catalyst was drying out or being fluffed up. In addition, the cyclone differential pressure may have been changing during those times and the plenum wall might have been slightly flexed (see Fig. 46).

Although the cause of this emission is not understood, it does appear that slight stressing occurred. Since no significant location data developed and the total number of events was low, the conclusion was that no active discontinuities were developing at this stage. Subsequent tests suggest that the very high attenuation on these vessels may also account for the low event totals.

The two pictures in Fig. 47 are segments of the total events versus time curves showing additional stages of the start-up procedure. Since temperature was fairly stable during this time period, the principal stress was thought to be pressure. Major changes in the differential pressure across the plenum can be seen in the preceding graph. The overpressure in the regenerator also increases dramatically as shown in Fig. 48.

Pressure stressing, like thermal stressing, is sufficient to promote discontinuity activity. Other sources of noise are also stimulated, so that detailed determination of discontinuity growth is not straightforward. The general conclusion is that no major discontinuities were detected and that acoustic emission activity from discontinuities was relatively low.

FIGURE 46. Graph of differential pressure across regenerator cyclones

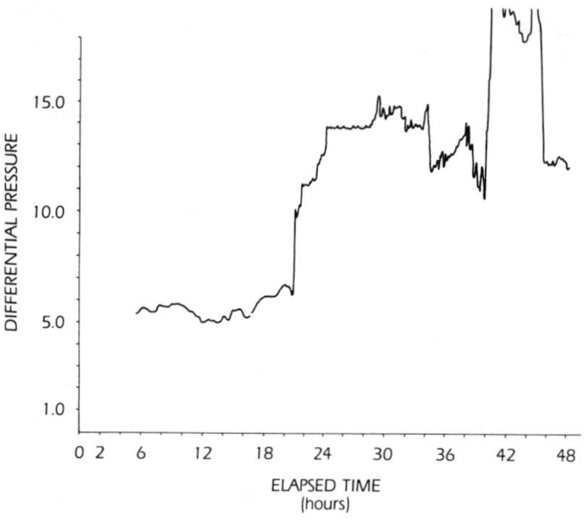

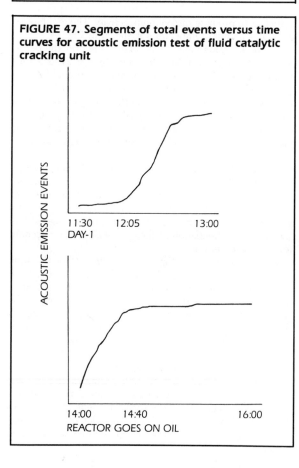

FIGURE 47. Segments of total events versus time curves for acoustic emission test of fluid catalytic cracking unit

TABLE 8. Sequence for acoustic emission testing of fluid catalytic cracking unit

Date	Time	Activity
August 24	18:30	Started test
		Checked calibration
		Began monitoring
August 25	01:00	Left system running unattended
	07:15	Returned to check on system
		No acoustic emission activity
		Performed system diagnostics (results OK)
	08:52	Restarted monitoring
	09:30	Left system running unattended
	17:30	Returned to check on system
		Computer had halted
		Performed diagnostics (results OK)
	18:52	Restarted system and began monitoring again
		Left system running unattended
		Hot air circulating in regenerator only
		No catalyst charged yet
August 26	07:30	Computer had halted overnight
		Reloaded system and restarted monitoring
		Operation now seems to be back to normal
	13:00	Dramatic increase in event rate
		Air flow increased
	17:00	Torch lit
August 27	03:00	Acoustic emission activity increases
		Catalyst being charged
	06:00	Acoustic emission activity increases
		Several sensors showing radiofrequency interference
		Decrease in sensitivity in hot area sensors noted
	11:00	Reactor begins to go on oil
	14:00	Reactor reaches significant feed processing
	16:00	Reactor stable on normal feed rate
	17:36	Monitoring stopped
		End of test

TABLE 9. Test instrumentation and setup for monitoring of regenerator upper shell (plenum weld area)

Material:	thought to be 2.25Cr-0.5Mo carbon steel
Thickness:	about 65 mm (2.5 in.)
Vessel diameter:	15 m (50 ft)
Temperature:	from ambient to about 760 °C (1,400 °F)
Pressure:	0 to about 170 kPa (25 psig)
Start-up conditions:	warm up, air circulation, catalyst charge, torch lit, catalyst circulation, on oil
Acoustic emission system:	multichannel computerized source location and characterization system
Number of channels used:	7
Array:	cylindrical, two-band of three sensors each with one dome sensor
Intersensor distance:	including waveguides, about 17 m (55 ft)
Sensor mounting:	
lower band	direct attachment
upper band	4 m (13 ft) carbon steel waveguides
dome	0.3 m (1 ft) waveguide
Difference of arrival time used:	4 ms (estimated)
Estimated velocity of sound:	1×10^6 mm·s^{-1} (4×10^5 in.·s^{-1})
frequency of sensors:	375 kHz
Gains:	
Channel 1	82 dB (40 dB preamp)
Channel 2	62 dB (40 dB preamp)
Channel 3	81 dB (40 dB preamp)
Channel 4	82 dB (40 dB preamp)
Channel 5	82 dB (40 dB preamp)
Channel 6	87 dB (40 dB preamp)
Channel 7	85 dB (40 dB preamp)
Acoustic attenuation:	about 1.6 to 3.2 dB·m^{-1} (0.04 to 0.08 dB· n.$^{-1}$)*
Sensor attachment:	
lower band	magnetic holders, direct contact, anti-seize high temperature couplant
upper band	waveguides, epoxy bonded
dome	waveguides, magnets and anti-seize high temperature couplant

*Could not be measured more accurately with instruments available. Cause was not known but was assumed to be a combination of refractory layer, hexmesh steel support layer and surface deposits of coke or paint.

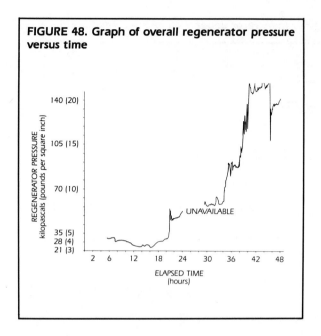

FIGURE 48. Graph of overall regenerator pressure versus time

PART 5
ACOUSTIC EMISSION TESTING OF METAL STORAGE TANKS

Tests of Flat Bottom Anhydrous Ammonia Storage Tanks

Figure 49 shows the dimensions of a 30,000 ton low temperature ammonia storage tank. Periodic in-service testing indicated the presence of discontinuities in the vessel wall. Intensity analysis indicated that the discontinuities were not severe but it was decided to empty, decontaminate and enter the vessel for internal inspection. This and other applications[20] help illustrate the correlation between acoustic emission test data and subsequent ultrasonic examination.

Ninety-six permanent sensors were mounted on the wall of the vessel and cables were run to a junction box at the base of the concrete secondary containment. The vessel was monitored during filling with anhydrous ammonia and during intermittent load holds. Acoustic activity was detected throughout the test but much of it was believed to be from boiling ammonia. Detailed data analysis was carried out to discriminate genuine from false acoustic emission and to determine the intensity of the discontinuities present. The analysis showed a number of areas with discontinuities but further indicated that they probably originated during construction and were not growing cracks.

Analysis of the acoustic emission data showed thirty-four sensors with activity suggesting the presence of discontinuities. Fourteen sensors were either dead or had high background noise of the type associated with bad cables or connections.

A 100 percent follow-up internal inspection of the shell welds was carried out using ultrasonics. All vertical and horizontal wall welds and the junction of the wall and floor were inspected. The ultrasonic inspection was independent of the acoustic emission results and the ultrasonic inspectors were not advised of the location of discontinuities found by acoustic emission. The ultrasonic inspection found fifty-five code-rejectable discontinuities.

The original construction discontinuities were incomplete penetration and slag in fusion lines between weld passes. No evidence was found to suggest that the discontinuities had grown during service or that new ones had developed. Repairs were carried out by grinding and rewelding about 8.2 m (27 ft) of shell plate weld.

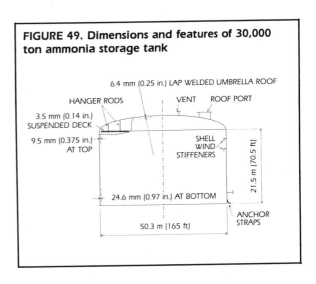

FIGURE 49. Dimensions and features of 30,000 ton ammonia storage tank

Comparison of Ultrasonic and Acoustic Emission Tests

Following start-up of the vessel, the ultrasonic and acoustic emission test results were compared. The comparison is shown in Table 10.

The agreement between the two test methods is excellent. Of the fifty-five discontinuities found by ultrasonics, forty-nine were originally found by acoustic emission. Three of the undetected discontinuities were in the zones of inactive sensors or sensors with high background noise. Only three of the ultrasonically detected discontinuities were near acoustic sensors that did not detect them.

In addition, five acoustic emission sensors indicated the presence of discontinuities that were not detected by ultrasonics. However, in the latter case, some of the active sensors were adjacent to sensor zones that had ultrasonically confirmed discontinuities, suggesting transmission of stress waves to an adjacent zone. Considering the strengths and weaknesses of ultrasonic and acoustic emission testing, agreement between the two methods is good. It is better than the agreement expected between other methods of nondestructive testing such as ultrasonics and radiography.

Tests of Horizontal Tanks

This case history[18] helps illustrate the correlation between acoustic emission data and discontinuities. A pressure vessel with various types of discontinuities was tested with acoustic emission and, as the pressure in the vessel was increased, an increasing number of discontinuities began to produce acoustic emission. It was found that crack-like discontinuities such as incomplete fusion and lack of penetration emitted at lower pressures than less structurally severe discontinuities, such as rust scale or weld porosity.

The vessel (see Fig. 50) was tested using a stepped increase in pressure from 69 kPa (10 psi) to 117, 138, 172 and 207 kPa (17, 20, 25 and 30 psi). The acoustic emission data were recorded during loading and load holds and the active sensors were noted. The data confirmed the presence of significant discontinuities and emission from many of the sensors exceeded the evaluation criteria. Following the acoustic emission test, the insulation was stripped from the vessel and thorough visual and ultrasonic examinations were conducted. All welds were scanned with ultrasonics and thickness readings were taken over the entire shell.

Correlation between acoustic emission evaluation and the visually and ultrasonically detected discontinuities is shown in Table 11. The table shows the test pressure at which the first evaluation criteria for a particular sensor was exceeded and the discontinuities within the zone of that sensor. Table 12 shows discontinuity data for acoustic emission sensors which do not exceed the acoustic emission evaluation criteria.

The data presented in Tables 11 and 12 provide an indication of the sensitivity of acoustic emission to different types of discontinuities. Acoustic emission is less sensitive to discontinuities which do not cause local yielding (for example, local reduction in shell thickness and weld porosity). Acoustic emission tests are more sensitive to crack-like discontinuities such as incomplete fusion and lack of penetration. It is significant that six areas of incomplete fusion are all detected by the time the pressure in the vessel reaches 138 kPa (20 psi). Acoustic emission can detect discontinuities such as rust scale, lack of penetration and undercut welds. However, the technique is less sensitive to this type of discontinuity than to the structurally severe crack-like discontinuities.

Of the first six sensors to become active, five have incomplete fusion in the primary zone. The sixth senor (number-1) is a special case that illustrates a phenomenon often observed when testing vessels. A 50 mm (2 in.) gouge with a depth about 33 percent of the vessel thickness was present 180 degrees around the vessel from the sensor. Frequently, a sensor directly opposite a discontinuity will detect that discontinuity. This is because of the waterborne signal traveling across the vessel.

Tests of Butane Storage Spheres

Acoustic emission testing has been found useful as a means for identifying the presence of discontinuities in vessels that could fail from brittle fracture during cold atmospheric operating periods. Critical periods can occur when

TABLE 10. Comparison of acoustic emission and ultrasonic test results following vessel start-up

Number of discontinuities found by ultrasonics:	55
Number of acoustic emission sensors indicating the presence of discontinuities:	34
Number of inactive sensors and sensors with high background noise:	14
Number of ultrasonically detected discontinuities in zones of acoustic emission sensors which indicated discontinuities:	49
Number of ultrasonically detected discontinuities in zones of dead or high background noise acoustic emission sensors:	3
Number of ultrasonically detected discontinuities in zones of acoustic emission sensors which do not indicate discontinuity:	3
Number of acoustic emission sensors indicating discontinuities with no ultrasonically detected code-rejectable discontinuity in zone:	5

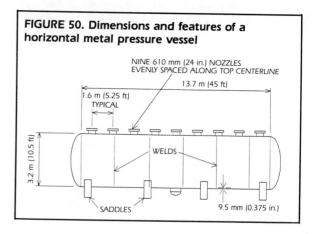

FIGURE 50. Dimensions and features of a horizontal metal pressure vessel

TABLE 11. Acoustic emission discontinuity correlation for sensors exceeding evaluation criteria in test of horizontal tank

Sensor with Activity Exceeding Evaluation Criteria	Pressure at Which First Evaluation Criterion Exceeded	Discontinuities Located with Ultrasonic and Visual Tests in Primary Zone of Acoustic Emission Sensor
2	69 kPa (10 psi)	Incomplete fusion Rust scale Gouge 50 mm (2 in.) deep or about 33 percent of thickness
13	117 kPa (17 psi)	Incomplete fusion
14	117 kPa (17 psi)	Incomplete fusion Pitting Rust scale Slag inclusion
16	117 kPa (17 psi)	Incomplete fusion
1	138 kPa (20 psi)	Gouge 180 degrees from 50 mm (2 in.) gouge at sensor-2
7	138 kPa (20 psi)	Incomplete fusion Rust scale
11	138 kPa (20 psi)	Lack of penetration
3	172 kPa (25 psi)	Undercut weld Reduction in thickness of 20 percent
4	172 kPa (25 psi)	Rust scale
8	172 kPa (25 psi)	Rust scale

filling vessels with warm, light hydrocarbon products during such cold periods. In order to more definitively indicate the presence of discontinuities under such low temperature, high pressure conditions, an acoustic emission hydrostatic test was applied to a series of eight storage spheres of the type described below.[21]

1. *Year built*: 1953
2. *Material*: A201B (A515, GR60)
3. *Design basis*: API/ASME
4. *Joint efficiency (design)*: 80 percent
5. *Diameter*: 15.5 m (51 ft)
6. *Capacity*: 2.4 × 10^6 L (5.35 × 10^5 gal or 1.24 × 10^{10} barrels)
7. *Thickness (design)*: 30 mm (1.18 in.)
8. *Operating pressure*: 275 to 385 kPa (40 to 56 psig)
9. *Operating temperature*: 13 to 38 °C (55 to 100 °F)
10. *Service*: butane
11. *Hydrostatic test pressure*: 724 kPa (105 psi)

TABLE 12. Acoustic emission sensors not exceeding evaluation criteria for test of horizontal tank

Sensor	Discontinuities Located with Ultrasonic and Visual Tests in Primary Zone of Acoustic Emission Sensor
0	None
5	Undercut weld
6	Porosity of 3.2 mm (0.125 in.)
9	Rust scale
10	Reduction in thickness of 15 percent
12	None
15	Pitting

Acoustic Emission Instrumentation for Testing Butane Spheres

A block diagram of the acoustic emission system is shown in Fig. 51. For these tests, twelve sensors and preamplifiers were positioned on the test vessel. Detected emission was preamplified by factors of 100 to 2,000 and sent by coaxial cables to a signal processing unit located 90 m (300 ft) away. In the conditioning unit, signals are filtered to reduce the masking effects of noise from bolt creep, flange movement or vibrational shock from nearby machinery. After filtering, the signals are further amplified up to 100 times and a single pulse is formed indicating time of arrival of the emission signal.

The pulses generated by three sensors are selected for input to the time analysis logic which in turn generates the time of arrival difference Δt used in discontinuity triangulation. The time analysis section also performs certain validity checks so that invalid data may be discarded. Valid Δt values are stored in memory and sorted by location. If a location assumes statistical significance, it is indicated on the digital printer.

An oscilloscope is used to display an isometric plot of the stored data so that an operator may make a real-time evaluation of accumulated emission sites. An additional feature permits display of selected Δt locations on the same oscilloscope screen in numeric form.

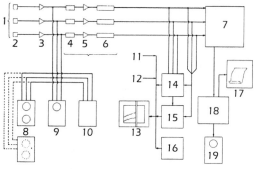

FIGURE 51. Block diagram of acoustic emission system used to test butane spheres; basic ten-channel system is shown, more channels can be added without redesign using the same memory

LEGEND
1. TEN CHANNELS
2. SENSORS
3. PREAMPLIFIER (100 TO 2,000) GAIN
4. SIGNAL CONDITIONING FILTER
5. SIGNAL CONDITIONING MAIN AMPLIFIER (1 TO 100 GAIN)
6. SIGNAL CONDITIONING NOISE REJECT AND PULSE FORMER
7. TIME ANALYSIS (TWO TO FOUR SENSORS)
8. FOURTEEN CHANNEL TAPE RECORDER
9. OSCILLOSCOPE
10. AUDIO
11. PRESSURE/FORCE
12. TIME BASE
13. XYY' RECORDER
14. PULSE ENERGY (RATE/TIME OR RATE/PRESSURE)
15. TOTALIZER
16. VOLUMETRIC FLOW
17. DIGITAL PRINTER
18. CORE MEMORY STORAGE (MAX 1,024 × 1,024 GRID)
19. OSCILLOSCOPE (REAL-TIME ARRAY DISPLAY)

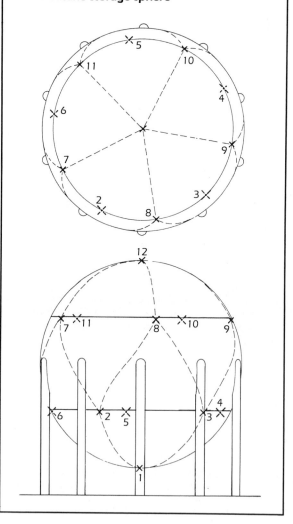

FIGURE 52. Sensor locations for acoustic emission test of butane storage sphere

The pulses from the signal conditioners are also used to derive total emission rate and energy emission rate. Emission rate may be given in emission per unit time or emission per pressure when a 0 to 1 V signal proportional to the parameter is available. This form of rate measurement evolved as variations in vessel pressurization rate were found to cause annoying distortions in emission rate data per unit time. The method extracts emission rate per unit stress which is the most meaningful data.

The emission rate, emission total or pressurizing fluid volumetric flow may be plotted versus time or pressure on an XYY' recorder and trends may be seen as they develop.

The raw data from the preamplifiers are stored on a fourteen-track magnetic tape recorder. These data may be replayed after the test for triangulation on a different set of sensors. The raw signals are also displayed on an oscilloscope and converted to the audible range through a loudspeaker. These two outputs are mainly signal quality checks and can pinpoint problems such as the development of leaks and sensor, preamplifier or cable grounding problems should they develop.

The preamplified signals are received by coaxial cables at the signal processing unit and put through an active filter to further reduce low frequency noise signals and also to eliminate interference from nearby radio transmitters.

The final step in the conditioning of the signal is determination of arrival time. The time analysis and storage section of the system (1) determines time differences between several sensors; (2) stores valid Δt pairs in memory; (3) prints pairs which are developing significance; and (4) creates a real-time oscilloscope display of all Δt pairs stored in memory. With this last information, an operator can qualitatively track results of the test as they accumulate. The display function generates a three-dimensional isometric visual view of a plane corresponding to 1,024 × 1,024 possible points that the Δt pair can occupy. Actual counts in a location appear to be elevated above the base plane.

Due to the statistical nature of signal detection, a discontinuity location will not correspond to a given point but rather to a narrow peak that is statistically centered at the Δt coordinates. Real-time triangulation may be performed while the data accumulate.

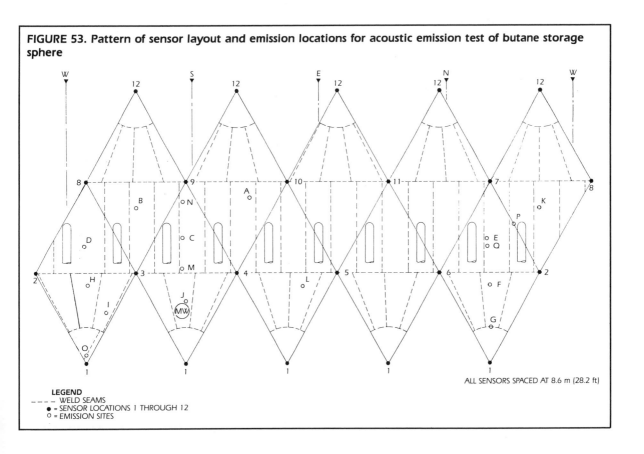

FIGURE 53. Pattern of sensor layout and emission locations for acoustic emission test of butane storage sphere

Setup for Acoustic Emission Tests of Butane Spheres

Preparation for acoustic emission testing consisted of locating and cleaning to bare metal twelve sensor stations, each about 50 mm (2 in.) in diameter, on the outside surface of the sphere. These stations were positioned to coincide with the nodes of an icosahedron inscribed on the 15.5 m (51 ft) diameter sphere. This layout (see Figs. 52 and 53) resulted in the equidistant and minimum adjacent spacing of 8.6 m (28.23 ft) between the twelve sensors used for this test.

After attaching the magnetically held sensors to their designated locations, individual coaxial cables were strung from the sensor preamplifiers to the acoustic emission mainframe components located 60 m (200 ft) away from the sphere. An electrical pulser, producing a signal similar to an acoustic emission, was placed adjacent to each sphere sensor station to verify the operability of the system.

Analysis of Acoustic Emission Response

Real-time analysis of acoustic emission produced during hydrostatic pressurization consisted of visually scanning the strip chart displays of emission rate, total count (Fig. 54) and Δt printouts. Evaluation of previous experimental results from acoustic emission testing of discontinuous specimens has shown that a pronounced and sustained increase in both

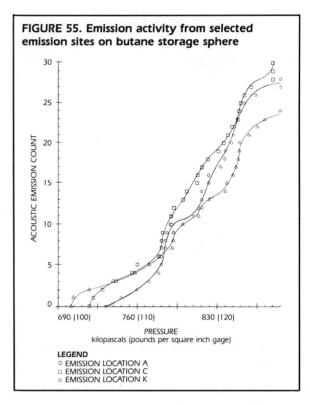

FIGURE 55. Emission activity from selected emission sites on butane storage sphere

LEGEND
○ EMISSION LOCATION A
□ EMISSION LOCATION C
△ EMISSION LOCATION K

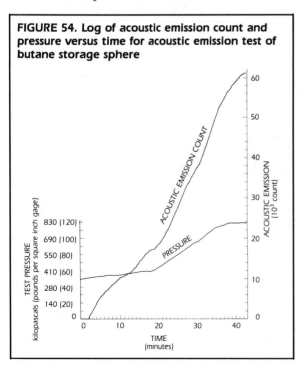

FIGURE 54. Log of acoustic emission count and pressure versus time for acoustic emission test of butane storage sphere

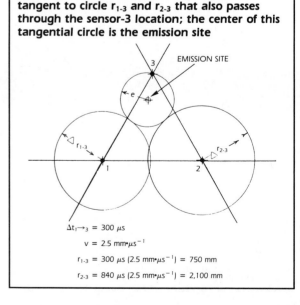

FIGURE 56. Diagram of triangulation using the Apollonian construction; e is the radius of a circle tangent to circle r_{1-3} and r_{2-3} that also passes through the sensor-3 location; the center of this tangential circle is the emission site

$\Delta t_1 \rightarrow_3 = 300 \ \mu s$

$v = 2.5 \ mm \cdot \mu s^{-1}$

$r_{1-3} = 300 \ \mu s \ (2.5 \ mm \cdot \mu s^{-1}) = 750 \ mm$

$r_{2-3} = 840 \ \mu s \ (2.5 \ mm \cdot \mu s^{-1}) = 2{,}100 \ mm$

emission rate and total count is observed from a growing discontinuity.

Emission rate and summation plots monitored during pressurization displayed a general but slow variation with time. Emission activity from all relevant emission sites showed only a slowly rising trend with pressure. The three plots in Fig. 55 (from emission locations A, C and K in Fig. 53) are typical and indicate a slow growth in discontinuities during vessel pressurization.

Emission Site Triangulation

The method used for locating emission sites employs the Apollonian construction discovered by Apollonius of Perga, a third century BC Greek mathematician. This construction locates the center of a circle that passes through a fixed point and is tangent to two given circles. Figure 56 (drawn to a microsecond per foot scale) illustrates this construction. The sign logic for Δt is as follows: the sign is positive (+) if acoustic emission is detected by sensor-1 *before* sensor-2; the sign of Δt is negative if emission is detected by sensor-1 *after* sensor-2.

In the illustration, sensor-1, 2 and 3 are spaced 3,050 mm (120 in.) apart. The measured velocity of sound during the test was 2.5 mm (0.1 in.) per microsecond. The measured transit time from sensor-1 to sensor-2 was 1,200 μs.

Therefore, the acoustic emission time differential from sensor-1 to sensor-2 ($\Delta t_{1\rightarrow 2}$) is +300 μs or a distance of 750 mm.

The $\Delta t_{1\rightarrow 3}$ value is +540 μs or a 1,350 mm distance. The $\Delta t_{2\rightarrow 3}$ can be calculated as $\Delta t_{1\rightarrow 2} + \Delta t_{1\rightarrow 3}$ or 300 μs + 540 μs = 840 μs or an equivalent of 2,100 mm. Locating the emission site using the construction in Fig. 56 requires (1) drawing a circle centered at sensor-1 with a radius of 540 μs; and (2) drawing a circle centered at sensor-2 with a radius of 840 μs. The emission site is found by locating the center of a circle tangent to the two previously drawn circles.

Verification of Test Results

Following the acoustic emission test, each of the seventeen relevant emission sites was located and its position marked on the sphere. Visual inspection indicated that several of the sources were probably weld undercuts and insufficient weld penetration. At other locations, sources were confirmed by ultrasonics examination as small weld cracks or plate laminations. At all of the seventeen sites, some type of weld or plate discontinuity was identified. All emission sites fell within a 76 mm (3 in.) diameter circle containing the probable discontinuity.

PART 6
ACOUSTIC EMISSION TESTING FOR LEAK DETECTION

Metal Pipeline Hydrostatic Testing

During hydrostatic testing of a newly constructed 114 mm (4.5 in.) diameter buried pipeline, a leak was detected but not located. Acoustic emission techniques were used to help locate the leak after conventional leak detection methods proved unsuccessful.

Eight bell holes were dug to access the pipe surface and to attach acoustic emission transducers. The holes were spaced at about 100 m (325 ft) intervals along the test section (see Fig. 57) and about 150 m (500 ft) of pipe was exposed. The entire test section of 1,080 m (3,500 ft) was scheduled to be excavated as the final alternative for locating the leak.

Hydrostatic Test-1

The acoustic emission level was measured at each consecutive bell hole to establish a reference for comparison of emission levels measured at higher line pressures. Test-1 line pressure was 4,100 kPa (595 psi).

A 30 kHz transducer was coupled with grease to the top of the pipe to measure the acoustic emission level. Figure 58 shows the acoustic emission system with the transducer coupled to the pipe. Emission level was measured for each of three filters (broadband, 30 kHz and 100 kHz) in the acoustic emission system. Data for the hydrostatic test are shown in Table 13.

A high level of emission was measured using the 100 kHz filter at hole-1 and relatively higher levels on all three filters at hole-6. Emission from water trickling into hole-8 was also measured.

Hydrostatic Test-2

Line pressure was increased to 11,200 kPa (1,625 psi) to increase the emission level from the leak. A higher level of emission was measured at hole-1 on the 100 kHz filter and the broadband filter. A trend showing an increase in emission level from hole-3 to a maximum at hole-6 was observed. The level decreased at hole-7 and hole-8. Data from hydrostatic test-2 are shown in Table 14.

FIGURE 57. Schematic diagram of test section for underground pipeline

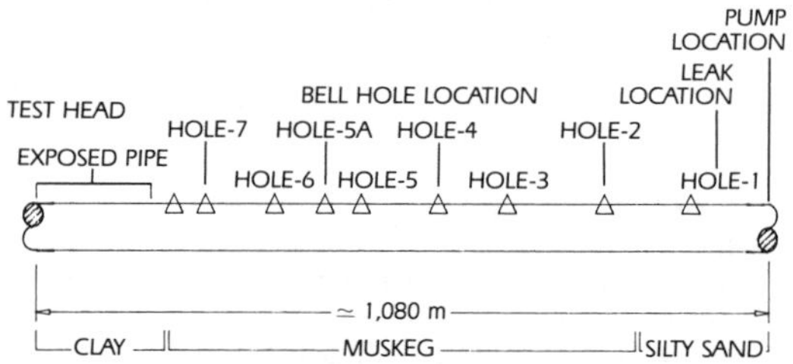

FIGURE 58. Acoustic emission leak detection system for pipelines (transducer coupled to pipe)

TABLE 13. Results of test-1: low pressure hydrostatic test of metal pipeline

Access Hole	Distance (meters)	Signal Amplitude (millivolts)			Instrument Gain	Comments
		Broadband	30 kHz	100 kHz		
Test head	0	—	—	—	—	—
1	120	77	70	210	1,000	Time: 8:50
2	250	63	60	60	1,000	
3	390	70	70	65	1,000	
4	490	70	70	66	1,000	Pressure: 4,100 kPa
5	600	68	70	65	1,000	
6	720	80	100	80	1,000	Water in hole
7	830	70	67	67	1,000	
8	880	76	76	70	1,000	Water in hole
Exposed	930 to 1,080	62	64	66	1,000	

TABLE 14. Results of test-2: high pressure hydrostatic test of metal pipeline

Access Hole	Signal Amplitude (millivolts)			Instrument Gain	Comments
	Broadband	30 kHz	100 kHz		
Test head	—	—	—	—	Pressure: 11,200 kPa
1	98	70	440	1,000	
2	64	64	68	1,000	
3	64	64	66	1,000	
4	62	68	67	1,000	Unsteady signals (30 kHz filter)
5	62	67	67	1,000	
6	70	80	70	1,000	Consistent ticking signal (like clock)
7	63	65	65	1,000	
8	60	60	65	1,000	
Exposed	58	58	64	1,000	Very steady signal

Hydrostatic Test-3

Hole-5A was dug between hole-5 and hole-6 to investigate a possible leak near hole-6. The results in Table 15 show a decrease in emission level in hole-5A.

Hydrostatic Test-4

Bell hole-1 was extended approximately 15 m (5 ft) toward hole-2 to investigate the higher emission level measured at hole-1 during test-2.

Acoustic emission measurements showed a rapid decrease in emission level toward hole-2 (Table 16). The leak was located 7.3 m (24 ft) in the direction away from hole-2 (toward the pump) in a circumferential weld (see Fig. 59). The attenuation curve in Fig. 60 shows a decrease in emission level from the maximum of 1,250 (full scale) to a background emission level of about 70 within 22 m (72 ft) of the leak.

FIGURE 59. Leaks detected by acoustic emission at circumferential weld; note spray from left side of weld

Interpretation of Hydrostatic Test Results

It was expected that the combined 30 kHz filter with the 30 kHz tuned transducer would be the most sensitive to leak emission. Measurements on the 30 kHz filter were therefore used to indicate a possible leak. Measurements on the broadband and 100 kHz filters were regarded as data to assist with interpretation. However, 100 kHz data eventually led to the leak detection.

High emission levels on the 100 kHz filter at hole-1 were discounted during test-1 because of the possibility that the emission was caused by equipment operating in the vicinity. However, test-2 results at hole-1 showed an emission level on the 100 kHz filter approximately double the level of test-1. This was strong evidence of a leak and in retrospect should have been investigated immediately after the higher emission level was measured. Based on test-1 and test-2 acoustic emission measurements, the leak could have been located within two hours. It was eventually located in only about five hours.

The rapid attenuation in emission level observed during test-4 was indicative of high frequency emission. Usually, the higher the frequency, the more rapid the attenuation. Increased sensitivity to high frequency emission could have increased the distance this emission was detected from the leak. The attenuation curve obtained from field data (Fig. 60) shows that emission levels above the background level would only have been measurable on the pipe for about 22 m (72 ft) on either side of the leak. This indicates the bell hole spacing of 100 m (328 ft) was too long.

To increase the potential for locating small leaks, transducers tuned to higher (greater than 30 kHz) frequencies are required.

Test-2 results showed a trend of increasing emission from hole-3 and hole-8 to hole-6. This evidence of a leak near hole-6 and neighboring holes was determined to be caused by the false indications. It was observed that as the water flow decreased (due to freezing), the emission level also decreased. When leak testing, high level emission (above background) should always be analyzed for possible environmental causes.

High level emission was expected at the pump location due to the exposed pipe. However, measured values during test-4 (Table 14) were low. It was observed during application of acoustic emission to the support leg that pipe was exposed to the environment. Low background level emission

TABLE 15. Results of test-3: hydrostatic test of metal pipeline

Access Hole	Signal Amplitude (millivolts)			Instrument Gain	Comments
	Broadband	30 kHz	100 kHz		
6	73	72	69	1,000	Consistent ticking signal (like clock)
5	63	67	67	1,000	
4	62	65	66	1,000	
5A*	65	64	68	1,000	
6	60	65	68	1,000	Water frozen, no more ticking

*HOLE DUG MIDWAY BETWEEN ACCESS HOLES 5 AND 6.

TABLE 16. Results of test-4: hydrostatic test of metal pipeline

Access Hole	Distance (meters)	Signal Amplitude (millivolts)			Instrument Gain	Comments
		Broadband	30 kHz	100 kHz		
1	120	108	70	450	1,000	
Test head	0	61	63	68	1,000	
2	250	62	62	69	1,000	
1	120	105	70	500	1,000	
1	120			550	1,000	
1	121.1			300	1,000	
1	126.4			180	1,000	
1	131.4			110	1,000	
1	117.7	500	490	950	1,000	
1	112.7	1,250	1,250	1,250	1,000	Leak Location Time: 13:45 Time: 14:15 Pressure: 9,800 kPa

was possibly due to the shelter provided by the trees on both sides of the right-of-way.

To reduce the possibility of high background emission, bell holes should be prepared according to the sketch in Fig. 61. The sloped sides of the hole should prevent soil slumping onto the exposed pipe. However, such a hole would have prevented easy access to the pipe because of ground water accumulation in the hole.

Ground water leakage required the holes to be deeper to allow water accumulation below the pipe (Fig. 62). This worked well for access to the pipe, but also resulted in the large exposed surface area, which led to false indications. It was however the most reasonable alternative to accessing the pipe in the muskeg and the method should be considered a reasonable alternate for acoustic emission leak detection in wet or muskeg areas.

Small leaks can be located using an acoustic emission leak detection system, but rapid, accurate interpretation of acoustic emission data is dependent on operator experience. Small leaks can produce emission frequencies higher than 30 kHz. Pipe buried in muskeg can result in higher levels of background noise than pipe buried in clay or silty sand.

Plastic Pipeline Pneumatic Testing

This test was conducted using a portable acoustic emission leak detector. Initially, an attenuation test was performed on 6.7 m (22 ft) of exposed pipe nearest the test fittings, valves and pressure gage. A total of five standard pipeline transducers were epoxy bonded on the line at distances shown in Fig. 63.

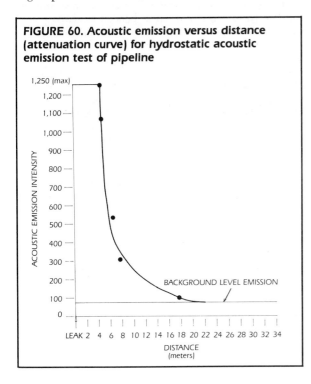

FIGURE 60. Acoustic emission versus distance (attenuation curve) for hydrostatic acoustic emission test of pipeline

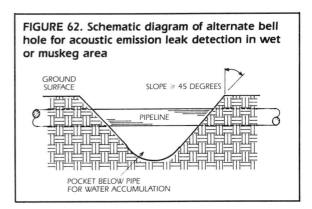

FIGURE 62. Schematic diagram of alternate bell hole for acoustic emission leak detection in wet or muskeg area

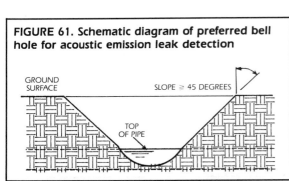

FIGURE 61. Schematic diagram of preferred bell hole for acoustic emission leak detection

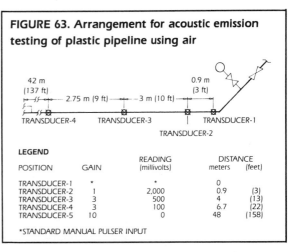

FIGURE 63. Arrangement for acoustic emission testing of plastic pipeline using air

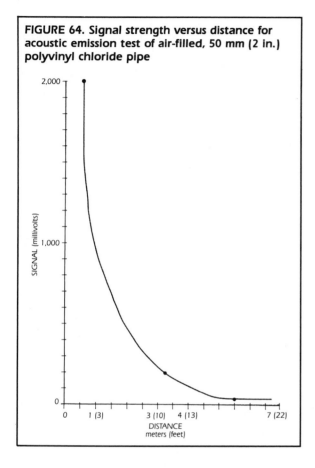

FIGURE 64. Signal strength versus distance for acoustic emission test of air-filled, 50 mm (2 in.) polyvinyl chloride pipe

Signals from a manual pulser were input at transducer-1 and readings were taken at each successive transducer on the line. Because of attenuation in the line, no signal was detected at transducer-5 at a distance of 48.5 m (159 ft) from the signal source. The resulting attenuation is shown in Fig. 64. Figure 65 shows a straight line approximation graph of signal ratio versus distance.

It was determined that the leakage rate during the hydrostatic test with water was not sufficient to generate acoustic noise detectable at the existing attenuation rates. Subsequent pressurization with air greatly increased the acoustic noise transmitted through the line.

After establishing attenuation curves for pressurization with air, transducers were epoxy bonded at access points along the line as indicated in Fig. 66. The maximum possible pressurization in the presence of the leak was about 350 kPa (51 psig).

Readings were taken and compared to background levels. Slightly higher readings were indicated at transducer-2 and transducer-3 as shown in Fig. 66. Comparing these readings on the signal ratio curves indicated that potential leakage was present 1.5 m (5 ft) from the centerline of the two transducers, at a position of 19.4 ± 1 m (63.5 ± 3 ft) from transducer-2.

Offshore Crude Oil Pipeline Testing

A section of 300 mm (12 in.) diameter pipe terminating on an offshore drilling platform was inspected for confirmation of a detected leak. The section of the line coming from the ocean floor to the drilling platform was monitored at an

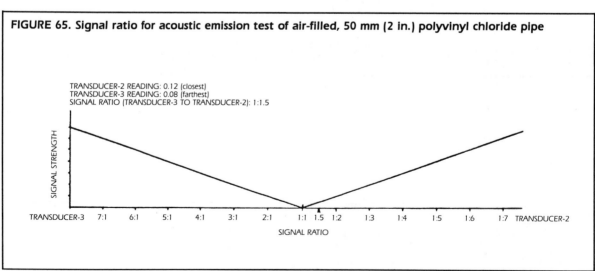

FIGURE 65. Signal ratio for acoustic emission test of air-filled, 50 mm (2 in.) polyvinyl chloride pipe

access point about 6 m (20 ft) above water level. Monitoring confirmed the presence of suspected leakage at a point below water level near the structure.

The structure under test was newly built and comprised 16 km (10 mi) of pipeline running from a shore installation to an offshore drilling platform. The pipeline was a 300 mm (12 in.) diameter steel pipe assembled on shore and towed into place in a continuous operation. The end of the pipe at sea was terminated in a welded flange located next to one of the support legs of the drilling platform.

Connections between the pipe and the riser on the drilling platform were made by fabricating a spool piece and then bolting it in place underwater. The lower end of the connecting spool piece was at a depth of about 64 m (210 ft). The riser section of pipe was fastened to support legs of the drilling platform by pipe clamps welded in place. The riser section terminated on the lower deck of the drilling platform.

During acceptance hydrotesting of the line it was noted that pressure decayed at about 0.4 $MPa \cdot s^{-1}$ (60 $psi \cdot h^{-1}$) starting at about 22 MPa (3,200 psig). The suspected source of leakage was at the spool piece flanges. After discovery of the leak in the pipeline, acoustic emission was used to determine if the leak was located in the spool piece.

Test Procedure for Pipeline Tests with Acoustic Emission

Signal level readings were taken on the 300 mm (12 in.) riser after pressure on the pipe was elevated to 22 MPa (3,200 psig). These signal readings were compared with readings taken on two adjacent pipes and on the nearest support leg for the structure (see Table 17). The additional readings were used to determine the amount of signal caused by sea motion and other interfering structural noise. The initial readings were taken with the platform in a shutdown condition and all construction workers on shore. The readings showed about a 50 percent increase in signal level on the leaking pipe as compared to the other two risers and the support leg. This indicated leakage close to the detection point and in effect verified that leakage was in the connecting spool piece flanges.

The confirmation prompted construction personnel to contact a diving company to tighten the flanges and to abort plans to drain the line and use other means of locating the leak. The following day, the diving company arrived on site and sent a diver to tighten the flanges. After completing the diving operation, pressure was once again applied to the pipe and acoustic emission data were taken to determine if the leak had been stopped. By the time the diver had returned to the surface, the acoustic emission inspector determined that the leak had been stopped. No further indications of leakage were detected either by mechanical means (pressure drop) or by acoustic emission.

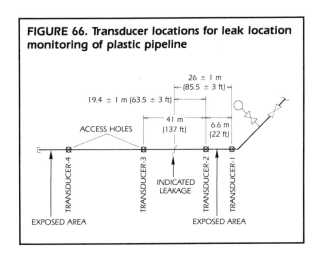

FIGURE 66. Transducer locations for leak location monitoring of plastic pipeline

TABLE 17. Data taken from leaking 305 mm (12 in.) pipeline and reference points on the platform during leak and after tightening spool piece flanges

Location	First-Day Reading	Comment
152 mm (6 in.) pipe riser	0.20 at 1,000 G	Reference
250 mm (10 in.) pipe riser	0.21 at 1,000 G	Reference
305 mm (12 in.) pipe riser	0.30 at 1,000 G	Leaking pipe
Corner support leg	0.21 at 1,000 G	Reference

Location	Second-Day Reading	Comment
152 mm (6 in.) pipe riser	0.20 at 1,000 G	Reference
250 mm (10 in.) pipe riser	0.20 at 1,000 G	Reference
305 mm (12 in.) pipe riser	0.20 at 1,000 G	Leak noise stopped
Corner support leg	0.21 at 1,000 G	Reference

Summation of Acoustic Emission Testing of Underwater Pipelines

The acoustic emission technique saved the pipeline construction company time and money by: (1) verifying the presence of a leak; (2) locating the leak in an underwater flange; (3) limiting contracted diver time for repair; and (4) verifying that the leak had in fact been stopped and thereby accelerating the test schedule.

Despite reservations, the acoustic emission method was practical for this application. The leak signals were detectable in the presence of noise caused by sea motion and at a distance of about 73 m (240 ft) from the leak. In addition, noises from normal platform activity and sea motion did not prevent or hinder the successful use of the technique.

Tests of Steam Condensate Lines

A steam condensate line network (designated AX, AY and AZ) was tested using acoustic emission inspection techniques and a portable acoustic emission leak detection system. The instrument was designed to detect high frequency acoustic noise created by fluids escaping from containment systems. These signals are amplified and displayed as a relative value on a digital readout device. The signal values are recorded and correlated to a leak location on the structure.

Before arrival of the test crew, access holes were dug to the steam condensate line at various locations. The positioning of the access holes was determined by lengths of line and by the installation geometry. In general, distances between access points were less than 107 m (350 ft).

An air compressor was attached at a valve station to supply 345 kPa (50 psig) pressure and create turbulent flow through leaks in the line during the test.

Signal amplitude readings were taken at each access point along the steam condensate line after the test pressure was attained. The access points are shown in Fig. 67 and the resulting leak data are presented in Table 18. The highest readings attained were at position-13 and position-14 where leaking valves were identified. The signals generated at these valves caused the level of position-12 to be high. The next highest readings were at position-4 through position-6, with the highest level detected at position-6. During the testing process, air was observed bubbling out of the ground between position-4 and position-2. However, the maximum leak-generated signals were between position-5 and position-6.

Interpretation of the data lead to the conclusion that there were numerous leaks in the line between access position-4, position-5 and position-6. A major leak was located between transducer-5 and transducer-6. The section of the steam condensate line between position-11 and position-6 appears to be free of detectable leaks.

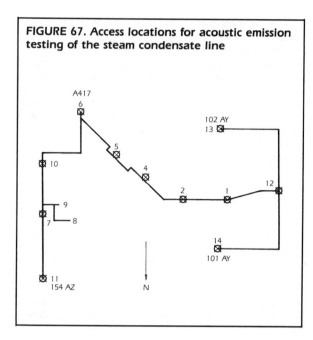

FIGURE 67. Access locations for acoustic emission testing of the steam condensate line

TABLE 18. Leak data for acoustic emission test of steam condensate line

	Access Point	Relative Leak Amplitude (megavolts at 1,000 G)
Confirmed leak	1	495
	2	750
	3 nonexistent	
	4	1,110
Leak area	4	1,110
	5	1,080
	6	1,650
	7 (compressor noise)	
	8	75
	9	110
	10	70
	11 (compressor noise)	
Leaking valves	12	1,350
	13	1,050
	14	900

Acoustic Emission Tests of Flat Bottom Tanks

Storage tanks for inflammable hydrocarbon liquids are inspected periodically in accordance with standard design specifications. These periodic inspections provide an overview of the mechanical condition of the storage tank and its tightness against leaks.

The conventional testing procedure is time consuming and expensive, often followed by difficulties in production planning for the refinery or tank farm. Depending on the type of tank and how much sludge and sediment are on the bottom, the storage tank can be out of service for two to sixteen weeks.

The determination of leakage by still gaging is limited because of the resolution of the readout devices. Leakage can be identified if a tank with diameter of about 57 m (190 ft) loses more than 2.5 $m^3 \cdot h^{-1}$ (89 $ft^3 \cdot h^{-1}$). For comparison, if a circular hole with an inside diameter of 1 mm is exposed to a differential pressure of 100 kPa, then a loss of 2.4 $L \cdot h^{-1}$ takes place.

Acoustic Emission Test Procedure for Flat Bottom Tanks

The detection of leakage by acoustic emission focuses on the noise generated by liquid or gas flow through a leak. These noises are caused not only by fluid leaks through material cracks or holes but also by liquid seepage through porous weldments and concrete walls.

Expanding gases generate a substantial increase in background noise called *continuous emission*. For liquids, depending on the type of leak and the differential pressure, an increase in background noise is mixed with small discrete pulses up to the the principal creation of burst signals.

These signals are detected by piezoelectric transducers. The main problem during the signal transfer is that the flat bottoms of storage tanks usually consist of overlapped metal sheets. The elastic waves are extremely attenuated or distorted at the joints. In open and buried pipelines, it is possible to attach the transducers on the metallic surface at intervals of up to 250 m (820 ft) and to detect the acoustic emission signals transmitted by the fluid media.

Sensors are chosen with a center frequency of 30 kHz for leak detection with low differential pressures. They are attached to the metallic wall about 250 mm (10 in.) above ground level. This is minimally higher than the sludge level inside the tanks. Typically, six sensors are spaced at equal distances around the periphery of the tank bottom.

Test results show that (1) sound velocities between 750 and 1,650 $m \cdot s^{-1}$ (2,500 and 5,400 $ft \cdot s^{-1}$) are obtained and (2) different hydrocarbon products have unique, highly stable velocities. For example, kerosene permits a sound velocity of 750 $m \cdot s^{-1}$ and, in a dispersion of hydrocarbon sludge with water, sound has a velocity of 1,650 $m \cdot s^{-1}$. In other hydrocarbon products, the sound velocity can be as low as 110 $m \cdot s^{-1}$.

During the zero test, all noises must be eliminated, especially those caused by heat exchangers, pumps and stirring rods. Leaks in neighboring valves must be avoided or equipment should be stopped. The storage tank can then be filled with its usual contents.

Flat Bottom Tank Test Results

A leak will raise the background noise and it is possible to detect and roughly locate the source by measuring the effective amplitude at different transducers.

If there are discrete pulse-like bursts, then the short rise time of the physical event is prolonged during sound travel in the liquid. Location accuracy is about five percent of the maximum transducer distance. By using only outside transducers, the accuracy is relative to the diameter of the storage tank. Using one or more inside transducers allows not only an increase in the accuracy of the location but makes it easier to determine if there is continuous acoustic emission.

The acoustic emission technique is somewhat limited in this application by unavoidable, disruptive noises, including rain, wind, snow and ice. An estimation of corrosion is not possible with this technique.

Advantages and Limitations of Acoustic Emission Testing for Flat Bottom Tanks

In 1982, an Austrian acoustic emission testing group inspected a refinery storage tank with three artificial and one natural discontinuity.[22] The test group had no data on the number or location of discontinuities. By finding the discontinuities, acoustic emission proved to be an efficient tool for detecting and locating leaks during hydrostatic tests of atmospheric storage tanks.

Since 1982, acoustic emission users have tested many storage tanks with various diameters up to 90 m (300 ft). The sensitivity of the technique is evident in one sample case where an area with high acoustic emission was located on a flat bottom tank. After opening the tank and visually inspecting the surface, pervasive discontinuities were discovered in a weld repair.

As with any nondestructive testing technique, interpretation of acoustic emission data is critical to the accuracy of the test. Another sample case illustrated this continuous noise during the tank test was at first thought to indicate a predominant leak. It was later determined that the actual acoustic source was an unwelded support leg on the heater coil structure (all the other support legs were properly welded to the bottom plates). This incident verified the need for a careful study of complete design documentation before testing.

One evaluation of the acoustic emission method's lowest

detectable leakage is 1.7×10^{-3} L·h^{-1} under optimal conditions. In relation to the hydrostatic test of a storage tank with a 57 m (185 ft) diameter, this is an improvement of $1:1.5 \times 10^6$. For tanks with a 5 m (16 ft) diameter, the detectability of acoustic emission is an improvement of 1:10.[4]

Acoustic emission has located leaks with diameters from 1 mm (0.04 in.), maximum differential pressure of 50 kPa (7 psi) and rates of 1.7 L·h^{-1} at the experimental tank. This is an improvement for storage tanks with 57 m (185 ft) and 5 m (16 ft) diameter of $1:1.5 \times 10^3$ and 1:10, respectively.

It is remarkable that a drill hole with a 1 mm (0.04 in.) diameter is *above* the lowest detectable size. A visual inspection cannot easily or consistently locate a discontinuity of this size when tank floor areas range from 20 m^2 (215 ft^2) to 2,500 m^2 (27,000 ft^2).

PART 7
ACOUSTIC EMISSION TESTS OF FIBER REINFORCED PLASTIC VESSELS

Testing Procedures for Pressure, Storage and Vacuum Vessels

Three organizations have accepted a procedure (with minor variations) for the examination of fiber reinforced plastic vessels. The originator of the procedure was the Society of the Plastics Industry (SPI), which developed procedures for glass reinforced vessels. Both the American Society for Testing Materials (ASTM) and the American Society of Mechanical Engineers (ASME) adopted similar examination procedures. In addition to fiberglass, the ASME procedure applies to vessels reinforced with other fibers.

The ASME Boiler and Pressure Vessel Code typically has state-passed legislative standing instead of serving as voluntarily applied consensus standards. Because of this, the easiest procedure to reference is probably ASME's Article 11, Section V which describes the requirements for applying acoustic emission examination of new and in-service fiber reinforced plastic (FRP) vessels under pressure, vacuum or other applied stress (see Table 19).

Figures 68 through 72 exemplify the stressing sequence and test algorithm flow chart. Table 20 lists the evaluation criteria.

Applications in the Chemical Industries

Tests of Composite Vessels

The use of acoustic emission for testing fiber reinforced plastic vessels has been extremely successful and the technique has become a standard test for fiber reinforced plastic equipment and piping.

In the chemical process industry, the performance of FRP pipe and vessels has been poor and many failures have been reported.[23] Apart from acoustic emission, there is no satisfactory test for determining the structural adequacy of fiber reinforced plastic equipment. Accordingly, the technique has filled an important need so that development and application of the technology has been rapid.

Cooperative research and development programs organized by the Society of the Plastics Industry have resulted in practical field test methods for tanks and piping.[24,25] Broad support from resin and glass suppliers, fiber reinforced plastic equipment fabricators, acoustic emission instrument manufacturers, fiber reinforced plastic designers and vessel owners was a major factor in the rapid development and ready acceptance of these test procedures. Initially, the research was performed to meet the needs of the chemical process industry. However, the technology has been adopted by other industries including aerospace, automotive, utility (lifts) and sports equipment manufacturers. Acoustic emission testing is accepted as the primary technique for evaluating most composites.

Figure 73 shows catastrophic fiber reinforced plastic tank failures of one major company during a twelve-year period.[26] The rate of more than two failures per year is clearly unacceptable. These major failures were accompanied by many minor failures including leaks, cracks at nozzles, breakage of holddown lugs and attachments, internal surface cracking and blistering. As a result, many tanks were unsuitable for their intended function.

The reasons for this poor performance are varied but they can be grouped into the following general categories: inadequate design; variability of fabrication quality; transportation, handling, storage and installation damage; in-service abuse; and corrosion attack. Compounding these problems, the lack of an adequate test method permitted discontinuities to go undetected until they reached serious proportions.

Some companies test all of their fiber reinforced plastic tanks as part of the initial purchase specification. The tanks are retested after installation and at intervals not exceeding two years. The dramatic reduction in catastrophic failures shown in Fig. 73 corresponds closely with implementation of acoustic emission testing in 1980. Other factors such as improved design methods, better control of fabrication quality and a greater understanding of the nature of composites have also contributed to the reduction in failures.

The failure in year-10 of Fig. 73 occurred in a tank that had been tested with acoustic emission. The test showed the tank to be unsafe for continued operation and a replacement tank was ordered. In the interim, the tank was operated at reduced levels and appropriate precautions were taken to

TABLE 19. Details of ASME's Section V *Nondestructive Examination*, Article 11

T-1110	Scope	90.13
T-1120	General	90.13
T-1121	Vessel Conditioning	90.13
T-1122	Vessel Stressing	90.13
T-1123	Vessel Support	90.14
T-1124	Environmental Conditions	90.14
T-1125	Noise Elimination	90.14
T-1126	Instrumentation Settings	90.14
T-1127	Sensors	90.14
T-1128	Written Procedure Requirements	90.14
T-1130	Equipment and Supplies	90.15
T-1140	Application Requirements	90.15
T-1141	Vessels	90.15
T-1142	Examination Procedure	90.15
T-1160	Calibration	90.16
T-1180	Evaluation	90.16
T-1181	Evaluation Criteria	90.16
T-1182	Emissions During Load Hold E_H	90.16
T-1183	Felicity Ratio Determination	90.16
T-1184	High Amplitude Events Criterion	90.16
T-1185	Total Counts Criterion	90.22
T-1190	Documentation	90.22
T-1191	Report	90.22
T-1192	Record	90.22
Figures		
T-1142(c)(1)(a)	Atmospheric Vessels Stressing Sequence	90.17
T-1142(c)(1)(b)	Vacuum Vessels Stressing Sequence	90.18
T-1142(c)(1)(c)	Test Algorithm — Flow Chart for Atmospheric Vessels	90.19
T-1142(c)(2)(a)	Pressure Vessel Stressing Sequence	90.20
T-1142(c)(2)(b)	Algorithm — Flow Chart for Pressure Vessels	90.21
Tables		
T-1121	Requirements for Reduced Operating Level Immediately Prior to Examination	90.14
T-1181	Evaluation Criteria	90.22
Mandatory Appendix I	Instrumentation Performance Requirements	90.23
I-1110	Acoustic Emission Sensors	90.23
I-1111	High Frequency Sensors	90.23
I-1112	Low Frequency Sensors	90.23
I-1120	Signal Cable	90.23
I-1130	Couplant	90.23
I-1140	Preamplifier	90.23
I-1150	Filters	90.23
I-1160	Power-Signal Cable	90.24
I-1161	Power Supply	90.24
I-1170	Main Amplifier	90.24
I-1180	Main Processor	90.24
I-1181	General	90.24
I-1182	Peak Amplitude Detection	90.24
I-1183	Signal Outputs and Recording	90.25

TABLE 19. continued

Mandatory Appendix II	Instrument Calibration	90.25
II-1110	General	90.25
II-1120	Threshold	90.25
II-1130	Reference Amplitude Threshold	90.25
II-1140	Count Criterion N_C and A_M Value	90.25
II-1150	Measurement of M	90.25
II-1160	Field Performance	90.25
Figure		
I-1183	Sample of Schematic of Acoustic Emission	
Instrumentation for Vessel Examination		90.26
Nonmandatory Appendix		
Appendix A	Sensor Placement Guidelines	90.28

protect against the possible failure. Failure predicted by the acoustic emission tests occurred during the period following the acoustic emission test and before installation of the replacement vessel.

Composite Pipe Testing Applications

In addition to the fiber reinforced plastic vessel tests, several hundred tests have been conducted on fiber reinforced plastic pipe sections. For both tanks and piping, discontinuities identified by acoustic emission have in all the stated cases been confirmed by subsequent visual or ultrasonic inspection. In addition, no structurally significant discontinuities escaped detection by acoustic emission.

The technical background for recommended practices in pipe testing is available in the literature.[27-31] A number of important features of the SPI practices are relevant as well for metal vessel testing.

Effect of Acoustic Emission Tests of Fiber Reinforced Plastic Structures

Figure 74 is a summary of discontinuities found by acoustic emission in one owner's installed fiber reinforced plastic tanks during a six-year period of testing. Not only has the percentage of discontinuities dropped sharply but the nature of each discontinuity is less serious. Because the acoustic emission test detects discontinuities at an early stage, repairs are possible before the discontinuity becomes a hazard or propagates beyond repair.

Information of the type shown in Fig. 74 and research work as reported in references 32 and 33 suggest that acoustic emission has an important role in establishing basic design criteria for fiber reinforced plastic structural components. Acoustic emission testing results in safer, better designed fiber reinforced plastic equipment. This in turn leads to reduced design factors and has produced valuable economies of material.

Zone Location in Fiber Reinforced Plastics

Discontinuity location in fiber reinforced plastics is accomplished with zone location methods. Zone location approximates the position of an event based on individual sensor activity. Figure 75 shows the sensor layout for a typical fiber reinforced plastic vessel, with the zones marked around each sensor. To ensure complete coverage, the zones of adjacent sensors overlap and in these areas acoustic emission will be detected more than once. The approximate location of the discontinuity can be determined by recording which sensors detect the emission. Conventional source location, using time of arrival and triangulation analyses, has been less than satisfactory for fiber reinforced plastics.

Problems with Time of Arrival Source Location

There are three main disadvantages to time of arrival source location. First, there is severe attenuation of stress waves as they travel through fiber reinforced plastic. Because of this, most events will not strike a sufficient number of transducers to permit triangulation. Placing the transducers closer together can reduce the effect of this problem. However, all transducers in a measurement array need to be within 0.6 m (2 ft) of an emission source to ensure location of all significant events. Such an arrangement would be prohibitively expensive for most applications.

A second problem arises because triangulation calculations assume that stress waves travel along the vessel wall at constant velocity. Most fiber reinforced plastic constructions are anisotropic and stress waves will travel much faster along the glass fibers than perpendicular to them. In theory, it might be possible to adjust for the differences in wave velocity. However, most practical constructions have at least two fiber orientations and the mathematics become extremely complicated. Time of arrival location problems are further compounded when stress waves reach a sensor by spiraling

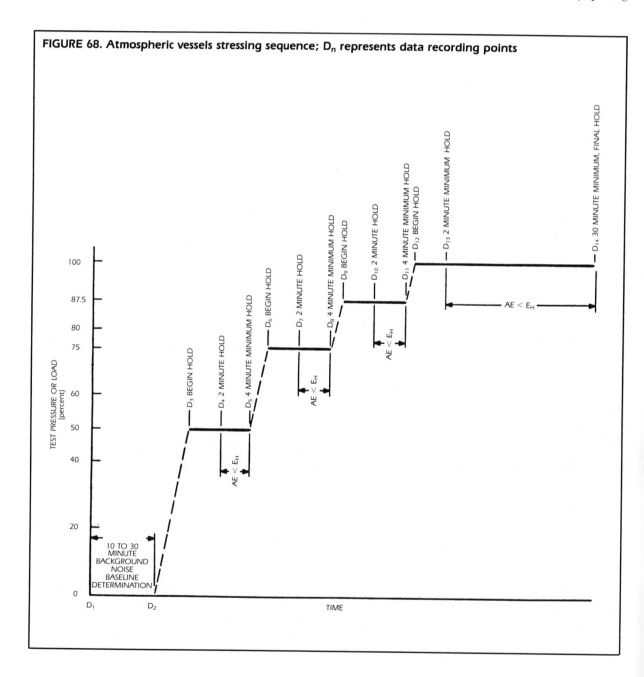

FIGURE 68. Atmospheric vessels stressing sequence; D_n represents data recording points

around the tank along the glass fibers but, because of attenuation, do not reach the same sensor by the shorter route along the tank wall.

Fiber reinforced plastic is an extremely good emitter and this leads to the third problem. A large number of events will be generated by a single source and each event will occur closely after the previous one. These events have a broad range of amplitudes and some will not reach all the sensors of a measurement array. Frequently, it becomes impossible to track which event is striking which sensor and the source location results become invalid.

Zone source location is by definition inexact. It does however limit the areas of a tank that are candidates for subsequent nondestructive testing with other methods. Normally, it is then a relatively simple matter to identify specific discontinuities within a zone.

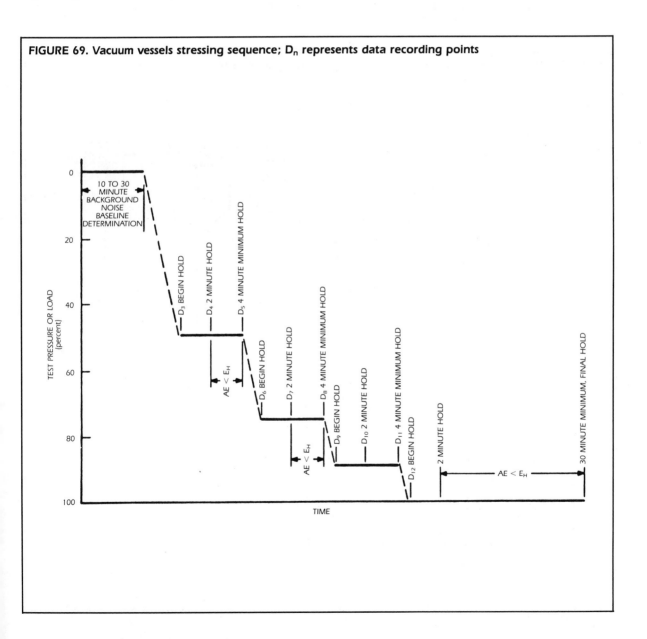

FIGURE 69. Vacuum vessels stressing sequence; D_n represents data recording points

Felicity Effect in Fiber Reinforced Plastics

The *felicity effect* is the presence of significant acoustic emission at stress levels below those previously applied. This phenomenon is a breakdown of the Kaiser effect. The felicity effect is an indication of discontinuities in fiber reinforced plastics. The *felicity ratio* is the stress at onset of emission divided by the previous maximum stress and serves as an indication of the severity of a discontinuity.

Acoustic Emission Signatures

Research has shown that specific types of composite discontinuities (dry glass, star cracking, internal blistering) have specific acoustic emission signatures.[34] Comparison of test data with standard discontinuity signatures permits identification of some types of discontinuity.

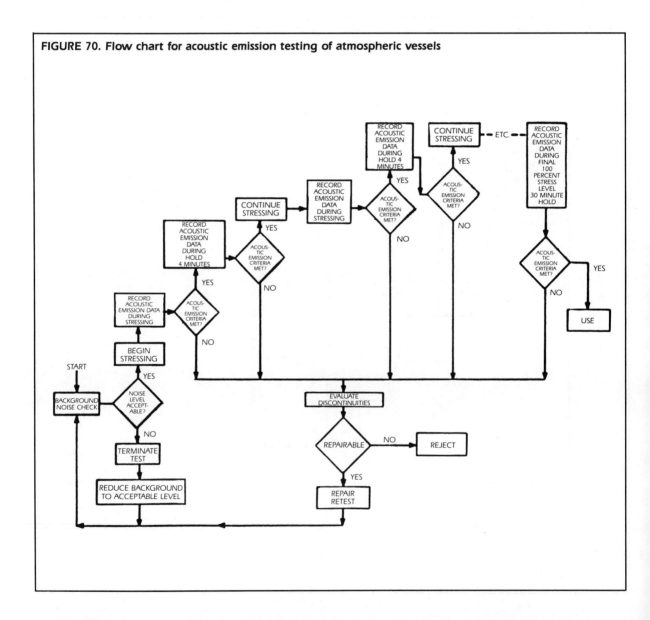

FIGURE 70. Flow chart for acoustic emission testing of atmospheric vessels

Criteria for Evaluation and Calibration

The SPI recommended practices define specific evaluation criteria and the most important is the one based on emission during load hold. Other criteria are based on the felicity ratio, total counts, number of large amplitude events and the acoustic emission energy parameter known as *DBC* (DBC tends to be sensitive to long duration events, such as those that accompany delamination). ASME uses the terms *M* or *signal value* instead of DBC.

Recommended calibration methods define instrument operating settings and evaluation criteria for different types of acoustic emission instruments. As specified, the calibration procedures require the use of a large lead sheet and a steel bar. This is clumsy to apply in practice. However, laboratory testing and field experience has shown that the procedures provide satisfactory cross-calibration between instruments.[35]

Acceptance of Acoustic Emission Techniques for Testing of Fiber Reinforced Plastics

The American Society for Testing Materials (Committees E7 and F18) has developed standards for acoustic emission

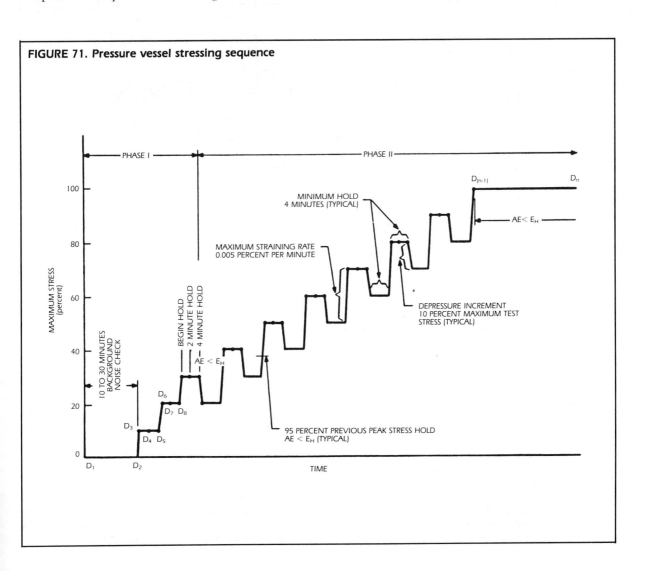

FIGURE 71. Pressure vessel stressing sequence

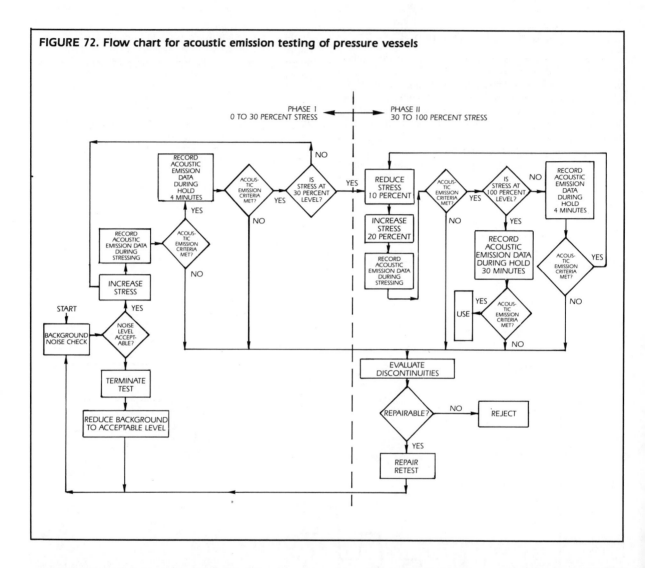

FIGURE 72. Flow chart for acoustic emission testing of pressure vessels

testing of tanks and piping and a standard for lift truck testing. The American Petroleum Institute is preparing a standard for acoustic emission testing of downhole tubing. The American Society of Mechanical Engineers (Boiler and Pressure Vessel Code committee) supports acoustic emission as a mandatory requirement for one-of-a-kind fiber reinforced plastic vessels.

Acoustic emission is accepted as the standard test method for glass fiber reinforced plastic equipment. New applications, such as advanced aerospace composites, carbon and graphite fiber reinforced plastics and high pressure piping, are regularly emerging and many are the subjects of intensive research and development programs. Acoustic emission is an important quality control tool and it is additionally becoming an important design tool for fiber reinforced plastic components.

Some interesting research[32,33] is directed toward the use of acoustic emission for defining the design strain and the limit of proportional linearity. Because acoustic emission is a sensitive method of detecting the onset of laminate microcracking, it may be suitable for experimentally determining the design strain value.

Further research is focused on extending the range and scope of signature definitions using techniques such as pattern recognition and trend analysis. A better understanding of the actual failure mechanisms of specific discontinuities has been a valuable secondary benefit of this work.

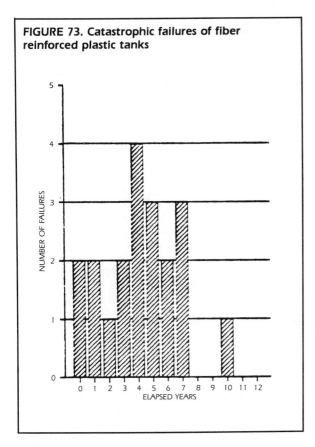

FIGURE 73. Catastrophic failures of fiber reinforced plastic tanks

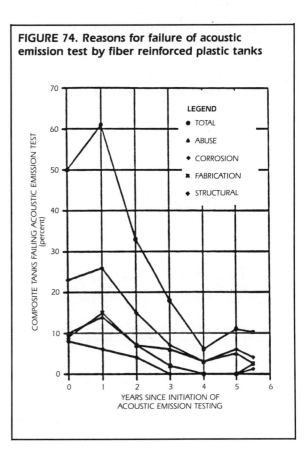

FIGURE 74. Reasons for failure of acoustic emission test by fiber reinforced plastic tanks

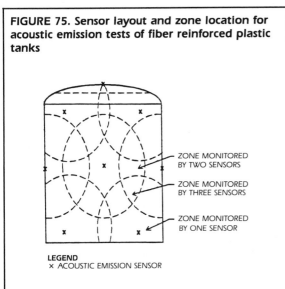

FIGURE 75. Sensor layout and zone location for acoustic emission tests of fiber reinforced plastic tanks

TABLE 20. Evaluation criteria for atmospheric (liquid head) and additional[1] superimposed pressures in fiber reinforced plastic vessels

	First Loading	Subsequent Loading	Comments
Emissions during hold	Less than E_H events beyond time T_H, none having an amplitude greater than A_M value [2]	Less than E_H events beyond time T_H	Measure of continuing permanent damage[3]
Felicity ratio	Greater than felicity ratio FA	Greater than felicity ratio FA	Measure of severity of previous induced damage
Total[4]	Not excessive[5]	Less than N_C total counts	Measure of overall damage during a load cycle
M[6]	No events with a duration greater than M	No events with a duration greater than M	Measure of delamination, adhesive bond failure and major crack growth
Number of events greater than reference amplitude threshold	Less than E_A events	Less than E_A events	Measure of high energy microstructure failures (this criterion is often associated with fiber breakage)

A_M, E_A, E_H, F_A, N_C and M are acceptance criteria values specified by the referencing code; T_H is specified hold time.

1. Above atmospheric
2. See Appendix II-1140 for definition of A_M.
3. Permanent damage can include microcracking, debonding and fiber pull out.
4. Varies with instrumentation manufacturer; see Appendix II for functional definition of N_C. Note that counts criterion N_C may be different for first and subsequent fillings.
5. Excessive counts are defined as a significant increase in the rate of emissions as a function of load. On a plot of counts against load, excessive counts will show as a departure from linearity.
6. If used, varies with instrumentation manufacturer; see Appendix II-1150 for functional definition.

PART 8
MACHINERY CONDITION MONITORING

Incipient Failure Detection

Acoustic emission techniques are particularly effective for machine condition monitoring. Such techniques are an extension of familiar vibration and sound monitoring. The presence of discontinuities in machinery is characterized by corresponding abnormalities and changes in the parameters being monitored and signature analysis can be applied to many different parameters.[36] For machinery, the most commonly used tests are vibration monitoring and ultrasonics; acoustic emission is the latest addition. All three technologies (acoustic emission, ultrasonics and vibration signature analysis) have been used for incipient failure detection (IFD).

Vibration analysis is limited in its ability to detect incipient problems since component vibration usually indicates that failure is imminent. This is particularly a problem for rigidly supported components such as short pump shafts on roller bearings. In addition, problems such as loss of lubrication often precede the initiation of vibration. This is where acoustic emission techniques serve an important role.

In relation to machinery condition monitoring, acoustic emission testing involves some concepts that are relatively unique. In the simplest sense, the technique is designed for direct discontinuity detection rather than component behavior analysis.

For early identification of failure, discontinuities must be detected when they are quite small and when detectable energy caused by their presence is correspondingly small. Because this energy extends to high frequencies while normal component related vibration is concentrated at low frequencies (Fig. 76), acoustic emission techniques can be used.

Friction, for example, is observable only with high frequency monitoring. Changes in friction may result from loss of lubrication, contamination and other factors, but the effects on vibration is usually negligible across the range of expected component frequencies. High frequency monitoring has proven sensitive even to variations in lubrication oil viscosity and thus can provide an excellent indicator of bearing film and surface condition.

The best frequency range for IFD monitoring has not been firmly established. Ranges between 10 kHz and 1 MHz have been tried, although the most successful attempt has been with the 90 kHz to 120 kHz band. Such a range is not the best for monitoring all machines or all discontinuities. An illustration of the frequency ranges for incipient failure detection are compared to conventional machine vibration analysis in Fig. 76b.

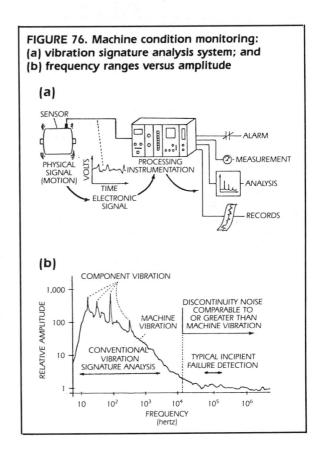

FIGURE 76. Machine condition monitoring: (a) vibration signature analysis system; and (b) frequency ranges versus amplitude

Root Mean Square Parameter

The rms parameter in incipient failure detection is the total effective amplitude of the measured signal across the frequency band of the IFD system. This is the same parameter used in vibration analysis. However, since the frequency is fixed, IFD analysis makes no attempt to relate frequency to component vibration characteristics. Instead, since the machine's characteristic vibration is assumed negligible in the IFD frequency band, what is measured is an indication of overall friction noise.

Signal Above Threshold

A second parameter used in IFD analysis has been called shock pulse emission or crest factor but is more often referred to as *signal above threshold* (SAT). The SAT value represents the quantity of signals that exceed a prescribed threshold. Where root mean square values are mostly a measure of random noise, SAT is a measurement of abnormal signals of high amplitude that have a specific cause: the sudden release of energy due to mechanical impact or fracture. The SAT measurement is often directly related to discontinuities present in the bearings and contact surfaces of rotating equipment.

Signal above threshold is unlike any parameter used in vibration signature analysis and has a direct relationship with acoustic emission monitoring. It represents a two-level amplitude distribution analysis that can distinguish between friction noise and metal-to-metal contact, the earliest indication of internal discontinuities. Where rms is a relative measurement, SAT is absolute (zero for normal operation).

Spectrum Analysis

The most significant feature of incipient failure detection monitoring is spectrum analysis. Such analysis may at first seem irrelevant since IFD monitoring is limited to a frequency band outside the range of normal component vibration frequencies. However, the detected signal may vary periodically and the frequency of that variation can be measured. Initially, it was thought that SAT bursts carried most of this information; the metal-to-metal impacts they represented would clearly recur at a rate related to rotational speed.[37] In addition, though, friction noise is also modulated by the periodically varying loading of a bearing and it too can be analyzed. The result is not unlike AM radio transmission and reception. Even though the radio signal is contained in a narrow frequency band above 500 kHz, it contains frequency information in the audible range.

The importance of this technique is that information directly related to the rotational frequency of a machine can be derived from a narrow band well above the frequency range of component vibration. As it turns out, spectrum analysis of signals from that quiet band is often more sensitive to incipient problems than analysis of changes in the full frequency, noisy, baseline vibration spectrum. A pictorial representation of these parameters is shown in Fig. 77.

Applications of Acoustic Emission in Machinery Monitoring

There are many types of instruments and systems available to measure these parameters, ranging from portable, single-channel ultrasonic (rms) probes to 500-channel computer-based trending systems. The choice of a particular instrument or system is based on the application, cost, potential benefits and the expertise of the user. The experience of the user is particularly crucial since monitoring instrumentation such as this ultimately requires some critical action by the user.

In small plants, for instance, potential benefits might be quite small because there is simply no one with the time or the inclination to evaluate machinery condition data. In addition, smaller plants often run simpler process schemes with less complex equipment that has proven to be very reliable. In such cases, the effort required to support new instrumentation, including training and maintenance, may quickly outweigh the benefits.

However, there are several factors involved that help justify this effort. Safety often can be improved simply because more is known about the operating conditions of the machinery. Acoustic data can also be helpful to understanding process conditions: turbulent flows, special reaction and extrusion processes and valve leakage can be monitored and have already been tried. Such data can be added to existing process records to better monitor plant operation.

Portable instruments are fairly simple and inexpensive. Even if a plant runs only a few pieces of rotating equipment,

FIGURE 77. Techniques for measuring information in a high frequency signal: (a) root mean square average energy; (b) signal above threshold; (c) envelope detection (demodulation); (d) low frequency pattern of envelope; and (e) spectrum analysis

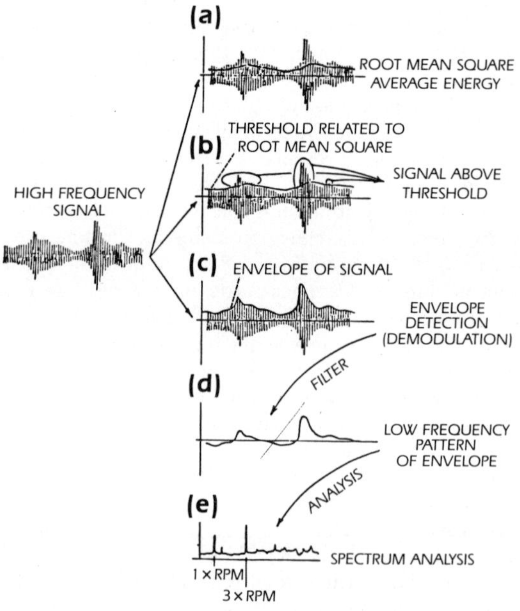

portable testing equipment has proven very valuable for quick checks of leakage in steam traps, safety valves and hydraulic systems. Incipient failure detection monitoring instrumentation can provide an additional effective technique, often without adding any significant training or labor cost.

In general, IFD techniques are particularly suited to computerized system methods. The added electronic complexity can be easily offset by the benefits of automatic, accurate and continuous monitoring and alarming. Such systems are particularly desirable for large arrays of machinery where it may prove impossible for an operator or an equipment specialist to pay close attention to performance. The installed cost of such systems can actually be low.[38] The benefits of a large data base on equipment condition can be quite high.

This is, in fact, a key point of high frequency acoustic monitoring: many measurements must be taken before accurate prediction of behavior can be made. Predictive maintenance can only be successful if a consistent effort is made to collect data on machine condition. For this reason, it is far better for a plant to purchase an automated system than a portable unit that is only rarely used.

Machinery Monitoring Case Histories

In actual application, IFD machinery condition monitoring is straightforward: changes in readings are the result of changes in operating condition. The critical decision is determining which changes are acceptable for normal operation and which are not. This need for evaluation and interpretation prevents IFD techniques from being foolproof. However, the quantitative condition data provided by IFD monitoring are sensitive and predictive enough to allow the time necessary to make an accurate, qualified decision.

The following examples illustrate these points and indicate some of the many practical applications of IFD techniques.

Simulated Discontinuities in a Pump

In a test conducted at a chemical company, a small forty horsepower centrifugal pump was subjected to cavitation and massive contamination of the oil sump. As shown in Fig. 78, the high frequency acoustic measurement proved to be

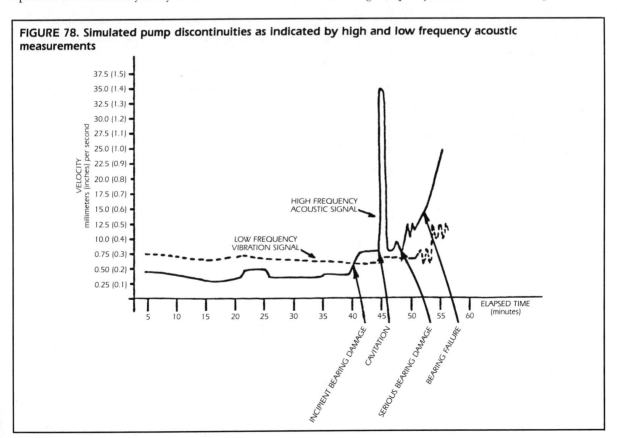

FIGURE 78. Simulated pump discontinuities as indicated by high and low frequency acoustic measurements

many times more sensitive to the degraded condition of the machine than the conventional low frequency vibration measurement.[39] Clearly, it was much easier to evaluate the machine's condition with the high frequency techniques. In fact, cavitation could not be detected with conventional vibration monitors.

Cavitation

Figure 79 is a further example of IFD monitoring's extreme sensitivity to cavitation. The high amplitude and random excursions are typical of measurements expected from a pump operating with barely sufficient net positive suction. Cavitation may be acceptable for short periods and some interpretive judgment is necessary. However, the long-term consequences of cavitation include impeller damage, seal failure and bearing damage. Often a modification in operation is essential. The ability to detect and control cavitation can therefore extend the lifetime of many varieties of process equipment.

Prevention of Bearing Failure

The trend graph shown in Fig. 80 was plotted from data taken on an extruder pelletizer. Although the extruder was not continuously running, the rms parameter had always been well below the one percent level.

The channel monitoring the inboard bearing housing began to run consistently at one percent and then jumped to two percent, creating an alarm on the IFD system. After a few more hours of operation, the trend continued and audible noise began.

Breaking down the pelletizer later revealed that the inboard bearing was indeed beginning to fail. The early warning allowed a normal shutdown and start-up of a sphere machine and probably prevented damage to the pelletizer knife

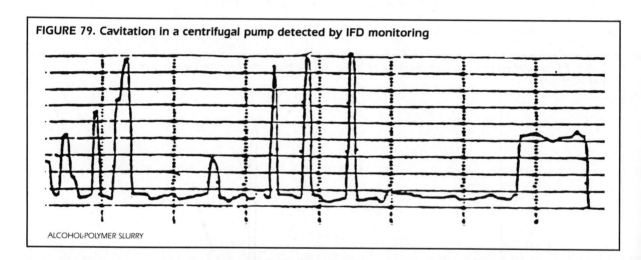

FIGURE 79. Cavitation in a centrifugal pump detected by IFD monitoring

ALCOHOL-POLYMER SLURRY

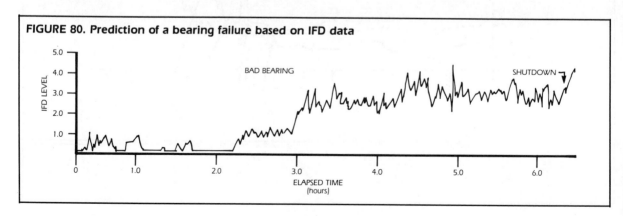

FIGURE 80. Prediction of a bearing failure based on IFD data

blades and die plate. Accurate assessment of savings is difficult, but it is certain that the failure was stopped in its early stages, well before any further, serious consequences occurred.

Process Noise Monitoring

A high frequency acoustic sensor on a process reactor circulation pump (Fig. 81) shows how acoustic emission data reveal the varying load and changing conditions experienced by the pump as the reaction cycle progresses. Although this is not directly related to machinery condition, it clearly indicates process condition and may have some potential for extending run lengths and improving quality control.

Detection of Seal Failure

Figure 82 shows a tabular data display that indicates an increasing trend in the IFD parameter (rms) being monitored. The initial alarms occurred when there were no machinery specialists available and no action was taken until the next morning. By that time, it was evident that the seal had failed and was leaking. The high friction noise (rms) was apparently due to loss of the seal facing material and was easily detected by high frequency acoustics.

Detection of Extruder Shaft Rubbing

In Fig. 83, a problem with a rubbing screw in an extruder was detectable as a high SAT parameter reading. The analysis was done after metal filings began appearing in the product. However after the unit was repaired, the high SAT reading reappeared. A spectrum measurement at this time (Fig. 84) helped to confirm the reoccurrence of the rubbing before filings again appeared in the product.

Tests of Gear Discontinuities

Many incipient gear failures are predictable with high frequency acoustic techniques even though gearboxes are often inherently noisy. The effect of a root crack in an intermediate stage gear on a large gearbox is shown in Fig. 85. Due to a series of alarms from an IFD system, an inspection was made during slow roll of the gear and the discontinuity was visually detected. Early warning clearly eliminated pumps as a potential cause, even though high fluid noise can almost completely mask bearing noise and sometimes make increasing flow look like a bearing failure.

Decision-Making as Part of Incipient Failure Detection

It is usually difficult to estimate time before failure even when the trend to failure has been recognized. The reason is that often the specific failure mechanism is not identifiable and it is only known that the machinery condition is degrading. For this reason, other techniques of inspection and analysis are brought in to determine the need for shutdown.

The IFD monitoring techniques have, in most cases, shown that failure can be anticipated and averted. It is the responsibility of plant management to take corrective steps as soon as possible. When plant operation and maintenance are done on a crisis basis, no monitoring program will pay off. When preventive steps are accepted and expected, incipient failure detection is the best single indicator of when to take them.

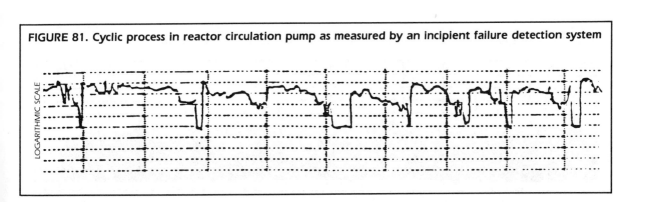

FIGURE 81. Cyclic process in reactor circulation pump as measured by an incipient failure detection system

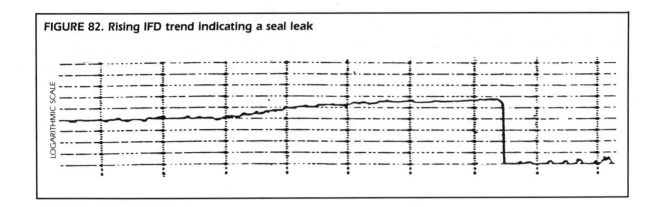

FIGURE 82. Rising IFD trend indicating a seal leak

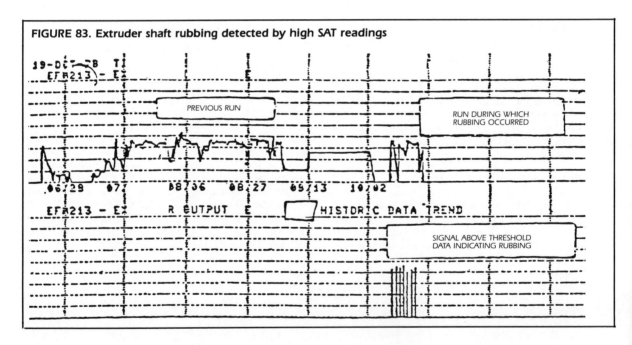

FIGURE 83. Extruder shaft rubbing detected by high SAT readings

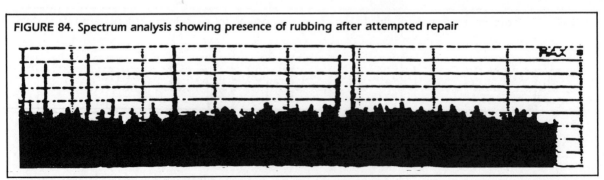

FIGURE 84. Spectrum analysis showing presence of rubbing after attempted repair

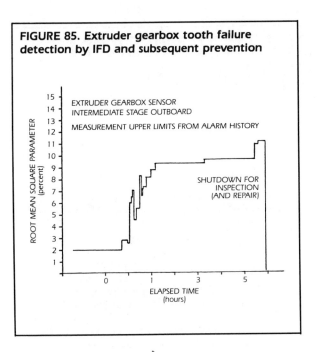

FIGURE 85. Extruder gearbox tooth failure detection by IFD and subsequent prevention

PART 9
ADVANTAGES OF ACOUSTIC EMISSION TESTING IN THE PETROCHEMICAL INDUSTRIES

Potential Cost Saving from the Use of Acoustic Emission Techniques

The dollar figures stated below are for comparison purposes and may not represent actual prices or savings.

For fiber reinforced plastic tanks, an estimated cost of one thousand dollars per acoustic emission test has been balanced against the costs of alternate inspection methods. These alternative methods required decontamination, entry into the tank and full vessel inspection at an estimated cost of twenty thousand dollars. This stark difference in cost allowed more frequent inspections and a correspondingly improved loss record.

For metal structures, the two cases noted below[40] are for a large cryogenic ammonia tank and for periodic in-line testing of a reaction vessel. The cost of shutdown, decontamination and full internal inspection of the cryogenic ammonia tank was estimated at one hundred fifty thousand dollars. Acoustic emission inspection was estimated at a cost of sixteen thousand dollars, nearly a 10:1 ratio. Additional savings would come from limiting plant downtime.

For the periodic in-line test of a reaction vessel, it is conservatively estimated that savings in additional product may be one and a quarter million dollars. The cost of alternative inspections was just over fifteen thousand dollars and the cost of the acoustic emission tests was fifteen thousand dollars. The potential saving here is very large and is unusual but not uncommon evidence of the actual value of acoustic emission tests.

Outline of Benefits for the Petrochemical Industries

Acoustic emission techniques have developed into reliable, valuable test methods for petroleum and chemical process industry equipment. The technique is widely used and recent research and development has led to very reliable test procedures. Acoustic emission has a number of advantages for use in petroleum and chemical process industries.

1. Equipment can be tested in-service, in many cases online and without a process shutdown. This has a number of advantages.
   a. Shutdown costs are eliminated.
   b. Equipment can be monitored on an as-needed basis, or on a regular cycle, often unconstrained by production and plant operating requirements.
   c. Because the test is carried out under operating conditions, the test provides an indication of discontinuities that are significant under these conditions.
   d. Internal inspection is often unnecessary. This reduces the need for decontamination and results in less personnel exposure.
   e. More frequent tests can be carried out and, in some cases, continuous monitoring becomes practical.
2. Acoustic emission provides information about the entire piece of equipment, rather than a small localized area. In this sense, it is complementary to other nondestructive test methods that provide high resolution of small areas.
3. The test monitors the growth of structural discontinuities. Discontinuities that are structurally insignificant are not normally detected with acoustic emission equipment.
4. The sensitivity of the test can be set so as to provide an early warning to discontinuity growth. Because of this, discontinuities can be detected before they become serious and appropriate corrective action taken.
5. Acoustic emission provides a measure of the structural severity of the discontinuity.
6. Signature analysis can be used to determine the type of discontinuity and thus allow implementation of measures to reduce the growth of the discontinuity or to prevent further deterioration.
7. Insulated vessels can be tested with only minor disturbance of the insulation.
8. Predetermined objective evaluation criteria have been developed. These criteria are less open to subjectivity or operator interpretation, a common problem with some other nondestructive examination methods.
9. Acoustic emission is a nonintrusive, passive inspection method. In comparison to other test methods, such as

radiography and ultrasonics, no local external energy input is required.

10. As with all nondestructive test methods, test practices are very important and data interpretation and analysis require particular care. The following is a list of warnings that apply to the use of acoustic emission for testing chemical process industry equipment.
    a. Nonstructural discontinuities, structurally unimportant discontinuities and passive discontinuities (those that do not grow under a particular set of load conditions) will not emit detectable stress waves.
    b. Successful acoustic emission tests require an increase in stress during a test. Nonstressed areas will not emit. For fiber reinforced plastic equipment, the test stresses must reach maximum operating values; for metal equipment, the stresses must exceed the maximum operating values.
    c. Equipment geometry, contents and material properties can obscure discontinuity emission.
    d. Background noise, such as radio interference, electromagnetic interference, fill noise, steam traps and air hoses, can interfere with the test and obscure genuine acoustic emission data.
    e. The test does not indicate the size of a discontinuity, even though it provides a measure of the discontinuity's structural severity.
    f. In metal vessels, it is not always possible to determine the specific type of discontinuity.
    g. Acoustic emission is a relatively new technology and there is a limit to the number of trained test operators.
    h. Few codes and standards cover testing of metal equipment.
    i. Normally, follow-up inspection, using other nondestructive examination methods, is required to determine the specific orientation of a discontinuity.
    j. Particular care is required with sensor spacing, use of waveguides, choice of source location technique and capture of all data.

Reference 40, in its conclusion, says that acoustic emission tests of metallic structures show more promise than they did originally and that the technique has benefitted from a growing maturity of its technology. With limited characterization of the discontinuity available, complementary investigation is often required. The authors caution that absence of acoustic emission may not always mean absence of harmful discontinuities and that the necessity to determine proper test conditions to maximize discontinuity detection is necessary.

For fiber reinforced plastic structures, it is acknowledged that the integrity of fiber reinforced plastic vessels, tanks and pipelines has been effectively assessed by acoustic emission analysis. Still to come is reliable discontinuity characterization.

The advantages and disadvantages of acoustic emission testing are listed in Table 21.

TABLE 21. Advantages and disadvantages of acoustic emission testing

Advantages:	Large area coverage (for steel, up to forty square meters per sensor)
	Discontinuities growing during test can be detected
	Location of discontinuities is possible
	Nondestructive
	Nonintrusive
	Inaccessible area inspection
	On-line application (even at high temperatures)
Disadvantages:	Not all discontinuities emit detectable acoustic emission
	Loading methods must be analyzed to ensure they promote detectable crack growth
	Many factors can obscure acoustic signals (geometry, materials, construction, noise)
	No standardized data interpretation for metals
	Discontinuity size not determinable

REFERENCES

1. Green, A.T. "Necessity: The Mother of Acoustic Emission Testing (1961-1972)." *Materials Evaluation*. Vol. 43, No. 5. Columbus, OH: The American Society for Nondestructive Testing (May 1985): pp 600-610. See also: *ASTM Standard Recommended Practice for Acoustic Emission Monitoring of Structures During Controlled Stimulation*. ASTM E569. Philadelphia, PA: American Society for Testing Materials (1976). See also: *Acoustic Emission Testing of Spherical Pressure Vessel Made of High Strength Steel and Classification of Test Results*. NDIS 2412. Tokyo, Japan: Japanese Society for NDI (July 1980). See also: *Recommended Practice for Acoustic Emission Testing of Fiberglass Tanks/Vessels*. New York, NY: Society of the Plastics Industry (1982).
2. Cross, N.O., et al. "Acoustic Emission Testing of Pressure Vessels for Petroleum Refineries and Chemical Plants." *Acoustic Emission*. ASTM STP 505. Philadelphia, PA: American Society for Testing Materials (May 1982): pp 270. See also: Blackburn, P.R. *Periodic Re-Testing of Trailer Tubes for Distribution of Industrial Gases*. Tonawanda, NY: Linde Division of Union Carbide Group (June 1986). See also: Craig, B.D. *Acoustic Emission Evaluation of Pressure Vessels in the Petroleum and Petrochemical Industry*. Englewood, CO: Metallurgical Consultants (July 1986).
3. Ingham, T., et al. "Acoustic Emission for Physical Examination of Metals." *International Metals Review*. Vol. 2, No. 41 (1980). See also: *Mobile Pipe Integrity Analysis Scoping Tests*. Jersey Nuclear Company (1971). See also: T. Ingham, A.L. Stott and A. Cowan. *Journal of Pressure Vessels Piping*. Vol. 2, No. 32 (1974). Reproduced in H.N.G. Wadley, C.B. Scruby and J.H. Speake. "Acoustic Emission for Physical Examination of Metals." *International Metals Review*. Vol. 2, No. 41 (1980).
4. Blackburn, P.R. and M.D. Rana. "Acoustic Emission Testing and Structural Evaluation of Seamless, Steel Tubes in Compressed Gas Service." *Transactions of the ASME: Journal of Pressure Vessel Technology*. Vol. 108 (May 1986): pp 234.
5. Nielsen, A. *Acoustic Emission Source Based on Pencil Lead Breaking*. The Danish Welding Institute (1980).
6. "Cylinder Service Life." *Seamless High Pressure Cylinders*. Pamphlet C-5. Arlington, VA: Compressed Gas Association.
7. *Monitoring Fracture in Pipeline Sections*. Pensacola, FL: Aerojet General (1967).
8. Private correspondence to *Acoustic Emission International* (1983).
9. "Use of Acoustic Emission Examination in Lieu of Radiography" (ASME case 1,968). *ASME Boiler and Pressure Vessel Code*. Sec. VIII, Div. 1. New York, NY: American Society for Mechanical Engineers (May 1985).
10. *Proposed Standard for Acoustic Emission Examination During Application of Pressure*. New York, NY: American Society for Mechanical Engineers (1975). See also: *Recommended Practice for Acoustic Emission Testing of Fiberglass Reinforced Plastic Piping Systems*. New York, NY: Society of the Plastics Industry (July 1983).
11. *Recommended Practice SNT TC-1A*. Columbus, OH: The American Society for Nondestructive Testing (1980).
12. "Specification for Wellhead Equipment." *Specification 6-A*. Washington, DC: American Petroleum Institute.
13. *Acoustic Emission of Blowout Preventers* (Code case BCB2-727). Sec. V. Acoustic Emission Working Group (1985).
14. Cross, N.O. "Acoustic Emission Test of High Pressure Cast Steel Gate Valves." *NDT Technology* (July 1986). See also: Cross, N.O. "An Acoustic Emission Test of a High Temperature Refractory During the Curing Cycle." *NDT Technology* (July 1986).
15. "Metal Vessel Test Procedure." *Procedure for Acoustic Emission Testing of Tanks/Vessels*. Issue 3. Monsanto Company. See also: "Testing Tubes the Sound Way." *Quality Magazine*. Vol. 25 (April 1986).
16. Adams, L. "Acoustic Emission Testing and Its Application to Ammonia Spheres and Stress Corrosion Cracking." *Proceedings of the Safety in Ammonia Plants and Related Facilities Symposium*. Los Angeles, CA (November 1982).
17. Berry, D.A. *Acoustic Emission Response of Hydroformer Reactor During Regeneration and Shutdown*. Sacramento, CA: Acoustic Emission Technology Corporation (August 1981).
18. Fowler, T.J. "Experiences with Acoustic Emission Monitoring of Chemical Process Industry Vessels" *Proceedings of the Eighth International Acoustic Emission Symposium*. Tokyo, Japan: The Japanese Society for Nondestructive Inspection (October 1986).
19. Finley, R.W. *Acoustic Emission Inspection During Start-up of Fluid Catalytic Cracking Unit, Upper Regeneration Section and Plenum*. Sacramento, CA: Acoustic Emission Technology (October 1980). See

also: "Acoustic Emission Examination of Fiber Reinforced Plastic Vessels." *ASME Boiler and Pressure Vessel Code.* Section V, Article 11. New York, NY: American Society for Mechanical Engineers (June 1985).
20. Bell, H.V. "Shutdown, Inspection and Start-up of a Cryogenic Ammonia Storage Tank." *Proceedings of the Safety in Ammonia Plants and Related Facilities Symposium.* Montreal, Canada (1981). See also: Fowler, T.J. "Acoustic Emission Testing of Process Industry Vessels and Piping." *Proceedings of the Fifth International Symposium: Loss Prevention and Safety Promotion in the Process Industries.* Cannes, France: European Federation of Chemical Engineering. Vol. II (September 1986).
21. Cross, N.O. "Acoustic Emission Test to Identify the Risk of Brittle Fracture of a Butane Storage Sphere." *NDT Technology* (July 1986). See also: Cross, N.O. "Acoustic Emission Test of Preflawed, Small Volume Pressure Vessel During Pressurization to Rupture." *NDT Technology* (July 1976).
22. Tscheliesnig, P. and H. Theiretzbacher. "Leakage Test by Acoustic Emission Testing (AET) on Flat Bottom Tanks." *Proceedings of the Second International Conference on Acoustic Emission.* Lake Tahoe, NV: The Acoustic Emission Working Group (October 1985).
23. Fowler, T.J. "Acoustic Emission Testing of Chemical Process Industry Vessels." *Proceedings of the Seventh International Symposium on Acoustic Emission.* Sendai, Japan (1984).
24. *Standard Practice for Acoustic Emission Examination of Fiberglass Reinforced Plastic Resin (RP) Tanks/Vessels.* E1067. Philadelphia, PA: American Society for Testing Materials (1985).
25. *Practice for Acoustic Emission Examination of Reinforced Thermosetting Resin Pipe (RTRP).* E1118. Philadelphia, PA: American Society for Testing Materials (1986.)
26. Fowler, T.J. "Acoustic Emission Testing of Fiber Reinforced Plastic Equipment." *Chemical Processing* (March 1984): pp 24-27.
27. Fowler, T.J. *Acoustic Emission of Fiber Reinforced Plastics.* Preprint 3092. New York, NY: American Society of Civil Engineers (1977).
28. Fowler, J.J. and E. Gray. *Development of an Acoustic Emission Test for FRP Equipment.* Preprint 3583. New York, NY: American Society of Civil Engineers (1979).
29. Fowler, T.J. and R.S. Scarpellini. "Acoustic Emission Testing of FRP Equipment." *Chemical Engineering.* Part I (October 1980) and Part II (November 1980).
30. Crump, T.N. and J. Teti. "Acceptance Criteria for Acoustic Emission Testing of FRP Tanks." *Emerging High Performance Structural Plastic Technology.* Andrew Green, ed. New York, NY: American Society of Civil Engineers (April 1982).
31. Fowler, T.J. and R.S. Scarpellini. "Acoustic Emission Testing of FRP Pipe." *Emerging High Performance Structural Plastic Technology.* Andrew Green, ed. New York, NY: American Society of Civil Engineers (April 1982).
32. Cortez, A., J.H. Enos, E. Francis and H. Heck. "Use of Acoustic Emission to Characterize Resin Performance in Laminates." *Proceedings of the First International Symposium on Acoustic Emission from Reinforced Composites.* New York, NY: Society of the Plastics Industry (July 1983).
33. Pickering, F.H. "Design of High Pressure Fiberglass Downhole Tubing: A Proposed New ASTM Specification." *Proceedings of the First International Symposium on Acoustic Emission from Reinforced Composites.* New York, NY: Society of the Plastics Industry (July 1983).
34. Scarpellini, R.S., T.L. Swanson and T.J. Fowler. "Acoustic Emission Signatures of FRP Defects " *Proceedings of the First International Symposium on Acoustic Emission from Reinforced Composites.* New York, NY: Society of the Plastics Industry (July 1983).
35. Fowler, T.J. "Calibration and Cross Calibration of Acoustic Emission Instrumentation Using the CARP Procedures." *Proceedings of the First International Symposium on Acoustic Emission from Reinforced Composites.* New York, NY: Society of the Plastics Industry (July 1983).
36. Lavoie, F.J. "Signature Analysis: Product Early-Warning System." *Machine Design* (January 23, 1969): pp 121-128.
37. Harting, D.R. "Demodulated Resonance Analysis — A Powerful Incipient Failure Detection Technique." *ISA Transactions.* Vol. 17, No. 1. Research Triangle Park, NC: Instrument Society of America.
38. Bloch, H.P. and R.W. Finley. "Using Modified Acoustic Emission Techniques for Machinery Condition Surveillance." *Proceedings of the Seventh Turbomachinery Symposium.* College Station, TX: Gas Turbine Laboratories of the Texas A&M University (1978).
39. Bloch, H.P. "Machinery Monitoring with High Frequency Acoustic Emission Measurements." *Hydrocarbon Processing* (May 1977).
40. "Guidance Notes on the Use of Acoustic Emission Testing in Process Plants." *Third Report of the International Study Group on Hydrocarbon Oxidation.* Rugby, Warwickshire, England: Institution of Chemical Engineers (1985).

SECTION 8

ACOUSTIC EMISSION APPLICATIONS IN THE NUCLEAR AND UTILITIES INDUSTRIES

Phillip H. Hutton, Pacific Northwest Laboratory, Richland, Washington

J.A. Baron, Ontario Hydro, Toronto, Canada
C.E. Coleman, Atomic Energy of Canada, Chalk River, Canada
Teruo Kishi, University of Tokyo, Tokyo, Japan
Hiroyasu Nakasa, Central Research Institute of Electric Power Industry, Tokyo, Japan
Peter Ying, Gilbert/Commonwealth, Reading, Pennsylvania

PART 1
CONTINUOUS IN-SERVICE ACOUSTIC EMISSION MONITORING OF NUCLEAR POWER PLANTS

The objective of the work described here is to develop and validate the use of acoustic emission methods for continuous surveillance of crack growth in reactor pressure boundaries. This technique will help to continuously assure physical integrity of reactor pressure boundaries.

The concept of applying acoustic emission technology to the continuous monitoring of nuclear reactor systems has received support *and* condemnation since the 1960s. Initially, it suffered from excessive enthusiasm on the part of proponents, resulting in some disappointing applications. There were also unsuccessful and premature efforts to incorporate acoustic emission methods for nuclear application into industry codes. The final result was a pessimistic assessment of acoustic emission techniques for continuous monitoring.[1,2] Critical reviews served a valuable function, however. For the most part, the critiques were well thought out and helped those supporting acoustic emission development to clarify what was needed to make the technique useful in continuous monitoring.

From 1977 to 1979, on-line acoustic emission monitoring was tested on Philadelphia Electric Company's Peach Bottom Nuclear Generating Station, Unit 3.[3] The work was supported by the US Department of Energy and provided some of the early evidence that instrumentation could be designed to endure the reactor environment for at least a year and that the background noise problem could be overcome. Subsequent work focused on applying acoustic emission monitoring for leak detection in selected areas of the plant[4] and proved that acoustic emission methods can be effective in detecting coolant leakage.

A carefully planned research program was conceived in 1976 to accomplish the technological development needed for achieving continuous on-line monitoring. The program was supported by the US Nuclear Regulatory Commission, Office of Nuclear Regulatory Research. The research was performed at Pacific Northwest Laboratory (PNL) operated by Battelle Memorial Institute. The program itself comprised three major phases designed to develop the needed technology in the laboratory; test it on a structure; perform necessary improvements; and retest to arrive at a reliable product. These phases included:

1. development of acoustic emission discontinuity relationships by laboratory specimen testing (a) to identify crack growth acoustic emission signals and (b) to use crack growth acoustic emission data to estimate crack severity;
2. evaluation and refinement of the technology developed in the laboratory by monitoring fatigue testing of a heavy section vessel under simulated reactor conditions; and
3. demonstration of continuous acoustic emission monitoring on a nuclear power reactor system.

Phase-1 produced an acoustic emission rate versus crack growth rate relationship for fatigue crack growth in A533B steel. Crack growth rates in the laboratory ranged from about 3.5×10^{-5} to 10^{-3} millimeters per second. Tests were conducted at room temperature and 290 °C (550 °F). Also developed were an instrumentation concept and statistical pattern recognition methods for identification of crack growth acoustic emission signals.

The NRC and PNL collaborated with the West German Materialpruefungsanstalt (MPA) Laboratory in a year-long test of an intermediate scale pressure vessel (ZB-1) to accomplish phase-2.[5] The vessel was 1.7 m (5.6 ft) outside diameter by 3 m (10 ft) long with a 120 mm (4.75 in.) wall thickness. It was fabricated from A508 steel with a 750 × 1,500 × 120 mm (29 × 59 × 4.75 in.) insert of A533B steel. Fatigue loading was supplied by cyclic internal pressurization with water. Test temperatures of 65 °C (150 °F) and 285 °C (545 °F) were used, and crack growth rates ranged from 2.5×10^{-6} to 1.3×10^{-4} millimeters per second. Test results demonstrated the ability to detect both crack growth from the known discontinuities and from an unanticipated crack that developed adjacent to a weld.

Phase-3 has begun at the Tennessee Valley Authority Watts Bar Unit 1 nuclear reactor. Acoustic emission sensors, permanent signal cables and an engineering prototype acoustic emission monitor system were installed for on-site test monitoring. Preservice tests (cold hydro and hot functional) have been monitored.[6,7]

Instrumentation

The fundamental requirements for this application include an acoustic emission instrument system capable of (1) operating continuously over a long period (one year minimum); (2) withstanding the reactor environment; and (3) providing mass data storage.

The acoustic emission system used with the ZB-1 vessel performed with very few problems during the one-year test and has continued in use on laboratory tests since that time. An engineering prototype system for installation at Watts Bar Unit 1 was based on the ZB-1 experience. The instrument concept is shown in Fig. 1.

The acquisition subsystem, which receives data from the sensors, is composed of 24 channels divided into six arrays. Arrays may be composed of 1, 2, 3 or 4 channels. Each channel has a maximum electronic amplification of 100 decibels (dB), which can be reduced by 15 dB with channel attenuators in 1 dB steps. A 20 dB gain preamplifier is used near the sensor, providing sensor impedance matching and cable driving. All amplifier stages are linear.

Data collected from each array consist of delta time for signal arrival, root mean square (rms) signal level, signal peak time (time from trigger to signal peak amplitude), peak signal amplitude and sensor hits. Delta times are used for source location and are measured with one microsecond resolution. The rms signal level is measured on each channel for leak detection analysis. Peak time and pulse height are measured on one channel of each array. Sensor hit totals are used to determine if the sensor is in functional condition.

All data are recorded on a 67 megabyte cartridge tape for long-term storage. The data are also passed on a high speed buss to a computer for real-time analysis.

The system is operated using a menu approach, where a video display shows function choices to the operator. Once a choice is made, a series of questions and prompts follows for operator guidance.

Data analysis is performed in a separate subsystem designed to maximize speed and information accessibility with a minimum of operator intervention. Analysis can be performed in either an on-line mode, with data directly from the acquisition subsystem, or in an off-line mode receiving data from cartridge tape.

High processing speeds are attained through the use of a 15 MHz clock digitizer, 32 bit central processing unit and more than two megabytes of random access memory. System software resides in battery backed, nonvolatile memory, eliminating the need for the slower mechanically based mass storage devices. The instrument system hardware is illustrated in Fig. 2.

A key element in the acoustic emission monitor system is a high temperature sensor. Two problems were experienced in attempts to develop acoustic emission sensors for mounting directly on a 245 to 300 °C (500 to 600 °F) surface: (1) the high temperature piezoelectric materials tend to have

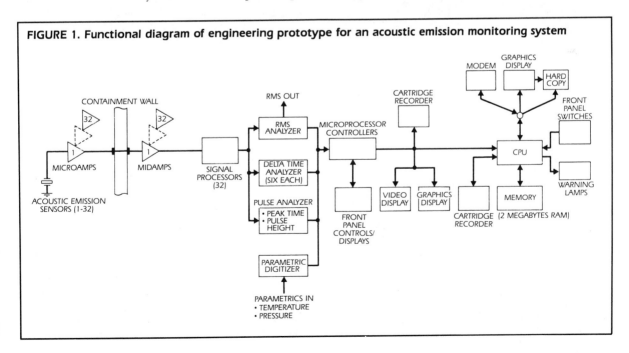

FIGURE 1. Functional diagram of engineering prototype for an acoustic emission monitoring system

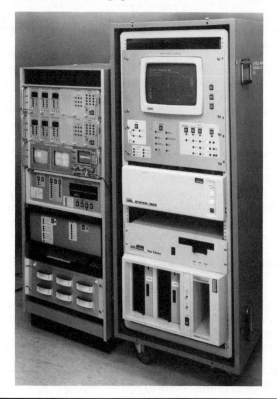

FIGURE 2. Engineering prototype acoustic emission monitoring system for Watts Bar reactor

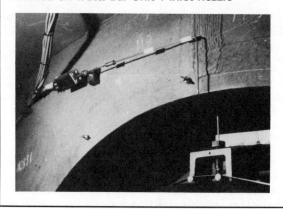

FIGURE 3. Waveguide acoustic emission sensor installed on Watts Bar Unit 1 inlet nozzle

Detection and Location of Discontinuities

In the ZB-1 vessel test,[5] fatigue crack growth from machined discontinuities was detected consistently during low temperature and high temperature testing. The most significant example of fatigue crack detection during the test, however, involved an unexpected discontinuity along an insert weld. Acoustic emission data associated with both the weld crack at the KS07R insert and the installed discontinuities are shown in Fig. 4. The illustration is a roll-out of the vessel cylinder with the match line running through the KS07R insert.

The presence of a growing fatigue crack in the weld region was identified early in high temperature testing (step 9) by the acoustic emission data. This was well ahead of confirmation by ultrasonic inspection. Another significant aspect of the test was the fact that the signals had to be received by sensors three meters from the source in order to be accepted and located by the monitor. This provided a calibration of the range of acoustic emission detection, for 400 kHz tuned waveguide sensors, pressure coupled to the vessel surface.

Stress corrosion cracking (SCC) was another cracking process of major interest. All of the information to date on detection of SCC by acoustic emission methods came from laboratory specimens ranging from 100 mm (4 in.) to 600 mm (24 in.) diameter austenitic stainless steel pipe. The SCC was successfully detected by acoustic emission in all of these tests. Figures 5 and 6 show the relatively small cracks that were detected. This information is from a test performed with 100 mm (4 in.) schedule 80 stainless steel pipe.

poor sensitivity; and (2) the sensors are very costly. Long-term acoustic coupling of these sensors to the structure surface can also be difficult.

These problems were overcome by using metal waveguides with a sensing crystal (PZT5A) and a preamplifier mounted on the outer end. Figure 3 shows one of these sensors mounted on a nuclear reactor vessel inlet nozzle. The sensor assembly was pressure coupled to the surface with a pressure of about 100 megapascals (MPa), which required a 133 newton (N) force with the tip design used. The acoustic waveguides are 3.5 mm in diameter. This diameter is a compromise between minimizing acoustic signal dispersion and retaining needed mechanical strength. These sensors were used effectively on both the ZB-1 vessel test and on hot functional testing at Watts Bar. The sensor assembly was exposed to 1.13 and 1.33 MeV gamma radiation at 17 sieverts (Sv) per hour or 1.7×10^3 roentgens (R) per hour, for a cumulative dose of 6,000 Sv (6×10^5 R) with no evident degradation.

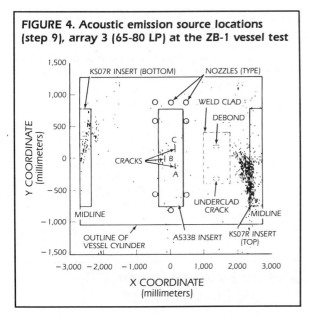

FIGURE 4. Acoustic emission source locations (step 9), array 3 (65-80 LP) at the ZB-1 vessel test

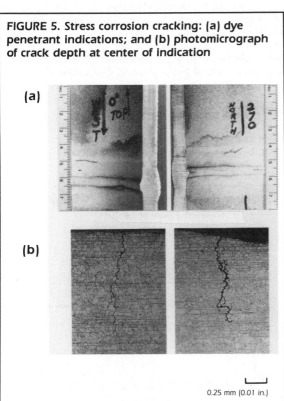

FIGURE 5. Stress corrosion cracking: (a) dye penetrant indications; and (b) photomicrograph of crack depth at center of indication

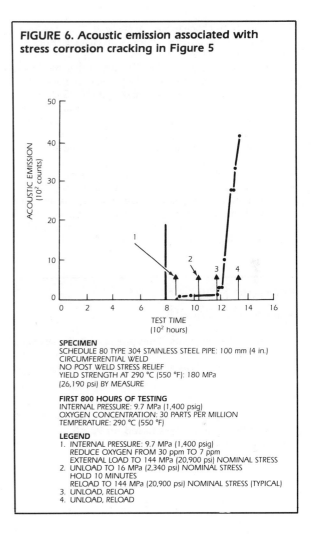

FIGURE 6. Acoustic emission associated with stress corrosion cracking in Figure 5

SPECIMEN
SCHEDULE 80 TYPE 304 STAINLESS STEEL PIPE: 100 mm (4 in.)
CIRCUMFERENTIAL WELD
NO POST WELD STRESS RELIEF
YIELD STRENGTH AT 290 °C (550 °F): 180 MPa (26,190 psi) BY MEASURE

FIRST 800 HOURS OF TESTING
INTERNAL PRESSURE: 9.7 MPa (1,400 psig)
OXYGEN CONCENTRATION: 30 PARTS PER MILLION
TEMPERATURE: 290 °C (550 °F)

LEGEND
1. INTERNAL PRESSURE: 9.7 MPa (1,400 psig)
   REDUCE OXYGEN FROM 30 ppm TO 7 ppm
   EXTERNAL LOAD TO 144 MPa (20,900 psi) NOMINAL STRESS
2. UNLOAD TO 16 MPa (2,340 psi) NOMINAL STRESS
   HOLD 10 MINUTES
   RELOAD TO 144 MPa (20,900 psi) NOMINAL STRESS (TYPICAL)
3. UNLOAD, RELOAD
4. UNLOAD, RELOAD

In summary, effective detection and location of fatigue crack growth under simulated reactor conditions was demonstrated. Fatigue cracking was detected by acoustic emission prior to detection or confirmation by ultrasonic inspection. Stress corrosion cracking was repeatedly detected in laboratory tests using techniques compatible with reactor monitoring.

Noise Interference

One of the major concerns with continuous acoustic emission monitoring on an operating reactor was the background noise from reactor coolant flow. Coolant flow noise for pressurized water reactors (PWR) and boiling water reactors

(BWR) has been characterized in the past.[8,9,10] This information was used to develop sensors with characteristics designed to prevent interference from flow noise. Results from hot functional testing[7] (see Fig. 7) showed that the tuned waveguide sensors do overcome the problem of noise interference at coolant temperatures above 120 °C (250 °F).

The sensors used on Watts Bar were the same as those used in the ZB-1 test (waveguide sensors tuned to 400 to 500 kHz and pressure coupled to the structure). At full cold flow conditions of 65 °C and 2.8 MPa (150 °F and 400 psig), even tuned sensors are ineffective. Increasing coolant temperature to about 120 °C (250 °F) causes a dramatic reduction in noise in the 400 to 500 kHz frequency range.

Except for the condition of full cold flow, the coolant flow noise problem can be overcome using high frequency, tuned waveguide sensors. These same sensors have shown the ability to detect acoustic emission at a distance of at least three meters.

Acoustic Emission Signal Identification

One of the key elements of this program was to develop a method of separating acoustic emission signals produced by crack growth from acoustic signals produced by innocuous sources. Signal pattern recognition was the technique chosen for this purpose.[11] Visual examination of recorded waveforms from the ZB-1 vessel test showed evidence of a consistent three-pulse signal pattern present in practically all waveforms originating near known and suspected regions of crack growth (Fig. 8).

It was determined experimentally that the three pulses are wave modes, traveling at different group velocities, and that separation of these wave modes takes place in the waveguide as a response to the short duration acoustic emission energy input pulse which is similar to a shock wave. Implementation of this waveguide response phenomenon to perform signal pattern recognition has produced correct classification rates of about 95 percent when applied to recorded waveforms from the ZB-1 vessel test.

Current emphasis is on testing the pattern recognition technique as part of the engineering prototype system to perform analysis on acoustic signals as they are received.

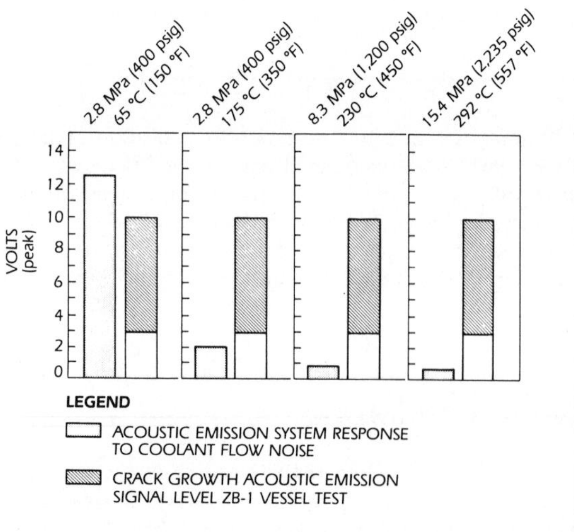

Severity Estimate from Acoustic Emission Data

Another major program objective was to develop a relationship for estimating the severity of discontinuities from acoustic emission data. The ZB-1 vessel test was the principal vehicle for evaluating a severity relationship developed in the laboratory. To test the relationship, acoustic emission data from each of three planned discontinuities (crack-A and crack-B were on the inside diameter, crack-C was on the outside diameter) were filtered with respect to source location, load position and signal amplitude by accepting only that data which fell within certain limits for each of these parameters. This was done to remove noise sources from the data. With the filtered information for each crack,

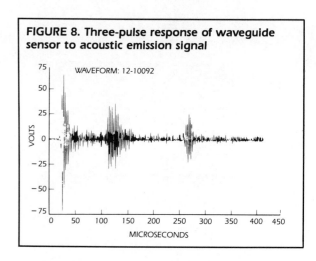

FIGURE 7. Acoustic emission system response to coolant flow noise; data based on 90 dB electronic amplification

FIGURE 8. Three-pulse response of waveguide sensor to acoustic emission signal

acoustic emission event rate values were established for each of the cyclic loading steps.

The rate for events per second was then used in conjunction with the lower bound severity relations developed from laboratory data to estimate crack growth rates in microinches per second. Figure 9 gives the experimental acoustic emission-to-fatigue crack growth rate relationship. Figure 10 shows the crack growth rates determined from the acoustic emission data for crack-A, B and C plotted against the crack growth rates determined from crack opening displacement and fracture surface measurements. The line in Fig. 10 shows where the two crack growth rates are equal to each other.

The results indicate that the severity relationship underpredicted the crack growth rates during the low temperature portion of the test. The underprediction may be a result of extrapolating the laboratory relationship to very low crack growth rates for which there are no laboratory data. There may also be a cycle-rate dependency on the acoustic emission data at low crack growth rates in a water environment. For the high temperature portion of the test, the relationship tended to overpredict the crack growth rate, which was expected since a lower bounds relationship from the laboratory data was used. Results from crack-B at 290 °C (550 °F) seem quite high relative to the other data in Fig. 10, but these results are within normal variations in the acoustic emission data, as can be seen in Fig. 9.

In summary, growth characteristics for three cracks were determined using nondestructive testing, crack opening displacement and fractographic measurements. This information was used to establish crack growth rates for comparison with predictions derived from the acoustic emission to crack severity relationship. A test of the relationship using data from crack-A, B and C indicates that the model gave reasonably good predictions.

Status and Future of the Technique

The feasibility of performing on-line acoustic emission monitoring to detect crack growth in reactor pressure boundaries has been demonstrated. Most of the necessary technology has been developed. The major areas requiring further attention are:

1. development of correlations between acoustic emission and intergranular stress corrosion cracking;

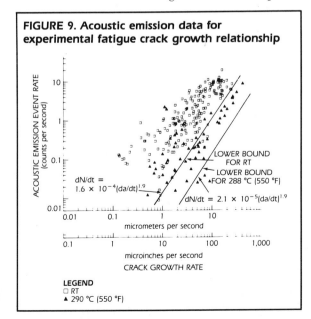

FIGURE 9. Acoustic emission data for experimental fatigue crack growth relationship

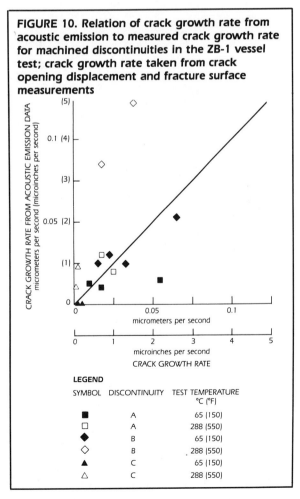

FIGURE 10. Relation of crack growth rate from acoustic emission to measured crack growth rate for machined discontinuities in the ZB-1 vessel test; crack growth rate taken from crack opening displacement and fracture surface measurements

2. upgrading the crack evaluation concept;
3. proving the acoustic emission signal identification approach through application testing;
4. generating field application data to substantiate the technology; and
5. gaining wider code acceptance of acoustic emission as a tool for assessing structural integrity.

It appears that acoustic emission testing will be applied initially to selected purposes such as pipe leak detection, valve operation monitoring, and extended service life for weld overlay pipe repairs. There is clear economic incentive in these cases and the technology is generally available. The key to realizing acoustic emission application lies in (1) field data gained under the umbrella of *test application* and (2) acceptance of continuous acoustic emission monitoring in codes and by regulatory bodies.

PART 2
APPLICATION OF ACOUSTIC EMISSION TECHNIQUES FOR DETECTION OF DISCONTINUITIES IN ELECTRIC POWER PLANT STRUCTURES

Diagnosis of structural integrity is the basis for correct treatment of and countermeasures against progressive structural abnormalities. An exact diagnosis is at present the most reliable means for determining the soundness of structures during power plant operation.

Acoustic emission technology has recently strengthened its application base, and improvements have been made in both instrumentation and practitioners' understanding of the technique's fundamentals. The practical use of acoustic emission technology, however, seems to be advancing slower than expected in such highly regulated industries as Japan's nuclear and utilities industries.

The reasons for this delay may be an overall slowdown in the economy, as well as the recent change in emphasis from *heavy, thick, long and big* to *light, thin, short and small*.[12] The following items, in particular, appear to contribute to the slowing of acoustic emission research and development activities in Japan's utility industry:

1. inverse demonstrations of the maturity of nuclear safety in analyses of past accidents;
2. economic concerns over cost increases for enhanced safety;
3. changing interests (focus on human errors and radiation protection);
4. improvements in manufacturing technology and structural integrity of the reactor pressure vessel and piping components;
5. lack of appropriate countermeasures after failure detection;
6. perceived reliability of other nondestructive testing methods for in-service inspection; and
7. doubt about the demanding needs of acoustic emission monitoring.

Apparently, there are two different attitudes that affect the increased use of acoustic emission field applications and these are based on (1) economics and (2) regulatory requirements. Mandatory codification may help increase acceptance of acoustic emission technology for quality assurance, but there is also the need for consensus among acoustic emission practitioners. Economically, the use of acoustic emission may be helped when it is acknowledged that structural integrity can be ensured by acoustic emission monitoring.

Nuclear Power Plant Applications

Reactor Pressure Vessels

For the purpose of investigating the dynamic behavior of fatigue cracks around the nozzle of reactor pressure vessels, fatigue tests of model pressure vessels were performed, and the relationship between acoustic emission characteristics and propagation behavior of fatigue cracks around nozzle corners was investigated using a multichannel acoustic emission source location system.[13,14] As shown in Figs. 11 and 12, the acoustic emission instrumentation system successfully traced fatigue crack propagation, but the signal-to-noise ratio was very low.

This occurred because (1) pressure loading alone generates little or no acoustic emission activity; and (2) acoustic emission generation often results from deformation and there was no bending load to cause deformation in this application.

Experimental approaches to continuous monitoring of specific critical areas in the primary coolant system were performed in a commercial nuclear power plant. Together with the back-up fatigue tests shown in Fig. 13,[15] acoustic emission activities were analyzed in various stages of reactor start-up, steady state operation and reactor shutdown.[14] Typical results are shown in Fig. 14, and the following conclusions were obtained.

1. Acoustic emission activities were very weak compared to environmental noise, and the detectability of acoustic emission signals depends on both the location monitored and the environmental conditions associated with reactor operation.

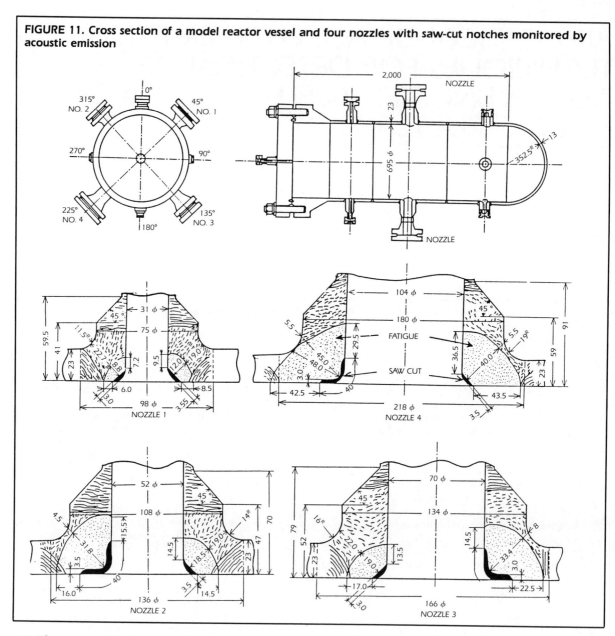

FIGURE 11. Cross section of a model reactor vessel and four nozzles with saw-cut notches monitored by acoustic emission

2. The event number [$N = \Sigma n(a)$], the derivative parameters [such as acoustic emission peak amplitude distribution $n(a)$, where a is the peak amplitude relative to noise] and average event energy [$E/N = \Sigma n(a)a^2/N$] serve as potent parameters for data evaluation.
3. Accumulation of acoustic emission data and experience is needed for the interpretation of the signature of acoustic signals detected.

To obtain acoustic emission characteristics under triaxial stress conditions similar to the operating plant, five kinds of small pressure vessels were manufactured from A508 steel, the same material used in the reactor pressure vessel nozzle.[16] The vessels contained artificial notches of different depths (see Fig. 15) and acoustic emission activities during crack growth were evaluated in hydrostatic fracture tests. The typical result is shown in Fig. 16, where total count

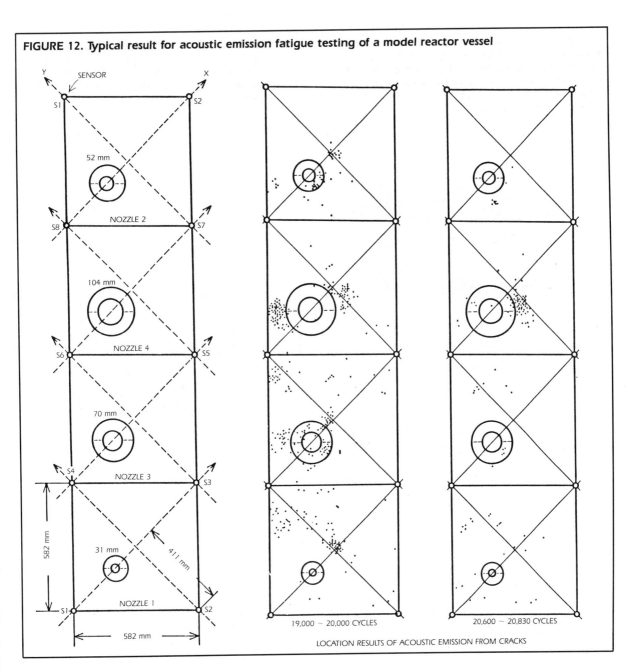

FIGURE 12. Typical result for acoustic emission fatigue testing of a model reactor vessel

curves are given on two nearly straight lines crossing at knuckle points. Each of these points was confirmed to correspond to the crack initiation time related to the initial notch depth.

Some research efforts were directed toward the use of acoustic emission in the evaluation of materials degradation, in a pressure vessel steel surveillance test. Acoustic emission characteristics were analyzed using neutron irradiated tensile and wedge opening loading (WOL) test pieces. The relationship to irradiation dose was evaluated using specimens exposed to higher degrees of radiation than was encountered on a reactor.[17,18] As shown in Fig. 17, peak amplitude

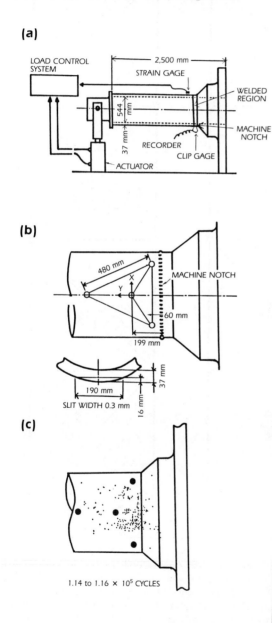

FIGURE 13. Typical result for fatigue testing of a model nozzle junction, showing: (a) the model nozzle junction; (b) the location of four acoustic emission sensors near a saw-cut notch; and (c) the result of acoustic emission signal source location at the last stage of the fatigue test

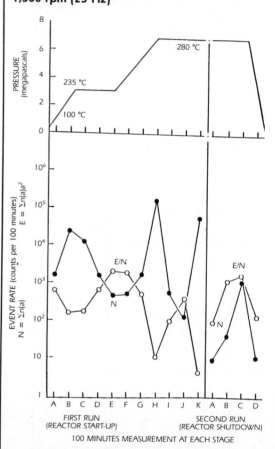

FIGURE 14. Acoustic emission activities obtained during on-line monitoring in a nuclear power plant; electric power output 0 to 70 MWe (maximum 760 MWe); steam flow rate 0 to 1,100 tons per hour; and turbine rotating speed 0 to 1,500 rpm (25 Hz)

distributions were strongly affected by irradiation hardening and embrittlement, in spite of short reactor operation time (one effective full power year). Such an application may become more important in the future when the reactor approaches the end of its service life.

Piping

Intergranular stress corrosion cracking (SCC) was experienced in the region adjacent to the heat affected zones in welds of Type 304 stainless steel piping on several components. To establish acoustic emission techniques for on-line monitoring of SCC and leak detection, several organizations

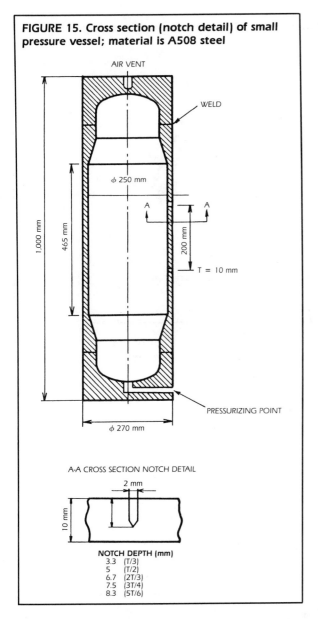

FIGURE 15. Cross section (notch detail) of small pressure vessel; material is A508 steel

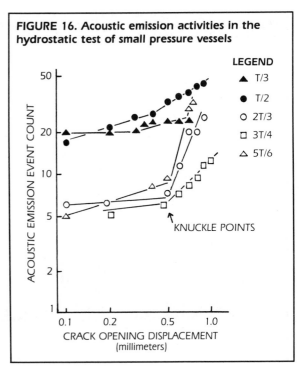

FIGURE 16. Acoustic emission activities in the hydrostatic test of small pressure vessels

in Japan performed cooperative research work. Their efforts proved the reliability of acoustic emission for detecting early stress corrosion cracking. The working group obtained the following results through basic research and accelerated pipe failure tests.[12]

1. Relatively high acoustic emission activities due to plastic deformation appear at the initial stage of loading, but these gradually decrease with work hardening. During SCC initiation, numerous small acoustic emission signals from the collapse of hydrogen bubbles can be detected, even during constant loading.
2. In the stage of SCC extension, relatively large signals can be located around the SCC damage area, when loading is applied to the pipe. Such an acoustic emission signal is mainly caused by secondary source mechanisms such as friction of crack surfaces and fracture of oxide layers.
3. Acoustic emission activities increase significantly if SCC develops rapidly at the final stage. Such a critical region as the SCC damage area showed high acoustic emission activities even at earlier stages.
4. Leakage noise is very large under reactor operating conditions but acoustic emission has a high sensitivity to leak detection and turbulent flow.

Continuous acoustic emission monitoring was conducted on a pipe test specimen with a single weld joint (as shown in Figs. 18 and 19) and good results (as shown in Figs. 20, 21 and 22) were obtained by a plant fabricator.[19] In the SCC test, the history of acoustic emission event counts reflected the early progression of corrosion damage and a step-like increase after the transition, corresponding to the propagation of intergranular SCC (Fig. 20). Many acoustic emission events were observed in load holding and unloading periods after the transition, contrary to a few events observed only in

FIGURE 17. Acoustic emission application for surveillance testing of irradiation embrittlement (fast neutron dose: 3×10^{22} n·m^{-2})

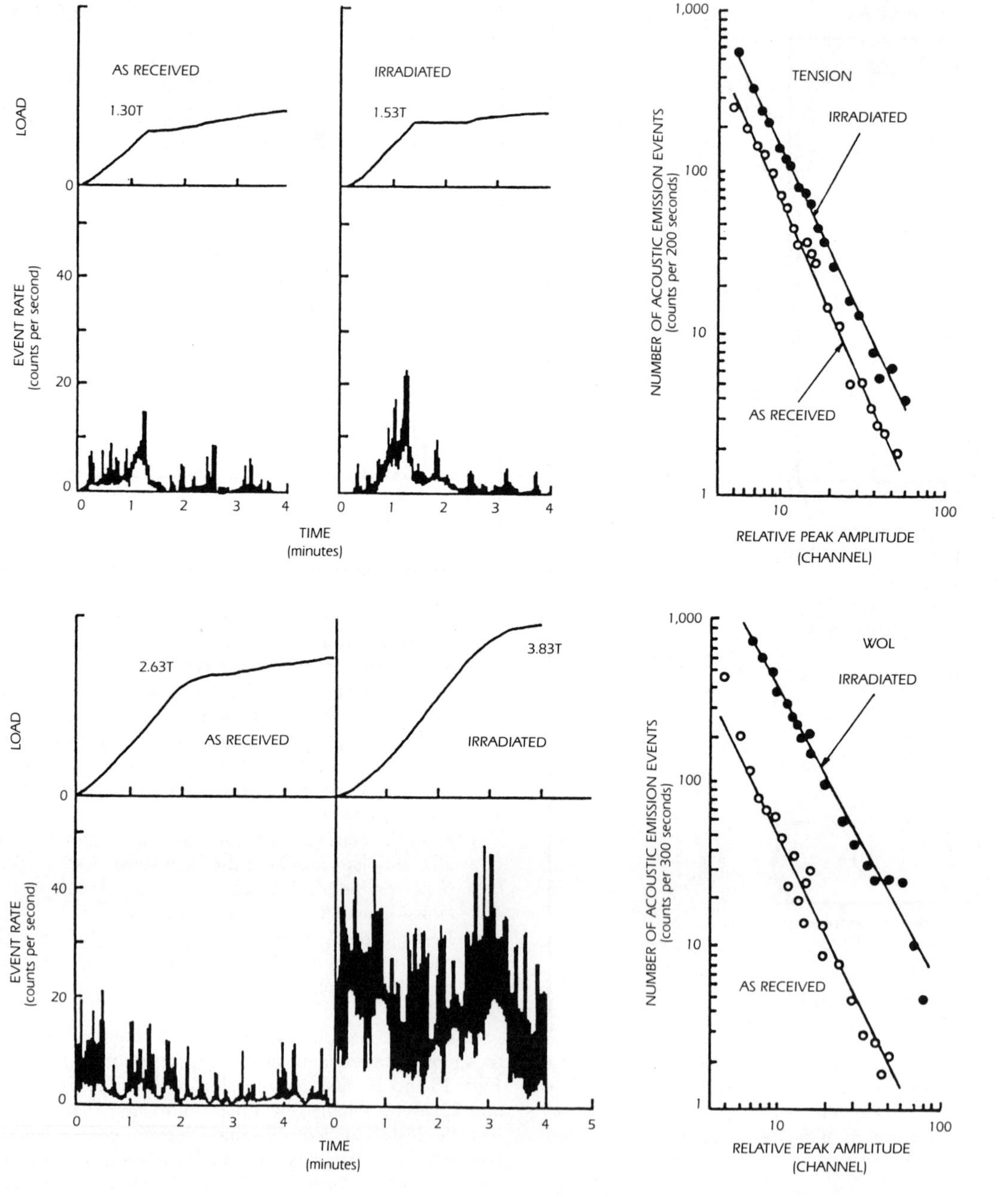

RT: ROOM TEMPERATURE ACOUSTIC EMISSION MEASUREMENT

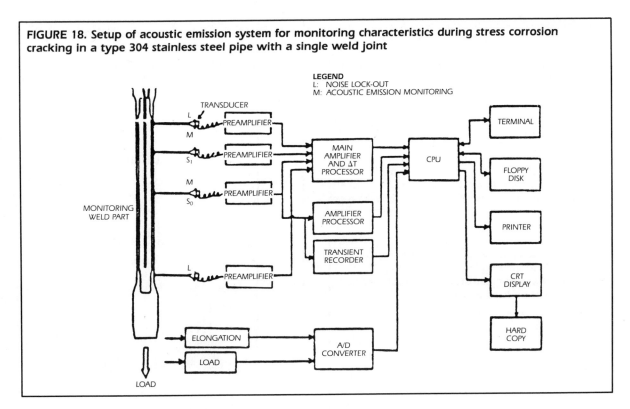

FIGURE 18. Setup of acoustic emission system for monitoring characteristics during stress corrosion cracking in a type 304 stainless steel pipe with a single weld joint

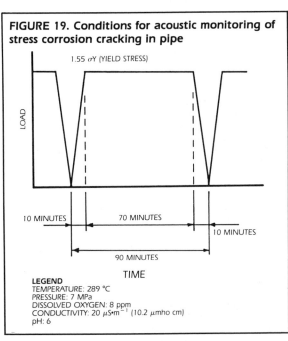

FIGURE 19. Conditions for acoustic monitoring of stress corrosion cracking in pipe

LEGEND
TEMPERATURE: 289 °C
PRESSURE: 7 MPa
DISSOLVED OXYGEN: 8 ppm
CONDUCTIVITY: 20 $\mu S \cdot m^{-1}$ (10.2 μmho cm)
pH: 6

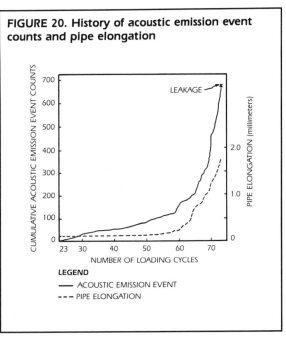

FIGURE 20. History of acoustic emission event counts and pipe elongation

LEGEND
— ACOUSTIC EMISSION EVENT
--- PIPE ELONGATION

the loading period before the transition (Fig. 21). These events showed high amplitudes and large released energies. The results of acoustic emission source location showed clear correspondence to the position of the damaged weld joint (Fig. 22).

A great number of fatigue tests, creep fatigue tests and thermal ratcheting tests were monitored by acoustic emission techniques for piping formations such as elbow, branch, reducer and nozzle components.[20,21] The results demonstrate the usefulness of acoustic emission techniques, not only for model testing but also for in-service monitoring of piping components. Typical tests were performed in conditions as near as possible to reactor operation. The results of acoustic emission monitoring in a 600 °C (1,100 °F) creep fatigue test carried out on a 300 mm (12 in.) diameter elbow

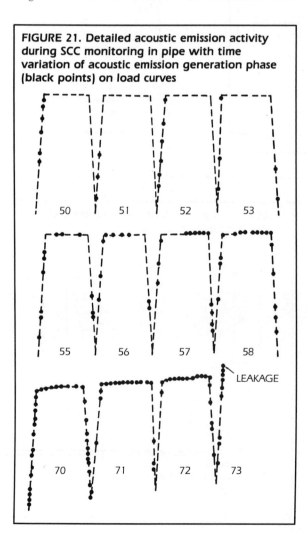

FIGURE 21. Detailed acoustic emission activity during SCC monitoring in pipe with time variation of acoustic emission generation phase (black points) on load curves

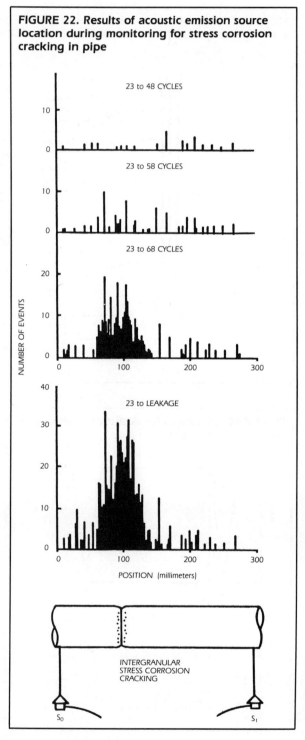

FIGURE 22. Results of acoustic emission source location during monitoring for stress corrosion cracking in pipe

pipe are shown in Figs. 23 and 24, where the acoustic emission event rate increases during both constant loading and dynamic loading, as creep fatigue cycles proceed.[21]

Figures 25, 26 and 27 show another result of acoustic emission monitoring in a thermal ratcheting test, using a liquid sodium piping test loop. The following characteristics were observed.[22]

1. The background noise generated from sodium flow was very low, and high acoustic emission activities due to thermal expansion and contraction were observed.
2. Such acoustic emission activities increased when axial stress rose, and they gradually decreased to a finite emission rate, which was proportional to the axial stress.
3. Such a proportional property disappeared after change of the failure mode from ratcheting to the occurrence of crack growth.

This phase of the research was followed by tests of the technique's ability to detect fatigue crack growth under changing parameters such as crack size, temperature, stress level and strain rate, as shown in Figs. 28 and 29.[23] The following conclusions were obtained.

1. Even in a straight stainless steel pipe specimen (in which detection of acoustic emission signals is normally difficult), large amplitude signals occurred in the active discontinuity area at elevated temperature and under bending load which causes deformation.

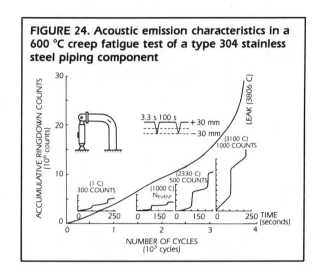

FIGURE 24. Acoustic emission characteristics in a 600 °C creep fatigue test of a type 304 stainless steel piping component

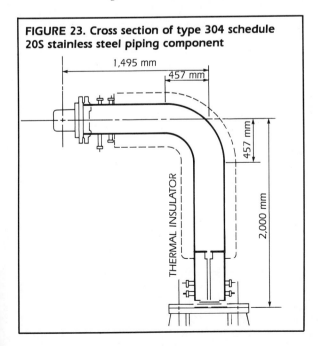

FIGURE 23. Cross section of type 304 schedule 20S stainless steel piping component

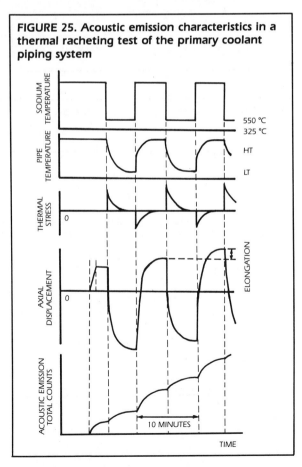

FIGURE 25. Acoustic emission characteristics in a thermal racheting test of the primary coolant piping system

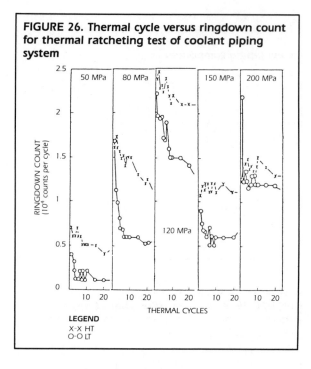

FIGURE 26. Thermal cycle versus ringdown count for thermal ratcheting test of coolant piping system

2. The existence of both a newly created fracture surface and a complicated stress field (such as shear force and torque) are needed for the acoustic emission generating mechanism to develop in the process of crack growth.
3. Acoustic emission signals originating from cracking of oxide and scale are useful for detection and location of pipe discontinuities such as abnormal deformation and cracks.

Basic research and applications studies were performed to evaluate structural integrity of the piping system by means of acoustic emission signals, as shown in Table 1.[24] The dependency of acoustic emission signals on loading phase and temperature was investigated from the viewpoint of acoustic emission source mechanism, using both small CT/CCT specimens and large, fast breeder reactor (FBR) piping components made of Type 304 stainless steel. It was found that acoustic emission techniques have a high capability for detecting and locating discontinuities in pipe. Many basic data were accumulated for the evaluation of structural integrity of plant components. In addition, acoustic emission signals generated during fatigue crack propagation were classified into the three following categories.

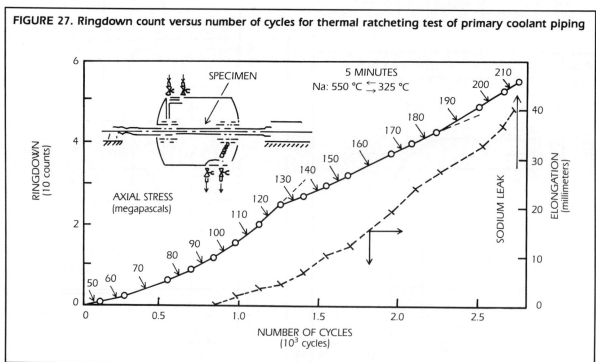

FIGURE 27. Ringdown count versus number of cycles for thermal ratcheting test of primary coolant piping

1. *Peak load acoustic emission* is generated during local crack growth at maximum applied load and is lower in activity than the opening and closure signals.
2. *Opening acoustic emission* is the signal generated during the increase of applied load and is related to local and abrupt crack opening.
3. *Closure acoustic emission* is the frictional sound occurring during crack closure.

Opening and closure acoustic emission signals are effective for detecting the existence of fatigue cracks because of their large amplitudes.

FIGURE 28. Cross section of pipe used in evaluation of acoustic emission applicability for detection and monitoring of pipe failure

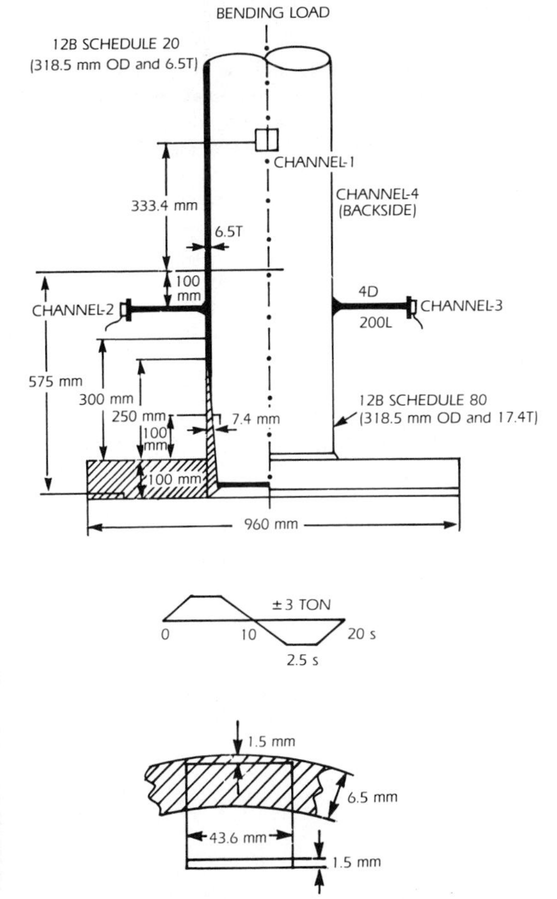

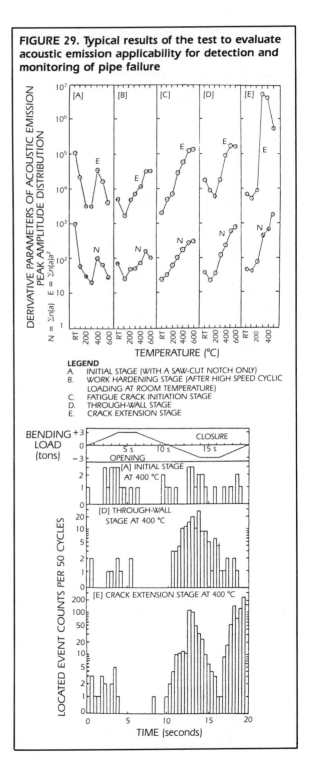

FIGURE 29. Typical results of the test to evaluate acoustic emission applicability for detection and monitoring of pipe failure

LEGEND
A. INITIAL STAGE (WITH A SAW-CUT NOTCH ONLY)
B. WORK HARDENING STAGE (AFTER HIGH SPEED CYCLIC LOADING AT ROOM TEMPERATURE)
C. FATIGUE CRACK INITIATION STAGE
D. THROUGH-WALL STAGE
E. CRACK EXTENSION STAGE

TABLE 1. Stages of an acoustic emission feasibility study

1. Development of the transducer.
2. Establishment of a calibration system.
3. Evaluation of waveguide characteristics and establishment of its handling method.
4. Development of instrumentation for the purpose of fatigue crack growth measurement (signal acquisition system based on cyclic loading phase).
5. Signal parameters available for evaluation of fatigue crack growth.
6. Acoustic emission activity and its source characterization during fatigue crack growth in materials testing.
7. Propagation, attenuation and mode transformation of acoustic emission waves in piping components.
8. Noise level measurement of liquid sodium piping components.
9. Location accuracy of fatigue cracks in mock-up piping components.
10. Acoustic emission event counts and amplitude distribution as a function of fatigue crack growth in mock-up testing.
11. Criterion for the evaluation of structural integrity based on the noise level in piping components, by means of the parameters such as event counts, amplitude, location and waveforms of acoustic emission signals.

Plant Construction

Acoustic emission research and development in the nuclear application fields were also directed beyond the area of steel structures. The Kaiser effect was utilized to provide information on geological stress to evaluate the strain energy level of rock foundations. A typical example of such an application is shown in Figs. 30, 31 and 32, where a geostress of 7.4 MPa was estimated from the Kaiser effect in the compression test of a core sample. The focus of the evaluation was a fault located beneath a nuclear power plant under construction.[14]

Fossil Fuel Power Plant Applications

Storage Tanks

The applicability of acoustic emission monitoring has also been evaluated for structures important to the safe operation of a fossil fuel power plant. To recognize and repair potentially dangerous structures at the earliest stage, before a catastrophic leakage accident or a large scale earthquake, many field tests were performed to monitor the structural integrity of oil storage tanks.

A series of field simulation tests were performed on 600 ton coolant water storage tanks in use at a steam power station.[25] The tanks were old and their reconstruction was being planned. Three kinds of water filling tests were carried out: (1) a normal operation mode test for an aged tank; (2) the failure mode test shown in Figs. 33 and 34; and (3) an initial loading test for a newly constructed tank. The following results were obtained.

Storage tank structures, made of relatively thin-walled steel, showed high acoustic emission activities during load change stages both in the preservice condition and in the aged state. The acoustic emission signals were caused mainly by such secondary emission sources as tearing and cracking of oxide or the corrosion product layer, and their occurrence depended on local bending deformation.

Relatively high acoustic emission activities in the preservice condition are a result of:

1. virgin loading of structural materials and weld joints;
2. generation of bending deformation related to the structural stability on the ground foundation; and
3. the presence of residual stress fields related to the fabrication condition without stress relief treatment.

The initial acoustic emission activities gradually decrease to zero following the Kaiser effect, as the tank goes into service. If high activities remain or occur during load change stages at the in-service condition, there may be concern for the presence of abnormal deformation or crack propagation.

Higher acoustic emission activities may be observed in the aged stages, where corrosion or degradation proceeds. Even in such stages, proper evaluation of structural integrity can be performed through the analysis of acoustic emission

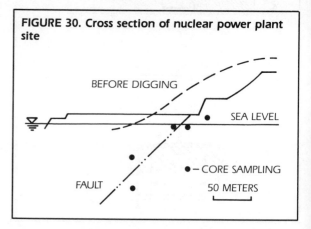

FIGURE 30. Cross section of nuclear power plant site

activities and source location on the basis of the reference data obtained here.

Acoustic emission monitoring methodology for the structural integrity assessment of cylindrical storage tanks includes:

1. inquiry into degrees of signal source concentration;
2. evaluation of acoustic emission using peak amplitude distribution; and
3. tracing the time variation of acoustic emission activities at designated areas, especially in the presence of the Kaiser effect.

The field simulation tests were followed by practical applications to oil storage tanks during hydrostatic tests after their repair.[26] As shown in Fig. 35 and Tables 2 and 3, several oil storage tanks were monitored by the same instrumentation system, and structural integrity was evaluated in comparison with the reference data mentioned above.

Pressurized Components

Acoustic emission monitoring of the superheater tube header (see Fig. 36) was performed during hydrostatic tests of steam power stations.[27] It was demonstrated that the structures tested were sound because of very low acoustic emission activities and that the technique was highly useful for structural integrity monitoring without undue interference from the environmental noise.

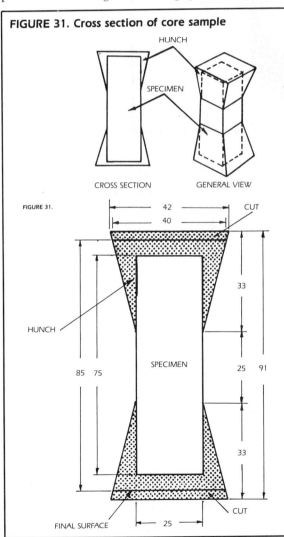

FIGURE 31. Cross section of core sample

Hydroelectric Power Plant Applications

Water Dam Gates

On-line monitoring during plant operation is one of the most desirable applications of acoustic emission testing, because early warning of abnormal conditions or incipient failure is very effective for increasing plant availability, improving performance and reducing outage time. Continuous or intermittent surveillance can be applicable to structures with good accessibility.

An acoustic emission application test to simulate on-line monitoring during operation was performed in a dam gate inspection, as shown in Figs. 37 to 39.[28] Four sensors were attached to the central part of the gate skin plate, the rack (right bank), and the rim (left and right bank). Acoustic activities during normal opening and closing of the dam gate showed a close relationship to such abnormal situations as cracks in the rack, twisting of the skin plate, and structural unbalance. Numerous large amplitude acoustic emission signals were observed from the sensor on the cracked right bank rack.

In the same gate inspection, another acoustic emission application was effectively performed for monitoring the safety of temporary partition steel walls. To make the visual inspection of the dam gate, water was drained between the partition walls and the dam gate. As water pressure was loaded to I-beams of the partition walls, acoustic emission signals from plastic deformation were observed by sensors on the I-beams. As shown in Fig. 39, however, acoustic emission activities decreased gradually with structural stabilization of the partition walls, and this helped to determine that there was little or no danger of cracking or deformation.

Turbine Rotors

Serious damage may occur when the rotor shaft of a large rotating machine breaks during operation. A wireless acoustic emission monitor was developed for the purpose of diagnosing abnormal conditions in such rotors.[29] This monitor

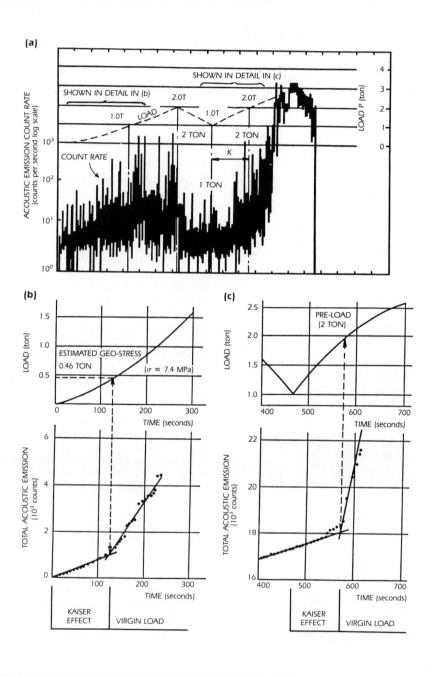

FIGURE 32. Results of geostress estimation using the Kaiser effect: (a) acoustic emission count rate; (b) estimation of geostress; and (c) influence of preloading on Kaiser effect

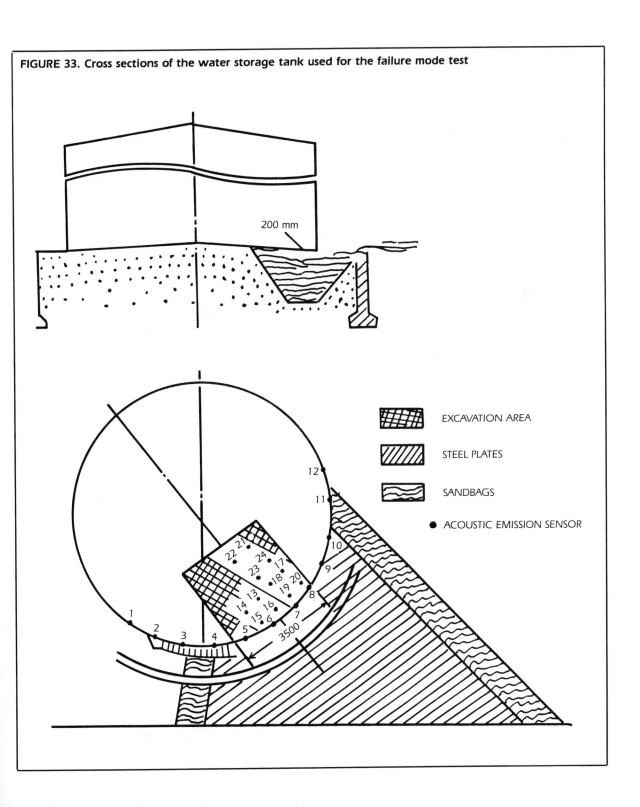

FIGURE 33. Cross sections of the water storage tank used for the failure mode test

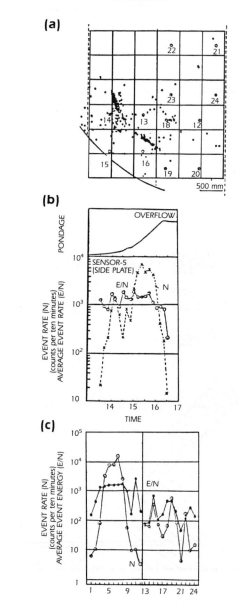

FIGURE 34. Failure mode test to simulate unequal subsidence using an aged water storage tank: (a) acoustic emission signal source location distribution (numbers correspond to numbered sensor positions in Figure 33); (b) time variation of pondage and acoustic emission activities of the sensor near the leak point; and (c) acoustic emission activities of all sensors

can detect and transmit acoustic emission signals from 50 kHz to 250 kHz, and can withstand temperatures from 0 to 100 °C (32 to 212 °F) and acceleration of 15,000 G during operation, without degradation in performance.

To evaluate this wireless monitor, a spinning breakdown test was conducted with a disk test specimen simulating the real generator rotor, as shown in Figs. 40 and 41. The disk was set on the rotating shaft and an acoustic emission sensor was inserted in a hole at the center of the disk. A transmitter module and battery were placed on the shaft and radiofrequency signals from a rotating antenna mounted on the shaft were transmitted to a stationary antenna mounted on a bearing housing. The following characteristics were observed.

1. Acoustic emission signals in the first test were generated by local plastic deformation at the points where balance weights and slots were in mutual contact. No discontinuities were discovered by a magnetic inspection after this test.
2. Acoustic emission activities in the second test were caused by plastic deformation at the root of teeth deformed by centrifugal force.
3. Acoustic emission activities in the third test (to fracture) showed higher amplitude levels caused by cracking.

These test results confirmed that the wireless acoustic emission monitor was highly sensitive and reliable. Burst signals caused by plastic deformation and cracking were observed during the increase in rotation speed of the shaft to the point of failure.

The main shaft in water power plants is subject to surface cracking. To detect such cracks and establish appropriate countermeasures, a test was conducted to monitor the main shaft, before and after its exchange, by means of the wireless monitor and other acoustic emission techniques.[30] The power plant used in the test produces 2,300 kW of power, has a water head of 13.4 m and a rotating speed of 240 rpm (4 Hz). Acoustic emission sensors were placed directly on the shaft surface and on the bearing metal of the shaft, as shown in Figs. 42 to 44. The test results were summarized as follows.

1. The shaft contained cracking before exchange and emitted numerous acoustic emission signals in the particular rotating phase.
2. The smaller cracks in the shaft were repaired and the acoustic emission activities decreased markedly.
3. Higher amplitude acoustic emission signals are expected from the impact of crack tips as cracks become larger. Acoustic emission monitoring, therefore, possesses a high capability for detection of shaft cracking.

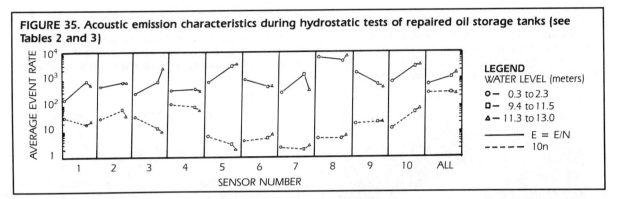

FIGURE 35. Acoustic emission characteristics during hydrostatic tests of repaired oil storage tanks (see Tables 2 and 3)

LEGEND
WATER LEVEL (meters)
○ — 0.3 to 2.3
□ — 9.4 to 11.5
△ — 11.3 to 13.0

——— E = E/N
----- 10n

TABLE 2. Conditions of oil storage tanks tested with acoustic emission

	TANK T-1	TANK T-2	TANK T-3	TANK T-4	TANK T-5
Tank type	cone roof	cone roof	floating roof	cone roof	cone roof
Diameter (millimeters)	11,640	15,520	44,600	32,900	43,400
Height (millimeters)	10,764	12,291	16,175	13,800	15,300
Capacity (cubic meters)	1,030	2,180	20,000	11,000	25,000
Content	gas oil	gas oil	naptha	kerosene	fuel oil
Shell plate material	S341	S341	HW50	S341	SM41 A
Bottom plate material	S341	S341	HW50	S341	S341
Construction date	1963	1967	1974	1964	1962
Repair	new bottom	new bottom	new annular plate	patched on bottom (overlayed)	patched on bottom (overlayed)

TABLE 3. Specifications of oil storage tanks tested with acoustic emission

Tank	Water Level (meters)	Averaged Rise Rate (millimeters per hour)	Measuring Time (minutes)	Number of Events	Event Rate (counts per minute)	Averaged Energy per Event
T-1	0.5 to 5.0		58	1,541	26.6	642
	5.7 to 10.2	710	56	1,100	19.6	574
	10.4 to 10.7		18	58	3.2	437
T-2	0.7 to 3.2		57	1,326	23.3	463
	6.1 to 8.0	420	63	1,164	18.5	449
	10.7 to 12.1		48	359	7.5	396
T-3	0.2 to 1.1		45	863	19.2	753
	7.3 to 8.0	170	60	828	13.8	1,121
	12.0 to 12.6		65	1,345	20.7	1,035
T-4	0.3 to 2.3		91	2,380	26.2	546
	9.6 to 11.5	400	94	2,522	26.3	1,044
	11.3 to 13.0		75	1,831	24.4	1,520
T-5	6.1 to 6.5		75	1,020	13.6	41
	11.1 to 11.9	260	56	434	7.8	133
	12.0 to 12.6		74	294	4.0	231

FIGURE 36. Acoustic emission application for hydrostatic tests of pressurized components in steam power plants: (a) superheater tube header monitored by acoustic emission; and (b) attachment of the acoustic emission sensor

(a)

(b)

FIGURE 37. Dam gate inspected with acoustic emission: (a) structural cross section; (b) gate positions during normal operation; and (c) positions during drainage for visual inspection

(a)
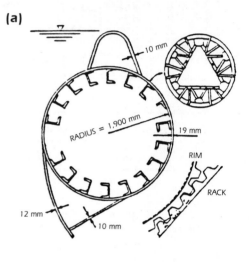

(b)
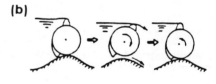

(c)
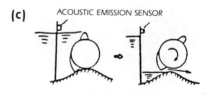

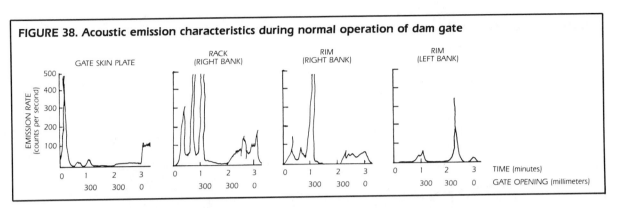

FIGURE 38. Acoustic emission characteristics during normal operation of dam gate

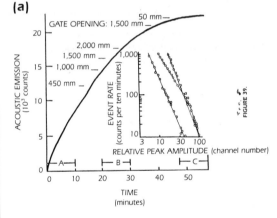

FIGURE 39. Acoustic emission characteristics used for safety monitoring of I-beam integrity during water drainage: (a) acoustic emission count versus time; and (b) event rate versus time

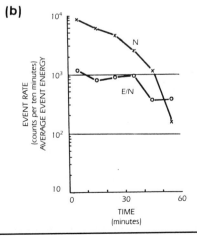

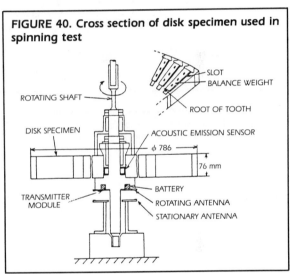

FIGURE 40. Cross section of disk specimen used in spinning test

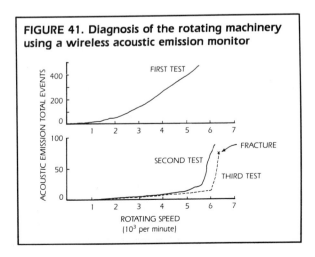

FIGURE 41. Diagnosis of the rotating machinery using a wireless acoustic emission monitor

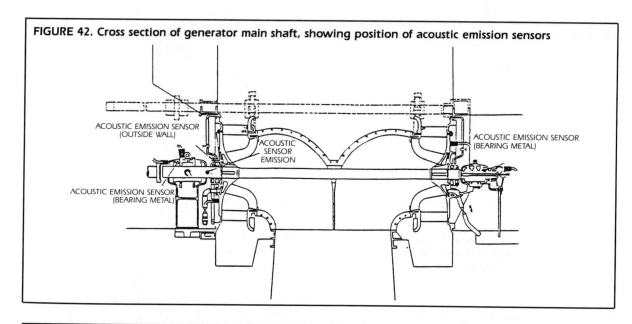

FIGURE 42. Cross section of generator main shaft, showing position of acoustic emission sensors

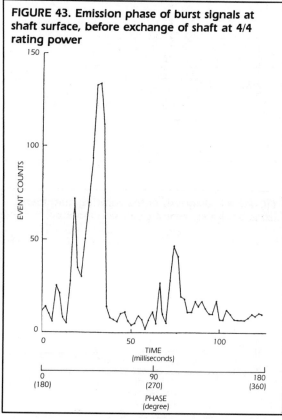

FIGURE 43. Emission phase of burst signals at shaft surface, before exchange of shaft at 4/4 rating power

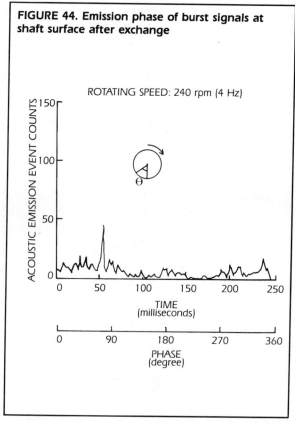

FIGURE 44. Emission phase of burst signals at shaft surface after exchange

PART 3
ACOUSTIC EMISSION LEAK DETECTION ON CANDU REACTORS

CANDU nuclear reactors are comprised of several hundred horizontal fuel channels. There are a number of advantages to this design, one being that refueling is performed with the reactor at full power by special fueling machines on each side of the reactor face (Fig. 45). The basic components of each fuel channel are two coaxial Zr-2.5Nb tubes. The inner tube (the pressure tube) is typically 6 m (20 ft) long with a 104 mm (4 in.) diameter and a 4 mm (0.16 in.) wall. The tube contains the nuclear fuel and coolant. The coolant is heavy water at a pressure of about 10 MPa at 250 to 300 °C. The outer tube (the calandria tube) is thinner than the pressure tube and is surrounded by the heavy water moderator which is under a static head equivalent to between 5 and 15 m at a temperature up to 70 °C.

The annular space between calandria and pressure tube is filled with a dry gas, either nitrogen or carbon dioxide at about atmospheric pressure. The annular gas spaces of all fuel channels are interconnected through small diameter tubing and a series of headers. The pressure tube is terminated at both ends by stainless steel endfittings. Each endfitting is 2.5 m (8 ft) long and protrudes through the biological shield and thermal insulation such that very little more than the extreme end is accessible for the fueling machine requirements.

Coolant enters at one endfitting through a connection at right angles to the fuel channel axis, and leaves at the other endfitting through an identical configuration (see Fig. 46). The feeder connections are in an insulated cabinet that is not normally accessible.

The pressure tube is attached to the endfitting by a rolled joint, made at site. The original installation procedures cause the rolling operation to extend beyond the area in the pressure tube that is restrained by the endfitting, causing in turn an area of high tensile residual stress. Hydrogen (present in the material from manufacture or added by ingress during service) precipitates as zirconium hydride when the reactor is cold.

The residual stress causes brittle hydride platelets to form in an unfavorable radial orientation and to concentrate at minute surface discontinuities. If the stress intensity is high enough, the hydride plates can crack spontaneously, and the resulting small discontinuities may extend by repetition of the process (delayed hydride cracking). In a few cases, this process has produced through-wall discontinuities. The subsequent leakage through the delayed hydride cracks was rapidly detected by dew point monitors on the annulus gas system, but these could not identify which of the several hundred pressure tubes were flawed. Typical leakage was 1 to 5 kg•h^{-1} at full pressure.

Acoustic Emission Identification of Delayed Hydride Cracks

Once the leak was detected through the annulus gas system, the problem was to identify the channels responsible. The first attempt involved forming ice plugs in the inlet and outlet feeders of a particular channel and then pressurizing the channel alone using the fueling machines. At a certain pressure, the channel fueling machine was valved closed and the rate of pressure decay was recorded. Unusually rapid fall in pressure would indicate a leak. Systematic use of this technique discovered the first leaking channel within ten days after reactor shutdown. Note that replacement energy costs for one shutdown of a Pickering reactor are about $250,000 per day.

The blocking of the single defective fuel channel did not stop liquid recovery from the annulus gas system, indicating more defective channels within the reactor. At this stage, acoustic emission was tried.

The acoustic emission system was very simple. A sensor was mounted onto a contoured shoe that in turn was affixed to a screw type hose clamp (see Fig. 47). The sensor happened to have a resonance at 500 kHz but others were also used. The implication here is that, in this application, the resonant frequency of the sensor is not of critical importance. In fact, as the system was developed toward completion, the sensor was changed to 150 kHz resonance with slight, but certainly not startling, improvement in performance. A 40 dB preamplifier, with about 1 m (3 ft) of cable between it and the sensor, boosted the signal to the main amplifier approximately 30 m (100 ft) away. The 200 to 800 kHz band-pass filtered signal was fed into a spectrum analyzer and then into a standard analog XY plotter.

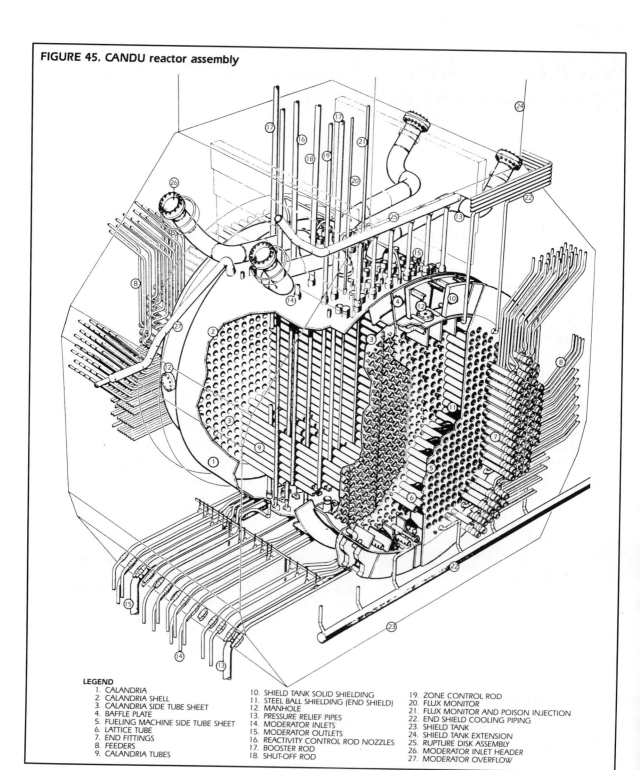

FIGURE 45. CANDU reactor assembly

LEGEND
1. CALANDRIA
2. CALANDRIA SHELL
3. CALANDRIA SIDE TUBE SHEET
4. BAFFLE PLATE
5. FUELING MACHINE SIDE TUBE SHEET
6. LATTICE TUBE
7. END FITTINGS
8. FEEDERS
9. CALANDRIA TUBES
10. SHIELD TANK SOLID SHIELDING
11. STEEL BALL SHIELDING (END SHIELD)
12. MANHOLE
13. PRESSURE RELIEF PIPES
14. MODERATOR INLETS
15. MODERATOR OUTLETS
16. REACTIVITY CONTROL ROD NOZZLES
17. BOOSTER ROD
18. SHUT-OFF ROD
19. ZONE CONTROL ROD
20. FLUX MONITOR
21. FLUX MONITOR AND POISON INJECTION
22. END SHIELD COOLING PIPING
23. SHIELD TANK
24. SHIELD TANK EXTENSION
25. RUPTURE DISK ASSEMBLY
26. MODERATOR INLET HEADER
27. MODERATOR OVERFLOW

An operator manually clamped the sensor to an endfitting at the reactor face. In this position, the sensor was about 2.5 m (8 ft) from the potential leak site. Operators working on the reactor vault floor recorded the frequency spectrum for that channel. The reactor face operator then unclamped the sensor, applied couplant (vacuum grease) and reclamped the sensor onto the next channel. Allowing for movement of the bridge carrying the reactor face operator, an inspection rate of less than two minutes per channel could be achieved. It was found that the most meaningful output occurred when a number of spectra were recorded on a single graph sheet (see Fig. 48). This process is described in reference 31.

With the reactor cold (pressurized but with minimal coolant flow), an entire reactor could be surveyed using two crews in a twelve hour shift. However, in order to gain insight into the value of the technique, a number of scans

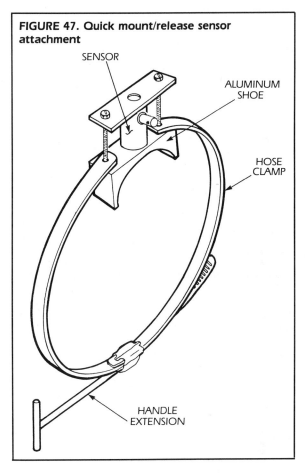

FIGURE 47. Quick mount/release sensor attachment

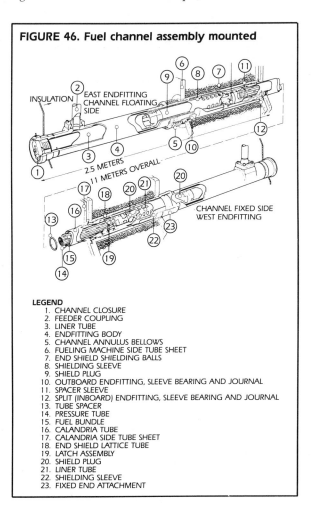

FIGURE 46. Fuel channel assembly mounted

LEGEND
1. CHANNEL CLOSURE
2. FEEDER COUPLING
3. LINER TUBE
4. ENDFITTING BODY
5. CHANNEL ANNULUS BELLOWS
6. FUELING MACHINE SIDE TUBE SHEET
7. END SHIELD SHIELDING BALLS
8. SHIELDING SLEEVE
9. SHIELD PLUG
10. OUTBOARD ENDFITTING, SLEEVE BEARING AND JOURNAL
11. SPACER SLEEVE
12. SPLIT (INBOARD) ENDFITTING, SLEEVE BEARING AND JOURNAL
13. TUBE SPACER
14. PRESSURE TUBE
15. FUEL BUNDLE
16. CALANDRIA TUBE
17. CALANDRIA SIDE TUBE SHEET
18. END SHIELD LATTICE TUBE
19. LATCH ASSEMBLY
20. SHIELD PLUG
21. LINER TUBE
22. SHIELDING SLEEVE
23. FIXED END ATTACHMENT

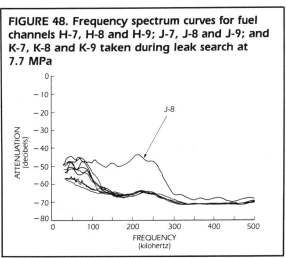

FIGURE 48. Frequency spectrum curves for fuel channels H-7, H-8 and H-9; J-7, J-8 and J-9; and K-7, K-8 and K-9 taken during leak search at 7.7 MPa

were completed with modulation of the basic reactor condition. Figure 49 illustrates the effect of reactor pressure on the acoustic emission signal for one particular leak. This manual method was initially used to identify nineteen leaking pressure tubes on two reactors. The system was then modified to allow remote operation and simplify data presentation.

Acoustic Emission Identification of Seal Leakage

During the replacement of defective pressure tubes, chemistry control of the primary heat transport coolant was less than optimum. This resulted in corrosion of the seal faces of the feeder endfitting connections in the channels that were replaced, and eventually leakage at the seals developed.

The seal leakage was quite different from the delayed hydride crack situation: (1) the leak was to atmosphere; (2) the main constraint was the economic penalty of having to upgrade the escaped heavy water rather than considerations of degradation of structural integrity; (3) the leak site was closer to the accessible face of the endfitting; and (4) because the repair was relatively straightforward and only affected very few channels, the overhead in terms of unavoidable shutdown time was proportionately higher.

Another consideration was that a leaky closure plug could generate an acoustic emission leak signal virtually identical to the seal leak situation. However, the closure plug leak could be eliminated simply by replacing the plug, using the fueling machine, with the reactor at full power. Repair of the seal leak would require reactor shutdown and special tooling. It was economically imperative that the two potential leak mechanisms be identified separately.

An unexpected poison outage occurred. This meant that for thirty-six hours the reactor would be inaccessible in terms of radiation field but operational in terms of temperature, pressure and coolant flow. A manual search was performed, duplicating the earlier acoustic emission technique, except that rms signal levels were recorded in addition to the frequency spectra. A number of seal leaks were identified but no action was taken before restart of the reactor. The total leakage rate remained at a few kilograms per hour. The rms readings confirmed that, in terms of leakage identification, there was no advantage in drawing spectra on an XY plotter.

It was decided to fabricate a leak identification system that could be carried over the reactor face using the fueling machine. This would allow surveillance with the reactor at full power. The experience under poison outage conditions (full coolant flow at 18 $kg \cdot s^{-1}$ per channel) showed that leakage of a very few kilograms an hour could be reliably detected even under conditions of full flow noise.

Acoustic Coupling and Insulation

Given that the fueling machine was going to carry the acoustic emission sensor around the reactor face, two potential problems were foreseen: (1) coupling the endfitting and the sensor; and (2) decoupling the largely hydraulic fueling machine noises from the sensor. Of these two problems, the former seemed to present the larger potential problem. Applying and removing couplant at each channel in an automatic mode would present significant difficulty. It is well known that dry coupling is usually less reliable than using the traditional couplant. In this case, the faces of the endfittings are a very fine machined finish and the fueling machine can exert an axial force of about 20 kN on the endfittings. Imposing such a force suggested rigid connection between the sensor, or sensor mounting device, and the acoustically noisy fueling machine. Figure 50 illustrates a simple experiment that was used to solve both problems. Figure 51 shows the success of the dry coupling technique and the quality of the insulation materials tested. Figure 52 is a schematic of the assembly carried by the fueling machine.

Note from Fig. 52 that a high frequency acoustic microphone was included. The function of the microphone was to detect airborne rather than structurally borne sound, which could then be used to discriminate between a benign closure plug leak and the more problematic feeder connection leak.

Experience with the Remote System

The first experience using rms measurements was with a digital voltmeter display. This was found to be less than optimum because the rapidly changing digits did not give the

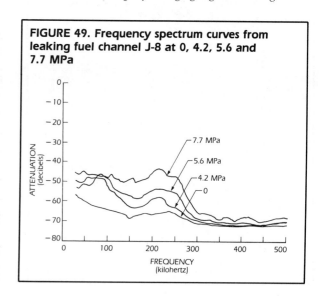

FIGURE 49. Frequency spectrum curves from leaking fuel channel J-8 at 0, 4.2, 5.6 and 7.7 MPa

operator the same feel achieved with an analog instrument. The remote instrumentation consisted of an acoustic emission amplifier, with a 100 to 400 kHz band-pass filter and rms output, and a large analog meter. Initially, the acoustic emission sensor was made from a PZT5A element, high temperature adhesives and high temperature cable. Later, this sensor was replaced by a commercially available sensor, with little difference in performance.

The mode of operation was to scan the test head across the reactor face with the fueling machine retracted. This allowed the sensing head to be within a few centimeters of each endfitting. A map of airborne noise was then drawn (see Fig. 53) and the fueling machine was directed to areas of higher than average readings. At selected channels, the fueling machine was then driven forward to press the acoustic emission sensor assembly against the endfitting, exerting a force of about 20 kN. Acoustic emission readings were then mapped, as shown in Fig. 54.

Data Processing

Reactor background noise has not been a significant problem for locating seal leaks but a processing algorithm was developed to aid interpretation of results. The flow in adjacent channels is in opposite directions, making alternate channels on one reactor face either an inlet or an outlet. It was found that the reactor noise field in both airborne and structure-borne systems varied between inlets and outlets and, in general terms, across the reactor face. Thus, each reading was evaluated against local channels of the same type. For example, an inlet would be compared with eight neighboring inlet channels.

Before making this local comparison, absolute measurements were at times ambiguous. The processing algorithm removed much of the ambiguity, as illustrated in Fig. 55, where the algorithm highlighted those channels in the area of J-15 having a seal leak. It should be emphasized that this is only a selection process to indicate general areas of high acoustic activity. Figure 54 shows no ambiguity in the acoustic emission evaluation. Most of the rms voltage figures indicated in Fig. 53 were taken at a single gain setting. Where the gain had to be changed to accommodate especially high or low readings, the recorded values were adjusted to accommodate the change.

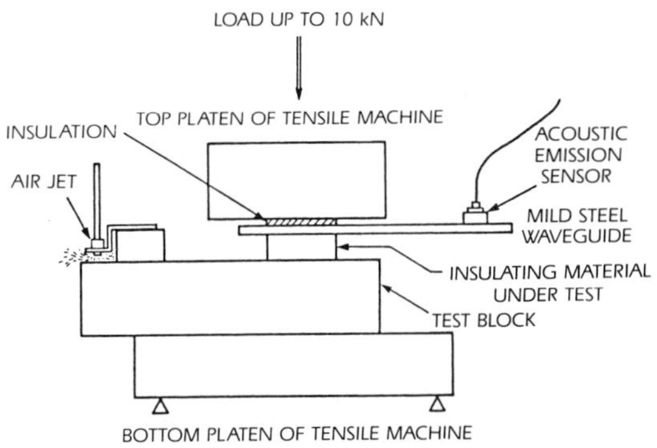

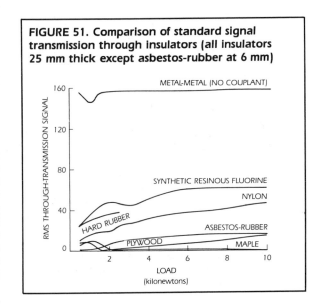

FIGURE 51. Comparison of standard signal transmission through insulators (all insulators 25 mm thick except asbestos-rubber at 6 mm)

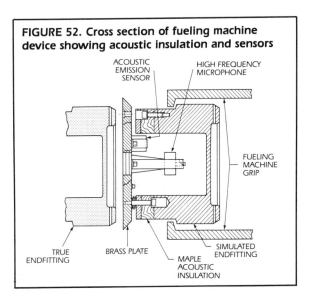

FIGURE 52. Cross section of fueling machine device showing acoustic insulation and sensors

FIGURE 50. Experiment to determine efficiency of couplant and acoustic insulators

FIGURE 53. Airborne root mean square noise scan with reactor at full power

	1	2	3	4	5	6	7	8	9	10	11	12	13	14	15	16	17	18	19	20	21	22
A								60	56	56	56	60	60	60	64							
B						60	56	56	60	56	60	60	64	64	64	68	68					
C					60	60	60	60	56	60	64	68	72	72	72	72	68					
D				56	56	52	56	56	56	60	50	60	60	60	60	105	85	80				
E			48	52	48	52	52	52	52	56	60	68	88	80	80	75	80	50	60	60		
F		52	52	52	52	52	52	52	52	56	64	70	90	95	115	100	90	76	64	56	60	
G		52	52	58	58	58	58	52	52	56	64	88	120	135	100	106	80	70	55	60	70	
H		58	56	58	56	52	56	52	52	52	60	75	66	94	180	173	125	92	68	60	60	
I	56	56	56	58	58	58	58	60	56	60	68	85	125	146	166	140	80	75	55	60	58	58
J	52	56	58	54	52	52	52	52	52	56	60	60	80	86	94	114	90	76	72	60	56	58
K	54	54	62	54	58	58	58	52	52	56	60	68	70	80	90	90	85	65	64	64	56	56
L	52	52	56	56	52	52	52	52	56	56	60	64	55	60	65	76	68	64	64	60	60	60
M	48	48	48	52	54	54	60	58	56	60	64	64	68	68	72	72	80	76	68	64	64	60
N	52	56	54	52	52	56	56	56	60	60	60	64	64	64	64	64	68	68	64	64	60	60
O	52	52	52	52	52	52	56	52	56	60	60	64	60	60	60	60	68	68	64	60	60	60
P		50	50	50	50	52	56	52	56	56	60	60	56	56	60	64	64	64	60	56		
Q		50	48	48	50	52	48	48	48	52	52	52	52	52	52	52	56	56	56	56	52	
R			52	46	48	46	44	48	48	44	44	44	44	44	48	48	48	48	52	52		
S				44	44	40	40	40	44	40	40	44	44	44	44	44	44	44	48			
T				47	47	47	47	47	47	47	47	47	47	47	47	33	40	44				
U					47	47	43	43	43	43	43	43	43	47	47	47						
V						43	43	43	43	43	32	43	47									

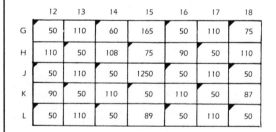

FIGURE 54. Contact acoustic emission rms readings; inlet channels indicated by triangle in upper left corners

	12	13	14	15	16	17	18
G	50	110	60	165	50	110	75
H	110	50	108	75	90	50	110
J	50	110	50	1250	50	110	50
K	90	50	110	50	110	50	87
L	50	110	50	89	50	110	50

Practical Experience

The remote system was first used when the heavy water vapor recovery from the reactor vault was at about twice normal, in the range 10 to 15 kg·h^{-1}.

A full reactor survey was completed, using both the airborne detector and localized structure-borne leak detection. This identified channel J-15 as leaking, and it was possible to eliminate both closure plug leakage and leakage through a delayed hydride crack, leaving the feeder seal as the leak source. The operation was performed with the reactor at full power. Repairs to the seal were delayed until all plans and tooling were ready, thus lost production time was minimized. The seal was repaired and the reactor returned to full operation.

The acoustic emission remote leak detection system's sensitivity to seal leaks exceeds the sensitivity of leak detection by observation of the heavy water recovery rates. For this reason, routine surveys of the reactor faces are performed so that small leaks can be detected and repaired at the most opportune time.

Conclusion

Acoustic emission has proven to be an extremely cost effective method of detecting fluid leakage across a pressure boundary. The instrumentation is simple and lends itself to remote and automatic operation. Background noise can be a problem if heavy reliance is put on absolute measurements but this was not necessary in the application discussed, and is probably not necessary in many applications.

The application described was unusual in that a single sensor was carried from place to place but the technique described would work equally well with multisensor arrays on single components such as pipes and pipelines. Comparison of readings to local background levels was better than relying on either absolute measurements or a global average.

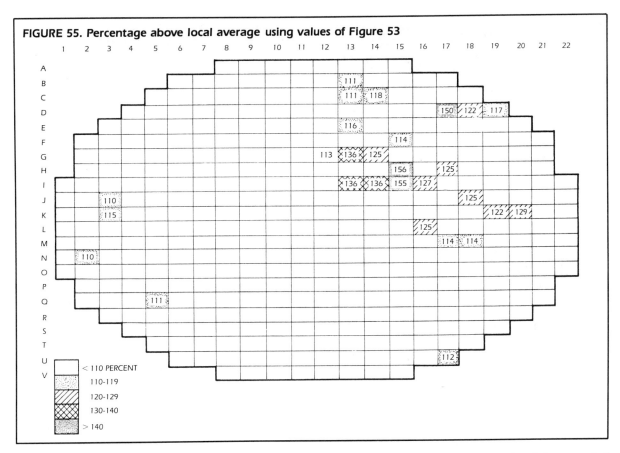

FIGURE 55. Percentage above local average using values of Figure 53

Comparison of a number of similar channels on a local basis is a better method for discrimination between leaking and not leaking because the background noise varies across the reactor face. Using a global (whole reactor) average does not enhance the information.

Acoustic emission is not only considered the prime leak location identification system for annunciated leakage on these reactors, but it is used in a routine survey mode to detect low level leakage at the endfitting feeder connection seals.

PART 4
ACOUSTIC EMISSION FROM ZIRCONIUM ALLOYS DURING MECHANICAL AND FRACTURE TESTING

Acoustic emission has been used extensively for monitoring deformation and cracking in zirconium alloys. Its nondestructive and remote sensing features make the technique ideal for laboratory applications where electrical and surface measurements or optical observations are impossible or too insensitive. Examples of the signals emitted by zirconium alloys during tensile and fatigue testing and various cracking processes are cited here to show the range and sensitivity of the acoustic emission technique. The limitations of the tests are discussed and future research is recommended.

Zirconium Alloys

The alpha phase of zirconium has a hexagonal, closely packed structure and, when fabricated into components, tends to be highly textured. As a result, strength properties and deformation are anisotropic. Zirconium has a great affinity for hydrogen and oxygen; brittle hydrides form at room temperature and the surfaces are never free from oxide. Zirconium alloys are also susceptible to stress corrosion cracking. Acoustic emission has been used to study the complex situations developed during mechanical and chemical interactions.

The two main alloys used in the nuclear industry are Zr-1.5Sn (called *Zircaloy*) and Zr-2.5Nb. The former is an alpha alloy used for fuel sheathing, pressure tubes, calandria tubes and other internal structures in the reactors; Zr-2.5Nb is an α/β alloy used for pressure tubes in CANDU reactors.

Plastic Deformation

The acoustic emission response from plastic deformation is studied (1) to separate the mechanisms of deformation; (2) to find how material variables affect acoustic emission; and (3) to provide data for fracture tests.

Acoustically, zirconium alloys behave similarly to other materials during tensile testing.[32-36] If slip is the predominant deformation mechanism, the majority of the acoustic emission is generated around the yield stress, starting at the proportional limit (see Fig. 56). At larger strains and up to ductile fracture, the material is silent.

This behavior has been interpreted in terms of Gilman's model[37] for the variation in density of mobile dislocations as a function of plastic strain. Twinning produces large bursts of noise and in large grained material, or when a specimen is pulled in a direction in which slip is difficult, the acoustic activity can continue up to fracture (see Figs. 57 and 58). After unloading and reloading at small strains, no acoustic emission is detected until the previous elongation is reached. This is the Kaiser effect which has been widely reported for other materials.

Neutron irradiation can reduce acoustic emission.[36] In experiments at room temperature on pure zirconium, twinning was suppressed by irradiation to 4×10^{23} neutrons per square meter at 470 K (197 °C or 387 °F). The specimens were deformed by dislocations clearing out the irradiation damage in channels. This latter process is very quiet. When the irradiation damage was removed by annealing at 1,020 K (747 °C or 1,377 °F), the original tensile properties were recovered, twinning occurred and copious acoustic emission was observed.

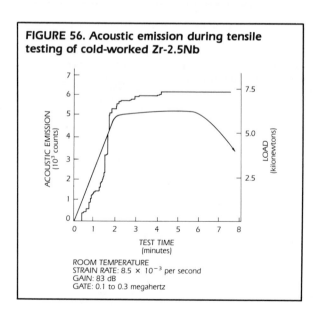

FIGURE 56. Acoustic emission during tensile testing of cold-worked Zr-2.5Nb

ROOM TEMPERATURE
STRAIN RATE: 8.5×10^{-3} per second
GAIN: 83 dB
GATE: 0.1 to 0.3 megahertz

The acoustic emission signal associated with twin formation has been studied in more detail at room temperature using hardness impressions in large grained pure zirconium.[38] The emission event took about 1 ms to complete but it consisted of about one thousand separate pulses. These pulses were interpreted as being caused by single or groups of twinning dislocations being accommodated by the undeformed zirconium crystal lattice.

Fatigue

To investigate the feasibility of on-line monitoring of fatigue, acoustic emission produced by the fracture and fatigue of heat treated Zr-2.5Nb was studied at room temperature.[35] For double and single edge notch specimens, the total acoustic emission count was plotted as a function of the square root of the notch edge opening displacement D which is proportional to K_I. The curves exhibited several stages (Fig. 58) that were interpreted in terms of the development of plastic zones.

When the plastic zone was small, the slope of the curve was usually greater than in later stages, when general yield was approached. Two fatigue tests, using simulated pressure cycling on tubes containing through-wall axial cracks, showed promise for continuous monitoring of cracks. The incremental acoustic emission energy per cycle (dE/dN) increased with the fatigue crack growth rate (da/dN) as a power law (see Fig. 59). No crack closure acoustic emission was detected. In the presence of water, the cracks provided signals that were up to 16 times more energetic, suggesting that fatigue would be easier to detect in the presence of liquids.

Cracking of Surface Oxide

Tensile specimens of Zr-1.5Sn coated in a thin oxide 1 to 3 μm thick behave acoustically like unoxidized specimens at room temperature, except for occasional energetic pulses at specimen deformations of 1 to 2 percent.[34] These results were surprising because the oxide was expected to be brittle, although other oxide cracking may have been lost in the early stages of deformation.

The expected acoustic emission from oxide cracking has been confirmed from ring tensile tests on Zr-1.5Sn fuel sheathing coated on the outside surface with thick oxides, up to 8 μm thick, produced by holding the specimen in steam at 1,270 K (997 °C or 1,826 °F) for times between 30 and 120 s.[39] A typical result (shown in Fig. 60) consisted of two maxima in count rate, one between 0.6 and 0.7 of the maximum load and the other at the maximum load. The mean and range of the maximum count rate for each peak (Table 4) were scattered but suggested that the rate at the first peak was independent of oxidation time while the rate at the second peak increased dramatically with oxidation time.

FIGURE 57. Profiles of stress-strain and acoustic emission rms voltage-strain in as-annealed zirconium; positive strain dependence of acoustic emission activity is characterized

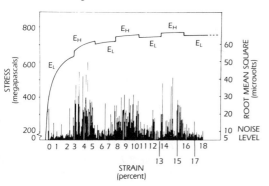

FIGURE 58. Typical least squares regression curves of acoustic emission total count versus the square root of notch edge opening displacement $(D)^{1/2}$ for an epoxy-bonded sensor

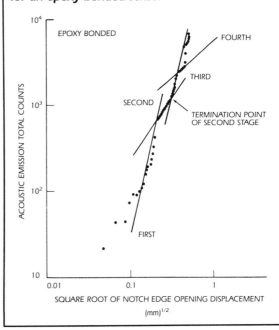

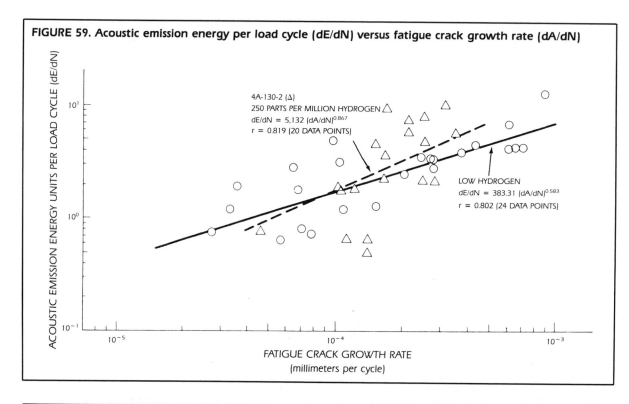

FIGURE 59. Acoustic emission energy per load cycle (dE/dN) versus fatigue crack growth rate (dA/dN)

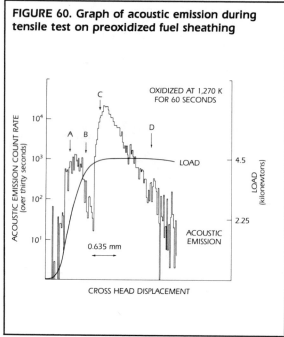

FIGURE 60. Graph of acoustic emission during tensile test on preoxidized fuel sheathing

TABLE 4. Maximum count rate at each peak during ring tensile tests on oxidized fuel sheathing (count time is 30 seconds)

Time at 1,270 K (seconds)	Peak 1 (count rate × 10³)	Peak 2
30	6.7 (4.1 to 9.8)	20.5 (9.5 to 30.3)
120	8.5	42.3
120 (vacuum)		2.2 (1.9 to 2.6)

Specimens with a very thin oxide (heated to 1,270 K in a vacuum for 120 s) had a peak count rate at maximum load but the total number of counts and the count rate were much smaller than with the oxidized specimens. Specimens oxidized for 60 s were loaded to point-A through point-D (Fig. 60) and examined by metallography and scanning electron microscope. At A, cracks with a spacing of about 30 μm were observed through the oxide and subsurface layer. At point-B, the spacing of the cracks was reduced to about 15 μm, but at point-C there was a little further cracking, spacing of 11 μm, but now the oxide was spalling off. Much spallation was observed at point-D. Acoustic emission from cracking in the oxide and oxygen rich subsurface layer was not distinguished.

Stress Corrosion Cracking

The first attempts at using acoustic emission to distinguish the stages in cracking processes were only partially successful.[32,40] Double cantilever beam and fuel cladding tube specimens of Zn-1.5Sn were placed in four environments: (1) methanol 1 percent iodine solution; (2) methanol 1 percent hydrochloric acid solution; (3) 5 percent aqueous sodium chloride solutions at room temperature; (4) and fused nitrate iodide salt mixture at 300 °C (570 °F). Acoustic emission was continuously recorded (see Fig. 61). Metallography was performed at stages during some tests and, after cracking, the fracture surfaces were examined to correlate the cracking mechanism with the acoustic emission.

Two characteristic acoustic signals were obtained: a continuous noise and a discontinuous noise. The fracture surfaces consisted of smooth cleavage facets separated by striations caused by ductile tearing and extensive slip and void formation. In the tests at 300 °C on fuel sheathing, grain boundary separation was observed. Often the fracture surfaces were etched by the solutions.

An incubation period was frequently observed before acoustic emission was detected. Continuous signals were associated with material dissolution, slip and the emission of gas bubbles (hydrogen if the specimen was cathodic and oxygen when it was anodic), as well as crack propagation. Discontinuous signals also indicated cracking but were produced by extra twinning or cracking of the protective oxide. Although several sources of noise were identified, no unique explanation of the signals was possible. The research continues for applications of acoustic emission to reliably monitor stress corrosion cracking in zirconium alloys.

Delayed Hydride Cracking

The most successful application of acoustic emission techniques to zirconium alloys has been with delayed hydride cracking (DHC).[41-50] This cracking has been responsible for failures of vessels in both the nuclear and chemical industries. The mechanism requires hydrogen to diffuse along a stress gradient and accumulate at a stress concentration where hydrides precipitate. If the stress is large enough, the hydrides crack. This sequence is then repeated and the crack grows in steps.

The inference from this description is that there should be an incubation time before cracking starts and then the crack should grow in discrete jumps. This was confirmed by acoustic emission measurements. Subsequent work established the relationship between the acoustic signals and delayed hydride cracking. Acoustic emission was then applied as a standard technique to study the important parameters of cracking.

The first experiment[4] used a notched tensile specimen of Zr-2.5Nb, loaded to 15.3 MPa at 420 K (147 °C or 297 °F) and allowed to crack. Elongation, representing crack opening, and resistivity change across the notch, as well as acoustic emission counts, were measured simultaneously as a function of time (Figs. 62 and 63). All three measurements showed the incubation time and changes associated with cracking and final failure.

The second experiment,[42] on a cantilever beam specimen of Zr-2.5Nb incorporated all the features exploited in later tests. The temperature and loading history were complicated but the acoustic emission was clearly detected and measured (see Fig. 64).

After cooling to 420 K, the first acoustic emission was observed after 250 hours (stage-1 of Fig. 64). Thereafter it was intermittent but the times between each acoustic burst were at least two orders of magnitude less than the crack initiation time. When the temperature was raised to 520 K, acoustic emission stopped (stage-2). It started again when the temperature was lowered back to 420 K (stage-3). On raising the temperature to 520 K, a few acoustic counts were detected (stage-4), but these were thought to be spurious.

Repetition of the thermal cycling around 520 K (stage-5 to stage-8) showed that if the specimen was cooled to 520 K it would emit noise. The effect of temperature change direction was repeatable up to 550 K. When the temperature was raised to 550 K, no acoustic emission was detected (stage-7), whereas when the specimen was cooled from 570 to 550 K, acoustic emission occurred (stage-10). Although acoustic emission occurred above 520 K, it could be stopped by first lowering the temperature (stage-11 and stage-12), then heating back up to 520 K (stage-13).

Reducing the stress resulted in incubation periods. In stage-14, the temperature was lowered to 420 K and the

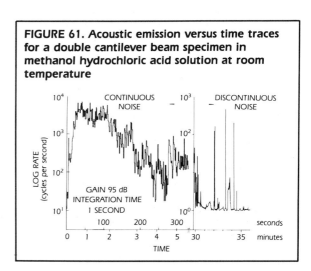

FIGURE 61. Acoustic emission versus time traces for a double cantilever beam specimen in methanol hydrochloric acid solution at room temperature

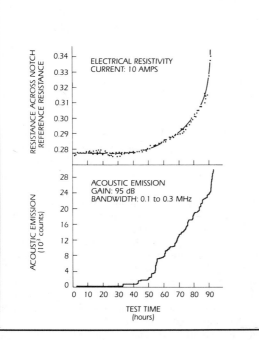

FIGURE 62. Comparison of resistivity change and acoustic emission during rupture test on round notched bar of cold-worked Zr-2.5Nb; hydrogen content 100 parts per million; test temperature 420 K; K_I at 15.3 MPa

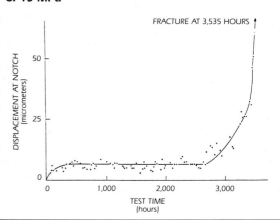

FIGURE 63. Displacement time curve for round notched bar of zircaloy-2 containing 100 parts per million hydrogen and tested at 350 K with K_I of 15 MPa

nominal stress lowered to 550 MPa. No acoustic emission was detected within three hours. When the stress was returned to 830 MPa, acoustic emission only started after incubation (stage-15). Acoustic emission stopped on reducing the stress to 690 MPa but started again after 17 hours (stage-16). Stage-17 shows that the incubation period was longer than 40 hours when the stress was lowered to 550 MPa. However, when the stress was raised back to 690 MPa, acoustic emission started immediately (stage-18). The ability to stop acoustic emission by raising the temperature to 520 K (stage-19), and restart it by lowering the temperature to 520 K (stage-24) was confirmed. After a new incubation period of 4.5 hours, induced by lowering the stress to 620 MPa, the specimen failed (stage-25).

The fracture surface associated with delayed hydrogen crack growth showed several colored bands corresponding to oxidation at higher temperatures. These could be directly related to the various stages of crack growth as indicated by acoustic emission.

In a third series of experiments,[48] cantilever beam specimens of Zr-2.5Nb were quenched at times less than and greater than the incubation time and examined metallographically. When the quench time was greater than incubation time, the hydrides were cracked. All of the above observations were the basis for accepting acoustic emission as a potent method for detecting delayed hydride cracking. Linear proportional correlations between acoustic emission counts and area of cracking have been established[45,47,49] and can be used to estimate crack lengths and velocities.

The constant of proportionality appears to be independent of K_I and temperature, although it varies slightly with each test. Thus, the usual practice was to break open the specimen after the test to confirm the constant. The constant varies with test apparatus, position of transducer and instrument settings. For example, a value of 2.4×10^3 counts·mm^{-2} was reported by one source[45] while another source[47] obtained 5×10^5 counts·mm^{-2}. Area of cracking is used because often the crack fronts are curved, making the crack longer in the midsection of the specimen. When translated into sensitivity, crack extensions of a few micrometers can be studied, thus many tests can be done on a single specimen. This leads to good testing efficiency. Four examples are listed below.

1. The determination of the threshold K_I, K_{IH}, for delayed hydride cracking by gradually reducing the load until acoustic emission and cracking stop (see Fig. 65).[42] This method can be repeated several times on one specimen and is less time-consuming than testing many specimens over a range of K_I values.
2. The determination of the effective solvus temperature T_s for hydrogen zirconium by finding the temperature above which cracking ceases.[43-48] The incubation time

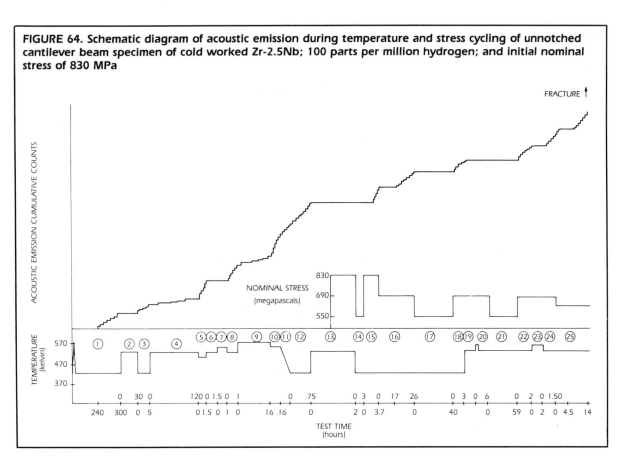

FIGURE 64. Schematic diagram of acoustic emission during temperature and stress cycling of unnotched cantilever beam specimen of cold worked Zr-2.5Nb; 100 parts per million hydrogen; and initial nominal stress of 830 MPa

was determined about the expected value of T_s. Below T_s the incubation time was short, while above T_s no acoustic emission was observed for very long times (Fig. 66).

3. The temperature dependence of crack velocity was determined for several metallurgical conditions using single specimens. Crack velocities were determined by allowing the crack to grow for short periods at one temperature then unloading, changing the temperature and reloading (see Fig. 67).[49]

4. Delayed hydride cracking in hydrogen gas is very rapid at room temperature but is sensitive to contamination.[50] Small amounts of oxygen can slow or suppress cracking. Slow cracking was detected by acoustic emission after 0.1 percent by volume of oxygen was added to the hydrogen atmosphere. This could not be observed visually through a microscope (see Fig. 68).

Acoustic emission has been used in attempts to establish the conditions for hydride cracking.[51,52] Tensile specimens of Zr-2.5Nb containing hydride platelets with normals parallel to the stressing direction produced acoustic emission at lower stresses and in larger quantities than specimens with hydrides containing normals perpendicular to the stressing direction. Notched specimens produced more acoustic emission than smooth specimens. In both studies, cracking associated with the hydrides was shown to be responsible for the acoustic emission, but the determination of the critical conditions for hydride cracking, which include stress, plastic strain and stress state, requires additional study.

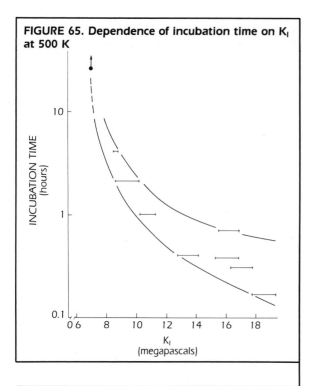

FIGURE 65. Dependence of incubation time on K_I at 500 K

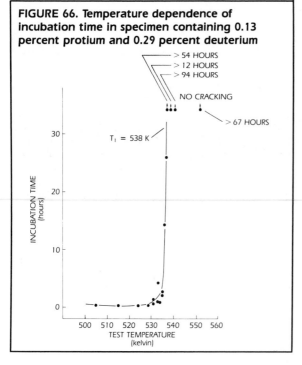

FIGURE 66. Temperature dependence of incubation time in specimen containing 0.13 percent protium and 0.29 percent deuterium

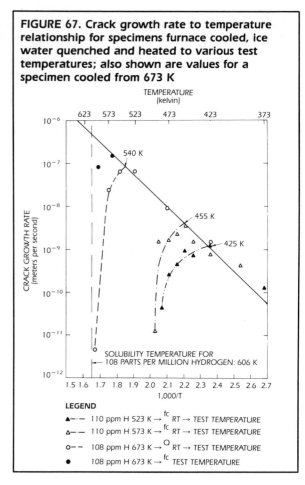

FIGURE 67. Crack growth rate to temperature relationship for specimens furnace cooled, ice water quenched and heated to various test temperatures; also shown are values for a specimen cooled from 673 K

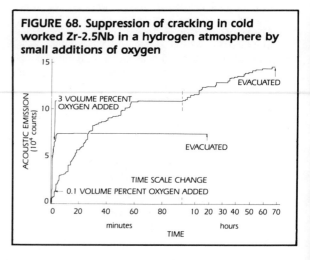

FIGURE 68. Suppression of cracking in cold worked Zr-2.5Nb in a hydrogen atmosphere by small additions of oxygen

PART 5
BUCKET TRUCK AND LIFT INSPECTION

If an aerial equipment or bucket truck fails mechanically or structurally it could result in injury or death to workers using the equipment. Many utilities have used regular test and inspection programs to ensure the safety of their line personnel and to minimize equipment repair costs consistent with safe operation.

Accidents, overloads and fatigue can occur during operation. Discontinuities in original materials also can cause problems. A thorough, regularly scheduled inspection can identify potential problems before they cause injuries or downtime. Furthermore, discontinuities discovered early are less expensive to repair than if left to develop.[57]

Insulated aerial devices can be divided into several categories: (1) extensible boom aerial personnel devices; (2) articulating boom aerial personnel devices; and (3) a combination of these two types. Figures 69 and 70 show the first two types of aerial devices or bucket trucks. Table 5 indicates which of their components may be inspected with acoustic emission methods.

Acoustic Emission Inspection Development

When acoustic emission techniques were being developed for metal and composite materials, several utilities, bucket truck manufacturers and independent testing companies began designing acoustic emission test methods. Acoustic emission from metals such as steel is a result of crack formation and propagation. On the other hand, acoustic emission from fiber reinforced plastics, used for platforms of bucket trucks, indicates matrix cracking or fiber breakage. The past work on acoustic emission inspection of bucket trucks can be found in the literature (see references 53 to 57).

The American Society for Testing and Materials (ASTM) issued standard F914 entitled *Standard Test Method for Acoustic Emission for Insulated Aerial Personnel Devices* in 1985.[58] The ASTM standard method, and most past applications, utilized the acoustic emission technique for *zone location* of discontinuities. Verification of emission sources may require the use of other nondestructive test methods such as radiography, ultrasonics, magnetic particle, liquid penetrant and visual inspection.

Described below are the acoustic emission methods most commonly used for bucket truck inspection. Some refined techniques for other applications may be valid as well.

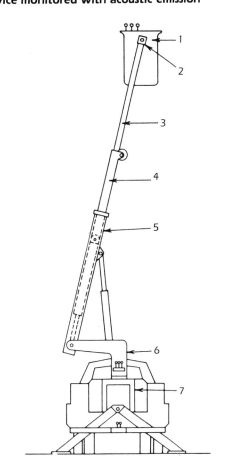

FIGURE 69. Components of an extensible aerial device monitored with acoustic emission

LEGEND
1. PLATFORM
2. PLATFORM ATTACHMENT
3. UPPER BOOM
4. INTERMEDIATE BOOM
5. LOWER BOOM
6. TURNTABLE AND ROTATION BEARING
7. PEDESTAL

FROM AMERICAN SOCIETY FOR TESTING MATERIALS (SEE REFERENCE 58). REPRINTED WITH PERMISSION.

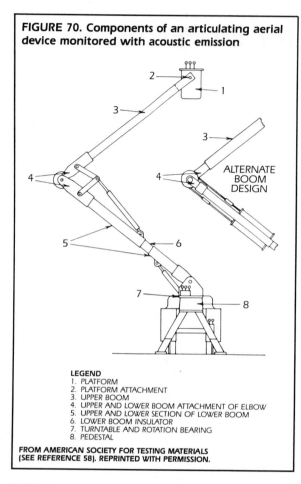

FIGURE 70. Components of an articulating aerial device monitored with acoustic emission

LEGEND
1. PLATFORM
2. PLATFORM ATTACHMENT
3. UPPER BOOM
4. UPPER AND LOWER BOOM ATTACHMENT OF ELBOW
5. UPPER AND LOWER SECTION OF LOWER BOOM
6. LOWER BOOM INSULATOR
7. TURNTABLE AND ROTATION BEARING
8. PEDESTAL

FROM AMERICAN SOCIETY FOR TESTING MATERIALS (SEE REFERENCE 58). REPRINTED WITH PERMISSION.

TABLE 5. Aerial device components monitored with acoustic emission techniques (see Figures 69 and 70 and reference 58)

Articulated Aerial Device Components	Extensible Aerial Device Components
Pedestal	Pedestal
Platform	Platform
Platform attachment	Platform attachment
Upper boom	Upper boom
Turntable and rotation bearing	Turntable and rotation bearing
	Intermediate boom
	Lower boom
Lower boom insulator	
Upper and lower boom attachment of elbow	
Upper and lower section of lower boom	

Instrumentation for Bucket Truck Inspection

Because sound attenuation for a given frequency in fiberglass composite material is higher than that in metal, the acoustic emission sensor frequency for the fiber reinforced plastic component of bucket trucks should be lower than that for use in metal. The transducer frequency for these components is usually in the range of 20 kHz to 75 kHz for acoustic emission waves that travel sufficiently long distances in the components. The frequency range from 100 kHz to 400 kHz is usually suitable for detection of acoustic emission in metal components for bucket truck inspections.

During proof load testing of aerial devices, a number of channels are needed to monitor individual components simultaneously. Typical acoustic emission monitoring systems have eight to sixteen independent channels. A typical sixteen-channel system will have ten to twelve high frequency channels for the metal components and four to six low frequency channels for the fiber reinforced plastic components.

A plot or display of acoustic emission events versus monitoring channels indicates which sensor is closest to the emission source. Most commercial instruments also have a first-hit circuit which registers the first channel when several sensors detect the same event.

According to the ASTM F914 standard, the amplitude of an acoustic emission signal at the output of a sensor is expressed in terms of decibels (dB) which are defined as:

$$\text{signal peak amplitude} = 20 \log_{10} \frac{A_1}{A_0} \quad \text{(Eq. 1)}$$

Where:

A_0 = one microvolt as the reference voltage; and
A_1 = peak voltage of the measured acoustic emission signal at the sensor output.

In order to distinguish significant events, an instrument usually has at least two threshold levels for the inspections. A high threshold level for 70 dB peak signals at the sensor output indicates serious damage such as fiber fracture and metal cracking. The ASTM standard also uses a 40 dB threshold level for the acceptance criteria in terms of acoustic emission counts.

Test Procedure for Bucket Truck Inspection

The test configurations for the proper load of articulated and extensible aerial devices are shown in Figs. 71 and 72.

Most aerial devices have two test configurations for complete monitoring of all potential damage.

A load measuring device such as an electronic load cell must be attached to the load application system, that in turn is attached to an adequate dead weight or anchor. The ASTM standard recommends the following method to provide sufficient load on the aerial lift.

1. The method of load application and attachment shall evenly distribute the load and shall not permanently deform the platform. The load system is attached to the bucket or platform so that the centerline of load application passes through the centerline of the bucket. On units with two platforms, the load must be distributed to both platforms. On units with platform rotators, the platform shall be rotated full around toward the end of the boom.
2. The load connection technique should simulate infield use as closely as possible.
3. The test load employed shall be two times the rated platform capacity.

The loading sequence is illustrated in Fig. 73. Prior to applying the load, possible background noise picked up by each acoustic emission sensor should be observed for two minutes. Potential background noise for bucket truck testing includes electromagnetic induction, close proximity to radio stations, improper grounding, rubbing interfaces and impact. If the background noise is excessive and cannot be eliminated, the test may be rescheduled or relocated.

Loading starts with a uniform rate, between 15 and 45 $N \cdot s^{-1}$ (3 and 10 $lbf \cdot s^{-1}$), until the maximum test load is attained. The load is held at maximum for four minutes, then removed at a constant rate, between 15 and 90 $N \cdot s^{-1}$ (3 and 20 $lbf \cdot s^{-1}$). After another four minute hold period at zero load, the load cycle is repeated.

During the loading, unloading and holding at maximum load and zero load, events, counts, amplitude distributions and location data for all metal and fiberglass channels are recorded versus time and load. If any acoustic emission test results indicate that damage could be occurring to the aerial device, the test should be stopped.

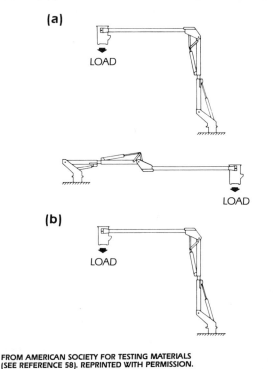

FIGURE 71. Test positions for articulated aerial devices: (a) over center model (upper boom travels past vertical); and (b) non-over center model (upper boom does not travel past vertical)

FROM AMERICAN SOCIETY FOR TESTING MATERIALS (SEE REFERENCE 58). REPRINTED WITH PERMISSION.

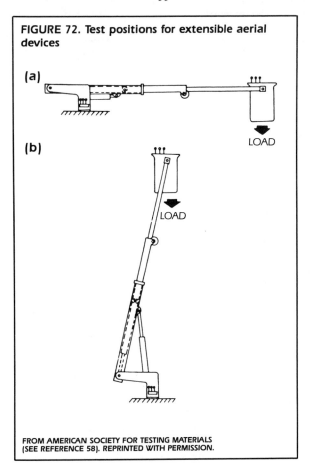

FIGURE 72. Test positions for extensible aerial devices

FROM AMERICAN SOCIETY FOR TESTING MATERIALS (SEE REFERENCE 58). REPRINTED WITH PERMISSION.

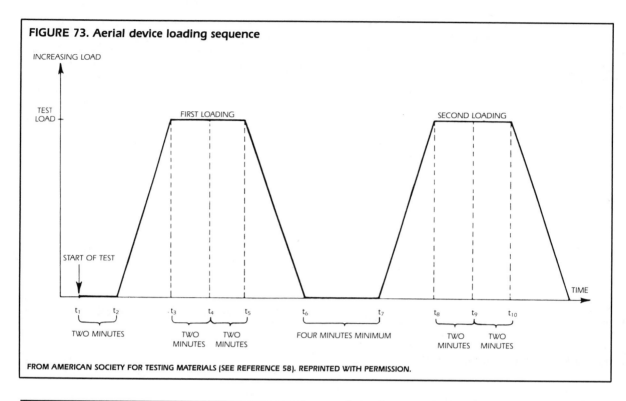

FIGURE 73. Aerial device loading sequence

FROM AMERICAN SOCIETY FOR TESTING MATERIALS (SEE REFERENCE 58). REPRINTED WITH PERMISSION.

Typical Test Data

Figure 74 shows typical sets of data that represent the accumulations of events in each channel.[56] The chart in Fig. 74a is obtained for those events that exceed the high amplitude threshold at 70 dB. Figure 74b is for the low threshold at 40 dB. The total events in each channel in Fig. 74b should be subtracted by the high amplitude events in the corresponding channel in Fig. 74a to obtain the low amplitude events. The high amplitude events indicate serious damage such as fiber fracture or metal cracking. The low amplitude events may result from other discontinuities such as insufficient lubricating and nonpropagating cracks. These low to medium amplitude events (40 to 70 dB) are detected during the loading periods but are quiet at a constant load.

The events shown in Fig. 74a indicate that high amplitude emissions were detected at channels 2, 4, 6 and 7. Since there were no high amplitude events detected by the sensors of channels 3 and 5, the sensors of channels 2, 4 and 6 must pick up those events independently. The events shown on channel 7 might be the acoustic emission pulses propagating from the component with the channel 6 sensor to the channel 7 sensor component through acoustic coupling of the two components. This interpretation can be confirmed through the first hit registration records.

After subtracting the high threshold events from the events in Fig. 74b, channels 1, 3, 5, 8, 9, 10, 12, 14 and 15 detected low amplitude emissions. Only channels 9, 12 and 15 definitely picked up low amplitude emissions independently, because channel 9 had more events than the adjacent channels 8 and 10; channel 12 had a significant amount of events and no event in either channels 11 and 13; and channel 15 had more events than channel 14. However, channels 1 and 3, 5 and 7 detected low amplitude emissions possibly coming from the same emission sources in the components with sensor 2 and 6, respectively.

There were two possible reasons for this: (1) the source that emitted high amplitude acoustic emission also emitted low amplitude emissions; and (2) the high amplitude emission pulses became weaker through the joints between two components. Therefore, the exact number of either high or low amplitude events generated in one component should be obtained in cooperation with the first hit registrations.

Figure 75 shows two amplitude distributions obtained from a metal component with a high frequency sensor, and a fiber reinforced plastic component with a low frequency sensor.[56] The histograms represent the number of events in every 5 dB interval, started at the 40 dB low threshold level. The total number of events in one histogram should equal the number of events in the corresponding channel in the events versus channels chart for the low threshold level.

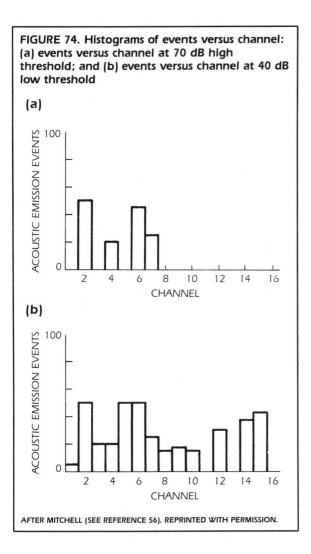

FIGURE 74. Histograms of events versus channel: (a) events versus channel at 70 dB high threshold; and (b) events versus channel at 40 dB low threshold

AFTER MITCHELL (SEE REFERENCE 56). REPRINTED WITH PERMISSION.

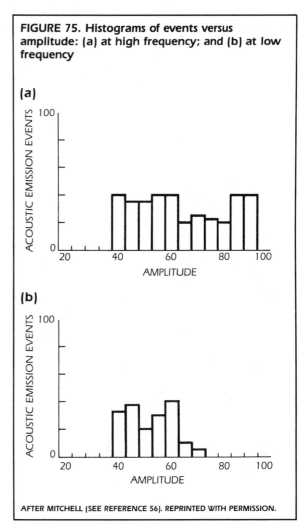

FIGURE 75. Histograms of events versus amplitude: (a) at high frequency; and (b) at low frequency

AFTER MITCHELL (SEE REFERENCE 56). REPRINTED WITH PERMISSION.

Count histograms of all channels for either high or low threshold are similar to those of event histograms as shown in Fig. 75.

Acceptance Criteria

The ASTM Standard F914-85 provides the following guidelines as the acceptance criteria for the acoustic emission testing of insulated aerial personnel devices:

1. less than 150 total events first hit, with amplitude greater than 70 dB on fiber reinforced plastic and metal channels combined;
2. less than 15,000 total counts first hit, with amplitude greater than 40 dB on fiber reinforced plastic and metal channels combined.

The above guidelines are only applied to an entire device in accordance with either total events or total counts in all test channels. There are no guidelines for a single component of either metal or fiber reinforced plastic.

REFERENCES

1. Stahlkopf, K.E. and G.J. Dau. "Acoustic Emission, A Critical Assessment." *Nuclear Safety*. Vol. 17, No. 1 (January-February 1976).
2. Bentley, P.G. "A Review of Acoustic Emission for Pressurized Water Reactor Applications." *NDT International*. Vol. 14, No. 6 (December 1981): pp 313-360.
3. McElroy, J.W. and W.F. Hartman. *An Experimental Study of Advanced Continuous Acoustic Emission Monitoring of BWR Components*. US Department of Energy Report No. C00-4592-2 (September 1980).
4. McElroy, J.W. *Acoustic Emission Leak Detection Methods for Light Water Reactor Application*. US Department of Energy Report No. C00-35208-3 (August 1982).
5. Hutton, P.H., et al. *Acoustic Emission Results Obtained from Testing the ZB-1 Intermediate Scale Vessel Pressure Vessel*. NUREG/CR-3915, PNL-5184. Richland, WA: Pacific Northwest Laboratory (August 1985).
6. Taylor, T.T., et al. *Acoustic Emission Monitoring of ASME Section III Hydrostatic Test, Watts Bar Unit 1 Nuclear Reactor*. NUREG/CR-2880, PNL-4307. Richland, WA: Pacific Northwest Laboratory (October 1982).
7. Hutton, P.H., et al. *Acoustic Emission Monitoring of Hot Functional Testing, Watts Bar Unit 1 Nuclear Reactor*. NUREG/CR-3693, PNL-5022. Richland, WA: Pacific Northwest Laboratory (June 1984).
8. Hutton, P.H. *Summary Report on Investigation of Background Noise at the San Onofre Nuclear Power Reactor as Related to Structural Flaw Detection by Acoustic Emission*. Southwest Research Institute Project 17-2440. Biannual Progress Report No. 6. Vol. I, Projects I through IV. Richland, WA: Pacific Northwest Laboratory (January 1972).
9. Smith, J.R., G.V. Rao and J. Craig. *Acoustic Monitoring Systems Tests at Indian Point Unit 1*. C00/2974-2, PC A06/MF A01. Pittsburgh, PA: Westinghouse Electric Corporation (December 1979).
10. Ying, S.P., J.E. Knight and C.C. Scott. "Background Noise for Acoustic Emission in a Boiling Water and a Pressurized Water Nuclear-Power Reactor." *The Journal of the Acoustical Society of America*. Vol. 53, No. 6 (1973).
11. Doctor, P.G., T.P. Harrington and P.H. Hutton. *Pattern Recognition Methods for Acoustic Emission Analysis*. NUREG/CR-0910, PNL-3052. Richland, WA: Pacific Northwest Laboratory (July 1979).
12. Nakasa, H. "Some Critical Remarks on Industrial Applications of Acoustic Emission." *Progress in Acoustic Emission II*. Tokyo, Japan: Japanese Society for Nondestructive Inspection (1984): pp 286-293.
13. Onoe, M., et al. "Multichannel AE Source Location System and Its Application to Fatigue Test of Model Reactor Vessel." *Proceedings of the Second Acoustic Emission Symposium*. Session 2. Tokyo, Japan (1974): pp 82-103.
14. Nakasa, H., H. Kusanagi and H. Ohno. "Review of Japanese Research and Development Activities on Acoustic Emission Applications in the Nuclear Industry." *Nondestructive Evaluation in the Nuclear Industry*. Metals Park, OH: American Society for Metals (1978): pp 206-220.
15. Uchida, K., et al. "AE Characteristics in a Fatigue Test of a BWR Model Pipe Joint." *Proceedings of the First National Acoustic Emission Conference*. Tokyo, Japan (1977): pp 37-42.
16. Shinomiya, Y., et al. "AE Characteristics in Hydrostatic Tests on Small Pressure Vessels." *Proceedings of the American Society for Nondestructive Testing Spring Conference*. Columbus, OH: American Society for Nondestructive Testing (1981).
17. Tokiwai, M., et al. "An Application of Acoustic Emission Techniques to the Study of Tensile Deformation Behaviors in Neutron Irradiated Zirconium and Other Nuclear Materials." *Proceedings of the Fourth Acoustic Emission Symposium*. Tokyo, Japan (1978): pp 62-88.
18. Nakasa, H. *AE Instrumentation Systems for Signal Analysis and Their Applications to Structural Integrity Diagnosis*. CRIEPI Report No. 203 (1979).
19. Komura, I., et al. "Acoustic Emission Characteristics of Stress Corrosion Cracking in Type 304 Stainless Steel." *Advances in Acoustic Emission*. Knoxville, TN: Dunhart Publishers (1981): pp 195-203.
20. Nakasa, H., S. Ueda and T. Nagata. "Acoustic Emission Analysis on Fatigue Failure of Fast Breeder Reactor Piping Components." *Proceedings of the Second Acoustic Emission Symposium*. Session 9. Tokyo, Japan (1974): pp 21-43.
21. Ohno, H., et al. *Application of Acoustic Emission Techniques to a Creep-Fatigue Test of a Type 304 Stainless Steel Elbow Component*. CRIEPI Report No. 77002 (1977).

22. Nakasa, H., et al. "AE Monitoring of a Thermal Ratcheting Test in the Sodium Piping Loop." *Journal of Nondestructive Inspection*. Vol. 24, No. 2. Japanese Society for Nondestructive Inspection. (1975): pp 82-83.
23. Nakasa, H. and T. Nagata. "Applicability of Acoustic Emission Techniques to Detection and Monitoring of Pipe Failure." *Proceedings of the Third International Symposium*. Japan Welding Society (1978): pp 215-220.
24. Sakakibara, Y., T. Kishi and K. Yamaguchi. "Evaluation of Structural Integrity of Piping Components for Fast Breeder Reactor by Acoustic Emission Signals." *Progress in Acoustic Emission II*. Japanese Society for Nondestructive Inspection (1984): pp 278-285.
25. Kawaguchi, K., et al. "Acoustic Emission Diagnostics for Integrity of Cylindrical Storage Tank Structures." *Proceedings of the Fifth Acoustic Emission Symposium*. Tokyo, Japan (1980): pp 521-535.
26. Sasaki, H. and M. Inagaki. "AE Characteristics During Hydrostatic Tests of Oil Storage Tanks Repaired." *Progress in Acoustic Emission II*. Japanese Society for Nondestructive Inspection (1984): pp 210-217.
27. *OHM*. Vol. 67, No. 6. Tokyo, Japan: The OHM Company (1980): pp 76.
28. Nakasa, H. and Y. Numazaki. "AE Applications in the Maintenance Management of Steel Structures Used in the Civil Engineering Field." *Journal of Nondestructive Testing*. Vol. 23, No. 9. Japanese Society for Nondestructive Inspection (1974): pp 500-501.
29. Sato, I., et al. "Development of a Wireless AE Monitor and Its Application to the Diagnosis of Rotating Machinery." *Proceedings of the Fifth Acoustic Emission Symposium*. Tokyo, Japan (1980): pp 454-464.
30. Imaeda, H., et al. "Acoustic Emission Monitoring of the Main Shaft in the Hydroelectric Power Plant." *National Conference on Acoustic Emission*. Hiroshima, Japan: Japanese Society for Nondestructive Inspection (1985): pp. 137-142.
31. Kupcis, O.A. "Nondestructive Inspection of Pressure Tubes at the Pickering Nuclear Generating Station." *Third Conference on Periodic Inspection of Pressurized Components*. Institution of Mechanical Engineers (1976).
32. Cox, B. "A Correlation Between Acoustic Emission During Stress Corrosion Cracking and Fractography Cracking of the Zircaloys." *Corrosion*. Vol. 30. Houston, TX: National Association of Corrosion Engineers (1974): pp 191-202.
33. Glygalo, V.N., V.V. Kirsanov, L.S. Kravtsova, and A.G. Leshchinskii. *Defektoskopiya*. Vol. 3 (1975): pp 140-142.
34. Cox, B. and C.E. Coleman. *Acoustic Emission During the Tensile Testing of Zirconium Alloys*. Atomic Energy of Canada Report AECL-5831 (May 1977).
35. Schwenk, E.B., P.H. Hutton and M. Igarashi. *Acoustic Emission During Room Temperature Fatigue Crack Growth and Fracture of Zr-2.5Cb Alloy Specimens*. ASTM STP 633. Philadelphia, PA: American Society for Testing and Materials (1977): pp 573-588.
36. Tokiwai, M., H. Kimura, H. Nakasa and H. Takaku. *Proceedings of the Fourth Acoustic Emission Symposium*. Tokyo, Japan (1978): pp 62-88.
37. Gilman, J.J. *Journal of Applied Physics*. Vol. 36 (1965): pp 2722.
38. Boiko, V.S., L.G. Ivanchenko and L.F. Krivenko. *Soviet Solid State Physics*. Vol. 26 (1984): pp 1,340.
39. Coleman, C.E. Unpublished data.
40. Cox, B. "Stress Corrosion Cracking of Zircaloy-2 in Neutral Aqueous Chloride Solutions at 25 °C." *Corrosion*. Vol. 29. Houston, TX: National Association of Engineers (1973): pp 157-166.
41. Coleman, C.E. and J.F.R. Ambler. *Susceptibility of Zirconium Alloys to Delayed Hydrogen Cracking*. ASTM STP 633. Philadelphia, PA: American Society for Testing and Materials (1977): pp 589-607.
42. Ambler, J.F.R. and C.E. Coleman. *Second International Congress on Hydrogen in Metals*. Paper 3C10. Paris, France (1977).
43. Coleman, C.E. and J.F.R. Ambler. *Canadian Metals Quarterly*. Vol. 17 (1978): pp 81-84.
44. Arora, A. and K. Tangri. *International Fracture Journal*. R93-R95. Vol. 15 (1979).
45. Arora, A. and K. Tangri. *Experimental Mechanics*. Vol. 21 (1981): pp 261-267.
46. Tangri, K. and X.Q. Yuan. *Fracture Problems and Solutions in the Energy Industry*. New York, NY: Pergamon Press (1982): pp 27-28.
47. Coleman, C.E. *Effect of Texture on Hydride Reorientation and Delayed Hydrogen Cracking in Cold-Worked Zr-2.5Nb*. ASTM STP 754. Philadelphia, PA: American Society for Testing and Materials (1982): pp 393-411.
48. Coleman, C.E. and J.F.R. Ambler. *Scripta Metallurgica*. Vol. 17 (1983): pp 77-82.
49. Ambler, J.F.R. *Effect of Direction of Approach to Temperature on the Delayed Hydrogen Cracking Behavior of Cold-Worked Zr-2.5Nb*. ASTM STP 824. Philadelphia, PA: American Society for Testing and Materials (1984): pp 653-674.
50. Coleman, C.E. and B. Cox. *Cracking Zirconium Alloys in Hydrogen*. ASTM STP 824. Philadelphia, PA: American Society for Testing and Materials (1984): pp 675-690.
51. Simpson, C.J., O.A. Kupcis and D.V. Leemans. ASTM STP 633 Philadelphia, PA: American Society for Testing and Materials (1977): pp 630-642.
52. Simpson, L.A. *Metallurgical Transactions*. Vol. 12A (1981): pp 2113-2124.
53. Pollock, A.A. and W.J. Cook. "Acoustic Emission Testing of Aerial Devices." New Orleans, LA: Southeastern

Electric Exchange. Engineering and Operating Division Annual Conference (April 1971).

54. Younger, W.O. and W.E. Swain. "Non-Destructive Testing Used on Bucket Trucks." *Transmission and Distribution.* (March 1980): pp 66-69.

55. LaRocca, P. "Acoustic Emission — The Bucket Stops Here." *Materials Evaluation.* Vol. 40, No. 5. Columbus, OH: The American Society for Nondestructive Testing (1982): pp 512-513.

56. Mitchell, J.R. and D.G. Taggart. "Acoustic Emission Testing of Insulated Bucket Trucks." *Electric Utility Fleet Mangement.* (October/November 1983): pp 41-50.

57. Moore, K. and C.A. Larson. "Aerial Equipment Requires Thorough Regular Inspection." *Transmission and Distribution.* (January 1984): pp 23-27.

58. "Standard Test Method for Acoustic Emission for Insulated Aerial Personnel Devices." *The 1985 Annual Book of ASTM Standards.* ASTM Standard F914-85. Philadelphia, PA: The American Society for Testing of Metals.

SECTION 9

ACOUSTIC EMISSION APPLICATIONS IN WELDING

Sotirios Vahaviolos, Physical Acoustics Corporation, Princeton, New Jersey

James Culp, Michigan Department of Transportation, East Lansing, Michigan
Phil Hutton, Battelle Pacific Northwest Laboratory, Richland Washington
R.F. Klein, Battelle Pacific Northwest Laboratory, Richland Washington
Charles McGogney, Federal Highway Administration, Washington, DC
Ronnie K. Miller, Physical Acoustics Corporation, Princeton, New Jersey
James Mitchell, Physical Acoustics Corporation, Princeton, New Jersey
Mansoor Saifi, Western Electric Company, Princeton, New Jersey
E.B. Schwenk, Battelle Pacific Northwest Laboratory, Richland Washington

PART 1
RESISTANCE SPOT WELD TESTING

Resistance Spot Welding

Direct current or capacitor discharge welding is a joining method that passes a high current, short time pulse across an interface to cause welding. Welding parameters pertinent to capacitor discharge welding of contacts have been extensively studied and characterized. It is more difficult to dynamically assess weld quality.

Although acoustic emission has been used to monitor the capacitor discharge welding process,[1] the relationship of the acoustic emission signal to the physical changes occurring in the weld zone needs to be clarified. Information from the melting and post-melting portions of the weld cycle needs to be extracted from the total acoustic emission signal, then analyzed and related to weld quality. The total welding cycle is apportioned into three parts: (1) premelting; (2) melting; and (3) post-melting.[2] By thermal analysis of a simplified body incorporating a change of phase, the melting portion of the cycle is analyzed for penetration depth (extent of melting).

Treating these periods separately is necessary to establish a correlation between welding and the detected acoustic emission signal from a transducer. From this fundamental information, it is possible to extract from the signal that portion believed to contribute most for determining an acoustic emission threshold value.[1] Note that the total detected acoustic emission signal is basically composed of three signal subsegments corresponding to the three aforementioned physical effects.

The premelting period helps determine the minimum required welder power, while the post-melting period gives the time needed to complete the welding process. As far as the welding itself is concerned, the interest here is primarily focused on the second period (melting). From this period comes the extent of melting and consequently a prediction of the thermal time constant for welding. The time constant is in turn used to set the test criterion in a nondestructive acoustic emission test for weld quality (under ideal conditions and with compatible materials).

Alternating current resistance spot welding can be treated as above by adding several half-cycles (direct current) into multiple cycle welding. Note that the energy of the half cycles (heat) can be electronically controlled and is denoted as a percent of heat.

Principles of Acoustic Emission Weld Monitoring

The resistance spot welding process consists of several stages: setdown of the electrodes; squeeze; current flow; forging; hold time; and lift-off. Various types of acoustic emission signals are produced during each of these stages. Often, these signals can be identified with their source. The individual signal elements may be greatly different or totally absent, in various materials or thicknesses. Figure 1 is a schematic representation showing typical signal elements that may be present in the acoustic emission signature from a specific spot weld.

Most of the depicted acoustic emission signal features can be related to factors of weld quality. The acoustic emission occurring during setdown and squeeze can often be related to the conditions of the electrodes and the surface of the parts. The large, often brief, signal at current initiation can be related to the initial resistance and the cleanliness of the part. For example, burning through certain oxide layers contributes to the acoustic emission response during this time.

During current flow, plastic deformation, nugget expansion, friction, melting and expulsions produce acoustic emission signals. The signals caused by expulsion (spitting, flashing or both), generally have large amplitudes and can be distinguished from the rest of the acoustic emission associated with nugget formation. Figure 2 shows typical acoustic emission response signals during current flow for both direct current and alternating current welding.

Following termination of the welding current, some materials exhibit appreciable acoustic emission during solidification. This can be related to nugget size and inclusions. As the nugget cools during the hold period, acoustic emission can result from solid-solid phase transformations and cracking.

During the lift-off stage, separation of the electrode from the part produces signals that can be related to the condition of the electrode as well as to the cosmetic condition of the weld.

Using time and amplitude or energy discrimination, the acoustic emission response corresponding to each stage can be separately detected and analyzed. Although the acoustic

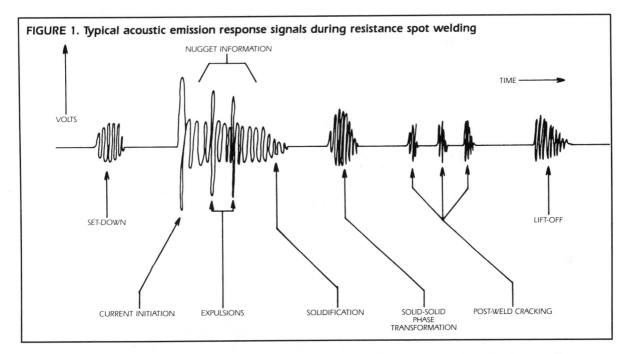

FIGURE 1. Typical acoustic emission response signals during resistance spot welding

emission associated with each stage of the spot welding process can be relevant to weld quality, this discussion only gives detailed consideration to the acoustic emission generated by nugget formation and expansion, expulsion and cracking.

Weld Quality Parameters that Produce Acoustic Emission

Acoustic Emission during Nugget Formation

When the material in the welding zone is heated, the pressure applied by the top electrode plastically deforms the material and acoustic emission is generated. The amplitude of the acoustic emission signals associated with this process is affected by both electrode pressure and the cleanliness of the part. These acoustic emission signals may be gated out, since they provide no useful information relative to the quality of the weld.

Further heating results in melting within the welding zone and growth of the nugget. Nugget formation and expansion produce acoustic emission signals that can be correlated with the strength of the weld. As soon as the welding current starts to decrease, the nugget begins to solidify and residual stresses occur in and around the weldment. If these residual stresses are severe, hot cracking occurs between cycles of alternating current welding.

By selecting a time and amplitude interval, measurements of the acoustic emission response for several samples may be correlated with weld strength or nugget size as determined from other tests. In this way, a minimum acceptance level of acoustic emission corresponding to acceptable weld quality may be established.

A storage scope or other device to record acoustic emission response should be used to verify the several stages of acoustic emission generation and detection shown for DC and AC spot welding in Fig. 2

Acoustic Emission from Expulsion

Expulsion occurs after a sufficient weld is formed. Within the weld period, it occurs sooner if the electrodes are clean or thinner material is welded. It occurs later if the welding conditions have deteriorated or thicker stock is welded with the same controller setting.

No matter when it occurs, expulsion is not desirable because it removes material from the weld nugget area. It may, however, be automatically kept to a minimum when expulsion signals are used to generate a feedback signal to control the welding process.

When a feedback control arrangement is not used, expulsion may be knowingly tolerated because of production considerations. In this case, several test coupons should be welded and the resulting weld strength or other quality parameter should be determined through destructive testing. Weld strength may be correlated with a suitable measure of

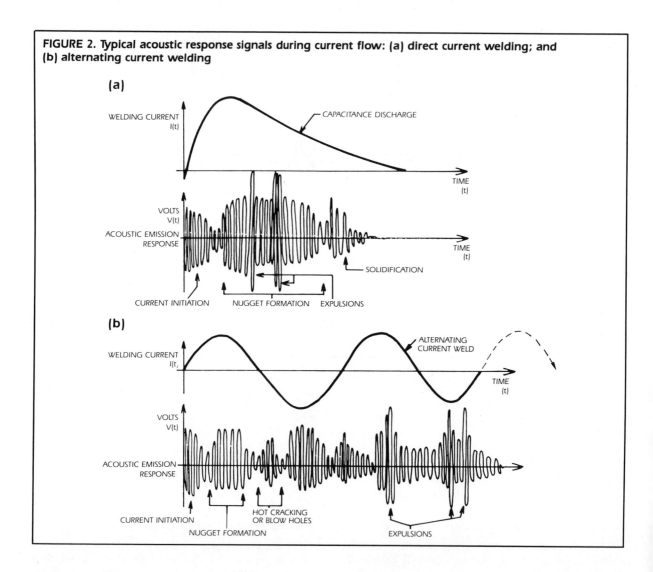

FIGURE 2. Typical acoustic response signals during current flow: (a) direct current welding; and (b) alternating current welding

the expulsion acoustic emission. In this way, a maximum acceptable level of expulsion acoustic emission can be determined and used to segregate unacceptable welds or welded parts.

Acoustic Emission from Phase Transformation

In certain carbon steel alloys, material in and near the spot weld undergoes martensitic phase transformation as the nugget cools. The total volume of material experiencing the transformation is related to both the nugget size and the area of fusion bonding. Therefore, the acoustic emission response during phase transformation may be related to the strength of the weld.

A waveform recorder should be used to isolate the phase transformation stage of the acoustic emission signal. By selecting a time and amplitude interval, a measure of the phase transformation acoustic emission for several samples may be correlated with the weld strength or size, as determined from other tests. In this way, a minimum acceptable level of acoustic emission corresponding to acceptable weld quality may be established.

Acoustic Emission from Cracking

The acoustic emission resulting from post-weld cracking can be monitored during the time interval between the current termination and lift-off (or, in some cases, between the phase transformation and lift-off). A waveform recorder or a storage oscilloscope may be used to identify the cracking interval by monitoring the acoustic emission response. For applications where crack size affects quality evaluation, acoustic emission monitoring should be supplemented with other test methods.

Acoustic Emission Instrumentation for Resistance Spot Welding

The acoustic emission system used for this kind of testing usually consists of a high frequency acoustic emission sensor attached to one of the spot welding electrodes. Matching amplifiers and filters with a center frequency of 0.5 MHz may be used to minimize the effects of mechanically induced low frequency signals. Detector and energy processor follow, producing a digital acoustic emission energy count proportional to the energy of the detected acoustic emission signal.

The initiation of each weld power impulse is sensed and a window is generated synchronous with the effective length of the impulse. The window gates the acoustic emission energy count into the counter, display and digital comparator. For multiple impulse welds, the energy counts are cumulative. The comparator output can activate a control circuit to terminate the weld cycle or if the anticipated preset energy count was not realized an alarm circuit can be activated. The energy count is displayed on a four decade light emitting diode readout. A simplified block diagram is shown in Fig. 3.

Sensor and Preamplifier Attachment

The sensor should be mounted to the lower (grounded) electrode or electrode holder. Liquid couplant may be used for periodic sampling of weld quality, provided that it is replenished and calibration of the system response is maintained. For sustained monitoring, such as on-line acoustic emission examination or control of each nugget, the sensor should be permanently mounted using an epoxy adhesive or a similar material.

A preamplifier is usually positioned near the sensor. However, when the instrumentation is located less than 1 m (3 ft) from the sensor, the gain otherwise supplied by the preamplifier may be incorporated into the main amplifier of the instrument.

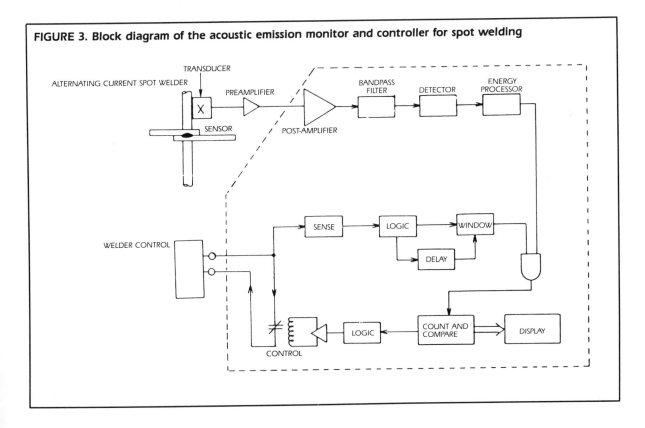

FIGURE 3. Block diagram of the acoustic emission monitor and controller for spot welding

Preliminary Measurements

The acoustic emission signal from a single spot weld should be displayed on a waveform recorder. A wire coil or Hall effect sensor positioned near an electrode can be used as a current sensor, thus providing a timing reference and trigger signal for viewing and measuring the acoustic emission.

This reference signal can also be obtained through an appropriate interconnection to the weld controller. Having established a typical acoustic emission trace, characteristic stages should be identified and one or more selected as an acoustic emission examination parameter. For example, weld quality indicators may be obtained from the acoustic emission response to nugget formation, expulsion or cracking.

Setup for Welding Applications

Preliminary measurements must be made to determine the instrument settings and the conditions for monitoring. The weld controller settings are determined from normal weld considerations. First, the complete acoustic emission response is observed on an oscilloscope. Next, the instrumentation gain is set to the maximum value where the acoustic emission signals (representing the selected examination parameter) do not saturate the amplifier. This step will ensure that the measurement is made with the best possible signal-to-noise ratio.

Next, the detection threshold level is established at a value slightly above (or below) the peaks of acoustic emission signals that are excluded from the measurement. The timing control is referenced to the onset of the weld current and consists of a delay and a time interval. The time interval is selected to catch relevant signals.

Finally, the count multiplier is set to a value that allows utilization of the maximum number of significant digits in the readout. The finalized settings of the weld controller and the acoustic emission instrumentation are recorded along with a photograph of the total acoustic emission signal. The counts obtained from individual welds may be kept on file for future reference.

Repeated Applications

If acoustic emission monitoring was previously applied to a particular controlling or monitoring problem, the purpose of the preliminary measurement is to reestablish the known and recorded original test conditions. The heat and the weld time settings on the weld controller. The gain, threshold level, timing and other settings on the acoustic emission instrumentation may be made identical to those on record.

A few sample welds are made to verify that the general appearance of the total acoustic emission signal exhibits the expected characteristic features and that the average count falls within the range of past measurements.

Typical Applications of the Acoustic Emission Method

Acoustic emission surveillance of spot welding can take two forms: monitoring and control. The purpose of on-line *monitoring* is to identify and segregate unacceptable welds for quality evaluation. On-line *control* uses the acoustic emission instrumentation to complete a feedback loop between the process and the source of welding current by automatically adjusting one or two process variables to compensate for deteriorating welding conditions.

Defective spot welds occur randomly or gradually in the process. Randomly occurring discontinuities are caused by extraneous changes in welding conditions (for example, changes in the composition, the geometry or the surface conditions of the welded parts). Gradual deterioration of spot welds are caused by electrode wear.

The heat applied is proportional to: (1) the electrical resistance and weld time; and (2) the square of the weld current. Of these, only the time and current are process variables. Typical constant heat curves are shown in Fig. 4. For example, if the initial settings of the controller were somewhere on the curve H_2, then after a large number of welds, the surface area of the electrodes may increase and cause the current to decrease. This is equivalent to a shift of curve H_2 to the position H_3 and requires compensation by adjusting either the weld time or the weld current or both.

On-Line Monitoring

The purpose of on-line monitoring a spot welding process by acoustic emission is to identify faulty welds or parts. This information can serve as a basis for acceptance or rejection of parts or for establishing a schedule for repair. Theoretically any combination of the weld quality parameters can be used for this purpose. Selection of the parameters is done by determining the most prevalent causes of defective welds in a particular welding application.

Excessive post-weld cracking can be caused by loose tolerances of the dimensions and surface conditions of the welded stock; burning; or excessive expulsion due to overwelding. Such cracking can be identified as randomly occurring discontinuities using corresponding acoustic emission responses.

If all the parameters are monitored simultaneously, an acceptable weld may be defined as: *a spot weld with adequate nugget volume that (1) is free of excessive expulsion and post-weld cracking; and (2) satisfies the average strength requirements*. The limits for a gradual deterioration of weld conditions, represented by a gradually increasing number of defective welds, can also be detected by monitoring the expulsion. This is the only examination parameter for certain applications.

Of the weld quality parameters described above, only nugget formation and expulsion occur during the welding process. Therefore, these parameters may be used for controlling weld process variables. The weld time and heat settings of the controllers usually produce an over-welded condition in industrial practice. This is done in order to cover variations of material, material dimensions and surface conditions and to allow some wear of the electrodes before the next maintenance period.

The equipment described previously should have the following inputs: (1) preamplified acoustic emission signal; and (2) pulse or step function from the weld controller which marks the onset of current synchronously with zero crossing point.

Acoustic emission feedback control automatically adjusts the required process variables. In practice, adjustment of the weld time is most common. After the acoustic emission weld parameters have been established as described previously, the acoustic emission monitor may be interfaced with an existing welder controller as shown in Fig. 5.

Nugget Formation as the Control Parameter

For a specific weld strength, the corresponding acoustic emission value is set on the acoustic emission controller. When a weld is made, a pulse signals the initiation of the gating section of the acoustic emission controller for acceptance of acoustic emission only during nugget formation.

If the acoustic emission value is reached, the controller sends a pulse to the welder in order to terminate the welding heat. The termination signal should be synchronous with the zero crossing of the weld current.

Expulsion as the Control Parameter[3]

In Fig. 4, if the curve H_1 represents the heat settings necessary to form a good nugget under initial and ideal welding conditions and if curve H_3 represents the actual controller settings, then the time difference between two corresponding points on the two curves represents the amount of over-welding. Under initial conditions, expulsion occurs close to H_1.

As the electrode conditions gradually deteriorate, the occurrence of expulsion shifts upward with constant current. Intermediate conditions can be represented by the curve H_2. A shift requires adjustment of the weld current, weld time or both. Acoustic emission feedback control automatically adjusts the required process variables. In practice, adjustment of the weld time is most common.

The input pulse is the timing reference for a single spot weld operation. The feedback controller produces a time delay corresponding to the time between points T and A in Fig. 4. From time A onward, acoustic emission is monitored and expulsion is detected when acoustic emission rises above the threshold level. Monitoring continues at least until the time corresponding to point B. If expulsion occurs earlier, the output of the feedback controller terminates the weld current (at C for example), thereby overriding controller setting B.

For both acoustic emission weld control parameters, it is possible that the maximum controller settings can be

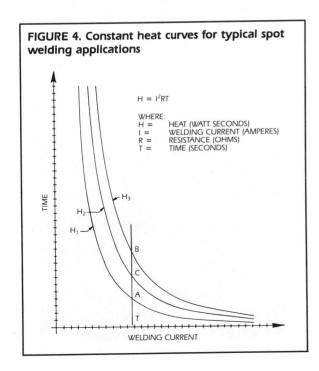

FIGURE 4. Constant heat curves for typical spot welding applications

$H = I^2RT$

WHERE:
H = HEAT (WATT SECONDS)
I = WELDING CURRENT (AMPERES)
R = RESISTANCE (OHMS)
T = TIME (SECONDS)

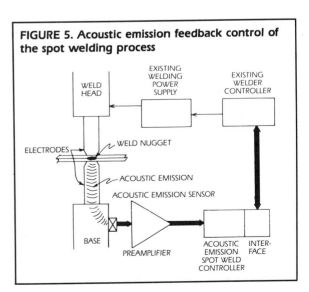

FIGURE 5. Acoustic emission feedback control of the spot welding process

reached without producing an acceptable weld. If this occurs, the electrodes have deteriorated to failure and should be changed.

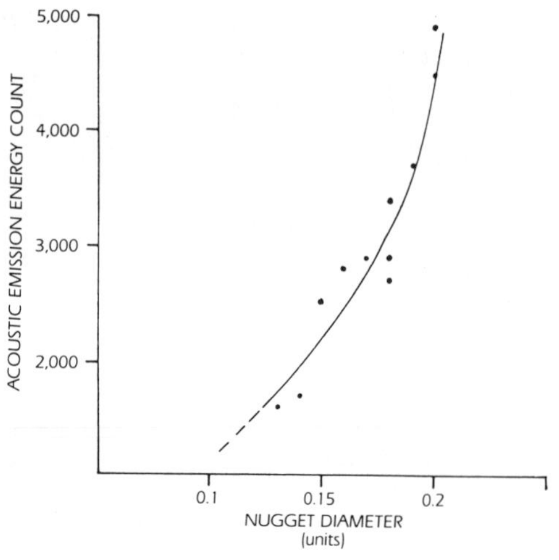

Monitoring Coated Steel AC Welds

Spot welding steel sheets coated with the zincrometal process expels the coating from the weld area in a violent manner, due to outgassing of the coating. With controlled welding parameters, the expulsion of the coating occurs during the first weld cycle. A one-cycle weld envelope from a zincrometal spot weld is shown in Fig. 6. The two initial bursts of acoustic emission are due to coating expulsion. Since this emission has little relation to true welding, it is blanked by preset logic in the acoustic emission weld analyzer, as indicated by the lack of windows during the first weld cycle. The following two weld cycles were processed in normal fashion.

Plotted in Fig. 7 is the weld energy, expressed as counts detected in a series of zincrometal spot welds versus nugget diameters. The curve would approach a straight line if the counts were plotted against nugget area. This curve is typical of the relation between weld energy count and spot weld size and is the basis for controlling the welding process with acoustic emission instrumentation.

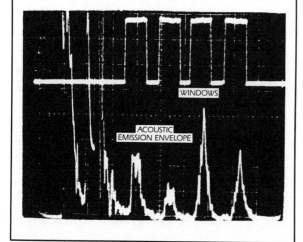

FIGURE 6. Acoustic emission from spot weld in zincro coated steel for a nugget diameter of 0.15 units; an acoustic emission energy count of 1,550; sweep of 5 milliseconds per centimeter; and a vertical of 0.5 volts per centimeter

AC Spot Welding Galvanized Steel

Because of the low melting point of zinc, acoustic emission signatures from spot welds in galvanized steels have two

FIGURE 7. Acoustic emission control of zincrometal spot welds

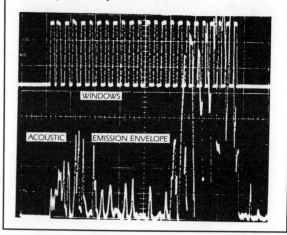

FIGURE 8. Acoustic emission envelope (1 volt per centimeter) detected during twelve-cycle spot weld in galvanized steel; acoustic emission energy count 21,200 for nugget diameter of 4.8 mm (0.185 in.)

features not encountered in ordinary steel welds. The coating is rapidly melted and expelled from the pressure zone beneath the electrodes. This mass flow of liquid zinc causes a slight to moderate increase in acoustic emission which subsides in a few cycles. The acoustic emission amplitude and energy count from galvanized welds are usually of greater magnitude than for steel, as shown in Fig. 8.

The second feature is apparent later in the weld interval. Since the temperature for welding steel is above 1,470 °C (2,678 °F), far above the boiling point for zinc, each power impulse is followed by residual zinc boil-off from the weld zone. This boiling is a considerable source of acoustic emission. By proper setting of delay and window for each power pulse, the acoustic emission from boiling can be largely rejected. If the zinc is not removed, the weld nugget is rich in iron zinc compounds and would be brittle. An extended trace showing the last four cycles of Fig. 8 is provided in Fig. 9, to more clearly illustrate the zinc boil off following each impulse while the window was closed.

Detecting the Size of Adjacent AC Welds

Shunting is caused by the presence of a previously made spot weld close to a weld being made and offering a shunt circuit for part of the weld power. The effect of shunting on weld nugget size becomes more pronounced with closer spacing between spot welds. To illustrate the effect, a series of spot welds were made in the uncontrolled mode. Welding parameters were held constant. The first welds made had some scatter of nugget size depicted by the acoustic emission instrument. These are referred to as initial welds.

Shunted spot welds were made from 0.25 units to 1 unit center-to-center and all were deficient in nugget size, as plotted in Fig. 10. All welds were made with three cycles. The welder control was reset to a maximum of five cycles and the acoustic emission spot weld controller was preset to turn off the welder after 1,100 weld energy counts. The results for a similar series of shunted welds was a uniform series of spot welds with 0.2 unit diameters requiring four to five cycles of weld power.

Control of Spot Weld Nugget Size

The principle that the size of the spot weld is proportional to the acoustic emission weld energy count is based on several measured parameters of the welding procedure. Figure 7, illustrating the relation between the two, is valid even for the extreme case of coated steels.

In Fig. 10, while the weld power was held constant, nugget size varied widely due to shunting. This effect was accurately evaluated by acoustic emission. When the acoustic emission energy count was used to turn off the welder, that was a constant energy count. The resulting nugget size was

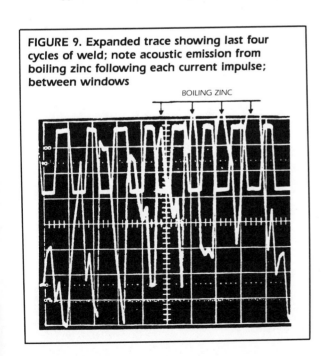

FIGURE 9. Expanded trace showing last four cycles of weld; note acoustic emission from boiling zinc following each current impulse; between windows

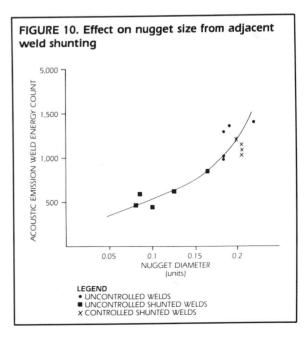

FIGURE 10. Effect on nugget size from adjacent weld shunting

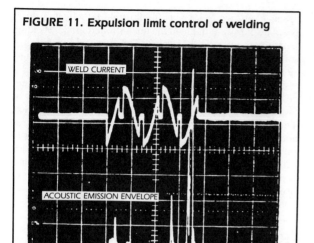

FIGURE 11. Expulsion limit control of welding

also constant in spite of wide variations in power input needed to compensate for shunting.[4]

Figure 11 is the acoustic emission record of a spot weld in steel using expulsion limit control.[3] A large spike of acoustic emission due to expulsion occurred during the first half of the third cycle. The expulsion limit of 100 counts preset into the instrument was exceeded and the instrument turned off the welder after 2.5 cycles. The instrument was working in expulsion mode. In the expulsion mode, a high level detector is activated. In this example, the expulsion bias was set at 3 V and only acoustic emission signals above that bias was detected and counted. The resulting spot weld was normal at 0.19 unit diameter with only a trace of expulsion.

The principal advantage of expulsion limit control is that expulsion of molten metal must occur after fusion. If the expulsion is detected late in the weld cycles, acoustic emission detection is sensitive enough to terminate welding before gross expulsion occurs. Thus, a reasonable control of welding is obtained with some assurance that there will be no stickers or underwelded spots.

Conclusions

Acoustic emission monitoring provides valuable insights into the welding process. Every burst of acoustic emission has a logical source and meaning.[5,6] Using acoustic emission for on-line evaluation of spot welding provides new applications for weld process control.

1. With instantaneous analysis, acoustic emission can fill in the missing link in a control loop including detection, feedback and control.
2. Acoustic emission monitoring can be automatic and integrated with numerical control (used with robotics).
3. Acoustic emission techniques are a unique diagnostic tool for evaluating weld parameters in real-time.
4. Over-welding can be reduced, improving weld efficiency.

Every mechanism occurring during a spot weld cycle results in acoustic emission that can be related to a specific source. The source mechanism may be beneficial or detrimental, with corresponding effects on weld integrity. Acoustic emission monitoring provides a tool for spot weld evaluation and control that is limited only by the initiative of the user. Other aspects might limit use, such as knowledge of the relationship between acoustic emission activity and the desired weld qualities.

PART 2
ELECTRON BEAM WELD MONITORING

Description of Electron Beam Welding

Electron beam welding[7] is a fusion joining process in which the workpiece is bombarded with a dense stream of high velocity electrons. Welding takes place in an evacuated chamber, which imposes certain limitations. However, it provides an advantage that outweighs all other considerations: in a pure and inert environment, metal may be welded without fear of chemical contamination.

The electron beam is capable of such intense local heating that it can almost instantly vaporize a hole through the entire joint thickness. The walls of the hole are molten and, as the hole is moved along the joint, more metal on the advancing side of the hole is melted. This flows around the bore of the hole and solidifies along the rear making the weld.

Heat input to the weld zone is controlled by four basic variables:

1. the number of electrons per second (beam current) impinging on the workpiece;
2. the electron speed at the moment of impact (accelerating potential);
3. the diameter of the beam at or within the workpiece; and
4. the speed of travel (welding speed).

Electron beam welding of small turbofan and turbojet engine parts is performed using a system of fixtures capable of presenting four parts in sequence to the welding head. It is a batch process, with considerable time and cost for each cycle. In general, a defective weld can be saved by making a second pass.

Discontinuities in Electron Beam Welding

In electron beam welding, as in all other welding processes, weld discontinuities can be divided into two major categories: (1) those that are open to or at the surface; and (2) those that occur below the surface. Surface discontinuities include: undercut, mismatch, underfill, reinforcement, cracks, missed seams and incomplete penetration. Figure 12 shows welds containing these discontinuities and a weld with none of them.

Surface discontinuities such as undercut, mismatch, reinforcement and underfill are macroscopic discontinuities related to the contour of the weld bead or the joint. As such, they may be detected visually or dimensionally. Large surface discontinuities such as cracks are detected visually by means of liquid penetrant inspection or magnetic particle inspection if the material is ferromagnetic.

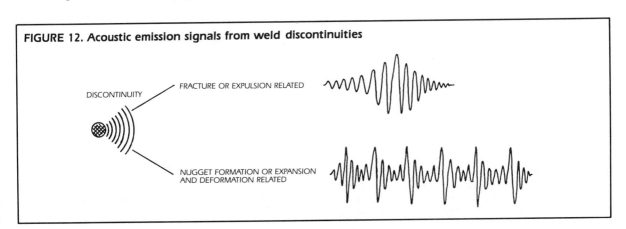

FIGURE 12. Acoustic emission signals from weld discontinuities

Occasionally, special inspection procedures must be employed to detect some types of surface discontinuities. Missed seams and incomplete penetration are often difficult to detect, because they are frequently associated with complex weld joints that prevent direct viewing of the affected surface.

Subsurface discontinuities are generally more difficult to detect than surface discontinuities because observation is indirect. Volume discontinuities such as porosity, voids and bursts are detected by radiographic or ultrasonic inspection, provided that their cross sections presented to the beam exceed 1 to 2 percent of the beam path in the metal. Discontinuities that present extremely thin cross sections to the beam path, such as cracks, missed seams and incomplete penetration, are detectable with X-rays only if they are viewed from the end along their planar dimensions.

Acoustic Emission Method of Testing Electron Beam Welds

Acoustic emission is the high frequency stress waves generated by the rapid release of strain energy that occurs within a material during crack growth, plastic deformation or phase transition. This energy may originate from stored elastic energy as in crack propagation or from chemical free energy as in phase transitions.[8]

Sources of acoustic emission that generate stress waves include local dynamic movements such as the initiation and propagation of cracks, twinning, slip, sudden reorientation of grain boundaries, bubble formation during boiling, martensitic phase transitions and nugget formation and expansion during welding. Most of these processes are found during electron beam welding.

Acoustic emission weld inspection detects and analyzes minute acoustic emission signals generated by discontinuities or nugget formation and expansion in materials under stress. Proper analysis of these signals can provide information on the location and structural significance of the discontinuities.

Acoustic emission monitoring enables necessary diagnosis to be made during each welding pass, so that remedial action can be taken at minimum cost, while the part is still mounted and under vacuum. Real-time discontinuity detection and characterization of welds is the major advantage of acoustic emission.

Electron beam welding used for turbine engine part fabrication is an excellent candidate for monitoring in real-time, using acoustic emission to achieve effective process control. Various welding parameters (such as beam current and position of beam focus) are amenable to feedback control during the weld pass. The principal goal of this work is to provide a diagnosis of the weld immediately after the pass or after a cool-down period, to indicate remedial action in the form of a second welding pass. This remedial action is applied immediately, without disrupting the batch processing.

The same acoustic emission criteria are again applied, with the result that in most cases a satisfactory weld is eventually obtained, without loss of time. In the original procedure, an unsatisfactory weld could not be identified until the part was removed from the vacuum chamber. The additional cost of a second pass included cycling the vacuum system, demounting, testing and remounting in the fixture.

The earliest acoustic emission applications to electron beam welding date to 1974 when acoustic emission was used in real-time monitoring of BeCu-BeCu welds in undersea repeaters.[9]

Advances in Acoustic Emission Testing of Welds

Important advances have been made in welding process control using acoustic emission monitoring. The technique enables the detection and quantification of physical processes, such as gas expulsion and microcracking, as they occur during the formation of a weld and during the cool-down. From observations of these processes, it has been possible in many cases to give an immediate pass/fail indication. It has even been possible, using observations during weld formation, to control welding parameters automatically during the process to save a weld that was in danger of being improperly formed. Successful acoustic emission monitoring has been performed on processes making use of resistance welding, electroslag welding and laser electron beam welding (all are detailed in this chapter).

Experimental results have shown that cracks and porosity in submerged arc, gas tungsten arc[10] and resistance welds[11] emit acoustic signals during and after fusion. In one instance, some weld discontinuities that had not been detected by radiography were detected by specially designed acoustic emission sensors. This was later verified by metallographic sectioning. It was found that the rate of acoustic emission from weld discontinuities was a function of weld temperature and that the acoustic emission count was related to the diameter of resistance spot welds. The quality of submerged arc welds was shown to be related to changes in the acoustic emission count rate recorded during welding.

Previous investigations of continuous welding methods using acoustic emission have generally made use of simple criteria for identifying different processes. High amplitude acoustic emission bursts resulting from expulsions or cracking are generally superimposed on an acoustic background caused by several processes, including solidification. The background varies in level as changes occur in such characteristics as weld depth. The acoustic emission bursts correlate with gross discontinuities such as cracks and voids.

In addition to crack monitoring, acoustic emission can be used to observe the signature of the welding process, which results from the summation of all emission processes. These processes can include liquid-solid and solid-liquid phase transitions, expulsion of weld metal droplets, cavitation and sloshing of the molten weld, arc noise, slag formation, cracking and deformation processes from thermal expansion and contraction near the weld. This emission varies with the welding parameters, so that the process can be evaluated using the acoustic emission signature.

In electron beam welding of nickel based alloys, it has been found that cracking during cooling produces high amplitude acoustic emission bursts easily distinguishable from a background of much lower amplitude.[12,13] Independent inspection after sectioning revealed cracks, detected with acoustic emission during welding, of types very difficult or impossible to detect by other nondestructive tests. The equipment used in this work was very simple, consisting of an acoustic emission transducer with a resonant frequency of 80 kHz, suitable amplification and filtering and tape recording of the waveform.

Analysis was a matter of observing the waveform during replay. While this method would not form the basis of any practical inspection system, the work made it clear that acoustic emission is likely to be the key to any real-time or near real-time electron beam welder process control.

Types of Acoustic Emission from Electron Beam Welds

Continuous Acoustic Emission from Welds

Basically, there are two types of acoustic emission: continuous and burst (Fig. 12). The waveform of continuous emission is similar to Gaussian random noise, but the amplitude varies with acoustic activity. In metals and alloys, this form of emission is thought to be associated with the dislocation movements in the grains. The energy of continuous emission is predominately in a low frequency range. For electron beam welding applications, the emission energy relates to the nugget (weld pass) size, depth of penetration and overall volume of the weld.

The stress that gives rise to an acoustic emission signal is proportional to the local strain jump caused by phase transition (nugget formation). It is also proportional to the propagation increment (nugget expansion) of this jump (Fig. 12). Suppose that the acoustic emission energy is proportional to

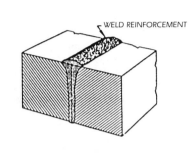
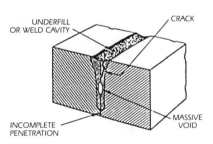
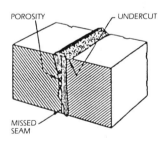
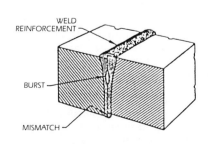

FIGURE 13. Electron beam welds showing (a) the absence of discontinuities in a good weld with reinforcement; and (b) discontinuities that can occur in poor welds

the rate of increase of nugget volume. That being the case, if welding parameters such as focus position, beam current and feed rate are held constant, a reduction in the detected rms signal is evidence of either (1) incomplete penetration or (2) underfill or weld cavity, two of the discontinuity types shown in Fig. 13.

Burst Acoustic Emission from Welds

Burst emission is short duration pulses (one microsecond to a few milliseconds in duration) associated with discrete releases of strain energy. Burst emission is generated by twinning, microyielding and the development of microcracks and macrocracks. This emission is of higher amplitude and higher frequency than continuous emission. Typical weld discontinuities detected from burst emission are:

1. crack initiation and propagation (during welding or cool-down);
2. excessive porosity; and
3. material expulsion and inclusions.

The different characteristics for these two types of acoustic emission, both carrying information about weld quality, make it possible to monitor them independently. This may be accomplished using a rolling differential transducer having two resonances: one in the frequency band of 140 to 200 kHz and one above 300 kHz, as detailed below.

Electron Beam Monitoring Instrumentation

Acoustic Emission Sensor

During the actual weld, the part executes a single revolution about its axis. This could be accommodated easily enough by almost any cable, so that sensors may be attached directly to the part. However, the technicalities of the fixture are such that each part executes a number of revolutions and the requirement for multiple passes (to correct bad welds) is a key part of the process. It is desirable to use a sensing method that avoids the problem of tangled leads.

Experience has been obtained in welding control using rolling dry contact transducers. These transducers consist of a stator, holding a piezoelectric crystal that always faces the contact surface and a rotor, that is essentially a rotating cylindrical case banded with silicone rubber or some other compliant compound to facilitate coupling to the surface. The rotor is filled with a liquid such as mineral oil for coupling to the stator.

This type of sensor can be mounted on the stationary part of the welder, rather than on the rotating table, in such a way as to make contact with each part when the part indexes into the welding position. A spring loaded plunger or elastic hinge mounting fixture will ensure consistent placement and acoustic coupling efficiency of the sensor. With this setup, the problem of tangled leads is eliminated.

To minimize the effect of radiofrequency interference and electromagnetic interference, a differential rolling transducer has been designed for a continuous welding process. Use of this component in electron beam welding applications requires only the ability to operate in a vacuum and at elevated temperature. Based on preliminary tests, no difficulties are anticipated unless the temperature of the part contacted by the rubber exceeds 150 °C (300 °F). Temperature of the body of the sensor is not expected to rise significantly, because of the poor thermal conductivity of the rubber and of the low pressure atmosphere.

It is not always necessary to make contact directly with a part in order to gather acoustic emission signals from it. A number of considerations can make it expedient to attach sensors to a clamp or fixture in good mechanical contact with the part. In this application, several different part geometries will be encountered, which means that sensor repositioning is necessary when changing from one part to another.

Because of this, it was proposed that the sensor make contact with one of the outer surfaces of the three-jaw chuck that holds the bottom of the part. If the rolling transducer rides on the surface of the chuck, no adjustment from part to part is required. The temperature of the chuck is not expected to rise significantly above ambient, giving the added benefit of removing temperature as a consideration. A spring loaded rolling sensor could be held in such a way that it would make contact with the chuck holding the part being welded and allow indexing of the table without any complicated system for placing and retracting the sensor. Another possibility is to attach rolling sensors to the live centers of the holding fixture and it is always possible to make rolling contact directly with the parts at the expense of added mechanical complexity, should this become necessary.

Signal Preamplifier

The rolling differential sensor is connected to a modified standard preamplifier. The twin axial cable from the sensor to the preamplifier is protected by means of a standard flexible armor sheath.

To avoid noise and acoustic emission signal loss, the preamplifier is mounted in the vacuum chamber. Its electrolytic capacitors are replaced with ceramic capacitors, since they will be subjected to vacuum. The preamplifier, having a bandwidth of 1 kHz to 1.2 MHz (which is further limited by the use of plug-in filters) is powered by the acoustic emission system through the shielded signal cable at 28 V DC.

Considerable care is taken with signal grounds and certain other precautions because of conditions inside the vacuum chamber. Careful routing of cables minimize problems with accumulation of deposited metal. Armored cables are used and the cabling is as short as possible to minimize the effects of electromagnetic interference.

A hermetically sealed BNC feed-through will be used for connection with equipment outside the vacuum chamber. It may be necessary to machine the feed-through from insulating material such as methyl methacrylate as part of the scheme for minimizing electromagnetic interference problems. This method has been used in vacuum systems previously with no difficulty.

Basic Acoustic Emission Instrumentation and Signal Analysis

Certain waveform characteristics have proved valuable in practical applications. These characteristics may be extracted from the acoustic emission signal at the front end of a dedicated microcomputer based system, allowing fast accumulation and analysis.

Many acoustic emission systems acquire a complete set of acoustic emission parameters. These are, for each acoustic emission hit, the peak amplitude, counts, energy, duration and rise time. The duration (microseconds) is determined from the times of the first and last threshold crossings. The peak amplitude (decibels relative to one microvolt at the sensor) is recorded along with the rise time (microseconds). The number of positive threshold crossings during the hit (at least one) is recorded as counts and the area under the signal envelope is the energy. In addition, time is recorded to a precision of 0.1 microseconds, digitizing channels are available for recording such parameters as applied load, temperature or pressure and the instantaneous average acoustic signal level can be recorded.

The basic configuration of a location analysis system for electron beam welding applications is shown in Fig. 14. To make the best use of the two types of acoustic emission, the signal is fed into both channels of an independent channel controller. A low frequency and high frequency configuration is used by means of plug-in filters. The output of the analog processed signal is then used by the digital section of the controller and the central processing unit (CPU) extracts the main features of the weld signal for further processing.

The CPU is also used for alarm processing and is capable of extracting go/no-go criteria. The main CPU is left alone to perform the tasks of graphing and storing the processed acoustic emission data. This type of parallel processing enables the user to perform real-time industrial weld quality measurement and control.

The real-time data acquisition system controls the accumulation, storage and display of acoustic emission data. The program is very general, allowing simultaneous compilation of up to nine graphs, showing any parameter plotted against any other, frequency distributions and a system of sequential graphical filters.

Acoustic emission during the weld is expected to consist of a relatively constant background, upon which is superimposed an occasional higher amplitude event, signifying a discontinuity such as a crack. Variations in the background are of interest in diagnosing certain characteristics of the weld, while the high amplitude events signify discrete welding faults. In this case, the system can be configured to monitor the background signal level and to perform feature extraction immediately on the higher amplitude events, making efficient use of two channels.

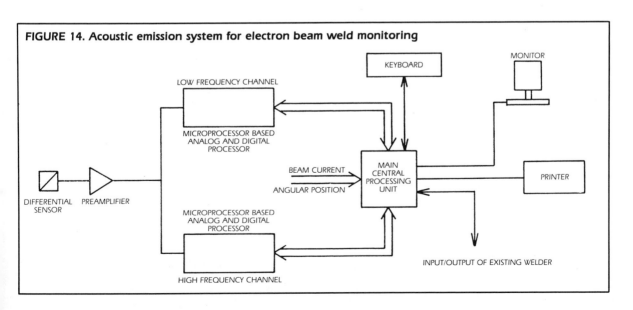

FIGURE 14. Acoustic emission system for electron beam weld monitoring

These signal characteristics can be examined for correlations with independently determined discontinuity type. During the first phase of evaluation, using simulated parts or weld coupons, the welds are examined by suitable independent methods to determine occurrence and severity of the various types of discontinuities. This information is then used to correlate the acoustic emission data. Location is determined based on the constant rotation rate of the part during the welding pass.

The raw acoustic signal can be recorded if necessary. The purpose of this is to allow replaying of recorded waveforms into the acoustic emission system in the process of optimizing system configuration. A tape recorder with 1 MHz bandwidth may be used.

Conclusion

Acoustic emission electron beam weld monitoring provides the user with a real-time technique for discontinuity detection and location, especially when rolling acoustic emission sensors are used. Crack detection and location, detection of incomplete penetration, and detection of burst/void and mismatch are possible. In addition, acoustic emission parameters can be used for process control.

When sufficient data and weld discontinuities are clearly understood, acoustic emission techniques can be used as a feedback element to adjust the electron beam current and to consistently produce excellent welds.

PART 3
ACOUSTIC EMISSION FOR DISCONTINUITY DETECTION IN ELECTROSLAG WELDS

Acoustic emission measurements have been made in-process and during cool-down of sample electroslag welds in 75 mm (3 in.) thick steel plate. The welds were fabricated to study the ability of various nondestructive testing techniques for finding and grading typical discontinuities. During in-process weld monitoring, acoustic emission techniques successfully identified excessive flux and shallow flux. During cool-down, cracks from copper, cold starts, hot starts and wet flux were detected and accurately located. Ultrasonic testing was used for confirmation.

The electroslag welding process for joining thick steel sections has been used over the past twenty years in the United States. The process is popular where the sections to be joined can be oriented in a vertical position and the savings in time compared to multipass methods make it economically feasible. Many state highway departments have used the electroslag process for butt welds in large plate girders and box sections for bridge structures.

The process usually involves one or more filler wires fed through a consumable guide tube into a molten bath of steel and flux confined by water cooled copper shoes. This is a very high heat input process and the finished weldment contains large coarse columnar grains. The finished grain structure is similar to that of a coarse grain casting, with low toughness value. Overall, the electroslag weld process is attractive from an economic point of view, however the quality of the weldments is highly dependent on operator skill and attentiveness.

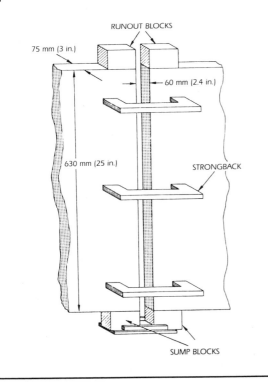

FIGURE 15. Specimen dimensions for acoustic emission testing of electroslag welds in steel plates

Test Specimens for Acoustic Emission Testing of Electroslag Welds

Steel plates of A36 and A588 were welded as test specimens and results of the A36 welds are reported below. Both steels are common in the bridge fabrication industry and

TABLE 1. Example of material and welding specifications for electroslag weld samples

Material specification:	ASTM A-36, ASTM A-588
Welding process:	electroslag, water cooled
Filler metal classification:	FES70-EMBK-EW
Filler metal classification:	FES70-EWT2
Flux:	PF 201
Specimen thickness:	75 mm (3 in.)
Number of passes:	single
Type of arc:	single guide, oscillating
Welding current:	500 ± 50 A
Preheat temperature:	none
Post heat temperature:	none
Welding voltage:	40 ± 2 V
Guide tube flux:	none
Guide tube specification:	AISI 1018
Guide tube diameter:	16 mm (0.63 in.)
Dwell:	two second dwell, three second travel
Cooling shoe:	water cooled, copper
Guide tube insulator:	201 flux, binder, fiberglass

were frequently welded with the electroslag process. The plates were fixed vertically as illustrated in Fig. 15.

Proper dimensions are maintained with steel strongbacks welded to both sides. Sump and runout blocks, that are removed after the weld is complete, are used to protect the finished product from start-up and stop irregularities in the weld.

The strongbacks are used to fix the position of the plates and also to hold the cooling shoes in place (see Fig. 16). The shoes are made of copper and cooled with water through internal passages, acting as a mold to contain the molten metal. Also indicated in Fig. 16 is the position of the acoustic emission transducers for in-process weld monitoring.

The Electroslag Welding Process

Electroslag welding can be described as a casting process. As illustrated in Fig. 17, molten metal supplied as filler wire mixes with molten parent metal in the weld pool. Above the weld pool is a flux pool that draws impurities from the weld pool and acts as a cover. The flux also surrounds the weld pools and forms a thin layer of slag between the weld and cooling shoes. A guide tube directs the filler wire into the weld pool and is gradually consumed by the weld. Further material and welding specifications are outlined in Table 1.

The undesirable characteristics of electroslag welding are the result of the tremendous amount of heat that must be used and the departure from strict adherence to welding procedures. Table 2 describes some of the discontinuities induced by deviations from correct welding procedure.

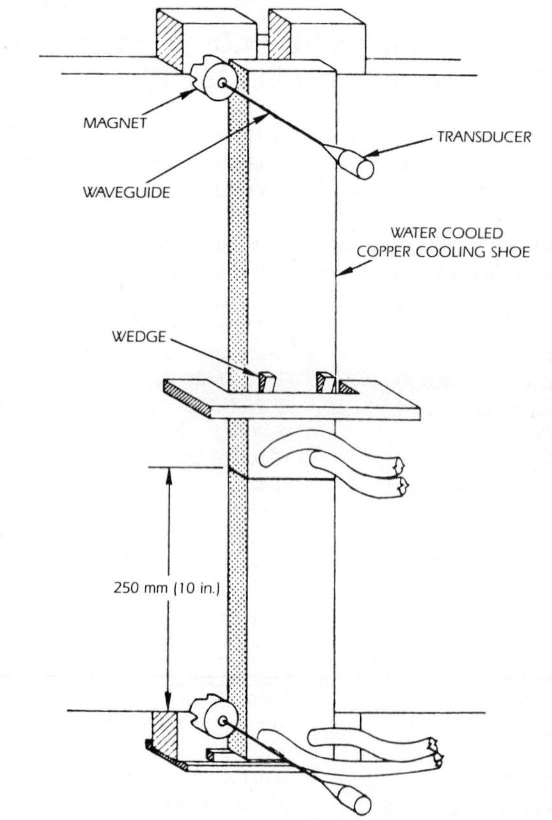

Acoustic Emission Equipment for Testing Electroslag Welds

The acoustic emission monitoring system[14] is made up of the standard modules listed in Table 3. A two-pen strip chart recorder and a storage oscilloscope were used to record the outputs.

FIGURE 16. Diagram of setup for acoustic emission testing of electroslag welds; note cooling shoes and transducer locations

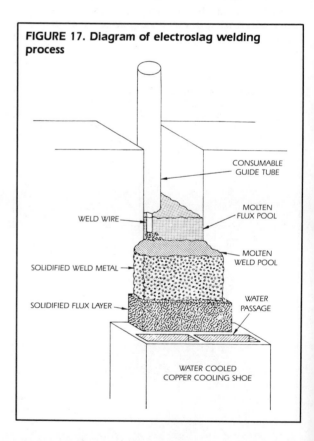

FIGURE 17. Diagram of electroslag welding process

TABLE 2. Typical discontinuities induced by deviations from acceptable electroslag welding procedure

Deviation	Description	Resulting Discontinuity
Wet flux	brings hydrogen into contact with molten metal	hydrogen cracking
Shallow flux pool [< than 25 mm (1 in.)]	decreases form factor, increases weld heat	intergranular cracking
Excessive flux pool [> than 50 mm (2 in.)]	increases form factor, cools weld pool and misaligns grain structure	small heat-affected zone, lack of fusion
Voltage drop	increases form factor, cools weld pool	small heat affected zone, lack of fusion
Hot restart	temporary halt of welding	
Cold restart	restart existing weld after cool down	small heat affected zone, lack of fusion
Copper contamination	copper resides in intergranular spaces	cracking

TABLE 3. Acoustic emission system components used for electroslag monitoring

Description	In-Process	Cool-Down
transducer: mounted on waveguide	88 mm (3.5 in.) of weld centerline top and bottom of weld	directly on weld top and bottom of weld
preamplifier, filter: 40 dB gain 190 kHz band-pass		
signal conditioner: 1 V preset threshold	40 dB gain, both channels	50 dB gain, both channels
counter: reset clock	count rate from traveling accept zone window and unprocessed count rate from one transducer	count rate in open window and unprocessed count rate from one transducer
distribution analyzer:	internal: locate spacing trim: 820 to 900 envelope: 10 ms log window manually adjusted as traveling accept zone	internal: locate spacing trim: 820 to 900 envelope: 10 ms log window 1 to 99
amplitude detector:	40 dB threshold	30 dB threshold
external memory:	external 1: amp envelope: 10 ms synchronous with window	external 1: amp envelope: 10 ms synchronous with window

The transducers were mounted on a waveguide with a magnetic base, as illustrated in Fig. 18. This arrangement allows the transducers to remain cool, provides very high contact pressure to the rough steel surface and eliminates the need for a viscous couplant. The transducers were a differential type designed to eliminate electromagnetic interference generated by the welding equipment.

Induced Discontinuities

Various discontinuities were induced during the weld process to produce weldments with known conditions. Aside from the acoustic emission monitoring, the induced discontinuities were studied with radiography and ultrasonic techniques to assess detectability. The discontinuities are then sectioned and micrographed. A list of induced discontinuities and their positions in the weld is given in Table 4.

Acoustic Emission Data Collection for Electroslag Welds

The electroslag welding process produces a variety of extraneous noise sources that are not indicative of weld quality. These include slag cracking, shoe rubbing, strongback weld failure and electromagnetic interference. Electromagnetic interference is eliminated by using differential transducers. The remaining noises are at least partially eliminated by using a traveling accept zone.

As illustrated in Fig. 19, solidified weld metal extends from the base of the weld pool to the beginning of the weld. The solidified metal extending behind the weld pool remains very hot and is subject to very high thermal stress. The concept of the traveling accept zone is to collect data from this region only while electronically gating out other sources. Acoustic emission count rate data are taken from within the traveling accept zone and are collected continuously during the welding process.

Other forms of acoustic emission data collected during the welding process include amplitude distribution of events within the traveling accept zone and linear locations of all events along the weld. Unprocessed count rate data from a single transducer are also collected. By switching count rate and distribution inputs, all the data are collected on the strip chart recorder.

After completing the weld, the slag is removed with a body grinder and chipping hammer in preparation for cool-down monitoring. Transducers are repositioned at the top and bottom of the weld. Data taken during the cool-down period include acoustic emission count rate from sources between the transducers, location distributions, amplitude distributions and count rate from a single transducer. Cool-down data are collected continuously for periods as long as 60 hours.

An audio monitor and storage oscilloscope are used in-process and during cool-down. Sliding of the shoes and other background noises are easily distinguishable from acoustic emission bursts (Fig. 20).

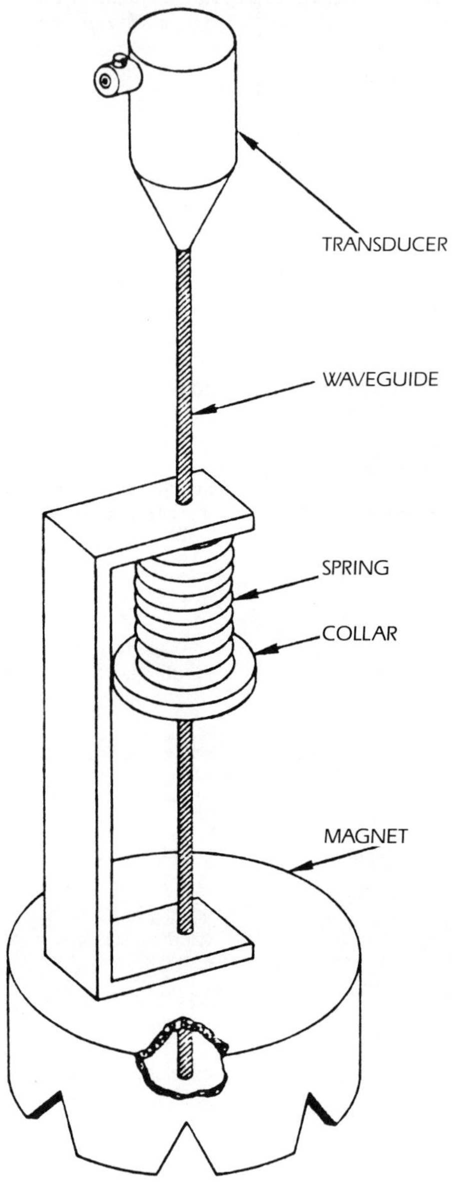

Results of Acoustic Emission Tests

The most revealing data are the linear location of events generated during cool-down after the slag had been removed. In almost every instance, the induced discontinuities were located along with some unexpected natural discontinuities.

During the welding process, count rate data are collected from the traveling accept zone behind the weld pool. This

FIGURE 18. Transducer mounted magnet and waveguide for testing of electroslag welds

technique is successful for identifying excessive flux type. Figure 21 illustrates the traveling accept zone data from four welds.

The control weld (36CP-6) shows very little activity except at the approximate times of the excessive flux pool and

TABLE 4. Position of induced discontinuities in sample electroslag welds

Weld Designation	Type of Discontinuity	Position in Weld
36F-1	cold restart (1.33 hour interval)	200 mm (8 in.)
	cold restart (14 hour interval)	400 mm (16 in.)
	hot restart (15 second interval)	475 mm (19 in.)
36F-2	excess flux	200 mm (8 in.)
	voltage drop 40-30 for 60 seconds	450 mm (18 in.)
36F-3	wet flux	0 to 175 mm (7 in.)
	steel shavings	300 mm (12 in.)
	copper	375 mm (15 in.)
	aluminum shavings	450 mm (18 in.)
	copper	525 mm (21 in.)
36CP-6	control weld (natural discontinuities)	
	excess flux	500 mm (20 in.)
	flux run out	175 mm (7 in.)

during flux runout between the shoes. In weld 36F-1, relatively little emission is detected until after the first cold restart. The emission from 36F-2 is very significant, beginning approximately 5 minutes after the depth of the flux pool was increased. The high count rate continued until the flux pool regained a normal depth about a half hour later. The 60 second voltage decrease is coincident with very high emission, but it is difficult to separate it from the excess flux emission. The wet flux and impurities thrown into 36F-3 produced no noticeable increase in count rate.

During cool-down (slag removed), the accept zone was fixed to include all locations along the weld except the sump and runout areas. Cool-down count rate data are presented in Fig. 22. Though the welds were monitored at different intervals and for different periods, it is evident that the control weld showed less activity than the others. Cool-down count rate data tend to diminish gradually over many hours after the weld is complete (60 hours in one case). It is not yet clear what time period is optimal for cool-down monitoring but judging from observed trends, one hour should be adequate, assuming that monitoring can begin within a half hour after the weld is finished.

Event locations were recorded during the cool-down period as illustrated in Figs. 23 to 26. Location data collected shortly after completing the weld and location data collected overnight are compared to the positions of known discontinuities. The control weld (36CP-6) had some very slight natural discontinuities that were not noticed until the acoustic emission event locations began to cluster. On close examination, it was apparent that excess flux had been used near the end of the weld. There was also evidence of flux runout between the cooling shoes.

FIGURE 19. Traveling accept zone for acoustic emission tests of electroslag welds

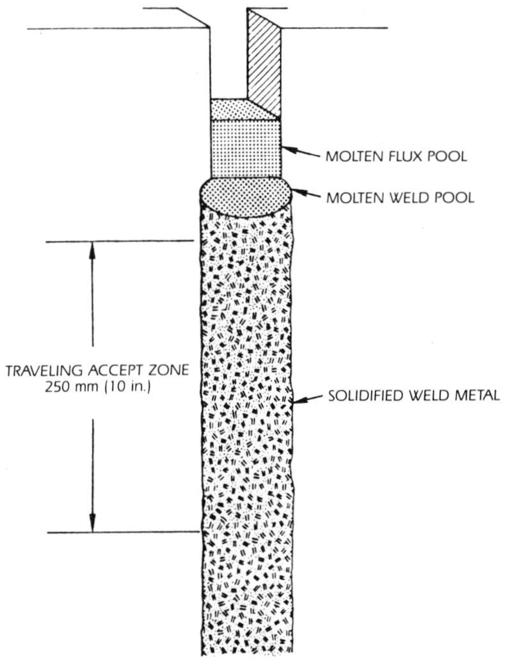

The cool-down data from 36F-1 were collected after the second cold start. The first 400 mm (16 in.) of weld including the first cold restart were 17 hours old. The location plot shows the positions of both cold restarts. The position of the hot start is close to the second cold restart and their event locations appear to be overlapped. A fourth active area slightly below the middle of the weld may again be flux runout from between cooling shoes, but this has not been confirmed.

Specimen 36F-2 produced event locations that correspond with the excessive flux pool and less prominently with the voltage drop. Of particular interest is the fact that one hour of monitoring, one hour after the weld was completed, revealed event locations very similar to the data collected over a twelve-hour period beginning three hours after the weld was complete.

Specimen 36F-3 was also monitored for a short period of time soon after welding and for a long, overnight period. The similarity of the locations plots indicate that the same discontinuities are being detected in both time segments. Event locations are suspiciously absent from the latter wet flux region. This suggests that wet flux is likely to induce hydrogen cracking early in the weld, but later as more heat is built up, the likelihood of hydrogen cracking diminishes.

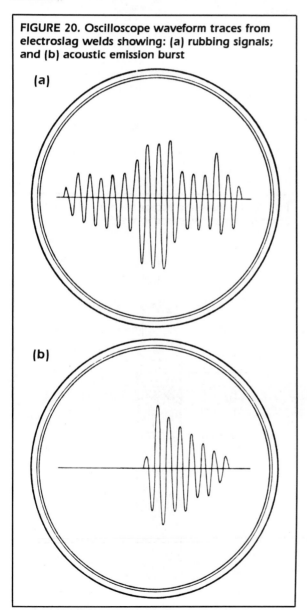

FIGURE 20. Oscilloscope waveform traces from electroslag welds showing: (a) rubbing signals; and (b) acoustic emission burst

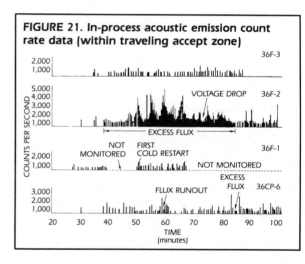

FIGURE 21. In-process acoustic emission count rate data (within traveling accept zone)

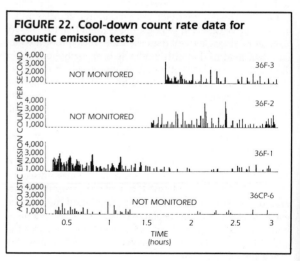

FIGURE 22. Cool-down count rate data for acoustic emission tests

Other data included single channel count rate, in-process location distributions, amplitude distributions and cool-down amplitude distributions. The count rate from a single transducer tends to show no correlation to good or bad welds.

Visual observation of event waveforms on an oscilloscope suggests that a significant amount of background noise is being collected. One potential solution for this is to use signal processing to eliminate events with incorrect waveform characteristics. A pulse width window would eliminate signals from rubbing. A higher frequency band is also recommended to help eliminate signals from slag popping.

From the test results, it is evident that poor quality welds can be identified either in-process or during cool-down. During the welding process, a pair of transducers and time gating techniques may be used to examine acoustic emission from below the weld pool, while eliminating other sources. Using this technique, deviations in the flux pool depth are immediately apparent.

After the weld is complete and the slag has been removed, the weld can be monitored under more favorable background conditions. During the cool-down period, linear location techniques accurately identify slag entrapment from cold starts, excess flux, copper induced cracking from wet flux and flux runout from poor fit-up of cooling shoes.

Acoustic Emission Techniques in Electroslag Welding Process Control

In the electroslag welding process, the long axis of the weld must be positioned vertically with the bottom end closed. Copper shoes are placed on both sides of the gap between the members to be joined. A consumable electrode is fed into the resulting channel. The weld is initiated by striking an arc between the electrode and the base metal to melt a quantity of dry flux at the bottom of the channel. From this point on, the resistivity of the molten flux pool is used to continuously produce the heat required to consume the electrode and melt the surface of the plates. The flux pool also maintains a protective layer over the molten metal beneath as it cools.

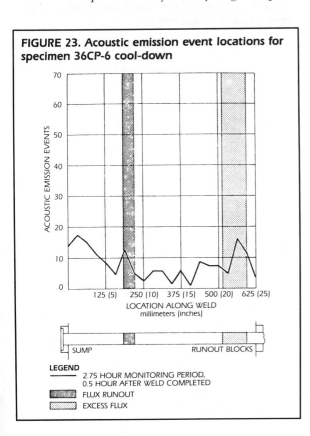

FIGURE 23. Acoustic emission event locations for specimen 36CP-6 cool-down

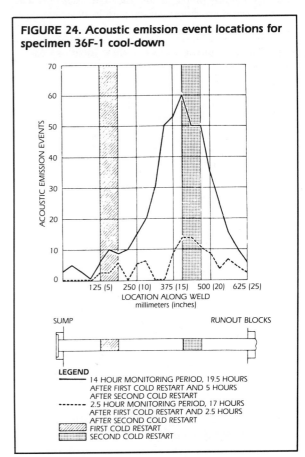

FIGURE 24. Acoustic emission event locations for specimen 36F-1 cool-down

Economy and weld joint quality are the primary advantages of the electroslag welding process. Surfaces to be welded are usually prepared by torch cutting the plates and they require no further preparation. The weld deposition is rapid when compared to the other processes. Because there is always a molten pool above the deposited metal, slag and gas have ample time to work their way to the surface, leaving a cleaner weld deposit. Preheating is generally not necessary because the process creates its own preheat.

There are also several problem areas. The nature of the process does not allow visual observation of the weld as it occurs. This prevents an operator from visually detecting possible weld discontinuities or process abnormalities during welding. The best means available for monitoring the weld in real-time has been the audible sound of the welding process. This relies on the experience, training and consistency of the operator. Thus, fabrication of a good electroslag weld is essentially dependent on specialized training and availability of the weld operator.

As with any welding process, discontinuities in the parent plate material may propagate into the weld. Lack of fusion, hot or cold cracking of the weld and porosity of various types are also likely problems. It is evident that methods to enhance process control and to rapidly detect, locate and analyze discontinuities as they form would be beneficial to developing expanded use of electroslag welding.

Acoustic Emission Test Procedure

The tests discussed here were approached with the assumptions listed below.

1. Changes in audible sound are used to control a portion of the electroslag welding process and this indicates that a more sophisticated method for applying acoustic analysis should produce more consistent and measurable control.
2. The ability to detect and locate some types of weld discontinuities by acoustic emission as they form during the more conventional welding processes (TIG, MIG, submerged arc and resistance) has been repeatedly demonstrated. The same techniques should be applicable to electroslag welding providing process noise is not too severe.

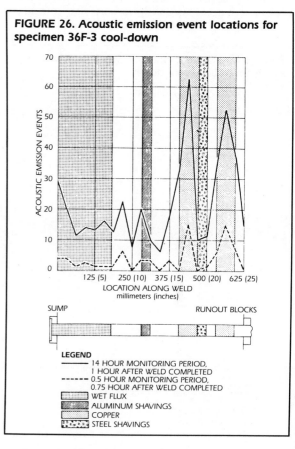

FIGURE 25. Acoustic emission event locations for specimen 36F-2 cool-down

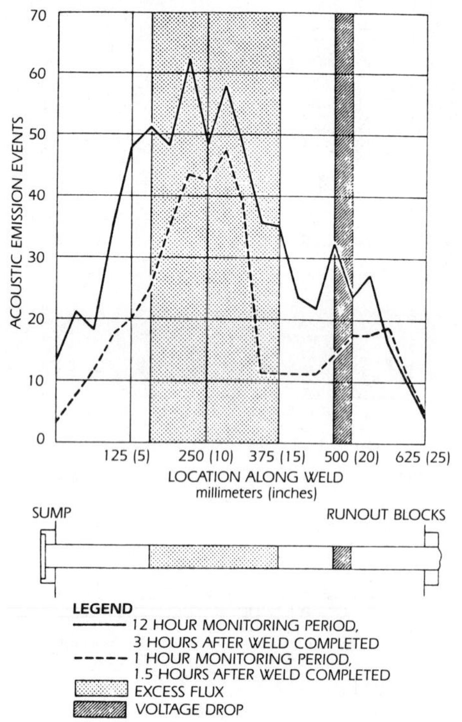

FIGURE 26. Acoustic emission event locations for specimen 36F-3 cool-down

Two separate tests were conducted to investigate the potential of putting the above premises into practice. The first was performed in collaboration with a structural fabrication shop. The intent of this work was to determine what electroslag weld parameter changes might produce corresponding acoustic emission changes.

Acoustic Emission Correlations to Weld Changes

Electroslag butt joint welds in ASTM Type A36 steel, 45 mm (1.75 in.) thick by 360 mm (14 in.) high, were acoustic emission monitored during welding. A constant voltage three-phase rectifier welder was used throughout the work. The guide tube was flux coated AISI 1020 seamless tubing 16 mm (0.6 in.) outside diameter by 3 mm (0.13 in.) inside diameter. The wire was 2.5 mm (0.1 in.) diameter. Stationary copper chill plates were used on all welds. Welding was conducted under field conditions at temperatures between -12 and -9 °C (10 to 15 °F).

Two acoustic emission monitor systems were used. One was a system with a broadband frequency response (100 kHz to 1 MHz) using a single sensor. The other was a commercial acoustic emission system using four sensors and four narrowband (550 to 750 kHz) preamplifiers. This system incorporated acoustic emission source location capability. A simplified schematic of the the test arrangement is shown in Fig. 27. Acoustic emission sensors were attached to the plate surface with air curing epoxy.

Normal and off-normal welds were made. The normal weld showed a relatively constant acoustic emission energy rate throughout the weld when the flux condition was adequate. When audible sounds indicated a need for additional flux, the acoustic emission energy rate had already increased. After fluxing, the acoustic emission energy increased slightly, then decreased back to the level that existed prior to fluxing. Acoustic emission count and energy plotted in Fig. 28 show a parallel response to weld conditions requiring flux addition. The source location system did not show any concentration of signal sources indicative of a discontinuity site.

Contrary to expectation, general welding process background noise was not a problem. The electroslag welding process proved to be quieter than some of the more conventional processes.

Typical results obtained from an off-normal electroslag weld are given in Figs. 29 and 30. In Fig. 29 as in Fig. 28, the acoustic emission rates generally follow flux additions. However, included in this test is the effect of increased wire feed with all other parameters held constant. Acoustic emission data show a direct and sizeable response to the change. It may be that the increased wire feed is indirectly producing the same condition as insufficient flux during normal welding. In other words, the normal flux pool becomes insufficient to accommodate the faster wire feed. It is significant that the acoustic emission returns to a more normal

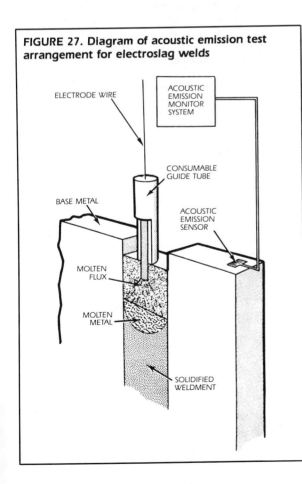

FIGURE 27. Diagram of acoustic emission test arrangement for electroslag welds

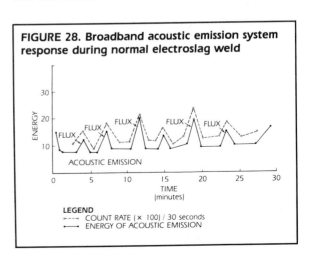

FIGURE 28. Broadband acoustic emission system response during normal electroslag weld

lower level as the wire feed rate is returned to normal. The wire feed rate's relationship to acoustic emission was also demonstrated in a second test in this series.

Reducing power supply voltage by as much as 25 percent did not produce any obvious acoustic emission response.

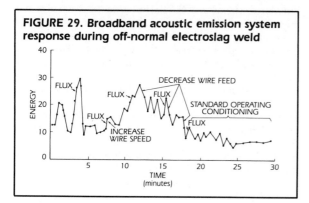

FIGURE 29. Broadband acoustic emission system response during off-normal electroslag weld

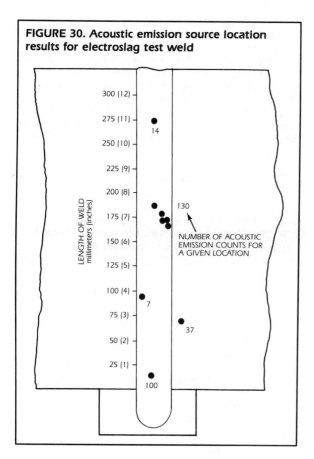

FIGURE 30. Acoustic emission source location results for electroslag test weld

Source Location with Acoustic Emission

Figure 30 shows acoustic emission source location results. Starting at the bottom of the weld, the first definite source point at about 13 mm (0.5 in.) showed 100 indications. From a radiograph, there appeared to be a crack in the starter weld just below the main plates. This is about 16 mm (0.63 in.) from the source point indicated by acoustic emission and within the accuracy expected from this type of source location.

The cause of the next two source indications of 37 counts at 75 mm (3 in.) and 7 counts at about 100 mm (4 in.) was not evident from the radiograph. These specimens were not sectioned to confirm acoustic emission sources. The next source indication of 130 counts at 180 to 200 mm (7 to 8 in.) appears on the radiograph to be from center line cracking. The 14 counts at 280 mm (11 in.) again relates to what appears on the radiograph to be some centerline cracking.

The source location method used in these tests indicates the source on the basis of a network of intersecting hyperbolas originating at the axes between two pairs of sensors. The hyperbola on each axis containing a given signal source is determined by measuring the difference in time of signal arrival (Δt) at two sensors in a pair. This is illustrated in Fig. 31. Such a location technique is often not as accurate as one that performs the hyperbolic calculations to directly identify the point source using composite Δt information from a multisensor array. The limited reduction in accuracy, however, is offset in many applications by the relative simplicity of instrumentation.

The source location results demonstrate that detection of discontinuities in an electroslag weld by acoustic emission is no more difficult than for more conventional welds.

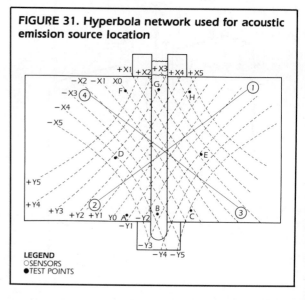

FIGURE 31. Hyperbola network used for acoustic emission source location

Acoustic Emission and Weld Flux

Research has been directed at examining the relation between acoustic emission and the weld flux parameter. Earlier work satisfactorily confirmed that discontinuity formation can be detected in electroslag welding by acoustic emission. Among the welding parameters, the acoustic emission flux relationship appeared to be the most significant. This is based on the fact that flux requirement is one parameter that cannot be directly measured. Welding voltage, current and wire feed rate can all be measured directly and maintained at a predetermined level.

In the test arrangement for this test, a constant voltage welding system was used. Guide tubes were AISI 1018, 13 mm (0.5 in.) diameter and the wire was AWS E70S-3. Plate material was ASTM type A36, 38 mm (1.5 in.) thick. Two plates 178 × 355 mm (7 × 14 in.) were joined on the long side.

The primary acoustic emission monitoring instrument was a digital acoustic emission recorder. This is a simplified instrument that uses solid state program read only memories (PROMs) to store digitized acoustic emission data. There are 256 storage locations in these memories, each of which can represent acoustic emission data accumulated over a preset time period ranging from a 0.75 second minimum up to a practically unlimited maximum. In this test, a 10 second update time was selected. Monitoring was performed over a broadband frequency range of 150 kHz to 1 MHz.

The finished weld appears to be good quality with no evident discontinuities. Figure 32 shows the acoustic emission response relative to flux additions during the course of the weld. The pattern may be compared to that from the earlier tests (Figs. 28 and 29). As the flux pool gets low, the acoustic emission level increases. When flux is added, the acoustic emission level generally decreases.

The data for the weld length from 250 to 300 mm (10 to 12 in.) are inconsistent with this pattern. This deviation may have been a result of allowing the flux to get very low. After what was intended to be an excess flux addition at 300 mm (12 in.), the acoustic emission level dropped. This gives some indication that the flux level was low to some degree through the total period.

These results are encouraging. Although the welding was done with a different system and acoustic emission results were measured with a different system, the acoustic emission versus flux pattern was generally similar to that observed in the earlier tests.

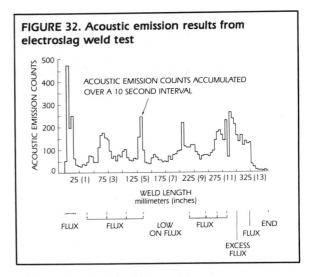

FIGURE 32. Acoustic emission results from electroslag weld test

Conclusions

It appears that acoustic emission techniques are beneficial to the electroslag welding process. Broadband (100 kHz to 1 MHz) information and particularly the lower end of that range is responsive to welding parameter changes. Higher frequency (600 to 700 kHz) shows identifiable response to weld discontinuity formation such as cracking.

Adequacy of the flux pool shows the most consistent reflection in level of acoustic emission data presented here. Changes in wire feed rate also produce a corresponding change in level of acoustic emission, but this may be reflective of the same weld conditions as low flux.

Dropping welding power supply voltage to 25 percent did not produce an identifiable change in acoustic pattern. Weld cracking can be detected and located acoustically as it occurs during or after the welding process.

This is good evidence that acoustic emission methods can be developed to provide a measurable control for at least one previously unmeasurable weld parameter (flux addition) and to detect weld discontinuity formation.

PART 4
ACOUSTIC EMISSION EVALUATION OF LASER SPOT WELDING[15]

Acoustic emission techniques have been used to predict the quality of laser welds in real-time. A criterion has been established for such a prediction and this criterion has been verified by resistance measurement, visual observation and, to a limited extent, mechanical pull testing. It is shown below that a nearly linear relation exists between the threshold crossings of an acoustic emission signal and the quality of the laser weld.

Laser Welding Procedures

Fusion welding is an important fabrication process where lasers are finding increasingly wide use. In early applications, the feasibility of laser welding was demonstrated, as was some of its advantages, including localized heat zone and the ability to weld dissimilar metals and dissimilar geometries. The application of pulsed laser welding to small electrical and electronic components has been discussed in the literature.[16-18]

Initially, laser welding was limited to the interconnection of cross-sectional thicknesses less than 0.5 mm. However, the development of multikilowatt CO_2 lasers allowed deeper penetration welding of large parts such as automobile bodies.[19]

Another important development in laser welding was the emergence of high repetition rates (up to 40 pulses per second), high power and efficient pulsed Nd-YAG lasers (infrared lasers in which the active material is neodymium ions in an yttrium aluminum garnet crystal). This made possible laser welding of small components at high production rates that were previously done with resistance and capacitance discharge welding.

In many electrical applications, it is necessary to join insulated copper wires to terminals. Such terminations often cannot be done by conventional welding techniques and in many cases require removal of insulation. This can be accomplished during a soldering operation by using special insulation such as polyurethane and a high temperature soldering gun. However, such methods are not amenable to mass soldering techniques such as wave soldering. Laser welding for such applications is very advantageous because the incident radiation can be used to remove insulation and cause a weld in a precise area.

To fully exploit the economic potential of this technique, there is a real need for on-line real-time weld evaluation techniques and this can be done with acoustic emission.

The first study of acoustic emission phenomena appears to have been done in the late 1940s and early 1950s in the United States[20] and Germany.[21] Efforts to develop the potential of acoustic emission for high resolution nondestructive surveillance of structures and pressure systems are of more recent origin. Welding geometry and material considerations are continuing studies that are detailed below, followed by a discussion of welding mechanics and analysis. The theoretical aspects of acoustic emission and its experimental correlation to laser welding end this chapter.

Laser Welding Geometry

Welding Insulated Wires to Terminals

The welding geometry to be considered is shown in Fig. 33. The copper wires are 310 μm (12.5 milliinches) in diameter and polyurethane insulated. The terminal shoulder under which the wire is wound protects the wire from direct irradiation from the laser beam. This geometry was found useful where it is desirable to protect the wire from severe necking in the welding process. However, this imposes certain restrictions on the choice of the terminal or the shoulder material (discussed later).

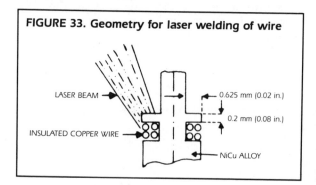

FIGURE 33. Geometry for laser welding of wire

In any welding process, the two materials to be welded must be in intimate contact. In many other welding techniques, this contact is accomplished by proper fixturing and application of mechanical pressure during the welding operation. In laser welding, this is not always possible because its high intensity incident radiation damages any external component used for fixturing. Proper winding and geometry considerations are required to keep the wire in contact with the shoulder terminal.

Material Considerations

Laser welding of insulated copper wires to terminal posts has been attempted and reported in the literature.[18] One common feature of these investigations has been that the terminals did not extend much beyond the termination point. This feature permits focusing of the laser beam on the terminal head, causing the head to flow over and weld to the wire. In the present application, the terminal geometry is more restrictive and extends much beyond the termination point for later wave soldering to a printed circuit board. Furthermore, the system geometry restricts the incident beam angle to within 15 degrees of the terminal axis.

For a wire completely protected by the terminal shoulder, as shown in Fig. 33, several terminal materials were investigated, including phosphor bronze, brass and a corrosion resistant NiCu alloy. For the particular geometry and laser parameters (pulse length of about 1 ms and energy at about 20 J), the most reliable welding results were obtained with the NiCu alloy. The analytical results and their experimental verification are discussed later. However, it is important to note here that the NiCu alloy melting and vaporization temperatures are higher than those of copper, whereas for phosphor bronze and brass the opposite is true. Moreover, copper and nickel are metallurgically compatible in the sense that they make solid solutions in almost all proportions.

The higher melting point of the NiCu alloy together with its excellent metallurgical compatibility with copper make this combination very useful for laser welding.

Effect of Insulation

The polyurethane insulation, while it can be removed by the incident welding pulse, does affect the weld quality and welding mechanics. First, even if the insulated wire is kept in contact with the shoulder, removal of its insulation during the welding cycle will create a gap between the wire and the shoulder. This can be avoided if the wire is wound with sufficient tension and the proper geometry is used to allow its movement towards the shoulder.

Second, there is some evidence that the optical properties of the polyurethane, particularly the absorption coefficient at the laser wavelength (1.06 μm), have a marked effect on welding mechanics. Even with an appropriate shoulder design, part of the wire is directly exposed to laser radiation and focused spot diameter should be carefully tailored to avoid direct exposure of the wire to the maximum beam intensity.

Laser Welding Mechanics

Analytical Results

In order to understand the welding process, consider a simplified one-dimensional model and the melting of a semi-infinite body (Fig. 34) with constant optical and thermal properties in both solid and molten regions. Another model based on Gaussian heat distribution of circular spot and finite thickness has been considered, but does not account for the molten region because its inclusion makes the governing equations nonlinear. For the spot size and pulse lengths considered here, other models give results very close to that of a semi-infinite body.

The modeling problem is further complicated by the fact that in the real welding geometry considered here, the beam is incident on two orthogonal axes (Fig. 33) and some surface vaporization cannot be avoided. The analysis carried out here is to support the qualitative arguments presented later in the discussion of welding process. These conclusions are corroborated with acoustic emission results and experiments carried out on welding of thin plates.

In the model, the first step is to calculate the time required for the surface to reach the melting (t_m) temperature and the vaporization (t_v) temperature. These times and the thermal properties of the material are then used to estimate

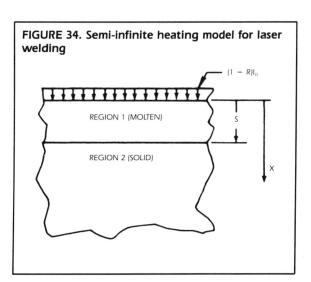

FIGURE 34. Semi-infinite heating model for laser welding

the depth of the molten zone. The optical and thermal properties of interest for the NiCu alloy and the laser welding parameters are shown in Table 5. The time (t_m) required for the NiCu alloy surface to reach melting point is given by:

$$t_m = \frac{\pi}{K}\left(\frac{kT_m}{2I_o}\right)^2 \quad \text{(Eq. 1)}$$

Where:

K = thermal diffusivity;
k = thermal conductivity;
I_o = absorbed radiation intensity; and
T_m = the NiCu alloy melting temperature.

Substituting the values from Table 5, t_m then equals 0.39 ms. From the curves in reference 17, the time for the surface to reach vaporization temperature is estimated to 1.66 ms and the depth of the molten zone is estimated to be 0.07 mm. Once the surface reaches boiling temperature, the molten zone could penetrate further, before heavy surface vaporization sets in. This is because the surface temperature stabilizes as the material takes its latent heat of vaporization.

Furthermore, the existence of vapor pressure causes the vaporization temperature to increase. The significance of the previous calculations lies in the fact that it predicts a minimum pulse length of about 1.6 ms. This prediction was experimentally confirmed by showing that small welds can be made with pulses of 2 ms, but not with pulses of 1.5 ms. At shorter pulse lengths and higher incident radiation intensity (constant energy per pulse) extensive evaporation occurs.

These data, and the fact that in the actual process heating is done from two sides, support the experimental results of producing good welds between the NiCu alloy shoulder and insulated copper wire with 3.5 ms duration laser pulses. The laser welding experiments were also carried out with shoulder terminals made out of phosphor bronze and brass. In both cases, it was difficult to obtain good welds to copper wire and in most cases heavy surface vaporization was observed.

TABLE 5. Thermal and optical properties of NiCu alloy and laser welding parameters

Thermal conductivity (k)	4.66 W•mm
Thermal diffusivity (K)	12.3 mm²•s⁻¹
Specific heat (C_p)	0.427 J•g⁻¹
Melting temperature (Tm)	~1,400 °C
Vaporization temperature (Tv)	~2,800 °C
Reflection coefficient (at 1.06 μm)	~0.6
Pulse length	3.5 ms
Pulse energy	8 to 10 J
Lens focal length	1,500 mm
Spot size (defocused)	1.25 mm

The analytical and experimental results suggest the welding model outlined below.

Experimental Results

Typically, the welds produced between the shoulder, the NiCu alloy terminal and the polyurethane insulated copper wires are made with a 3.5 ms duration pulse focused to about a 1.25 mm diameter spot.

Because of the excellent solid solubility of nickel and copper, the weld region indicates good material mixing. There is also an absence of voids and residual polyurethane insulation. Several of these welds were pull tested to determine the weld strength and the strength of the wire in the vicinity of the weld region. The tensile strength of the welded wire was found to lie between 50 and 75 percent of its parent value. However, in no case was it possible to break the weld joint itself. The welds were also subjected to an electrical test by passing current through several series connected welded terminals. In all cases, wire melted and broke without any deleterious effects or heating of the weld region itself. This result is perhaps due to the large thermal mass of the terminal itself.

Several studies report satisfactory laser welding of nickel and copper alloys. In this experiment, the higher melting temperature of the NiCu alloy was used to produce good welds with insulated copper wire. The proposed welding mechanics are as follows. The wire interface should reach insulation vaporization temperature (600 °C or 1,110 °F) prior to the time when the NiCu alloy shoulder surface begins to vaporize. This is estimated by assuming a constant surface temperature and then calculating the temperature at the wire/NiCu alloy interface from the expression below.[22]

$$T(x, t_m) = T_s \, \text{erfc} \, \frac{x}{2\sqrt{kT_m}} \quad \text{(Eq. 2)}$$

Where:

T_s = surface temperature (about 3,000 °C); and
x = shoulder thickness (centimeters).

If the wire is wound under tension, the copper surface comes in the intimate contact with the NiCu alloy. When the NiCu alloy at the interface approaches its melting point, copper, because of its lower melting point, melts and begins to develop a fusion zone.

Because of its lower melting point, the copper melts and begins to develop a fusion zone. The higher thermal conductivity of the copper inhibits heavy melting and destruction of the wire. This and the fact that the fusion zone melting temperature is higher than that of copper but lower than that of the NiCu alloy, tends to stabilize the process.

Real-Time Evaluation Theory

In many applications, it is important to determine the weld quality during the weld cycle. This is particularly important in order to realize the economic potential of high speed spot welding possible with high repetition rate pulsed lasers.

Generation of Acoustic Emission

Acoustic emission is a phenomenon arising from energy released as a solid material undergoes (elastic) plastic deformation and fracture. A laser spot welding operation releases acoustic emission energy and thus it constitutes an acoustic emission source. The released energy is converted to elastic waves that propagate through the material and can be detected at the material surface using highly sensitive sensors. A detection technique of these waves is shown in Fig. 35.

During the laser welding cycle and the acoustic emission detection time, losses due to reflection at the specimen boundaries can be neglected if the acoustic impedance of the surrounding medium approaches that of the specimen. The amount of energy transmitted to the adjacent medium is equal to the ratio of acoustic impedances:

$$U_E' = \frac{P_s v_s}{P_m v_m} \times 10^2 \qquad (Eq.\ 3)$$

Where:

U_E' = percent of energy transmitted;
P_m = density of the medium;
P_s = density of the specimen;
v_m = wave propagation velocity in the medium; and
v_s = wave propagation velocity in the specimen.

In this experiment, the medium is steel and the sensor is a PZT transducer. A U' value of about 80 percent of acoustic emission energy transfer is expected. Since acoustic emission pulses emanate from point sources, they can be considered as spherical waves (Fig. 35) whose sound pressure decreases inversely with distance from the source. Geometrical losses are independent of frequency and material properties, and are of the form below.[23]

$$p = p_o \frac{x_o}{x} \qquad (Eq.\ 4)$$

Where:

p_o = the sound pressure (pascals);
x_o = an arbitrary point (usually the sensor location).

When frequency and material dependent attenuation is considered, then:

$$p = p_o e^{-\alpha x} \qquad (Eq.\ 5)$$

where α is the attenuation coefficient in decibels per unit distance (frequency dependent).

It was determined that for a transducer frequency of 500 kHz and small intertransducer distances, signal losses should be attributed to geometrical losses. This is shown in the acoustic emission detection of Fig. 36.

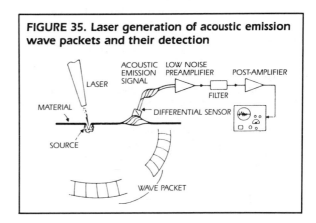

FIGURE 35. Laser generation of acoustic emission wave packets and their detection

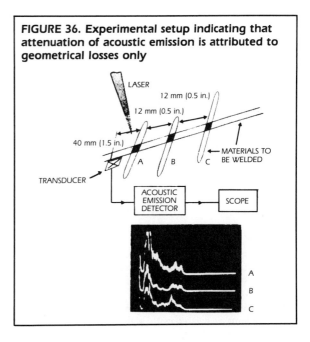

FIGURE 36. Experimental setup indicating that attenuation of acoustic emission is attributed to geometrical losses only

An important point should be made: the attenuation introduced does not affect the proportionality of the transducer output to initial displacement of the lattice other than to introduce a dissipative factor. To obtain optimum performance from a piezoelectric device, the circuit to which it is connected must have high input impedance and frequency match. Hence, depending on the application, some crystals are more desirable than others.

Detection of Acoustic Emission

When the laser beam is absorbed by the surface of the material under consideration, a stress wave packet is generated due to the elastic behavior of the material as well as the thermal stresses generated. As the wave packet travels it excites a differential piezoelectric crystal that in turn generates an electrical acoustic emission signal. This signal is then amplified to obtain usable voltage levels.

Two outstanding features of acoustic emission as an integrity monitoring device are: (1) the ability to detect melting, plastic deformation, resolidification and crack formation or movement at the time it occurs; and (2) this detection is accomplished remotely. Also simultaneous interrogation of multiple sensors provides the necessary information to locate the source of the signals by triangulation.

For the static case of a pressure wave at normal incidence to a crystal that deforms only in the thickness dimension, the maximum peak voltage of the transducer may be shown to be:[24]

$$V = g_{33} d p \quad \text{(Eq. 6)}$$

Where:

- p = the applied pressure (pascals);
- d = the thickness of the transducer element (millimeters);
- g_{33} = the transducer sensitivity constant (V•m/N).

Assuming a U'_E of 100 percent, the minimum theoretical detectable displacement of a material could be calculated from Eq. 7:

$$\delta_{min} = \frac{V_n}{g_{33} \cdot Y_{33}^E} \quad \text{(Eq. 7)}$$

Where:

- V_n = system electrical noise (volts); and
- Y_{33}^E = transducer modulus constant (newtons per square meter).

The theoretical δ_{min} value is equal to about 9.5×10^{-8} mm. Problems associated with high transducer sensitivity and ringing are shown in Fig. 37. In Fig. 37a, the saturated transducer masks signals associated with information concerning the quality of the weld (second distinct burst). Waveforms of the form shown in Fig. 37b are preferable. These curves indicate that extreme sensitivity might be undesirable. This is based on the finite time constant associated with the ringdown of a piezoelectric transducer, according to the equation:

$$V_T(t) = V[f(t) + e^{-\beta t} \sin \omega_o t] \quad \text{(Eq. 8)}$$

FIGURE 37. Detected acoustic emission from laser weld indicating that oversensitivity is not acceptable in detecting acoustic emission:
(a) transducer attached to the welded part; and
(b) transducer attached in proximity to the welded part

(a)
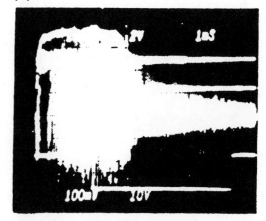

(b)
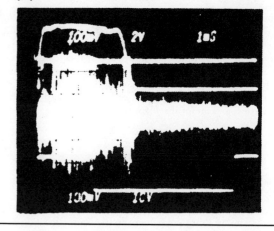

Where:

- V_t = voltages as a function of time;
- V = initial voltage;
- $f(t)$ = a function of time (associated to Rayleigh waves);
- β = time constant of the transducer; and
- ω_o = angular resonant frequency of the transducer.

Equation 8 defines a single acoustic emission event. The acoustic emission detected during laser welding could be thought of as a summation of events resulting from a series of discrete thermal stress sources similar to the shot effect of a semiconductor device. Also, the elastic behavior and the material expulsion resulting from the laser beam could contribute to the overall source of acoustic emission signals. These discrete sources can be characterized because they result in mechanical waves governed by a wave equation having these properties: (1) randomness in time; and (2) stress displacement relation behaving as a function of the form:

$$u(x) = \psi f(x) e^{-\gamma x} \qquad \text{(Eq. 9)}$$

Where:

- $u(x)$ = displacement function;
- ψ = the random amplitude of the displacement;
- γ = the displacement coefficient; and
- $f(x)$ = the displacement function.

If the width of the displacement signal caused by the laser remains constant, the following wave, shown in Fig. 38 arrives at the transducer:

$$u(x) = \sum_{i=0}^{N} \psi_i f(x - x_i) \exp -\gamma(x - x_i) \cdot \mu(x - x_i) \qquad \text{(Eq. 10)}$$

Where:

- N = the number of acoustic emission events; and
- $\mu(x - x_i)$ = the unit step function.

The output of the transducer will consist of a function similar to Eq. 8, but will be random in time and amplitude. Therefore:

$$V_T(t) = \sum_{i=0}^{N} V_i(t - t_i) e \exp -\beta_i(t - t_i) \\ \sin \omega_o(t - t_i) \mu(t - t_i) \qquad \text{(Eq. 11)}$$

where $\mu(t - t_i)$ is the unit time step function.

This is illustrated in Fig. 38b. For simplicity, the envelope of the detected acoustic emission signal is also displayed. It is this envelope that enables the evaluation of the laser spot weld. The widths of these envelopes define such parameters as melting time, cooling time and particulate material removal time. For low ringing transducers, the acoustic emission signal could be characterized as a composite Poisson process.

Experimental Correlation for Laser Spot Welding

Electronic Setup

The electronic setup used for real-time evaluation of the laser welds is shown in Fig. 39. Whenever a weld is made, acoustic emission is generated by the fusion of the material. This mechanical stress signal is then detected by a differential piezoelectric transducer whose output is amplified and filtered.

The output of the filter is subjected to a threshold crossing and counting quantization procedure based on the assumption of having an amplitude and time invariant transducer time constant. Thus, the larger the V_o, the greater the

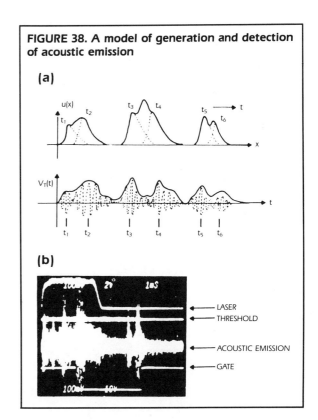

FIGURE 38. A model of generation and detection of acoustic emission

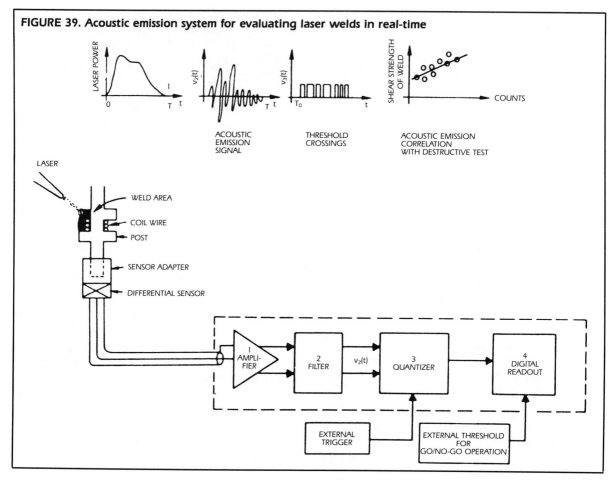

FIGURE 39. Acoustic emission system for evaluating laser welds in real-time

number of threshold crossings. It has been shown[24] that an upper limit of the number of crossings exists, assuming $f(t) = 1$ and can be calculated from:

$$N = \frac{f_o}{\beta_o} \ln \frac{V_o}{V_t} \qquad \text{(Eq. 12)}$$

where N is the number of crossings at threshold V_t.

Experimental Results

Because signals detected from a laser welding operation are multiple events, the derived number should be multiplied by the number of events. Note that these events are only counted during the proper external triggering that is set by the laser intensity window.

Whenever a correlation is necessary, the previously given count is compared to a destructive weld specimen and, as indicated in Fig. 39, the results favor an almost linear relation. The relative quality of the weld can be obtained by reading the digital readout of the acoustic emission evaluation system. The relation can also be depicted graphically.

Before that it is important to characterize the envelope of the signals detected during welding. As shown in Fig. 40, the laser striking a terminal generates the signal of Fig. 40a. When burning the wire insulation, the laser generates the signal in Fig. 40b.

The second event generated in Fig. 41 is attributed to the fusion of two materials and is indicative of weld quality. When the welds were subjected to visual inspection, it was found that they were in full agreement with the illustrations. A go/no-go signal can be generated using a comparator to compare the magnitude of the weld envelope to a variable threshold externally adjusted and indicative to an acceptable weld.

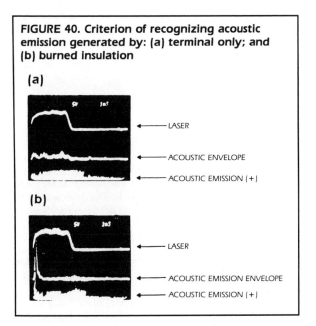

FIGURE 40. Criterion of recognizing acoustic emission generated by: (a) terminal only; and (b) burned insulation

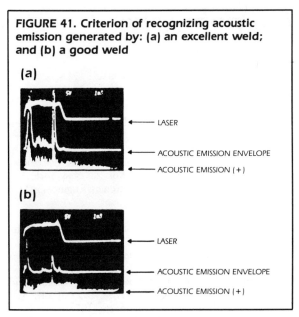

FIGURE 41. Criterion of recognizing acoustic emission generated by: (a) an excellent weld; and (b) a good weld

Conclusions

The evaluation of these laser welds was done nondestructively in real-time using acoustic emission techniques.

This evaluation method is an effective tool for real-time analysis of laser weld quality. Prediction of the failure mechanisms (insulation burning and striking high or low on the terminal) has been done successfully.

Data indicate that an almost linear relation exists between the number of the threshold crossings of an acoustic emission signal and the pull strength of the laser weld. A go/no-go signal can be generated by comparing the magnitude of the laser weld envelope signal to a variable threshold externally adjusted and indicative of an acceptable weld.

REFERENCES

1. Vahaviolos, S.J., V.C. Paek and G.E. Kleinedler. "Network Approach to a Welding Process and Its Experimental Verification with Stress Wave Emission Techniques." *IEEE Transactions on Industrial Electronics and Control Instrumentation*. Vol. 23. New York, NY: Institute of Electrical and Electronics Engineers (May 1976): pp 123-128.
2. Paek, V.C., S.J. Vahaviolos and G.E. Kleinedler. "Thermal Analysis of Capacitor Discharge Welding and Its Correlation with Observed Stress Wave Emission." *Journal of Engineering Materials and Testing* (October 1977): pp 379-386.
3. Havens, J.R. "Controlling Spot Welding Quality and Expulsion." Technical paper AD76-279. Dearborn, MI: Society of Manufacturing Engineers (1976).
4. Vahaviolos, S.J. *Method and Apparatus for the Real-Time Evaluation of Welds Emitted Stress Waves*. US Patent 3,965,726 (June 1976).
5. Notvest, K.R. "Effects of Thermal Cycles in Welding D6aC Steel." *The Welding Journal*. Vol. 45. Miami, FL: American Welding Society (April 1966): pp 173-178.
6. Notvest, K.R. *Acoustic Emission Spot Welding Controller*. US Patent 3,824,377 (July 1974).
7. *Modern Joining Processes*. A.L. Phillips, ed. Miami, FL: American Welding Society (1976).
8. "Nondestructive Inspection and Quality Control." *Metals Handbook*, eighth edition. Vol. 11. Metals Park, OH: American Society for Metals (1976).
9. Vahaviolos, S.J. *Method and Apparatus for the Real-Time Monitoring of a Continuous Weld Using Stress-Wave Emission Techniques*. US Patent 3,986,391 (October 1976).
10. Drouillard, T.F. "Crack Detection in PIGMA Welding of Beryllium." *Proceedings of the Seventh Acoustic Emission Working Group Meeting* (May 1971).
11. Jon, M.C., C.A. Keskimaki and S.J. Vahaviolos. "Testing Resistance Spot Welds using Stress Wave Emission Techniques." *Materials Evaluation*. Vol. 36, No. 4. Columbus, OH: The American Society for Nondestructive Testing (March 1978): pp 41-44.
12. Dickhaut, E. and J. Eisenblatter. "Acoustic Emission Measurements during Electron Beam Welding of Nickel-Based Alloys." ASME Paper 74-GT-45. *ASME Transactions: Journal of Engineering for Power*. Vol. 97, No. 1. New York, NY: American Society for Mechanical Engineers, (January 1975): pp 47-52.
13. Bolotin, Y.I. and V.M. Belov. "Welding Quality Control by Acoustic Emission during Electron-Beam Welding." *Welding Production*. Vol. 23, No. 4 (April 1976): pp 56.
14. Mitchell, J.R., C.H. McGogney and J. Gulp. "Acoustic Emission for Flaw Detection in Electroslag Welds." *Advances in Acoustic Emission*. H.L. Dunegan and W.F. Hartman, eds. Columbus, OH: The American Society for Nondestructive Testing (1979): pp 336.
15. Vahaviolos, S.J. and Mansoor Saifi. "Laser Spot Welding and Real-Time Evaluation." *IEEE Journal of Quantum Electronics*. Vol. QE-12, No. 2. Copyright © 1976 Institute of Electrical and Electronics Engineers (February 1976).
16. Gagliano, F.P. and D.H. Lockhart. "Pulsed Laser Welding of Transistor and Other Electronic Component Parts." *IEEE Digest*. 69C-14. New York, NY: Institute of Electrical and Electronics Engineers (1969): pp 150-151.
17. Cohen, M.I., F.J. Mainwaving and T.G. Melone. "Laser Interconnection of Wires." *Welding Journal*. Vol. 48 (1969): pp 191-197.
18. Cohen, M.I. and J.P. Epperson. "Application of Laser to Microelectronic Fabrication." *Advances in Electronics and Electron Physics*. A.B. El-Kareh, ed. New York, NY: Academic Press (1968).
19. Yessik, M. and D.J. Schmatz. "Laser Processing in the Automotive Industry." *Digest of Technical Papers: IEEE/OSA Conference on Laser Engineering and Applications*. New York, NY: Institute of Electrical and Electronics Engineers (1975).
20. Mason, W.P., H.J. McSkimm and W. Shockley. "Ultrasonic Observations of Twinning in Tin." *Physical Review*. Vol. 73 (May 1948): pp 1,213-1,214.
21. Kaiser, J. "Information and Conclusions from the Measurements of Noise in Tensile Stressing of Metallic Materials." *Arch. Eisenhuetten*. Vol. 24 (1953): pp 43-45.
22. Carslaw, H.S. and J.C. Jaeger. *Conduction of Heat Solids*. New York, NY: Oxford Press (1959).
23. Mason, W.P. and H.J. McSkimin. "Attenuation and Scattering of High-Frequency Sound Waves in Metals and Glasses." *Journal of the Acoustical Society of America*. Vol. 19 (1947): pp 464-473.
24. Vahaviolos, S.J. "Real-Time Detection of Microcracks in Brittle Materials Using Stress Wave Emission (SWE)." *IEEE Transactions on Parts, Hybrids, Packaging*. Vol. PHP-10 (September 1974): pp 152-159.

SECTION 10

ACOUSTIC EMISSION APPLICATIONS IN CIVIL ENGINEERING

Theodore Hopwood, Kentucky Transportation Research Program, Lexington, Kentucky

Charles McGogney, Federal Highway Administration, Washington, DC

PART 1
ACOUSTIC EMISSION GEOTECHNICAL APPLICATIONS

Historical Background

In 1939, a suspension bridge at Portsmouth, Ohio experienced stress corrosion cracking of the main cable wires at anchorage points located at each end of the bridge. Watchmen were placed in the anchor chambers where the fractures had been detected. Subsequently, they reported hearing the sounds of further wire breakage on quiet nights. When this was reported, a decision was made to recable the bridge.[1] That was one of the earliest documented instances for the use of acoustic emission phenomena in a structural application.

Also in the late 1930s, L. Obert and W.I. Duval at the US Bureau of Mines were performing sonic tests on rock mines. They were surprised to find that stressed rock pillars emitted microlevel sounds.[2] Those noises were later termed *rock talk*.

In contrast to the early acoustic emission structural monitoring at Portsmouth, the rock talk phenomena has been the subject of continuing geotechnical research since the time of its discovery.

Over the years, much progress has been made in civil engineering applications using acoustic emission testing. However, most of those applications are still in the developmental stage. Also, some of the past research is contradictory. Therefore, the potential acoustic emission user should perform preliminary tests to ascertain the viability of the intended acoustic emission procedure. Both laboratory and field tests should be performed under controlled conditions to ensure the applicability and usefulness of the test method before it is employed in-service. While this approach is expensive, subsequent cost savings that may be derived by employing acoustic emission testing in-service, compared to the costs of other nondestructive testing methods, usually justifies the expenditures.

The following text details the primary applications of acoustic emission in civil engineering: (1) geotechnical; (2) structures; and (3) special component testing. The reader is referred to the acoustic emission literature for references to more specific applications.[3,4]

Geotechnical Applications

The acoustic emission phenomena can be applied to a variety of geotechnical materials. Those materials include but are not limited to soils (sandy and clayey), rocks (igneous and metamorphic), fossilized deposits (coal) and ice. Acoustic emission monitoring can be conducted on foundations, mines, tunnels, cuts and fills, embankments, tied back and retaining walls and dams. Acoustic emission monitoring may be useful in nearly every geotechnical application where subsurface deformation can be anticipated.

In the application of the acoustic emission test method, instability must exist in the material before meaningful results can be obtained. Instability creates in the disposition of the material a change that may be detected by the acoustic emission sensors. The instability may be caused by the removal of material (as encountered in tunnels and mines or in cuts); by the addition of material (as encountered in a fill, retaining wall or earthen dam); by the imposition of a load (as in a foundation); or by the interaction of the material with water or chemicals (as in ground water seepage). Physical sources of acoustic emission activity in a material include shear movement, fracture and, in a liquid medium, particle flow. Acoustic emission is detected either as stress wave packets traveling through the material or as moving particles contacting a sensor or waveguide.

Diverse acoustic emission source mechanisms operate in geotechnical testing and the resulting stress waves are often transmitted through heterogeneous mediums. This has led to a variety of testing techniques and systems. For acoustic emission geotechnical studies as shown in Fig. 1, frequencies range from the low audible (25 Hz) to the ultrasonic (50 kHz).

Acoustic emission signals have three important characteristics that must be taken into consideration.[5] Those are signal strength, frequency and attenuation. An approximate relationship has been derived for maximum acceleration of the signal at its source (for soils).[6]

$$a_{max} = \frac{\sqrt{\pi f^2 e^2 VRc}}{2(\Delta t)r} \quad \text{(Eq. 1)}$$

Where:

- a = acceleration of the elastic wave at r;
- f = frequency of the elastic wave at r;
- e = elastic strain released;
- V = volume of soil involved in the process;
- R = radiation of coefficient (efficiency of radiating elastic waves);
- c = wave velocity;
- Δt = time increment for release of elastic strain e; and
- r = distance from source.

Acoustic emission events contain a spectrum of different frequencies. The acoustic emission spectrum usually contains higher frequencies at the emitting source than at some distance from the source. This is due to the greater attenuation (see below) of high frequency portions of the acoustic emission as it propagates through the material (Fig. 2). In the example, an acoustic emission sensor is sensitive only to stress waves with frequencies greater than 100 kHz and the maximum range of acoustic emission detection is 2 m. Frequencies measured during acoustic emission events are affected by: (1) the source of the acoustic activity; (2) the acoustic properties of the transmitting material; and (3) the frequency response characteristics of the sensor and acoustic emission system.

Acoustic emission response is directly affected by the loss of acoustic emission energy as waves travel through a material. Loss of energy is called *attenuation*. Attenuation may be measured by placing several transducers at different distances from an acoustic emission source. Attenuation can be calculated as:

$$\gamma = \frac{20}{x} \log \frac{A_1}{A_2} \quad \text{(Eq. 2)}$$

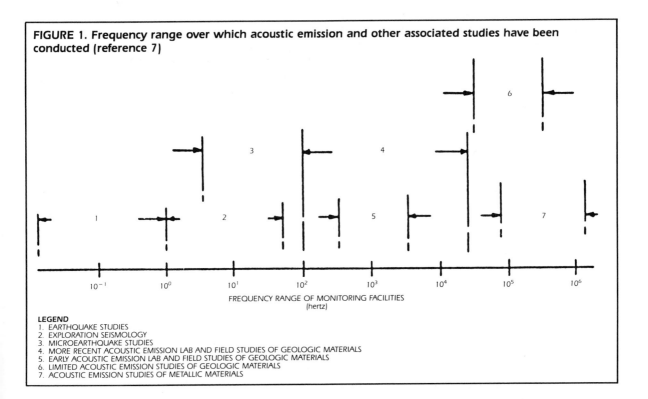

FIGURE 1. Frequency range over which acoustic emission and other associated studies have been conducted (reference 7)

LEGEND
1. EARTHQUAKE STUDIES
2. EXPLORATION SEISMOLOGY
3. MICROEARTHQUAKE STUDIES
4. MORE RECENT ACOUSTIC EMISSION LAB AND FIELD STUDIES OF GEOLOGIC MATERIALS
5. EARLY ACOUSTIC EMISSION LAB AND FIELD STUDIES OF GEOLOGIC MATERIALS
6. LIMITED ACOUSTIC EMISSION STUDIES OF GEOLOGIC MATERIALS
7. ACOUSTIC EMISSION STUDIES OF METALLIC MATERIALS

Where:

γ = attenuation coefficient (decibels over distance);
x = distance between pickup points;
A_1 = amplitude of signal at point-1; and
A_2 = amplitude of signal at point-2.

Attenuation is a function of the frequency content of the acoustic emission waves and the medium through which they travel. As shown in Fig. 3, soils have higher attenuation responses than rocks or coal.

Laboratory Behavior of Acoustic Emission from Geotechnical Materials

A good understanding of the acoustic emission response of geotechnical materials is necessary in order to design acoustic emission field tests. Field situations may not always duplicate laboratory results especially if circumstances require simplified laboratory designs. However, laboratory tests will provide valuable insights into the responses of geotechnical materials and may be useful in preparing feasibility studies before expensive field testing is performed.

Earlier studies of acoustic emission laboratory compressive tests on rock[7] indicated that, as the applied stress in a material approached failure levels, the acoustic emission rate increased (Fig. 4). In another study, the acoustic emission activity rate decreased after structural failure occurred in a rock mine.[8] Such observations indicated that acoustic emission rate is a function of the degree of rock stability.

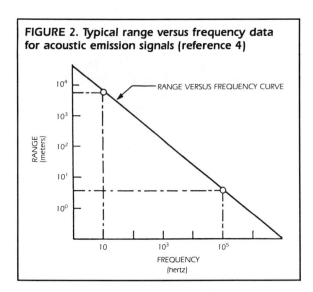

FIGURE 2. Typical range versus frequency data for acoustic emission signals (reference 4)

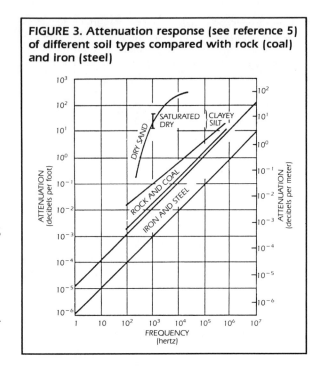

FIGURE 3. Attenuation response (see reference 5) of different soil types compared with rock (coal) and iron (steel)

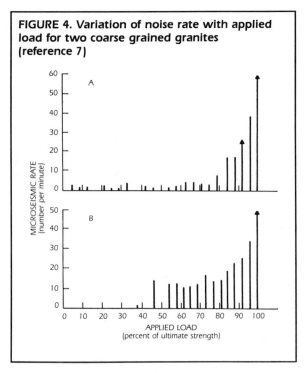

FIGURE 4. Variation of noise rate with applied load for two coarse grained granites (reference 7)

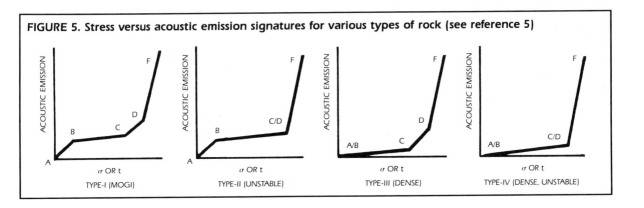

FIGURE 5. Stress versus acoustic emission signatures for various types of rock (see reference 5)

The repeated compressive loading and unloading of rock samples has also been studied.[9] Above a certain high stress level, the acoustic emission event rate increased significantly. If the rock was subjected to stress cycling below that level, the material became less emissive with each succeeding stress cycle. When the rock was subjected to stress cycling above that level, it became noisier with the application of successive stress cycles. Rock was also found to be emissive during the unloading portion of each stress cycle.

The basic acoustic emission behavior of rock subjected to compressive stress is shown in Fig. 5.[10] Cumulative acoustic emission count is plotted as a function of compressive stress. Four different zones of acoustic emission behavior may be identified. They are related to crack closure (zone-AB); linear elastic deformation (zone-BC); stable fracture propagation (zone-CD); and unstable fracture (zone-DF).

Experimental work has revealed other types of acoustic emission signatures. Those were discerned from tests of ten different types of rocks including: dolomite (type-I); shale (type-III); mica schist from Philadelphia, Pennsylvania (type-III); horneblende (type-II); marble (type-II); dolomitic marble (type-I); granite (type-I); shale (type-II); mica schist from Germantown, Pennsylvania (type-III); and limestone (type-IV). Type-II rocks lacked stable fracture behavior; type-III rocks lacked the crack closure region; and type-IV rocks lacked both crack closure and stable fracture regions. Differences in acoustic emission response with stress indicate the need to have a good understanding of the type of material existing in the field prior to monitoring.

The state of stress also affects the acoustic emission response of rock. Split tensile tests of the same rocks produced only type-III and type-IV behavior. In tension, type-I and type-II signatures would not be expected.

Direct shear tests of rock produced either quick or slow failure.[11] Slow failure was characterized by initial slow shear movement of mating rock faces followed by sudden failure. Both failure modes produced high amplitude acoustic emission activity.

Direct shear acoustic emission monitoring tests of soils were performed on three types of soils: beach sand, clayey silt and kaolinite clay.[12] The resulting shear stress and accumulated acoustic emission count versus strain curves indicate that the soil producing the most total acoustic emission activity during loading (sand) showed the least strain and that the soil producing the least total acoustic emission activity during loading (kaolinite clay) showed the most strain (Fig. 6). Other laboratory tests[13] indicate that increased amounts of water in soils will result in less acoustic emission activity when strained.

Laboratory tests also showed that soils produce relatively large amounts of acoustic emission activity when they fail. A series of triaxial compression tests was conducted on clayey, silty soils using confining pressures of 35, 70 and 105 kPa (5, 10 and 15 psi) to simulate soils at depths of 2.5, 5 and 7.5 m (8, 16 and 24 ft) respectively.[14] The test results shown in Fig. 7 show a good correlation between total acoustic emission activity and strain. Based on those results, both acoustic emission and strain can be used to evaluate soil stresses and therefore failure. This type of correlation exists for many soils.

The frequency content of soil's acoustic emission activity depends on many variables, including: the distance between the source and acoustic emission sensor; the nature of the acoustic emission sensor; and the type of sensor pickup used to monitor acoustic emission activity. Figure 8 shows the frequencies obtained from clayey, silty and sandy soils tested under identical conditions. While such conditions may be monitored in the laboratory, difficulties may be encountered when attempts are made to perform frequency spectral analysis in the field without due consideration of the aforementioned variables.

The amplitude curve shown in Fig. 9 reflects the relatively weak acoustic emission emissivity of cohesive (clayey) soils compared to granular soils (sands). Increasing acoustic emission amplitudes are produced by both types of soils with increasing stresses. As expected from Fig. 6, granular

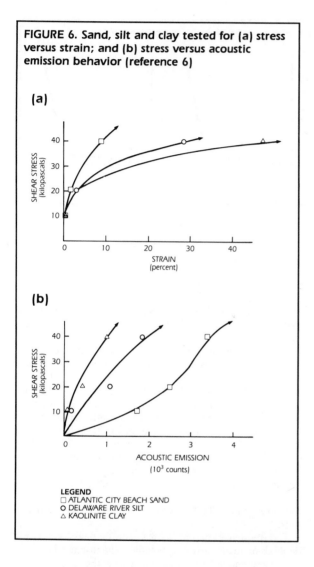

FIGURE 6. Sand, silt and clay tested for (a) stress versus strain; and (b) stress versus acoustic emission behavior (reference 6)

FIGURE 7. Response of a clayey silt soil with 15 percent water content tested in triaxial shear at confining pressures of 105, 70 and 35 kPa (15, 10 and 5 psi) with accelerometer embedded in the sample: (a) stress versus strain; and (b) stress versus acoustic emission (reference 13)

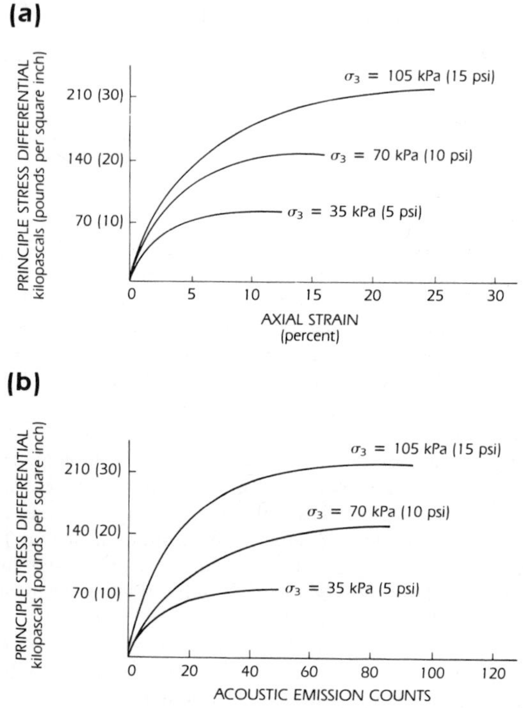

soils produce higher amplitude acoustic emission activity than cohesive soils. Also, there are differences in acoustic emission amplitude behaviors of cohesive and granular soils as the failure stresses of these two different materials are approached. Cohesive soils show a decrease in acoustic emission amplitudes above 70 percent of the failure stress whereas the acoustic emission amplitudes in granular soils increase until failure. Such behavior indicates that acoustic emission amplitude analysis may be better suited for granular soils than for cohesive soils.

Laboratory acoustic emission tests have been conducted to determine if water seepage could be detected.[15] A hydrophone was embedded in a column of gravel and both clear and turbid water were passed across the hydrophone. The turbid water could be detected at a lower flow rate than the clear water (10 mL·s^{-1} versus 45 mL·s^{-1}). The acoustic emission rate activity for both specimens was found to increase exponentially with flow rates (Fig. 10).

Past laboratory tests of rocks and soils have shown that acoustic emission activity is an indicator of material instability. Unstable material conditions may be associated with a number of failure mechanisms encountered in geotechnical structures. Based on initial laboratory tests, the only geotechnical materials that appear to present a problem in acoustic emission testing are clayey (cohesive) soils. High monitoring sensitivities are necessary to detect failure of clayey soils.

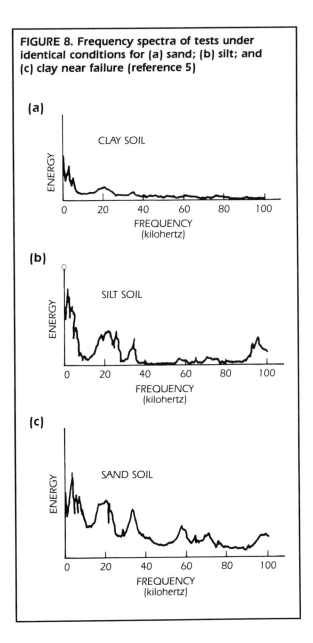

FIGURE 8. Frequency spectra of tests under identical conditions for (a) sand; (b) silt; and (c) clay near failure (reference 5)

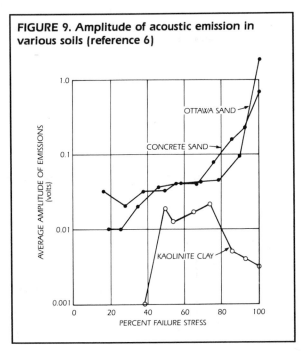

FIGURE 9. Amplitude of acoustic emission in various soils (reference 6)

consists of: (1) optional waveguide; (2) sensor or transducer; (3) preamplifier; (4) signal cable; (5) filters; (6) amplifier; (7) counter; and (8) output recording devices, typically an oscilloscope, tape recorder, XY or chart recorder.

Geotechnical acoustic emission equipment is typical of instruments used in other acoustic emission applications except that it usually monitors lower frequencies and in some cases employs volumetric acoustic emission source location.

A wide variety of sensors are employed in geotechnical work, including: (1) accelerometers; (2) piezoelectric transducers; (3) geophones; and (4) hydrophones (see Fig. 12).

The wide range of geotechnical acoustic emission test frequencies shown in Fig. 1 requires resonant frequency responses of the sensors in these ranges. Geophones and hydrophones are used in tests involving frequencies between 1 and 2,000 Hz. At frequencies greater than 2,000 Hz, accelerometers and piezoelectric transducers are used.

As geophones and hydrophones monitor lower frequency acoustic emission activity, they are capable of detecting high amplitude acoustic emission events that typically occur at those frequencies. As noted in Fig. 3, high frequency waves are subject to attenuation as they travel through a material. The greater attenuation problems encountered in high frequency acoustic emission monitoring can be offset by mounting accelerometers or piezoelectric transducers near the source of material instability or by using waveguides that penetrate through the unstable material. Higher frequency

Geotechnical Acoustic Emission Equipment

A schematic diagram of a simple, single-channel acoustic emission monitoring system is shown in Fig. 11. The system

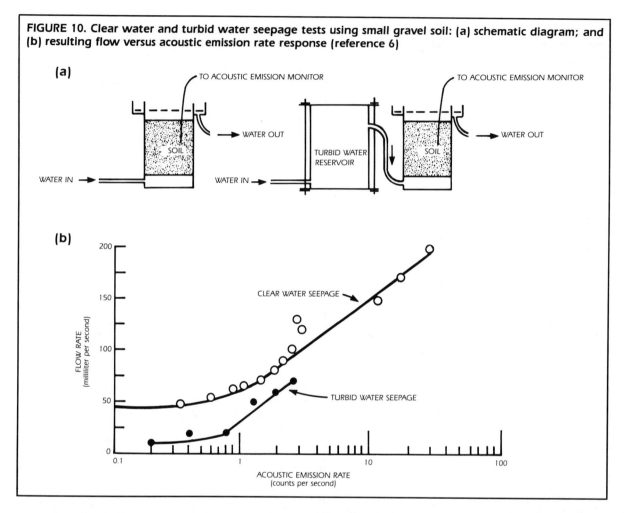

FIGURE 10. Clear water and turbid water seepage tests using small gravel soil: (a) schematic diagram; and (b) resulting flow versus acoustic emission rate response (reference 6)

transducers offer the advantage of being more resistant to ambient noise sources around the test site.

As shown in Fig. 13 (see reference 16), geotechnical acoustic emission monitoring can be conducted at a number of locations. For an underground geological structure (ST), the transducers can be located on the surface of the structure or at holes drilled outward from it (B). If access is not convenient, the substructure can be monitored from transducers installed from the ground surface (C1-C4).

Mounting Techniques

A number of localized mounting techniques have been used (Fig. 14) for underground mines and tunnels. Where surface installation is desirable, a number of installation techniques have been implemented (Fig. 15). Surface installation includes the borehole technique (Fig. 15d). In this method the borehole is filled to a specified level with water (which acts as a coupling medium) and a hydrophone is then placed in the water. In monitoring deeply buried structures or structures with deep overburdens, special deep burial techniques (Fig. 16) have been used.[16]

The indirect mounting technique uses metal waveguides placed in the earth or sidewall of a mine (Fig. 17). A transducer or accelerometer is bonded on the exposed end of the waveguide. One application used 13 mm (0.5 in.) diameter steel rods with threaded ends for waveguides. A section of the rod is driven into the ground until only the threaded end is exposed (Fig. 18). Another section of rod is coupled to the buried rod and they are driven further into the ground. Once the desired depth is reached, the transducer is bonded to the end of the exposed rod.

Another technique makes use of the well casing as the waveguide. The transducer is mounted to the wellhead (see

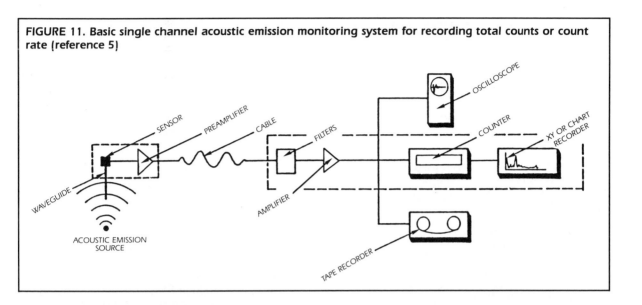

FIGURE 11. Basic single channel acoustic emission monitoring system for recording total counts or count rate (reference 5)

Fig. 17). Any material instability or acoustic emission waves occurring along the length of the waveguide will transmit the acoustic emission wave activity from the soil to the waveguide. The acoustic emission waves travel along the waveguide and activate signals by the acoustic emission transducer.

There are several advantages and disadvantages in the direct and indirect transducer mounting techniques. Below about 3 m (10 ft), the direct surface mounting techniques are usually more difficult and expensive to install. Also, the mounting depth of the acoustic emission sensor must be close enough to the material instability to allow the transducer to detect the resulting acoustic emission activity.

Waveguides are easier to install than the direct mounting methods. Also, since the entire length of the waveguide can be a receptor of acoustic emission activity, it can be placed sufficiently deep in the ground to ensure that it will be near a location of material instability such as a plane of shear failure. However, it has been noted[17] that the depth of acoustic emission activity cannot be determined using waveguides and that the possibility exists for waveguides to interact with the surroundings and create acoustic emission activity. Therefore, caution should be exercised when using waveguides.

Monitoring Devices

Geotechnical acoustic emission monitors are usually either simple single-channel acoustic emission devices or complex multichannel units capable of volumetric instability location.

Single-channel acoustic emission monitoring devices are usually portable and are normally mounted in shock resistant, weatherproof cases (Fig. 19). The units are often battery powered and can be left in the field for remote monitoring for periods up to 48 hours. These devices are suitable for general condition monitoring. This means that one single-channel acoustic emission monitor is dedicated to one acoustic emission transducer or, in effect, one test location.

Many practical test locations require concurrent acoustic emission monitoring at several test sites. Due to their relatively low cost, multiple sites can be economically monitored using a number of single-channel acoustic emission systems. Proper placement of acoustic emission sensors (and possibly waveguides) and the high attenuation encountered in many geotechnical applications should provide acoustic emission isolation between the different test sites. In employing such acoustic emission systems, the user must be confident where to locate his sensors, to ensure detection of any acoustic emission activity.

For more complex monitoring tasks, multichannel acoustic emission systems are employed and are capable of acoustic emission source location. For source location along a horizontal plane, a four-channel system is required. For volumetric (three-dimensional) location a minimum of five channels is required. A five-channel system is shown in Fig. 20, with a source located at point P_0. Computerized methods for determining source location normally utilize a least squares iteration technique that searches for the best average solution for a set of equations that contain the transducer coordinates, X_i, Y_i and Z_i ($i = 1, 2, 3 \ldots n$) and the corresponding acoustic emission arrival times and velocities

t_i and V_i. The method involves the solution of the following set of i equations:

$$(X_i - X_o)^2 + (Y_i - Y_o)^2 \\ + (Z_i - Z_o)^2 = V_i^2(t_i - T_o)^2 \quad \text{(Eq. 3)}$$

where X_o, Y_o and Z_o are the true coordinates of the source; and T_o is the true origin time.

Such analyses require interfacing the acoustic emission system with a computer, though a number of the newer commercial acoustic emission systems contain microprocessor circuitry for source location. A number of computer programs have been prepared for source location. The Pennsylvania State University program (IBMSL) assumes that the acoustic emission source and monitoring transducers are located in a homogeneous medium with provision for anisotropic velocity characteristics. The US Geological Survey has prepared a program for source location in bedded strata.

The data output from the single-channel and multichannel acoustic emission systems can include: total acoustic emission counts; acoustic emission event rate; acoustic emission amplitude; signal rise time; signal-to-noise background ratio; acoustic emission energy; acoustic emission energy rate; and acoustic emission event duration. Multichannel systems will also furnish coordinates of the acoustic emission sources.

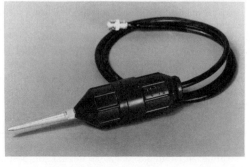

FIGURE 12. Typical geophones used as acoustic emission sensors

FROM PENNSYLVANIA STATE UNIVERSITY (H.R. HARDY). REPRINTED WITH PERMISSION.

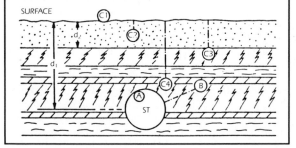

FIGURE 13. Simplified field situation illustrating various possible locations for acoustic emission transducers (reference 4)

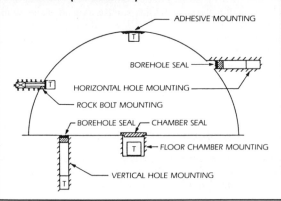

FIGURE 14. Various methods employed for underground mounting of acoustic emission transducers (reference 4)

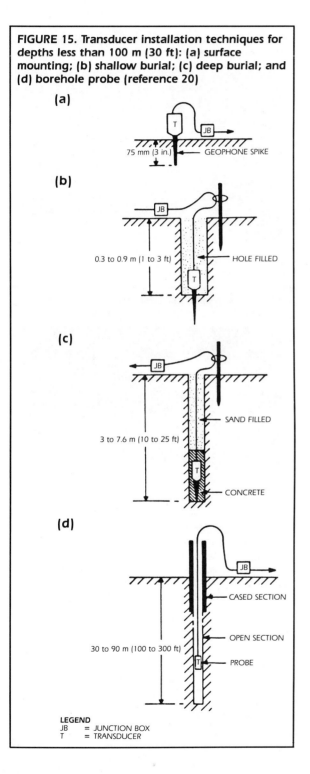

FIGURE 15. Transducer installation techniques for depths less than 100 m (30 ft): (a) surface mounting; (b) shallow burial; (c) deep burial; and (d) borehole probe (reference 20)

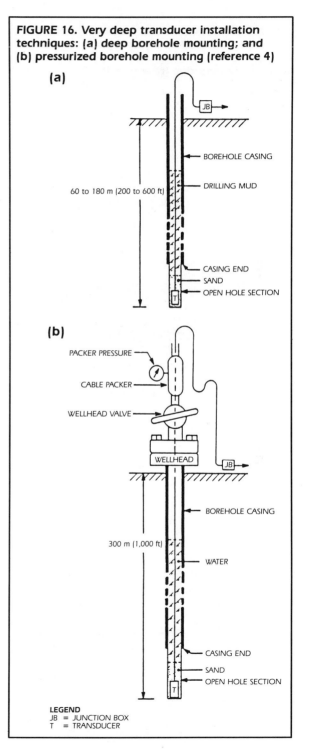

FIGURE 16. Very deep transducer installation techniques: (a) deep borehole mounting; and (b) pressurized borehole mounting (reference 4)

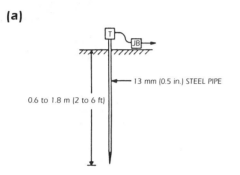
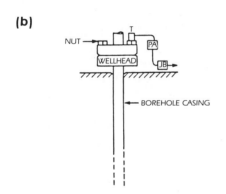
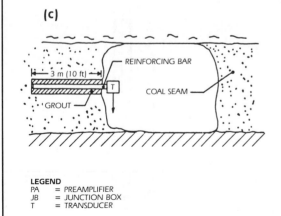

FIGURE 17. Various indirect methods for transducer installation: (a) waveguide mounting; (b) wellhead mounting; and (c) modified waveguide mounting (reference 4)

LEGEND
PA = PREAMPLIFIER
JB = JUNCTION BOX
T = TRANSDUCER

FIGURE 18. Accelerometers attached to threaded steel rods (waveguides)

FROM DREXEL UNIVERSITY (R.M. KOERNER and A.E. LORD). REPRINTED WITH PERMISSION.

FIGURE 19. Typical portable acoustic emission monitors used in geotechnical applications

FROM DREXEL UNIVERSITY (R.M. KOERNER and A.E. LORD). REPRINTED WITH PERMISSION.

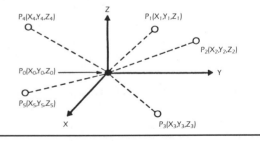

FIGURE 20. Geometry of a typical acoustic emission transducer array; transducers are located at points P_1 through P_5 with the acoustic emission source at point P_0 (reference 4)

Acoustic Emission Geotechnical Field Applications

Foundation Monitoring

The potential for application of the acoustic emission method to foundation monitoring, especially for shallow foundations such as spread footings, appears good. A laboratory bearing capacity test involving settlement of a footing has also been conducted (see Fig. 21).[18] The transition between linear deflection and failure is clearly indicated by both the footing deflection and total acoustic emission activity load diagrams.

The two types of shallow foundation failures, punching settlement and tilting settlement due to a bearing capacity failure, should be relatively easy to monitor. Transducers or guide rods should usually be placed close to or beneath the foundation footing and should be located close to the anticipated failure shear planes. The soil shear planes in punching failures should occur symmetrically around the footing. In the case of bearing capacity tilting failures, a portion of the substructure shear activity is located on the opposite side of tilt.

The largest portion of foundation settlement usually occurs during construction. Generally, after the structure is completed, foundation settlement (secondary compression) is relatively small, or in certain cases may be inconsequential. This settlement is very dependent on time. Therefore, it is desirable to perform acoustic emission monitoring on foundations or footings during construction. Simultaneous monitoring of several, similar foundation structures to determine the comparative acoustic emission activity is also useful. The acoustic emission activity could also be correlated with the amount of settlement observed at the various sites. The acoustic emission activity during a given time period could also be compared with the amount of settlement or tilt that has occurred at the foundation over that time interval.

If the foundation settlement or tilt stabilized with time, then periodic acoustic emission testing could be used to determine the onset of failure. If nearby excavation or tunneling work was being performed, acoustic emission testing would be useful to detect any unfavorable subsurface interaction between that work and the foundation that would result in further foundation movement.

Tunnel Monitoring

Application of acoustic emission monitoring to detect instability of tunnels appears promising. Tunnels embedded in grouted soil, plain soil and rock have been monitored.[19] Sensors used to monitor acoustic emission activity included hydrophones and accelerometers. The hydrophones were used in downhole installations partially filled with water for the

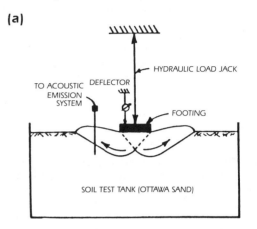
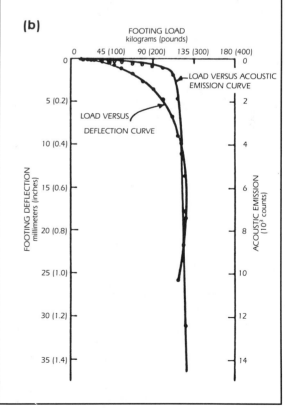

FIGURE 21. Bearing capacity failure of a shallow foundation: (a) schematic diagram of test; footing is 100 × 200 mm (4 × 8 in.); (b) resulting load versus deflection and load versus acoustic emission response curves (reference 6)

tunnel through-soil tests. In those tests, the holes were placed adjacent to tunnel shafts.

The tunnel through-rock tests were conducted from within the tunnel. Accelerometers were attached on threaded studs epoxied to the rock wall, clamped on a spiling bar and threaded on a piton (steel wedge) driven into a tight joint. Noise from construction activity interfered with the responses and meaningful acoustic emission data could not be obtained during construction hours. However, one acoustic emission test conducted during off-hours gave good indication of material instability that was manifested as a minor roof fall. Further tunnel monitoring experience is required to correlate acoustic emission activity with incipient failures.

Embankment Monitoring

The acoustic emission monitoring of soil and rock embankments and cut sections may provide warnings of slope failures. Numerous experiments have been conducted to detect slope instability of embankments and cuts consisting of rocks and soils.[20] Previous tests indicate (1) that slope and cut instability will produce acoustic emission activity and (2) that acoustic emission activity can provide warnings of slope instability prior to visible signs of movement and subsequent failure. To detect these failures, acoustic emission sensors should be placed on top of a berm (embankment) or cut (to detect tension cracks) or at the toe of an embankment (to detect shear movement).

If failure or severe shear is possible, waveguides should also be used to safeguard the sensor. Waveguides should probably be used in soft soils as well. Due to the low amplitude acoustic emission from soft soils, it may be necessary to have waveguides penetrating directly through areas of possible soil shear movement.

Monitoring Mines

A large amount of research has been conducted on mines,[21] including studies of rock bursts, column failure, roof falls and outbursts in coal. As found in other applications of acoustic emission monitoring, collapse or material instability is usually preceded or accompanied by large increases in acoustic emission activity. One interesting analytic technique consists of monitoring daily acoustic emission activity at various locations in the mines. Cumulative plots of acoustic emission activity are compiled as a function of time at various locations. Areas with persistent and increasing levels of acoustic emission activity are candidates for potential failures.

Most acoustic emission tests in mines require long-term monitoring. In many cases, the sensors have been located inside the mine shafts and the signal wires are routed up the mine shaft to the acoustic emission monitoring system. Another method used by researchers at Pennsylvania State University involves surface monitoring.[22] In one case, geophones were placed in boreholes at various depths over and in advance of a long wall coal mine. The tests were bandpass filtered between 0 and 1,000 Hz using a system gain of 70 to 90 dB. High amplitude acoustic emission events were subjected to source location analysis. Viable results were obtained.

Other Applications

Earthen dams are also a good prospect for acoustic emission monitoring to detect soil movement or water seepage. Waveguides can be mounted either vertically or horizontally in areas where soil shear is likely to occur (Fig. 22). Acoustic emission sensors and possibly waveguides could be placed at the toe of a dam to monitor water seepage. Acoustic emission activity in this area of an earthen dam is often a precursor to subsequent failure.

Many acoustic emission tests have been performed on subterranean storage areas. Those areas are usually old caverns or mines converted to storage of materials such as oil, natural gas or nuclear waste. Monitoring techniques are

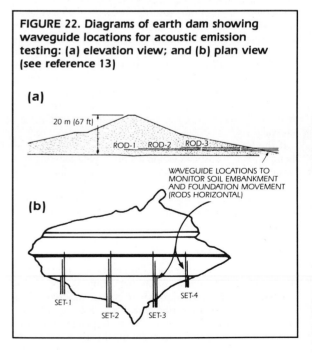

FIGURE 22. Diagrams of earth dam showing waveguide locations for acoustic emission testing: (a) elevation view; and (b) plan view (see reference 13)

similar to those employed for mines. Acoustic emission methodology has been used to monitor the structural integrity of caverns and mines. In the case of pressurized natural gas storage, acoustic emission leak detection methodology is possible. Work has also been done on acoustic emission activity of salt dissolution by water and the results are applicable to some salt mine storage problems.[23]

Some research indicates that acoustic emission can be used to determine prestress in soils and rock.[24] In general, the technique consists of coring a hole in the soil or rock mass and placing a pressurizing device and an acoustic emission sensor in the hole. The acoustic emission sensor must be in contact with the material as the pressurizing device is actuated. The soil or rock will apparently produce acoustic emission activity after its previous high stress is exceeded (the Kaiser effect). This technique promises to provide rapid on-site determinations of prestress. However, the method is still in the research stage.

PART 2
ACOUSTIC EMISSION MONITORING OF STRUCTURES

Acoustic emission appears to be most useful for the nondestructive inspection of structures made from steel (plates or rolled shapes) or prestressed concrete. Structures made of other materials such as conventional reinforced concrete, masonry, wood or aluminum may also be inspected by acoustic emission testing. However, there has been little acoustic emission work performed on such structures.

Prestressed and post-tensioned concrete structures have proven especially difficult to inspect by conventional NDT methods.[25] The primary area of concern on those structures are steel wire strands used as the tension ligaments. The strands are made from high strength steel wires that are susceptible to corrosion cracking in some environments. Cracking of the strand wires can be associated with live loading (fatigue or corrosion fatigue) or in a static environment due to stress corrosion. The utility of acoustic emission testing on prestressed concrete structures has yet to be demonstrated. Some review of acoustic emission experiments related to corrosion of reinforced concrete is contained in Part 3 of this Section.

A major potential civil engineering application of acoustic emission monitoring is to inspect large steel structures such as bridges, dams, penstocks, towers, buildings, cranes and oil platforms. Those structures are usually subjected to some cyclic loads that can promote fatigue cracking. The potential for fracture and the severity of such events have increased over the past forty years due to widespread use of welding as a fabrication method.

Welded structures are often monolithic. When fractures occur in welded structural members, the potential exists for their complete fracture. If the structure is nonredundant, it can catastrophically collapse. Even if the structure is redundant, it can be severely debilitated by loss of a structural element.

Structural failure of steel members usually occurs at connections such as welded splices and bolted or riveted joints (Fig. 23). Welds are often sites of stress concentrations and discontinuities. They are initiation sites for fatigue cracks in the presence of cyclic stresses. Less frequently, fatigue cracks can nucleate from bolt hole or rivet hole areas into structural members. Occasionally, some structural steels will exhibit brittle behavior at low temperatures. Structural elements made from such steel may fail at very low stresses.

The potential service problems offer good opportunities for acoustic emission monitoring. Any acoustic emission testing plans should be formulated by a multidisciplinary team consisting of structural engineers, metallurgists and nondestructive testing engineers to obtain the best results. Acoustic emission testing may involve the application of service loads (dead loads and live loads) or proof loads. The magnitudes of those loads should be anticipated and, at least in initial acoustic emission tests, those loads should be measured.

Acoustic Emission Activity in Structural Steel Members

Three possible acoustic emission sources have been proposed for steel.[26] Those are Luders band propagation, carbide fracture and decohesion fracture of inclusions. These are believed to be responsible for the copious acoustic emission activity that usually accompanies the stressing of steel. Much of that activity occurs over short time intervals (bursts). Acoustic emission activity is produced by both gross deformation and by fracture processes of a wide variety of ductile and brittle structural materials (Table 1).[27-31]

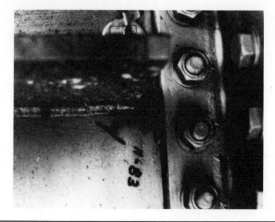

FIGURE 23. Acoustic emission transducer partially visible above stiffener; note crack at the end of the welded stiffener

Crack Propagation

When cracks propagate across a steel member, a plastic zone is generated at the tip of the crack. It is probable that the major source of acoustic emission activity is the plastic work done in generating the crack. For metals, the plastic work in generating a crack is usually 50 to 1,000 times greater than the work generated creating new crack surface area. In two-dimensional perspective, the plastic zone can be assumed to be circular. The radius of the plastic zone at yield r_y has been related to fracture toughness by the following equation.

$$r_{y(max)} = \frac{1}{A}\left(\frac{K_c}{\sigma_{ys}}\right)^2 \qquad \text{(Eq. 4)}$$

Where:

A = 2π for plane stress;
    = 6π for plane strain;
K_c = critical stress intensity of a crack; and
σ_{ys} = yield stress.

As the stress intensity increases (for example, $K \rightarrow K_c$), the plastic zone size increases. One relation[32] compared the amount of acoustic emission activity on the count basis N with stress intensity K by the equation:

$$N = \frac{Da}{\frac{16\sigma_{ys}^{4}}{K^4}} \qquad \text{(Eq. 5)}$$

Where:

N = number of ringdown counts;
a = crack length for a penny-shaped crack;
D = proportionality constant between plastic volume and acoustic emission counts;
K = stress intensity of a crack; and
σ_{ys} = yield stress.

The growth of the plastic zone and subsequent acoustic emission activity of a crack, is in some ways reflective of the acoustic emission response of tensile specimens. This is because some of the same acoustic emission source events may be active during the crack growth process. However, the magnitude of the acoustic emission activity and the general stress level in the remainder of a discontinuous structure may differ markedly.

The acoustic emission wave packet released from a small, slowly growing discontinuity can be pictured as propagating through the material as an expanding sphere (Fig. 24). When the wave contacts a bounded surface, the wave mode changes from a body wave to a Raleigh wave or a plate wave.

TABLE 1. Factors that influence acoustic emission detectability

Factors Resulting in Higher Amplitude Signals	Factors Resulting in Lower Amplitude Signals
High strength	Low strength
High strain rate	Low strain rate
Anisotropy	Isotropy
Homogeneity	Nonhomogeneity
Thick section	Thin section
Twinning materials	Nontwinning materials
Cleavage fracture	Shear deformation
Low temperatures	High temperatures
Presence of discontinuities	No discontinuities
Martinsitic phase transformations	Diffusion controlled transformations
Crack propagation	Plastic deformation
Cast structure	Wrought structure
Large grain size	Small grain size

AFTER DUNEGAN AND GREEN (REFERENCE 31). REPRINTED WITH PERMISSION.

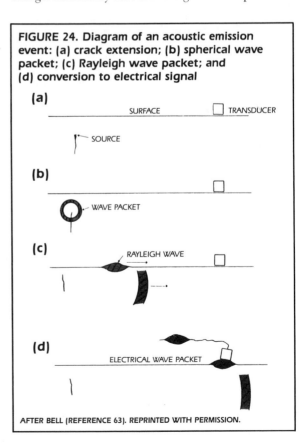

FIGURE 24. Diagram of an acoustic emission event: (a) crack extension; (b) spherical wave packet; (c) Rayleigh wave packet; and (d) conversion to electrical signal

AFTER BELL (REFERENCE 63). REPRINTED WITH PERMISSION.

Those waves expand on the bounded surface as concentric circles radiating from the epicenter of an acoustic emission source event. While body waves may propagate through the material at higher velocity than surface waves, they are more readily attenuated than surface waves. The velocity of plate waves in steel is about that of shear waves: 3.3×10^6 mm·s^{-1} (1.3×10^5 in.·s^{-1}).

The Nature of Acoustic Emission in Structures

Acoustic emission may be defined as transient elastic waves generated by the rapid release of strain energy in a material. In metals such as steel, the energy can be furnished from a number of possible sources, including dislocation motion and crack propagation. Only a small percentage of the actual energy released is available for acoustic emission.

While measurement of this acoustic emission energy might provide insights about the initial process, the acoustic emission signal is affected by (1) the rate of energy release at the source; (2) the properties of the material adjacent to the source; (3) the distance of the source from the transducer; and (4) the shape of the structural member. The resulting wave must also undergo other changes when being converted from a mechanical wave to an electrical signal at the transducer.

One approach is to consider the dynamic effect of the acoustic emission event in terms of rise time of waves detected at the surface of a body, produced by the rapid release of strain energy inside the material. However, much theoretical work needs to be done to relate rapid transient events such as force step functions to specific sources.

Considering the source of acoustic emission as a point, the newly generated waves propagate through an infinite body as a series of expanding, spherical, surface waves. There are several factors that weaken the sound pressure from an acoustic emission source as it propagates through the material to the acoustic emission sensor. Those factors include scattering; true absorption; true attenuation; and retransmission through a different material. Many typical civil engineering materials are isotropic, with the notable exception of fiber composites. The sound transmission properties for isotropic materials are not affected by material orientation.

Scattering occurs because transmission through a body is affected by inhomogeneities. In steel these include cementite, inclusions, pores and grain boundaries. True absorption is loss of sonic pressure due to the conversion of mechanical energy (wave oscillations) to heat. This process is called *damping*. True attenuation is caused by spreading of the spherical wave as it travels away from the acoustic emission source. Higher frequencies are subject to greater amplitude attenuation losses than lower frequencies.

When a sound wave hits a boundary, the wave will be reflected if the bounding surface is smooth. If the surface is rough, the wave is partially reflected and partially scattered. The transmission of acoustic emission waves through a couplant from the source material to the acoustic emission sensor also offers distortion of the acoustic emission source event. If a solid couplant such as paste or glue is employed, both shear and transverse waves can be detected. If fluid couplants are used, the transmission of shear waves will be attenuated.

Use of Transducers in Tests of Structures

The distance of the transducer from the epicenter of the event affects the type of wave measured. In an unbounded material, the waves arriving at the acoustic emission sensors usually bear good resemblance to the source event. However, in a bounded material such as structural steel plate, the acoustic emission waves reaching the sensors may have undergone multiple reflections, interferences and mode conversions. When the distance between the acoustic emission source and the transducer is large in a bounded material, the surface waveform dominates and it becomes difficult to use dynamic modeling to predict acoustic emission sources. The surface waves then might bear less resemblance to the waveform generated by the source than to the effects of specimen geometry.

There have been a number of attempts to relate acoustic emission source properties to acoustic emission waveforms in bounded specimens. However, there is some doubt about the practical application of those methods. All of the aforementioned problems reflect the difficulties in relating available information from small finite tensile type or compact fracture specimens to the nearly infinite situations encountered in the field.

Acoustic emission transducers detect surface displacement waves and convert the mechanical vibrations of the structure to electrical signals that are then processed to provide information about the acoustic emission source. The ideal transducer would measure both horizontal and vertical displacement (or velocity) and convert these linearly into electrical signals over a bandwidth up to 100 MHz. Acoustic emission signals can be expected to be generated in steel in frequencies up to and exceeding 10 MHz. Unfortunately, most of the existing wide band transducers do not have the sensitivity to measure small amplitude displacements below 10^{-10} meters. Some capacitive transducers are displacement sensitive over a frequency range from direct current to 50 MHz. However, these are less sensitive to surface displacements than the narrow band piezoelectric transducers that are widely used in acoustic emission tests.

Piezoelectric transducers are able to measure displacements down to 10^{-14} meters. However, they respond over a narrow band about their resonant frequency which usually gives a response range of from 50 to 1,000 kHz. This type of transducer cannot cover the full spectrum of monitorable acoustic emission waves, but is good at detecting and locating the position of weak emission sources. Piezoelectric transducers have been employed in amplitude and energy distribution analyses, but they can only sample emissions from a small spectrum of frequencies.

Signal Processing

The most common means of processing acoustic emission signals are: (1) ringdown counting; (2) energy analysis; (3) amplitude analysis; and (4) frequency analysis.

Ringdown Counting

Counting is a technique that records the number of times a signal amplitude (the cumulative ringdown count) or its time derivative (the ringdown count rate), exceeds a threshold. This method has been the most common means of displaying acoustic emission results. A less common type of counting is recording the number of acoustic emission events by eliminating the successive ringdown waves with a time delay before the following signals exceeding threshold can be counted. This method provides the least information about an acoustic emission event but it is the easiest method to record, especially when limited data storage capacity is available. Standard ringdown counting data are strongly influenced by test variables, including the test specimen, detection threshold and equipment variables. It is also difficult to relate these data to an acoustic emission source function, especially when measuring only total counts.

Energy Analysis

Acoustic emission energy is assumed to be proportional to the integral of the square of the transducer voltage. The commonly measured root mean square (rms) voltage is closely related to energy rate or acoustic emission power. Energy analysis is usually measured after amplification of 40 to 100 dB and a bandwidth of about 1 MHz.

For several reasons, it is difficult to relate acoustic emission event energy to detected acoustic wave energy. Two of these are (1) an uncertainty of the transducer's mode of operation and (2) the partial coverage of the source bandwidth by the detection system. The advantage of rms measurement is that it gives continuous measurement of a parameter that can be standardized and used for comparative experiments.

Amplitude Analysis

In amplitude analysis, the amplitudes of voltage signals from a piezoelectric transducer are plotted as a distribution and then compared. It has commonly been found that the signals exceeding a specified level can be expressed as a power law:[33]

$$N(a) = \left(\frac{a}{a_o}\right)^{-b} \qquad \text{(Eq. 6)}$$

Where:

a = voltage of the signal;
a_o = threshold voltage of the detection system;
N = normalized amplitude distribution function for the signal; and
b = the cumulative distribution slope.

With this representation, b is independent of system gain so that attenuation in the structure has no effect when the acoustic emission source is remotely located.

Frequency Analysis

Frequency analysis has the potential to yield information on the source rise time and fracture type. However, signal processing is extremely complex. This is usually accomplished by passing the amplified wave through a transient recorder for digitizing and processing the wave using Fourier transform routines in a small digital computer. This restricts the upper limit of frequency analysis to 50 MHz. Most experimental frequency analysis has been done with an upper limit of about 5 MHz. The characteristics of the acoustic emission signal are then analyzed in terms of power spectra, frequency bandwidth and phase data. Frequency analysis has been performed[34] on emissions produced during the deformation and fracture of pressure vessel steel.

Discontinuity Location

Not only is acoustic emission testing capable of detecting discontinuities in structures, but it is also able to locate or isolate them. When attempting to monitor acoustic emission from a known source, several techniques apply. One technique employs one channel of a multichannel system to actively monitor emissions only from the discontinuity. The active channel acoustic emission transducer is located adjacent to the discontinuity. Several other channels of the system would be used as guards. Acoustic emission transducers for these channels are placed on the structure more distant from the discontinuity.

Acoustic emission from the discontinuity is expected to strike the active transducer first. If one of the guard transducers is activated first that indicates that the acoustic emission events detected by the active transducer (in the time interval required for a sound wave to travel from the guard to the active transducer) is probably extraneous noise. This is often prevented by the acoustic emission system circuitry which may be designed to ignore any acoustic emission signals from the active transducer for that time period. A second prevention method is related to planar discontinuity location and will be discussed shortly.

Another type of acoustic emission monitoring occurs when an acoustic emission source occurs at some unknown location between two acoustic emission transducers located a fixed distance apart. This technique is called *linear discontinuity location*. When one transducer is struck by an acoustic emission burst, a clock in the acoustic emission system is started. When the second acoustic emission transducer is struck by that burst, the clock is stopped. The difference in the times of arrival of the acoustic emission wave at the two transducers can be used to locate the discontinuity between the two transducers.

If the two acoustic emission transducers lie in a plane (steel plate surface), the loci of all possible sources having the same time of arrival difference would be a pair of hyperbolas symmetrical to the bisector of the line drawn between the two transducers (Fig. 25). The correct hyperbola (the one containing the acoustic emission source) would be the one closest to the transducer that first received the acoustic emission event. To eliminate acoustic emission sources that are not on the line between the two active transducers, two or more guard transducers can be employed. If the guard is struck first by acoustic emission sources transverse to the active array, the subsequent signals from the active transducers are not processed by the acoustic emission system.

If the boundaries of the active acoustic emission region are narrow, such as in a weld or a row of fasteners and the two transducers lie close to that boundary, the source can be accurately located with only the two transducers. If the sound sources lie in an extended area, then it is necessary to have at least one additional transducer to determine a second hyperbola (Fig. 26). The line between the second array is usually normal to that of the first array. The source will then lie at the intersection of the two appropriate hyperbolas. This technique is called *planar discontinuity location*.

A second planar method called *planar source isolation* was employed in the digital acoustic emission monitor (DAEM) developed by Battelle Pacific Northwest for the Federal Highway Administration. In that system, planar source isolation was achieved by using two transducer pairs. The transducers shown in Fig. 27 were grouped into set-A (sensor-1 and sensor-2) and set-B (sensor-1 and sensor-3). Data accept windows were preset in the device, allowing only acoustic emission events with predetermined times of arrival to be accepted.

The time of arrival accept limits shown in Fig. 27 are also equivalent to the hyperbolae that are created on the planar surface and that define the data accept zone. The overlap of the two hyperbola accept intervals creates an acceptance

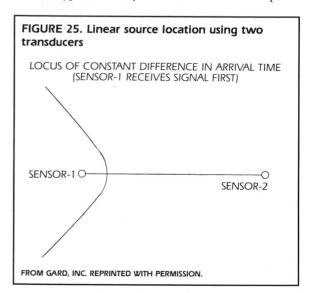

FIGURE 25. Linear source location using two transducers

LOCUS OF CONSTANT DIFFERENCE IN ARRIVAL TIME (SENSOR-1 RECEIVES SIGNAL FIRST)

SENSOR-1 SENSOR-2

FROM GARD, INC. REPRINTED WITH PERMISSION.

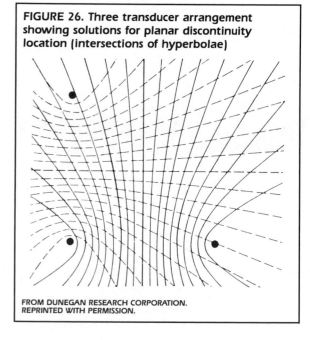

FIGURE 26. Three transducer arrangement showing solutions for planar discontinuity location (intersections of hyperbolae)

FROM DUNEGAN RESEARCH CORPORATION. REPRINTED WITH PERMISSION.

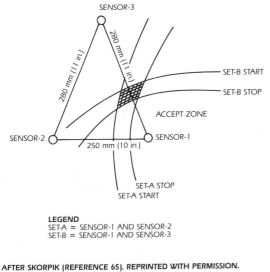

FIGURE 27. Acoustic emission source isolation array with the following physical requirements: discontinuity nearer sensor-1 than sensor-2; discontinuity nearer sensor-1 than sensor-3; and discontinuity inside sensor triangle

AFTER SKORPIK (REFERENCE 65). REPRINTED WITH PERMISSION.

zone of fixed size and location relative to the transducer arrays. Only acoustic emission events emitted from that zone are expected to be accepted by the system.

Source Activation

The performance of successful field monitoring of structures depends on being able to activate deleterious acoustic emission sources, detect and define acoustic emission events and locate these acoustic emission sources. Discontinuity location and noise rejection capabilities are basic requirements for acoustic emission tests in the field. There are two principal types of acoustic emission field tests: proof testing and service monitoring. Both methods have certain operational advantages and disadvantages.

Proof Testing

Proof testing requires that a structure or structural element be deliberately stressed to a level above the maximum anticipated service stress, but usually below the yield stress of the structure (or element). The proof test has several advantages over service monitoring. The application of high stress increases the chances of acoustic emission source activation. Weak sources of emission may be more readily detectable when proof stressed. Critical discontinuities that were not subject to subcritical crack growth can also be detected. Acoustic emission monitoring with this method can be performed in real time, eliminating some cause and effect questions. The tests can be completed in a shorter time than required for service monitoring.

The proof test also has some disadvantages, compared to service monitoring. Proof tests require multiple personnel. Special techniques and equipment must be developed to proof test large structures. To test complex structures, multichannel acoustic emission sensor arrays must be employed. Proof testing will activate more acoustic emission sources, making data analysis more difficult. Those tests must be performed at relatively high ambient temperatures to ensure maximum material toughness. Also some structures may not be adaptable to proof testing techniques because of design limitations.

Service Monitoring

Service monitoring has several corresponding advantages over proof testing. The test can be performed with relatively simple acoustic emission monitoring devices. The tests can be set up and left, requiring less field personnel. The tests can be conducted over a period of time, giving some idea of the activity of subcritical crack mechanisms. Few coincidental acoustic emission source mechanisms will be activated, simplifying data analysis. Due to their design, some structures are stressed to significant levels in service and those structures may not require proof testing. Older structures will be more prone to subcritical crack growth than to catastrophic failure by large or severe discontinuities. Service monitoring can be performed over a wide range of temperatures and can be safely performed on all members of a structure.

Service monitoring techniques do have several limitations. There may be times in the growth of a subcritical discontinuity when no acoustic emission sources are active. If they are active, the sources may be very weak. Structural loads are generally not measured in relation to acoustic emission activity and the service monitoring tests may need to be run over long periods of time. That ties up equipment and makes it difficult to relate acoustic emission activity to events on the structure. Also, due to the duration of these tests, the equipment may need to be left unattended, exposing it to vandalism and theft, with coincident replacement or insurance costs. If the equipment is to be fully portable, the data handling capabilities will be limited, especially if battery power is required.

Proof testing may be a more desirable method for testing newer structures and those of limited size. Service monitoring may be appropriate on older structures and on larger, more complex structures. In either case, the number of structural members to be monitored can be limited by several practical considerations. Members should be critical to the structural integrity and should have tensile loading components. The most critical areas of a structural member are points of geometric discontinuity or of fabrication (welding).

Types of Sources

The plausible sources of detectable acoustic activity on steel structures include: crack initiation; crack growth; crack closure; plastic deformation; elastic deformation; paint decohesion and oxide fracture; rubbing noises; hydraulic noises; and electrical noises.

It is highly unlikely that any acoustic emission monitoring would be fortunate enough to detect the initiation of fatigue cracks. That early stage of fatigue crack growth in steel involves surface shear which may be an energy process too weak to be practically detected by a commercial acoustic emission system. However, it is highly likely that normal fatigue crack growth could be detected by acoustic emission monitoring. However, most fatigue cracks of concern are those that grow from large preexistent discontinuities that arise from material, fabrication and erection problems.

Crack closure is a valid acoustic emission location mechanism. Stress reversal or complete tensile stress relief is not required. Cracks usually corrode, the corrosion product expands and fills the crack opening. Forces acting on the crack displace and impact the corroded crack faces creating detectable noise.

The plastic deformation processes detected in the field can be considered anelastic events. Unless high stresses are imposed during the acoustic emission tests, it will be difficult to explain these emissions. Acoustic emission events can also be expected from elastic strains. However, as with most of the plastic strain emissions, these may occur randomly along the stressed structural member. Crack activity may be indicated when many emissions of high energy content are addressable to one location in the absence of a detectable geometric discontinuity.

Acoustic Emission Noise

Paint decohesion and surface oxide fractures are possible sources of acoustic activity especially when high stresses are imposed on structural members. These activities are also likely when acoustic emission monitoring is conducted at temperature extremes. Most of this activity can be anticipated on older structures with built-up paint, cracked or spalled paint and general corrosion.

Rubbing or fretting noises present the greatest problems in performing acoustic emission monitoring. This is a detriment because one of the areas of highest concern is the joints of structural members. Joints are usually the noisiest areas on a member. It is extremely difficult to use wide band spectrum analysis or discontinuity location methods at joints. Low pass filtering and transducer characteristics can eliminate low frequency (audible) noise. Higher frequency noise must be eliminated by signal processing techniques.

Air flow is not a problem, because the acoustic amplitude falls off rapidly with increasing frequency. Hydraulic noise may be a problem if it cannot be isolated from the test area. Tests of offshore platforms have shown that wave noise is not a serious impediment to acoustic emission monitoring.

Electric noise problems can severely affect the performance of an acoustic emission detection system. Electric noise can be dealt with in several ways. Differential (anticoincident) transducers can attenuate some electric noise. Electric isolation of the transducer and signal cable from the structure is also necessary. It is often effective to use high pass filters to attenuate signals with frequencies greater than 1 MHz in the main acoustic emission system.

Electric noise tends to exist in the form of voltage spikes of short duration. An instrumentation acceptance window should require valid acoustic emission signals to have a predetermined duration and should eliminate the acceptance of short duration electrical transients (voltage spikes). Most acoustic emission detection systems require a line voltage power supply. In the field, a transient voltage suppressor should be placed on-line between the power cable and the acoustic emission device to reduce problems with voltage spikes.

Properly shielded connectors and signal cables will also help preclude electrical noise problems. Placement of transducers and signal cables away from vehicular traffic may ease the situation. The use of discontinuity location and noise rejection techniques will also lessen any anticipated problems from electrical noise.

Acoustic Emission Structural Instrumentation

Acoustic emission instruments used for structural monitoring are usually multichannel systems capable of discontinuity location (linear or planar), source isolation or noise rejection. Acoustic emission instrumentation ranges from simple battery powered units to complex multichannel systems capable of monitoring many locations simultaneously.

Most of the acoustic emission structural monitoring systems are capable of detecting acoustic emission activity in the 100 to 500 kHz range. They usually store analog or digitized test data for record keeping or post-test processing. Some acoustic emission systems provide real-time discontinuity detection and location.

Some of the newer acoustic emission systems are capable of pattern recognition data processing to distinguish between discontinuity related acoustic emission activity and noise. Parameters that are commonly analyzed include:

acoustic emission ringdown counts; acoustic emission amplitude; acoustic emission signal rise time; acoustic emission event duration; acoustic emission location data; acoustic emission frequency content; and external load or strain data.

Usually, the detected acoustic emission signals are front-end filtered and the temporal properties are extracted in digital form and stored on disks. This greatly reduces data storage requirements compared to storing recordings of raw acoustic emission data. Most of the new acoustic emission systems can be used to post-process the digitized data. This allows selection of acoustic emission discontinuity activity patterns and scanning of the stored digitized data to see if preselected patterns are present.

Instrumenting a Structure

Instrumenting and wiring a large structure can be a difficult task. This is especially true if long-term monitoring is performed. Often it is difficult to achieve the desired acoustic emission sensor locations because of the design of the structure being monitored. Routing signal wires from remote sensor locations is often troublesome and most structures require considerations not encountered in the laboratory.

Wires must be protected from wind vibration, abrasion, snagging and excessive heat. Wires used to connect the transducer (and preamplifiers) to the acoustic emission monitoring systems should be shielded. For wire runs exceeding 30 m (100 ft), coaxial cable such as RG58 has proven satisfactory. This is especially true if power transmission towers are in the vicinity. For shorter signal wire runs, cable such as RG174 is acceptable; it is smaller and can be obtained in multiple coaxial cable assemblies. These are convenient when using a multiple transducer array to monitor a specific location. For electrical field connections, BNC or TNC connectors are used. Lighter duty electrical connections are usually too frail for field applications.

In most field tests, preamplifiers are required to transmit a sufficiently strong signal from the transducers to the acoustic emission system. Care should be taken to electrically isolate the preamplifiers and transducers from the structure. Usually magnetss or tape are used to mount transducers on the structure. Lead wires from the transducers to the preamplifiers are usually less than 3 m (10 ft) long so that preamplifiers must also be mounted to the structure. Integral preamplifier and transducers eliminate the need for preamplifier mounting.

The transducers should be firmly coupled to the structure. Most standard transducers have flat wear plate faces. Care should be taken to ensure that the face is mounted to a relatively flat, clean surface on the structure. If the structure's surface is not flat, the transducer wear plate can be machined or sanded to conform to the structure contour. Sensors can tilt and loose contact with the structure but clamping devices can be used to eliminate such problems.

Long-term coupling can be achieved by gluing the transducer to the structure. If the transducer mounting is exposed to the elements, the bonded area should be protected with a caulk. Short-term coupling can be achieved with a silicon grease or thick resin. Care should be taken to prevent grit from contaminating the coupled area as it may interfere with sonic coupling. Poor coupling can also be caused by loose paint or rust on the structure.

Once the transducers are mounted, it is desirable to check for sonic coupling. This can be achieved using an ultrasonic or spark pulser or by breaking pencil leads or glass capillaries (Fig. 28). This will serve to: (1) determine the acoustic emission system function; (2) check coupling between the transducer and the structure; and (3) calibrate or check the discontinuity location (or noise rejection).

Acoustic Emission Monitoring of Structures

One of the first documented uses of acoustic emission testing for civil engineering structures was in 1971.[35] Acoustic emission testing was conducted on buildings that were water-loaded.

Also in 1971, acoustic emission tests were conducted on a portable military bridge being proof tested by the British army.[36] During the test, one bridge girder was instrumented with seven transducers including several pairs used for linear discontinuity location. Analysis of acoustic emission ringdown counts was conducted on-line during load periods,

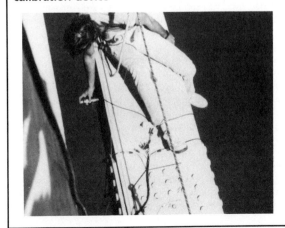

FIGURE 28. Calibrating a transducer array on a bridge member using a hand held spark bar calibration device

hold periods and repeat tests. Post-test analysis yielded further information on acoustic emission amplitude distributions and source locations. The acoustic emission sources were attributed to locations where plastic deformation had occurred.

In 1972, acoustic emission tests were performed on eight cables of the Dumbarton lift bridge near San Francisco, California. That work is discussed later. Also in 1972, acoustic emission sound measurements were made by the Argonne National Laboratory on a steel bridge located on Interstate 80 in Illinois.[37]

The Kentucky Transportation Research Program performed an acoustic emission monitoring test on a continuous eyebar truss bridge in 1973.[38] A single-channel acoustic emission device was used for this study. The test was run using a 140 kHz resonant transducer with a system gain of 80 dB and a threshold of 1 V. The test revealed that mechanical noise was a serious problem for acoustic emission testing of bridges. It was also found that good sound transmission existed between the pinned eyebars.

The next notable acoustic emission testing on steel bridges was performed by Battelle Pacific Northwest for the Federal Highway Administration. This project consisted of developing and demonstrating an acoustic emission system for inspecting in-service bridges.[39,40] Initially, work was aimed at determining the acoustic spectrum of bridges and developing an acoustic emission system for centralized acoustic emission signature analysis.

The first field tests were conducted on the Interstate 5 bridge, a tied arch structure over the Toutle River in Washington State. Cracks had been detected on the floor beams of that bridge and single transducers were attached to floor beams exhibiting varying degrees of cracking. Transducers with different resonant frequencies were used in the range from 5 kHz to 1 MHz. The bridge was subject to normal traffic during the monitoring period. Amplified acoustic emission activity was stored on a video recorder. Later the recorded tape was replayed through a spectrum analyzer.

The tests revealed that traffic responses did not seem to be present above 250 kHz. Attempts to detect a crack signature based on frequency content, as determined from a single transducer, proved futile.

A second test was performed on girders of a prestressed concrete bridge, the Colorado Street overpass on Route 12 near Richland, Washington. The bridge possessed some broken strands in a tendon where the bridge was damaged by impact with an oversized vehicle. A single transducer was attached to both good girders and the damaged girder. Spectral analysis was used on the recorded acoustic emission data. Both piezoelectric transducers and accelerometers were used as sensors. The traffic excitation was centered about frequencies in the 15 to 40 kHz range. Test personnel believed that the concrete attenuated the higher frequencies. The spectral analyses did not distinguish the damaged girder from those that were in good condition. On this structure, the traffic noise disappeared below 100 kHz.

Battelle personnel then monitored the Route 224 bridge over the Yakima River near West Richland, Washington. That work was limited to attenuation measurements. Those tests indicated that frequencies above about 500 kHz were almost completely attenuated by the joints in the structure.

Initial work indicated the feasibility of developing a central monitoring system to perform signature analysis on acoustic emission data from transducers located about the bridge. It was believed that such work would need to be conducted with monitoring frequencies below 50 kHz but the program was not pursued. Instead, Battelle concentrated on developing a small self-contained acoustic emission discontinuity monitor.

The original instrumentation used a 1 MHz high pass filter for noise rejection. The system was battery operated and capable of unattended acoustic emission monitoring for limited periods. Thereafter, the device was taken to a laboratory. It contained an erasable programmable read only memory (EPROM) chip on which the acoustic emission field data were recorded. Such chips store data in a formatted, readable form and, in this application, the data were acoustic emission event counts taken over fixed time intervals. The chip was read and subsequently erased by exposure to ultraviolet light for future reuse.

The instrumentation was tested in conjunction with an acoustic emission system that recorded three channels of concurrent acoustic emission data on a magnetic tape recorder. The data for that system came from a triangular three-transducer array enclosing the single transducer of the self-contained acoustic emission unit.

The self-contained system was used on the Interstate 80 Carquinez bridge near Vallejo, California. Eyebars on the upper chord of the truss bridge were to be removed. Several parallel eyebars were monitored prior to removal. The eyebars monitored by the acoustic emission system during this operation assumed the load of the eyebar that was removed. The acoustic emission system monitoring frequency was restricted to 400 kHz and above. Single transducers were epoxy bonded to two eyebars adjacent to the eyebar pin.

The acoustic emission data were very low, indicating the absence of discontinuities in the eyebars. The low acoustic emission data peaks that were present generally corresponded to load transfer as the eyebars were removed.

The acoustic emission system was subsequently placed on a cracked floor beam of the Toutle River bridge. The three-transducer system was placed near an extending crack and the single transducer of the self-contained system was located inside that array some 75 mm (3 in.) from the crack tip. The system monitored the bridge for several periods of about fifteen days each.

Taped data from the three-transducer system were replayed through a commercial acoustic emission discontinuity location system. The results indicated that possible acoustic emission sources could be (1) along the bridge deck floor beam interface; (2) the riveted connection between the floor beam and the tie girder; and (3) the crack. The majority of the acoustic emission sources were from locations other than the crack.

The results indicated that the self-contained acoustic emission system did not have sufficient discontinuity discrimination capabilities. Battelle personnel felt that discontinuity location and source isolation capabilities would be required. The portable, self-contained acoustic emission system was upgraded with a three transducer, two-linear array system. That system used adjustable time accept limits for each linear array to define a set of hyperbolas. The acoustic emission data falling outside the set of hyperbolas were rejected. Only data that met the time of arrival of the two hyperbola sets were accepted. That created an accept zone defined as the area bounded by the overlapping hyperbola boundaries.

The modified digital memory acoustic emission monitor (DAEM) was tested on the floor beams of the Toutle River bridge and suspect areas of a girder weld on the airport overpass bridge at Walla Walla, Washington. The system was also used to monitor edge fillet welds on box girders at a structural steel plant in Clinton, Tennessee.

The system was used to monitor a floor beam crack on the Toutle River bridge for a period of 10.5 days in 1976. The three-transducer array was placed and the time of arrival accept limits were set to provide a valid zone that was located at the crack tip. The zone was defined by breaking pencil leads and defining the valid zone on the surface of the floor beam web about the crack. The unit employed two EPROM memories. One recorded all acoustic emission event count data from sensor-3. The other EPROM was used to record acoustic emission events from the valid zone.

The EPROM data showed that the total acoustic emission counts varied between 300 and 3,000 counts (the data from the sensor-3). The valid acoustic emission data ranged from 0.5 to 3 percent of the total acoustic emission. As part of a second one-day test, a crack-free zone in the web was monitored to determine the effectiveness of the source isolation system. The acoustic emission level at that location was found to be less than 0.5 percent of the level at the first test site.. Some crack growth was believed to have occurred during the 10.5 day monitoring period.

A third test for twenty-two unattended days was performed on the Toutle River bridge floor beams in 1977. The 1976 test was conducted using a detection frequency of 600 kHz. The 1977 test used a detection frequency of 400 kHz.

The resulting data indicated that the frequency change increased the amounts of both valid and total acoustic emission data. It was difficult to correlate the valid acoustic emission data to crack growth during this monitoring period. The data showed that the bridge loading was irregular during the test, weekdays being more active than weekends.

During the testing period, the valid acoustic emission data were much less than the total acoustic emission data, less than ten percent for most of the twenty-two day monitoring period. In several instances, no data were recorded on the valid acoustic emission EPROM, while large numbers of events were recorded on the total acoustic emission EPROM. This indicated that the source isolation technique was effectively rejecting noise.

A weld on the airport overpass bridge revealed high acoustic emission activity when tested earlier by another acoustic emission monitoring system. The DAEM was placed on the bridge with the transducer array and accept limit set to form an acoustic emission accept or valid zone. The weld was monitored for a sixteen-day period using one-hour updates.

Subsequent data analysis indicated a period of high total acoustic emission activity. During several periods, the valid acoustic emission data equaled the total acoustic emission data from the single transducer. However, the event counts during those periods were less than fifty counts per hour.

From August 1980 to July 1982, the Kentucky Transportation Research Program conducted tests on the Interstate 471 bridge over the Ohio River at Newport, Kentucky. The DAEM system was used to monitor a tie chord butt weld that contained discontinuities detected with ultrasonics (Fig. 29). The DAEM was stored in the tie chord box and the EPROMs were replaced on regular intervals over a 1.5 year period.

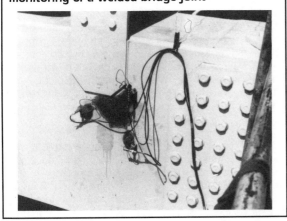

FIGURE 29. Three transducers in an acoustic emission source isolation array for DAEM monitoring of a welded bridge joint

Those tests showed that high acoustic emission event activity could be detected during peak traffic hours over the bridge. Rainfall also produced high acoustic emission rates. However, comparative acoustic emission tests between the weld area and a similar weld found to contain no ultrasonic discontinuity indications proved inconclusive.

From March 1982 to January 1983, the West Virginia Department of Highways monitored acoustic emission activity on the Interstate 64 Dunbar bridge over the Kanahwa River near Charleston, West Virginia. A commercial acoustic emission monitor was mounted on a pier of the bridge.[41] Eight weld locations that contained subsurface acoustic emission indications were instrumented. The resulting acoustic emission data were transmitted over telephone lines and placed on a digital tape. Planar source location was subsequently performed using a copy of the data tape.

The transducer arrays were of particular interest. Special angle beam 500 kHz transducers were placed along the weld lines. These were found to be 20 dB more sensitive to signals from acoustic emission activity originating in the weld line than to sources approaching from the sides. Additionally, conventional guard transducers were placed offset of the midpoints of the weld lines. The guards acted to lockout the angle beam transducers when first struck by acoustic emission activity. The transducers were cemented to the steel girders.

The planar discontinuity location system required at least three of the transducers to be struck for an acoustic emission event to be considered valid. One array location produced 12,560 such events. This was almost 1,000 times greater than the least active array (fourteen planar valid events) and about ten times more active than the second most active location (1,461 planar valid events).

In the period of 1982 to 1984, a broad banded piezoelectric transducer was developed for the Federal Highway Administration.[42] The transducer was a point contact type with a conical piezoelectric element. Laboratory tests showed the transducer had flat, continuous wave response between 100 kHz and 1 MHz. The transducer was intended for use with broad band instrumentation and signal processing to provide signal characterization as a means of differentiating between noise and acoustic emission activity from cracks in steel bridges. The transducer offered the possibility of an advanced approach to discontinuity evaluation.

Acoustic Emission Weld Monitor

Since 1982, the Kentucky Transportation Research Program has used an acoustic emission weld monitor (AEWM) system to detect crack activity on steel bridges.[43,44] The system was evaluated under a research program sponsored by the Federal Highway Administration and was originally developed to monitor in-process welding operations.

Analog electronics are used to acquire and preprocess acoustic emission activity. This includes the use of analog signal amplification and band-pass filtering from signals produced by standard resonant transducers. The conventional time of arrival (Δt) technique is employed for linear discontinuity location using two active transducers. The AEWM possesses a microprocessor based multiparametric filtering program that analyzes the acoustic emission data, rejects noise and locates and characterizes discontinuities in real-time.

Consecutive acoustic emission events are subjected to a three-step sequential test or acoustic emission pattern recognition filtering program (Fig. 30). First, the analog preprocessing circuitry computes the ringdown count and the time of arrival. Then the microprocessor portion of the system tests the collected analog information for each event.

As the first step in the filtering program, the ringdown count must lie within fixed limits. If this is satisfied, the second filtering step is imposed wherein the acoustic emission event must occur within a predetermined minimum event rate with other acoustic emission events preceding or following it (which have also passed the ringdown test). The third step determines whether all the events passing the first two filtering tests were located by time of arrival from within a right locational tolerance. All acoustic emission event data that fail to pass any one of the tests are discarded. Additionally, the frequency content of each acoustic emission event is analyzed using a comb filter. Valid acoustic emission events having high frequency biases are classified as cracks. Other data that satisfy the model are characterized as unclassified discontinuities.

The AEWM can continuously process large numbers of acoustic emission events occurring at rates too fast for an operator to analyze. The microprocessor circuitry also determines when valid discontinuity activity occurs. The operator is informed of discontinuity related events by an indicating lamp and by a light emitting diode (LED) panel that displays the relative location of the discontinuity between the two active transducers. The unit is also capable of data storage by disk and direct hard copy output subsequent to a test.

Over four years, thirteen field tests were conducted on nine different bridges in four states. The bridges inspected include: the Interstate 24 bridge over the Ohio River at Metropolis, Illinois (Illinois Department of Transportation); the Interstate 24 bridge over the Tennessee River; the Interstate 471 bridge over the Ohio River at Newport, Kentucky; the Interstate 75 bridge over the Ohio River at Cincinnati, Ohio; the Interstate 64 bridge over the Ohio River at Louisville, Kentucky; the US 25 bridge over the Rockcastle River (Kentucky Department of Highways); the US 18 bridge over the Mississippi River at Prairie du Chien, Wisconsin; the Interstate 94 Oklahoma Avenue overpass in Milwaukee, Wisconsin (Wisconsin Department of Transportation); and the Interstate 310 bridge over the Mississippi River at Luling, Louisiana (Louisiana Department of Transportation and Development).

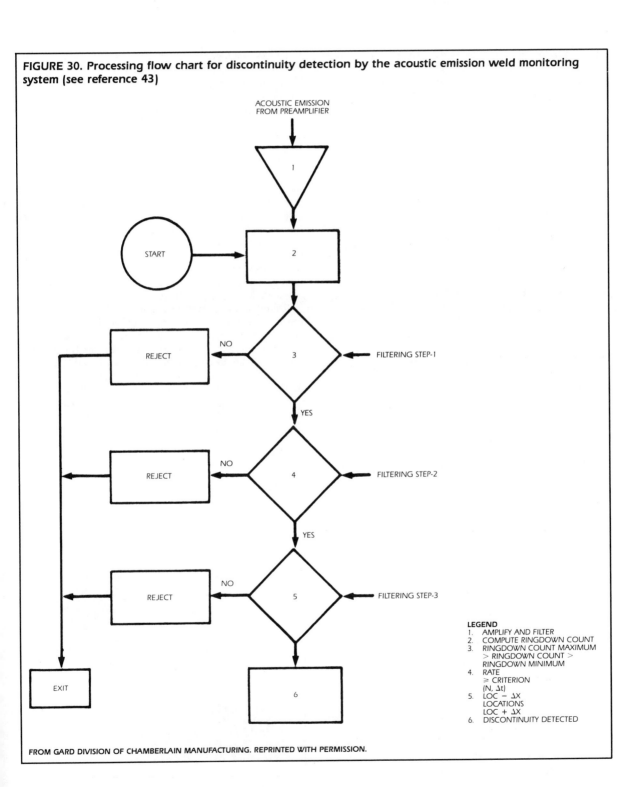

FIGURE 30. Processing flow chart for discontinuity detection by the acoustic emission weld monitoring system (see reference 43)

During these tests, the AEWM was used to monitor visible cracks, ultrasonic subsurface discontinuity indications, stress intensifying weld details and elements containing riveted or bolted connections. The system detected acoustic emission discontinuity activity emitted from visible cracks on three occasions. Once, it detected acoustic emission discontinuity activity from a subsurface ultrasonic indication. On those occasions, the AEWM interpreted discontinuity activity that was located at the crack, in relation to placement of the two transducers that comprised the linear array.

In every instance except for the Interstate 310 bridge at Luling, Louisiana, the bridges were subject to normal traffic. Heavy proof loads were employed on the former structure. Typically, the tests have used system gains from 70 to 80 dB and linear array spacings from 0.5 to 2 m (20 to 80 in.). Standard 175 kHz resonant transducers were used.

Preliminary data indicated a correlation between fatigue crack activity, structural vehicular loading and acoustic emission discontinuity activity. The most active discontinuity monitored was a 150 mm (6 in.) crack in the floor beam of an interstate bridge. The bridge was subjected to frequent truck loads. Monitoring that crack produced acoustic emission discontinuity activity about every fifteen minutes. The acoustic emission discontinuity activity corresponded to the passage of one or more heavy vehicles over the bridge. Some bridge cracks have been monitored for extended periods (up to 128 hours) without detecting acoustic emission activity. In those cases, even with large surface cracks, long-term visual inspection has revealed no fatigue growth (indicating that those cracks are benign).

Strain gage work was done in conjunction with the AEWM on a bridge with visible large cracks. On that bridge, even heavy truck loading did not produce high stress fields about the cracks. Therefore, even though the cracks were large, the vehicular loads were insufficient to cause fatigue crack growth. The cracks were created prior to erection and at the time of the acoustic emission tests they were inactive. In some cases, fatigue cracks are created and propagate to low stress areas and become benign. At least two such instances were monitored using the AEWM. The ability to determine if a discontinuity is harmful is one of the strong abilities of the acoustic emission test method. It is now believed that acoustic emission monitoring of bridges heavily traveled by trucks does *not* have to be performed for extended periods to detect fatigue cracks.

Tests with the AEWM have been successful in rejecting the large amounts of mechanical noise present on bridges. The noise rejection model will possibly exclude some valid acoustic emission discontinuity data. However, past testing indicates that amount is not significant. Typically, on a bridge member with an active fatigue crack, the AEWM will detect 1,000 to 3,000 discrete acoustic emission events per hour. The device may consider only three or four of those to be discontinuity related. This indicates the difficulty of attempts to use long-term acoustic emission monitoring of structures to integrate acoustic emission discontinuity activity from background noise. Similar locations on bridges have been found to produce widely varying amounts of background noise.

The amount of acoustic emission noise is dependent on the type of structural detail near the acoustic emission transducer array and the amount and characteristics of vehicular loading on the bridge. Members that are primarily in tension have proved to be very quiet. A good example would be the tie chords of a tied arch bridge (see Fig. 31). Bridge members directly abutting a concrete deck or near some types of bolted connections tend to be very noisy. Typical examples of those are floor beams, stringers and floor beam-to-girder bolted connections (see Fig. 32). Acoustic emission noise not only depends on the volume of traffic on the bridge, but also on the vehicle weight and speed. Heavy vehicles traveling at high speeds created large noise backgrounds on bridges, but also are best at triggering acoustic emission discontinuity activity.

Testing of the AEWM has indicated the occasional need to use guard transducers to prevent false hits on the linear array. A typical example of such situations is encountered in monitoring welded cover plates on steel beam bridges. The cover plates are located on the lower flange of the beam. To monitor such areas, it is best to mount the transducer array on the lower flange. However, fretting noise from the concrete deck-to-upper flange interface can create false signals that can trigger false indications which appear to come from the center of the array. One solution would be to shift the array along the lower flange in relation to the cover plate and ignore any discontinuity indications from the center of the array.

The second and better choice is to employ a guard system that will preclude the analysis of acoustic emission events

FIGURE 31. Transducer array on the tie chord of an arch bridge

FIGURE 32. Linear transducer array adjacent to a bolted angle splice on a bridge

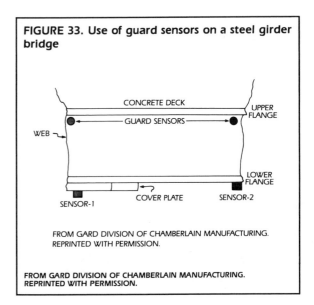

FIGURE 33. Use of guard sensors on a steel girder bridge

FROM GARD DIVISION OF CHAMBERLAIN MANUFACTURING. REPRINTED WITH PERMISSION.

that first strike one or more guard transducers (Fig. 33). In this case, two guard transducers are placed on the upper flange and are located about halfway between the two active transducers when viewed from the ground. The guards must be located so that they are at least twice as far from the array as the array spacing itself. Otherwise, they will be struck first (even by some sources within the active transducer array) and this prevents reception of any acoustic emission signals from sources along the line between the two active transducers.

Acoustic emission testing of large structures is not restricted to bridges. Recent acoustic emission tests were performed on a semisubmersible offshore platform.[45] The rig contained a 115 mm (4.5 in.) through-wall crack in a column leg. The crack was surrounded by four resonant transducers in a rectangular array. The array spacing was varied close to the crack to a maximum transducer-to-crack distance of 1.5 m (5 ft).

The crack was loaded by sea waves. When the closed array was used, the wave loading was about one-half of that at the time of maximum transducer spacing. At the closest transducer spacing, wave stressing produced 5 to 10 acoustic emission events per minute that were addressable to the crack tips. At an intermediate transducer array spacing, the wave loading increased with a resulting increase of crack related acoustic emission events to 99 per minute. At the greatest transducer array spacing and highest wave loading, the acoustic emission rate was still high (about 95 acoustic emission events per minute). Most of the events occurred during the rising load. However, some occurred at the peak load.

Those results indicate that offshore platforms may be good structures to consider for acoustic emission inspection while subject to in-service wave loading.

PART 3
COMPONENT TESTING

Much civil engineering effort is directed toward the fabrication of structural components especially by welding. Newer structures take advantage of the high tensile strength of steel wires. Unfortunately, both welded and steel wire structural components pose unique problems.

Welding can induce discontinuities in a structural component and these can lead to eventual failure of the completed structure or, at best, an expensive repair of a completed weld. Structural components containing steel wire are more susceptible to corrosion damage than normal steel beams or girders. The increased use of prestressed concrete beams containing wire reinforcement presents many future inspection problems.

Testing of Welded Structures

The earliest use of acoustic emission monitoring of welds was for post-weld monitoring of high strength steel.[46] Later acoustic emission tests of in-process slagless welding also showed promise.[47,48] However, most welding of steel structures required the use of slag welding methods. That limited early acoustic emission efforts to post-weld monitoring.[49,50]

One slag method, electroslag welding is relatively quiet and several successful acoustic emission tests have been employed on in-process welding operations and in post-weld monitoring.[51,52] In both cases, researchers were able to determine if the flux pool was satisfactory. During post-weld monitoring using linear discontinuity location, researchers were also able to detect typical electroslag discontinuities, including cracking.

Early tests of acoustic emission monitoring of slag welding operations[53] led to the evaluation of the acoustic emission weld monitoring system described earlier. The Federal Highway Administration then contracted to test the typical submerged arc butt welding operations.[54] A bulk of the laboratory tests were performed on grooved plates of typical highway bridge steels. Discontinuities (including cracks, porosity, slag inclusions and lack of fusion) were deliberately embedded in the welds as they were deposited. The system successfully detected all twenty-four of the planar discontinuities (cracks and lack of fusion), all six of the slag inclusions and seven out of eight of the porosities. Following those tests, field tests were conducted at three fabrication shops (Figs. 34, 35 and 36).

During those tests, the system monitored 210 m (690 ft) of in-process welds. Six minor discontinuities were detected and all of those were visually confirmed. No other discontinuities were detected by the AEWM or by inspection with radiography. In 1985, the system was demonstrated to highway personnel from twenty-one states. The Federal Highway Administration initiated extensive tests of the device by the Kentucky Transportation Research Program.

Testing of Steel Wire Rope

Early laboratory tests of wire rope showed that wire fracture was readily detectable.[55,56] Continuous monitoring of cyclic acoustic emission tests provided warning of fatigue failure as early as half the lifetime. Proof testing and acoustic

FIGURE 34. Acoustic emission monitoring of semiautomatic submerged arc butt welds (note magnetic transducer hold-down under the welding operator's elbow)

FIGURE 35. Acoustic emission monitoring of a large complex weldment using automatic submerged arc welding; the transducers are attached to the backside of the flange; the acoustic emission system is to the left of the weldment

FIGURE 36. Acoustic emission monitoring of concurrent welding operations; note transducers attached to the plate on the backside of the welding operation

emission monitoring of cyclically loaded wire rope also provided warning of impending fatigue failure.

In 1972, acoustic emission tests were performed on the cables of Dumbarton lift bridge near San Francisco, California.[57] The old bridge showed wear on the cables and connectors. To prevent high sound attenuation, radiator hose clamps were placed around the wire ropes and tightened to consolidate the strands. The 150 kHz transducers used in the tests were attached to the strands by radiator clamps. The cables were proof loaded by providing a transverse load with a hand winch. The load was applied and held for ten minutes. Transducers were also placed on the wire rope connectors for continuous acoustic emission monitoring over a twenty-four hour period.

Several of the cables showed more continuous acoustic emission activity than others. However, the acoustic emission proof load tests showed no sign of serious deterioration.

Testing of Other Components

Early acoustic emission tests on a model prestressed concrete reactor vessel[58] showed that acoustic emission testing could detect: (1) the onset of failure; and (2) the progression of failure processes; (3) previous loading levels; and (4) locations of failure.

Acoustic emission techniques have also been used for in-service monitoring of concrete bridges, precast reinforced roofing panels, for mixing concrete and for grading aggregate. Acoustic emission testing has often been used as an indicator or warning preceding the use of more comprehensive inspection techniques.[59]

Researchers at Florida Atlantic University conducted a series of laboratory experiments on the corroding of steel reinforcement embedded in concrete.[60-62] Some of those tests were conducted at a natural corrosion rate and others were accelerated by impressing an anodic direct current on the reinforcing steel (usually in the presence of a saline solution or sea water). The tests revealed that acoustic emission activity from the corroding specimens was much greater than from specimens in a noncorrosive environment.

Acoustic emission activity was associated with the creation of a corrosion product, an expansive ferrous oxide on the reinforcing steel surface. The corrosion induces internal hoop stresses in the concrete which leads progressively to concrete fracture.

Corrosion activity in reinforced concrete was studied using amplitude distribution analysis and the simple power law (see Eq. 6). Amplitudes were determined by counting the number of acoustic emission events detected at discrete thresholds. A commercially available pulse height analyzer was used for that task. The b value was determined by dividing by 20 the number of decibels associated with a one-decade drop in the number of events. Changes in that value within the linear portions of the power law (or cumulative distribution) are indicative of changes in the relative contributions of any of several mechanisms related to acoustic emission.

Figure 37 shows the acoustic emission and b slope behaviors for a reinforced concrete specimen corroded to surface

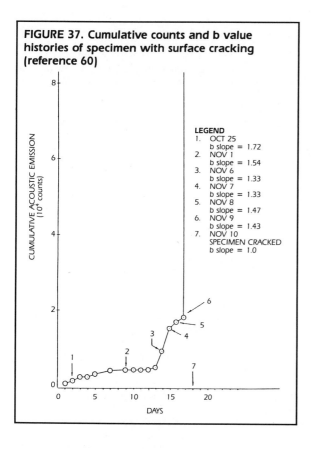

FIGURE 37. Cumulative counts and b value histories of specimen with surface cracking (reference 60)

cracking of the concrete. During days when the number of acoustic emission counts increased, the b values were closer to 1.0. Those values indicated that (1) more cracks which released higher amplitude signals were activated; and (2) since the value of b was 1.0 when surface cracking occurred, it was determined that lower b values corresponded to the growth of large cracks.

The concrete was found to be highly attenuative of acoustic emission signals in 100 to 300 kHz range with 500 mm (20 in.) considered the maximum useful detection distance.

Field tests were conducted on a concrete cap and walk at Delray Beach, Florida. The site was near the sea and several locations along the walkway exhibited surface cracking of the concrete with rust stains seeping from the steel reinforcement. A commercial acoustic emission system with source isolation was used to monitor three locations. Two of those sites possessed cracks and the third was crack-free. The sites were monitored for several months.

The reference area with no visible distress emitted only low level emissions. One of the active sites with a surface crack produced numerous acoustic emission events of moderate amplitude. The second active site which exhibited only a small amount of surface cracking produced many high amplitude acoustic emission events. The tests demonstrated that the acoustic emission pattern varied with the corrosion process. The area with the larger crack probably produced lower acoustic emission activity as the crack acted as a vent for the corrosion product formed on the reinforcing steel. At the second active site, the expanding corrosion product was probably developing its cracking process.

REFERENCES

1. "Bridge Recabled after Twelve Years Service." *Engineering News Record* (April 1984): pp 63-67.
2. Obert, L. *Measurement of Pressures on Rock Pillars in Underground Mines — Part I*. RI 3444. US Bureau of Mines (1939).
3. Drouillard, T.F. *Acoustic Emissions: A Bibliography with Abstracts*. New York, NY: Plenum Press (1979).
4. Hardy, H.R. "Applications of Acoustic Emission Techniques to Rock and Rock Structures: a State of the Art Review." *Acoustic Emissions in Geotechnical Engineering Practice*. ASTM STP 750. Philadelphia, PA: American Society for Testing and Materials (1981): pp 4-92.
5. Koerner, R.M. and A.E. Lord, Jr. *Use of Acoustic Emissions to Predict Ground Stability*. Report No. FHWA/RD-82/052. Washington, DC: Federal Highway Administration (December 1982): pp 16.
6. Koerner, R.M., W.M. McCabe and A.E. Lord, Jr. "Acoustic Emission Behavior and Monitoring of Soils." *Acoustic Emissions in Geotechnical Engineering Practice*. ASTM STP 750. Philadelphia, PA: American Society for Testing and Materials (1981): pp 101.
7. Hardy, H.R. "Applications of Acoustic Emission Techniques to Rock Mechanics Research." *Acoustic Emission*. STP 505. Philadephia, PA: American Society for Testing Materials: pp 41-83.
8. Op. cit. 4: pp 57.
9. Op. cit. 7: pp. 48.
10. Mogi, K. "Study of Elastic Shocks Caused by Fractures of Heterogeneous Materials and Its Relation to Earthquake Phenomena." *Bulletin of the Earthquake Research Institute*. TPJKA. Vol. 74 (1963): pp 487-490.
11. Op. cit. 5: pp 69-75.
12. Op. cit. 6: pp 120.
13. Koerner, R.M. and A.E. Lord, Jr. *Earth Dam Warning System to Prevent Hazardous Material Spills*. Philadelphia, PA: Drexel University: pp 6
14. Koerner, R.M. and A.E. Lord, Jr. "Acoustic Emission Testing of Soils." *Proceedings of QualTest-3*. Paper IQ 84-675. Columbus, OH: The American Society for Nondestructive Testing (1984): pp 7.
15. Op. cit. 6: pp 119.
16. Op. cit. 4: pp 25.
17. Op. cit. 4: pp 26.
18. Op. cit. 15: pp 118.
19. Op. cit. 5: pp 98-128.
20. Hardy, H.R., Jr. "Evaluating the Stability of Geologic Structures Using Acoustic Emissions." *Monitoring Structural Integrity by Acoustic Emission*. ASTM 571. Philadelphia, PA: American Society for Testing and Materials (1975): pp 86.
21. Op. cit. 4: pp 50-64.
22. Op. cit. 20: pp 86.
23. Hardy, H.R., Jr. "Acoustic Emission During Dissolution of Salt." *Journal of Acoustic Emission*. Vol. 4, No. 213 (April-September 1985): pp 19-20.
24. Koerner, R.M. and A.E. Lord. *AE Detection of Prestress in Soil and Rock*. Philadelphia, PA: Drexel University.
25. Dunn, S.E. *Discussion of the Application of Acoustic Emission Methods for Field Monitoring of Corrosion Induced Structural Deterioration of Reinforced and Prestressed Concrete Structures*. Boca Raton, FL: Florida Atlantic University (January 1985).
26. Wadley, H.N.G., C.B. Scruby and J.H. Speake. "Acoustic Emission for the Physical Examination of Metals." *International Metals Review* (February 1980): pp 41-64.
27. Dunegan, H.L., D.O. Harris and A.S. Tetelman. "Detection of Fatigue Crack Growth by Acoustic Emission Techniques." *Materials Evaluation*. Vol. 28, No. 10. Columbus, OH: The American Society for Nondestructive Testing (October 1970): pp 221-227.
28. Ritchie, R.O. and J.F. Knott. "Mechanisms of Fatigue Crack Growth in Low Alloy Steel." *Acta Metallurgica* (May 1973): pp 639-648.
29. Frederick, J.R. "Acoustic Emission as a Technique for Nondestructive Testing." *Materials Evaluation*. Vol. 28, No. 2. Columbus, OH: The American Society for Nondestructive Testing (February 1970): pp 43-47.
30. Harris, D.O. and H.L. Dunegan. *Continuous Monitoring of Fatigue Crack Growth by Acoustic Emission Techniques*. Livermore, CA: Dunegan Research Corporation (February 1973).
31. Dunegan, H.L. and A.T. Green. "Factors Affecting Acoustic Emission Response from Materials." *Materials Research and Standards* (March 1971): pp 1-4.
32. Harris, D.O., H.L. Dunegan and A.S. Tetelman. *Prediction of Fatigue Lifetime by Combined Fracture Mechanics and Acoustic Emission Techniques*. Report DRC-105. Livermore, CA: Dunegan Research Corporation (1969): pp 11.
33. Pollock, A.A. "Acoustic Emission Amplitudes — Acoustic Emission 2." *Nondestructive Testing* (October 1973): pp 264-269.
34. Graham, L.J. and G.A. Ahlers. "Spectrum Analysis of Acoustic Emission in A533-B Steel." *Materials Evaluation*. Vol. 32, No. 2. Columbus, OH: The American Society for Nondestructive Testing (1974): pp 34-37.
35. Liptai, R.G., P.O. Harris, R.B. Engle and C.A. Tatro. "Acoustic Emission Techniques in Materials Research." *Journal of Nondestructive Testing* (May 1971): pp 263-264.

36. Pollock, A.A. and B. Smith. "Stress-Wave Emission Monitoring of a Military Bridge." *Nondestructive Testing* (December 1972): pp 348-353.
37. *Formal Proposal for Acoustic Emission of Steel Highway Bridges*. Argonne, IL: Argonne National Laboratory. Submitted to the National Science Foundation (April 1973).
38. Hopwood, T. *Acoustic Emission, Fatigue, and Crack Propagation*. Report No. 457. Kentucky Bureau of Highways (October 1976).
39. Hutton, P.H. and J.R. Skorpik. *Acoustic Emission Methods for Flaw Detection in Steel in Highway Bridges, Phase I*. Federal Highway Administration Report No. FHWA-RD-78-97. Richland, WA: Battelle Pacific Northwest (March 1975).
40. Hutton, P.H. and J.R. Skorpik. *Acoustic Emission Methods for Flaw Detection in Steel In Highway Bridges, Phase II*. Federal Highway Administration Report No. FHWA-RD-78-98. Richland, WA: Battelle Pacific Northwest (March 1978).
41. "Acoustic Emission Monitoring of Electroslag and Butt Welds on Dunbar Bridge, West Virginia (June 1, 1980 to April 30, 1985)." *Final Report: West Virginia Department of Highways Project 64*. San Juan Capistrano, CA: Dunegan Corporation (1985).
42. Miller, R.K., H.I. Ringermacher, R.S. Williams and P.E. Zwicke. *Characterization of Acoustic Emission Signals*. Report No. R83-996043-2. East Hartford, CT: United Technologies Research Center (1983).
43. Prine, D.W. and T. Hopwood. "Improved Structural Monitoring with Acoustic Emission Pattern Recognition." *Proceedings of the Fourteenth Symposium on Nondestructive Evaluation*. San Antonio, TX (April 1985).
44. Hopwood, T. and D.W. Prine. "Acoustic Emission Structural Monitoring in Noisy Environments Using Event Based Processing." *Proceedings of the International Conference on Fatigue, Corrosion Cracking, Fracture Mechanics and Failure Analysis* (Salt Lake City, UT). Metals Park, OH: American Society of Metals (December 1985): pp 277-282.
45. Lovass, S. "Acoustic Emission of Offshore Structures: Attenuation; Noise; Crack Monitoring." *Journal of Acoustic Emission* (April-September 1985): pp A161-A164.
46. Notvest, K. "Effect of Thermal Cycles in Welding D6Ac Steel." *The Welding Journal*. Vol. 45. Miami, FL: American Welding Society (1966): pp 73-78.
47. Jolly, W.D. "Acoustic Emission Exposes Cracks During Welding." *The Welding Journal*. Miami, FL: American Welding Society (1969): pp 21-27.
48. Jolly, W.D. "The Application of Acoustic Emission to In-Process Inspection of Welds." *Materials Evaluation*. Vol. 28, No. 6. Columbus, OH: The American Society for Nondestructive Testing (June 1970): pp 135-144.
49. Hartbower, C.E. "Application of SWAT to the Nondestructive Inspection of Welds." *The Welding Journal*. Miami, FL: American Welding Society (February 1970): pp 59s-60s.
50. Hopwood, T. and J.H. Havens. "Acoustic Emission of Weldments." *Journal of Testing and Evaluation* (April 1979): pp 216-222.
51. Hutton, P.H., R.F. Klein and E.B. Schnenk. *Potential for Applying Acoustic Emission Techniques to Electroslag Welding for Process Control and Flaw Detection*. Richland, WA: Battelle Pacific Northwest.
52. Mitchell, J.R., C.H. McGogney and J. Culp. "Acoustic Emission for Flaw Detection in Electro-Slag Welds." *Proceedings of the First International Conference on Acoustic Emission*. Knoxville, TN: Dunhart Publishing (1980): pp 336-349.
53. Prine, D.W. *NDT of Welds by Acoustic Emission*. Report No. DE-73X. Livermore, CA: Dunegan Endevco Corporation (1973).
54. Prine, D.W., et al. *Improved Fabrication and Inspection of Welded Connections in Bridge Structures*. Federal Highway Administration Report No. FHWA-RD-83/006. Niles, IL: GARD, Inc. (October 1984).
55. Communication from J.R. Matthews (Department of the National Defense — Canada) to Charles McGogney (US Federal Highway Administration): June 1985.
56. Harris, D.O. and H.L. Dunegan. *Acoustic Emission Testing of Wire Rope*. Report No. DE-72-3A. Livermore, CA: Dunegan Corporation (October 1972).
57. Harris, D.O. *Acoustic Emission Monitoring of Lift Span Cables on Dunbarton Bridge*. Technical memorandum DC-72-TM11. Livermore, CA: Dunegan Corporation (December 1972).
58. Green, A.T. *Stress Wave Emission and Fracture of Prestressed Concrete Reactor Materials*. Report No. DRC-71-3. Livermore, CA: Dunegan Research Corporation (May 1971).
59. Muenow, R.A. "Large Scale Applications of Acoustic Emission." *IEEE Transactions on Sonics and Ultrasonics*. Vol. SU-20. New York, NY: Institute of Electrical and Electronics Engineers (January 1973).
60. Dunn, S.E., J.D. Young and W.H. Hartt. "Acoustic Detection of Corrosion Cracking in Reinforced Steel Concrete (Part 1)." *Characterization of Acoustic Emissions Related to Cracking of Concrete*. Boca Raton, FL: Florida Atlantic University (April 1983).
61. Dunn, S.E., R.M. Marshall and W.H. Hartt. "Acoustic Emissions Associated with Embedded Metal Corrosion Cracking of a Concrete Slab (Part 2)." *Acoustic Detection of Corrosion Cracking in Reinforced Concrete*. Boca Raton, FL: Florida Atlantic University (April 1983).
62. Dunn, S.E., R. Babakian, A.M. Ward and W.H. Hartt. "Applications of Acoustic Emission to Inspections of Corrosion Induced Cracking of Concrete in the Field

(Part 3)." *Acoustic Detection of Corrosion Cracking in Reinforced Concrete.* Boca Raton, FL: Florida Atlantic University (April 1983).
63. Bell, R.L. *Acoustic Emission Transducer Calibration — Transient Pulse Method.* Livermore, CA: Dunegan Corporation (1973).
64. Skorpik, J.R. *Acoustic Emission Methods for Flaw Detection in Steel in Highway Bridges.* Richland, WA: Battelle Pacific Northwest (1978): pp 23.

SECTION 11

ACOUSTIC EMISSION APPLICATIONS IN THE ELECTRONICS INDUSTRY

George Harman, National Bureau of Standards, Washington, DC (Part 4, 5 and 9)

Christy Bailey, Dunegan Corporation, Irvine, California (Part 8)
Robert Checkaneck, AT&T, Springfield, New Jersey (Part 1)
J.R. Dale, Signetics Corporation (Part 6)
S.A. Gee, Signetics Corporation (Part 6)
Y. Iwasa, Massachusetts Institute of Technology, Cambridge, Massachusetts (Part 7)
Min-Chung Jon, AT&T, Princeton, New Jersey (Part 2)
Sherwin Kahn, AT&T, Princeton, New Jersey (Part 1 and 3)
Dennis Miller, AT&T (Part 3)
John Stapleton, AT&T (Part 2)
C.G.M. van Kessel, Philips Research Laboratories, Sunnyvale, California (Part 6)
Governor Ware, Teletype Corporation, Little Rock, Arkansas (Part 2)

INTRODUCTION

Acoustic emission is used in the electronics industry for production line control and product evaluation but these applications are relatively new. Because of the small size of electronic components and semiconductor chips, it is more difficult to implement this technology than in most other applications.

The wavelength of acoustic emission in the usual detection frequency range is typically on the order of 10 millimeters. In the case of nuclear pressure vessels and other large structures, this wavelength offers sufficient resolution for triangulation to locate the acoustic emission source. However, for objects the size of microelectronics, multiple internal reflections inevitably mask the position or nature of the acoustic emission source.

When testing electronic components with acoustic emission techniques, investigators are often restricted to simply attaching a transducer (that may be larger than the specimen) in whatever manner possible (such as using tapered acoustic waveguides) and then analyzing the received signal. Waveform signatures of frequency and amplitude are recorded and empirically correlated with appropriate thermal or mechanical stress tests (such as destructive bond pull tests). Because of the circumstances listed above, the interpretation of acoustic emission signals from typical electronics applications may never be amenable to clear mathematical solution.

Equipment is another major difference that distinguishes electronics industry applications of acoustic emission from more conventional usage. Conventional usage generally employs off-the-shelf components and signal processing equipment from the variety of commercial suppliers. However, much of the acoustic emission equipment (as well as the techniques used for electronics production control) must at present be developed or modified in-house for each electronics application.

Caution is required before implementing acoustic emission techniques for new applications in electronic production process control. This is not a simple application; the operator must know and understand all of the failure modes and all the mechanisms that will be encountered by the applied stresses before acoustic emission development work commences. Talented and experienced personnel are required for implementation.

PART 1
ACOUSTIC EMISSION TESTING OF MULTILAYER CERAMIC CAPACITORS[1]

The reduction in size of electronic equipment, achieved through integrated circuit technology, has dictated the development and use of small capacitors with sufficiently large capacitance to serve in coupling and bypass applications. Capacitors using ferroelectric ceramic as the dielectric material can achieve a very high capacitance per unit volume because of the high permittivity of this material ($15 < \epsilon < 9{,}000$). Moreover, ceramic capacitors can be manufactured very economically in a wide range of capacitance values and working voltages.

An acoustic emission proof testing technique has the ability to detect defective capacitors early in manufacture. If these are removed before further processing, 50 to 74 percent of the cost of a finished capacitor may be saved.

If a capacitor meets all electrical specifications but contains hidden physical discontinuities, the long term reliability of the component is affected, thus increasing system service and maintenance costs. Acoustic emission techniques can be used to locate such capacitors, and additional savings are realized by preventing their use in electronic systems.

The Ceramic Chip Capacitor and Its Failure Modes

Ceramic chip capacitors are made by alternating layers of metal and ceramic in a simple parallel plate arrangement. The capacitance is a function of the number and size of the electrodes, their separation and the permittivity of the ceramic dielectric. Typical ceramic formulations are *NPO* ($15 < \epsilon < 80$), *X7R* ($1{,}200 < \epsilon < 2{,}800$) and *Z5U* ($400 < \epsilon < 9{,}000$).

The allowable working voltage is dictated by the electrode separation and the electrical breakdown strength of the ceramic formulation.

It is likely that all ceramic capacitors have small physical discontinuities that enter the product because of undesirable variations in manufacturing conditions but that do not degrade performance in any way. Such devices are usually termed *defect free*. However, when discontinuities are sufficiently large or numerous, the long term reliability of the capacitor is affected. Failure due to low insulation resistance or short circuiting has been linked to three types of structural discontinuities: cracks, voids and delaminations (Fig. 1).

Accelerated-life testing of ceramic chip capacitors[2] has shown that parts with large delaminations (greater than 40 percent of capacitor length) or a number of smaller delaminations (five or more at less than 40 percent capacitor length) fail at a significantly higher rate than parts having no delaminations or only minor delaminations. In one such study, most of those capacitors that failed during a 1,000 hour test exhibited severely degraded insulation resistance while the capacitance and dissipation factor remained within tolerance. A small number of devices failed catastrophically by short circuiting.

Discontinuity Detection Methods

Techniques for detecting mechanical defects in ceramic chip capacitors are limited to destructive sample lot evaluation, certain types of acoustic microscopy and, to a limited extent, radiography. However, a nondestructive acoustic emission technique has been developed for detecting the presence of physical discontinuities in multilayer ceramic capacitors.[3]

The acoustic emission signal obtained during application of a mechanical proof load may be quantified, providing an

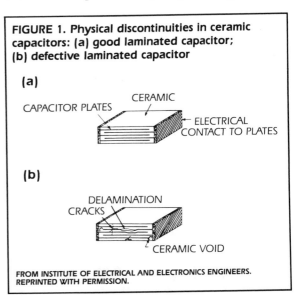

FIGURE 1. Physical discontinuities in ceramic capacitors: (a) good laminated capacitor; (b) defective laminated capacitor

FROM INSTITUTE OF ELECTRICAL AND ELECTRONICS ENGINEERS. REPRINTED WITH PERMISSION.

indication of the severity of physical discontinuities. Unlike other inspection methods, this technique can be used to economically screen capacitors in production quantities.

The technique makes use of the great mechanical strength exhibited by ceramics when loaded in compression. A perfectly monolithic ceramic chip capacitor containing no voids, cracks or delaminations will support a large, uniform compressive stress on its faces with no physical degradation and no release of acoustic emission. The presence of physical discontinuities, however, reduces the ability of the device to withstand such a load. If a sufficiently large stress is applied, discontinuities will propagate with the emission of detectable acoustic energy (see Fig. 2).

A capacitor is placed on a compliant base so that the plane of its laminations is parallel to that surface. Using a compliant ram, a linearly increasing load is applied to the free face of the capacitor in a direction orthogonal to the plane of its laminations. Acoustic emission from the capacitor is detected by a piezoelectric device whose signal is processed by suitable electronic circuitry to yield a number indicative of the presence and severity of discontinuities.

The test configuration (shown in Fig. 3) allows the application of sufficiently large loads to obtain good sensitivity to small but significant discontinuities without generating bending stresses that might fracture or otherwise damage good capacitors. There are several advantages to using this geometry: (1) no special fixturing is needed to hold the test capacitor in a critical preferred position; (2) the acoustic path is precisely reproducible; and (3) discontinuities are, at most, one capacitor's thickness from the surface of load application.

A total of 2,159 capacitors were screened in one test using this acoustic emission procedure, yielding the count distribution shown in Fig. 4. Although the observed acoustic emission activity ranged from 1 to 138,779 counts, 84.3 percent of the test sample exhibited fewer than 500 counts. This fact, along with the knowledge that typical production yields are in the 80 to 85 percent range, strongly suggests that capacitors below the 500 to 750 count range have a low likelihood of containing physical discontinuities. Units with higher counts are likely to have increasingly severe discontinuities.

It should be noted that the activity count closely follows a log normal distribution (Fig. 5) over the low count range and begins to deviate significantly at approximately 435 counts, the defined dividing line between good and bad devices. The physical significance of the occurrence of a log normal distribution is not known.

Conclusions

The ability of acoustic emission procedures to detect capacitors having internal physical discontinuities is clearly supported by destructive physical analysis. Moreover, correlation of data from two accelerated-life tests show, with a

FIGURE 2. Acoustic emission behavior of acceptable and unacceptable capacitors

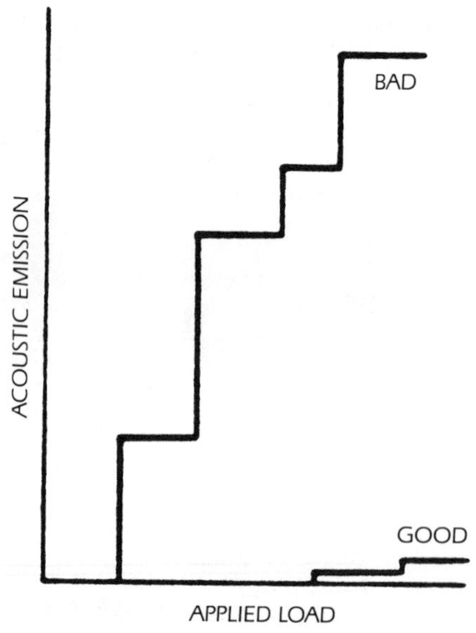

FROM INSTITUTE OF ELECTRICAL AND ELECTRONICS ENGINEERS. REPRINTED WITH PERMISSION.

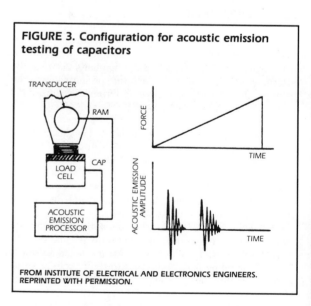

FIGURE 3. Configuration for acoustic emission testing of capacitors

FROM INSTITUTE OF ELECTRICAL AND ELECTRONICS ENGINEERS. REPRINTED WITH PERMISSION.

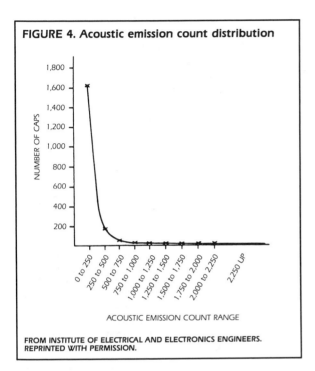

FIGURE 4. Acoustic emission count distribution

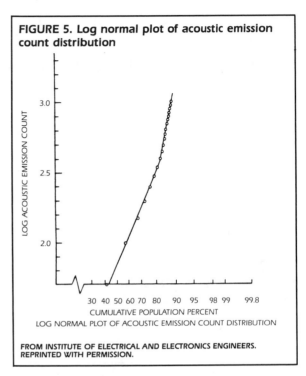

FIGURE 5. Log normal plot of acoustic emission count distribution

high degree of confidence, that such devices have abnormally poor reliability. Not only do these capacitors fail at a higher rate, but it has been shown that they exhibit a high probability of being intermittent, due to a self-healing phenomenon. The acoustic emission proof test provides the capability of increasing ceramic capacitor reliability by detecting physical discontinuities.

The acoustic emission test may be used in at least four distinct ways. First, as a lot checking technique, a representative ceramic sample is rapidly evaluated immediately after kiln firing. This yields feedback on the physical integrity of the capacitors in the lot and allows better control of the firing process.

A second mode of operation is for the detection of discontinuities in very high reliability capacitors where no internal physical discontinuities can be tolerated. In this application, cost is of secondary importance, and devices having any acoustic emission activity during proof testing are discarded. Radiographic examination, as presently used, might be eliminated or made more cost effective by radiographing only those devices that have passed the acoustic emission test.

A third mode of operation for the acoustic emission proof test is to evaluate every part in a lot known to contain a high proportion of grossly defective capacitors. By removing the most defective devices, the lot may be returned to processing without the economic penalty of finishing a large number of parts that will fail acceptance tests somewhere along the path to completion.

The final mode of operation is to check every part in a normal lot as it leaves the firing kiln. This acoustic emission test is used by at least two large capacitor manufacturers and represents one of the few acoustic emission tests in high production within the microelectronics industry.

PART 2
ACOUSTIC EMISSION APPLICATIONS IN UNDERSEA REPEATER MANUFACTURE[4]

For an undersea electronic repeater, lying on the ocean floor and boosting transmission signals in an undersea cable, maintenance is impossible and replacement is extremely expensive. Consequently, during the manufacture of these repeaters, the device's circuitry and the waterproof closure of its cylindrical protective housing are subjected to extremely rigid quality controls.

Application of acoustic emission technology helps ensure the structural integrity of the blocking capacitor in an undersea repeater's circuitry.

High Voltage Capacitor in the Repeater Circuitry Unit

The high voltage DC blocking capacitor separates the AC voice signals from the DC power supply in the cable. It is an oil-filled paper capacitor housed in a cylindrical ceramic casing with metal end caps, as seen in Fig. 6. The paper capacitor unit is wound on a ceramic core to allow a coaxial cable to pass through its axis. Both the casing and the core have a metallized band on each end for solder sealing (see Fig. 6).

During the rotational soldering operation (Fig. 7), a poor quality ceramic casing or capacitor core could crack because of microscopic discontinuities or deviations from optimal processing conditions. While external cracks can be detected visually, it is not possible to detect internal hairline cracks in the ceramic using visual or optical methods, because both ends of the capacitor's casing and core are enclosed by the soldering operation. Even though the subsequent life test might identify some defective capacitors by revealing an oil leakage, this test is costly and extremely time consuming. Moreover, optical inspection and life-testing will not reveal cracks covered by the solder.

FIGURE 6. At left, the capacitor's ceramic casing is positioned between two metal covers that fit over its ends; the covers are then soldered to the ceramic casing (right); the ceramic core, also susceptible to cracking during the soldering operation, is shown projecting from the end cap in the assembled capacitor (right)

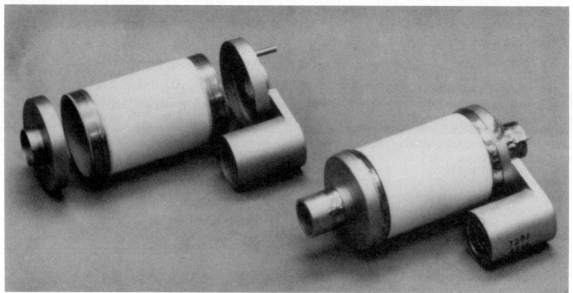

FROM BELL LABORATORIES. REPRINTED WITH PERMISSION.

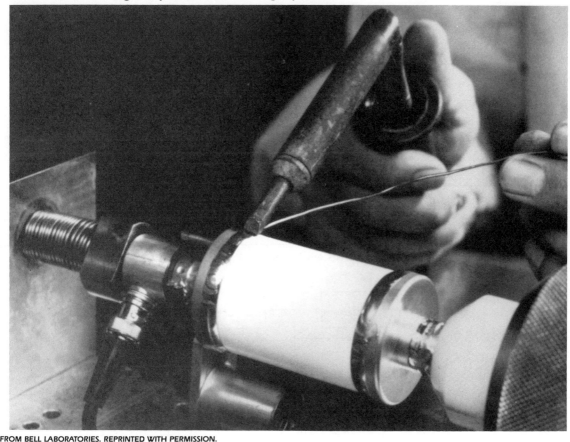

FIGURE 7. The transducer that detects acoustic emission originating in the ceramic during soldering is seen at left, also serving as a pivot for the rotating capacitor

FROM BELL LABORATORIES. REPRINTED WITH PERMISSION.

Once a capacitor is operating within the repeater housing, a hairline crack in a casing or core could propagate, releasing a seepage of oil. Besides contaminating the repeater, this seepage could cause dielectric breakdown in the high voltage section of the cable. A shorted capacitor would render the repeater inoperable, requiring costly replacement. Hence, a system that can detect cracks as they occur during manufacture would help ensure the quality of the capacitor.

Instrumentation and Analysis

Ceramic Capacitor Crack Detector

Because of the continuous nature of the soldering process, a window or gate period could not be defined and existing crack detectors could not be used effectively for the undersea repeater's capacitor soldering operation. The required system would have to monitor cracking during the entire processing cycle. The design used to overcome this problem was also made to distinguish cracking from noise utilizing a signature recognition scheme.

Figure 8 is a block diagram of the crack detection system used in conjunction with the rotational soldering operation that joins the metal caps to the ends of the ceramic capacitor case described above. The incoming acoustic emission signal from the preamplifier (Fig. 9) is filtered and further amplified. This analog signal is then passed to both a threshold detector and an envelope detector. The output of the threshold detector is a wave train of pulses corresponding to each threshold crossing of the filtered acoustic emission signal. The output of the envelope detector is a voltage following the peak input excursions of the acoustic emission signal.

The acoustic emission envelope is then passed to both a

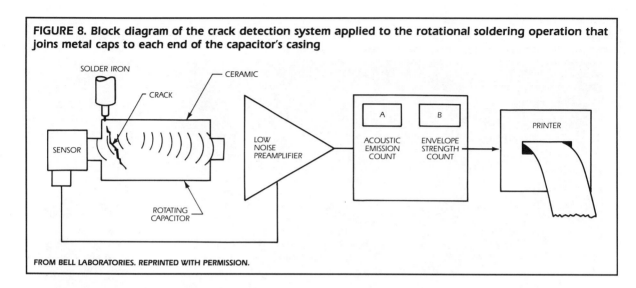

FIGURE 8. Block diagram of the crack detection system applied to the rotational soldering operation that joins metal caps to each end of the capacitor's casing

FROM BELL LABORATORIES. REPRINTED WITH PERMISSION.

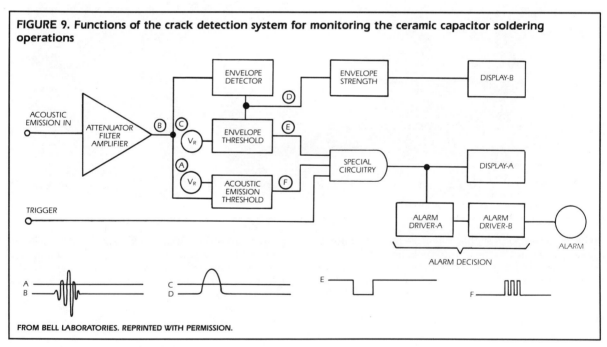

FIGURE 9. Functions of the crack detection system for monitoring the ceramic capacitor soldering operations

FROM BELL LABORATORIES. REPRINTED WITH PERMISSION.

threshold detection circuit and an envelope strength circuit. A window opens when the envelope surpasses a threshold and persists as long as the envelope remains above that threshold. The envelope strength circuit converts the acoustic emission envelope into pulses whose number is proportional to the area (amplitude and length) of the envelope curve.

As seen in Fig. 9, the acoustic emission counts (display-*A*) and envelope strength counts (display-*B*) are accumulated during the envelope generated window. At the end of the window, the accumulated counts are displayed in light emitting diode (LED) counters. The counts are also sent to a printer where the values are stored and then printed at the unit's earliest convenience.

ACOUSTIC EMISSION APPLICATIONS IN THE ELECTRONICS INDUSTRY

The printer's storage element consists of a parallel recirculating stack of 128 bytes that is capable of storing information from acoustic emission bursts occurring less than 10 microseconds apart; priority is given to storage rather than printing. The recirculating concept actually expands the memory capability to more than 128 bytes. Once the information in each memory location is printed, it is no longer needed and can be written over during subsequent cycles. The electronics feed acoustic emission burst data to the printer until all the information that is stored is printed.

The printout allows the operator to further analyze relevant acoustic emission bursts for indications of cracking in the ceramic. This is done by determining the ratio of the envelope strength counts (display-B) to the acoustic emission counts (display-A).

Experimental results show the following: if the count ratio B/A falls within a specified range (experimentally determined to be 1.6 to 2.9), the burst is considered a signal caused by cracking in the ceramic.

After successful work in the laboratory, the ceramic capacitor crack detector was installed in the production line. The system included a calibrator that generated simulated acoustic emission signals to check the normal functions of the transducer and the acoustic emission electronics. Installation was accomplished by replacing a standard nylon pivot on the soldering station with a specially modified transducer (Fig. 7) that also acted as a pivot for the rotating capacitor. This was the only change required for transfer of the acoustic emission system from the laboratory to the manufacturing environment.

Tubulation Pinchweld on the Repeater Housing

The housing of the entire undersea repeater is designed to resist more than 75 MPa (11,000 psi) of water pressure on the ocean floor. The tubulation pinchweld seals off the pressurized nitrogen gas in the repeater housing and a poor quality weld could cause premature failure of the circuitry by admitting salt water into the unit.

In the past, the quality of the pinchweld was determined with a radioisotope test, trying to force isotopes through cracks or voids that might exist in the pinchweld area. Water

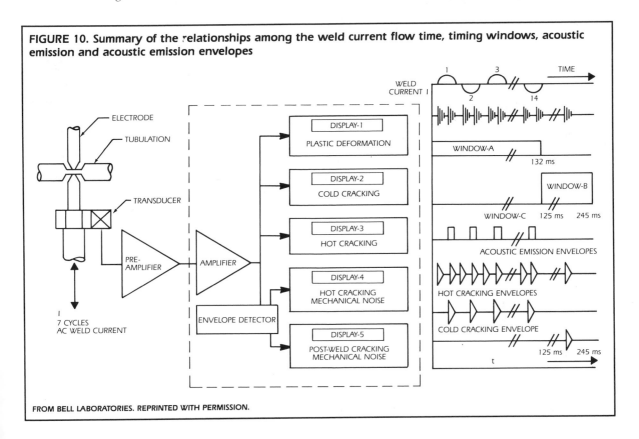

FIGURE 10. Summary of the relationships among the weld current flow time, timing windows, acoustic emission and acoustic emission envelopes

FROM BELL LABORATORIES. REPRINTED WITH PERMISSION.

containing isotopes of cesium was used as the medium. After the pinchweld area was subjected to this high pressure water for a specific time period, a Geiger counter was passed over the repeater housing to check for leakage. The time consumed, the complexity of the facilities and the difficulties associated with this test made its replacement desirable.

Pinchweld Analyzer

The pinchweld analyzer differs from the conventional acoustic emission resistance spot weld analyzer in that it can perform the following functions:

1. measuring the degree of plastic deformation related to pinchweld thickness;
2. detecting hot cracking during the welding cycle and cold cracking during the post-weld period; and
3. recognizing mechanical noise generated during the welding operation.

The pinchweld analyzer has five LED counter displays; three show acoustic emission counts and two show envelope counts. The five displays are controlled by three time windows. The time windows are generated externally by the weld current; that is, a time window is formed whenever the weld current surpasses a threshold, and persists as long as the current remains above the threshold. Figure 10 summarizes the relationships among the weld current flow time, the timing windows, the acoustic emission and the acoustic emission envelopes. A calibrator generates simulated acoustic emission signals used to check the normal functions of the transducer and pinchweld analyzer.

The first window (window-A in Fig. 10) is opened during the time of welding current flow. This detects the number of acoustic emission counts associated with material plastic deformation and material expulsion. Acoustic emission counts collected in this window are displayed in the first LED counter. Upper and lower acoustic emission count control limits for this window have been established experimentally (shown in Fig. 11). The lower limit corresponds to insufficient weld melting and the upper limit represents excessive melting, possible material expulsion, and therefore porosity formation.

To monitor post-weld cold cracking, a window opens after the weld period is completed (window-B). This acoustic emission count is displayed in the second LED counter. A higher count corresponds to the occurrence of cold cracking and therefore a bad weld. An upper count limit is established empirically; if the post-weld cracking count exceeds this value, the weld is rejected.

Hot cracking is monitored by window-C which opens during the time that each half-cycle of weld current is decaying. Acoustic emission signals generated during this time are related to residual stress buildup occurring during the cooling period as the weld current decreases. This acoustic emission count is displayed in the third LED counter. If the count exceeds an experimentally predetermined maximum value, the weld is rejected because hot cracking has occurred.

An additional capability of the pinchweld analyzer is that it can distinguish weld cracking from mechanical noise. Mechanical noise releases energy very slowly while the acoustic emission signals generated by weld cracking are a rapid energy release phenomenon. Weld cracking is identified by

FIGURE 11. Upper and lower control limits: (a) for acoustic emission counts of window-A in Figure 10; (b) lower limit corresponds to insufficient weld melting; (c) upper limit represents excessive melting and possible material expulsion, indicated by the arrow

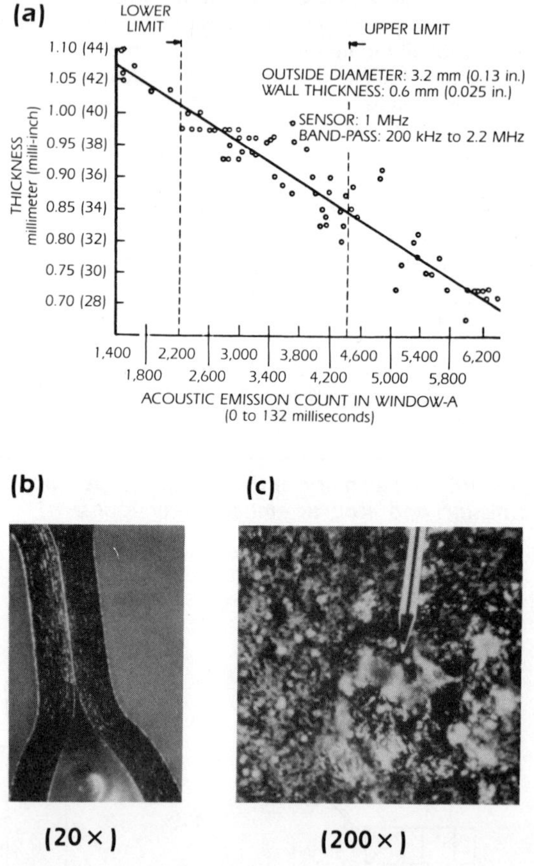

FROM BELL LABORATORIES. REPRINTED WITH PERMISSION.

FIGURE 12. Acoustic emission crack detector used during the soldering of ceramic capacitors

FROM BELL LABORATORIES. REPRINTED WITH PERMISSION.

taking the ratio of the respective acoustic emission cracking count (value in LED counter two or three) to the corresponding envelope count occurring during the same period (fifth LED counter for cold cracking and fourth LED counter for hot cracking). If the ratio is less than an experimentally predetermined value, it is attributed to cracking. For a ratio higher than the predetermined value, it is attributed to mechanical noise.

Over a six month field trial of the pinchweld analyzer, the acoustic emission results compared favorably with those of the existing isotope test method, which was subsequently discontinued. The ceramic capacitor crack detector is shown in production line operation in Fig. 12.

The detector provides a means of checking for cracking during the soldering operation. Previously, only a visual inspection was made after the capacitor was soldered. As a result of using the acoustic emission technique, in-process quality analysis of the soldering operation has been achieved and the reliability of the capacitor has been enhanced.

In the case of the pinchweld analyzer, its ability to identify poor quality welds by detecting amounts of plastic deformation, material expulsion, hot and cold cracking, and mechanical noise interference has made it an excellent system for the undersea repeater pinchweld application.

The cost of adding the acoustic emission weld analyzer to the pinchweld production line is minimal and is far outweighed by eliminating the previous isotope test. The earlier test facility occupied a lead lined room containing six test sites, at which six repeaters simultaneously underwent lengthy tests in high pressure water. This space is now available for other purposes. In addition, the radioactive testing materials, requiring strict adherence to Federal government regulations and inspection procedures, have also been eliminated.

PART 3
ACOUSTIC EMISSION DETECTION OF CERAMIC SUBSTRATE CRACKING DURING THERMOCOMPRESSION BONDING[5]

Cracks may form in the ceramic material during thermocompression bonding of integrated circuit lead frames to thin film circuits on ceramic substrates. Such microscopic cracks, invisible to the unaided eye, may not be discovered until after the circuit is in service when their growth leads to costly failure.

The formation and growth of cracks in a brittle material are accompanied by detectable acoustic emission.[6] An acoustic emission system is used to monitor the bonding of integrated circuit lead frames to brittle ceramic substrates by a thermocompression bonding process.

The Bonding Process

Thermocompression bonding is used extensively in the manufacture of plastic molded dual in-line integrated circuit packages (DIP). An integrated circuit chip is first bonded to a small, highly pure ceramic substrate bearing thin film circuits. These circuits are formed by evaporation of titanium and palladium, followed by the selective electroplating of a layer of gold.

The gold plated ends of the leads, part of a punched copper lead frame, are then attached to the ceramic substrate (Figs. 13 and 14) by a bonding machine that simultaneously forces them against thin film circuits on the ceramic.

Proper time, temperature and pressure are necessary to achieve a solid state bond between the two gold surfaces. Typically, during a one-second bonding time at thermode and pedestal temperatures of 350 °C (660 °F), about 50 percent deformation takes place. These temperatures and deformations are achieved by pressing a thermode against the top of the lead frame, while the prepositioned ceramic substrate is supported by a heated base (pedestal).

Though thermocompression bonding is a reliable technique for forming interconnections, defective products can occur. A worn pedestal or thermode may crack the ceramic when pressure is applied. Poor quality ceramic may crack even under optimal bonding conditions. The degree of cracking may vary from microscopic hairline cracks, an easily visible macroscopic crack or complete fracture of the substrate.

Hairline cracks are of greatest concern. They may go undetected, leading to failure during the manufacturing process or after a complete integrated circuit assembly has been placed in service. Since delayed failure caused by cracks can be costly, it is desirable to detect any ceramic cracking during the bonding process. Development of an acoustic emission detection system was undertaken for this purpose.

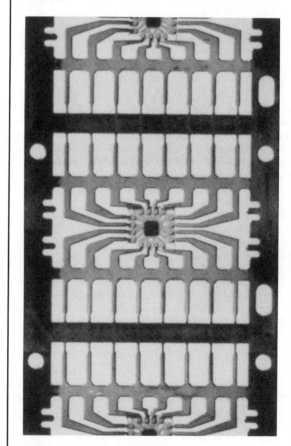

FIGURE 13. Bonding of the lead frame to the ceramic substrate (center) takes place after the integrated circuit chip has been bonded to the substrate

FROM BELL LABORATORIES. REPRINTED WITH PERMISSION.

Acoustic Emission System for Detection of Bonding Cracks

Acoustic Emission Propagation Path in Ceramics

In thermocompression bonding, cracking of a ceramic substrate may occur when the thermode makes physical contact with the device being bonded. During the bonding process, a good mechanical contact (and thus a good acoustic path) exists between the ceramic and the thermode, and between the ceramic and its supporting pedestal. Acoustic energy emitted during cracking of the ceramic may propagate outward through the thermode structure or the pedestal. A piezoelectric transducer placed on either the thermode or pedestal will ensure a short propagation path and allow observation of the acoustic emission signals associated with cracking.

However, the high temperatures of the thermode and the pedestal can cause depoling (loss of piezoelectric properties) in all but very special piezoelectric materials. Moreover, the thermocompression bonding machine uses four dual positions that are sequentially loaded and rotated under the bond head. This motion further complicates the placement of a transducer.

As in most cases where acoustic emission systems must be retrofitted to an existing machine, a compromise was made when determining the location of the transducer. While it is desirable to place the transducer as close as possible to the source of emission, this is not always practical. In this particular case, the transducer was mounted on a flat surface of the bond head (Fig. 15), a water cooled structure that houses the thermode. The acoustic emission signal that reaches the transducer is modified by passage through the thermode and the insulating material that isolates the thermode from the bond head. The rest of the bonding machine also affects the signal that is observed, because the mechanical structure of the machine is acoustically coupled to the ceramic through the thermode. Thus any characteristic resonances of the free, unconstrained ceramic will be modified or completely damped by the loading effect of the mechanical structure.

Bond Head Transducer

The acoustic emission signal that propagates to the surface of the bond head is detected by a piezoelectric transducer that is coupled to the surface of the bond head (Fig. 15) by a thin layer of epoxy adhesive. The transducer is designed for resonant operation in the thickness mode with a resonant frequency of 650 kHz. A resonant transducer is used to obtain greater sensitivity and also some rejection of machine vibration, which is stronger at lower frequencies.

The acoustic emission crack detectors modify a transducer's differential output with a charge amplifier having a 40 decibel (dB) gain, and a noise level of approximately 100 microvolts. After this stage of preamplification, the signal is further amplified by a high stability amplifier providing an additional 20 dB gain.

It has been observed that in certain restricted bandwidths, the amplitude of an acoustic emission signal originating from a propagating crack in the ceramic exceeds the signals arising from the impact and vibration of the bonding machine. Therefore, the amplified signal from the transducer is filtered by a twin-T active filter that provides a sharp band-pass response centered at 650 kHz, the transducer resonant frequency. The resulting signal is passed through a buffer amplifier before signal processing.

Signal Processing and Decision Making

The amplified and filtered transducer output signal contains, in its amplitude, the information necessary for deciding whether a crack has occurred. This information is extracted by using a simple signal processing technique known as peak detection. The use of a precision clamping circuit followed by a peak detector yields a DC voltage proportional to the maximum peak-to-peak amplitude of the acoustic emission burst. The system compares the level with a preset DC level (threshold) which represents the highest signal expected from machine noise. If the peak-detected signal exceeds this level, an audible alarm circuit is activated to indicate cracking. To minimize the effects of extraneous noise, the alarm circuit is activated only during the time the thermode is in contact with the ceramic (the only time cracking might occur). This is accomplished by using the motion of the bond head to provide a synchronizing signal to the alarm circuit.

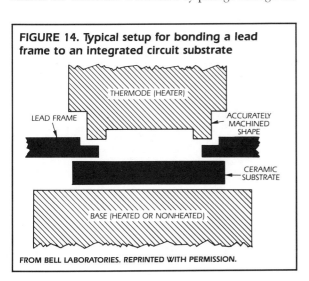

FIGURE 14. Typical setup for bonding a lead frame to an integrated circuit substrate

FROM BELL LABORATORIES. REPRINTED WITH PERMISSION.

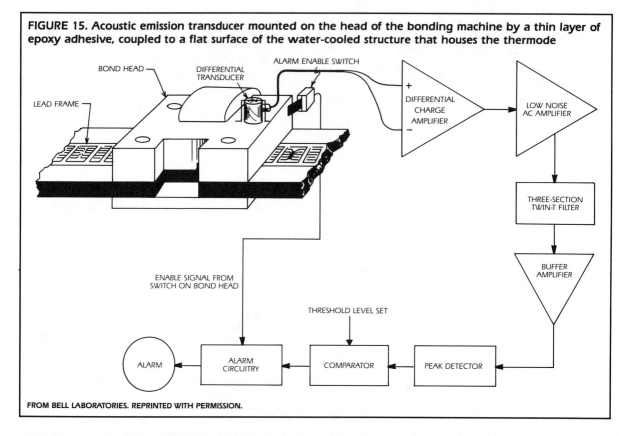

FIGURE 15. Acoustic emission transducer mounted on the head of the bonding machine by a thin layer of epoxy adhesive, coupled to a flat surface of the water-cooled structure that houses the thermode

FROM BELL LABORATORIES. REPRINTED WITH PERMISSION.

The alarm has been augmented with additional circuitry to operate a pen that marks the defective assembly. This reduces the need for constant operator attention.

Conclusions

The acoustic emission crack detector is a useful means for identifying cracked substrates in real-time. However, it is most valuable as a process monitor, indicating when a large number of substrates are being cracked because of worn pedestals or thermodes, or misadjustment of the bonder. Without the acoustic emission system, a quantity of bad product might be turned out before visual inspection could discover a problem.

Though the crack detectors used in thermocompression bonding are quite reliable, they are subject to inaccuracy because of the variability of machine related noise. It has been found that the noise background attributable to the bonder is extremely variable, depending on the machine's operating speed, mechanical adjustment and its state of lubrication. This variability necessitates periodic resetting of the threshold amplitude level used to make the crack/no crack decision, because upward drift in machine noise could result in false crack indications.

PART 4
ACOUSTIC EMISSION TEST FOR MICROELECTRONIC TAPE AUTOMATED BONDING

An acoustic emission test method for tape automated bonding (TAB) integrity has been developed, consisting of a precision testing machine that simultaneously applies (1) a clamping force on the semiconductor chip and (2) a lifting force on the electrical interconnecting leads. Acoustic emission signals are transmitted through a waveguide to the detector. The system also permanently forms (raises) the leads, which is a normal requirement for TAB devices. Large acoustic emission signals occur when a lead breaks, a bump lifts or a weld crack propagates. Appropriate acoustic emission signal processing is also employed.

An advanced technique of interconnection between a semiconductor chip and its package, tape automated bonding (TAB)[7] is accomplished by increasing the thickness of aluminum bonding pads on the chip with copper or gold bumps, using plating or other techniques. A special metallized polyimide tape with individual extended metal leads is then bonded to the bumps by a thermocompression or a solder process. This is called *inner-lead bonding*. Figure 16 shows two examples of chips bonded to such tape. The tape in Fig. 16a is called *testable tape*. Here the individual insulated leads terminate on pads that may be probed so that the device can be electrically tested after bonding. This tape system is used in acoustic emission tests.

The bond system consists of gold bumps on the chip and tin-plated copper leads on the tape. During the bonding process, the tin melts, forming various gold-tin intermetallic compounds. These form a strong but brittle bond which under certain conditions can result in acoustic emission under applied force. The tape in Fig. 16b is of all-metal construction and the device cannot be tested before packaging. This type of tape is usually used for inexpensive, low lead count integrated circuits where no mechanical or electrical testing is justified at this stage of production. All tests described below were performed on the tape shown in Fig. 16a.

Once the chips are attached, the tape is usually wound onto reels and stored until final assembly into either an integrated circuit package or a hybrid. In the case of some calculators and watches, the lead's outer end is excised from the tape and reflow soldered onto a printed circuit board. The chip is then protected by a drop of epoxy.

When working with any microelectronic interconnection bonding system, it is necessary to evaluate the bond quality as well as the bonding machine setup parameters with some form of stress test. This is normally done with a type of pull tester for both wire and tape bonds. This tester is also used to confirm acoustic emission test data.

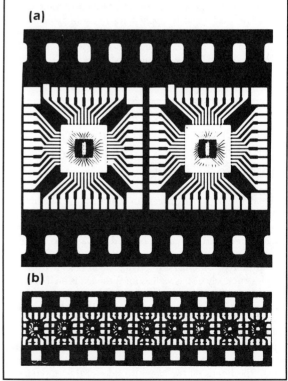

FIGURE 16. TAB integrated circuits: (a) three-layer 35 mm polyimide tape; (b) single layer 11 mm all-copper tape; the 35 mm tape has tin-plated copper leads and is melt-alloy bonded to gold bumps on the chip; copper leads of the 11 mm tape were thermocompression bonded to gold-plated bumps on chip

A diagram of the pull test is shown in Fig. 17. The forces applied to the TAB-chip bond during this test can be calculated from the resolution of forces using Eqs. 1 and 2. The variables are defined in Fig. 17.

The tensile pull force along the lead is:

$$F_L = F_{app}(1 - \epsilon)\sqrt{\frac{1 + \epsilon^2 d^2}{h^2}} \quad \text{(Eq. 1)}$$

Where:

F_L = tensile pull force along the lead;
F_{app} = applied force;
ϵ = distance ratio; and
d = distance from chip to tape.

The peel force (vertical component of F_L) applied to the bond interface at the bond heel is:

$$F_p = F_{app}(1 - \epsilon) \quad \text{(Eq. 2)}$$

These pull test equations are also used for the nondestructive acoustic emission push-up testing of tape bonds described later. The equations assume that the leads extend perpendicularly from the edge of the chip. For the most effective acoustic emission testing of TAB bonds, force probes should be positioned to apply a relatively high peel force, which promotes crack propagation along the interface of weak bonds or bumps. A high peel force is achieved by pulling or lifting the leads as near as practical to the edge of the chip, as can be seen from Eq. 2.

An Acoustic Emission Nondestructive Bond Quality Tester and Lead Forming System for Tape Bonded Integrated Circuits

Lead forming is a process in which the leads are permanently bent upward, away from the chip, to increase the chip-to-lead clearance and prevent shorting. Forming is usually required on tape bonded devices for face-up mounting.

A prototype automatic tester has been designed to both nondestructively test as well as form the leads on TAB inner lead bonds. Bond failure is indicated by acoustic emission monitoring. This tester is shown in Fig. 18, with closeups of the test stage, without the TAB device, in Figs. 19 and 20. The adjustable force wedges are shown in Fig. 19e. In operation, the 35 mm tape is advanced until the chip is centered in the fixture. A tool clamps the chip by pressing down on top of the bonds on two opposite sides of the chip, leaving

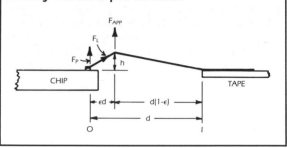

FIGURE 17. Resolution of forces diagram used to identify TAB bond pull variables

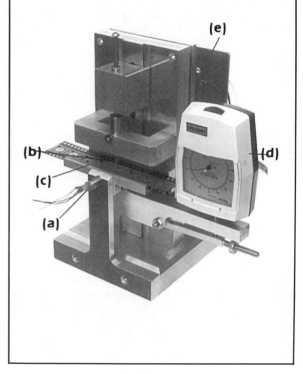

FIGURE 18. Prototype of acoustic emission monitored TAB bond integrity tester: (a) acoustic emission detector; (b) clamping tool to hold one-half of the leads fixed while the other half are free to be tested and formed; (c) 35 mm tape with TAB semiconductor device in center under clamping tool; (d) force gage used to adjust and read force applied to TAB leads; and (e) drive motor panel

FIGURE 19. Closeup of TAB lead testing portion of Figure 18 without tape, shown in portion of peak force application: (a) clamping tool; (b) nondestructive testing and lead forming wedges; the lead contacting surfaces are ground flat to a width of approximately twice the TAB lead width (~0.15 mm) and then the edges of the face are rounded; (c) anvil to hold chip and conduct acoustic emission signal to detector; (d) acoustic emission detector and 40 dB preamplifier in a package; and (e) enlargement of wedges from (b).

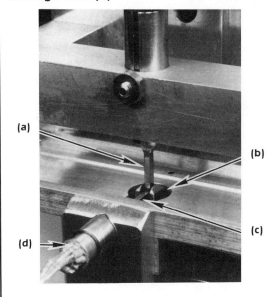

FIGURE 20. Cross section of the active portion of the TAB testing apparatus in Figure 19: (a) semiconductor device (chip); (b) bumps on chip; (c) tin-plated copper leads (approximately 0.08 mm wide by 0.02 mm thick); and (d) polyimide tape

FIGURE 21. Block diagram of acoustic emission detection and signal processing equipment; in various experiments, only totalizer and printer or transient recorder (with or without computer) may be used to record acoustic emission data

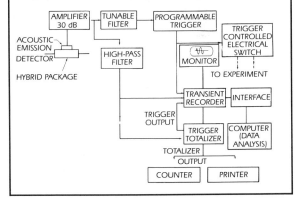

the leads on the other two sides free to be tested. Another tool, consisting of two adjustable wedges (Fig. 19e) rises from below, exerts a vertical testing and lifting force against the unclamped leads, and bends them upward.

This force is determined for each device type or lead design (based on destructive pull tests and the amount of lead forming desired) but is usually on the order of 100 millinewtons (mN) per lead. When a lead breaks, a crack propagates, or a bump lifts, the acoustic emission signal travels through the semiconductor chip to the anvil (Fig. 19c) and then to the acoustic emission detector (Fig. 19d). The acoustic emission signal may be attenuated at the chip-to-anvil interface unless care is exercised. Tape bonded chips sometimes have tiny bits of plastic left on the back side, from the adhesive applied to hold the wafer during sawing, leaving an uneven surface. Normal acoustic emission couplant cannot be used to increase transmission because it could contaminate the chip. A plastic film having a little compliance is bonded to the anvil and generally provides adequate contact.

Both the acoustic emission trigger and the totalizer (Fig. 21) are used to assist in the interpretation of failures. The acoustic emission amplifier may be force gated (or time gated) to be sensitive only after a force of approximately 10 or 20 mN per lead is applied, in order to avoid interference noise from contact or scraping of the wedges against the leads. The gate window would be closed after maximum force is applied. With the present equipment, gating was found unnecessary for slow application of force (about 20 mN per second). However, it was essential when the rate of force application was increased by a factor of 5, as required for practical production testing.

Representative data selected from four of these sample tests are given in Fig. 22. The data are plotted as simple cumulative acoustic emission counts (with no other data processing) in order to show the actual data obtained from the tests. Curves A and B represent cases where the lead-to-bump and bump-to-chip interfaces were strong. Points on curve-A are the result of cumulative acoustic emission noise. This is the ideal case, where no gold-tin intermetallic cracking occurred. Curve-B results from leads having strong bond interfaces, but reveals gold-tin intermetallic cracking at the bond heel during the test. Curves C and D are the result of bumps separating from the chip (the lead-to-bump interface was strong in these samples).

It is apparent from curves C and D that the same type and number of lead failures can produce widely different cumulative acoustic emission counts (in this case, 4,000 and 9,000 counts). Such differences may result from the particular form and velocity of crack propagation. However, the separation of weak bonds (curves C and D) from strong bonds (curves A and B) can be improved if signal processing methods such as multiple amplitude thresholds and squaring signal amplitude are implemented. Examination of the acoustic emission waveforms (from the transient recorder output in Fig. 21) showed that far higher peak amplitudes occurred during interface failures (curves C and D) than during intermetallic microcracking (in curve-B), making separation relatively easy. Separation was also greater when the force was applied faster.

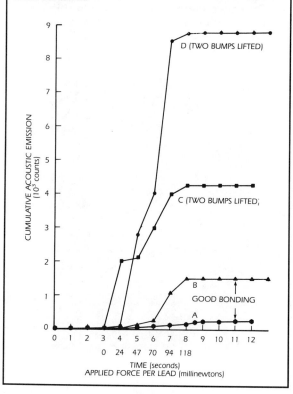

FIGURE 22. Cumulative acoustic emission data from 35 mm TAB devices tested on the prototype bond integrity tester; curve-A and curve-B are from devices with strong bonds; curve-C and curve-D are from devices having weak bump-to-chip bonds; force peaked at 118 mN (8 seconds), then released

Certain problems related to quiet, wear resistant organic surface coatings for the lead contacting tools have not been completely solved. A polyester film coating over the steel wedges has been used for this purpose, but it would be desirable to make the entire wedge out of a hard plastic such as nylon. However, the small top-flat dimensions of the wedges (0.15 mm) make this difficult to machine and plastic molding processes were not available. The present prototype only tests half the leads on a device. If 100 percent testing and lead forming were required, a second test stage oriented at right angles to the first, that applies force to the other leads, would be necessary.

PART 5
ACOUSTIC EMISSION THERMAL SHOCK TEST FOR HYBRID MICROCIRCUIT PACKAGES

An acoustic emission test for hybrid microcircuit package integrity was developed and consists of a hot stage operating at 400 °C (750 °F), a special water cooled acoustic emission detector mount, and the appropriate signal recording equipment. In use, the detector is coupled to the back of a hybrid package and both are set on the hot stage for approximately 30 seconds. Any acoustic emission signals indicate thermal excursion damage to the glass-to-metal seals. The acoustic emission signals are correlated with both room temperature and 125 °C (260 °F) leak test of the packages.

Hybrid microcircuit packages are generally made of a low expansion metal and have very small glass-to-metal seals permitting electrical leads to enter the sealed hermetic interior. A typical hybrid package is shown in Fig. 23. Such packages may have as many as one hundred glass-to-metal seals that together must be hermetic to at least 10^{-7} Pa•m^{3}s^{-1}. The processes and materials used to manufacture these packages have enough variations to occasionally produce packages that develop hermetic leaks during later handling and hybrid circuit assembly. Various thermal shock tests have been employed (followed by a hermeticity test) to reveal the weak, failure-prone seals. Some manufacturers use expensive military-specified liquid-to-liquid shock for 10 to 15 cycles, followed by a room temperature leak test. Others use a simple hot plate shock test, also followed by a leak test, unless one of the military tests is specified and paid for by the purchaser.

The hot plate thermal shock test is performed as follows. A hybrid package containing glass-to-metal seals is placed on a hot plate preheated to a temperature in the range of 350 to 400 °C (660 to 750 °F). The package temperature is allowed to reach equilibrium (about 30 seconds). Following removal and cooling, the glass seals are examined and leak tested for damage. Because of the frequent use of the shock test as a reliability screen, it is desirable to instrument it with acoustic emission detection equipment to obtain rapid information on the glass-to-metal seal integrity that may not be revealed in a visual inspection or even in a room temperature leak test.[8] Such an acoustic emission test is faster than a liquid-to-liquid shock test followed by a leak test and can also give information related to the high temperature reliability of the package.

Characterization of the Hot Stage Shock Test

An acoustic emission hot plate shock test requires that the packages be put on the hot plate in a face-down configuration with the acoustic emission detector on top (attached to the normal bottom of the package). The apparatus consists of a temperature controlled hot stage with a separate thermocouple. For this purpose, 0.08 mm (0.003 in.) diameter thermocouple wires are welded to the inside wall of a hybrid package adjacent to and at the same level as the lead

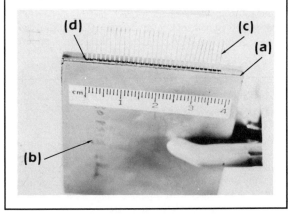

FIGURE 23. A large hybrid package made of an alloy having poor thermal conductivity but excellent glass-sealing characteristics: (a) the side walls, rim or perimeter are usually about 1.5 mm (0.06 in.) thick and 2.5 mm (0.1 in.) high; (b) the lids, on the top and bottom, are about 0.4 mm (0.015 in.) thick and are welded to the rim; (c) the leads are 0.25 mm (0.01 in.) thick, 0.4 mm (0.015 in.) wide and pass through glass seals; and (d) glass seals are about 1 mm (0.04 in.) in diameter

throughs. The hot plate surface is flattened and smoothed with a fine file. This smoothing also helps protect the package weld seal plated surface.

Figure 24 shows the temperature rise curves for a hybrid package. Curve-A gives the average temperature time curve of a package, open face up, on the 400 °C (750 °F) hot stage for a gold plated package of 32 × 32 × 7.5 mm (1.25 × 1.25 × 0.3 in.) dimensions. Each point is the average of three runs. Factors that may lead to varying temperature rise versus time at the seals are: (1) package design (height of seals above bottom surface and size of package); (2) surface plating material, its reflectivity and particles, contaminants or irregularities on the surface of the hot stage; and (3) possible ambient air motion (drafts) across its surface.

After characterizing the normal hot stage thermal shock test, it is necessary to obtain equivalent data on the face-down package configuration with a convection radiation shield (on the hot stage under the central part of the hybrid) and a cooled detector attached to the bottom of the hybrid. After several runs, it was determined that the equivalent rate of temperature rise in the initial 20 to 200 °C (70 to 400 °F) range (the temperature range that produces maximum stress on the perimeter of glass-metal seals in the hybrid configuration) was obtained with the stage temperature increased to about 25 to 425 °C (80 to 800 °F), as shown in Fig. 24, curve-B. However, because this temperature exceeds the one normally employed, a maximum temperature of 400 °C (750 °F) was set for further tests. This corresponds to a face-up shock temperature of about 375 °C (700 °F) which is well within the range normally used. In general, it took about five seconds to increase the temperature of the seals from room temperature to 200 °C (390 °F). Thus, acoustic emission instrumentation can be time delayed for two to three seconds to eliminate mechanical noise generated during placement of the package and detector on the hot stage.

Different organizations may use different hot stages, ranging from small substrate heaters to large hot plates, as well as a large variety of package types. Therefore, each particular setup should be characterized, and the surface of the stage smoothed and kept clean in order to obtain consistent results.

Acoustic Emission Monitoring of the Shock Test

Most commercial hot plates generate copious acoustical noise during operation. For purposes of acoustic emission monitoring tests, a quiet hot plate is required. It was found that units heated by insert type elements meet this requirement, whereas ones with wrapped coils, as are typical of most commercial hot plates, are extremely noisy.

Since the package rests on a stage that is maintained at about 400 °C (750 °F), some thermal protection or isolation is required for the acoustic emission detector. This was achieved by putting the detector in a water cooled jacket and spring loading it to contact the thin alloy bottom of the hybrid package.

The acoustic emission detector was protected from the high temperatures of the hot stage by a small water cooled mount, a thermal couplant to prevent heat conducting into the detector from the package walls. The detector and mount were attached to the thin back side of the package with vacuum grease. This served as an acoustic emission couplant and as an adhesive that held the lightweight package to the detector mount during placement on, and removal from, the hot stage. After the test, the vacuum grease was removed with a solvent. A photograph of the cooled detector mount, hybrid package, and hot stage is shown in Fig. 25. An expanded cross section of the acoustic emission detector mount is given in Fig. 26.

A block diagram of the acoustic emission equipment is given in Fig. 27. Acoustic emission output during the shock tests was found to extend over a period of up to 30 seconds,

FIGURE 24. Temperature rise curves for a hybrid package, at the position of the glass seals, on the hot stage of Figure 25: curve-A represents the package face up, hot stage at 400 °C; curve-B represents the package face down, hot stage at 425 °C

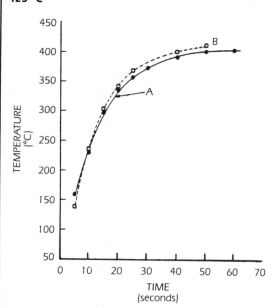

which was expected from the temperature rise characteristics given in Fig. 24. For this work, the signal processing was simple, consisting of the totalizing system and the printer portion of Fig. 27. Each acoustic emission cycle or pulse (count) above a set threshold was recorded and the summation of these was printed each second.

Hot stage shock tests on high quality hybrid packages show that they emitted essentially no acoustic emission during the test, so data from one of these were used as a control to define a good package. The accumulated counts versus time on the hot stage, for the control sample, are plotted as curve-D in Fig. 28. The small, essentially linear increase in signal with time is due to the cumulative system noise above the set threshold and is not due to glass seal degradation.

This package and others were tested for leaks at room temperature and at 120 °C (250 °F). The control sample leak rate was less than 10^{-8} Pa•m^{3}s^{-1} at both temperatures.

Most of the forty-eight packages tested in this manner produced acoustic emission versus time curves similar to that of curve-D in Fig. 28. However, a small lot of four packages (from one manufacturer) that were initially hermetic at both room temperature and at 120 °C (250 °F) emitted significant acoustic emission when shock tested. Plots of the data from three of these are given as curves A through C in Fig. 28. The delay of acoustic emission onset for a few seconds after placing the packages on the hot stage was expected, since it takes several seconds to reach the 150 to 200 °C (300 to 390 °F) required to put the seals in tension. A retest of the high acoustic emission producing samples resulted in very low acoustic emission output, similar to and only slightly greater than that of the control sample. This phenomenon is an example of the Kaiser effect[9] in which little or no acoustic emission occurs until previously applied stress levels are exceeded.

The packages that produced large acoustic emission signals were leak tested at both room temperature and 120 °C (250 °F). All of them were hermetic (less than 10^{-9} Pa•m^{3}s^{-1}) at room temperature. However, two of these packages (Fig. 28, curve-A and B) became fine leakers (in the 10^{-6} to 10^{-7} Pa•m^{3}s^{-1} range) at 120 °C (250 °F). The seals that leaked were identified by carefully spraying helium over the glass beads during the leak test. There was no visual damage to identify the leaking seals. This temperature

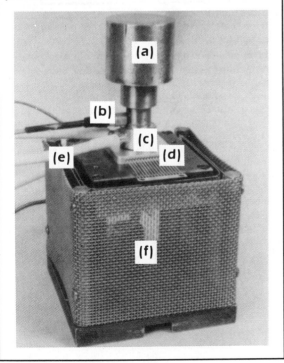

FIGURE 25. Thermal shock hot stage with a 25 by 25 mm (1 by 1 in.) hybrid package and a water cooled acoustic emission detector on top: (a) brass weight to hold the spring loaded detector against the package; (b) part of the lead and connector from the acoustic emission detector; (c) water cooling jacket around the detector; (d) hybrid package shown face down on the hot stage; (e) silicone rubber coolant tubing; and (f) hot stage surrounded by a protective screen

FIGURE 26. Expanded portion of acoustic emission detector in Figure 25 plus its cooler: (a) spring to press detector against hybrid package; (b) weight to hold entire detector assembly against package; (c) high temperature grease to establish thermal contact between cooled assembly and thin bottom of package and to serve as a couplant to the detector

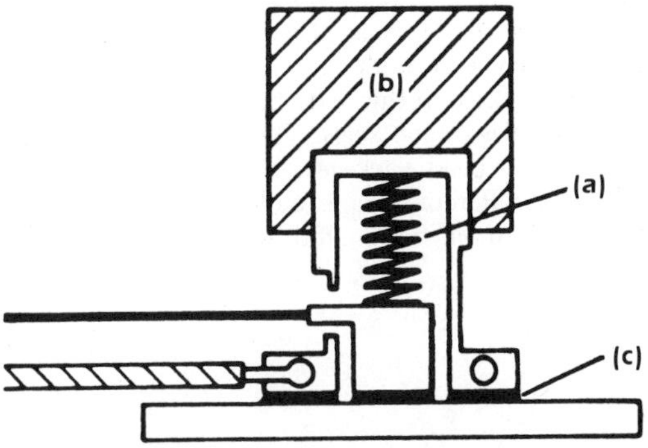

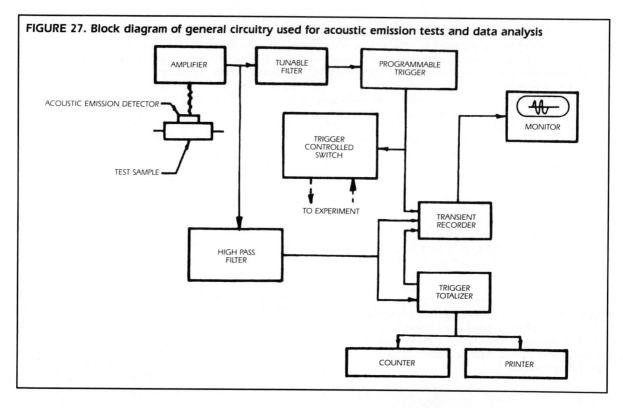

FIGURE 27. Block diagram of general circuitry used for acoustic emission tests and data analysis

is about the minimum at which the forces in the perimeter of the glass seals change from compression to tension. The available high temperature leak testing apparatus in its current configuration cannot go to a higher temperature. However, since the acoustic emission test detected seal damage in all of these packages, it is speculated that others would also leak if the test temperature could have been raised. Even if the others remained hermetic, a 50 percent detection rate for marginal packages is adequate to reveal seal problems on a lot basis.

The total number of packages tested in this experiment was relatively small. Forty-eight packages from three manufacturers involving six production lots were available for the work. Each package in this and other tests had 30 or more independent glass-metal seals. Thus, the total number of leads tested on good packages, where each seal may be considered a different sample under test, was statistically significant. However, the limited number of poor packages screened and found good was small and each one may have had only one or two potentially weak seals. A larger study, involving direct manufacturer participation, should be undertaken to verify results and to develop practical procedures and standards for test implementation.

This acoustic emission shock test is most effectively performed at the manufacturer's plant, on a lot sample basis, to ensure delivery of high quality lots. The verified Kaiser effect for this test renders results, obtained from an incoming inspection at a user's facility, somewhat less definitive unless the thermal shock history of the packages is supplied by the manufacturer. Without such information, a more time consuming high temperature leak test or a liquid-to-liquid shock test followed by a room temperature leak test must be used. The acoustic emission shock test can be performed more rapidly and more economically than either of the other tests. The ability of the acoustic emission test method to reveal marginal packages appears to be unique.

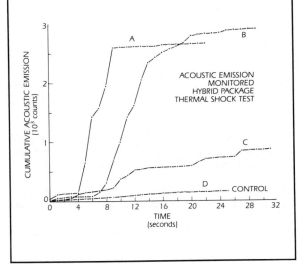

FIGURE 28. Cumulative acoustic emission counts versus time for the acoustic emission hot stage shock test; curves A, B and C are data from testing packages with weak seals; curve-D data are from a good control package and represent cumulative noise counts (typical 6 to 8 per second)

PART 6
EVALUATION OF FRACTURE IN INTEGRATED CIRCUITS WITH ACOUSTIC EMISSION

The long-term mechanical integrity of integrated circuits is a major concern in achieving device reliability. Because integrated circuits are composed of brittle materials such as silicon, glass and silicon nitride, discontinuities measuring a few micrometers can reduce mechanical strengths below tolerable levels. Discontinuities of characteristic dimensions are introduced by specific manufacturing processes and, when subjected to subsequent manufacturing stresses, have been correlated with particular device failure modes. For example, wafer dicing and the wafer backdamage treatment are two manufacturing processes that have been correlated with device failures.[10]

Failure mechanisms range from macroscopic fracture running completely through the device (die cracking)[10-12] microscopic cracks in one micrometer thick dielectric passivation layers.[13-15] The ability to evaluate specific manufacturing processes and their correlation with device reliability depends on mechanical tests that adequately simulate the stresses responsible for device field failures. A number of possible test configurations are explored here, with acoustic emission being used to monitor the specimen response to stress.

Acoustic emission has been applied to several manufacturing processes in the semiconductor and microelectronics industry. Examples include work on acoustic emission in alumina under compression loading.[16] These results led to an assembly monitoring system for the *in situ* detection of crack formation during thermocompression bonding of beam lead devices.[17] The difference between the acoustic waveforms for crack formation as opposed to the waveforms for slip has also been investigated for thermocompression bonding to GaAs single crystal substrates.[18-19] Crack and void detection in multilayer ceramic capacitors has been described in a paper concerning acoustic monitoring as part of a compression proof testing scheme.[20] During the soldering of larger capacitors, acoustic emission has also been used as an *in situ* monitor of the assembly process.[21] Additionally, several nondestructive tests have been developed involving acoustic evaluation of integrated circuit assembly steps (wire bonding, tab beam lead bonding and solder joints in flip chip integrated circuits).[22-25] Recent studies involving acoustic monitoring of electromigration tests have successfully correlated passivation crack formation with the extrusion of migrating aluminum into the passivation layer.[15]

The acoustic response of test specimens to three different stressing configurations are described below. Interpretation of crack initiation loads, crack propagation and final failure loads were facilitated by the evaluation of acoustic emission data. In the first configuration, the effect of various wafer processing steps on wafer strength is evaluated in biaxial bending tests of complete wafers. Some of the wafer processing steps analyzed were: the wafer backside damage process; thermal oxidation; and the deposition of phosphorus glass and silicon nitride films. Comparison of cumulative ringdown counts as a function of test load for various processing conditions gave insights into particular strength reduction mechanisms. Ideally, these plots of acoustic emission versus load could be used to characterize wafer fabrication processes with respect to mechanical integrity.

In a second experiment, localized stressing of 20×20 μm areas on one micrometer thick passivation layers was accomplished through Knoop, Vickers and spherical indentation with an ultramicrohardness microscope. In these experiments, the onset of acoustic emission was used to determine the resistance of various dielectric layers and passivation structures to crack initiation and propagation. This type of experiment proved to be useful in evaluating the fracture sensitivity of passivation structures, a property that is important in avoiding cracks in these structures due to plastic encapsulation stresses.[13-14]

A final experiment involved acoustic monitoring of the die shear test.[26] In these experiments, cumulative acoustic ringdown counts were used to monitor crack growth in the die attachment and interfacial layers. Acoustic response coupled with dye penetrant tests showed that failure in adhesively attached dice occurred through a stable crack growth mechanism, the die shear failure load being determined by the last phase of die separation.

Acoustic emission techniques can be applied to virtually any mechanical strength test. In the described experiments, acoustic emission data provided insights into deformation and fracture processes that would have otherwise gone unnoticed. The additional effort involved in collecting acoustic emission data is small. However, because of the sensitivity of the technique, the major difficulty is in the interpretation of results and the identification of acoustic emission sources.

Acoustic Emission Electronics

The experiments described below were done in controlled laboratory environments with a minimal amount of mechanical and electrical noise. A block diagram of the acoustic detection system is shown in Fig. 29. Evaluation of a range of piezoelectric transducers and high pass and bandpass filters eventually led to the use of a 100 kHz high pass filter and high sensitivity piezoelectric transducer for optimum signal sensitivity. In this configuration, the frequency response was peaked in the 100 kHz to 500 kHz range. Fixturing and acoustic isolation of the test sample were accomplished through the placement of high impedance materials (tape and silicon rubber pads) at the contact surfaces. Additionally, precautionary cleaning of sample debris was done prior to mounting of fresh test samples.

A standard, commercial viscous liquid coupling medium was used to decrease the acoustic impedance at contact boundaries in the path from test sample to transducer. Transducer output voltages in excess of 30 microvolts triggered the system and were processed in terms of cumulative ringdown counts. The relative magnitudes of individual acoustic events were also processed in terms of the maximum event amplitude. The event amplitude is defined as the transducer output voltage in decibels above a 7 μV reference level.

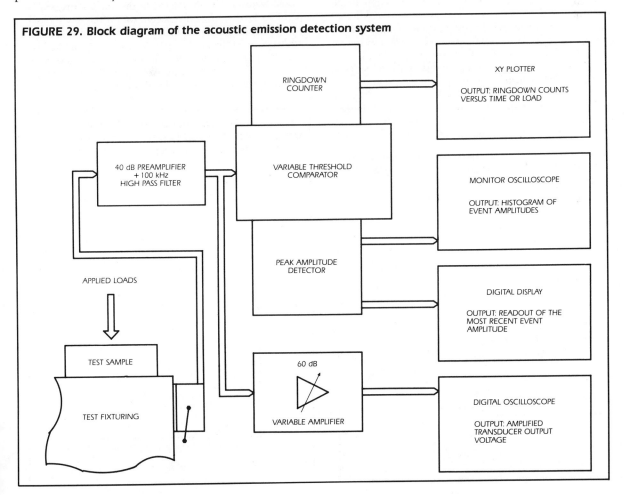

FIGURE 29. Block diagram of the acoustic emission detection system

Biaxial Wafer Bending Tests

Biaxial bend tests on 75 mm (3 in.) diameter wafers were performed with a circular stainless steel waveguide, to which piezoelectric transducers could be attached (see Fig. 30). The waveguide was placed, via an acoustic damping pad of silicone rubber, on top of the load cell of a commercial mechanical test system. Wafers 400 μm in thickness were placed face down on the waveguide and loaded from the top surface with a steel ball, 25 mm in diameter. Tape 75 μm in thickness was placed at the ball-to-silicon contact. Using a crosshead speed of 0.1 mm per minute, the duration of a test was typically 20 minutes.

In this configuration, the lower wafer surface was stressed in tension and the upper surface in compression. Because of this, fracture initiation occurs on the bottom surface of the wafer where tensile stresses are maximum. In Fig. 31, the effects of surface preparation on wafer fracture strength are demonstrated for unprocessed wafers from four different vendors. Wafers with the largest amount of back surface damage failed at the lowest loads. These wafers were also stressed with the polished wafer surface in tension. Due to the considerably smaller discontinuity sizes on this surface, the wafer fracture load was correspondingly larger.

The amount of acoustic emission during a wafer bending test can vary by several orders of magnitude as is demonstrated by the example of Fig. 32. The only difference between the wafers is a 50 nm sandwich layer of thermally grown SiO_2. The upper family of curves came from wafers in which a planar, 1 μm thick layer of plasma silicon nitride was deposited directly on silicon. The lower family of curves represent the standard process, in which a thin oxide is grown before deposition of the silicon nitride. Although no evidence of acoustic emission sources could be detected with either optical or scanning electron microscopy, the extensive acoustic emission in samples without the SiO_2 layer are believed to result from interfacial delamination. The sandwich SiO_2 layer promotes the adhesion of the silicon nitride and, during its growth at high temperature, residual stresses in the silicon are relieved by a simultaneous annealing. The noisy samples processed without the thin oxide layer failed at an average load of 120 ± 40 N, while the quiet samples failed at 230 ± 140 N.

In these and other biaxial bending experiments, final fracture occurs almost instantaneously due to the large amount of elastic energy stored in the wafer prior to failure. For this reason, acoustic activity related to the crack formation responsible for final fracture occurs within the few milliseconds prior to catastrophic failure. This behavior reflects the brittle nature of silicon. The statistical scatter in failure loads results from the distribution of natural discontinuities in the test sample.[27] Due to the nature of this final fracture, acoustic activity up until the final milliseconds prior to failure

FIGURE 30. Schematic diagram of the biaxial wafer bending experiment

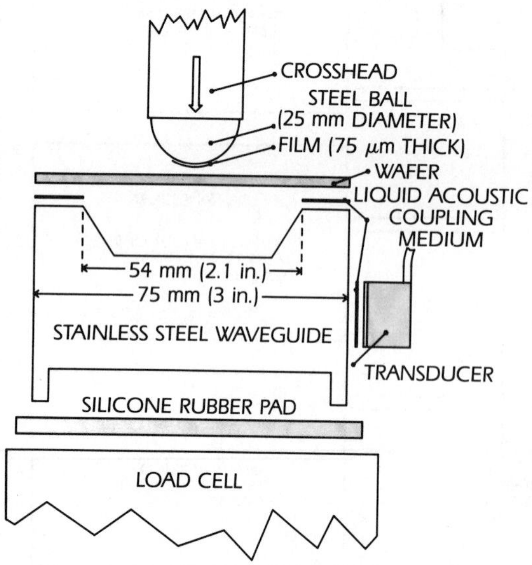

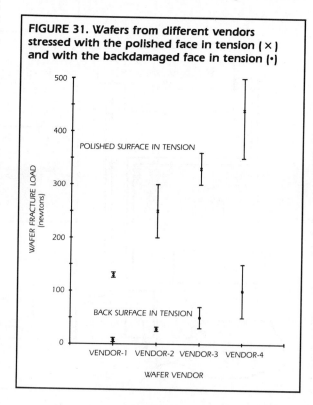

FIGURE 31. Wafers from different vendors stressed with the polished face in tension (×) and with the backdamaged face in tension (•)

could not be used to predict wafer fracture loads; the cumulative sum of acoustic ringdown counts within a particular wafer group typically varied by an order of magnitude but bore no relationship to the ultimate wafer fracture load. Acoustic response was, however, dependent on past loading history, always becoming less noisy after preloading.[28]

Wafer fracture experiments on unprocessed silicon wafers and on fully processed wafers showed that there was an approximately 70 percent drop in wafer fracture strength after completed device fabrication. In order to pinpoint the processing steps responsible for this reduction in wafer strength, individual processing steps were evaluated. The results of these tests are presented graphically in Fig. 33. In this figure, each data point represents the average wafer fracture strength and standard deviation for a minimum of four wafer bending tests. In order to isolate key fabrication processes, wafers were processed exactly as indicated. Particular wafer processing steps are listed in the order that they generally occur during actual device fabrication.

Wafer Backdamage Process

The wafer backdamage process is a process whereby discontinuities are created on the wafer's back surface, generating a surface layer with high residual stresses. This back surface layer attracts impurities such as iron, nickel, chromium and carbon which are mobile at temperatures in excess of 500 °C (930 °F).[29] The wafer backdamage process is necessary in the formation of bipolar devices and usually involves mechanical processes such as grinding or sandblasting. The mechanical strength of a brittle material like silicon depends strongly on surface finish, so a wafer strength test was used to evaluate the mechanical effects of backdamage processing. Figure 34 shows the biaxial bending strengths of backdamaged wafers obtained from four different vendors. Testing involved placing the wafer back surface face down on the bending fixture so as to stress the back surface in tension. The backdamage treatments involved sandblasting, the removal of debris and a light silicon etch. Although the backdamage processes were similar, profilometer measurements of surface roughness showed that backdamage grooove depths varied from vendor to vendor.

In an attempt to develop a more reproducible wafer back surface finish, a study was initiated to determine the feasibility of using laser beam irradiation for backdamage processing.[30,31] Within this program, the biaxial bending test was used to evaluate the effects of the laser energy density on wafer fracture strength. The back surfaces of 100 silicon wafers were irradiated with a scanning, repetitively Q-switched Nd-YAG laser system. Figure 35 shows the acoustic emission response versus load curves for wafers irradiated at 15, 20, 24 and 35 J·cm^{-2}. From this figure, it is evident that the wafer fracture strength is independent of laser energy density within the range examined. This is in contrast to acoustic emission response which is strongly dependent on the laser energy density. Scanning electron micrographs of cleaved cross sections from laser irradiated wafers showed evidence of microcracks down to a depth of 20 to 40 μm.[31] Increased acoustic activity at 24 J·cm^{-2} corresponds to the laser energy required for the evaporation of silicon.[31] After a 15 minute heat treatment at 1,050 °C (1,920 °F) in either O_2 or N_2, evidence of the healing of microcracks through a surface diffusion mechanism was found. The effect of a heat treatment on the wafer strength of irradiated and standard material is illustrated in Fig. 36. Laser irradiation leads to a significant reduction in the wafer fracture strength. However, after heat treatment, the healing of microcracks results in a recovery of the wafer fracture strength. In particular, heat treatment in an oxidizing ambient resulted in a complete recovery of the wafer fracture strength.

Stepped Passivation Layers

Biaxial bend testing of fully processed wafers has shown about a 70 percent drop in wafer fracture strength with respect to the starting material. Unlike backdamage evaluations, wafer processing strength tests were carried out with the device surface of the wafer in tension. Evaluation of the

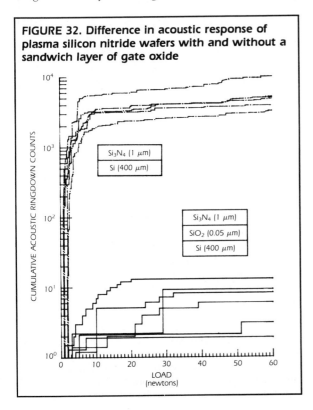

FIGURE 32. Difference in acoustic response of plasma silicon nitride wafers with and without a sandwich layer of gate oxide

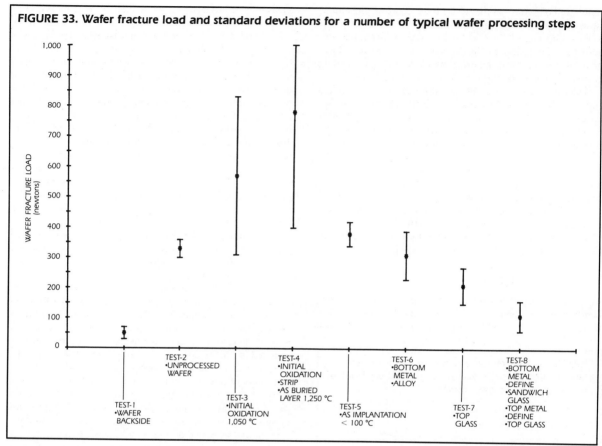

FIGURE 33. Wafer fracture load and standard deviations for a number of typical wafer processing steps

individual fabrication processes has implicated stepped passivation layers over metal traces as the primary factor in strength reduction. In fact, processing steps such as ion implantation, diffusions, thermal oxidations and planar CVD phosphorus glass (P-glass) or metal depositions resulted in either general increases or no change in the wafer fracture strength. It is not until the passivation layer (P-glass or silicon nitride) is deposited over metal traces that the wafer strength drops to the levels found in fully processed wafers. This acute drop in fracture strength is demonstrated by comparing test-8 with test-2 in Fig. 33. Similar results on single and double layer metal structures with P-glass or silicon nitride passivations are shown in Fig. 37. Optical and scanning electron microscope inspection of the fracture surfaces of tested samples suggest that crack initiation occurs along the edges of metal traces. Finite element calculations of the stresses in such stepped metal and passivation layer structures[14,32] have indicated that regions of high tensile stress exist in the layers adjacent to the corners of metal traces (see Fig. 38). At these stress concentrations, microcracks may occur in the dielectric layers in the vicinity of the metal trace. This results in inferior passivation and possible electrical shorts between metal traces. To investigate the fracture sensitivity of dielectric layers, a microscopic technique of stress testing was developed and is described in the next section.

Diamond Indentation Experiment

To determine the fracture sensitivity of dielectric films used for device passivation, diamond indentation was chosen as the method of stressing, because it can be applied selectively to specific device features. The resulting cracks are localized, stable and convenient to study. The indentations were made with a microscope equipped with a microhardness tester. Figure 39 shows the Knoop, Vickers and spherical (radius 12.5 μm) indenters used with this system. Test samples were mounted on top of the transducer (Fig. 40) with a thin layer of coupling fluid to give a low acoustic impedance and a simple attachment. Site selection and

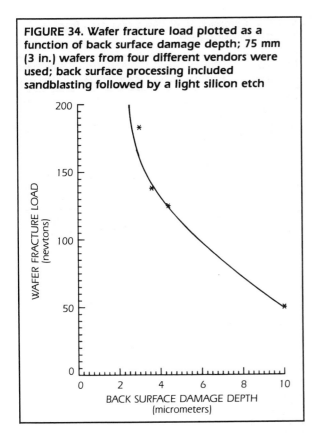

FIGURE 34. Wafer fracture load plotted as a function of back surface damage depth; 75 mm (3 in.) wafers from four different vendors were used; back surface processing included sandblasting followed by a light silicon etch

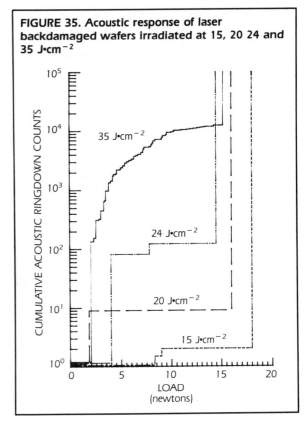

FIGURE 35. Acoustic response of laser backdamaged wafers irradiated at 15, 20 24 and 35 J·cm^{-2}

length measurements were done at an optical magnification of 760×. Test loads ranged from 10 to 500 mN (1 to 50 g) and were controlled manually with a gear reduction mechanism connected to the fine focus knob of the microscope. The acoustic instrumentation was configured to generate plots of cumulative counts as a function of either load or time.

Knoop Indentation

Knoop indentation was done on 0.7 μm silicon nitride films deposited under varying conditions of temperature, silane ratio and atmospheric pressure. The indentation site was an area of the integrated circuit where the nitride layer was deposited on top of 1 μm of P-glass, 0.5 μm of thermal oxide and the substrate of silicon. The typical acoustic response of silicon nitride to loading with the Knoop indenter is shown in Fig. 41.

The critical stress for crack formation is built up during the initially quiet part of the loading curve and culminates in crack formation as detected by a single high amplitude pulse of 50 to 60 dB. Subsequent pulses are believed to be related to the propagation of this crack and were generally of a lower amplitude (30 to 40 dB). The critical load at which this initial, high amplitude acoustic pulse occurred was a systematically repeatable feature of these experiments and appeared to depend on the silicon nitride deposition temperature (see Fig. 42). For films deposited at 280, 340 and 425 °C (540, 640 and 800 °F), penetration of the Knoop indenter into the silicon nitride film at the moment of crack initiation was approximately 25 percent, 20 percent and 17 percent of the film thickness, respectively. Within the process range examined, silane ratio and gas pressure appear to have a negligible effect on the fracture initiation load. Measurements of the as-deposited stress level in these films indicate that all films were compressive. The magnitude of this compressive stress was found to rise with increasing deposition temperature and decreasing silane ratio.[33]

Vickers Indentation

The crack made by indentation with a Knoop diamond is a vertical subsurface crack along the main diagonal.[34] Microscopic inspection of this crack is difficult because of crack closure in the plastic deformation region directly underneath the indenter. In a second series of experiments, a

Vickers indenter was used to enable crack length measurements. With the Vickers indenter, radial surface cracks at the indentation corners are directly visible and have been correlated with acoustic emission in experiments on glass, silicon and a number of other materials.[35,36]

Vickers indentations were made on three sample structures: bulk silicon, silicon nitride films and P-glass films. The silicon nitride and glass samples were chemical vapor deposition (CVD) films deposited directly on unprocessed silicon wafers. In the bare silicon and the silicon nitride on silicon samples, the amplitude of initial pulse could not be distinguished from other pulses; in loading up to 20 grams, all pulse amplitudes remained in the 30 to 40 dB range. The general shape of the acoustic emission curves for the P-glass samples were similar to those of the Knoop indentations; the initial pulse had the largest amplitude ($\sim$ 57 dB) and was followed by 30 to 40 dB pulses. The critical load at which the initial acoustic pulse occurred, the amplitude range of this acoustic pulse and the measured stresses in various thin films are shown in Table 1. The film stress data were obtained by calculating the wafer curvature from Fizeau fringes.

For the P-glass samples with 13 percent, 15 percent, 17 percent and 19 percent P_2O_5, the measured compressive film stress showed a decrease with increasing P_2O_5 concentration. Diamond penetration into the glass films at the moment of crack initiation was about 70 percent of the film thickness. The threshold load of 80 mN (8 g) for crack formation in bare silicon is in reasonable agreement with the value of 50 mN (5 g) reported elsewhere.[35] The presence of a 0.7 μm nitride film reduces this value to 20 mN (2 g). This is similar to the Knoop indentation results of Fig. 42. The silicon nitride film in the present case was deposited at 425 °C (800 °F).

Both the Knoop and Vickers indentation experiments show, to some extent, the tendency of a reduction in acoustic threshold load with increasing compressive stress in the top surface film. At present the mechanisms of crack initiation are not well understood. Recent studies on soda lime glass[37,38] describe subsurface deformation and the interaction of shear flow near indenter edges as the nucleation mechanism, but more work needs to be done to determine if a similar mechanism occurs in the thin films tested here. While crack initiation seemed to occur more readily with a compressive stress in the top layer film, the K_{Ic} measurements described below indicate that the load for crack extension in the same samples increased for compressive film stresses.

Once a microcrack is formed, its extension under an applied stress is well described for brittle materials by means of fracture toughness K_{Ic}. The fracture toughness is used to determine the relationship between crack size c and stress level σ_c for which crack extension will occur.[27] This relationship generally has the form:

TABLE 1. Comparison of film stress data from acoustic emission tests

Thin Film on Silicon	Critical Load (millinewtons)	Acoustic Amplitude	Film Stress (MN/m^2)
13% P_2O_5	121 ± 16	53 ± 2	−130
15% P_2O_5	142 ± 7	56 ± 1	−60
17% P_2O_5	135 ± 11	58 ± 1	>−40
19% P_2O_5	134 ± 4	59 ± 1	$\simeq$0
Si_3N_4	20 ± 5	35 ± 5	−300
Bare silicon	86 ± 23	35 ± 5	

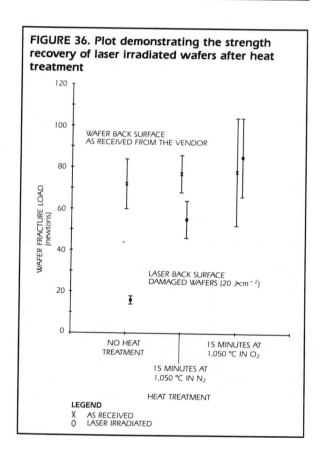

FIGURE 36. Plot demonstrating the strength recovery of laser irradiated wafers after heat treatment

$$\sigma_c = A \frac{K_{Ic}}{\sqrt{\pi c}} \quad \text{(Eq. 3)}$$

where A is a geometrical factor that accounts for crack shape and the type of applied stress field. The fracture toughness is a material property of dimensions $N \cdot m^{-3/2}$ obtained from

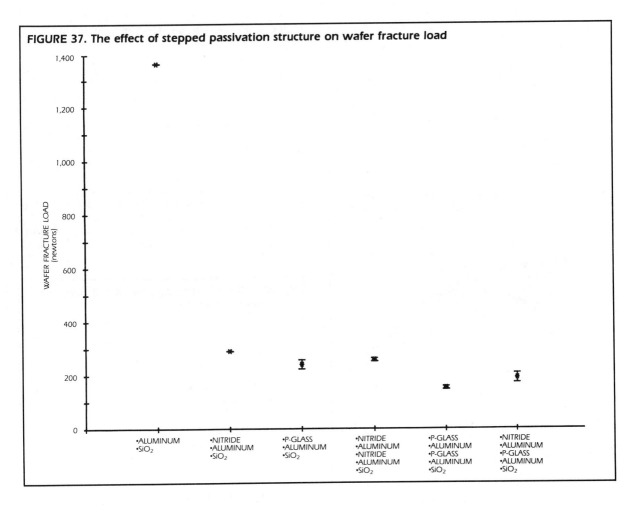

FIGURE 37. The effect of stepped passivation structure on wafer fracture load

a controlled fracture experiment.[27] To obtain K_{Ic} from an indentation experiment, the lengths of the radial cracks at indentation corners are measured as a function of applied load.[39-41] To this end, each sample type was indented with a load of 100, 200, 300, 400 mN (10, 20, 30, 40 g) and, in some cases, 500 mN (50 g). Usually, at these low loads, crack extension was not visible until a load of 300 mN (30 g). At the low loads used, cracks at the four corners of the indentation developed independently and were of different lengths. This suggests that they were shallow Palmquist cracks[41] that had not yet developed into radial cracks.

The first step in the determination of K_{Ic} involves plotting the 3/2 power of the crack length as a function of the indentation load. This is shown in Fig. 43 where c is the crack radius (micrometers) and P is the indentation load (millinewtons). From the inverse slope $(P/c^{3/2})$, K_{Ic} can be determined from Eq. 4.[39,41]

FIGURE 38. Regions of high tensile stress exist at the edges of metal traces; these areas are denoted by the arrows

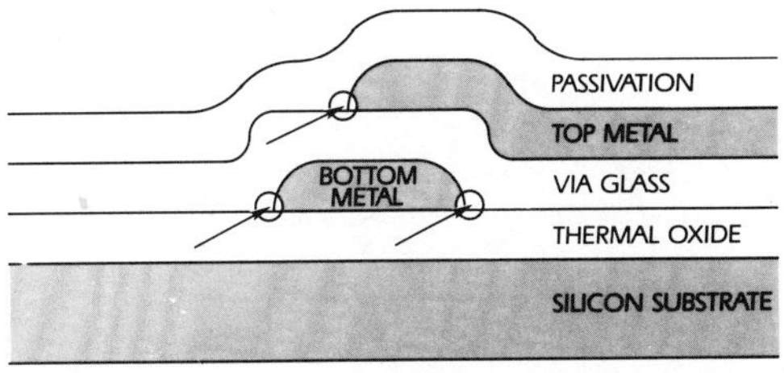

$$\frac{K_{Ic}\phi}{H\sqrt{a}}\left(\frac{H}{E\phi}\right)^{0.4} = 0.142\left(\frac{c}{a}\right)^{-3/2} \qquad \text{(Eq. 4)}$$

Where:

H = Vickers hardness (newtons per meter squared);
E = the elastic modulus (newtons per meter squared);
ϕ = a geometric factor; and
a = the indentation diagonal length (meters).

FIGURE 39. The geometry of the Knoop, Vickers and spherical diamond indenters

LEGEND
d = diagonal
t = indentation depth

For the Vickers indenter, $\phi \sim 3$, $H = 0.046\ P/a^2$ and, in $N{\bullet}m^{-3/2}$, Eq. 4 becomes:

$$K_{Ic} = 3.38 \times 10^{-1}\left(\frac{E}{H}\right)^{0.4}\frac{P}{c^{3/2}} \qquad \text{(Eq. 5)}$$

Because the elastic stress field extends far into the silicon substrate, the elastic modulus of silicon ($E = 169 \times 10^9$ $N{\bullet}m^{-2}$) can be used. Experimentally determined values for the hardness, the thin film stress, and the critical stress intensity factor are shown in Table 2. The measured value of K_{Ic} for silicon is comparable to values quoted in the literature.[40,42-45]

During crack extension experiments, the diamond indenter penetrates the silicon substrate and crack extension depends both on the film and substrate properties. The thin films studied here were all in a state of compressive stress. The CVD glass films with phosphorus doping are tensile directly after deposition but become compressive due to the

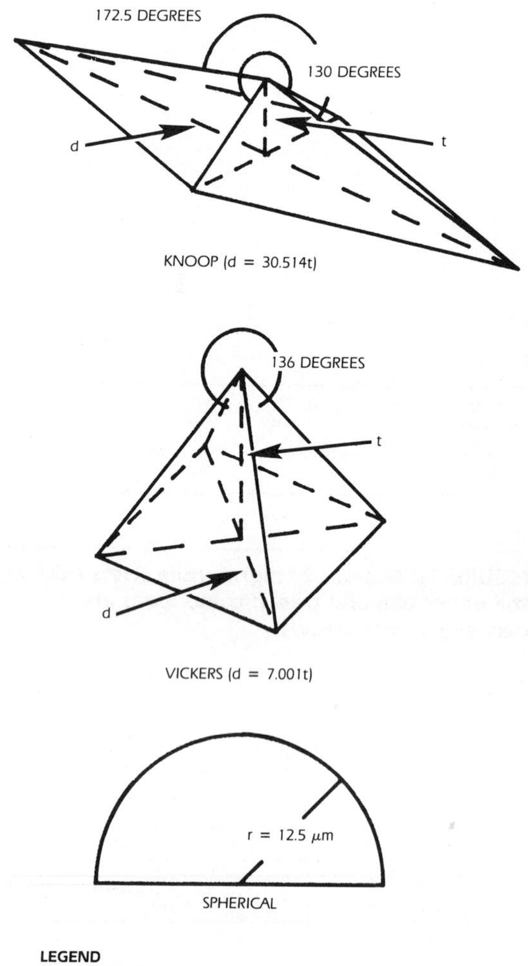

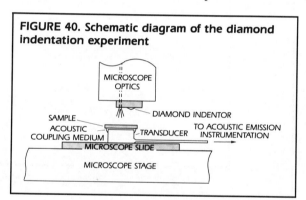

FIGURE 40. Schematic diagram of the diamond indentation experiment

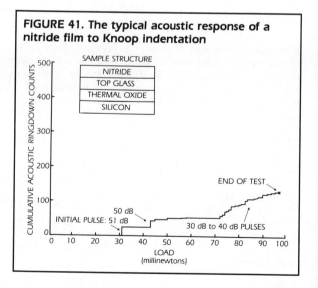

FIGURE 41. The typical acoustic response of a nitride film to Knoop indentation

absorption of water from the atmosphere.[46,47] During diamond indentation, the induced tensile stress field is partially compensated for by the compressive stress in the top layer film. As a result, a higher indentation load is needed for crack extension. Both stress fields may be superimposed to yield the relation[48] shown in Eq. 6.

$$\frac{\sigma p}{c^{3/2}} = K_{Ic} - 2\psi\sigma_f \sqrt{d_f} \qquad \text{(Eq. 6)}$$

In this equation, σ and ψ are geometric constants that depend on the indenter and the crack shapes, respectively, and d_f is the film thickness. The correction to fracture toughness, resulting from stress in the thin film, σ_f, is taken into account in the second term on the right. This equation is

TABLE 2. Comparative data for stress tests using the wafer curvature method

Thin Film on Silicon	Knoop Hardness (GN/m²)	Film Stress (MN/m²)	Film Stress (MN/m^{3/2})
13% P_2O_5	10.5	−130	0.54
15% P_2O_5	12.1	−60	0.59
17% P_2O_5	11.2	>−40	0.47
19% P_2O_5	10.9	≃0	0.42
Si_3N_4	12.2	−300	0.64
Bare silicon	10.7*		0.49**

* Literature data: 9.0 to 11.0 MN/m² (reference 40 and 42)
** Literature data: 0.5 to 0.7 MN/m^{3/2} (reference 40 and 42 to 45)

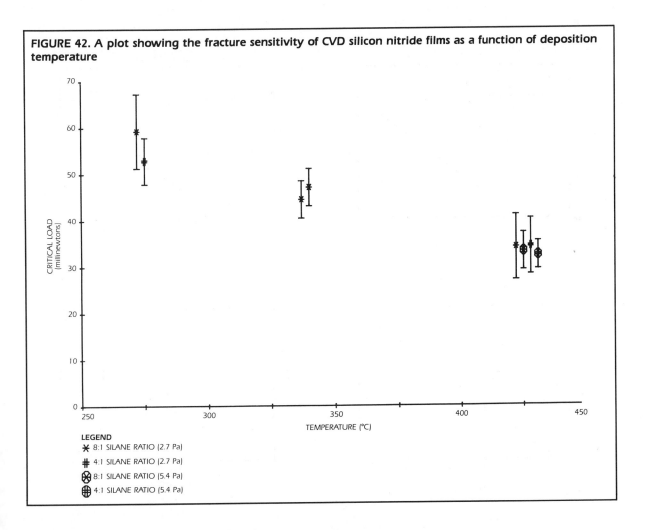

FIGURE 42. A plot showing the fracture sensitivity of CVD silicon nitride films as a function of deposition temperature

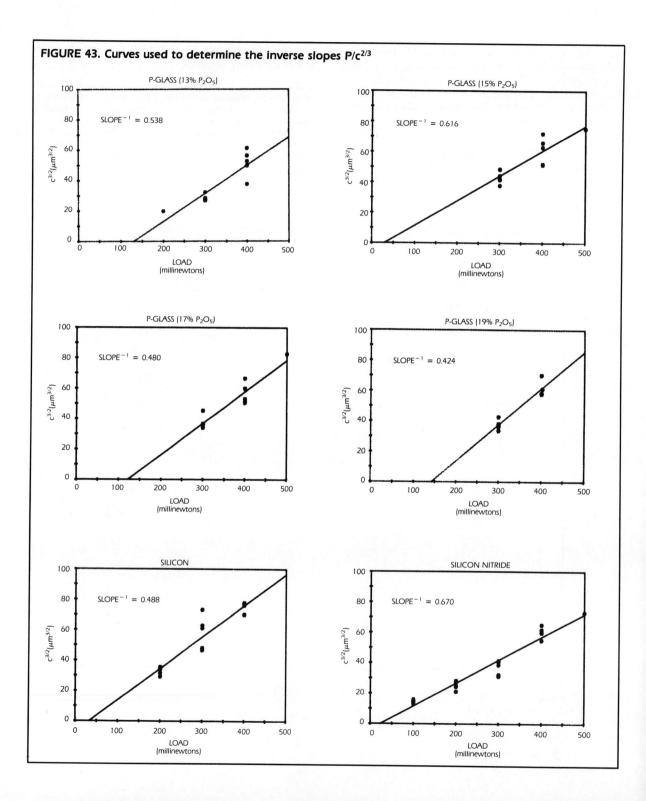

FIGURE 43. Curves used to determine the inverse slopes $P/c^{2/3}$

valid when the thin film is small compared with the crack length. The equation could be used to calculate thin film stresses without resorting to the wafer curvature method used for the stress data in Table 2. However, at the low loads used, cracks at the four corners of the indentation varied in length, so the present set of data is too limited to match this relation accurately. Still, the trend in measured K_{Ic} values demonstrates the influence of compressive film stress on crack resistance.

Spherical Indentation of Passivation/Metal Layer Structures

To test the fracture resistance of passivation films on top of single or double layer metallization, the spherical diamond indenter was chosen. With the spherical indenter, a compressive stress field is set up to a depth on the order of the contact diameter. Large tensile stresses occur along the outer perimeter of the contact circle and have a maximum value at the surface of the sample.[49] A spherical indentation test is therefore more sensitive to surface discontinuities than the Vickers or Knoop tests. For these experiments, a diamond profilometer stylus with a radius of 12.5 μm replaced the normal diamond indenter in the microhardness microscope. For a 50 mN load on a glass substrate, appropriate equations for Hertzian loading[49] result in an expected contact diameter of 3.5 μm. This gives a suitable compression loading of the approximately one micrometer thick dielectric layers covering the single and double layer metal structures examined.

Initial experiments indicate that the acoustic emission response during Hertzian loading could be used to evaluate the relative fracture sensitivity of various passivation films on top of ductile metal layers. Passivation structures were ranked by the load at which the acoustic pulse occurred with the maximum amplitude. Unlike Knoop and Vickers tests where this quantity was typically the initial pulse, numerous 30 to 40 dB pulses occurred prior to the solitary high amplitude pulse in the 50 dB range. Examples of a typical high and low amplitude pulse are shown in Fig. 44.

To verify that acoustic emission is related to cracking in the passivation layer, samples were chemically etched and inspected with a high power interference microscope. Figure 45 shows representative acoustic emission plots and the corresponding etching features of an acoustically active and an acoustically quiet sample. Drops of diluted hydrofluoric acid (ten parts water to one part hydrofluoric acid) were placed at indentation sites for ten minutes. This resulted in the delineation of microcracks in the nitride layer by etching in the underlying P-glass layer.

In another sample structure (P-glass on top of aluminum), crack delineation was accomplished with an aluminum etch of 15 minute exposure to a solution of five parts water to one part sodium hydroxide. Figure 46 shows the cumulative acoustic ringdown counts for this structure at a load of 20 mN (2 g) for eight indentations. For all indentations in which cracking was visible, acoustic emission was generated.

There were indentations, however, for which cracking could not be found, and yet limited acoustic emission was detected. Presumably, crack arrest occurred within the top glass layer, so that the chemical etchant could not reach the aluminum layer. In the instances where etching did occur, the extent of visible cracking was found to be related to the cumulative sum of acoustic ringdown counts for samples taken to the same load. However, at relatively higher load levels, where cracking can be seen without resorting to chemical etching, it was observed that the indentation load was a more accurate measure of the extent of cracking than the cumulative sum of acoustic ringdown counts. This is believed to be due to reduced levels of acoustic energy generated during stable crack extension in the latter stages of spherical indentation loading.

In Fig. 47 various passivation layer structures are ranked in terms of their maximum amplitude load. This method of ranking suggests that: (1) CVD glass layers are more resistant to cracking than silicon nitride layers; (2) increasing the hardness of the metallization by alloying the aluminum with copper inhibits fracture in adjacent dielectric layers; and (3)

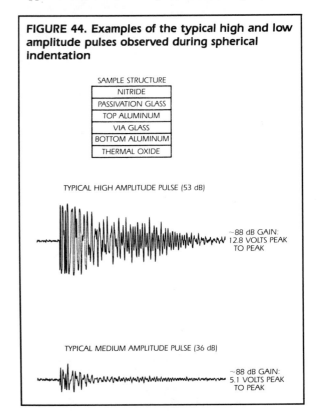

FIGURE 44. Examples of the typical high and low amplitude pulses observed during spherical indentation

addition of dielectric layers (increasing the dielectric thickness) leads to an increase in the maximum amplitude load. These three factors suggest that the low yield stress of the metallization is the main cause for differences in the maximum amplitude load. Spherical indentation on similar structures without the underlying soft metal layer have much higher critical loads (greater than 1 N). Spherical indentations were also done on device features that had double layer metallization structures. It was found that only the dielectric layers above the top metal layer influenced the maximum amplitude load. This point is demonstrated in Fig. 48 where similar double and single metal structures are grouped together for comparison.

Stable Crack Growth During Die Shear Testing

The final mechanical test described here is the die shear test. In the assembly of integrated circuits, die attachment is the step of mounting the silicon device (die) to a substrate. For plastic encapsulated devices, adhesives are commonly used for die attachment to metal substrates (lead frames). For ceramic packages, a eutectic solder, usually AuSi, is used to mount the die to a ceramic substrate. The die shear test is an extensively used industry monitor[50] whose purpose is to check the mechanical integrity of the die attachment layer (see Fig. 49). Using acoustic emission techniques, the onset and propagation of fracture during the die shear test was analyzed.[26]

The shear test fixture was designed around a linear translation stage (see Fig. 50) and universal testing machine. Anvil displacements were measured with a transducer and acoustic emission was detected with a piezoelectric transducer mounted on the back plane of the cylindrical sample

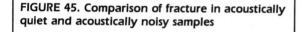

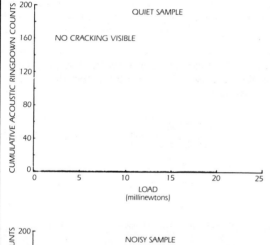
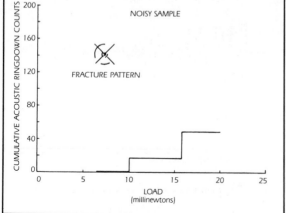

FIGURE 45. Comparison of fracture in acoustically quiet and acoustically noisy samples

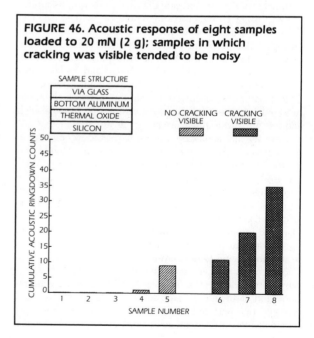

FIGURE 46. Acoustic response of eight samples loaded to 20 mN (2 g); samples in which cracking was visible tended to be noisy

support mount. Cumulative acoustic ringdown counts, shear force and anvil displacement were plotted during the test and stored on a desktop computer. The die shear tests involved 3 × 3 mm² dice, bonded with a silver loaded polyimide adhesive, dried at 160 °C (320 °F) for 10 minutes then cured at 270 °C (520 °F) for 10 minutes. Due to the fact that bonding was to alloy-42 instead of copper lead frames, die shear strengths were significantly increased (by about a factor of four). This is due to the relatively smaller thermal expansion mismatch between silicon and alloy-42 which leads to a smaller prestress of the die attachment layer. In addition to increased failure loads, the silicon to alloy-42 adhesive die bond also exhibited increased acoustic activity when compared with the silicon-to-copper adhesive die bond.

Features typically observed in the shear force versus anvil displacement plots are shown in Fig. 51. Settling of the fixturing and the plastic compression of the tape on the shear anvil are responsible for the slow loading evident during the early stages of the die shear test (0 to 20 N). Thereafter, the loading rate steadily increases until final (or catastrophic) fracture. The decrease in loading rate, beginning at 145 N, is indicative of yielding in the adhesive layer just prior to catastrophic die shear failure.

A more sensitive measure of fracture in the adhesive bond appears to be contained in the acoustic ringdown counts curve. Three acoustic bursts at approximately 10 N are probably related to the formation of a minute crack in the adhesive layer, just underneath the die, at the anvil contact

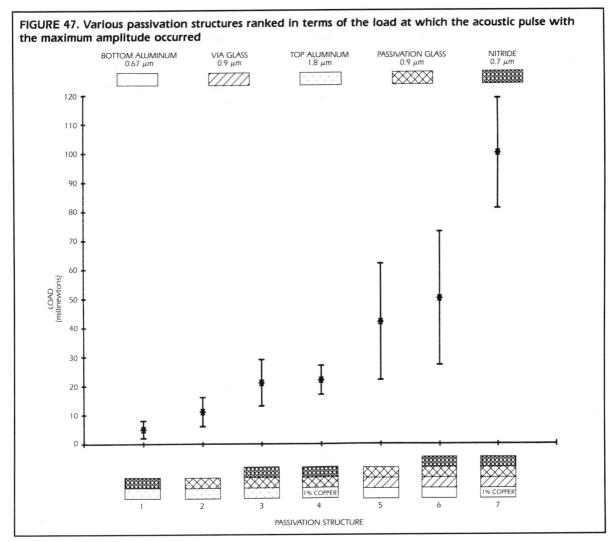

FIGURE 47. Various passivation structures ranked in terms of the load at which the acoustic pulse with the maximum amplitude occurred

edge. Propagation of this minute discontinuity does not occur until approximately 110 N, when high amplitude continuous acoustic emission begins.

To determine whether acoustic emission really was an indication of die shear fracture, dye penetrant experiments were done on samples that were loaded but not taken completely to failure. Figure 52 shows the shear force and acoustic emission curves for one of these samples. Each sample was soaked in dye penetrant for half an hour then lightly rinsed in acetone. The sample was then reloaded and taken to complete failure in a second test. Examination of the resulting fracture surfaces under ultraviolet light is shown in Fig. 53. The fluorescent regions indicate that 30 percent of the adhesive had fractured during the first loading cycle. This sample emitted 2,714 acoustic ringdown counts prior to unloading. In contrast, another sample (Figs. 54 and 55) emitted only 60 counts and showed only minute cracking. Interestingly, the quiet sample was loaded to 260 N whereas the partially fractured sample was only loaded to 148 N. Both of these samples showed yielding in the load displacement curve prior to unloading. These results indicate that stable cracking is the mechanism by which the polyimide bonds failed.

Die shear failure load plotted against adhesive thickness at the anvil contact edge is given in Fig. 56. While there is a significant scatter in the data, there appears to be a trend

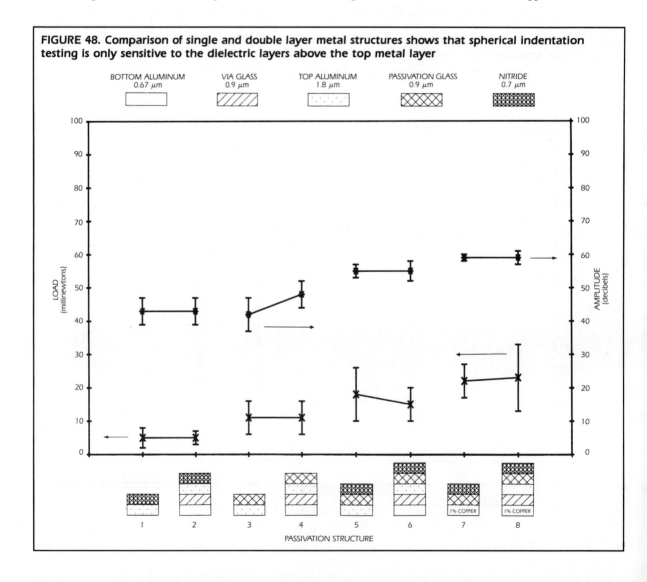

FIGURE 48. Comparison of single and double layer metal structures shows that spherical indentation testing is only sensitive to the dielectric layers above the top metal layer

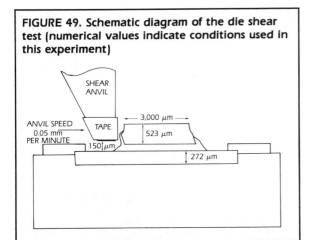

FIGURE 49. Schematic diagram of the die shear test (numerical values indicate conditions used in this experiment)

toward increasing shear strength with increasing adhesive thickness. Figure 57 is a repeat of Fig. 56 except that the starred points indicate the loads at which 500 acoustic emission counts occurred. The scatter from linearity in the acoustically defined failure load appears to be less than that for the normally defined failure load. In the four samples, where failure occurred at loads higher than might be expected, acoustic activity begins at loads more normal for that thickness. This suggests that adhesive thickness is an important parameter in die shear crack propagation. In cooling from curing temperatures, thick adhesive layers should be in a state of lower stress than thin layers because of increased viscoelastic creep. This may, to some extent, explain some of the strength differences between thick and thin samples. Another factor is that the stress concentration at the adhesive anvil edge might be reduced for thicker adhesive layers. This behavior is known to occur in lap shear joints.[51,52]

FIGURE 50. Photograph of the shear test fixture; for clarity the shear anvil and sample support have been repositioned for better visibility

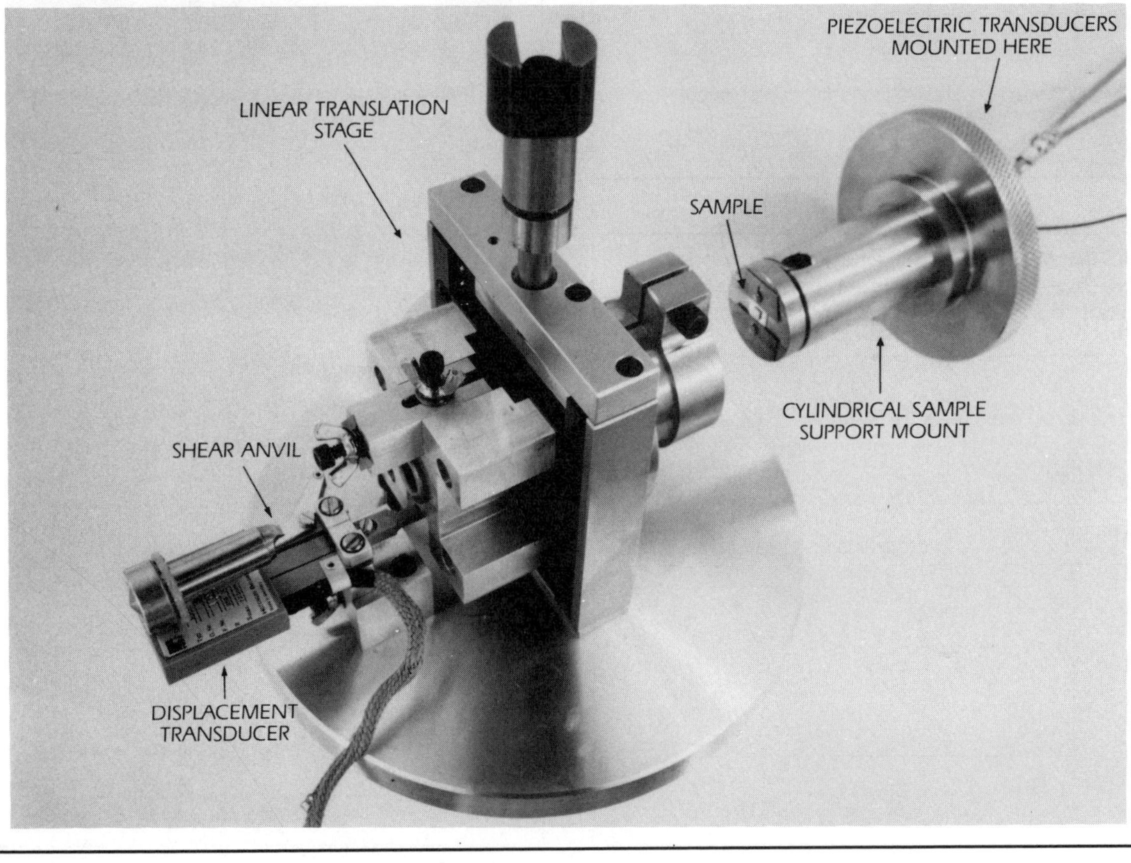

Summary

Acoustic emission has been used to detect crack initiation and propagation in a variety of strength tests applicable to the manufacture of integrated circuits. In wafer level biaxial bending tests, the effects of wafer processing steps on wafer fracture strength were examined. It was found that stress concentrations in the vicinity of passivation/metal edges were responsible for a large drop in the wafer fracture strength. Wafer back surface damage processing was also found to be detrimental to the wafer fracture strength.

FIGURE 51. Typical load displacement and acoustic response curves for the silver-loaded polyimide alloy-42 lead frame system

FIGURE 52. Load displacement and acoustic response curves for a sample loaded but not taken to failure (acoustically noisy sample)

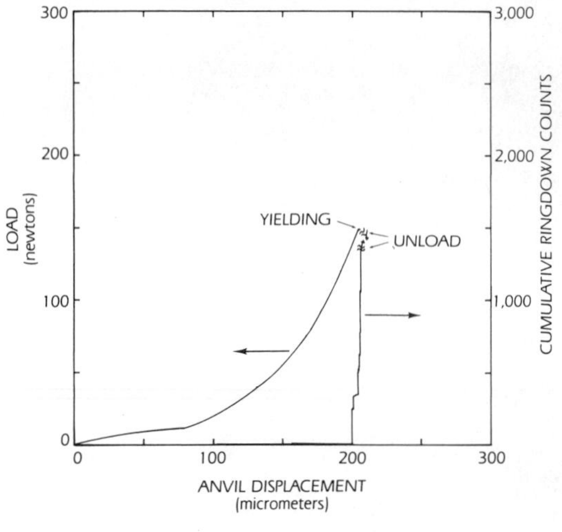

FIGURE 53. Dye penetrant results on an acoustically noisy sample taken to 145 N before unloading

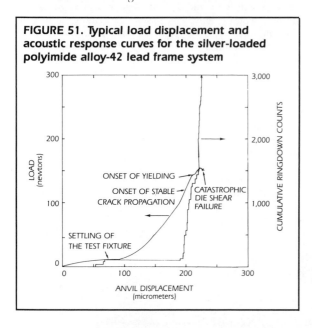

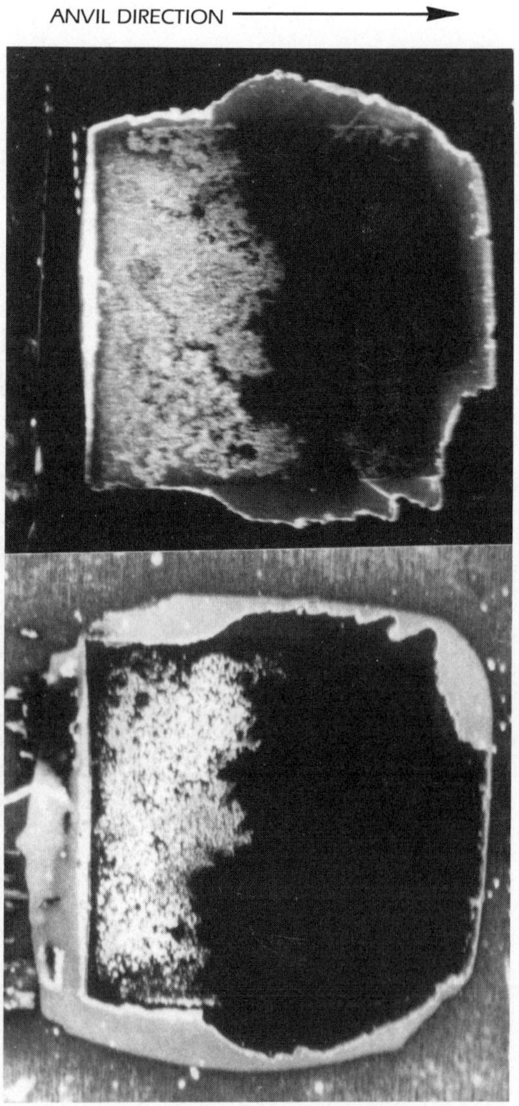

Processes such as arsenic implantation or thermal oxidation resulted in a compressive top layer in the silicon wafer and were found to increase the wafer strength. Substantial strength recovery also resulted from the annealing of microdamage during high temperature processing. Generally, acoustic emission occurred as a sequence of random bursts that (except for the few milliseconds prior to failure) bore no relationship to the ultimate wafer fracture load.

In diamond indentation tests on thin dielectric films, acoustic emission was used to determine crack initiation loads. In Knoop diamond indentation experiments, increasing the processing temperature from 280 to 425 °C (540 to

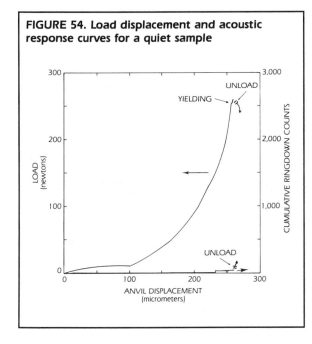

FIGURE 54. Load displacement and acoustic response curves for a quiet sample

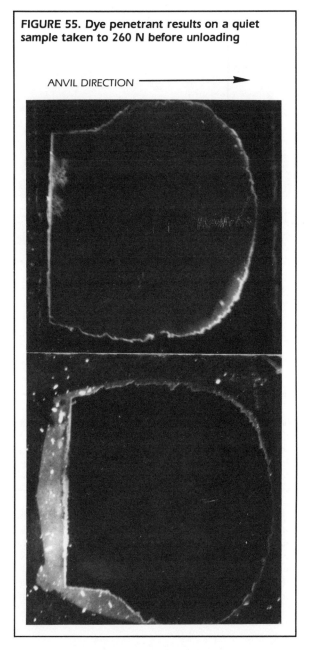

FIGURE 55. Dye penetrant results on a quiet sample taken to 260 N before unloading

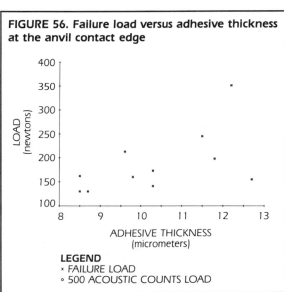

FIGURE 56. Failure load versus adhesive thickness at the anvil contact edge

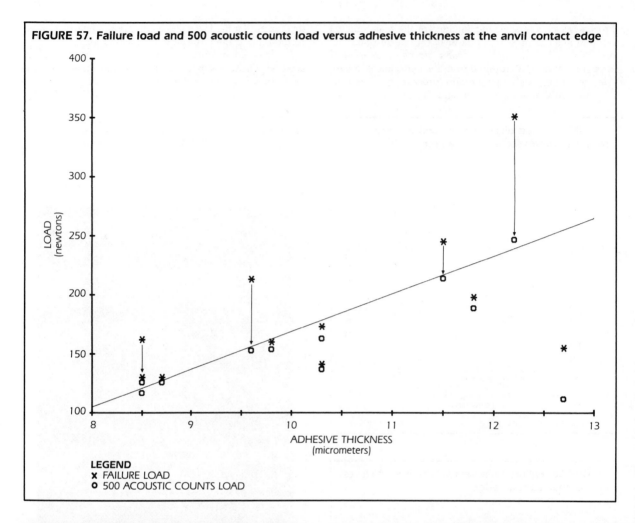

FIGURE 57. Failure load and 500 acoustic counts load versus adhesive thickness at the anvil contact edge

LEGEND
× FAILURE LOAD
○ 500 ACOUSTIC COUNTS LOAD

800 °F) in nitride films resulted in an apparent decrease in the crack initiation load. Vickers indentation experiments showed that increasing the P_2O_5 concentration in CVD glass films resulted in a slight increase in the crack initiation load. Measurements of the thin film stress indicate that, in both cases, an increase in compressive stress level corresponded to reduced crack initiation loads.

In contrast to this, after crack initiation, the compressive top layer inhibited crack propagation at higher diamond indentation loads. In spherical diamond indentation experiments on various passivation layer structures, it was shown that increased dielectric structure thickness and increased metal hardness resulted in an improved resistance to fracture. In these spherical indentation experiments, P-glass layers were also shown to be more resistant to fracture than silicon nitride layers.

Acoustic monitoring of the die shear test was used to show that shear failure in the silicon to alloy-42 adhesive bond involved a stable crack mechanism. In these experiments, it was shown that acoustic emission could be used to assess the extent of bond fracture prior to catastrophic die shear failure.

To conclude, acoustic emission provided additional information about the progress of failure in three mechanical test methods described. In full wafer experiments, all sources of acoustic emission were not identified but the test method differentiated wafers with differences in processing history. By using acoustic emission for crack initiation detection, indentation fracture offers a new method to test fracture sensitivity of thin dielectric layers. In nonbrittle tests such as the die shear test, acoustic emission can be implemented to monitor crack growth.

PART 7
ACOUSTIC EMISSION DIAGNOSTIC AND MONITORING TECHNIQUES FOR SUPERCONDUCTING MAGNETS

The application of acoustic emission technology to monitoring and diagnostic techniques for superconducting magnets includes: (1) identification of quench-causing mechanical disturbances; (2) quantitative relationship between motion-induced acoustic activity and an analytical model based on frictional motion in the magnet winding; (3) normal zone detection; and (4) energy quantification. In addition, acoustic emission sensors specifically developed for magnet use in the cryogenic environment are discussed below.

The Phenomenon of Superconductivity

The reduction of electrical resistivity with a decrease in temperature is a property of metals that has long been known. As the temperature of a metallic crystal is lowered, the thermal vibrations are reduced, causing the conduction electrons to become less scattered, thus allowing current to pass through the material with less resistance. This effect is linear down to about one-third the Debye temperature for a given material, at which point the electron scattering by impurities becomes dominant. Further cooling results in a less rapid reduction of electrical resistivity.

Speculation about the low temperature resistivity of metals has for many years led researchers to perform experiments at the lowest temperatures technically achievable. The successful liquefaction of helium by Kamerlingh Onnes in 1908 allowed him to perform experiments at a few kelvins above absolute zero. In 1911, while investigating the electrical behavior of pure mercury, Onnes discovered a complete and sudden loss of resistivity as the material was cooled to about 4 K. Further research has shown that dozens of elements and many alloys exhibit this phenomenon when cooled to the critical temperature T_c specific to each material.

Superconductivity was the name coined for the special behavior exhibited by many materials when they are cooled to extremely low temperatures. Such materials are called *superconductors* and are said to be superconducting when below the transition state and are said to be normal when above the transition. Although zero resistance is the best known property of superconductors, other striking electrical and magnetic properties distinguish a superconductor from a material which is simply very conductive.

Hopes of applying the newly found property of zero resistance to the construction of large electromagnets were soon dashed with the discovery that the property disappeared in the presence of magnetic fields. Just as there is a maximum tolerable temperature at which the phenomenon occurs, there is also a maximum critical field H_c to which materials may be subjected while maintaining zero resistance. Not until the development in the early 1960s of alloys with sufficiently high critical fields could superconductors be used to generate high fields in large volume.

An upper limit also exists on the current carrying capacity of a superconducting material. This critical current density J_c results in part from the self-field generated by current flowing within the conductor and hence is related to H_c. The three critical parameters (temperature, magnetic field and current density) define a critical surface, within which the material is in the superconducting state and outside of which the material is in the normal state.

The only superconducting materials that are widely used and commercially available are an alloy of niobium and titanium (NbTi) and an intermetallic compound of niobium and tin (Nb_3Sn). Their approximate critical temperature values are 10 K and 18 K, respectively. In 1987, new types of superconductors with critical temperatures above 90 K were discovered. It is estimated that at least ten years of development work will be required before these materials can be used as magnets, if indeed all the material properties can be resolved.

Applications for Superconductors

Superconductors are extraordinary in their tremendous capacity for carrying current while dissipating little or no power. Conventional copper conductors may carry a current density on the order of 10^7 A•m^{-2} while commercial superconducting alloys have the capacity to carry a current density over 10^9 A•m^{-2}. The use of superconductors has enabled the manufacture of many types of large electromagnetic devices that would not have been feasible with normal conductors because of their requirements for prohibitive amounts of power.[53]

Present applications of superconductors include high field research magnets (1) for nuclear magnetic resonance; (2) for experimental work focused on investigating the magnetic properties of light, bioorganisms and semiconductors; and (3) for many diverse areas of solid state physics. Superconducting magnets are now being used in high energy physics research, capitalizing on their low power consumption and high field strength. Accelerators at the Fermi National Accelerator Laboratory employ many superconducting magnets. A collider proposed by the US physics community, the superconducting super collider (SSC), might employ 10,000 superconducting dipole and quadrupole magnets arranged in a ring about 100 km in diameter.

Today, commercial uses of superconducting magnets also include magnetic resonance (MR) imaging of the human body. This technique, which promises much better resolution than differential X-ray absorption with less harmful side effects, was recently made possible by an adaptation of the nuclear magnetic resonance (NMR) methods long used in physics research. Superconducting magnets are also employed in a test facility for magnetically levitated transport systems in Japan.

Potentially the most important role that superconducting magnets will play is in the generation of large volume, high strength magnetic fields required in new energy conversion and storage schemes such as fusion, magnetohydrodynamic cogeneration and magnetic energy storage. By far, the largest research efforts have been aimed at achieving fusion reactor feasibility. These multinational efforts focus on magnetic confinement of ionized hydrogen gas (plasma) either in a bottle-like apparatus with a magnetic mirror at each end (tandem mirror), or in a toroidal configuration with the plasma constrained by three complex, interacting magnetic fields. Increased sophistication in superconducting magnet technology is necessary before it is possible to build or efficiently operate successive generations of research machines or commercial reactors.

Development of superconducting materials for electrical conductors has followed a lengthy and arduous course. The high resistance of superconducting alloys in the normal state requires the fabrication of composite conductors, adding to the superconductor a good conventional conductor such as copper or aluminum, which will share the current in the event of temperature increases. Initially, large filaments of superconducting NbTi were embedded in a matrix of copper. It was soon found that flux jumping (spontaneous redistributions of magnetic flux within the superconducting filaments) dissipated enough energy to raise the temperature of the composite conductor beyond T_c, thus causing a transition to the normal state. Degraded performance due to flux jumping was alleviated by reducing filament diameter and twisting the filaments. The cross section of a typical multifilamentary composite superconductor is shown in Fig. 58. This strand has a 2.3 mm diameter and contains 1,069 filaments, each with a 50 μm diameter. The strand can carry a current of about 20,000 A in a magnetic field of 8 T at 4.2 K.

FIGURE 58. Cross section of NbTi copper multifilamentary composite; conductor diameter: 2.3 mm (0.1 in.); filament diameter: 50 μm (0.002 in.); number of filaments: 1,069

Cryostable Versus Adiabatic Magnets: Stability to Quench

The simplest design for a superconducting magnet utilizes coils wound from multifilamentary composites and immersed in a bath of liquid helium. The operating point of the magnet must remain below the critical surface. Thermal disturbances, generated by friction (wire motion) or by electrical dissipation (flux jumps, eddy currents and magnetic hysteresis) are removed by nucleate boiling heat transfer to maintain the operating point. Should the conductor at any location become nonsuperconducting (or normal), the resistance of that section rises dramatically. The current then must flow through the copper matrix, which produces joule heating in the conductor. A heat balance expression can be written per unit volume of the normal zone, as shown in Eq. 7.

$$\frac{d}{dt}(CT) = \nabla(k\nabla T) - h(T - T_\infty) + pj^2 + g(t) \quad \text{(Eq. 7)}$$

The left side of the equation represents the change in stored energy of the element. The first term on the right side indicates the heat flow via conduction, the second term the heat convected out to the liquid helium, the third term the heat generated by joule heating, and the last term represents other heat inputs, electrical and mechanical. If more heat is removed from the normal zone by conduction and boiling than is generated by joule heating, the $d(CT)/dt$ term becomes negative and the normal zone will shrink and disappear. A magnet design that depends heavily on the cooling term is *cryostable*. If a normal zone, once started, continues to grow, it will become necessary to shut down the magnet. This is called a *quench*. Large scale superconducting magnets with a magnetic energy in excess of 100 MJ (such as those required for fusion devices)[54,55] are designed to be cryostable; they should not quench under normal operation.

High performance magnets, such as those used in MR spectrometers and in high energy accelerators, are designed for maximum overall current density, and are therefore operated just below the critical surface. Furthermore, in order to achieve the highest possible current density, these magnets are designed without space for liquid helium. Although this practice allows more conductor to be jammed into the magnet, it also eliminates the possibility for convective cooling — they are virtually adiabatic. Adiabatic magnets are much more susceptible to quench by minute heat input induced principally by mechanical disturbances within the winding.[56]

Sources of Acoustic Emission in Superconducting Magnets

Acoustic emission monitoring of superconducting magnets was first done in the late 1970s.[57-61] It has now been firmly established that acoustic signals in superconducting magnets are emitted principally by mechanical events such as conductor strain,[62] conductor motion,[63] frictional motion,[64] and epoxy cracking.[56,65-67] Despite earlier suggestions,[61,68,69] flux motion, except during flux jumping, does not appear to be an important source of acoustic emission signals in superconducting magnets. These mechanical disturbances are transitory, each generating a packet of signals that can be triangulated with strategically located sensors. Source triangulation has been achieved with a few sensors in simple cases[70-72] and with many sensors in a large magnet.[73,74]

The performance of large cryostable magnets, on the other hand, is virtually unaffected by mechanical disturbances. For these magnets, acoustic emission should be incorporated in a long-range, early warning monitoring system to detect structural discontinuities and incipient failure.[75] An impending failure is likely to be preceded by an increase in mechanical activity; that is, a mechanical failure will follow an increase in acoustic emission activity over and above that which is normal. Large cryostable magnets, therefore, can be treated like other mechanical systems presently monitored by acoustic emission techniques.

Quench Identification

In ordinary magnets wound with multifilamentary conductors, acoustic emission signals are chiefly generated by mechanical disturbances (conductor motion and cracking of the impregnant material). Because each event generates an acoustic emission signal, monitoring of both signal and magnet voltage makes it possible to identify and localize the source of an event responsible for a premature quench. Generally, quench events in epoxy impregnated superconducting magnets may be divided into three classes.[76]

Class-1: a quench preceded by a voltage spike and an acoustic emission signal. Because sudden conductor motion generates an acoustic emission signal and induces a voltage spike across the coil, the simultaneous occurrence of a voltage spike and an acoustic emission signal indicates that such a quench is induced by conductor motion. Figure 59 presents an oscillogram showing an example of a quench event induced by conductor motion and observed in a superconducting dipole 1.75 m long. The top two traces are acoustic emission signals and the bottom trace is the terminal voltage of the coil.

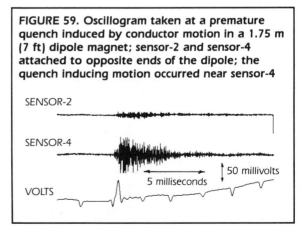

FIGURE 59. Oscillogram taken at a premature quench induced by conductor motion in a 1.75 m (7 ft) dipole magnet; sensor-2 and sensor-4 attached to opposite ends of the dipole; the quench inducing motion occurred near sensor-4

Sensor-2 was located at one end of the dipole and sensor-4 was located at the other end. The voltage pulse that precedes the slowly rising resistive voltage was probably induced by microscopic conductor motion (microslip).[64] The event generated an acoustic emission signal that was captured by sensor-4 and after a delay of 0.3 ms the event was detected by sensor-2. The propagation speed for the acoustic wave is about 5,000 m·s^{-1} and a delay of 0.3 ms suggests a wave travel distance of about 1.5 m. This suggests that the event did indeed take place near one end of the dipole.

Another example of class-1 quench was recorded in a Nb$_3$Sn solenoid operated in superfluid helium at 1.8 K (see Fig. 60).[77] Both the acoustic signal and the voltage spike at the start of the quench occurred at 520 A.

Class-2: a quench with an acoustic emission signal but no voltage spike. This quench is triggered by events, such as epoxy cracking, that are not related to conductor motion. Figure 61 illustrates an epoxy cracking event triggering a quench in a small test coil.[66] The top oscillogram (Fig. 61a) is similar to the oscillogram of Fig. 59 except there is no motion-induced voltage spike (bottom trace). Note that a resistive voltage grows about 16 ms after the cracking event.

The two acoustic emission traces shown in an expanded time scale (about 25 μs per division) in Fig. 61b clearly indicate that the event actually occurred near the top flange of the coil. The acoustic emission sensor for the top trace was mounted on the top flange and the sensor for the bottom trace was mounted on the bottom flange. The coil length, or the distance between the flanges, was 109 mm. A time delay of nearly 50 μs (two divisions) gives rise to an expected propagation speed of about 2,000 m·s^{-1} for the acoustic emission wave through the coil winding.

Class-3: a quench with neither a voltage spike nor an acoustic emission signal. This quench results from pure joule heating as the conductor reaches critical current and enters into the normal (nonsuperconducting) state. Because the

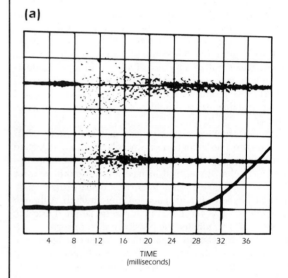

FIGURE 61. Oscillograms taken at an epoxy crack induced quench: (a) acoustic emission signals (top two traces) and coil voltage (bottom trace); and (b) acoustic emission signals

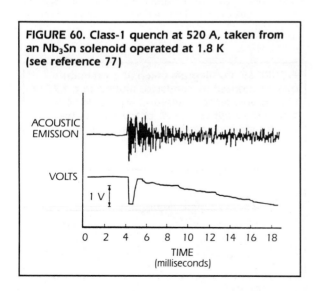

FIGURE 60. Class-1 quench at 520 A, taken from an Nb$_3$Sn solenoid operated at 1.8 K (see reference 77)

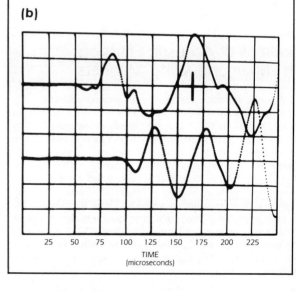

quench is induced by pure heating rather than by a mechanical event, no voltage spike or acoustic emission signal appears at the start of the quench. Figure 62 presents an example of class-3 quench (at 540 A) recorded in the Nb_3Sn solenoid discussed with Fig. 60. Note the absence of both an acoustic emission signal and a voltage spike.

This acoustic emission voltage technique has been applied successfully to identify, and sometimes localize, quench sources in many magnets.[70-74,76-82]

Acoustic Emission Signals Induced by Conductor Motion in a Superconducting Dipole

Acoustic emission results obtained in a superconducting dipole at the Fermi National Accelerator Laboratory are presented below.[72] The data have shown that most acoustic emission signals in these adiabatic superconducting dipoles (and quadrupoles) are induced by conductor motion taking place in the winding. The motion-induced acoustic emission data are correlated with a simple friction model. Agreement between data and theory is excellent.

To monitor specifically superconducting dipoles and quadrupoles of high energy accelerators, an instrumentation system capable of (1) real-time location of genuine acoustic emission events from several sensors, and (2) real-time extraction of acoustic emission energy was developed at the Francis Bitter National Magnet Laboratory and Plasma Fusion Center at MIT.[72,83] The system was field tested to monitor a model dipole at Fermi National Accelerator Laboratory (FNAL). The coils for this particular model were wound from a special alloy of niobium, titanium, and tantalum (NbTiTa) with a copper-to-superconductor ratio of 1.8.

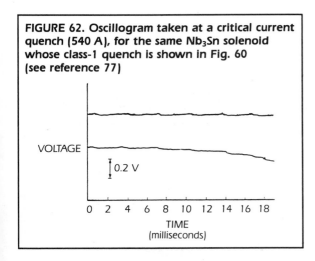

FIGURE 62. Oscillogram taken at a critical current quench (540 A), for the same Nb_3Sn solenoid whose class-1 quench is shown in Fig. 60 (see reference 77)

The insulation was fiberglass. Stainless steel collars restrain the coils with a nominal aperture of 76 mm. Other parameters are the same as previously discussed.

Experimental Procedure

The sensors were the differential type. Each sensor contained two piezoelectric elements cut from one circular disk, one element oriented positively with respect to ground and the other negatively; in practice, one crystal was simply flipped over. The differential sensor has higher sensitivity and a higher signal-to-noise ratio than the single ended type. The two halves were enclosed in a copper jacket that shielded them from electromagnetic interference. The sensor was then mounted on the magnet collars with a G-10 piece fixed on either end by studs welded to the collars and nuts. The sensors were further protected from physical abuse by diecast aluminum guard boxes. Belleville washers acted as stiff springs preventing the guard boxes from loosening on cool down.

Amplification was accomplished in two stages: first, with a commercial preamplifier that had good common mode rejection and wide bandwidth (2 to 600 kHz). The preamplifier gain was selectable at 40 or 60 dB. Further amplification was done in the secondary amplifier (part of the signal processing system). At FNAL, the preamplifiers were located immediately outside the magnet test rig and connected to the sensors by 3 m coaxial cables. The amplified signals were carried over 30 m of shielded cable to the signal processing system located in the control room. The signals usually picked up noise (mainly low frequency power supply noise or its higher harmonics) along the path to the control room. Active high pass filters with cutoff frequency of 10 kHz were inserted between the preamplifier outputs and the signal processor.

The experimental arrangement is shown in Fig. 63. There are four sensors in all, one each at the top, middle and bottom of the magnet. The fourth sensor is a *guard*, mounted on the current leads to detect extraneous noise. The signal processor uses the *first hit* method of event detection. Event count, energy count, and time of arrival are digitized and transferred to the computer for storage and further processing.

Results

Only those results relevant to the friction model are presented here. Quench identification and localization results are not included. From the principal set of data (cumulative record of acoustic emission events, energy and transport current history), only cumulative energy count versus transport current I data are presented and compared with the prediction of the friction model theory. Given a motion-induced acoustic emission signal, $v_{AE}(t)$, acoustic energy E_{AE} is defined as:

$$E_{AE} = \int v_{AE}^2 dt \qquad \text{(Eq. 8)}$$

For each motion-induced acoustic emission signal arriving *first* at one of the three sensors, E_{AE} was computed and accumulated for each sensor as the transport current in the dipole was energized from zero to 5,000 A.

Of several results obtained, it was recorded that there was substantially more acoustic emission activity (and thus greater E_{AE}) in the virgin run than in any subsequent runs. The virgin run occurs when the dipole is energized for the first time; the virgin run returns when the dipole is driven normal, heated substantially and subsequently returns superconducting. Later runs, on the other hand, did not differ appreciably from each other. Figure 64 presents E_{AE} versus I plots recorded by the bottom sensor for the virgin run (trace 1) and nonvirgin runs (traces 2, 3, 4). The dotted lines are theoretical predictions based on the frictional conductor motion model presented below.

Frictional Conductor Motion Model

Analytical expressions for E_{AE} versus I may be developed, based on a simple frictional mechanism controlling conductor motion within the dipole winding.[63] Assume that the conductor is subjected to three main forces: spring, electromagnetic and friction. The force balance for a unit conductor is then:

$$kx = f_e - f_c \qquad \text{(Eq. 9)}$$

Where:

x = displacement from the stationary position;
k = the spring constant;
f_e = the electromagnetic force; and
f_c = the static friction.

If $f_e > kx$, then static friction f_c may be expressed as:

$$f_c = f_s \qquad \text{(Eq. 10)}$$

or

$$f_c = -f_s$$

if $f_e < kx$. The term f_s is a positive value. The sign change occurs because friction tends to oppose motion. This point is illustrated in Fig. 65 with a mass spring model.

The conductor begins in the virgin state with zero transport current (point A in Fig. 66). As the conductor transport current I is increased, f_e increases. Initially, the whole conductor remains still, emitting no acoustic signals. This corresponds to the vertical trajectory beginning at point A. When f_e reaches a threshold value f_o, small sections of the conductor (where f_s is least) begin to move. This is point B on the $f - x$ space.

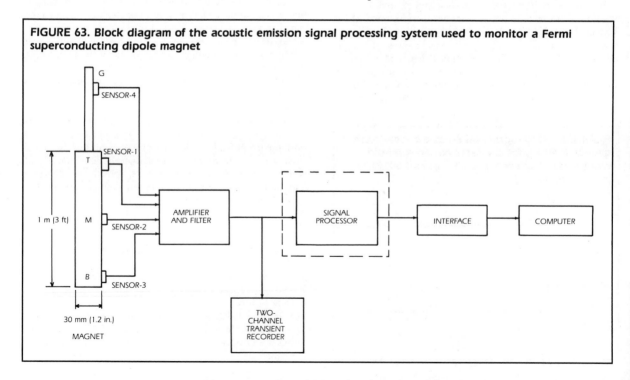

FIGURE 63. Block diagram of the acoustic emission signal processing system used to monitor a Fermi superconducting dipole magnet

Beyond point B, the energy released (δE) when the conductor moves a distance δx due to excess electromagnetic force over friction $f_e - f_o$ is given by:

$$\delta E = (f_e - f_o)\delta x \quad \text{(Eq. 11)}$$

Substituting $\delta f = k\delta x$ and $\delta f = \delta f_e$ (since f_o is constant):

$$\delta E = \frac{f_e - f_o}{k}\delta f \quad \text{(Eq. 12)}$$

Integrating Eq. 12, we obtain an expression for the total energy E released as f_e is increased from f_o to f_e:

$$E = \int_{f_e}^{f_o}\left(\frac{f_e - f_o}{k}\right)df_e \quad \text{(Eq. 13)}$$

and

$$E = \frac{(f_e - f_o)^2}{2k} \quad \text{(Eq. 14)}$$

Assuming a linear system, the energy picked up at the sensor will be:

$$E = \frac{\beta(f_e - f_o)^2}{2k} \propto E_{AE} \quad \text{(Eq. 15)}$$

The β term is a proportionality constant that depends on the transfer properties of the medium, the sensitivity of the sensors, and the calibration factor of the energy processor.

This equation holds until point D when I begins to decrease. So far the conductor has moved outward. When I

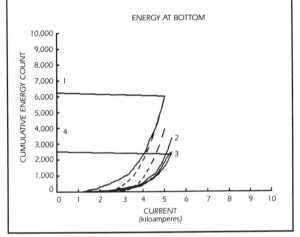

FIGURE 64. Cumulative energy versus current plots recorded by the bottom sensor for the virgin run (trace-1) and nonvirgin runs (trace-2, 3 and 4); dashed lines are predicted by Eqs. 14 and 15

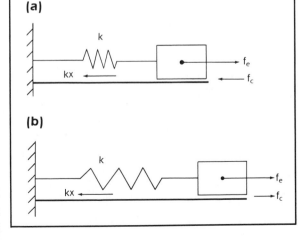

FIGURE 65. Mass spring model with friction force f_c: (a) corresponds to a sweep-up mode; and (b) corresponds to a sweep-down mode

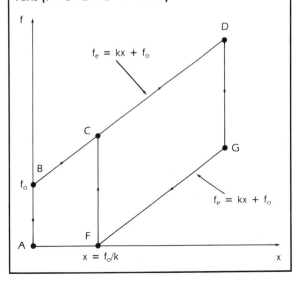

FIGURE 66. Force displacement traces for the virgin run (A→B→C→D→G→F) and nonvirgin runs (F→C→D→G→F→C ...)

begins to decrease, the conductor will tend to move inward. However, the frictional force reverses direction such that the conductor remains still, emitting no acoustic signals until the conductor again begins to move over sections with the least friction (point G). The value of f_e at point G is $f_m - 2f_o$, where f_m is the value of f_e at point D. Using similar reasoning to that above, the energy released during sweep down as f_e decreases from $f_m - 2f_o$ is:

$$E = \int_{f_e}^{f_o} \frac{f_e - (f_m - 2f_o)}{k} df_e \quad \text{(Eq. 16)}$$

and thus:

$$E = \frac{\beta[f_e - (f_m - 2f_o)]^2}{2k} \quad \text{(Eq. 17)}$$

This equation holds until the current reaches zero at point F where f_e becomes zero and the conductor stops moving. The displacement at this point is $x_o = f_o/k$.

During nonvirgin runs, the trajectory starts from F. The conductor remains still and no acoustic emission occurs until C when $f_e = 2f_o$. Thereafter, the conductor moves, generating acoustic signals. Following the same steps as before, the total energy released as f_e is found to increase from $2f_o$ and is given by:

$$E = \frac{\beta(f_e - 2f_o)^2}{2k} \quad \text{(Eq. 18)}$$

In the return passage, the trajectory follows path DGF, as in the virgin run. During nonvirgin runs, the trajectory always follows path FCDGF with reproducible acoustic characteristics.

Theoretical Predictions

The electromagnetic force on a noninductive winding with a transport current I is proportional to I^2:

$$f_e = \alpha I^2 \quad \text{(Eq. 19)}$$

where α is a constant. Substituting appropriately into Eqs. 15, 17, and 18, the following expressions are obtained for the predicted acoustic emission energy as a function of I for the virgin run (Eq. 20) and the nonvirgin run (Eq. 21):

$$E = 0 \quad \text{if } 0 < I < I_o \quad \text{(Eq. 20)}$$

$$E = \eta (I^2 - I_o^2)^2 \quad \text{if } I_o < I < I_m$$

$$E = \eta (I_m^2 - I_o^2)^2 \quad \text{if } I_m > I > \sqrt{I_m^2 - 2I_o^2}$$

and

$$E = \eta [I^2 - (I_m^2 - 2I_o^2)]^2 \quad \text{if } \sqrt{I_m^2 - 2I_o^2} > I > 0$$

$$E = 0 \quad \text{if } 0 < I < \sqrt{2} I_o \quad \text{(Eq. 21)}$$

$$E = \eta (I^2 - 2I_o^2)^2 \quad \text{if } \sqrt{2} I_o < I < I_m$$

$$E = \eta (I_m^2 - 2I_o^2)^2 \quad \text{if } I_m > I > \sqrt{I_m^2 - 2I_o^2}$$

and

$$E = \eta [I^2 - (I_m^2 - 2I_o^2)]^2 \quad \text{if } \sqrt{I_m^2 - 2I_o^2} > I > 0$$

The current at which the coil begins to move in the virgin run is $I_o = (f_o/\alpha)^{1/2}$ and:

$$\eta = \frac{\beta \alpha^2}{2k} \quad \text{(Eq. 22)}$$

Note that Eqs. 20 and 21 are the same as those obtained by the previous analysis,[63] where it was assumed (1) that the conductor moved in discrete steps, each step corresponding to one acoustic emission event, and (2) that the number of oscillatory signals above a given threshold was proportional to the distance traveled during the step motion. In deriving Eqs. 20 and 21, however, the only assumption made was that the system was linear. The theoretical predictions of Eqs. 20 and 21 (using the values of I_m and I_o obtained from the experiments) are plotted by the dotted lines in Fig. 64.

Figure 67 is an example of data taken while the magnet was ramped continually up and down, well below its quench

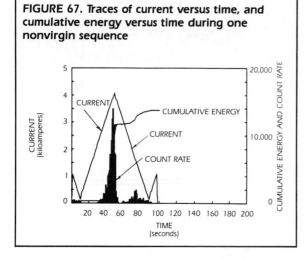

FIGURE 67. Traces of current versus time, and cumulative energy versus time during one nonvirgin sequence

TABLE 3. Comparison of Theoretical and Experimental Current Peak

	Middle Sensor (amperes)	Bottom Sensor (amperes)
Theoretical value	1,400	1,800
Experimental value	1,600	1,500

current. As predicted by theory, there are initially no acoustic signals until a certain current level is reached, at which point acoustic emission (E_{AE} in Fig. 67) begins to occur, the rate rising to a peak coincidentally with the transport current. During sweep down, a small local peak occurs in the E_{AE} rate. When the current stops increasing and starts decreasing, the static friction reverses direction, temporarily stopping the coil from moving. The acoustic emission rate will remain zero for a while after the current starts decreasing, until static friction is overcome. The acoustic emission rate must also be zero at the end of the sweep down when there is no current in the conductor. In between, the acoustic emission rate goes through nonzero values and must therefore have at least one local peak. This result may be predicted from the theory (the acoustic emission energy during sweep down is given by Eq. 21):

$$E = \eta\,[I^2 - (I_m^2 - 2\,I_o^2)]^2 \qquad \text{(Eq. 23)}$$

A peak in the acoustic emission rate will occur during sweep down at current I_p:

$$\left(\frac{d^2E}{dI^2}\right)_{I_p} = 0 \qquad \text{(Eq. 24)}$$

The corresponding current is then given by:

$$I_p = \sqrt{\frac{I_m^2 - 2\,I_o^2}{3}} \qquad \text{(Eq. 25)}$$

Values of I_p predicted by Eq. 25 are compared with experimental data in Table 3.

Estimation of Conductor Motion

The fact that most of the quenches were diagnosed as conductor motion induced further strengthens the belief that conductor motion is the dominant mechanical disturbance in the dipole.

The extent of the motion may be estimated as follows. When a conductor of length ρ moves perpendicularly to a magnetic field B at a velocity v, the motion-induced voltage V across the conductor terminals is given by:

$$V = B\rho v \qquad \text{(Eq. 26)}$$

or

$$V = B\rho\,\frac{dx}{dt} \sim B\rho\,\frac{\Delta x}{\Delta t} \qquad \text{(Eq. 27)}$$

where Δx is the distance moved in time Δt.

From measurements made during the experiments, typically $V \sim 10$ mV and $\Delta t \sim 40$ μV. With a peak field of 5 T, and taking $\rho \sim 50$ mm, then from Eq. 27, $\Delta x \sim 2$ μm, which is on the same order of magnitude as microslip distances measured in earlier experiments.[64] Microslips do not induce quenches at currents much below critical current and, irrespective of the current level, not every microslip leads to a quench: its size may be too small, the length of affected conductor may be short, or its location may have extra cooling. None of the voltage transients observed, except perhaps at unusually high charge rate (1,000 A•s^{-1}), could be characterized as macroscopic slips. These seem to have been successfully eliminated from the Fermi coils.

Estimation of Frictional Power Dissipation

One useful set of information that acoustic emission data may provide is the amount of power dissipated from the magnet by mechanical disturbances. Acoustic emission data may be used to estimate the power dissipated by friction during magnet energization. The strong agreement between theory and experiment in acoustic emission history as presented above justifies this approach.

The energy released during one slip event lasting an interval t_1 is given by:

$$e = \int_{t1}^{t1+\Delta t} i(t)v(t)\,dt \qquad \text{(Eq. 28)}$$

Where:

$i(t)$ = the transport current; and
$v(t)$ = the voltage across the conductor.

The current is practically constant over the time interval Δt so that Eq. 28 may be rewritten:

$$e = I\int_{t1}^{t1+\Delta t} v(t)\,dt \qquad \text{(Eq. 29)}$$

where $I = i(t) \sim$ constant during a motion event.

The voltage pulses observed during the experiments were typically about 5 mV in amplitude and about 40 μs in duration. Putting these numbers into Eq. 29 gives an estimate of the energy released during a slip event: ($e \sim 1$ mJ). This compares well with values calculated for slip events in previous experiments.[63] The average power dissipation over a complete charging sequence may be obtained directly from Eq. 30.

$$P_{avg} = e \frac{dN}{dt} \quad \text{(Eq. 30)}$$

Here dN/dt is the average acoustic emission event rate during the sequence.

From raw data, dN/dt is about 20 events per second. Substitution into Eq. 30 gives: $P_{avg} \sim 20$ mW.

With a total winding volume of 2,000 cm^3, the average power dissipation per unit volume is $P_{avg} \sim 10\ \mu\text{W}\cdot\text{cm}^{-3}$.

Note that in quantifying a disturbance such as frictional slip, both the size and duration of the disturbance are important because cooling, which mainly counterbalances frictional dissipation in superconducting magnets, is time dependent. For short durations, cooling is dominated by transient phenomena while for long durations, steady state phenomena prevail.

The calculation above was performed on the assumption, based on experimental data, that most acoustic emission events were generated by frictional motion and each event released the same amount of energy. In practice, microslips do not occur uniformly and some acoustic emission will always be generated by sources other than motion. Note, too, the assumption that all the energy released is dissipated; actually, some fraction may be recovered. Nonetheless these results are not trivial, for the heat capacities of materials at 4.2 K are extremely low. In high current density superconducting magnets, microslips may generate sufficient heat to induce quench, as these and earlier experiments have shown.

It should also be noted that, although the computed frictional dissipation is small, there are other important forms of power dissipation such as AC losses and magnetic hysteresis losses that are not included in this analysis.

Normal Zone Detection in Superconducting Magnets

When a superconducting magnet is driven normal, partially or wholly, a nonuniform temperature distribution is created within the winding, which in turn creates a nonuniform stress distribution within the winding. The nonuniform stress distribution causes cracking in structural materials and may even cause conductor motion, both dominant sources of acoustic emission signals. Acoustic emission signals can therefore be used to detect the presence of a localized or global normal state in superconducting magnets. This basic concept was demonstrated successfully in an experiment with small test coils impregnated with epoxy resin.[84] Each test coil, instrumented with thermocouples and acoustic emission sensors, contained a heater wire in the winding to create a nonuniform temperature distribution within the winding. The experimental observations are summarized below.

1. In the epoxy impregnated superconducting coil driven normal, acoustic emission signals are generated by the normal zone induced change in the temperature distribution within the winding. Furthermore, the presence of acoustic emission signals signifies that the maximum winding temperature has reached at least 30 K. Thermal expansion coefficients of winding material are very small below 20 K and become significant above 30 K.[85]
2. Acoustic emission activity disappears when the magnet returns to the superconducting state.

As an illustration of the normal zone detection technique, recent results[86] obtained in the testing of a pulse superconducting quadrupole[87] are described below.

HEP Quadrupole

A high energy physics (HEP) quadrupole was wound with NbTi cables with B-stage epoxy impregnated glass tapes cured under compression. The winding was permeable to liquid helium. The quadrupole was energized, and in the sequence shown in Fig. 68, it quenched at 3,600 A. Both acoustic emission count rate and current traces are presented in Fig. 68. As inferred from the acoustic emission signals (which persisted over four minutes after the

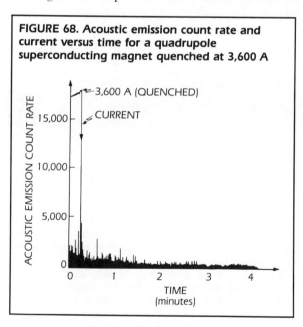

FIGURE 68. Acoustic emission count rate and current versus time for a quadrupole superconducting magnet quenched at 3,600 A

quench), the quench apparently drove a large section of the quadrupole normal.

Acoustic emission signals disappeared when the quadrupole again became superconducting, which was confirmed by measuring the magnet's resistance. Figure 69 presents a similar set of data when the current (at 3,880 A) was discharged at a much slower rate through a dump resistor, allowing the quadrupole to remain superconducting throughout the discharge sequence. Note the complete absence of acoustic emission signals immediately after the completion of discharge.

Disturbance Energy Quantification

Because of the extremely small specific heat of materials at 4.2 K (about 0.1 percent of that at room temperature), a minuscule amount of energy dissipated by an event such as conductor motion is sufficient to drive the conductor normal. At 4.2 K, motion often consists of a series of microslips, each a sudden slippage covering a distance of 1 to 10 μm.[64] Epoxy cracking is another localized event that releases dissipative energy quickly and gives rise to localized heating in the winding. Because each of these events emits an acoustic emission signal $v_{AE}(t)$, it is possible to use an energy parameter E_{AE} to quantify event energy, as was assumed earlier in connection with the friction motion model (see Eq. 8).

In a series of experiments,[88] microscopic dissipative energies of frictional events and cracking were measured by a custom energy transducer and compared with measured input mechanical energies and E_{AE}. As seen from results presented below, agreement between measured energies and E_{AE} is excellent.

Figure 70 presents dissipative energy data taken for a copper-copper frictional interface immersed in liquid helium. In the figure, each $F_t \Delta x$ represents the mechanical work supplied to the interface by the experimental rig and dissipated as heat during one microslip event; E_p is the corresponding dissipation energy measured *directly* by an energy transducer. Note that there is an excellent one-to-one (45 degree) correspondence between $F_t \Delta x$ and E_p. Presented in the same figure are E_{AE} data, computed from v_{AE} recorded for each microslip event. Although E_{AE} cannot provide energy information on an absolute basis, the 45 degree line drawn through E_{AE} data suggests E_{AE} to be a useful parameter in quantifying dissipation energies.

Cryogenic Temperature Acoustic Emission Sensors

The acoustic emission sensors used at 4.2 K for superconducting magnets were designed and constructed at MIT. A schematic of this sensor is shown in Fig. 71. The piezoelectric transducer is cut in half, and its halves are mounted in the sensor casing with reverse polarity.[89] When used in conjunction with a differential amplifier, the voltages due to stress waves add, while those due to electrical noise subtract. Since the electrical noise is similar on each side, it tends to cancel, leaving mostly the stress signal. This is the principle of the differential sensor. Past acoustic emission projects using nondifferential (single ended) sensors were

FIGURE 69. Acoustic emission count rate and current versus time for the same quadrupole superconducting magnet discharged slowly at 3,880 A

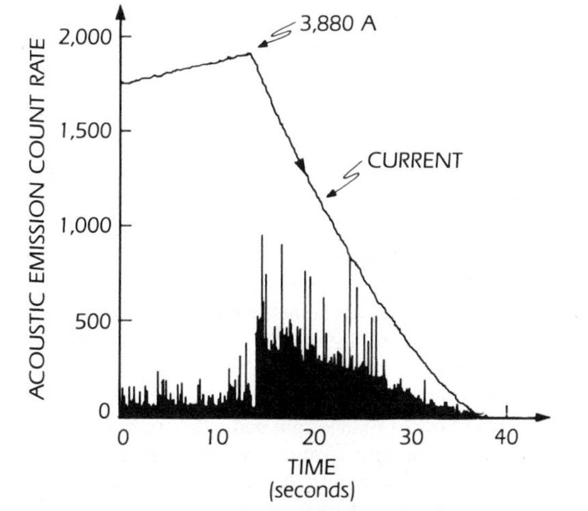

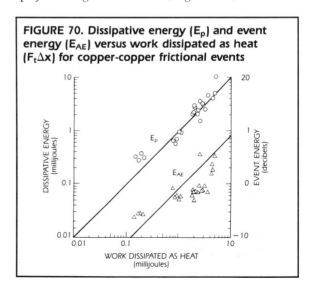

FIGURE 70. Dissipative energy (E_p) and event energy (E_{AE}) versus work dissipated as heat ($F_t \Delta x$) for copper-copper frictional events

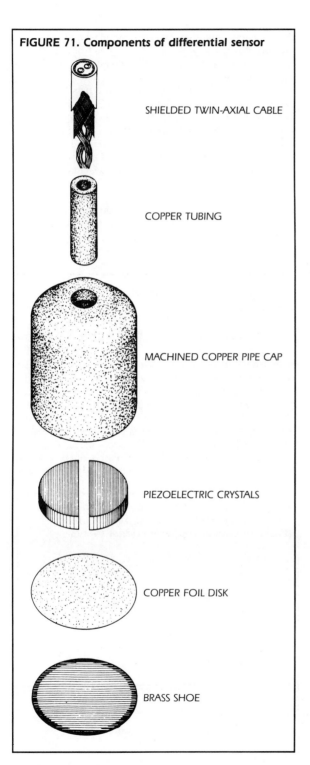

FIGURE 71. Components of differential sensor

plagued by noise.[73] The differential sensor has been very successful in eliminating electrical noise while retaining the extremely high sensitivity of the piezoelectric crystal. Although not presented here, another important acoustic emission application in superconducting magnets is for detection of electric arcs in high voltage operation of pulse magnets.[86,90] Because an electric arc generates both sound waves and electromagnetic waves, differential sensors have proven very effective in suppressing electromagnetic waves.[83,86]

The sensors used on superconducting magnets must also be able to withstand extreme variations in temperature. To prevent thermal stresses from debonding the crystals from their supporting structure, the sensors have been mounted on a thin piece of copper foil. Thermal stresses in the brass plate have little effect on the crystals.

Conclusion

The progress toward acceptance of acoustic emission as a standard monitoring technique in superconducting magnets has followed steps similar to other acoustic emission applications,[86] including those below.

1. An initial stage of exaggerated, and sometimes exotic, claims.
2. Genuine successes accompanied with some failure. As with any technology, there is always a temptation to apply the method beyond its current state of technical development or analytical understanding.
3. Realization for the need to undertake basic studies.
4. Accumulation of field data and experience.
5. Development of techniques to enhance reliability and accuracy.
6. General acceptance of standard techniques.

Some acoustic emission techniques for superconducting magnets are approaching stage-6, while others are still at stage-2 or 3. At the present time, reliability is the most critical quality needed in acoustic emission technology before it can reach the final stage as an accepted monitoring technique for superconducting magnets.[90]

To achieve this goal, our understanding needs to advance in the following areas: (1) sensor sensitivity, particularly at cryogenic temperature; (2) acoustic emission wave propagation in the magnet winding media; and (3) quantitative correlation between acoustic emission amplitude and physical quantities such as displacement, velocity, force and energy. It is also important to realize that acoustic emission technology serves best as a complementary method rather than as the only means of monitoring. Combined with voltage measurement, for example, it has proven almost indispensable.

Acknowledgments

Thanks to many colleagues, visiting scientists and graduate students without whose productive work this article could not have been written. The author is particularly indebted to Osami Tsukamoto of the Yokohama National University, Yokohama, Japan, who has been a constant and steady collaborator.

PART 8
PARTICLE IMPACT NOISE DETECTION AS A NONDESTRUCTIVE TEST FOR DETERMINING INTEGRITY OF ELECTRONIC COMPONENTS

Much of the equipment and many of the measurement techniques used in acoustic emission testing are shared by PIND testing. The text below is included here not because PIND is considered an acoustic emission test, but for the completeness of this chapter's information on acoustic tests of electronic components.

Contaminant particles within cavities of electronic components can pose a significant integrity problem in high reliability electronics applications.

As early as 1960, government space organizations, electronic component manufacturers and end users employed PIND testing as a nondestructive test technique to determine the integrity of electronic components.

The use of PIND testing affords the quality assurance engineer the ability to screen electronic components by administering a series of mechanical shocks and vibrations to the device under test, allowing contaminant particles to move freely within the component cavity, generating acoustic signals and thereby revealing themselves.

In this section, an overview of PIND testing, its theory of operation and the components of various research studies will be discussed to illustrate the value of PIND testing as a nondestructive test technique for ensuring the integrity of electronic components.

Definition of PIND

Particle impact noise detection (PIND) is used to determine the integrity of electronic components by monitoring loose particles inside the cavities of electronic components such as transistors, integrated circuits, hybrids, diodes, relays and switches. The technique is not an acoustic emission method, but it is a nondestructive technique based on important acoustic principles.

The PIND test simulates dynamic environments such as aircraft landings and launchings by administering a series of mechanical shocks and vibrations to the device under test. These shocks and vibrations free particles adhering to component cavity walls. The high frequency acoustic noise from the resulting impacts between the particles and the package interior are detected by a transducer on which the test component is mounted.

The importance of PIND testing can be most dramatically illustrated with the US Space Shuttle Columbia. In November 1983, the Columbia's landing was delayed eight hours with speculation about indefinite suspension in space because of loose particles in integrated circuits (which had not been PIND tested) that were housed in the guidance and navigation computers.[91,92]

Dollar figure losses due to loose particles, upwards of $32 million on Delta launches and $300 thousand on Nimbus retrofits, are also significant reasons to consider the importance of PIND testing.[93]

PIND Testing History

Particle impact noise detection, once known as acoustical loose particle detection (ALPD) began to emerge in the early 1960s with the widespread use and acceptance of transistors in applications such as missiles and satellites.[94]

The problem of loose particles was identified when an analysis of telemetry data from a failed satellite mission showed the cause of failure to be a short in a transistor. The cause of the short was due to a small conducting particle encapsulated within the device cavity.

Texas Instruments began developmental work on equipment to detect loose particles inside semiconductors in the early 1960s. Their first PIND tester consisted of a sinusoidal vibration shaker, an accelerometer with a high output voltage, a passive low frequency filter and oscilloscope.[95] In 1966, both Lockheed (Sunnyvale) and General Electric (Valley Forge) experienced problems with particulate contamination in relays. Lockheed attempted to duplicate GE's system for PIND testing which consisted of a large vibration shaker and a very low G force.

During Lockheed's attempt, problems in fixturing were experienced. Lockheed borrowed an ultrasonic frequency translator and used a smaller shaker and their first PIND system for relays was born.[96] Around 1976, PIND testing was also used in electromechanical relays during a Delta launch vehicle countdown.[97] A part failure was traced to a loose wire and a million dollar retrofit ensued. McDonnell Douglas was contracted by NASA to update and improve the equipment system and technique. As a result of McDonnell Douglas's work and the SPWG, Military Standard 883C Method 2020 evolved.

A commercial, mechanical PIND system was built in 1970 by Dunegan Corporation. This system consisted of a vibration shaker, acoustic emission transducer, high frequency amplifier, active low noise filter, oscilloscope and audio speaker. The system has evolved into a sophisticated microprocessor based unit with software tailored to military specification requirements.

Military Standards Governing PIND Testing

Particle impact noise detection of completely fabricated electronic components is most often employed because it is generally accepted as the most economical and, technically, the best test for a finished device.[98] This is supported by the fact that PIND testing has been incorporated into the most widely used PIND Military Standard 883C Method 2020, Military Standard 750C Method 2052 and Military Standard 202F Method 217 and is required for Class-S microcircuits and semiconductors.

Military Standard 883C Method 2020 falls under the auspices of the Air Force and defines testing of integrated circuits and hybrids. Devices are tested in one of two categories; condition-A or condition-B. Components tested in condition-A are usually high reliability flying hardware type components and are tested at 20 G vibration at frequency ranges of 40 to 250 Hz. Components tested in condition-B are usually ground based hardware and are tested at less stringent levels of 10 G at 60 Hz.

Military Standard 750C Method 2052 falls under the auspices of the Air Force and defines testing of discrete devices while Military Standard 202F Method 217 falls under the auspices of the US Army and defines testing of relays.

Theory and Instrumentation for PIND Testing

Instrumentation Overview

The PIND tests are performed with instrumentation similar in function to acoustic emission testing instruments. The acoustic signals produced by particle impacts are converted to electrical signals with a piezoelectric transducer. The very small electrical signals produced by this transducer are amplified by a factor of 1,000 to increase their amplitude for processing. Three different methods are used for detection of impacts. They are (1) audio detection; (2) oscilloscope (visual); and (3) threshold detection (electronic).

Instrumentation Specifications

Military Standard 883C Method 2020 defines the major specifications of the instrumentation (see Fig. 72) used for PIND testing.

The *transducer* is specified as having a nominal peak sensitivity of 77.5 dB referred to one volt per microbar. Restated in other terms, a transducer with this sensitivity produces 133 microvolts of electrical signal from a pressure of 1 microbar on the transducer face. The sensor is used at a frequency that corresponds to one of the crystal resonances, and that resonance must lie between 150 and 160 kHz.

Typically, the *amplifier* assembly consists of a low noise preamplifier, followed by a filter to limit the noise bandwidth and a second amplifier to bring the total amplification to 1,000. In one commercial system, the preamplifier is located on the shaker head itself. The amplifier must contribute no more internal noise than the equivalent of a ten microvolt transducer signal.

Audio detection uses electronic circuits to shift the ultrasonic acoustic signal down to the audible frequency range. It is amplified and applied to either a loudspeaker or headphones. The operator listens for the characteristic, slightly musical click or clicks from particle impacts.

Oscilloscope detection applies the amplified ultrasonic signal to the vertical axis of an oscilloscope having a sensitivity of 20 millivolts per vertical division. Impacts appear as sinusoidal bursts having a fast rising edge and a longer (approximately exponentially decaying) trailing edge. The exact nature of the oscilloscope display depends on how the horizontal axis of the oscilloscope is driven. Military Standard 883C Method 2020 allows the horizontal beam position to be a function of the shaker excitation or semiconductor acceleration or a linear triggered sweep.

Threshold detection compares the amplified ultrasonic signal to a precision threshold that is set to 5 mV above peak system noise. When the amplifier output exceeds this threshold, it typically lights an indicator lamp that must be manually reset.

Nature of Particles

It is important to point out that particle mass, shape, size, composition and the type of device the particle is encapsulated in, have direct bearing on whether the particle is detectable, whether the PIND test results are repeatable and whether the particle is capable of compromising device integrity.

A variety of examples of contaminant loose particles have emerged from various PIND test studies. These include textile fibers, silicon splinters, gold wire, glass beads, aluminum wire, weld splatter, potting particles, acoustic ceiling fibers, hair, ceramic splinters and even a contact lens.

In the early years of PIND testing, before the advent of more sophisticated sensor devices, the first PIND systems were capable of reliably detecting particles with masses of 30 micrograms or greater.[99] Detection of particles with masses of 7 micrograms was sometimes possible. Today, most commercial PIND systems are capable of detecting particles as small as 0.16 micrograms (using a gold ball as a standard).

An excellent study was conducted where high speed movies were made to view the behavior of particles at frequency ranges of 60 to 2,000 Hz at levels of 7 to 35 G and power spectral density levels of 0.0065 to 0.49 G^2 per hertz.[98] The device types used were specially fabricated glass cans mounted on headers. Particle sizes and compositions included: 0.05 and 0.1 mm (0.002 and 0.004 in.) lead spheres, 0.05 and 0.1 mm (0.002 and 0.004 in.) gold flakes and 0.18 × 0.1 mm (0.007 × 0.004 in.) gold wire. The following observations were made.

1. At 2,000 Hz, vibration of 35 G, only the 0.1 mm (0.004 in.) sphere and gold flakes exhibited activity. The sphere moves along the cavity floor but the flakes' bounce height is significant enough to allow shorting in some devices.
2. At 1,000 Hz, vibration of 35 G, particles exhibit virtually the same behavior as 2,000 Hz. Both masses are capable of causing component failure under this condition.
3. At 300 Hz, vibration of 35 G, vigorous activity is displayed by all particles, allowing frequent contact of critical component elements both vertically and laterally.
4. At 60 Hz, vibration of 7 G, only the spheres are active, bouncing off cavity floor and ceiling but with little lateral movement.
5. At 150 Hz to 2,000 Hz random vibration at 0.49 G^2 per hertz, all particles are very active vertically and laterally. All particles make contact with critical component elements.
6. At 150 Hz to 2,000 Hz random vibration at 0.0065 to 0.49 G^2 per hertz), activity is very limited until a *break away* point is reached.

Additional studies were also conducted. Figures 73, 74 and 75 further illustrate particle behavior.

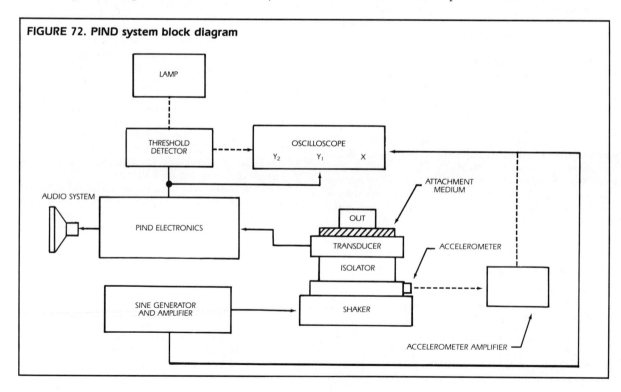

FIGURE 72. PIND system block diagram

In Fig. 73, the comparison of output amplitudes of particles at various levels of acceleration, shows that in all instances output amplitudes increased slightly with increased acceleration. Additionally, the greater the mass, the greater the output amplitude, independent of the shape of the loose particle.[95]

Figure 74 shows that an indication of size can be inferred by the magnitude of output amplitude.[100] The observed range of maximum and minimum amplitudes is due to the shape of the particle and its orientation at impact. The closer the particle is to a sphere, the greater the chances the output will be at a uniform level. This is because a sphere's effective mass (the mass actually 'seen' by the impact surface) is independent of orientation. With an odd shaped piece of wire, effective mass will be different depending on the wire's orientation at impact.

The comparison of vibration frequency versus amplitude in Fig. 75 illustrates again that the greater the mass, the greater the output amplitude. Additionally, the output amplitude levels decrease with increasing frequency. The optimum test frequency is depending on device configuration but generally falls in the 30 to 90 Hz range. Lower frequencies do not provide enough signals per unit time; higher frequencies cause the particle displacement to be reduced, providing less signal amplitude on impact.

As mentioned earlier, there are a variety of examples of contaminant loose particles. It should be noted that conductive particles (gold, aluminum, silicon and solder) are

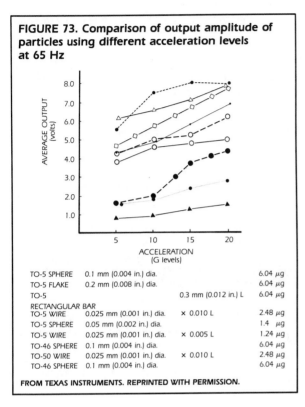

FIGURE 73. Comparison of output amplitude of particles using different acceleration levels at 65 Hz

FROM TEXAS INSTRUMENTS. REPRINTED WITH PERMISSION.

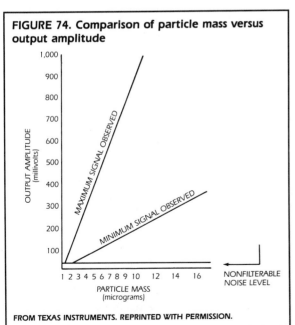

FIGURE 74. Comparison of particle mass versus output amplitude

FROM TEXAS INSTRUMENTS. REPRINTED WITH PERMISSION.

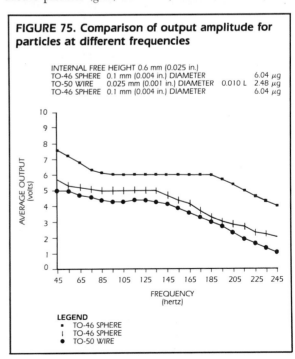

FIGURE 75. Comparison of output amplitude for particles at different frequencies

the only particles capable of causing short circuits in components. Additionally, the greater the activity of the loose particle, the greater the probability it will impact on a short-causing location within the device. Nonconductive particles do present problems during vibration by striking of other component elements. In components such as relays, nonconductive particles can create as great a threat as conductive particles because they can hold relay contacts apart.

Another property of loose particles is their ability to adhere to component cavity walls. This *latch-up* phenomenon can be caused either electrostatically or mechanically when the particles become trapped in cavity seams or crevices. This can hamper repeatability of PIND test results.

Methods for Elimination of Particles Prior to PIND Testing

There are several methods used to eliminate particles before the device becomes completely fabricated, including: proper manufacturing process controls; cleanliness in the manufacturing environment; visual inspection of the device prior to capping; and source surveillance.

An alternative for eliminating the effects of particles is the passivation of the device cavity using a conformal coating. Conformal coatings are most commonly employed in commercial applications and are dispensed in various ways: vapor deposited; brushed on; dipped; sprayed or applied with a hypodermic syringe. Use of conformal coatings is an effective method for immobilizing particles, however, implementing the coating process can create problems because of a general lack of information on the process, limited availability of production equipment, limited expertise for implementing application techniques, very high costs of equipment and handling losses associated with application.

PIND Test Effectivity

There have been numerous PIND studies performed to determine the effectiveness of the PIND test. For the purposes of this discussion two such studies will be detailed.

PIND test effectiveness is subject to the following test variables and they must be remembered when reviewing PIND test data:

1. equipment used to conduct PIND test and its operational status and calibration status;
2. test environment (noise conditions, etc.);
3. skill, experience and fatigue level of test operator;
4. test method and procedure;
5. particle composition, shape, size and mass;
6. type of device being tested; and
7. proper mounting of devices under test.

In 1979, the NASA Goddard Space Flight Center conducted a study to determine the effectiveness of PIND testing.[98]

The study was performed within the aerospace industry and comprised fifty-four companies including semiconductor manufacturers, test labs and users testing 297 various package styles with various seeding materials. Each company was asked to test parts twice, once according to Military Standard 883C Method 2020 condition-A and again at condition-B. Some companies preferred to use variations to both condition-A and condition-B as determined by experience, equipment limitations and contractual requirements. Their test result inputs were categorized as test condition-C.

The test results showed that a 44 percent average detection score for all test conditions, all device types, and all seeding materials. This prompted the conclusion that "consideration be given to adding PIND to the screening tests performed on Class-B Military specification microcircuits, to JANTXV transistors (PIND is already required on S levels for both) and to similar quality high reliability source controlled drawings."

It should be noted that 44 percent is not necessarily a high percentage detection score. However, the reason for this could be the *memory effect*. Memory effect is the theory that the more times a device is subject to PIND testing, the easier and faster it is for particles to latch up or adhere to cavity walls either electrostatically or mechanically, and the harder it is to detect them.

In analyzing the three test conditions, results showed that the average detection score for condition-A was about 6 percent higher than the average score for condition-B. Additionally, out of only 35 company tests, the average score for condition-C was 24 percent. Therefore, test condition-A was found to be only slightly superior to the other two test conditions.

Results also showed that particle detection is very sensitive to package style; metal lid TO styles, for example, have greater detectability over all the ceramic body package styles.

One of the most important results of this study was the wide variation from company to company in their ability to detect particles. Extensive training of PIND operators and qualification of testing companies and equipment is imperative.

Another study was conducted at Crouzet in 1982 at Valence, France to examine the effectiveness of the PIND test. This study was conducted over a period of three years on a sample of 1,930 devices. The sample represented six types of devices from a number of manufacturers.

The test sequences used during the Crouzet study included:

1. one shock at 1,500 G;
   vibration at 10 G for five seconds;
   frequency set by the instrument according to internal package;
   height of device under test; and
   sequence repeated four times.
2. one shock at 1,500 G;
   vibration of 20 G for five seconds;
   frequency set by instrument; and
   sequence repeated four times.

Of the 1,930 components tested, 4.15 percent failed at condition-*A* (20 *G*) and 1.66 percent failed at condition-*B* (10 *G*). As in the NASA study, condition-*A* was shown superior to condition-*B*.

Radiographic inspection and internal visual inspection were performed on the devices identified as defective; 33.8 percent of the failed devices had loose particles confirmed by X-ray tests.

The limitations of X-ray testing must also be considered: some particles are invisible to X-rays; particles smaller than 0.025 mm (0.001 in.) are difficult to image; and there can be masking by attached materials.

Additionally, when the identified failed components were opened for visual inspection, loose particles were found in all of them. No loose particle was found in any device that passed the PIND test. The quality of visual inspection is dependent on the skill and precision of the test operator unsealing the device.

It is important to note that conductive particles capable of causing serious malfunction were found in 64.1 percent of the components that failed the PIND test.

The authors of the Crouzet study concluded that "as a means of selecting and screening components, the PIND test remains the most effective way of detecting loose particles inside such devices."

Using PIND Testing to Conduct Failure Analysis

The PIND test can be used for failure analysis by isolating and capturing particles. A simplified explanation of the procedure is to first determine (by signature) that a particle is present inside a cavity and to then carefully open the package, reidentify and capture the particle to determine its source.

Once a particle has been detected, the next step is to lap the package lid down to a thickness of 0.025 to 0.04 mm (0.001 to 0.0015 in.).[101] Lapping of the package lid must be done very carefully because if the lid is perforated, particles may be introduced into the cavity. Conversely, the lapping must be performed to correct thinness or excessive force will be required to puncture the package lid, resulting in particle introduction into the cavity.

The lid is punctured using a 0.8 mm (0.032 in.) diameter mandrel with a specially prepared point[102] and the puncture is covered with cellophane tape. The package is then mounted on a PIND transducer, shocked and vibrated until the particle is captured on tape.

The signal should be carefully observed to compare with initial signature previously observed. If the signal levels are significantly different, it is reasonable to assume that additional shock pulses are necessary to knock loose the particle that was originally observed.

After no additional signals are observed, the tape can be carefully peeled from the device and a microscopic examination made to see if particulate matter is present. When used in conjunction with a scanning electron microscope, a conclusive determination of composition, shape and dimensions of captured particles can be made.

Alternate Tests for Loose Particle Detection

Vibration Testing

In addition to the PIND test, there are two other procedures for conducting electrical monitoring during vibration for particle detection. These procedures are (1) monitored vibration with power applied to the device, and (2) sinusoidal vibration with mechanical shocks applied at various intervals.[93]

The monitored vibration method is usually specified for testing power transistors and diodes. The technique uses a sinusoidal vibration either at a fixed frequency or a swept range of frequencies. Power is applied to the device under test and a latching circuit is used as a means of detecting and displaying electrical malfunctions.

This test is not considered completely effective because in order to be detected a particle must occupy one of many specific locations at a specific time. Additionally, the test is lengthy and expensive.

The sinusoidal vibration test is sometimes termed the *autonetics test*. It is similar to monitored vibration tests except that it subjects the device under test to shocks at frequent intervals.[93] These shocks serve to dislodge particles that may be adhering to cavity walls and allows them to freely move about the cavity.

As with the monitored vibration test, the sinusoidal vibration test is considered ineffective; correlation of data on devices identified as containing particles and the results of analysis on the devices were poor.[93] Analysis of devices that passed sinusoidal vibration tests indicated that many contained particles. Additionally, the imposed mechanical stresses are believed to generate particles. This test is also considered impractical due to high cost, limitation of available test equipment, and the time and cost of designing and instrumenting the driving and monitoring circuitry.

Radiographic Testing

Radiographic tests can be effective in detecting large loose or affixed particles as well as manufacturing process deficiencies. It is inexpensive and very easily performed. Some of the drawbacks of X-ray testing of electronic components include: some conductive particles are invisible to X-rays due to composition and size; particles smaller than 0.025 mm (0.001 in.) are not detectable and some small particles can be masked by silicon-gold die attached material.

A variation on radiographic testing combined with vibration is conducted by imaging the device and making a radiograph. The device is then subjected to vibration and radiographed again. The two exposures are compared to locate particles that have moved during vibration.

Conclusion

Loose particle detection has been a problem since the early 1960s. Since then, PIND systems have evolved from custom mechanical systems to the more sophisticated microprocessor based systems available today. PIND system detection capability has dramatically increased from 30 micrograms to 0.16 micrograms.

In comparison to alternate test methods, and because the PIND test has been incorporated into Military Specifications 883C, 750C and 202F and is required for Class-S microcircuits and semiconductors, it is generally accepted as the most economical and technically advanced test for finished electronic devices.

It is important to remember that particle mass, shape, size, composition and the type of device the particle is encapsulated in have direct bearing on whether the particle is detectable, whether the PIND test is repeatable and whether the particle is capable of compromising device integrity.

Care should be taken to eliminate particles prior to completed fabrication: proper manufacturing process controls; cleanliness in the manufacturing environment; visual inspection of the device prior to capping; and source surveillance are necessary.

To ensure the highest effectiveness of the PIND test, care should be taken to guarantee that equipment is operating at peak performance levels; that the test environment is as clean and noise free as possible; that the test operator is highly skilled and thoroughly trained; and that proper observation procedures are followed.

PART 9
DESIGNING MICROELECTRONIC WELDS FOR ACOUSTIC EMISSION TESTABILITY

Nature of Welds in Microelectronics

In developing tests to detect poor welds in microelectronic components, the failure modes must be understood and this is done by understanding the characteristics of unacceptable welds. An unacceptable weld may be defined here as one having a small percentage of its interface welded, either because of an underbond weld schedule or contamination in the interface. In either case, the weld interface is actually joined together by isolated microwelds.

Thermocompression welds made between contaminated gold interfaces using normal bonding parameters (see Fig. 76) have been found to contain numerous isolated microwelds with dimensions of a few micrometers.[103] Underwelding tends to produce fewer but somewhat larger microwelds.[104] As a weld between clean interfaces matures, the microwelds increase in size and join together, resulting in a high percentage of the interface (more than 50 percent) being welded.

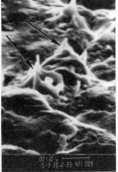

FIGURE 76. Scanning electron microscope photographs of chrome gold substrate with thermocompression bonded gold beam lead; substrate metallization intentionally contaminated to inhibit bonding (arrows indicate broken microwelds): (a) substrate gold pulled off and adhering to beam lead; and (b) gold lead microweld breaks on substrate

(a)   (b)

Griffith Cracking

Most weldments used in electronic assembly involve copper, gold, aluminum and other ductile materials. Cracks in the bulk of such polycrystalline materials can be expected to propagate erratically, with considerable yielding of the metal. However, surface cracks in brittle materials such as glass propagate in relatively straight lines with velocities about 0.3 times the velocity of sound in that material. This is known as *Griffith crack propagation*.[105] As crack length increases, the stress necessary to keep the crack growing *decreases*. Thus, once started by the application of a given stress, complete fracture usually ensues.

The edge of an unacceptable weld serves as a notch or crack for stress buildup, and the interface serves as a channel through which the extending crack must propagate. This generally occurs rapidly without major deformation or yielding of the weldments. This type of break between ductile weldments, therefore, has many similarities to the Griffith break.

Because the density of the microwelds in an unacceptable weld may vary from point to point, a propagating crack could meet increased resistance and stop until a higher force is applied, as in the brittle cleavage of polycrystalline materials when a crack encounters a grain boundary. This is equivalent to adding a microweld density function to the theoretical breaking stress of the Griffith equation.

Such a function could be position dependent within a given weld, could vary from one weld to the next and would result in changing the specific surface energy. Establishing an effective surface energy was the approach taken by investigators using the basic Griffith equation to explain crack propagation through polycrystalline zinc and iron.[106] The breaking stress is then proportional to the square root of the microweld density. The stress σ_b at which the crack propagates is:

$$\sigma_b = \sqrt{\frac{D}{a}\frac{E\rho}{4c}} \qquad \text{(Eq. 31)}$$

Where:

D = the effective surface energy;
a = the interatomic distance;
E = Young's modulus;
ρ = the radius of curvature at the crack leading edge; and
$2c$ = the crack length.

Considering the half-power relation between the breaking stress and the microweld density, it is likely that the majority of such welds would fail quickly with a single acoustic emission burst or with a series of bursts within a few milliseconds and not be preceded by detectable cracks at significantly lower stress levels. This has been observed in large numbers of acoustic emission monitored microelectronic wire-bond pull tests and in TAB pull tests. However, some welds (perhaps 10 to 20 percent) can and do fail by two or more partial lifts when there is a large change in the number of microwelds across the weld interface.

Designing Welds for Testability

It would be desirable to have all welds fail by a series of small cracks as the stress level is increased, thus giving a warning burst of acoustic emission at a low stress level. This could then be related to impending failure if stresses were increased.

Typical unacceptable welds break rapidly without acoustic emission warning. Changing this characteristic by changing the design of the weld interface would allow for more efficient acoustic emission testing.

The most likely design change applicable to electronic leads is to shape the welding tool to apply lower force to the bond heel region. In a normal strong weld, this results in the heel region being slightly weaker than the bulk of the weld. Underbonded or contaminated welds made with the same tool would have a heel region significantly weaker than the bulk of the weld and would begin to peel, emitting acoustic emission at a low test force. Low stress detection of unacceptable welds would then be possible.

Designing welds, other joints and entire structures for testability is an important development. In the field of microelectronics, such design considerations improve the reliability of acoustic emission testing as a post-production control technique.

REFERENCES

1. Kahn, Sherwin R. and Robert W. Checkaneck. "Acoustic Emission Testing of Multilayer Ceramic Capacitors." *IEEE Transactions on Components, Hybrids and Manufacturing Technology*. Vol. CHMT-6, No. 4. New York, NY: Institute of Electrical and Electronics Engineers (December 1983).
2. Spriggs, R.S. and A.H. Cronshagen. *Insulation Circuits* (February 1978): pp 59-62.
3. US Patent 4344326 (August 1982).
4. Jon, Min-Chung, Governor Ware and John Stapleton. "Applications of AE in Undersea Repeater Manufacture." *The Western Electric Engineer* (October 1979): pp 31-37.
5. Kahn, Sherwin and Dennis Miller. "Detecting Ceramic Substrate Cracking During Thermocompression Bonding." *The Western Electric Engineer* (October 1979): pp 15-19.
6. Vahaviolos, Sotirios J. "Real-Time Detection of Microcracks in Brittle Materials Using Stress Wave Emission (SWE)." *IEEE Transactions on Parts, Hybrids and Packaging*. Vol. PHP-10, No. 3. New York, NY: Institute of Electrical and Electronics Engineers (September 1974).
7. Harman, G.G. "Acoustic Emission-Monitored Tests for TAB Inner Lead Bond Quality." *IEEE Transactions on Components, Hybrids and Manufacturing Technology*. Vol. CHMT-5. New York, NY: Institute of Electrical and Electronics Engineers (1982): pp 445-453.
8. Harman, G.G. and K.A. Harmison. "The Assessment of Hybrid Package Glass-Metal Seal Reliability Using Acoustic-Emission Measurement Techniques." *International Journal of Hybrid Microelectronics*. Vol. 5 (November 1982): pp 248-259.
9. Harman, G.G. *Semiconductor Measurement Technology: The Use of Acoustic Emission to Determine the Integrity of Large Kovar Glass-Sealed Microelectronic Packages*. NBS Special Publication. Gaithersburg, MD: National Bureau of Standards (May 1982): pp 400-470.
10. van Kessel, C. and S. Gee. "The Use of Fractography in the Failure Analysis of Die Cracking." *International Symposium for Testing and Failure Analysis*. Los Angeles, CA (October 1984): pp 258.
11. Taylor, T.C. and F.L. Yuan. "Thermal Stress and Fracture in Shear-Constrained Semiconductor Device Structures." *Transactions of Electrical Developments*. No. ED-9. Institute of Radio Engineers (1962): pp 303.
12. van Kessel, C., S. Gee and J. Murphy. "The Quality of Die-Attachment and Its Relationship to Stresses and Vertical Die Cracking." *Proceedings of the Thirty-Third Electronic Components Conference* (1983): pp 237.
13. Isagawa, M. "Deformation of Al Metallization Caused by Thermal Shock." *IEEE Reliability Physics Symposium*. New York, NY: Institute of Electrical and Electronics Engineers (1980): pp 1980.
14. Okikawa, S., M. Sakimoto, M. Tanaka, T. Sato, T. Toya and Y. Hara. "Stress Analysis of Passivation Film Crack for Plastic Molded LSI Caused by Thermal Stress." *International Symposium for Testing and Failure Analysis* (1983): pp 275.
15. Severn, H. Huston and J. Lloyd. "Acoustic Emission Study of Electromigration Damage in Al-Cu Thin Film Conductor Stripes." *IEEE Reliability Physics Symposium*. New York, NY: Institute of Electrical and Electronics Engineers (1984): pp 256.
16. Vahaviolos, S. "Real Time Detection of Microcracks in Brittle Materials Using Stress Wave Emission (SWE)." *IEEE Transactions on Parts, Hybrids and Packaging*. Vol. PHP-10. New York, NY: Institute of Electrical and Electronics Engineers (1974): pp 152.
17. Kahn, S. and D. Miller. "Acoustic Emission Detection: Part II, Detecting Ceramic Substrate Cracking During Thermocompression Bonding." *The Western Electric Engineer*. Vol. 23 (October 1979): pp 15.
18. Ikoma, T., M. Ogura and Y. Adachi. "Acoustic Emission from Single Crystals of Gallium Arsenide." *Third Acoustical Emission Symposium*. Tokyo, Japan (1976): pp 329.
19. Ogura, M., Y. Adachi and T. Ikoma. "Acoustic Emission from Gallium Arsenide Single Crystals During Deformation." *Journal of Applied Physics*. Vol. 50 (1979): pp 6745.
20. Kahn, S. and R. Checkaneck. "Acoustic Emission Testing of Multi-Layer Ceramic Capacitors." *Proceedings of the Thirty-Third Electronic Components Conference* (1983): pp 77.
21. Carlos, M. and M. Jon. "Detection of Cracking During Rotational Soldering of a High Reliability and Voltage Ceramic Capacitor." *Proceedings of the Twenty-Eighth Electronic Components Conference* (1978): pp 336.
22. Harman, G. "Non-Destructive Tests Used to Insure the Integrity of Semiconductor Devices with Emphasis on Passive Acoustic Techniques." *Non-Destructive Evaluation of Semiconductor Materials and Devices*.

J. Zemel, ed. New York, NY: Plenum Press (1979): pp 677.
23. Harman, G. "The Use of Acoustic Emission as a Test Method for Beam Lead Bond Integrity." *IEEE Reliability Physics Symposium*. New York, NY: Institute of Electrical and Electronics Engineers (1976): pp 86.
24. Harman, G. *The Use of Acoustic Emission as a Test Method for Electronic Interconnections and Joints, Soft Soldering and Welding in Electronics and Precision Mechanics*. DVS meeting. Munich, Federal Republic of Germany (November 1981): pp 104.
25. Harman, G. "Acoustic Emission Monitored Tests for TAB Inner Lead Bond Quality." *IEEE Transactions, Components, Hybrids and Manufacturing Technology*. Report CHMT-5. New York, NY: Institute of Electrical and Electronics Engineers (1982): pp 445.
26. Gee, S., C. van Kessel, J. Murphy and N. Panousis. "Acoustic Emissions Monitoring of Die Shear Testing." *International Symposium for Hybrid Microelectronics*. Philadelphia, PA (October 1983): pp 410.
27. Lawn, B. and T. Wilshaw. *Fracture of Brittle Solids*. Cambridge, UK: Cambridge University Press (1975): pp 24.
28. Spanner, J. *Acoustic Emission: Techniques and Applications*. Evanston, IL: Intex Publishing Company (1974): pp 3.
29. Vengurlekar, A. "Model of Backsurface Gettering of Metal Impurities in Silicon." *Applied Physics Letters*. Vol. 41, No. 9 (November 1982).
30. Eggermont, G., D. Allison, S. Gee, K. Ritz, R. Falster and J. Gibbons. "Characterization of Laser Induced Backside Damage for Gettering Purposes." *Laser and Electron Beam Interactions with Solids*. B. Appleton and G. Ceeller, eds. North-Holland, NY (1982): pp 615.
31. Eggermont, G., S. Gee, C. van Kessel, R. Falster and J. Gibbons. "Thermal Annealing of Microcracks Produced by Backside Laser Irradiation of Silicon." *Applied Physics Letter*. Vol. 41 (December 1982): pp 1133.
32. Huang, D. Private correspondence. Signetics (PRLS).
33. Ritz, K. Private correspondence. Signetics (PRLS).
34. Lawn, B. and R. Wilshaw. "Indentation Fracture: Principles and Applications." *Journal of Material Sciences*. Vol. 10 (1975): pp 1049.
35. Lankford, J. and D. Davidson. "The Crack-Initiation Threshold in Ceramic Materials Subject to Elastic/Plastic Indentation." *Journal of Material Sciences*. Vol. 14 (1975): pp 1662.
36. Lee, L. and H. Kim. "Acoustic Emission During Indentation Fracture of Soda-Lime Glass." *Journal of Material Sciences*. Vol. 3 (1984): pp 907.
37. Hagan, J. "Micromechanics of Crack Nucleation During Indentation." *Journal of Material Sciences*. Vol. 14 (1979): pp 2975.
38. Lawn, B., T. Dabbs and C. Fairbanks. "Kinetics of Shear-Activated Indentation Crack Initiation in Soda-Lime Glass." *Journal of Material Sciences* (1983): pp 2785.
39. Evans, A. and E. Charles. "Fracture Toughness Determinations by Indentation." *Journal of the American Ceramic Society*. Vol. 59. Columbus, OH: American Ceramic Society (1976): pp 371.
40. Anstis, G., P. Chantikul, B. Lawn and D. Marshall. "A Critical Evaluation of Indentation Techniques for Measuring Fracture Toughness: Direct Crack Measurements." *Journal of the American Ceramic Society*. Vol. 64. Columbus, OH: American Ceramic Society (1985): pp 533.
41. Lankford, J. "Indentation Microfracture in the Palmqvist Crack Regime: Implications for Fracture Toughness Evaluation by the Indentation Analysis." *Journal of the American Ceramic Society*. Vol. 62. Columbus, OH: American Ceramic Society (1979): pp 347.
42. Lawn, B. and D. Marshall. "Hardness, Toughness, and Brittleness: An Indentation Analysis." *Journal American Ceramic Society*. Vol. 62. Columbus, OH: American Ceramic Society (1979) pp 347.
43. St. John, C. "The Brittle-to-Ductile Transition in Pre-Cleaved Silicon Single Crystals." *Philadelphia Magazine*. Vol. 32 (1975): pp 1193.
44. Chen, C. and M. Leipold. "Fracture Toughness of Silicon." *American Ceramic Society Bulletin*. Vol. 59. Columbus, OH: American Ceramic Society (1980): pp 469.
45. Lawn, B., D. Marshall and P. Chantikul. "Mechanics of Strength-Degrading Contact Flaws in Silicon." *Journal of Material Sciences*. Vol. 16 (1981): pp 1769.
46. Shintani, A., S. Sugaki and H. Nakashima. "Temperature Dependence of Stresses in Chemical Vapor Deposited Vitreous Films." *Journal of Applied Physics*. Vol. 51 (1980): pp 4197.
47. Blech, I. and U. Cohen. "Effects of Humidity on Stress in Thin Silicon Dioxide Films." *Journal of Applied Physics*. Vol. 53 (1982): pp 4202.
48. Lawn, B. and E. Fuller. "Measurement of Thin-Layer Surface Stresses by Indentation Fracture." *Journal of Material Sciences*. Vol. 19 (1984): pp 4061.
49. Lawn, B. "Hertzian Fracture in Single Crystals with the Diamond Structure." *Journal of Applied Physics* (1968): pp 4828.
50. "Method 2019.1: Die Shear Strength." *Military Standard Test Method and Procedures for Microelectronics*. MIL-STD-883B. Washington, DC: Department of Defense (August 1977).
51. Wu, S. *Polymer Interface and Adhesion*. New York, NY: Marcel Dekker, Inc. (1981): pp 504.
52. Cherry, B. *Polymer Surfaces*. Cambridge, UK: Cambridge University Press (1981): pp 90.

53. Wilson, Martin N. *Superconducting Magnets*. Oxford, UK: Clarendon Press (1983).
54. Kozman, T.A., et al. "Magnets for the Mirror Fusion Test Facility: Testing of the First Yin-Yang and the Design and Development of Other Magnets." *IEEE Transactions on Magnetics*. Report MAG-19. New York, NY: Institute of Electrical and Electronics Engineers (1983): pp 859.
55. Haubenreich, P.N. "Superconducting Magnets for Toroidal Fusion Reactors." *IEEE Transactions on Magnetics*. Report MAG-17. New York, NY: Institute of Electrical and Electronics Engineers (1981): pp 31.
56. Iwasa, Y. "Experimental and Theoretical Investigation of Mechanical Disturbances in Epoxy-Impregnated Superconducting Coils." *Cryogenics*. Vol. 25 (1985): pp 304.
57. Nomura, H., K. Takahisa, K. Koyama and T. Sakai. "Acoustic Emission from Superconducting Magnets." *Cryogenics*. Vol. 17 (1977): pp 471.
58. Schmidt, Curt and Gabriel Pasztor. "Superconducting Under Dynamic Mechanical Stress." *IEEE Transactions on Magnetics*. Vol. MAG-13. New York, NY: Institute of Electrical and Electronics Engineers (1977): pp 116.
59. Turowski, P. "Acoustic Emission and Flux Jump Phenomena during Training of Superconducting Magnets." *Proceedings of the Sixth International Conference on Magnetics Technology*. Report MT-6 (1978): pp 648.
60. Pasztor, G. and C. Schmidt. "Dynamic Stress Effects in Technical Superconductors and the 'Training Problem' of Superconducting Magnets." *Journal of Applied Physics*. Vol. 49 (1978): pp 886.
61. Pasztor, G. and C. Schmidt. "Acoustic Emission from NbTi Superconductors during Flux Jump." *Cryogenics*. Vol. 19 (1979): pp 608.
62. Pasztor, G. and C. Schmidt. "Dynamic Stress Effects in Technical Superconductors and 'Training' Problem of Superconducting Magnets." *Journal of Materials Science*. Vol. 16 (1981): pp 2154.
63. Tsukamoto, O. and Y. Iwasa. "Sources of Acoustic Emission in Superconducting Magnets." *Journal of Applied Physics*. Vol. 54 (1983): pp 997.
64. Maeda, H., O. Tsukamoto and Y. Iwasa. "The Mechanism of Friction Motion and its Effect at 4.2 K in Superconducting Magnet Winding Models." *Cryogenics*. Vol. 22 (1982): pp 287.
65. Bobrov, E.S., J.E.C. Williams and Y. Iwasa. "Experimental and Theoretical Investigation of Mechanical Disturbances in Epoxy-Impregnated Superconducting Coils. Shear-Stress Induced Epoxy Fracture as the Principal Source of Premature Quenches and Training — Theoretical Analysis." *Cryogenics*. Vol. 25 (1985): pp 307.
66. Iwasa, Y., E.S. Bobrov, O. Tsukamoto, T. Takaghi and H. Fujita. "Experimental and Theoretical Investigation of Mechanical Disturbances in Epoxy-Impregnated Superconducting Coils. Fracture-Induced Premature Quenches." *Cryogenics*. Vol. 25 (1985): pp 317.
67. Fujita, H., T. Takaghi and Y. Iwasa. "Experimental and Theoretical Investigation of Mechanical Disturbances in Epoxy-Impregnated Superconducting Coils. Prequench Cracks and Frictional Motion." *Cryogenics*. Vol. 25 (1985): pp 323.
68. Nomura, H., M.W. Sinclair and Y. Iwasa. "Acoustic Emission in a Composite Copper NbTi Conductor." *Cryogenics*. Vol. 20 (1980): pp 283.
69. Pappe, Markus. "Acoustic Emission in Superconducting Magnets." *IEEE Transactions on Magnetics*. Report MAG-17. New York, NY: Institute of Electrical and Electronics Engineers (1981): pp 2082.
70. Tsukamoto, O., M.R. Steinhoff and Y. Iwasa. "Acoustic Emission Triangulation of Mechanical Disturbances in Superconducting Magnets." *Proceedings of the Ninth Symposium on Engineering Problems of Fusion Research*. IEEE Publication 81CH1715-2 NPS. New York, NY: Institute of Electrical and Electronics Engineers (1981): pp 309.
71. Tsukamoto, O. and Y. Iwasa. "Acoustic Emission Triangulation of Disturbances and Quenches in a Superconductor and a Superconducting Magnet." *Applied Physics Letters*. Vol. 40 (1982): pp 538.
72. Ige, O.O., A.D. McInturff and Y. Iwasa. "Acoustic Emission Monitoring Results from a Fermi Dipole." *Cryogenics*. Vol. 26 (1986): pp 131. See also: Ige, O.O. *Acoustic Emission Monitoring Instrumentation and Results from a High-Performance Superconducting Dipole*. Master of Science thesis. Department of Mechanical Engineering, Massachusetts Institute of Technology. Unpublished (January 1985).
73. Lore, J., N. Tamada, O. Tsukamoto and Y. Iwasa. "Acoustic Emission Monitoring Results from the MFTF Magnets." *Cryogenics*. Vol. 24 (1984): pp 201. See also: Lore, J. *Monitoring Disturbances in Large Superconducting Magnets Using an Acoustic Emission Technique*. Master of Science thesis. Department of Mechanical Engineering, Massachusetts Institute of Technology. Unpublished (January 1983).
74. Cogswell, F.J., Y. Iwasa, J.W. Lue and J.N. Luton. "Acoustic Emission (AE) Techniques for Large Superconducting Magnets." *Advanced Cryogenic Engineering*. Vol. 31 (1986): pp 269. See also: Cogswell, F.J. *Monitoring Large Superconducting Magnets Using Acoustic Emission Technology*. Master of Science thesis. Department of Mechanical Engineering, Massachusetts Institute of Technology. Unpublished (September 1985).

75. Iwasa, Y. and M.W. Sinclair. "Acoustic Emission in Superconductors and Superconducting Magnets and Its Diagnostic Potential." *Mechanics of Superconducting Structures*. ASME AMD-41. New York, NY: American Society of Mechanical Engineers (1980): pp 109.
76. Tsukamoto, O., J.F. Maguire, E.S. Bobrov and Y. Iwasa. "Identification of Quench Origins in a Superconductor with Acoustic Emission and Voltage Measurements." *Applied Physics Letters*. Vol. 39 (1981): pp 172.
77. Maeda, H. "Application of Acoustic Emission Technique to a Multifilamentary 15.1 tesla Superconducting Magnet System." *Advances in Cryogenic Engineering*. Vol. 32 (1986): pp 293.
78. Nishijima, S., K. Shibata, T. Okada, K. Matsumoto, M. Hamada and T. Horiuchi. "An Attempt to Reduce Training Using Filled Epoxy as an Impregnating Material." *IEEE Transactions on Magnetics*. MAG-19. New York, NY: Institute of Electrical and Electronic Engineers (1983): pp 216.
79. Caspi, S. and W.V. Hassenzahl. "Source, Origin and Propagation of Quenches Measured in Superconducting Dipole Magnets." *IEEE Transactions on Magnetics*. MAG-19. New York, NY: Institute of Electrical and Electronic Engineers (1983): pp 692.
80. Maeda, H. "Mechanical Disturbances for a Cable-In-Substructure Superconductor. *Cryogenics*. Vol. 24 (1984): pp 208.
81. Yoshida, K., et al. "Acoustic Emission Measurement on Large Coils at JAERI." *Advanced Cryogenic Engineering*. Vol. 31 (1986): pp 277.
82. H. Iwasaki, S. Nijishima and T. Okada. "Application of Acoustic Emission Method to the Monitoring System of Superconducting Magnet." *Proceedings of the Ninth International Conference on Magnet Technology*. Swiss Institute for Nuclear Research (1985): pp 830.
83. Ige, O.O., H. Fujita and Y. Iwasa. "Acoustic Emission Instrumentation for High-Performance Superconducting Magnets." *Advanced Cryogenic Engineering*. Vol. 31 (1986): pp 303.
84. Tsukamoto, O. and Y. Iwasa. "Correlation of Acoustic Emission with Normal Zone Occurrence in Epoxy-Impregnated Windings: an Application of Acoustic Emission Diagnostic Technique to Pulse Superconducting Magnets." *Applied Physics Letter*. Vol. 24 (1984): pp 922.
85. Hartwig, G. "Low Temperature Properties of Potting and Structural Materials for Superconducting Magnets." *IEEE Transactions on Magnetics*. MAG-11. New York, NY: Institute of Electrical and Electronic Engineers (1975): pp 536.
86. Tsukamoto, O. and Y. Iwasa. "Acoustic Emission Diagnostic and Monitoring Techniques for Superconducting Magnets." *Advanced Cryogenic Engineering*. Vol. 31 (1986): pp 259.
87. Tsuchiya, K., et al. "A Prototype Superconducting Insertion Quadrupole Magnet for the Tristan Main Ring." *Advanced Cryogenic Engineering*. Vol. 31 (1986): pp 173.
88. Fujita, Hiroyuki and Yukikazu Iwasa. "High-Resolution Experimental Techniques for Cryomechanics — A Study of Mechanical Behavior of Materials at 4.2 K." *Experimental Mechanics*. Vol. 26 (1986): pp 128.
89. Jacob, K., Y. Iwasa, J.W. Lue and J.R. Miller. "Acoustic Emission Monitoring of LCP Superconducting Magnets." *Proceedings of the Tenth Symposium on Engineering Problems of Fusion Research*. IEEE Catalog No. 83 CH1916-6NPS. New York, NY: Institute of Electrical and Electronic Engineers (1983): pp 754. See also: Jacob, K.D. *Acoustic Emission Instrumentation System for Monitoring Large Superconducting Magnets*. Master of Science thesis. Department of Mechanical Engineering, Massachusetts Institute of Technology. Unpublished (January 1984).
90. Shen, S.S., C.T. Wilson and J.N. Luto, Jr. "Acoustic Emission Measurements for Locating High-Voltage Breakdowns in Large Superconducting Magnet Systems." *Advanced Cryogenic Engineering*. Vol. 31 (1986).
91. Golden, Frederic. "Those Balky Computers Again." *Time Magazine* (December 19, 1983).
92. Dembart, Lee. "Computer Flaws on Shuttle Tied to Tiny Objects." *Los Angeles Times* (December 22, 1983).
93. Adolphsen, John, William Kagdis and Albert Timmins. *A Survey of Particle Contamination in Electronic Devices*. Greenbelt, MD: Goddard Space Flight Center (December 1976).
94. Clark, L.D., J.C. Burrus and R.D. Clark. "Evaluation of Selected Methods for Detecting Contaminants within Semiconductors." *Proceedings of the Institute of Environmental Sciences Annual Technical Meetings* (1965).
95. Reynolds, Willy. *PIND Test Training Report*. Report No. 03-060-WRR. Texas Instruments (May 1985).
96. Hurd, Walter. Personal correspondence. Burbank, CA: Lockheed Corporation (March 1984).
97. McGuiness, N. *PIND Testing Workshop Report*. Report TOR 0082 (2902-04)-4 (March 1982): pp 14.
98. Adolphsen, John W. *The Effectivity of PIND Testing*. Greenbelt, MD: NASA Goddard Space Flight Center (1979).
99. McCullough, Ralph E. "Screening Techniques for Intermittent Shorts." *IEEE Reliability Physics Symposium*. New York, NY: Institute of Electrical and Electronics Engineers (1972): pp 19.

100. McCullough, Ralph E. "Hermeticity and Particle Impact Noise Test Techniques." *IEEE Reliability Physics Symposium*. New York, NY: Institute of Electrical and Electronics Engineers (1976): pp 256.

101. McCullough, Ralph E., James C. Burrus and Willy R. Reynolds "PIND's Role as a Failure Analysis Tool." *Proceedings of the ATFA Conference* (1979).

102. Chesne, T. and B. Lemaire. "PIND Test — Summary of Studies Conducted at Crouzet." *Proceedings of Third International Conference on Reliability and Maintainability*. Crouzet, France (October 1982).

103. Harman, G.G. *IEEE Transactions on Parts, Hybrids and Packaging*. Vol. PHP-13, No. 2. New York, NY: Institute of Electrical and Electronics Engineers (September 1977): pp 116-127.

104. Harman, G.G. and K.O. Leedy. *Proceedings of the Tenth Annual IEEE Reliability Physics Symposium*. New York, NY: Institute of Electrical and Electronics Engineers (April 1972): pp 49-56.

105. Anderson, O.L. *The Griffith Criterion for Glass Fracture*. New York, NY: John Wiley and Sons Publishing (1959).

106. Greenwood, G.W. and A.G. Quarrell. *Journal of the Institute of Metals*. Vol. 82 (1954): pp 551-560.

SECTION 12

ACOUSTIC EMISSION APPLICATIONS IN THE AEROSPACE INDUSTRY

Stuart McBride, Royal Military College of Canada, Kingston, Ontario

William Boyce, Sikorsky Aircraft, Stratford, Connecticut
J. Philip Perschbacher, Sikorsky Aircraft, Stratford, Connecticut
William Pless, Lockheed Georgia Company, Marietta, Georgia

PART 1
ACOUSTIC EMISSION TESTING OF AEROSPACE METAL ALLOYS

The most common metal alloys currently used in aircraft structures are AlCu, AlMgZn, AlTi and high strength steel. These materials are used in a variety of component types but the structural use of steel is normally restricted to attachment fittings and bushings where very high yield strength or wear properties are required.

Acoustic emission from airframe metal alloys may result from:

1. plastic deformation of the matrix material;
2. local microfracture processes during slow, stable crack growth (less than one micrometer per load cycle);
3. unstable crack advance due to excessive loading near component failure (greater than one micrometer advance during load application); and
4. rubbing and fretting noises (either component or crack face friction effects).

The maximum sizes of representative signals measured at the acoustic emission sensor output before amplification are shown in Table 1.

The general background noise during careful laboratory acoustic emission measurements on metal alloys would typically be in the range of 10 to 100 μV peak. In airframe structural applications, it could become as high as 1 mV. For in-flight monitoring applications, signal-to-noise considerations would restrict acoustic emission detection to type-2, 3 and 4 (listed above) for the aluminum alloys, and type-3 and type-4 for high strength steels. In this practical application, most of the accessible acoustic emission data relating to crack presence and size are contained in rubbing and fretting noises except near component failure when extremely large fracture related signals occur in most structural materials. Attempts have been made to separate the different types of noise sources in laboratory specimens using multiple classifier, adaptive learning techniques.[1,2] An external parameter (applied stress at the time of occurrence of the emission) has been used as a single signal classifier.[3,4] In airframe structures, signal rise time is an important parameter for distinguishing remote noise sources from sources originating in the vicinity of a crack.

Type-2 and type-4 are the important acoustic emission sources for in-flight monitoring of airframe structural components with respect to crack initiation and growth. While type-4 produces abundant data, it has generally been regarded by experimenters as a nuisance during studies of type-2 and has not been quantitatively documented except in applications to rotating machinery. Type-4 should be regarded as an essentially untapped source of information about crack presence, providing data that are not dependent on fracture mechanisms. In high strength steels, type-4 sources may provide the only detectable acoustic emission information relating to crack presence. There is, however, an important limitation to the efficacy of such data: the effect of lubricants can radically change or even eliminate the acoustic activity from rubbing and fretting.

TABLE 1. Approximate maximum signal peak amplitudes for several types of acoustic emission processes observed using a commercial piezoelectric sensor

	Source Mechanism			
Material	Plastic Deformation	Slow Crack Growth	Crack Overload	Rubbing and Fretting
4340 steel	100 μV	100 μV	1 V	100 mV
Aluminum alloys	100 μV	10 mV	1 V	100 mV

FROM DUNEGAN CORPORATION. REPRINTED WITH PERMISSION.

Quantitative Observations of Acoustic Emission from Crack Growth in Aircraft Aluminum Alloys

Acoustic emission during crack growth in aluminum alloys can be observed by cyclically loading the specimen in a conventional fatigue testing system using conventional acoustic emission apparatus. Recording of the load history, occurrence of acoustic emission signals, crack length and acoustic emission signal amplitude is desirable. All of these parameters can be recorded using a dedicated digital data acquisition system and simultaneously analyzed in the form of parametric and distribution analysis. If noise from spatially separated locations is present, a zone isolation system involving more than one acoustic emission sensor would be included to restrict data acceptance to a chosen location.

Data obtained in this manner for crack growth in commercial 7075-T6 aluminum plate material show the general behavior depicted in Figs. 1 and 2. Figure 1 shows a uniform increase in the cumulative number of acoustic emission events with increasing crack length. The source mechanism for this emission has been shown to be the fracture of intermetallic inclusions.[5,6] The size distribution can be directly related to the area size distribution of Mg_2Si inclusions as shown in Fig. 2.

This effect permits the prediction of the acoustic emission size distribution from an inclusion size distribution measurement on a polished metallographic specimen which characterizes the inclusion content of a given component or batch of material. Inclusion free and annealed material (7075-0) show no fracture related acoustic emission during

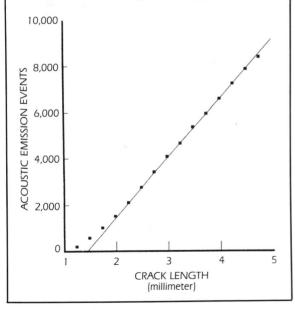

FIGURE 1. Cumulative number of acoustic emission signals with amplitudes greater than a selected threshold (−30 dB relative to 1 mV at the sensor output) and resulting from crack advance in a 7075-T6 aluminum specimen (single-edged notch) with dimensions 4.7 mm × 25 mm × 300 mm; note the uniform increase in number of events and increasing crack length

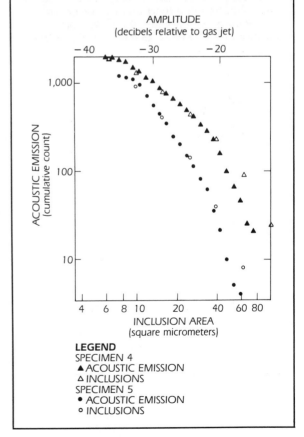

FIGURE 2. Amplitude distribution of acoustic emission voltage signals and the area size distribution of inclusions measured from a polished fracture face for two specimens of 7075-T6 aluminum; crack propagation carried out at room temperature; the selected fit suggests that the acoustic emission voltage signals have a size of −33 dB relative to the helium gas jet, corresponding to an inclusion area of 10 μm^2

crack advance for the amplitude range shown here. Similar behavior would be expected in other aluminum alloys containing inclusions.

Effect of Temperature

Airframe materials may be subjected to considerable temperature variations during use. Such temperature variations have associated variations in the bulk mechanical properties of the material, including yield strength and toughness.[7] In commercial aluminum alloys, this can affect the probability of inclusion fracture and hence the acoustic emission activity resulting from crack growth.

Figure 3 shows the measured temperature variation for specimens of 7075-T651 material. To obtain these results, the single-edged notch (SEN) specimen was chosen to permit fatigue crack growth at the chosen ambient temperature while the sensor was maintained at room temperature. This eliminates the effects of temperature on the sensor or couplant during the experiment. The misfit stresses developed at the inclusion/matrix interface are a result of differential thermal contraction resulting from the heat treatment.

Effect of Heat Treatment

The bulk mechanical properties of aircraft aluminum alloys are affected by heat treatment. This changes the probability of inclusion fracture, which in turn results in changes in the acoustic emission activity. Figure 4 shows the observed variation of acoustic emission activity as a function of material yield strength for crack growth in overaged 7075-T651 aluminum. Note that there is a linear relation between the number of events above threshold and the yield strength for the overaged material. This effect can be explained in terms of the misfit stresses at the inclusion/matrix interface that develop as a result of differential thermal contraction from the heat treatment.

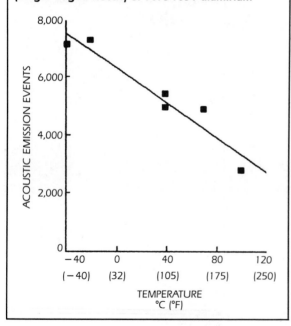

FIGURE 3. Variation of the number of acoustic emission events greater than a selected threshold (−30 dB relative to 1 mV at the sensor output) plotted as a function of crack growth temperature; data are for 10 μm^2 of crack advance in a 4.7 mm × 25 mm × 1 m specimen (single-edged notch) of 7075-T651 aluminum

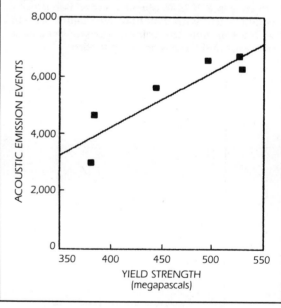

FIGURE 4. Variation of the number of acoustic emission events greater than a specified threshold (−30 dB relative to 1 mV at the sensor output) plotted as a function of bulk yield strength; data are for 10 μm^2 of crack advance in overaged 7075-T651 aluminum specimens (single-edged notch) 4.7 mm × 25 mm × 1 m

PART 2
ACOUSTIC EMISSION MONITORING OF AIRCRAFT STRUCTURES

Acoustic emission testing of aircraft structures is a challenging and difficult problem. The structures involve bolts, fasteners and plates, all of which move relative to one another due to differential structural loading during flight. The complex geometry of the airframe results in multiple mode conversions of acoustic emission source signals which compound the difficulty of relating source event to detected signal. Further, the most common discontinuity of interest is a crack initiating at a bolt hole. Detection of size change directly from micromechanical fracture sources would require unambiguous recognition of the fracture source in the presence of crack face and bolt hole rubbing when all sources differ in location by distances on the order of an acoustic wavelength.

To solve this problem, it is necessary to focus on the location in question, forsaking the most desirable feature of the acoustic emission technique: its use for surveillance of large areas or volumes of structure. Sources such as crack face rubbing (considered noise in a crack advance measurement) can be potential indicators of crack presence and size due to the frequent changes in component loading that occur during flight.

Several studies have investigated the feasibility of using acoustic emission techniques to monitor the integrity of aircraft components during flight.[8-14] It has been demonstrated that crack face rubbing noises can provide an indication of crack size during flight.[13] It has also been shown that crack growth is detectable during flight.[14] These observations are not necessarily applicable to an arbitrary location in a randomly selected airframe type, but could only be relevant to a particular component when the noise at this location is found to be suitable.

The techniques and apparatus required to perform acoustic emission tests in aircraft structures are similar to those used in laboratory and other field testing. When measurements are made during flight, however, there is the need (1) to strengthen the apparatus to withstand substantial vibrations (up to approximately ±9 G to 2 kHz) and (2) to provide protection against electromagnetic interference from avionics (radar and communications) systems, relays and possible bus switching between AC and battery power sources. It is advantageous to have the apparatus battery powered to minimize interaction with avionics apparatus and to prevent the loss of data due to power interruption or shutdown.

The actual apparatus normally involves several channels to permit zone isolation and source location and simultaneously records data to document the aircraft maneuver (accelerometers) or state of stress of the component or area of the airframe under investigation (strain gages). Because of the complex structural geometry of the airframe, impulse mapping of the region of interest normally is necessary for interpretation of the signal parameters and optimization of sensor location. This could be accomplished using either a piezoelectric pulser, capillary fracture or graphite fracture or other convenient impulse sources. For complex structures such as airframes, the repetitive capability of the piezoelectric pulser is a considerable advantage.

Principal findings on airframe acoustic emission for frequencies greater than about 100 kHz are listed below.

1. Most of the data detected by the acoustic emission system result from airframe rubbing and fretting.
2. Airframe noise is abundant when the airframe is experiencing a change in loading. This effect is substantial during low altitude, turbulent flight. Like all bolted structures when local loading is constant and high enough, the structure is locally quiet.
3. Noise sources remote from the sensor due to the presence of interfaces or airframe geometry are characterized by a very long rise time and predominantly low frequency content.
4. Aircraft jet engines do not produce appreciable noise above 300 kHz.
5. At threshold levels appropriate for the detection of slow crack growth in aircraft aluminum alloys (less than one micrometer per load cycle), thousands of airframe noise signals per flying hour are to be expected. Slow crack growth typically results in one to ten signals per flying hour.

Experimental Observations during Flight

Acoustic Emission Signal Amplitudes for Airframe Noises and Crack Growth

Using a single-channel recording system, acoustic emission data from several different airframe locations and different airframe types have been recorded during flight.[12]

For a particularly noisy component (CF100 wing trunnions), it was found that the transient burst signals varied over at least two orders of magnitude in amplitude and in number of events per flying hour. The largest signals were about twenty times larger in amplitude than those attributed to inclusion fracture emission during fatigue crack growth in airframe aluminum alloys.

The same signals are, by about a factor of five, smaller than those associated with rapid (greater than one micrometer per load cycle) crack propagation near failure. The particular component studied was attached to the fuselage and wing by several bolts and was expected to exhibit considerable friction noise from rubbing at the bolt holes. These conclusions have been confirmed for several locations involving different airframe types.

Effect of Crack Presence on Airframe Noise

Crack presence can be expected to result in friction noises from crack face rubbing and fretting during changes in component loading. Simultaneous acoustic emission data recording from similar components in symmetrical airframe locations would therefore be expected to exhibit different activity if one of the components contains a crack.

The validity of this was established using a dual-channel system to simultaneously record data from two symmetrically located wing attachment fittings.[13] The apparatus was portable and was used to study a fleet of aircraft, each with an identified crack in one of the components under investigation. One hour test flights were used and, in all cases studied, only one component of the symmetric pair was cracked. It was found that the ratio of the number of events above a selected threshold in the cracked component to the number of events above the same threshold in the symmetrically located uncracked component (N_c/N_{uc}) was greater than unity and increased progressively with increasing crack size (Fig. 5). No detectable crack advance occurred during these flights.

Relation of Component Noise to Airframe Loading

Acoustic emission activity results from changes in airframe loading during flight. Figure 6 shows the number of events above a specified threshold as a function of time for a cracked and an uncracked component. Figure 7 shows the occurrence of these same events as a function of amplitude, time and aircraft load for the cracked component. The data from these two components were simultaneously recorded during the same test flight.

The general form of the two curves in Fig. 6 is similar and results from the specific flight profile, yet the cracked component produced a larger number of events. Further, as seen in Fig. 7, high constant loads (+4 G aircraft acceleration in this case) produced a decrease in the maximum amplitude of events by a factor of about 300, resulting in a noise level similar to that experienced in fatigue tests using a screw driven testing machine. High acoustic activity resulted from changing loads during flight.

Acoustic Emission of Crack Advance during Flight

Crack advance has been shown to be detectable during flight using a precracked laboratory test specimen mounted in a structure across the inner fuselage of a CC130 aircraft.[14] In series with this specimen was an uncracked specimen and a load cell. Crack length was monitored by an eddy current

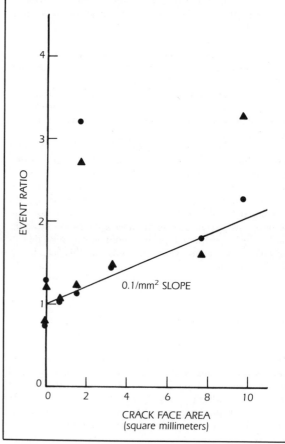

FIGURE 5. The ratio of the number of in-flight recorded acoustic emission signals from symmetrically located cracked and uncracked upper forward wing trunnions (N_c/N_{uc}) versus the crack face area in the cracked trunnion; each point is for a different airframe; circle gives the ratio through the controlled gravity maneuver portion of the flight; triangle gives the ratio for the entire flight

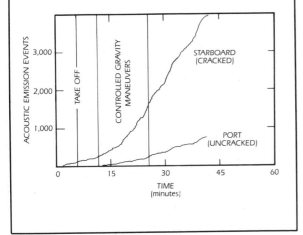

FIGURE 6. The in-flight acoustic emission cumulative event versus time plot for two symmetrically located upper forward wing trunnions on the same airframe during the same flight; port trunnion is uncracked and starboard trunnion contains a crack with a crack face area of 9 mm^2

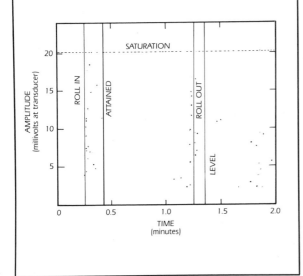

FIGURE 7. Amplitude versus time plot for in-flight recorded acoustic emission events occurring in an uncracked upper forward wing trunnion during a 4 G turn

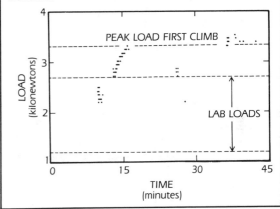

FIGURE 8. The occurrence of acoustic emission signals as a function of load and time during the first forty-five minutes of flight; also indicated is the load range used to crack the 7075-T6 specimen in the laboratory (1.2 to 2.7 kN); signals not attributable to crack advance occur at 10 minutes and 26 minutes; these can be distinguished by observation of the voltage-time behavior

probe attached to the cracked specimen. The load on the specimen was varied by changing aircraft altitude thus causing expansion of the inner fuselage across which the test specimen was attached.

Figure 8 shows the occurrence of acoustic emission data as a function of load and time during a test flight. Note the occurrence of signals each time the previous maximum load is exceeded, indicating fracture related acoustic emission activity. After the test flight, the specimen was fatigued to failure in the laboratory and examined in the scanning electron microscope. This revealed that the crack advance resulting from the two in-flight overloads was 20 μm (0.8 milli-in.). The experiment conclusively demonstrates the detection of crack advance by acoustic emission activity during flight.

Acoustic Emission Signal Characteristics Observed during Flight

Envelope-followed acoustic emission signals recorded during flight are shown in Fig. 9. Crack advance, crack face rubbing and airframe noises can be distinguished using simple parameters such as signal rise time and pulse length. For the cases shown, the sensor was placed 65 mm (2.5 in.) from the crack and airframe noises traveled distances greater than 1 m (3 ft) through the complex airframe structure. The degradation of rise time with propagation path is the feature most important to in-flight acoustic emission applications and is also evidenced by a progressive reduction in high frequency information with propagation distance.

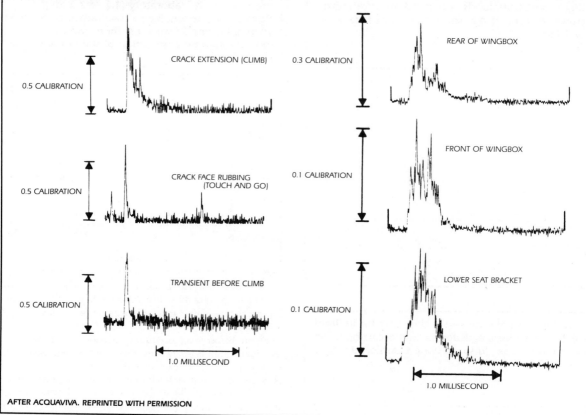

FIGURE 9. Comparison of the voltage-time behavior of typical recorded acoustic emission signals resulting from crack extension, crack face rubbing and noise signals in three airframe locations; note the long rise times of 0.2 milliseconds compared to the crack growth signals; the amplitude scale indicated is in units of the helium gas jet calibration signal

AFTER ACQUAVIVA. REPRINTED WITH PERMISSION

PART 3
AEROSPACE ORGANIC COMPOSITES

A composite is an engineered material consisting of two or more dissimilar materials that, in combination, provide desired properties and synergistic effects not exhibited by either of the materials acting alone. Composites are heterogeneous, anisotropic materials designed for high efficiency. The composite consists of one or more high strength, high modulus reinforcement materials embedded in a relatively low strength, low modulus matrix. There is a definite interface between the constituents. The reinforcement member usually dominates the mechanical and thermomechanical properties of the composite.

Organic Composites

An organic composite is one whose matrix is an organic nonmetallic material and the reinforcement fibers are either organic or metallic. A common aerospace organic composite consists of graphite fibers in a cured epoxy matrix. Advantageous properties of aerospace composites (compared to their conventional metallic counterparts) include lightness, high strength, high stiffness, directional stress distribution, corrosion resistance and high thermomechanical stability. In some structures, the composite may also be compound, bonded to a metallic part or containing interlaminar metallic shims. Selection of the constituent materials and the composite manufacturing technique is governed by the product's function, performance requirements and service environment.

Reinforcement fibers can be long continuous strands, short whiskers or particulates. Many aerospace organic composites are fabricated from sheets (plies) of continuous, resin impregnated collimated filaments or woven yarns. Manufacturing techniques include the winding of single continuous filaments, sheets or yarns on a mandrel, or spacing of oriented sheets or yarns to obtain a desired thickness and shape. These are then cured at specified temperature, pressure and time profiles. The most common method for achieving the cure conditions is the use of an autoclave. Fibers used as reinforcement in aerospace resin matrix composites are commonly made from the following materials: graphite, aramid, silicon carbide, boron tungsten, aluminum, quartz or fiberglass.

Aerospace Composite Applications

The use of organic composites in military and civilian aerospace applications is increasing rapidly. Several small all-composite aircraft have been produced and filament-wound graphite epoxy motor cases for the Space Shuttle solid rocket boosters have been fabricated and tested. Additional applications include: (1) wing skin panels and leading edges; (2) control surfaces such as ailerons and rudders; (3) complete wing box structures; (4) pressure chambers; (5) rocket motor cases and rocket nozzles; (6) ablators and nose cones; (7) radomes; (8) propellers and helicopter rotor blades; and (9) turbine blades, inlet ducts and cowlings.

Organic composites produce copious amounts of acoustic emission when stressed to deformation. Acoustic emission monitoring fulfills the need to apply a passive, nondestructive, noninterfering, real-time test method for evaluation of damage initiation and discontinuity growth. In a sound composite, acoustic emission may initiate at 40 to 60 percent of the ultimate failure load, or as low as 20 percent, if significant damage already exists. The acoustic emission waveform characteristics are influenced by the composite's physical characteristics, the failure mechanisms, the loading and environmental history of the structure. Therefore, acoustic emission waveforms carry potentially useful information about the condition of the material.

It is important for the researcher using acoustic emission techniques to understand not only acoustic emission but also to know a great deal about composites and how acoustic emission is generated within them. Acoustic emission monitoring is particularly applicable to composite materials testing, especially in view of the limitations of other nondestructive testing methods for these materials.

Applications for Monitoring Organic Composites with Acoustic Emission

Some purposes for monitoring organic composite specimens and structures include:

1. determining the onset of damage due to induced strains;
2. detecting damage propagation;
3. predicting burst pressure or failure load;
4. predicting service life;
5. assessing relative quality or progressive degradation; and
6. identifying failure mechanisms and sequence of failure.

Applications of acoustic emission techniques to aerospace materials research include three broad areas: (1) characterization of material behavior under mechanical, thermal and environmental stresses; (2) quality control of materials and structures; and (3) in-service surveillance. Much research has been done to establish relationships between composite failure mechanisms and acoustic emission events. The objective most often is twofold: (1) to relate recorded acoustic emission signal attributes to specific failure events such as matrix cracking, fiber-to-matrix disbonds, fiber pullout and fiber fracture; and (2) to establish the ability of acoustic emission techniques to display the sequence of failure events.

The underlying assumption is that each type of failure event releases stress energy packets that differ in amplitude, frequency content or other waveform characteristics and that the acoustic emission detection system can distinguish these differences. The first part of this assumption is probably well founded, but present acoustic emission sensors, particularly the piezoelectric variety, have not exhibited an adequate ability to identify separate failure events. Research is progressing in optical laser sensors that may overcome this difficulty.

In-Service Applications

Such applications will certainly expand greatly. Acoustic emission applications to components in-service will be concerned with detection of damage and long-term degradation. In-service applications require the use of (1) specified loading to produce acoustic emission and (2) standardized acoustic emission monitoring techniques and analysis methods. Much of the technology for in-service applications will utilize research data bases.

Degradation is the irreversible accumulation of damage that alters important mechanical properties (such as stiffness and interlaminar shear strength) and that may lead to impairment of the structure. Acoustic emission has been proven through research to reliably detect the presence of damage such as fiber fractures and impact damage at loads as low as 20 percent of limit load using sensitive source location techniques.[15] In addition, the felicity effect (deviation from the Kaiser effect at increased loads) seems to indicate the presence of damage in organic composites.[16] The Kaiser and felicity effects are illustrated in Fig. 10.

The feasibility of using acoustic emission to detect damage, degradation and poor quality in service structures may be achieved with the following approaches: (1) correlation of the felicity effect to damage and degradation; (2) correlation of changes in the felicity locus to long term degradation; and (3) correlating excessive amounts of acoustic emission at given proof loads to decreasing stiffness or shear strength.

In-Flight Composite Monitoring Systems

In-flight acoustic emission monitoring of experimental or aging organic composite primary structures may be desirable for detecting and locating damage induced by the flight loads and environments. In-flight monitoring may be accomplished when formally specified by military or Federal Aviation Administration requirements. Damage detection rather than characterization will likely be the primary objective of early in-flight monitoring systems, to pinpoint areas of the structure for subsequent testing to verify and define the damage.

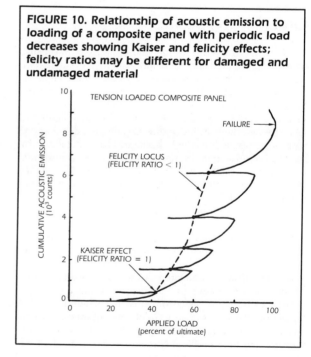

FIGURE 10. Relationship of acoustic emission to loading of a composite panel with periodic load decreases showing Kaiser and felicity effects; felicity ratios may be different for damaged and undamaged material

The technology for this application is comparatively advanced, so that system design and optimization for composites are the major effort. Some important concerns for an in-flight system include: noise rejection or suppression; a method of processing, analyzing and interpreting the relatively high volume of acoustic emission data coming from the stressed composite; in-flight stress correlation and consideration of the Kaiser effect; and flight-hardened sensors, selected for detection (damped narrow band types).

Composite Testing and Failure Mechanisms

Characterization Testing

Because of the impact on design and application, materials scientists are highly concerned with how organic composites diminish in strength, stiffness and durability as a result of repeated loading and exposure to adverse environments. Laboratory testing is a means of learning these things under simulated conditions. In the materials development and characterization stage, composite specimens may be subjected to mechanical tests such as those listed in Table 2. Some of these tests may be employed to determine certain service properties listed in the table.

Prior to or during these tests, the materials may be further conditioned through heat, moisture or ultraviolet exposure, or simulated damage representing impact, cracks and delaminations. Acoustic emission monitoring can be applied to detect real-time deformations during these tests. Figure 11 shows a graphite epoxy test panel instrumented with acoustic emission transducers and strain gages to monitor the growth of impact damage at the panel's center.

Failure Modes and Failure Mechanisms

It is now well known that ultimate failure of continuous fiber composites is preceded by a sequence of microstructural damage, debonding, interlaminar cracking, edge delamination and fiber fracture. The accumulation of damage may substantially reduce the material's strength and stiffness during the life of the component. Failure mechanisms that may occur in organic composites during testing are listed in Table 3. Fiber breaks are particularly significant when they occur in contiguous fibers or as bundles. This is the critical event in the component failure process.

This kind of discontinuity controls failure and its critical size appears to be insensitive to the presence of matrix cracking and delaminations. The matrix damage does affect the load under which the discontinuity propagates unstably. Impact damage, overstress fractures, delaminations and general degradation are the most likely failure modes for in-service organic composite structures.

TABLE 2. Acoustic emission used to evaluate damage in composites

Type of Test	Properties
Static tension	Ultimate strength
Static compression	Residual strength
Cyclic tension-tension	Damage tolerance
Cyclic tension-compression	Durability
Block or random spectrum loading	Fracture toughness
Pressure	Rupture strength
Three-point bending	Shear strength
Cantilever beam	Stiffness, flexure
Delamination tests	Interlaminar shear strength
Impact	Impact resistance
Thermal shock	Deformation

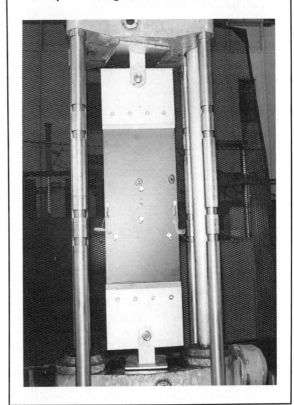

FIGURE 11. A graphite epoxy panel installed in a test machine for tension loading to failure; the panel is instrumented with a four-transducer triangular array for monitoring acoustic emission from impact damage in the center of the panel

TABLE 3. Failure mechanisms in organic composites

Tension	Compression
Fibers	Fibers
Debonding from matrix	Debonding from matrix
Fiber coating brittle fracture	Fiber coating brittle fracture
Fiber coating debonding from core filament	Fiber coating debonding from core filament
Fiber breaks (single)	Fiber kinking
Fiber bundle fracture	Fiber microbuckling
Fiber pullout	Laminar buckling
Matrix	Matrix
Microcracks	Shear cracking
Transverse cracking	In-ply splits parallel to fibers
In-ply splits parallel to fibers	Ply delaminations in buckling
Ply separations (delaminations)	Plastic flow

Acoustic Emission Relationships to Composite Mechanisms

Acoustic Emission Waveform Characteristics

The acoustic emission stress wave in the composite material is a pulse wave derived from a step displacement created by deformation. A primary characteristic of the stress wave is its spectral energy. When this energy is incident upon the sensor, the acoustic emission system detects the stress wave and presents a waveform that can be characterized by measurable factors, including: spectral energy, as modified by system response; also total energy; amplitude; damped oscillations and counts; rise time; and duration.

Any acoustic emission waveform characteristic can potentially form the basis of a correlation to failure mechanisms in the composite.

Propagation of Acoustic Emission in Organic Composites

This application is concerned with acoustic emission as a broadband sound impulse generated by disparate failure mechanisms and propagating through a nonhomogeneous medium. The amplitude, frequency and duration of the impulse is strongly influenced by (1) whether the failure takes place in the matrix or in the fibers; (2) the size of the failure; and (3) the energy released by the failure. Thus, information about the failure event is inherently carried in the impulse.

The interaction of the impulse with the material controls other characteristics that will be evident in the detected waveforms. These include velocity of the wave, attenuation through scattering by the fibers and microvoids, frequency dependent attenuation through dispersion, mode conversions, internal reflections from discontinuities, and external damping mechanisms.

When triangulation techniques are used to locate damage sites in composite structures, it is necessary to know the predominant wave velocity in the material. Because most aerospace organic composites are heterogeneous layered materials, they may exhibit several wave velocities, depending on direction of the wave with respect to fiber direction. Velocities for boron epoxy and graphite epoxy composites are commonly available in the literature.[17]

Through superposition of waves, considerable confusion and error can occur in acoustic emission location data. Quite often one mode and velocity predominates and this is usually the low velocity flexural mode. This can be determined empirically by monitoring simulated acoustic emission impulses caused by fracture of a 2H, 0.5 mm diameter graphite load on the part surface (Hsu method).

Attenuation of acoustic emission can be relatively high in organic composites due to absorption of sound energy by the matrix. High frequency components may be selectively absorbed by the matrix and scattered by the filaments and other small discontinuities. Degradation of the composite due to imposed loading and environmental effects over a period of time will cause time dependent changes in attenuation and dispersion. The acoustic emission wave, when compared to previous baseline wave data, can reveal this degradation because the information is impressed on waveforms as they propagate through the material. This is a potentially important area for research because the operational use of composites will demand effective in-service inspection methods to evaluate material integrity.

The effect of internal reflections is totally dependent on (1) the geometry of the material; (2) location of acoustic emission sources; and (3) location of the acoustic emission sensors. Small variations in any of these can produce large changes in the waveform at the sensor because of wave superposition effects.

Research in correlating acoustic emission waveform data to specific composite failure mechanisms is hampered by the effects of nonuniform absorption, dispersion, scattering and reflections, in addition to the self-response of the acoustic emission sensor and detection system.

Because of the information that may be potentially extracted from an acoustic emission waveform, it is desirable to use a broadband sensor and as low a frequency cutoff in the preamplifier as is practical. Noise, which is usually more prevalent at lower frequencies, limits the practicality of observing acoustic emission at low frequencies. In the absence of noise, the low frequency system response may be selected as 30 to 60 kilohertz. Noise generally is present, however, necessitating the use of low frequency cutoff filters at 100 kilohertz or higher. This is a matter for experimentation.

Correlations of Acoustic Emission and Composite Failure Mechanisms

Acoustic Emission Profiles

The materials scientist is often interested in determining when damage initiates and propagates under applied loads. Acoustic emission sensors can be applied to specimens under test to acquire a profile of acoustic emission versus load (or time, strain or some other parameter). This profile can reveal the load at which the sequence of failure events occurs. Such an acoustic emission load profile is shown in Fig. 12.

It should be noted that most of the acoustic emission in many organic composites originates in the matrix and at the fiber/matrix interface. Acoustic emission begins when the load is sufficient to cause microcracking in the matrix, the onset of damage. As the load is increased to a limit load or failure, the profile shows dramatic increases in acoustic emission as more significant matrix cracks, delaminations, splits, and finally, fiber fractures occur. The parameter that controls acoustic emission is the stress intensity factor rather than load, stress or crack length alone. Through careful building of a data base, the occurrence of the failure events can be recognized in profiles or even in the characteristics of individual signals.

The data base correlations of acoustic emission to failure mechanisms (and to onset and propagation of damage during the loading cycle) must be firmly verified from destructive analyses such as fractography or other established methods.[18] In these methods, composite specimens are tested to some percentage of fracture load, radiographed using enhancement penetrants, marking the damage with a permanent penetrant, then deplying the specimens using a pyrolysis technique. Each laminae can be examined in detail and the damage measured and characterized. The total damage can then be related to the acoustic emission profiles and other data.

Acoustic emission activity is affected by the same factors that directly affect individual failure events and ultimate failure, so that there is potentially good correlation between acoustic emission and failure. These factors include:

1. type of fibers;
2. fiber orientations and ply stacking sequence;
3. type of resin or matrix;
4. fiber to resin ratio;
5. resin cure conditions;
6. material conditioning;
7. previous load history;
8. the presence of discontinuities or damage; and
9. overall quality.

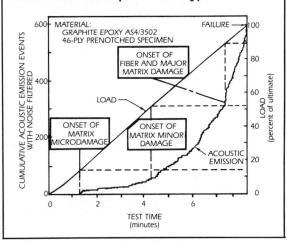

FIGURE 12. Acoustic emission and load profiles for a precracked multidirectional graphite epoxy specimen loaded in tension to failure; the loads for onset of certain types of damage or severity can be identified in plots of this type

To fully understand and utilize acoustic emission profiles, it is important that the investigator completely document the material design, processing and loading histories as well as the mechanical test and acoustic emission parameters. This documentation and all the acoustic emission profiles obtained on that material become the data on which further development is based.

Acoustic Emission as an Indicator of Discontinuity Significance

The identification of failure mechanisms and their sequencing is important in material development because of its importance to operational service requirements. Therefore, there is great interest in developing acoustic emission detection capabilities whereby stress induced failures in composites can be located and identified as matrix, fiber or interfacial failures. Identification of failure mechanisms using acoustic emission is based on the potential for the mechanism to imprint its signature on one or more of the acoustic emission waveform characteristics described earlier. Reference 19 contains a bibliography on the subject of significance of discontinuities in organic composites.

Noise Suppression

In order to extract the greatest information possible from an acoustic emission waveform, the sensor should be very broadband, and the low frequency cutoff should be as low as practical. This means that noise must be suppressed. Electromagnetic interference, caused by electrical relays, motor start-ups and sparking, can be minimized by good shielding and grounding practices. Mechanical noise poses a tougher problem.

In the laboratory, loads are applied using mechanical test machines that are a major source of noise. Options for suppressing mechanical noise include:

1. spatial discrimination or coordinate gating;
2. frequency discrimination or band-pass filtering;
3. gating on portions of the load signal produced by discontinuities;
4. gating on acoustic emission pulse rise time;
5. gating on acoustic emission amplitude, counts or duration;
6. acoustic emission signal threshold and gain settings;
7. specimen tabs; or
8. rubber damping materials.

All but the last two of these are accomplished within the instrumentation; the final two are accomplished on the test material.

When using the threshold and gain parameter to suppress noise, it should be remembered that the input electronic white noise level of most acoustic emission systems is six to ten microvolts. This noise voltage will be amplified through all stages of amplification. For example, at 80 dB of signal gain, the electronic noise will be amplified to about 0.1 volt. Therefore, a practical threshold setting to suppress electronic white noise (and higher level shot noise) should be 0.3 volt or higher.

Load Transfer Interface Noise

Considerable noise can be introduced at the load transfer interface and steps should be taken to prevent it. In particular, the load injection mechanism should not make direct contact with the test material, such as when the grip jaws bite directly into the grip areas of the specimen. Grip tabs made from plastic, fiberglass epoxy or aluminum can be bonded to the grip areas of specimens and small structures to provide a low noise interface.

For large composite structures, rubber-faced loading pads can be used. Because noise limits the usefulness of acoustic emission data and makes an additional level of complexity necessary in detection and processing systems, the signal-to-noise ratio should always be maximized through use of appropriate noise suppression techniques.

Standardizing Acoustic Emission Tests

An effective method for standardizing acoustic emission monitoring equipment involves the controlled fracture of a short section (~1.5 mm) of 2H graphite lead 0.5 mm in diameter on the surface of the test piece. The resulting stress wave very closely resembles an acoustic emission in pulse, rise time and amplitude.

When the lead is broken at a convenient location with respect to the sensor, the processed pulse can be analyzed for amplitude, counts, energy or duration, and one of these parameters can be selected for reference. The system gain, threshold and other parameters are adjusted to obtain a practical reference level. The acoustic emission from failure processes in the material during an actual test can then be compared to this reference level for screening and analysis purposes.

Instrumentation for Organic Composite Tests

Acoustic emission instrumentation for use in organic composite tests is, in major respects, similar to that used in other applications. From the standpoint of optimization for composites, however, certain capabilities should be present, including broadband sensor detection capabilities and selection of the lowest practical low frequency cutoff. A block diagram of a generic acoustic emission system for monitoring composites is shown in Fig. 13. Many good systems suitable for composite monitoring are available commercially. Some are specifically designed for use on composites. Many investigators prefer to assemble their own systems from suitable modular equipment. A computer for processing and analyzing the data in real-time is a necessity.

A good acoustic emission system for use on composites should be able to acquire, process, compute and store the data; analyze and display processed data; and correlate factors (time, load, strain, temperature, pressure and others) with acoustic emission events. For maximum flexibility and effectiveness in composites research, the instrumentation should be capable of providing all or most of the following acoustic emission processing parameters:

1. total acoustic emission events and event rate;
2. total threshold crossing counts and count rate;
3. amplitude;
4. integrated event energy;
5. pulse rise time;
6. event duration;
7. waveform; and
8. spectral amplitude analysis.

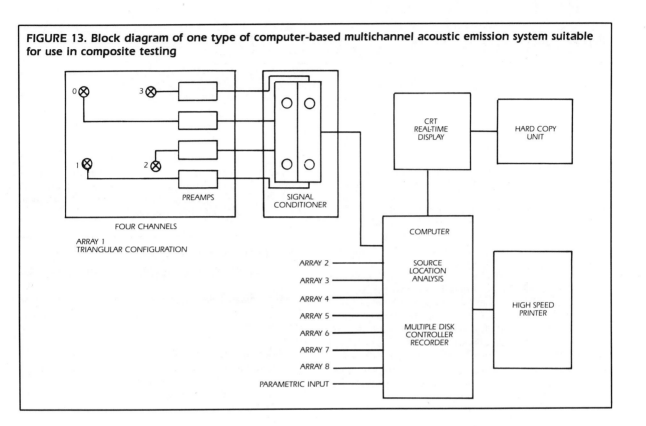

FIGURE 13. Block diagram of one type of computer-based multichannel acoustic emission system suitable for use in composite testing

PART 4
METAL MATRIX COMPOSITES

The discussions of acoustic emission monitoring applications with organic composites are also applicable to metal matrix composites (MMC). This discussion addresses areas that are unique to MMC or areas where important differences exist between the two types of composites.

Metal matrix composites are precisely engineered materials whose matrix is made from various metallic alloys to which is added high strength, high modulus fibers. This provides even stronger, stiffer, more dimensionally stable metallic parts. The reinforcement fibers in MMC materials may be continuous, chopped (whiskers) or particulate.

Continuous fiber metal matrix composites have a layered structure and directional mechanical and thermomechanical properties. One method of producing MMC materials is by a process of winding continuous filaments on a metal foil substrate, flame spraying powdered metal after each wound layer, cutting and laminating, then consolidating the molds in an autoclave at elevated temperatures and pressures, followed by aging.

The discontinuous reinforcement types of MMC are made by blending a specified ratio of whiskers or particulates with powdered metal, pouring into billet molds, and consolidating in a press, followed by secondary processing. Discontinuous MMC materials can be rolled, extruded or forged. The continuous fiber types must be molded near to their final shape.

Some common types of aerospace metal matrix composites use aluminum, titanium, magnesium or nickel matrix alloys. Aluminum and magnesium composites may use graphite, silicon carbide on carbon, boron carbide on tungsten or boron on tungsten as filament materials. Titanium and nickel composites may use silicon carbide or boron carbide on boron. Silicon carbide (ceramic) whiskers and particulate fibers are also used in aluminum and magnesium MMC matrices. Note that some fibers have composite structures themselves.

Some probable aerospace applications of metal matrix composites include the following, where high performance and mechanical, thermal and dimensional stability are required: primary airframe structures; load bearing skin and webs; airfoil control surfaces; landing gear components; spacecraft airframes; spacecraft hatches and doors; rocket engine nozzles and thrust chambers.

Organic composites are prolific emitters of acoustic emission, arising predominantly from matrix and matrix fiber failure mechanisms. Metal matrix composites are not nearly so prolific because of the ductile metal matrix. Acoustic emission is emitted primarily from the fibers themselves, their interfaces with the matrix, or from intermetallic inclusions when adequate strains are induced.

Applications of Acoustic Emission Testing to Metal Matrix Composites

Depending on the broader area of application, the objectives for monitoring acoustic emission during MMC testing include:

1. determining onset of stress induced damage;
2. determining onset of specific failure mechanisms;
3. identifying failure propagation;
4. detecting damage propagation;
5. evaluating dynamic deformation under thermomechanical loading; and
6. predicting yield, shear or fracture strength.

Acoustic emission monitoring methods can potentially be used during MMC material characterization and structural verification testing to detect discontinuity initiation and growth and to assess relative material quality. Acoustic emission is emitted during dynamic deformation processes, even when microscopic failures occur in the matrix, in the fibers or at the fiber/matrix interface. Some possible MMC failure modes and mechanisms include:

1. microcracking in the matrix;
2. transverse cracking of the matrix;
3. in-plane matrix crack propagation between fiber layers;
4. matrix crack propagation between the fibers and through the thickness;
5. debonding at the fiber/matrix interface;
6. cracking, shearing or splitting of the fiber outer case;
7. fiber pullout;
8. breaking of the fiber or core filament;
9. buckling or shear fracture in compression;
10. thermomechanical deformation; and
11. corrosive reactions at the matrix/filament interfaces.

Acoustic emission is generated by any of these processes which are themselves dependent on the type of imposed loads and strains. The types of mechanical and environmental characterization tests performed on MMC specimens and structures are standard and are virtually the same as those used for organic composites.

Failure mechanisms involving only the metal matrix will not usually produce high acoustic emission activity in any of the common types of aerospace MMC materials. Acoustic emission amplitudes are relatively low from ductile matrix sources. When a specimen of aluminum matrix material not containing fiber reinforcement is pulled in tension to failure, acoustic emission initiates just as the material experiences the nonlinear stress-strain region. The acoustic emission rate builds to a maximum just prior to yield, then decreases sharply beyond the yield point. Time correlated plots of acoustic emission and applied loads are shown in Fig. 14.

Curve-A in Fig. 14 is the plot of acoustic emission for nonreinforced aluminum. Curve-C is the load curve. Curve-B is a plot of cumulative acoustic emission for a silicon carbide whisker reinforced aluminum matrix specimen (otherwise identical to the curve-A specimens). The whiskers make a pronounced difference in the load dependent acoustic emission profile, which is probably due to debonding at the fiber/matrix interface. Whisker restraint on matrix yielding probably helps keep the acoustic emission rate low. The total acoustic emission is also considerably lower relative to comparable organic composite specimens. Such plots produced for metal matrix composites aid the materials scientist in characterizing failure mechanisms and overall behavior of the material under stress.

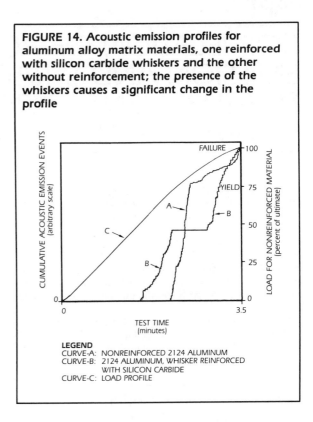

FIGURE 14. Acoustic emission profiles for aluminum alloy matrix materials, one reinforced with silicon carbide whiskers and the other without reinforcement; the presence of the whiskers causes a significant change in the profile

Proof Load Testing of Metal Matrix Composite Structures

It is often necessary to submit in-service components for proof tests or limit load tests that verify the material's ability to support service stresses. Acoustic emission monitoring is a potentially valuable nondestructive testing method for detecting discontinuity initiation and growth that may occur during proof tests. Prior damage or corrosion that exists in the material at the time of the test may extend in response to the test load and will emit acoustic emission as it does so.

A most effective technique is the use of a multichannel array, in which the transducers are strategically placed in the vicinity of a suspect area. The resulting acoustic emission signals are computer processed using a program designed to provide source coordinates through triangulation. Acoustic emission signal amplitude, counts or duration and delta times are recorded coincident with load, strain gage or pressure signals. This technique provides both location and severity information, and a basis for determining whether the material is potentially defective. Other nondestructive testing methods are then used to verify and characterize suspected discontinuities.

Instrumentation for Testing Metal Matrix Composite Materials

The instrumentation discussed for organic composites is applicable to metal matrix composites. Significant differences are likely to exist in the way the equipment is set up for testing, particularly with respect to signal gain, threshold and low frequency cutoff filtering. These factors must be established experimentally for the type of matrix and fiber combinations. Since a metal matrix composite produces lower levels of acoustic emission, signal gains of 80 dB or higher in combination with low threshold (greater than 0.3 V) and low frequency cutoff (40 kHz) are typically satisfactory. When significant noise is introduced into the structure from loading or pressurization apparatus, it may be advisable to use higher threshold and low frequency cutoff settings if the noise cannot be suppressed or sufficiently discriminated.

PART 5
CORRELATION OF ACOUSTIC EMISSION AND STRAIN DEFLECTION MEASUREMENT DURING COMPOSITE AIRFRAME STATIC TESTS

Acoustic emission has been used to improve detection, location and characterization of damage during ultimate load testing of composite airframe structures. The fundamentals of acoustic emission testing, system hardware and software are reviewed below, along with the criteria used to determine the placement of the transducers. More specifically, the ability of an acoustic emission system to detect, locate and to some extent characterize acoustic emission events in composites is shown with specific examples from airframe static test.

Acoustic emission count rates are shown to correlate with load strain and load deflection nonlinearities during significant structural damage such as cracks and delaminations.

Detecting Damage During Airframe Structural Tests

Strain and deflection linearity are among the traditional methods for detecting damage during testing of airframe structures. Unfortunately these methods are often imprecise and can be misleading. Composite structures permit efficient design by allowing a component's internal composition to be tailored to the load requirements, however this additionally complicates the determination of structural strength limits by increasing the types and number of potential origins for structural breakdown. Improper assessments of test events can lead to unnecessary reinforcement of incorrectly identified parts or may allow understrength parts to be overlooked. Acoustic emission techniques provide a valuable tool for dealing with composite airframe testing. The following topics are detailed in this discussion:

1. the advantages found in using acoustic emission for several different composite airframe tests;
2. the observed correlations between levels of acoustic activity and structural damage, as evidenced by strain and deflection measurements and post-test visual analysis of airframe structures; and
3. suggestions for improving acoustic emission hardware and software.

Effect of Airframe Structure on Testing

Airframe tests are complicated by the complexity of the structure. Most significant structural parts are located in the interior of the airframe and are hidden from view. Only external structural damage or events are detectable with traditional visual inspection. Events within the structure's interior are detectable only after severe damage has occurred and even then they are difficult to locate. Composite structures increase these problems by creating subsurface (often invisible) origins for structural discontinuities.

Strain measurements are useful during airframe tests but the size of the structure limits the number of strain gages available for monitoring all of the critical areas. Additionally, strain rosettes or single active strain gages (if correctly aligned to the principal strain axis) provide only local information.

Airframe deflection measurements are difficult to make, especially on interior structures. Rigid body rotation at mounting points (often a large percentage of total motion) must be subtracted and this significantly reduces the accuracy and usefulness of the technique.

The use of acoustic emission monitoring techniques alleviates these problems. The method can accurately detect and locate significant structural events, often at lower load levels than strain and deflection measurements. In fact, one of the major problems with acoustic emission monitoring is determining at what point the emission indicates significant structural deformation. Presented below is a correlation between the acoustic emission level and significant structural activity as monitored by strain and deflection measurements.

Analyzing Acoustic Emission Signals

Acoustic emission is the name given both to the elastic waves generated within a solid as a consequence of deformation and fracture processes and to the techniques employed for their measurement (Fig. 15). Acoustic emission

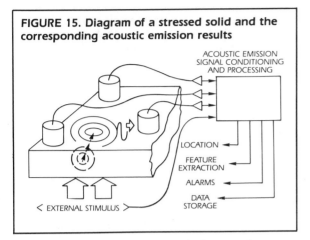

FIGURE 15. Diagram of a stressed solid and the corresponding acoustic emission results

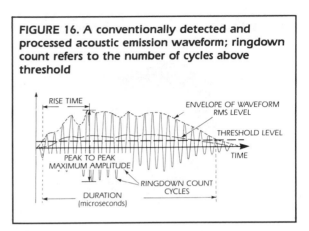

FIGURE 16. A conventionally detected and processed acoustic emission waveform; ringdown count refers to the number of cycles above threshold

has an advantage over ultrasonic methods in nondestructive testing of composites because it is a global, passive technique. Acoustic emission sensors are mounted on the test object during the entire test phase, allowing the structure to be continuously monitored.

There are several procedures for processing acoustic emission signals. The most prevalent procedure is to record events as a function of time and analyze parameters such as spectral content. An acoustic emission event consists of many ringdown counts recorded by a transducer attached to the structure.

For example, if a steel structure is struck with a hammer, it rings audibly. These oscillations decay and are detected by the transducer. The detected signals are passed through a counter with a set threshold (Fig. 16). The number of threshold crossings is called the *event count* or *ringdown count*. In practice, these parameters or their time rate of change are recorded versus time or versus an external stimulus, such as load or temperature. It can be shown, both experimentally and theoretically that the total number of ringdown counts per event is a good indication of that event's energy.[20-23]

The rate at which the counts occur is indicative of the extent of structural degradation. However, care must be taken in comparing rates since they are dependent on the acoustic emission system's collection time interval. The data reported here were evaluated in a one-second interval.

Another procedure used by a number of acoustic emission instrument manufacturers is rms voltage measurement. The detected acoustic emission signal voltage is squared, then the area under the voltage squared versus time curve is measured. This technique has been reported to be superior to ringdown counting for characterizing high energy events.[24,25]

The amplitude distribution of a series of acoustic emission events may also be measured. The relative success of amplitude distribution analysis is highly dependent on the dynamic range of the instrumentation.[26,27] It is shown below that the amplitude distribution can also be used to indicate the type of degradation.

Considerations for Acoustic Emission Sensors

Most acoustic emission applications use a damped piezoelectric sensor to detect the emission event. The sensor, generally imperfectly damped, superimposes its own resonant response onto the true emission signal and thereby distorts it. These transducers are velocity sensitive and typically have large contact areas that further complicate signal analysis.

An ideal transducer would be (1) broadband from DC to several megahertz (permitting choice of bandwidth to fit the situation); (2) displacement sensitive; (3) point contacting; with (4) a large signal output. Such a device would generate a true wave amplitude signal at a surface point and this would be easily interpreted, mainly because it eliminates phase interference artifacts arising at the higher frequencies from large contact areas.

As part of a National Bureau of Standards acoustic emission standards program,[28-30] a conical element, piezoelectric broadband transducer was developed. It has a frequency response that is flat to within 5 dB from a few kilohertz to approximately 1.5 MHz. This transducer was designed to serve as a laboratory standard, not for field use. Based on the NBS design, a new transducer was developed for practical applications. This nonresonant, point contact, self contained transducer is shown in Fig. 17.[31]

A comparison of the transducer response to a simulated event shows the unique qualities of the point contact sensor.[32] The reference event for acoustic emission calibration (known as the *seismic surface pulse*) is the breaking of a small glass capillary tube or a fine pencil lead (0.3 to 0.5 mm

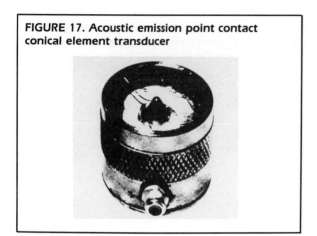

FIGURE 17. Acoustic emission point contact conical element transducer

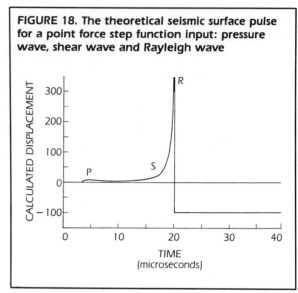

FIGURE 18. The theoretical seismic surface pulse for a point force step function input: pressure wave, shear wave and Rayleigh wave

diameter) on the surface of a polished steel block. The resulting waveform is the response to a unit step function force applied at a point to an infinite surface. The theoretical waveform for such an event[33] is shown in Fig. 18. The initial, small amplitude bump arises from the pressure or P wave arrival. The pressure wave is a longitudinal wave causing the particles of the structure to oscillate back and forth parallel to the direction of wave motion.

The knee or inflection in the curve just before the steep rise is the shear or S wave, coming at a later time because it travels more slowly. The shear wave is a transverse wave causing particles of the structure to oscillate at right angles to the direction of wave propagation. As pressure and shear waves reflect off of (or hit) the free surface of the structure, they generate very intense Rayleigh or R waves.

The response of the NBS transducer to a glass capillary break is shown in Fig. 19. Comparison to the theoretical response is excellent, clearly showing the pressure wave, shear wave and Rayleigh wave components. The point contact transducer response to a glass capillary break on a steel block is shown in Fig. 20. The shear, pressure and Rayleigh waves clearly stand out in proper proportion, indicating a good high frequency response.

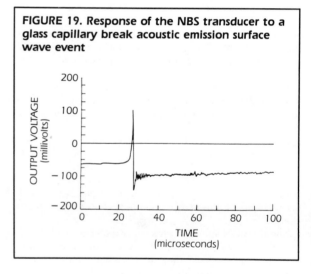

FIGURE 19. Response of the NBS transducer to a glass capillary break acoustic emission surface wave event

Acoustic Emission Setup for Tests of Airframes

An eight-channel microcomputer acoustic emission system was used to acquire and process acoustic emission data. Point contact transducers were used with preamplifiers (20 kHz high pass filter).

Global, single transducers were located as closely as possible to analytically critical structure sites. Because discontinuous, interwoven composite layers attenuate the transmission of sound waves, two transducers were joined in a linear array only on structures of predominantly unidirectional material. The transducers were attached with tape or pliable silicone adhesives. For each test the individual channel threshold was set between 35 and 40 dB. No other filtering technique was used.

Calibration of the structure, in those cases where source location was being used, was accomplished by making pencil lead breaks at different points on the structure. A 0.5 mm graphite was broken on the composite structures in the vicinity of each sensor location to study wave propagation patterns through the complex composite structure.

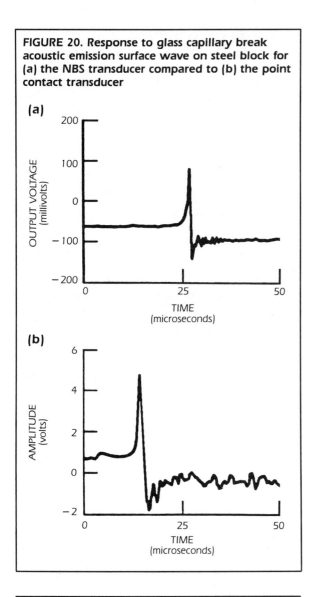

FIGURE 20. Response to glass capillary break acoustic emission surface wave on steel block for (a) the NBS transducer compared to (b) the point contact transducer

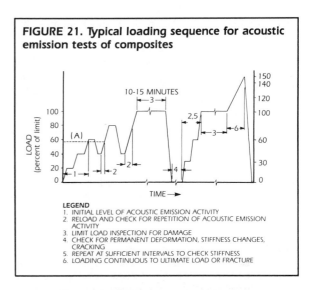

FIGURE 21. Typical loading sequence for acoustic emission tests of composites

LEGEND
1. INITIAL LEVEL OF ACOUSTIC EMISSION ACTIVITY
2. RELOAD AND CHECK FOR REPETITION OF ACOUSTIC EMISSION ACTIVITY
3. LIMIT LOAD INSPECTION FOR DAMAGE
4. CHECK FOR PERMANENT DEFORMATION, STIFFNESS CHANGES, CRACKING
5. REPEAT AT SUFFICIENT INTERVALS TO CHECK STIFFNESS
6. LOADING CONTINUOUS TO ULTIMATE LOAD OR FRACTURE

Limit Load Test Techniques

Composite structures produce acoustic activity when first traversing a load level. This may not indicate significant structural damage but may represent the release of built-in fabrication prestresses. This phenomenon, along with the Kaiser effect,[34] dictates a special approach to the application of test loads. Figure 21 illustrates the approach to a load sequence.

Satisfactory limit load performance requires that the application of design limit loads shall not result in: (1) damage or permanent deformation that is detectable on inspection; or (2) detrimental deformation under load. Detrimental deformation inhibits or degrades the mechanical operation or aerodynamic characteristics to the extent that flying qualities are significantly impaired.

Traditional techniques for evaluating the limit load performance of ductile metal structures depend on the ability to detect damage or detrimental deformation. With composite structures, there are a variety of potential damage modes, including matrix cracking, fiber fracture, disbonding and buckling. In addition, composites exhibit characteristics such as high stiffness and small deflections, little or no plasticity prior to fracture and creep aftereffects. All of these composite properties make detection and evaluation of damage or permanent deformation more difficult.

High stiffness and small deflection characteristics can present problems in determining if degradation has occurred. Deflections may be too small to measure reliably or may not reflect matrix degradation. The structure may exhibit a time delay in assuming its at-rest shape, giving indications of permanent deformation where none exists. In these situations, monitoring of acoustic emission activity can supplement traditional load versus deflection and load versus strain monitoring and locate areas of potential damage.

Load Procedure

The following procedures were used to evaluate performance during limit load testing. Limit loads were applied in approximately 20 percent increments to full limit load. Each load increment was held for a sufficient time to permit a thorough visual inspection as well as monitoring of acoustic activity and acquisition of pertinent load versus deflection and load versus strain data. Critical structural areas and areas of suspect damage or significant acoustic activity were identified and marked for post-test evaluation.

If significant damage occurred (matrix cracks or delaminations, fiber fractures, bonded joint separations, or buckling of primary structural beams, longerons, stringers, or frames), testing was suspended until further evaluations could be performed and, if necessary, changes could be incorporated. Elastic buckling of reinforced skin panels (diagonal tension behavior) was permitted, provided that the buckling was reversible and that there was no indication of delamination in the buckled areas. Load versus strain and load versus deflection data were recorded at each increment and then reviewed for nonlinear or divergent trends that could indicate damage or significant deformation.

If acoustic emission activity was detected, load levels were maintained to check for continued acoustic emission activity and to locate the acoustic emission source for visual inspection. The loads were then reduced and reapplied to test the significance of the activity. Qualitatively, acoustic activity occurring during initial loading was discounted.

Activity recurring during reloading was used as an indicator of possible structural changes. If activity recurred on reloading, the load levels were maintained for an additional ten to fifteen minutes to check for increasing deflections and continued acoustic emission activity. At limit, load was maintained for a ten to fifteen minute interval to permit visual observations and checks for possible permanent damage and continuing acoustic activity.

Post-test activities included a complete visual inspection. Special emphasis was placed on areas that exhibited buckling or acoustic emission activity. In these areas, visual inspections were supplemented by ultrasonic inspections to ensure that damage had not occurred. A sufficient rest period was permitted to allow the structure to assume its at-rest zero readings. Load versus deflection and load versus strain data were compared and correlated with acoustic activity and reviewed for nonlinearity, divergences and evidence of permanent damage.

Results of Acoustic Emission Tests of Composite Airframes

The results of composite airframe static tests highlight the correlation betweeen damage detection using acoustic emission and traditional strain and deflection techniques. Tests of four different structures are described: (1) a graphite auxiliary support beam manufactured using a resin transfer process; (2) a graphite curved frame; (3) a complex aramid graphite roof structure; and (4) an airframe transition section (shell structure).

Auxiliary Support Beam Test

The support beam was a graphite structure formed by molding instead of traditional hand lay-up techniques.

The beam has two integrally attached titanium lug boxes and was supported at four stations simulating frame attachments. Four static conditions incorporating a variety of normal, in-plane and side loads were applied to ultimate or failing levels. Strain, deflection and acoustic emission transducers were installed to monitor the system (see Fig. 22).

Time constraints prevented the completion of limit load for all conditions prior to ultimate load testing. It was therefore critical to the success of the test that catastrophic failure be avoided. The first test was successfully completed to ultimate load (150 percent limit load) with no failures and generally benign acoustic emission outputs.

The second condition (test data shown in Fig. 23) progresses without incident until 60 percent limit load. At this level nonlinearities appeared in the deflection of mounting arms and significant acoustic emission was recorded (count rates greater than 600 counts per second and amplitudes greater than 70 dB).

Load was removed to evaluate the Kaiser effect and to check for permanent strain or deflection. No permanent deformation was indicated. When reloading from 60 percent to 80 percent limit load, acoustic emission count rates above 600 counts per second were recorded. Nonlinearities again occurred in the deflection and also appeared in the strain measurements. At 80 percent, nonlinearities and acoustic emission activity also occurred at the fitting that was later found to be the eventual fracture origin. Testing was interrupted at this level. The part was inspected ultrasonically but only minor discrepancies were noted. The completion of this test was postponed, based primarily on the acoustic emission activity.

The remaining two test conditions were completed to ultimate load with no indicated deflection or strain irregularities and with only minor acoustic emission outputs.

The second condition was rerun and the beam failed just above 90 percent limit load at the intersection of the arm and the central channel section of the beam (Fig. 22). The failure was predominantly delamination with some fiber breakage (the acoustic emission amplitude distribution is shown in Fig. 23). The use of acoustic emission techniques had avoided a potentially serious test failure and had predicted the cracking origins.

Graphite Curved Frame Tests

The curved frame specimen (see Fig. 24) is representative of thin flange, light gage airframe construction and was made up of a web of woven graphite epoxy and two back-to-back channels.[35] The web is structurally reinforced by beaded stiffeners. The frame caps (flanges) are graphite

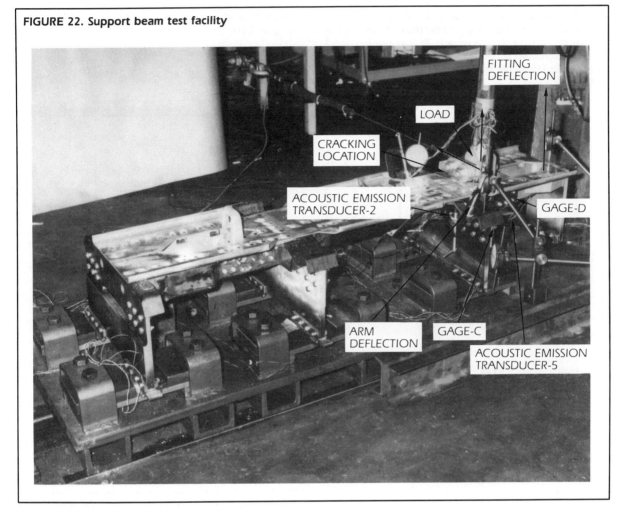

FIGURE 22. Support beam test facility

epoxy using tape and woven material in addition to the woven plies of the web.

The curved frame was installed and tested as shown in Fig. 25. The test load duplicated frame bending, shear and axial forces with a single point load application. Test specimens were not directly attached to a test facility but were free standing with the hydraulic cylinder. Specimens were supported by the loading arms and associated hardware. Strain gages were installed on the cap and beaded stiffener. Acoustic emission transducers were installed on the cap flange as shown in Fig. 24.

The curved frame was then incrementally loaded. At 80 percent of design load it fractured. A preliminary review of the data plotted during the test indicates that the fracture occurred when the inner (compression) cap and web beaded stiffeners became unstable. This instability was indicated by the strain data of gages TS-5, TS-2, TS-3 and was also detected by the acoustic emission activity as the fracture load was approached. A cross plot of acoustic emission data and gages TS-5 and TS-6 is shown in Fig. 26. The location of the fracture region on the curved frame correlates with the maximum region of acoustic emission output shown in Fig. 27.

Tests of Complex Aramid and Graphite Roof Structure

The aramid and graphite roof structure is shown in Fig. 28. It is composed primarily of two graphite beams running the length of the specimen. Additional support near the

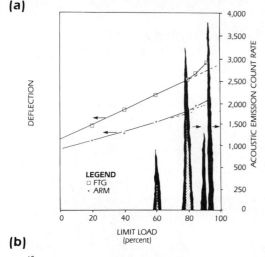
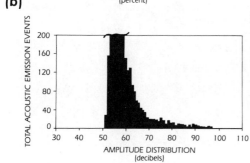
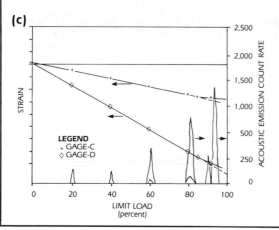

FIGURE 23. Acoustic emission test data for graphite support beam: (a) deflection versus load versus count rate for transducer-5; (b) count versus amplitude distribution for transducer-5; and (c) strain versus load versus count rate for transducer-2

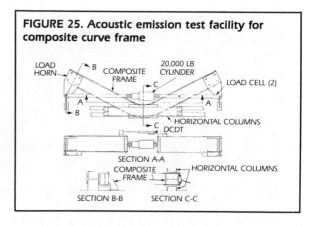

FIGURE 24. Composite curved frame strain gage and acoustic emission transducer locations

FIGURE 25. Acoustic emission test facility for composite curve frame

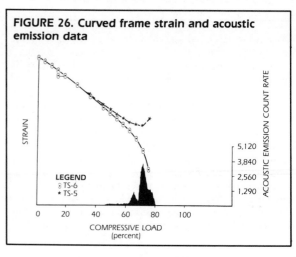

FIGURE 26. Curved frame strain and acoustic emission data

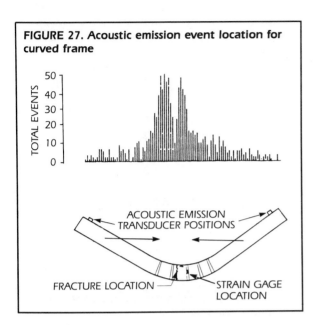

FIGURE 27. Acoustic emission event location for curved frame

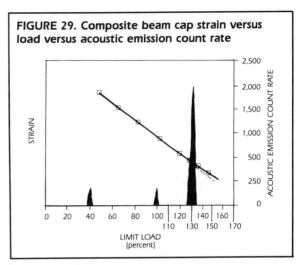

FIGURE 29. Composite beam cap strain versus load versus acoustic emission count rate

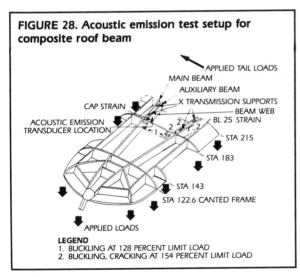

FIGURE 28. Acoustic emission test setup for composite roof beam

mounts is provided by auxiliary beams and metal fittings. The framing members are either aramid or graphite. The roof skin is also a mixture of aramid and graphite areas and the roof was part of a complete airframe specimen.

The specimen was tested to a simulated flight maneuver by introducing representative air and inertial loads through the frames. The load condition was critical for the tail section, not the roof. The specimen was instrumented with both strain and acoustic emission transducers as shown in Fig. 28.

No significant events were noted during limit load testing. During ultimate load testing, while loading to 128 percent limit, audible events were recorded. Load was removed and a review of the acoustic emission data showed the most significant events occurring in the central roof area. Visual inspection found buckling and cracking of a large frame of two-ply graphite web at its intersections with the main roof beams (see Fig. 28).

This area was not directly instrumented for either strain or acoustic emission but an acoustic emission transducer about 150 mm (6 in.) away did pick up the event, as did the main transmission beam cap strain transducer whose output is shown in Fig. 29. The damage was repaired and the area was reinforced prior to continuing. Repairs and reinforcements were made on the beams in the assembly, saving significant calendar time.

Unfortunately, the acoustic emission system was not available when the reinforced structure was retested. On reloading, audible events were almost continuous above 120 percent limit load. Visual inspection was made, but no significant origin could be found, and loading proceeded cautiously thereafter.

At 154 percent limit load, a load crack was heard. The load was dropped to 80 percent and another visual inspection detected cracks in the left main and auxiliary roof support beams as shown in Fig. 28. Due to limited instrumentation capacity, the strain gages on the left side beams had been dropped in favor of gages in the predicted critical tail cone area. The right-hand beam gages shown in Fig. 30 did not detect or predict the problem.

Tests of Airframe Transition Section

The composite fuselage transition section shown in Fig. 31 was a monocoque structure, made principally of aramid

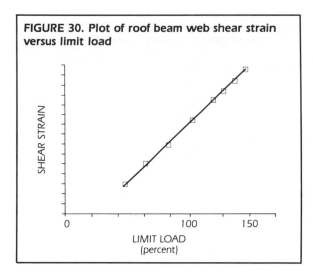

FIGURE 30. Plot of roof beam web shear strain versus limit load

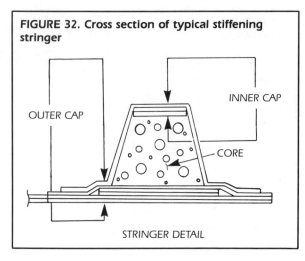

FIGURE 32. Cross section of typical stiffening stringer

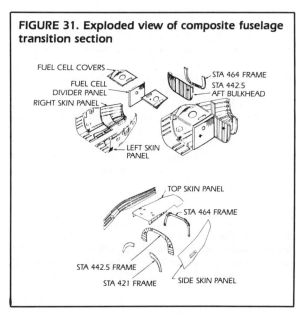

FIGURE 31. Exploded view of composite fuselage transition section

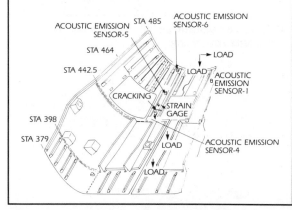

FIGURE 33. Setup for acoustic emission test of composite transition section

epoxy skins, with graphite epoxy beams and frames, and with graphite epoxy foam stiffening stringers. Figure 32 shows a typical stringer cross section. The shell tapers from a large cabin frame to a small tailcone frame in about 3 m (9 ft). The skin has numerous fuel ports and holes for access panels. The specimen was instrumented with strain gage rosettes and during ultimate load testing, acoustic emission transducers were added as shown in Fig. 33.

The fuselage section was tested to a severe, combined lateral and vertical bending, simulated maneuver. Limit load testing conducted without acoustic emission was successfully completed with no indication of permanent deformation. During ultimate load testing, apparently significant acoustic emission activity was encountered at 80 percent limit load and verified by strain nonlinearity as shown in Fig. 34.

When load was dropped and reapplied, the count rate did not repeat. At 130 percent limit load, significant acoustic emission activity did occur as corroborated by the corresponding strain discontinuity. In addition, skin buckling was visually detected. Acoustic emission activity continued at higher load levels but the fuselage section reacted to ultimate load without failure. Post-test inspection revealed two buckled and cracked stringers (see Fig. 32). The amplitude distribution at failure is shown in Fig. 35.

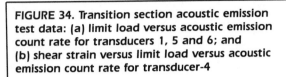

FIGURE 34. Transition section acoustic emission test data: (a) limit load versus acoustic emission count rate for transducers 1, 5 and 6; and (b) shear strain versus limit load versus acoustic emission count rate for transducer-4

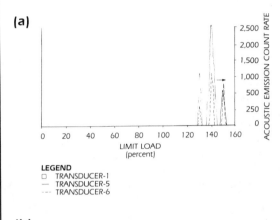

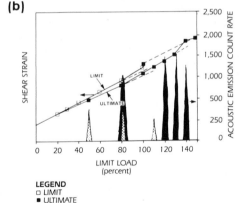

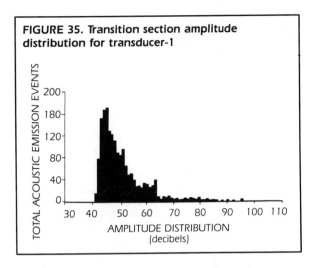

FIGURE 35. Transition section amplitude distribution for transducer-1

Conclusion

The following conclusions can be drawn as a result of these tests.

1. Acoustic emission activity for thin composite fuselage structures, with count rates above 600 counts per second, is indicative of significant structural damage.

2. In general, acoustic emission techniques warned of and located impending failures, providing the opportunity for repair and reinforcement of the test structure.

3. Because of the complex nature of the cracking modes, strict acoustic emission amplitude correlation with delamination or fiber breakage was not possible. However, as a general guide, significant structural activity was typically associated with amplitudes above 80 dB and these were assumed to be fiber breaks.

For acoustic emission to serve as a quantitative structural monitoring system, reliable classification of emission source and severity must be provided. However, most existing acoustic emission instrumentation and analysis methods are designed for detection rather than characterization. A quantitative acoustic emission monitoring system should have the following features:

1. increased channel capacity to more effectively cover a large structure;
2. faster rates of data acquisition and digitizing acoustic emission waveforms;
3. advanced signal processing algorithms that evaluate waveforms to track the propagation of structural damage and analyze the digitized acoustic emission waveforms to provide a real-time determination of structural integrity.

REFERENCES

1. Graham, L.J. and R.K. Elsley. "AE Source Identification by Frequency Spectral Analysis for an Aircraft Monitoring Application." *Journal of Acoustic Emission.* Vol. 2, No. 1/2 (1983): pp 47-56.
2. Hay, D.R., R.W.Y. Chan, D. Sharp and K.J. Siddiqui. "Classification of Acoustic Emission Signals from Deformation Mechanisms in Aluminum Alloys." *Journal of Acoustic Emission.* Vol. 3, No. 3 (1984): pp 118-129.
3. McBride, S.L. and J.W. Maclachlan. "Effect of Crack Presence on In-Flight Airframe Noises in a Wing Attachment Component." *Journal of Acoustic Emission.* Vol. 1, No. 4 (1982): pp 229-236.
4. McBride, S.L. and J.W. Maclachlan. "Acoustic Emission Due to Crack Growth, Crack Face Rubbing and Structural Noise in the CC-130 Hercules Aircraft." *Journal of Acoustic Emission.* Vol. 3, No. 1 (1984): pp 1-10.
5. Acquaviva, S., et al. "Interlaboratory Comparisons of Acoustic Emission Spectra." *NDT International* (1980): pp 230.
6. McBride, S.L., J.W. Maclachlan and B.P. Paradis. "Acoustic Emission and Inclusion Fracture in 7075 Aluminum Alloys." *Journal of Nondestrustive Evaluation.* Vol. 2, No. 1 (1981): pp 35-41.
7. "Properties, Physical Metallurgy and Phase Diagrams." *Aluminum.* Vol. 1. Philadelphia, PA: American Society for Testing and Materials (1967).
8. Bailey, C.D. "Acoustic Emission for In-flight Monitoring on Aircraft Structures." *Materials Evaluation.* Vol. 34, No. 8. Columbus, OH: The American Society for Nondestructive Testing (1976): pp 165-171.
9. Bailey, C.D. and W.M. Pless. "Acoustic Emission Structure-Borne Noise Measurements on Aircraft During Flight." *Materials Evaluation.* Vol. 34, No. 9. Columbus, OH: The American Society for Nondestructive Testing (1976): pp 189-195, 201.
10. Hutton, P.H., J.R. Skorpic and D.K. Lemon. "Fatigue Crack Monitoring In-Flight using Acoustic Emission — Hardware, Technique and Testing." *Proceedings of the Nondestructive Testing Forum.* Washington, DC: Air Transportation Association (1981).
11. McBride, S.L. "Canadian Forces In-Flight Acoustic Emission Monitoring Programme." *Proceedings of the ARPA/AFML Review of Progress in Quantitative NDE* (July 1978).
12. McBride, S.L. and J.W. Maclachlan. "In-Flight Acoustic Emission Monitoring of a Wing Attachment Component." *Journal of Acoustic Emission.* Vol. 1, No. 4 (1982): pp 223-228.
13. McBride, S.L. and J.W. Maclachlan. "Effect of Crack Presence on In-Flight Airframe Noises in a Wing Attachment Component." *Journal of Acoustic Emission.* Vol. 1, No. 4 (1982): pp 229-235.
14. McBride, S.L. and J.W. Maclachlan. "Identification of Acoustic Emission Signals due to Crack Growth, Crack Face Rubbing and Structural Noise in the CC-130 Hercules Aircraft." *Journal of Acoustic Emission.* Vol. 3, No. 1 (1984): pp 1-11.
15. Bailey, C.D., J.M. Hamilton, Jr. and W.M. Pless. "Acoustic Emission of Impact Damaged Graphite-Epoxy Composites." *Materials Evaluation.* Vol. 37, No. 5. Columbus, OH: The American Society for Nondestructive Testing (1979): pp 43.
16. Fowler, T.J. and E. Gray. "Development of an Acoustic Emission Test for FRP Equipment." *Proceedings of the ASCE Convention and Exposition.* Boston, MA (April 1979).
17. Daniel, I.M. and T. Liber. *Wave Propagation in Fiber Composite Laminates.* NASA CR-135086 final report, part II. Contract NAS3-16766.
18. Freeman, S.F. "Characterization of Lamina and Interlamina Damage in Graphite-Epoxy Composites by the Deply Technique." *Composite Materials Testing and Design.* ASTM STP 787. Philadelphia, PA: American Society for Testing Materials (May 1981): pp 50.
19. Chatterjee, S.N. and R.B. Pipes. *Composite Defect Significance.* NAPC 80048-80 (August 1981).
20. Fowler, T.J. *Acoustic Emission Testing of Chemical Process Industry Vessels.* St. Louis, MO: Monsanto Company (1985).
21. Williams, R.S. "Acoustic Emission Monitoring of Degradation in Monolithic Refractories." *Proceedings of the Conference on Environmental Degradation of Engineering Materials* (October 1977): pp 675.
22. Awerbuch, J. *Acoustic Emission During Damage Progression in Composites.* Princeton, NJ: Physical Acoustics Corporation (May 1985).

23. Otsuka, H. and H.A. Scarton. "Variations in Acoustic Emission Between Graphite and Glass-Epoxy Composites." *Journal of Composite Materials.* Vol. 5 (November 1981): pp 591.
24. Beattie, A.G. and R.A. Jaramillo. "The Measurement of Energy in Acoustic Emission." *Review of Scientific Instruments.* Vol. 45 (1974): pp 352-357.
25. Beattie, A.G. "Energy Analysis in Acoustic Emisision." *Materials Evaluation.* Vol. 34, No. 4. Columbus, OH: The American Society for Nondestructive Testing (1976): pp 73-78.
26. Pollock, A.A. "Acoustic Emission Amplitudes." *Nondestructive Testing.* Vol. 6 (1973): pp 265-270.
27. Ono, K. "Amplitude Distribution Analysis of Acoustic Emission Signals." *Materials Evaluation.* Vol. 26, No. 8. Columbus, OH: The American Society for Nondestructive Tesing (1976): pp. 177-181.
28. Breckenridge, R.R., C.E. Tschiegg and M. Greenspan. "Acoustic Emission — Some Applications of Lamb's Problem." *Journal of the Acoustic Society of America.* Vol. 57 (1975): pp 626-631.
29. Nelson, N. Hsu and S.C. Hardy. "Experiments in Acoustic Emission Waveform Analysis for Characterization of Acoustic Emission Sources, Sensors and Structures." *Proceedings of the ASME Conference on Elastic Waves and NDE of Materials.* New York, NY: American Society of Mechanical Engineers. (December 1978): pp 85-106.
30. *The NBS Conical Transducer for Acoustic Emission.* Gaithersburg, MD: National Bureau of Standards (December 1981).
31. Boyce, W.C. and R.K. Miller. "Assessment of Structural Damage in Composites Utilizing Acoustic Emission Technology." *Proceedings of the Forty-first American Helicopter Society Conference* (May 1985).
32. Miller, R.K., P.E. Zwicke, R.S. Williams and H.I. Ringermacher. *Characterization of Acoustic Emission Signals.* UTRC R83-996043-2. Washington, DC: Federal Highway Administration (1983).
33. Pekeris, C.L. "The Seismic Surface Pulse." *Geophysics.* Vol. 41 (1955): pp 469-480.
34. Kaiser, J. *Untersuchungen uber das Auftreten Gerauschen Beim Zuguersuch.* Ph.D. thesis. Munchen, Germany: Technische Hoschschule (1950).
35. Lowry, D.W., N.E. Krebs and A.C. Dobyns. *Design, Fabrication and Test of Composite Curved Frames for Helicopter Fuselage Structure.* NASA Contractor Report 172438 (October 1984).

SECTION 13

ACOUSTIC EMISSION RESEARCH IN BIOMEDICAL ENGINEERING

Timothy M. Wright, The Hospital for Special Surgery, New York, New York

Roberto Binazzi, The Hospital for Special Surgery, New York, New York
R.W. Bright, Foundation for Skeletal Research, Baltimore, Maryland
James Carr, The Hospital for Special Surgery, New York, New York
Kjell Eriksson, Karolinska Institutet, Danderyd, Sweden
John Insall, The Hospital for Special Surgery, New York, New York
Ulf Jonsson, Karolinska Institutet, Danderyd, Sweden
J. Lawrence Katz, Rensselaer Polytechnic Institute, Troy, New York
S.J. Poliakoff, Foundation for Skeletal Research, Baltimore, Maryland
Thomas Santner, Cornell University, Ithaca, New York
Michael Soudry, The Hospital for Special Surgery, New York, New York
David Wypij, Cornell University, Ithaca, New York
Hyo Sub Yoon, Rensselaer Polytechnic Institute, Troy, New York

INTRODUCTION

Biomedical engineers have employed acoustic emission technology as a research tool with potential diagnostic applications. Their efforts have centered mostly on the musculoskeletal system. In their function as load bearing and protective structures, the skeleton (see Fig. 1) and its prosthetic replacement parts are subject to mechanical damage. The use of acoustic emission affords the opportunity to investigate damage mechanisms and to develop nondestructive testing techniques for diagnosing conditions where the amount of such damage reaches a detrimental level. The use of acoustic emission is quite appealing because the other techniques currently used to diagnose such conditions result in significant risk to the patient.

In this section, an overview of acoustic emission research in biomedical engineering is presented. Comprised of several research projects, this review illustrates the potential of acoustic emission to contribute to the study of a diverse range of orthopedic problems.

PART 1
ACOUSTIC EMISSION FOR STUDIES OF THE MECHANICAL PROPERTIES OF ENTIRE BONES

The mechanical properties of the long bones of the skeleton have been studied for many years. A large amount of experimental data has been gathered by testing entire bones, as well as standardized specimens of bone tissue machined from whole bones. The load-to-deformation relationship for both entire bones and machined specimens of bone material is initially linear with a subsequent nonlinear region prior to fracture. It is generally agreed that the deformation within the linear range is elastic.

Many investigators[1,2] have suggested that the deformation within the nonlinear range is elastic-plastic. This means that a bone loaded into the nonlinear range should be permanently deformed. It is, however, known from clinical practice that parts from a fractured bone are not permanently deformed; that is, exact reduction of the fracture is generally possible. It has been suggested that the nonlinear deformation is due to microcracking in the bone material.[3] By using acoustic emission techniques, strong evidence was found for such a damage mechanism.[4] Acoustic emission has proven most useful for gaining insights into the nature of the nonlinear load deformation of whole bones prior to fracture.

Methods for Bone Studies

A commercial acoustic emission system was used for registration of preamplified acoustic emission signals from bone. High pass filters eliminated extraneous noise. The total amplification needed was 88 to 95 decibels (dB). Sound propagation in cortical bone is severely dampened, especially perpendicular to the long axis of the bone. Adequate amplification of the test arrangement was assured using breakage of lead from an ordinary pencil to propagate an appropriate acoustic emission signal toward the area of interest on the bone.

Most whole bones used for experiments are large in comparison to ordinary transducers and the bone surface is not smooth, therefore it is advantageous to maximize the amount of contact between the transducer and the bone surface. This can be accomplished by using paste or heavy grease as a couplant. If the ends of the test bones are attached to the test equipment by grips, the transducer may also be attached to the grips and still record an adequate acoustic emission signal. Pinpoint waveguides also have proven efficient for transmitting acoustic emission signals between the bone surface and the transducer. Another advantage with the pinpoint waveguide is that slipping is avoided between transducer and bone during testing.

Results and Discussion of Bone Studies

Distinct bursts of high amplitude signals, typical for formation or growth of cracks, were found within the nonlinear range, but not within the linear portion of the corresponding load deformation curve for the bone. This finding strongly suggests that the nonlinear deformation is caused by microcracking and not by some mechanism of plastic deformation.[4] The final fracture is preceded by a multitude of burst signals indicating microcrack coalescence.

Acoustic emission signals of burst type were found in load and unload tests of whole bones, not only during loading within the nonlinear deformation range, but also during the initial part of the unloading.[5] The signals during unloading cannot be distinguished from those observed during loading. It follows that cracking probably also occurs during initial unloading in the nonlinear range. Microcracking during unloading may occur through growth of previously formed cracks and formation of new ones. The growth of cracks during unloading must be due to a time dependent crack growth resistance, which decreases with time, and/or to an increase of the stress intensity around crack tips through local stress redistribution. Both these effects require viscoelastic material properties. Viscoelastic residual deformation was found after unloading from the nonlinear range.[4] Other results in the literature indicate that bone tissue possesses viscoelastic properties, especially at high deformation rates.[6]

Acoustic emission testing has been valuable for the studies of mechanical properties of whole bones, showing that the amount of microcracking increases during the nonlinear deformation range until final fracture occurs through crack coalescence.

PART 2
DIAGNOSIS OF BONE FRACTURES BY A NONINVASIVE AND NONTRAUMATIC ACOUSTIC TECHNIQUE

Diagnosis of bone fractures and diseases is done with various techniques, including conventional radiography,[7] X-ray spectrophotometry,[8] direct photon absorptiometry (single[9] or dual beam[10]) and X-ray computed tomography (CT).[11] These techniques are all based on the determination of bone mineral content and employ ionizing radiation, a risk to the patient. Furthermore, the techniques can be expensive and imprecise. Nuclear magnetic resonance (NMR) imaging[12] and ultrasonic computed tomography[13] are useful for detecting soft tissue abnormalities, but their resolution for imaging bone is currently poor.

Based on previous acoustic emission studies on fresh animal bones,[14-16] a noninvasive and nontraumatic acoustic emission technique has been developed and applied to animal and patient studies. Because applying loads to the bone under test would introduce additional trauma to the patient, simulated low intensity ultrasonic pulses are injected to the bone through an acoustic emission transducer. Another transducer is used to receive the signal, which is then processed by usual acoustic emission techniques. Acoustic emission terminology has been adapted to describe experimental results, even though the technique is a combination of acoustic emission and ultrasonic methods.

The new acoustic emission technique was first applied to nonbony or nonbiological specimens, such as wood, plastics and metals. When these well defined material systems were found to yield reproducible results, animal bones were investigated both *in vitro* and *in situ*. Then, *in vivo* animal and human studies followed. To aid in the presentation of these studies, the bones in the human skeleton are shown in Fig. 1. The bones of other mammals are qualitatively similar.

In the *in vivo* animal study, a surgical defect was introduced in the right mid-tibia of a dog, while the left leg served as the contralateral control. Subsequently, healing processes were monitored both with X-rays and acoustic emission.[17-19] The human study involved fourteen volunteers, primarily runners with histories of stress fractures in their tibiae. Acoustic emission data from these patients provided a consistent picture of normal versus injured bones. It is known that stress fractures are not detectable by conventional X-ray techniques until they have begun to heal.

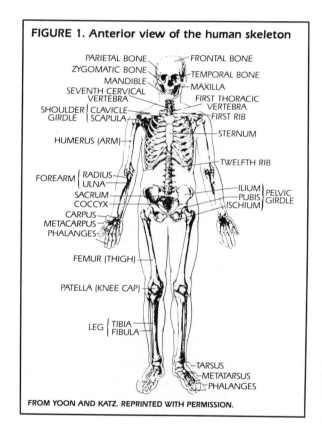

FIGURE 1. Anterior view of the human skeleton

FROM YOON AND KATZ. REPRINTED WITH PERMISSION.

Historical Brief of Medical Applications

To place the new acoustic emission techniques in perspective, it is helpful to briefly review the use of auscultation and acoustic emission in biomedical applications (Table 1). The thoracic percussion of Auenbrugger (1761)[20] and the stethoscope of Laennec (1819)[21] in combination may be considered a primitive form of acoustic emission.

TABLE 1. Review of biomedical acoustic applications*

Early Biomedical Applications
  Direct [Immediate] Auscultation
    Hippocrates (c. 400 BC)   Succession
    Harvey (1628)   Heard heartbeats
    Auenbrugger (1761)   Invented thoracic percussion

  Indirect [Mediate] Auscultation
    Hooke (1705)   Suggested the use of sound for diagnosis
    Laennec (1819)   Invented the stethoscope
    Lisfranc (1823)   Applied the stethoscope for diagnosis of fractures[25]

Recent Biomedical Applications
  Georgia Institute of Technology (1973)[26]
  University of Akron (1974)[27]
  Rensselaer Polytechnic Institute (1977)[14]
  The Hospital for Special Surgery (1978)[26]

*Reference 29 reviews resonant or vibrational techniques applied to bone but not covered in this table.

Prior to these inventions, Hooke appears to have been the first to discuss help for the audition of heartbeats,[22] although he was certainly not the first to have heard them. Harvey, in his classic treatise of 1628 on the circulation of blood, drew a suggestive analogy in referring to these sounds.[23] It is noteworthy that Hooke's suggestion to use sound as a diagnostic tool had been ignored by physicians who studied the heart until Laennec's stethoscope became the wanted help.[24] Thereafter, another century or more passed before sound, now known as acoustic emission or stress wave emission, found its use in fields such as seismology, mining and materials science.

Experimental Procedure

For *in vitro* animal studies, fresh bovine bones were obtained from a local slaughterhouse and tested to failure in bending while monitored for acoustic emission. For the *in vivo* animal study, a stabilized (thirty-day quarantine) beagle of approximately 16 kg (35 lb) was purchased from a laboratory animal supplier. The right tibia was partially sawed surgically to simulate a fracture, while the left hind leg served as the contralateral control. Subsequently, healing processes were monitored both with X-rays and acoustic emission for up to 41 days. In addition, the acoustic emission technique was applied at various times prior to surgery. During both the acoustic emission measurements and radiographic examinations of the hind legs, the dog was anesthetized and lying sideways on a surgical table. The dog's leg was either laid down on the table or held up during acoustic emission measurements.

The tests performed on the fourteen volunteers used in the human study are summarized in Table 2. The number preceding a hyphen in the first column specifies each individual; the number following it indicates different experimental conditions for the same patient. Patient-1 through patient-6 are Dr. L.J. Micheli's patients in the Children's Hospital Medical Center, Boston, Massachusetts, of whom the first five were diagnosed as having had tibial stress fractures, based on their radiographs and other clinical information. Patient-6 was injured in an automobile accident and sustained a fracture of the left tibia, clearly evident in the corresponding radiograph. Patient-7 through patient-12 were participants in the 1980 Boston Marathon and were all tested one day before the race. Two of these (patient-9 and patient-11) were retested after the race. Patients 0 and 00 were normal subjects and were not active runners. The ages of the volunteers, male and female, ranged from fourteen to fifty-one.

The new acoustic emission technique consists of an ultrasonic (pulse) sending system and an acoustic emission receiving or processing system. The sending transducer introduces a repetitive series of well defined ultrasonic pulses into the material under examination, and each of these pulses produces simulated stress waves that mimic acoustic emission events. The receiving transducer intercepts some of the simulated stress wave energy. The sending and receiving transducers are mounted at both ends of the specimen across a known or suspected fracture site.

Figure 2 shows a block diagram of the experimental setup. A pulser driver/pulser synchronizer combination works as an ultrasonic sending system. A conventional acoustic emission setup serves as a receiving system to collect the acoustic emission data. Since a commercial high sensitivity acoustic emission sensor was found to be sensitive to both bulk (longitudinal) and surface waves, it is used both as a transmitter and as a receiver. Both transducers are mounted on the specimen using tape or an elastic bandage and a water soluble couplant gel. During the *in situ* and *in vivo* animal acoustic emission measurements, the transducers were bonded on the smoothly shaved skin of the leg. For the human study, the patient was seated comfortably in a chair with both feet resting on the floor. For the upper extremity test, the forearm was resting on a table.

Results and Discussion

A typical result from the *in vitro* animal study is shown in Fig. 3, in which cumulative counts are plotted versus loading time for a bovine femur loaded in bending. Although

TABLE 2. Acoustic emission experiments on human patients*

No.	Type of Bone	Right/Left	Limb Status	Transducers Location	Distance (millimeters)	X-Rays Taken
0-1	ulna/radius	L	normal	medial	140	no
0-2	tibia	L	normal	m-a/a-1/p	190	no
0-3	tibia	L/L/R	normal	medial	190	no
0-4	tibia	R	normal	medial	varied	no
0-5	tibia	L	normal	medial	223	no
0-6	tibia	L	normal	medial	varied	no
0-7	tibia	L	normal	medial	225 (remount)	no
0-8	tibia	L	normal	m-XMTR/p-RCVR	190	no
0-9	tibia/femur	L	normal	above knee-XMTR/ankle-RCVR		no
0-10**	tibia	L	normal	medial	220	no
0-11**	tibia	R	normal (light motion)	medial	225	no
00-1**	tibia	R/L	normal	medial	218/225	no
1	tibia	R/L	stress fractured/normal	medial	290/285	yes
2-1	tibia	R/L	both stress fractured	medial	280/275	yes
2-2	tibia	R/L	both stress fractured	lateral	300/309	yes
3	tibia	R/L	both stress fractured	medial	330/330	yes
4-1	tibia	R/L	normal/stress fractured	lateral	310/330	yes
4-2	tibia	R/L	normal/stress fractured	medial	320/330	yes
5-1	tibia	R/L	stress fractured/normal	lateral	275/285	yes
5-2	tibia	R/L	stress fractured/normal	medial	265/250	yes
6-1	tibia	R/L	normal/fractured	lateral	290/285	yes
6-2	tibia	R/L	normal/fractured	medial	280/250	yes
7	tibia	R/L	stress fractured/normal	medial	333/328	no
8	tibia	R/L	stress fractured/normal	medial	253/260	no
9-1	tibia	R/L	stress fractured/normal	medial	265/260	no
9-2**	tibia	R/L	stress fractured/normal	medial	265/260	no
10	tibia	R/L	normal/stress fractured	medial	375/405	no
11-1	tibia	R/L	normal/normal	medial	300/300	no
11-2**	tibia	R/L	normal/normal	medial	280/260	no
12	tibia	R/L	normal/normal	medial	280/280	no

* In each case all the five types of acoustic emission data were recorded unless indicated otherwise.
**In this case only the three types of acoustic emission data C/T, C, A were obtained.

such a conventional acoustic emission measurement is useful to distinguish between initial low loadings and fracture, it requires a destructive, invasive test. Note that the bovine femur exhibits a Kaiser effect (acoustic emission is not generated during the reloading until the stress exceeds its previous maximum level).

Using the new acoustic emission technique, five characteristic types of acoustic emission data were typically recorded: (1) cumulative counts versus scanning time (up to one minute), C/T; (2) number of events versus count, C; (3) number of events versus peak amplitude (in decibels), A; (4) number of events versus energy (in V^2 seconds), E; and (5) number of events versus pulse width (in milliseconds), PW. Since the pulse rate was kept constant at five pulses per second, only the slope has meaning in the plot of cumulative counts versus time.

Acoustic emission data from the right tibia of the dog monitored *in vivo* are shown in Figs. 4 through 8 for C/T, C, A, E and PW, respectively, just prior to surgery, and eight and twenty-four days thereafter. The corresponding radiographs of the right leg are given in Fig. 9, which shows the healing process of the fracture up to 41 days. Unlike the human experiments, it was difficult to consistently mount the transducers at the same locations and at the same distance apart on the tibia from one test to the next. Simply lifting up the leg often changed the transducer distance due to the

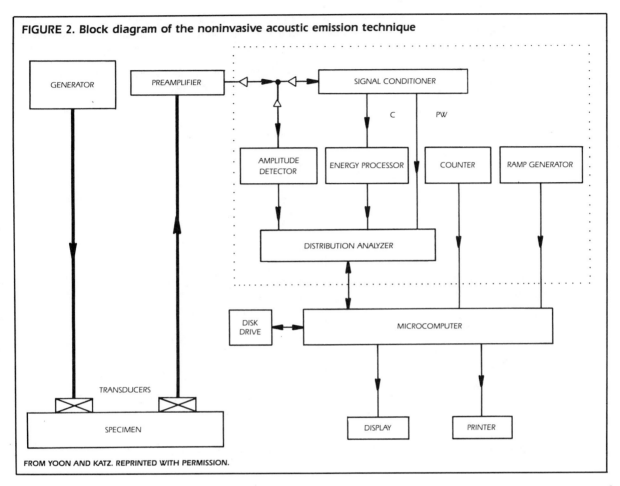

FIGURE 2. Block diagram of the noninvasive acoustic emission technique

FROM YOON AND KATZ. REPRINTED WITH PERMISSION.

sliding motion of the thick surrounding soft tissue layer near the knee. Even so, eight days after surgery all of the acoustic emission parameters differ significantly from the values before surgery, but return to these values by twenty-four days.

The same trend was observed in the human study, although the time scales were different. For example, patient-6 was tested three months after sustaining a traumatic tibial fracture (the cast was removed the day the acoustic emission tests were conducted). The radiograph taken on the same day is shown in Fig. 10. All the acoustic emission parameters of the left tibia were consistently higher than the corresponding values of the uninjured right tibia. The same results were obtained with the transducers mounted on the medial aspect of the tibia as when they were mounted on the lateral aspect.

Figure 11 shows an example of acoustic emission data from patient-5, who suffered a stress fracture in her right leg from running. Testing was performed eighteen months after her initial injury because she was still experiencing pain. All of the acoustic emission parameters measured from her tibiae were clearly distinguishable, although, as expected, the differences were less pronounced than those of patient-6. Again, all the acoustic emission parameters of the injured tibia were higher than those of the normal tibia, even though the radiographic appearance of the injured tibia was normal. Similar results were obtained for patient-4, whose left leg was also injured from running about a year before testing. This contrasts with a comparison of two normal tibiae from the same individual, such as patient-11, who was tested before running the marathon (see Fig. 12).

Patient-1 had stress fractures only in her right tibia, which appeared to have healed at one year, with subsequent occasional pain. The acoustic emission data on the tibiae of this patient were similar to each other. All parameters for the left tibia were slightly higher than those for the right tibia, except for the energy distribution, where the difference was somewhat greater. Since both patient-2 and patient-3 incurred injuries to both their tibiae, no specific implications could be derived from the acoustic emission data. Similarly consistent acoustic emission results were obtained for patients 7, 8, 9, 10 and 12.

From these human and animal studies, it appears that, with time, callus formation and possibly increased mineralization around a fracture site contribute to decreased attenuation of the acoustic emission signal compared with normal bone. This explanation is consistent with known fracture healing processes of bone.

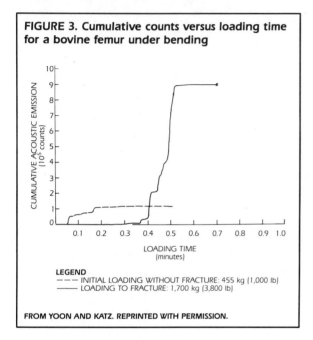

FIGURE 3. Cumulative counts versus loading time for a bovine femur under bending

LEGEND
--- INITIAL LOADING WITHOUT FRACTURE: 455 kg (1,000 lb)
— LOADING TO FRACTURE: 1,700 kg (3,800 lb)

FROM YOON AND KATZ. REPRINTED WITH PERMISSION.

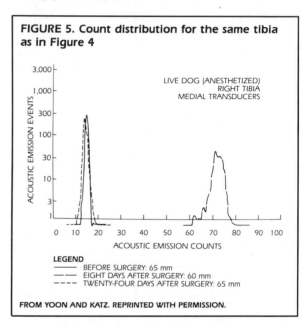

FIGURE 5. Count distribution for the same tibia as in Figure 4

LEGEND
— BEFORE SURGERY: 65 mm
--- EIGHT DAYS AFTER SURGERY: 60 mm
---- TWENTY-FOUR DAYS AFTER SURGERY: 65 mm

FROM YOON AND KATZ. REPRINTED WITH PERMISSION.

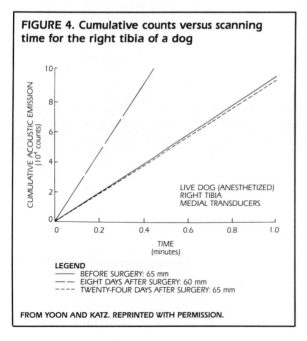

FIGURE 4. Cumulative counts versus scanning time for the right tibia of a dog

LEGEND
— BEFORE SURGERY: 65 mm
--- EIGHT DAYS AFTER SURGERY: 60 mm
---- TWENTY-FOUR DAYS AFTER SURGERY: 65 mm

FROM YOON AND KATZ. REPRINTED WITH PERMISSION.

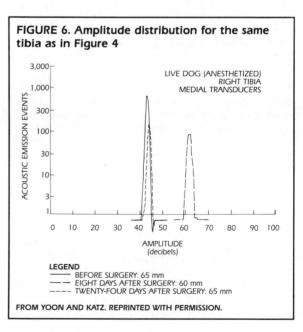

FIGURE 6. Amplitude distribution for the same tibia as in Figure 4

LEGEND
— BEFORE SURGERY: 65 mm
--- EIGHT DAYS AFTER SURGERY: 60 mm
---- TWENTY-FOUR DAYS AFTER SURGERY: 65 mm

FROM YOON AND KATZ. REPRINTED WITH PERMISSION.

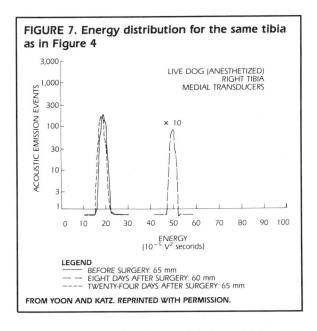

FIGURE 7. Energy distribution for the same tibia as in Figure 4

LEGEND
— BEFORE SURGERY: 65 mm
— — EIGHT DAYS AFTER SURGERY: 60 mm
- - - - TWENTY-FOUR DAYS AFTER SURGERY: 65 mm

FROM YOON AND KATZ. REPRINTED WITH PERMISSION.

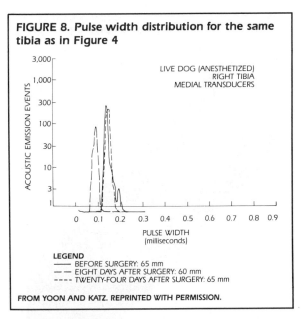

FIGURE 8. Pulse width distribution for the same tibia as in Figure 4

LEGEND
— BEFORE SURGERY: 65 mm
— — EIGHT DAYS AFTER SURGERY: 60 mm
- - - - TWENTY-FOUR DAYS AFTER SURGERY: 65 mm

FROM YOON AND KATZ. REPRINTED WITH PERMISSION.

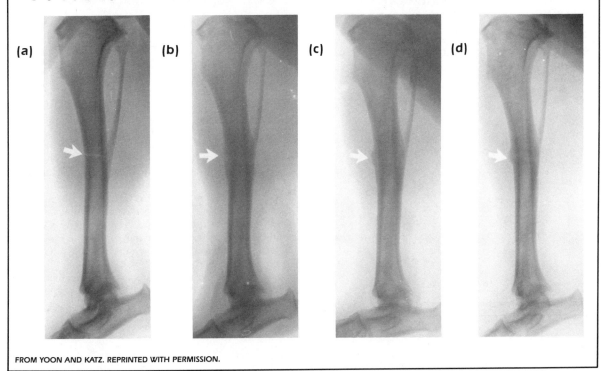

FIGURE 9. Radiographs of the right tibia and fibula of the same dog used for Figure 4: (a) 8 days after surgery; (b) 15 days after surgery; (c) 24 days after surgery; and (d) 41 days after surgery

FROM YOON AND KATZ. REPRINTED WITH PERMISSION.

FIGURE 10. Radiograph of the fractured left tibia of patient-6

FROM YOON AND KATZ. REPRINTED WITH PERMISSION.

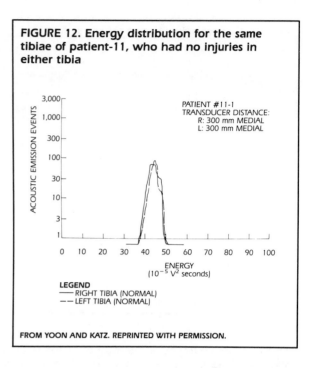

FIGURE 12. Energy distribution for the same tibiae of patient-11, who had no injuries in either tibia

FROM YOON AND KATZ. REPRINTED WITH PERMISSION.

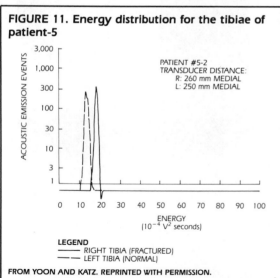

FIGURE 11. Energy distribution for the tibiae of patient-5

FROM YOON AND KATZ. REPRINTED WITH PERMISSION.

Conclusions

The acoustic emission data from both human and animal experiments have been extremely encouraging. The acoustic emission technique distinguished between normal and stress fractured legs in human patients, consistent with similar results obtained when the technique was applied to the fracture healing process of a laboratory animal. The importance of acoustic emission data is that it provides a measure of bone strength when failure is being initiated. To monitor fracture healing, or detect stress fractures or diagnose bone pathologies such as osteoporosis, a measure of the risk of fracture or refracture is critical. The relationship between density, stiffness (or rigidity) and strength in bone (more than in most other materials) is varied and difficult to determine. Acoustic emission holds great promise for medical diagnoses of skeletal defects over conventional techniques that rely to some extent on such a relationship. Even new vibratory techniques, such as modal analysis,[31] will not detect a bone's risk of failure.

PART 3
ACOUSTIC EMISSION TECHNOLOGY IN EXPERIMENTAL AND CLINICAL PEDIATRIC ORTHOPEDICS

Long bones in the body grow in length by the growth of cartilage cells in the epiphyseal (or growth) plate (see Figs. 13 and 14). Acoustic emission technology offers new and exciting possibilities for understanding the fracture mechanics of the growth plate and for the development of clinical methods of assessing and treating growth disorders.

Leg length discrepancy of 25 mm (1 in.) or more can be a significant problem in children and adults, leading to curvature of the spine, back pain, alterations in the normal biomechanics of the lower extremities and early painful degenerative arthritis. Methods of limb equalization include stopping the growth of the longer side, shortening the longer side or lengthening the shorter side. Procedures on the longer side usually require operating upon the unaffected side while present procedures to lengthen the shorter side require multiple major operations and prolonged postoperative care extending over many months.

Recently surgeons have attempted to lengthen the shorter limb by pulling across the growth plate of skeletally immature children at empirically chosen rates. The technique requires a relatively small operation and very little hospitalization time. Unfortunately, colleagues from different parts of the world report completely different results after lengthening through the growth plate. Because closure of the growth plate has frequently occurred subsequent to this procedure, clinical applications have been limited to adolescents approaching skeletal maturation. However, significant leg length discrepancy occurs in younger children as well, and if techniques could be developed to allow limb

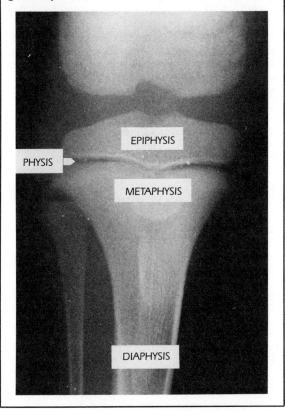

FIGURE 13. Regions of a skeletally immature canine tibia; growth occurs from the physis or growth plate

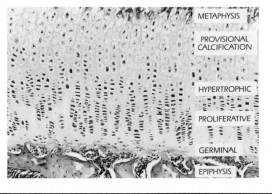

FIGURE 14. Regions of a skeletally immature rat proximal tibial physis (see Figure 13)

lengthening by distraction across the growth plate without causing its closure, then this population of younger children might also be able to benefit from this technique.

Rate dependent distraction at 1 mm a day will cause separation of the epiphysis from the growth plate (epiphysiolysis), but Italian research suggests that a slower rate (0.25 mm twice a day) will allow lengthening without epiphysiolysis.[32] If a method could be developed to detect impending separation through the growth plate, such as the release of acoustic signals from initial microcrack formation, then conceivably the distraction process could be stopped (before causing epiphysiolysis) and resumed on the following day. To examine this theory, acoustic parameters that correlate with the initiation of structural damage to the physis during distraction must first be determined. The *in vitro* studies detailed below will report the results of these preliminary investigations.

In Vitro Model

Method for the In Vitro Model

Skeletally immature rats were sacrificed and their tibiae dissected. Each rat tibial shaft was securely placed into a testing jig while the proximal epiphysis was placed between two steel pins of a yoke and attached to a loading cam. The bones were placed in a horizontal position to subject the growth plate to bending loads or in a vertical position to apply tension forces on the growth plate. An acoustic emission sensor was affixed to one of the yoke pins, which acted as a waveguide. A load cell measured the total force applied to the specimen by the loading cam and a displacement transducer, attached in parallel to the loading armature, recorded displacement during force application.

Throughout the test, acoustic data were collected using a commercial acoustic emission system. Load versus deformation curves were recorded on a storage oscilloscope and photographed for later analysis. Some specimens were loaded to complete failure while monitoring acoustic events. Other tibiae were loaded only until specific acoustic emission amplitudes were recorded and then the load was released. After testing, the specimens were removed from the testing jig and placed in formalin prior to decalcification. Specimens were then serially sectioned and examined microscopically for any histological alterations or microcracks.

Results of the In Vitro Model

Histologic evaluation of the tibiae loaded to failure showed complete separation through the growth plate. A vast array of acoustic signals were recorded for these specimens. However, in those specimens that were only partially loaded and loading was stopped prior to the initiation of the failure crack, microfractures were noted in a variety of layers within the growth plate (see Fig. 15).

These microfractures were noted when either bending moments or tension forces were applied to the rat tibia. When the plate was loaded in tension, microfractures appeared more commonly in the germinal layer of the growth plate. These cracks represent the earliest signs of significant damage[33] within the substance of the growth plate and occurred in specimens that had emitted acoustic events with amplitudes of 50 dB or greater during partial loading.

Significance of the In Vitro Model

Acoustic emission monitoring appears capable of detecting the earliest signs of structural damage within the growth plate (the formation of microcracks in the germinal layer) during tension loading. Such microcracks can disrupt the blood supply to the cells of the germinal layer which may result in the death of germinal layer cells and closure of the growth plate.[34,35] Monitoring acoustic emission from the growth plate provides a technique for early detection of damage in the germinal layer when the physis is subjected to tension loading. What is not clear at this point, however, is how extensive microfractures have to be before irreversible damage is done to the future growth potential of the physis.

Therefore, an *in vivo* model has been developed to determine whether the growth plate can tolerate daily increments of distraction which are needed to achieve mechanical bone lengthening. It has been suggested that the growth plate can respond to the distraction process by accelerating the rate of new bone formation.[36] Such a theory requires distraction of

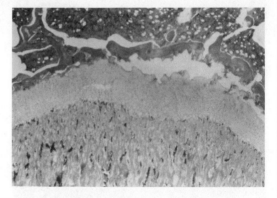

FIGURE 15. Microcracks in growth plate cartilage with partial loading; separations are noticeable in the upper proliferative layer and in between the epiphysis and the growth plate

FROM POLIAKOFF AND BRIGHT. REPRINTED WITH PERMISSION.

the physis without causing irreversible damage to the growth plate. Thus, some distraction may serve as a stimulus for new bone formation, while excessive distraction could result in complete cessation of growth. To test the possibility of lengthening at a slow enough rate to not alter future growth, the following *in vivo* biochemical studies were performed while monitoring the growth plate during lengthening.

In Vivo Model

Method for the In Vivo Model

Skeletally immature puppies underwent femoral lengthening using an Orthofix external fixator in which heavy stainless steel pins were inserted into the regions of the distal femur above and below the growth plate and then connected to a rigid external frame. Small threaded stainless steel wires were also inserted above and below the growth plate to act as radiographic markers so that longitudinal growth could be quantitated on sequential X-rays. Lengthening was accomplished at a fixed rate through a distractor that was connected to the external fixator. Three rates of lengthening were employed: (1) a daily lengthening of 1 mm; (2) a daily lengthening of 0.5 mm or less if a single acoustic event with an amplitude of 50 dB or greater was recorded; and (3) every other day lengthening of 0.5 mm or less if a single acoustic event with an amplitude of 50 dB or greater occurred.

During distraction a sensor was attached to one of the epiphyseal pins below the distal femoral growth plate and acoustic events were recorded. Preliminary tests showed that the pins were excellent waveguides and that the signal was best received at the flat end of the pin.

Results of the In Vivo Model

In developing the animal model, it was found that anesthesia in addition to sedation was necessary to avoid spurious acoustic events arising from animal movement. Pins needed to be wrapped securely with gauze and treated with unpleasant tasting substances to prevent the animals from removing the dressings and exposing the percutaneous pin wounds to kennel pathogens. Appropriate antibiotic prophylaxis was also required during surgery and while the fixator remained in place.

Growth continued in the puppies after lengthening was completed if distraction had been performed using either of the 0.5 mm protocols (see Fig. 16). Epiphysiolysis did not occur in any of these animals. Acoustic parameters monitored included duration, counts, energy and amplitude. Frequently, the greatest magnitude of a given acoustic parameter obtained during the distraction process was obtained on

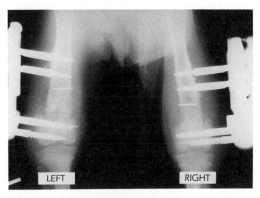

FIGURE 16. Radiograph of the hind limbs of a dog with bilateral distal femoral distractors; left side shows epiphyseal distraction with separation through the growth plate after 14 days of 1 mm distraction per day; the right side was distracted 0.5 mm per day for 14 days and did not separate through the physis

FROM POLIAKOFF AND BRIGHT. REPRINTED WITH PERMISSION.

the fifth or sixth day of distraction. This time frame corresponded to those reported from clinical studies of children in England[37] undergoing physeal distraction where maximum forces across the growth plate were obtained on about the fifth to seventh day of lengthening. Afterwards, lesser values, denoting mechanical failure of the growth plate, were recorded and epiphysiolysis occurred.

Significance of the In Vivo Model

It is known that epiphysiolysis can cause growth to stop and hence distraction across a physis must be stopped prior to the occurrence of epiphysiolysis. In this study, distraction, limited and monitored by acoustic emission technology, could successfully be performed without causing epiphysiolysis. These acoustic events may be forewarnings of an impending separation of the growth plate and can thus provide information about when the distraction process should be discontinued to prevent such a catastrophic event.

Summary of Acoustic Emission Application

Much more research will be required before clinical applications can become routine. However, these early investigations have been encouraging. Acoustic emission technology does seem to offer a very sensitive and reproducible method for determining the earliest indication of growth plate damage (microcracks).

In addition, acoustic emission technology may also be useful in other aspects of limb lengthening. For example, children and adults whose growth plates have closed and who require limb lengthening must undergo distraction through an osteotomy (surgical division) of the shaft of the bone. Although distraction was previously started at the time of the osteotomy, clinicians have more recently awaited the formation of healing callus before beginning distraction.

In this way, the surgically created osteotomy has newly formed callus already bridging the gap between the cut bone surfaces during distraction and there is less chance that the bone will fail to unite. Once the bone has been lengthened, the external fixator must be left in place for an empirically chosen period of time so that the bone which forms will be able to withstand both the compressive muscular forces within the limb and the compressive and bending forces of weight bearing after the fixator is removed. By using a transducer to transmit a pulse wave across the osteotomy gap with a sensor transducer on the other side, acoustic emission technology may also offer a safe and more objective method of assessing the quality and strength of the new bone which is formed.

PART 4
DIAGNOSIS OF LOOSE OR DAMAGED TOTAL JOINT REPLACEMENT COMPONENTS

The surgical replacement of diseased and damaged joints by prosthetic components is an established treatment for relieving pain and restoring function. Though as many as 90 percent of total joint replacement patients have a successful outcome, many patients suffer serious complications such as infection and loosening of the prosthetic components from their surrounding bony attachment. Early, reliable diagnosis of these complications is necessary to prevent pain and subsequent loss of bone stock, making revision surgery to replace the components more difficult.

The techniques currently being used to diagnose complications such as loosening include radiography, arthrography and radionuclide scanning. Unfortunately, these techniques are often unreliable in detecting the early stages of loosening and present significant risk to the patient through exposure to X-rays or injection of radiopaque or radioactive material. Another problem is that these techniques provide only a static view of a dynamic process. The need exists for a noninvasive technique that can provide early diagnosis of loosening and mechanical damage with no risk to the patient.

Studies have been conducted to develop acoustic emission as a diagnostic technique for detection of loosening and damage. The aim of this study was to examine the efficacy of acoustic emission monitoring to predict failure in total joint replacements of the knee. In a preliminary study, the attenuation of acoustic emission signals by soft tissue was measured[38] using a transducer placed on the skin. It was established that minimal attenuation would occur through the soft tissue covering the knee joint. It has also been previously demonstrated that signals received from damaged total knee replacements were significantly different from signals received from controlled, well functioning total knee replacements.[39]

Methods for Diagnosis

Patients were monitored during their follow-up visit to the Total Knee Replacement Clinic at the Hospital for Special Surgery. Monitoring was performed by attaching an acoustic emission transducer over the tibial tubercle at the lower front of the knee. The transducer was attached to the skin using surgical tape and sterile jelly as a couplant. Each patient was asked to perform two tasks: rocking back and forth on the affected limb and stepping up onto a step stool, leading with the affected limb. Each task was repeated three times with the resulting acoustic emission data collected on a commercial acoustic emission system. Total gain of the system was 69 dB.

Pertinent clinical information, such as original diagnosis, weight, sex and length of time the total joint replacement was implanted, were obtained from the patient's medical records. Standard radiographs taken during the same clinic visit were examined and used to determine the condition of the total knee replacement (see Fig. 17). The knee replacement was considered a failure if the radiographs demonstrated a radiolucent region around the entire border between the knee replacement and the surrounding bone or if, upon examining earlier radiographs, a radiolucent region progressing in length with time was noted.[40]

Patient Population

For this study, 182 total knee replacements in 127 patients were monitored with acoustic emission. There were 55 patients with bilateral knee replacements and both joints were monitored in each of these patients. Eighty-three percent of the patients were female. The average age of the patients was 71 years, with a range of 41 to 91 years. Patient weight ranged from 47 to 140 kg (104 to 309 lb), with an average of 79 kg (174 lb).

The diagnosis at implantation was osteoarthritis in 137 knees (75 percent of the cases), rheumatoid arthritis in nine knees (5 percent) and osteonecrosis in nine knees (5 percent). Twenty knee replacements (11 percent) had been revised and the diagnoses for the remaining seven knees (4 percent) were unknown. There were 103 right knee replacements and 79 left knee replacements. The average follow-up at the time of acoustic emission monitoring was 49 months, with a range from three to 107 months.

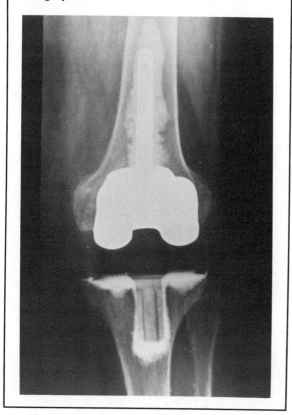

FIGURE 17. Radiograph showing the anterior view of a knee with a total knee replacement; the thigh bone (femur) has a metallic component implanted, while the shin bone (tibia) has a polyethylene component which can only be seen in outline against the surrounding radiopaque acrylic cement; the cement is used as a grout to secure the components to the surrounding bone; no radiolucent lines are visible on this radiograph

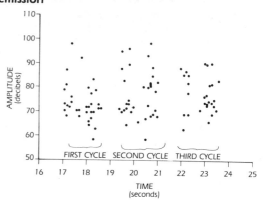

FIGURE 18. Amplitude plotted against real-time for three repetitions of the step-up task by a total knee patient monitored with acoustic emission

Results of the Acoustic Emission Tests

The acoustic emission data for each knee replacement included the total number of counts, the total number of events, and the total energy determined from the postprocessed acoustic emission signals recorded during the rocking and step-up tasks. Typical data recorded from one patient during the three repetitions of the step-up task are shown in Fig. 18. The number of events with an amplitude equal to or greater than 80 dB was also included as part of the acoustic emission data. This latter variable is a component of the total number of events and was chosen as a separate variable based on results from previous work,[39] in which high amplitude events were shown to be an important part of the acoustic emission signal arising from loosened total knee replacements. Table 3 summarizes the acoustic emission data for the 182 total knee replacements monitored during the study.

Review of the patients' radiographs revealed 35 total knee replacements with complete or progressive radiolucency. These were considered failures, with the remaining 147 replacements taken as normal or functioning well.

A statistical analysis was performed combining the demographic and acoustic emission data to determine if such data could be used to predict total knee joint failure (as determined from the presence of progressive or total radiolucency). The analytical method used is an extension of logistic regression analysis. The standard logistic regression model is a mathematical description of how the probability of failure depends on demographic and acoustic emission variables. It assumes all responses are independent of one another.

In the present study, much of the data were from bilateral total knee replacements. The responses from knee replacements on the *same* patient cannot be considered independent. A correlated logistic model was used to accommodate the dependent binary responses. A series of regression models were fitted to screen the demographic and acoustic emission variables. The coefficients of each variable were estimated, as well as their standard errors. For each model, the ratio of the estimated coefficient to its standard error was used to judge whether the variable added significantly to the predictive capability of the model.

TABLE 3. Acoustic emission data from 182 total knee replacements

	Rocking Task average (range)	Step-Up Task average (range)
Counts	172 (0 to 5,400)	375 (0 to 2,805)
Events	7 (0 to 60)	15 (0 to 94)
Events greater than 80 dB	4 (0 to 44)	10 (0 to 82)
Energy	149 (0 to 5,510)	286 (0 to 3,466)

TABLE 4. Hypothesis tests for model variables

Variable	Ratio of Estimated Parameter to Its Standard Error
Side	−0.43
Weight	1.08
Age	−1.30
Sex	−1.20
Follow-up time	0.23
Counts (rocking)	0.98
Energy (rocking)	0.99
Counts (step-up)	2.43
Energy (step-up)	1.90
Events (rocking)	2.69
Events greater than 80 dB (rocking)	2.73
Events (step-up)	3.78
Events greater than 80 dB (step-up)	3.64

Results of the regression analysis are summarized in Table 4. None of the demographic variables examined had a significant effect on the predictive probability of failure. The counts and energy variables of the acoustic emission signals collected during the rocking and step-up tasks also showed no significant effect.

The only variables showing evidence of predictive capability were the acoustic emission events (total events and events with an amplitude above 80 dB) obtained during the rocking and step-up tasks. Both acoustic emission event variables measured from the step-up task had larger standardized coefficients than the corresponding events measured from the rocking tasks. This indicates the step-up tasks have greater predictive ability to determine progressive or total radiolucency and is consistent with the higher values shown in Table 3 for the step-up task.

Multivariate models were fitted, based on several exploratory variables. This effort showed that simple single-variable models based either on total events or events with amplitude above 80 dB fit as well as more complicated models.

The single-variable model was used to predict the outcome (success or failure) for the total knee replacements monitored in this study. If the model predicted, with a probability greater than 0.5, that a knee replacement would be a failure, it was considered a failure. Otherwise, it was considered a success. Overall, the model correctly predicted the outcome for 83 percent of the knee replacements. All but one of the knee replacements with no radiographic evidence of failure were predicted as successes. However, only six of 35 knee replacements considered as failures radiographically were predicted to be failures.

Discussion of the Acoustic Emission Technique

The results of this study further demonstrate that significant acoustic emission signals emanate from damaged total knee replacements under load. As in previous work,[39] the amplitude of these signals appears to be the most appropriate diagnostic variable. Comparison of the acoustic emission data from the rocking and step-up tasks reveals that with the higher load (4.4 times body weight)[41] applied during step-up, more signals of greater magnitude are emitted than with the lower load (2.8 times body weight)[41] incurred during the rocking task. Therefore, patients should be tested using the step-up task whenever possible. The rocking task can be employed when pain or deformity make the step-up task impossible.

The inaccuracy in predicting total joint failure based on acoustic emission data should not be considered a final evaluation of the technique. Several potential solutions to improve the accuracy must be examined. One step is to expand the data base to include more patients. The population in the present study had few documented clinical failures. This required the adoption of a less precise radiographic definition for failure, based on total or progressive radiolucency around the prosthetic components.

As more patients are added and as the patients already being studied are continually monitored during successive clinic visits, a more precise definition of failure may emerge. Furthermore, time can be included as a variable, so that changes in the acoustic emission signals from the same total knee replacement can be examined to see if they accurately predict early damage occurring in the joint.

Finally, other means of including acoustic emission data into the predictive model might prove fruitful. One potential method is the use of correlation plots, in which various acoustic emission parameters are plotted against one another to examine possible significant correlations. These correlations might be mechanism specific and might therefore have predictive capability.

REFERENCES

1. Piekarski, K. "Fracture of Bone." *Journal of Applied Physics*. Vol. 41 (1970): pp 215-223.
2. Currey, J.D. "Changes in the Impact Energy Absorption of the Bone with Age." *Journal of Biomechanics*. Vol. 12 (1979): pp 459-469.
3. Netz, P., K. Eriksson and L. Stromberg. "Nonlinear Properties of Diaphyseal Bone. An Experimental Study on Dogs." *Acta Orthopaedica Scandinavica*. Vol. 50 (1979): pp 139-143.
4. Netz, P., K. Eriksson and L. Stromberg. "Material Reaction of Diaphyseal Bone Under Torsion. An Experimental Study on Dogs." *Acta Orthopaedica Scandinavica*. Vol. 51 (1980): pp 223-229.
5. Jonsson, U. and K. Eriksson. "Microcracking in Dog Bone Under Load. A Biomechanical Study of Bone Visco-Elasticity." *Acta Orthopaedica Scandinavica*. Vol. 55 (1984): pp 441-445.
6. Sammarco, G.J., A.H. Burstein, W.L. Davis and V.H. Frankel. "The Biomechanics of Torsional Fractures. The Effects of Loading on Ultimate Properties." *Journal of Biomechanics*. Vol. 4 (1971): pp 113-117.
7. Lachman, E. "Osteoporosis: The Potentialities and Limitations of Its Roentgenologic Diagnosis." *American Journal of Roentgenology, Radium, Thermal and Nuclear Medicine* (now *American Journal of Roentgenology*). Vol. 74 (1955): pp 712-715.
8. Gustafsson, L., B. Jacobson and L. Kusoffsky. "X-Ray Spectrophotometry for Bone-Mineral Determinations." *Medical and Biological Engineering* (now *Medical and Biological Engineering and Computing*). Vol. 12 (1974): pp 113-119.
9. Cameron, J.R., R.B. Mazess and J.A. Sorenson. "Precision and Accuracy of Bone Mineral Determination by Direct Photon Absorptiometry." *Investigative Radiology*. Vol. 3 (1968): pp 141-150.
10. Mazess, R.B., M. Ort, P.F. Judy and M. Mather. *Proceedings of the Bone Measurement Conference*. J.R. Cameron, ed. USAEC Conference 700515 (1970): pp 308-312.
11. Brooks, R.A. and G. Di Chiro. "Principles of Computer Assisted Tomography (CAT) in Radiographic and Radioisotopic Imaging." *Physics in Medicine and Biology*. Vol. 21 (1976): pp 689-732.
12. Brownell, G.L., T.F. Budinger, P.C. Lauterbur and P.L. McGeer. "Positron Tomography and Nuclear Magnetic Resonance Imaging." *Science*. Vol. 215 (1982): pp 619-626.
13. "Special Issue on Computerized Medical Imaging." *IEEE Transactions on Biomedical Engineering*. A.C. Kak, ed. Vol. BME-28, No. 2 (1981): pp 49-234.
14. Thomas, R.A., H.S. Yoon and J.L. Katz. "Acoustic Emission from Fresh Bovine Femora." *Ultrasonics Symposium Proceedings*. J. deKlerk and B.R. McAvoy, eds. New York, NY: Institute of Electrical and Electronics Engineers (1977): pp 237-241.
15. Yoon, H.S., B. Caraco and J.L. Katz. "Further Studies on the Acoustic Emission of Fresh Mammalian Bones." *Ultrasonics Syposium Proceedings*. J. de Klerk and B.R. McAvoy, eds. New York, NY: Institute of Electrical and Electronics Engineers (1979): pp 399-404.
16. Yoon, H.S. "Applications of a New Acoustic Emission Technique to Head Injuries." *First Year Progress Report to the Whitaker Foundation* (1979): pp 94.
17. Yoon, H.S., B. Caraco, H. Kaur and J.L. Katz. "Clinical Application of Acoustic Emission Techniques to Bone Abnormalities." *Ultrasonics Symposium Proceedings*. B.R. McAvoy, ed. New York, NY: Institute of Electrical and Electronics Engineers (1980): pp 1,067-1,072.
18. Yoon, H.S., B. Caraco, H. Kaur, J.L. Katz and L.J. Micheli. "Noninvasive and Non-Traumatic Diagnosis of Bone Abnormalities by AE Techniques." *Proceedings of the International Conference on Acoustic Emission and Photo-Acoustic Spectroscopy* (1981): pp 4.
19. Yoon, H.S. "Applications of a New Acoustic Emission Technique to Head Injuries." *Final Report to the Whitaker Foundation* (1983): pp 76.
20. Auenbrugger, L. *Inventum Novum ex Percussione Thoracis Humani*. Vienna: Trattner (1761). *On Percussion of the Chest* (English translation by J. Forbes): London (1824).
21. Laennec, R.T.H. *De l'Auscultation Médiate, ou Traité du Diagnostic des Maladies des Poumons et du Coeur*. two volumes. Paris: Brosson et Chaudé (1819). *A Treatise on the Diseases of the Chest and on Mediate Auscultation* (English translation by J. Forbes): London (1821) first edition; (1834) fourth edition.
22. Hooke, R. *The Posthumous Works of Robert Hooke*. R. Waller, ed. London: Smith and Walford (1705): pp 39-40.
23. Harvey, W. *Exercitatio Anatomica de Motu Cordis et Sanguinis in Animalibus*. Frankfurt: Fitzeri (1628). *The Works of William Harvey* (English translation by R. Willis). London: Sydenham Society (1847); Johnson reprint (1965): pp 9-86.

24. Hunt, F.V. *Origins in Acoustics*. New Haven: Yale University Press (1978): pp 137.
25. Lisfranc, J. *Mémoire sur de Nouvelles Applications du Stéthoscope de M. Le Professeur Laënnec*. Paris: Gabon et Cie (1823).
26. Hanagud, S., R. G. Clinton and J.P. Lopez. "Acoustic Emission in Bone Substance." *Proceedings of the Biomechanics Symposium*. Y.C. Fung and J.A. Brighton, eds. New York, NY: American Society of Mechanical Engineers (1973): pp 79-81.
27. Chu, M.L., I.A. Gradisar, M.R. Railey and G.F. Bowling. "An Acoustical Technique for the Evaluation of Knee Joint Cartilage." *Proceedings of the 27th Annual Conference on Engineering in Medicine and Biology*. Vol. 16 (1974): pp 294.
28. Wright, T.M., A.H. Burstein and S.P. Arnoczky. "In Vivo Monitoring of Ligament Damage in the Canine Knee by Acoustic Emission." *Proceedings of the ASNT National Fall Conference* (1978): pp 122-125.
29. Nokes, L.D.M., H.J. Williams, J.A. Fairclough, W.J. Mintowt-Czyz and I.G. Mackie. "A Literature Review of Vibrational Analysis of Human Limbs." *IEEE Transactions in Biomedical Engineering*. Vol. BME-31 (1984): pp 187-192.
30. King, B.G. and M.J. Showers. *Human Anatomy and Physiology*. Philadelphia, PA: Saunders (1969): pp 163.
31. Van der Perre, G., R. Van Audekercke, J. Vandecasteele, M. Martens and J.C. Mulier. "Modal Analysis of Human Tibiae." *Proceedings of the Biomechanics Symposium*. W.C. Van Buskirk and S.L.Y. Woo, eds. New York, NY: American Society of Mechanical Engineers (1981): pp 173-176.
32. DeBastiani, G., P. Armotti, L.R. Brivio and R. Aldegheri. *Chondrodiatasis: An Experimental Study on Orthofix Paediatric Applications*. England: Electro-Biology International (UK) Ltd.
33. Bright, R.W. and S.M. Elmore. "Physical Properties of Epiphyseal Plate Cartilage." *Surgeons Forum*. Vol. 19 (1968): pp 463-464.
34. Trueta, J. and J.D. Morgan. "The Vascular Contribution to Osteogenesis, I. Studies by the Injection Method." *JBJS*. Vol. 42B, No. 1 (February 1960): pp 97.
35. Trueta, J. and V.P. Amato. "The Vascular Contribution to Osteogenesis, III. Changes in the Growth Cartilage Caused by Experimentally Induced Ischaemia." *JBJS*. Vol. 42B, No. 3 (August, 1969): pp 571.
36. DeBastiani, G., R. Aldegheri, L.R. Brivio and G. Trivella. *Controlled Symmetrical Distraction of the Epiphyseal Plate (Chondrodiatasis) to Achieve Limb Lengthening in Children* (work in progress, personal communication).
37. Jones, C. et al. *Epiphyseal Limb Lengthening*. Personal communication to Dr. Robert Bright (December 1984).
38. Wright, T.M., and J.M. Carr. "Soft Tissue Attenuation of Acoustic Emission Pulses." *Journal of Biomedical Engineering*. Vol. 105 (1983): pp 20-23.
39. Wright, T.M., R.W. Hood and W.J. Flynn. "Acoustic Emission Monitoring in the Diagnosis of Loosening in Total Knee Arthroplasty." *Proceedings of the Biomechanics Symposium*. W.C. Van Buskirk and S.L.Y. Woo, ed. New York, NY: American Society of Mechanical Engineers (1981).
40. Insall, J.N., P.F. Lachiewicz and A.H. Burstein. "The Posterior Stabilized Condylar Prosthesis: A Modificaton of the Total Condylar Design." *Journal of Bone and Joint Surgery*. Vol. 64-A, No. 9 (1982): pp 1317-1323.
41. Walker, P.S. *Human Joints and Their Artificial Replacements*. Springfield, IL: Charles C. Thomas (1977).

SECTION 14

PROCESS MONITORING WITH ACOUSTIC EMISSION

Ronnie K. Miller, Physical Acoustics Corporation, Princeton, New Jersey

David Dornfeld, University of California, Berkeley, California
Anthony Hamilton, LTV Aerospace and Defense Company, Dallas, Texas
James Mitchell, Physical Acoustics Corporation, Princeton, New Jersey
Ken Yee, National Bureau of Standards, Gaithersburg, Maryland

PARTS 2, 3 AND 4 FROM SOCIETY OF MANUFACTURING ENGINEERS. ADAPTED AND REPRINTED WITH PERMISSION.
PART 5 FROM AMERICAN SOCIETY FOR METALS (SOCIETY FOR CARBIDE AND TOOL ENGINEERS). ADAPTED AND REPRINTED WITH PERMISSION.

INTRODUCTION

The use of acoustic emission techniques for process monitoring has the potential of ensuring high product quality while minimizing the total cost of a product. Processes such as cutting, grinding, forming, joining and curing all generate acoustic emission for reasons unique to each process. In many cases, this emission can be monitored to: (1) characterize the process; (2) detect discontinuities or process abnormalities *in situ*; and (3) detect discontinuities shortly after the process has been terminated.

For processes that can be characterized in real-time, it is possible to generate a feedback signal that relates to one of two conditions: (1) that the process is on track and the process variables are ideal; or (2) that the process is not on track and the process variables need to be changed.

Even when the process is optimized and controlled, it is still possible to experience abnormalities and have discontinuities in the product. Knowledge of these events can be provided with acoustic emission techniques, thus avoiding unnecessary manufacturing steps and saving considerable expense.

A separate section in this *NDT Handbook* volume is dedicated to weld monitoring, a joining process.

PART 1
ACOUSTIC EMISSION MONITORING OF AN AUTOMOTIVE COINING APPLICATION

During the fabrication of fuel injectors in the automotive industry, it is common to use a joining technique called *coining* or *peening* to fasten the tip of the fuel injector to the nozzle. A cutaway view of one such product is shown in Fig. 1, with the nozzle fully supported. The coining head is repeatedly dropped onto this assembly, until the collar of material surrounding the tip deforms, thus holding the tip securely in place (Fig. 2). In production, the number of times the coining head strikes the nozzle is fixed at three.

Through experience, it was determined that defective products were a result of three basic problems.

1. There was a potential for crack formation in the tips due to improper heat treatment of the tips or existing cracks that were impossible to detect before coining.
2. Whenever a nozzle or tip was improperly mounted, the potential for eccentric coining existed. This situation resulted in material being scraped off the collar of the nozzle resulting in a weak joint.
3. If the coining head applied more force than was programmed, there was a potential for excessive deformation or cracking.

Sources of Acoustic Emission

Given the description of the problem and an understanding of the process, five major acoustic emission sources were identified: (1) coining impact; (2) deformation of the nozzle walls; (3) cracking of the nozzle walls; (4) scraping of the nozzle walls; and (5) cracking of the tip. The first two sources, impact and deformation, are expected during a normal process. The last three are not. When they are present, the part is considered rejectable. In addition, if the force of the coining head strike is higher than programmed, the deformation should also deviate from the norm.

Development Monitoring

A coining machine was modified by replacing one of the nozzle supports with a special assembly. This assembly, shown in Fig. 1, consists of a spring, a 150 kHz resonant transducer and a waveguide. The waveguide was permanently bonded to the transducer on one end and established a dry mechanical coupling to the nozzle whenever a part was inserted. It was determined that the spring supplied sufficient coupling force to ensure repeatable results from part to part.

During the coining operation, the acoustic emission associated with the process would propagate down the nozzle to the transducer via the waveguide. The output signal from the transducer was then amplified by 40 dB using a preamplifier and then band-pass filtered between 100 and 300 kHz. An additional gain of 40 dB was provided using a commercial system and the signal was then compared to a 1 V

FIGURE 1. Coining machine assembly, with fuel injector in position and nozzle fully supported

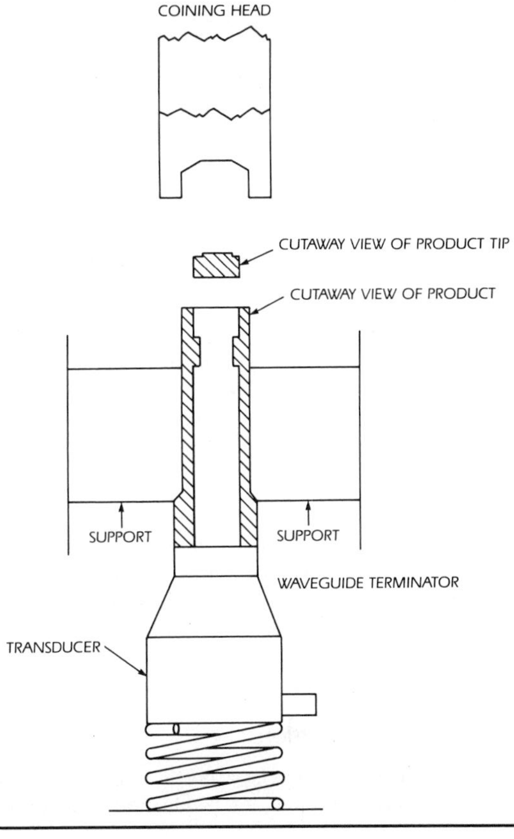

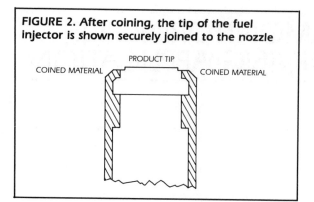

FIGURE 2. After coining, the tip of the fuel injector is shown securely joined to the nozzle

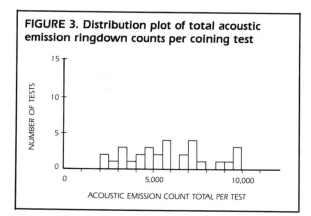

FIGURE 3. Distribution plot of total acoustic emission ringdown counts per coining test

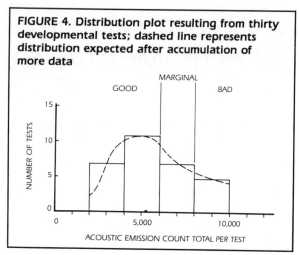

FIGURE 4. Distribution plot resulting from thirty developmental tests; dashed line represents distribution expected after accumulation of more data

threshold. The resulting ringdown counts for each coining cycle were totaled and recorded (see Fig. 3).

Results of the first thirty tests were used to develop this application and a production coining machine was instrumented in the same fashion. It was assumed that the distribution of ringdown counts per test would grow as shown in Fig. 4. This figure shows the results of the same thirty tests done in the development stage but plotted less discretely. The dashed line overlayed on this plot represents the shape of the distribution that would be expected after more test values had been acquired.

More data were acquired through production monitoring and a skewed distribution similar to the one shown in Fig. 4 was obtained. Also, accept/reject criteria were established based on total ringdown counts per coining cycle. In the production development, the area shown as marginal in Fig. 4 was redefined rejectable.

Production Instrumentation

A single-channel, commercial acoustic emission system was purchased and modified for production monitoring.

The system had an adjustable alarm threshold for ringdown counts and an alarm output. The counter was reset from an external source, in this case, the foot pedal that actuated the coining machine. Figure 5 shows a diagram of this setup.

During operation, the counter would reset to zero whenever the foot pedal was pressed to start the coining cycle. As a nozzle was coined, the system would sum the ringdown counts and compare that number with the alarm threshold. If the number stayed below the alarm threshold, the operator was allowed to continue on to the next part. If not, the alarm would trigger, the coining machine would lock and an audio alarm would sound. The operator was required to use a special key to release the coining machine and reset the audio alarm.

Benefits of Coining Monitoring with Acoustic Emission

The initial intent of using acoustic emission to monitor coining was to detect and reject parts with cracked tips. A fuel injector failure in the field could potentially lead to a costly liability. Although this objective was met, the real benefit came from reducing the alarm threshold to the marginal level illustrated in Fig. 4. This approach provided a small percentage of parts, about four percent of a production lot, that could be subjected to destructive fatigue testing. The service life of the part was determined by this postcoining test which was already a production requirement. If the service life was less than engineering specifications, the entire production lot could be held back for additional testing and the coining machine examined for proper operation.

In summary, it can be said that the overall objective of monitoring the process for the detection and rejection of

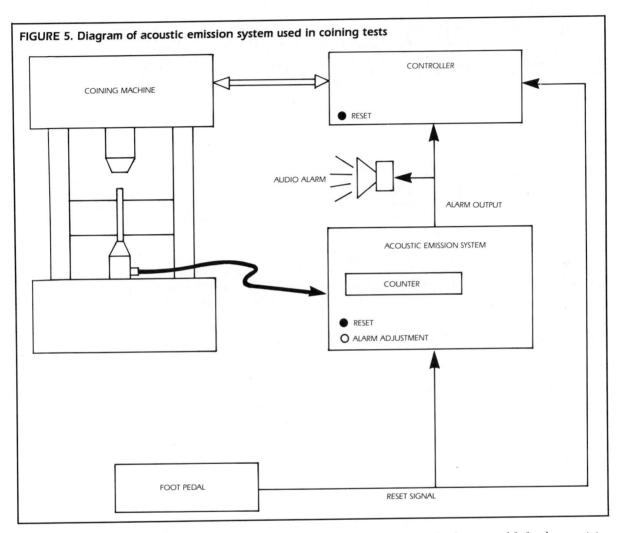

FIGURE 5. Diagram of acoustic emission system used in coining tests

discontinuous parts was met. An additional benefit was seen by reducing the alarm threshold to the marginal level to provide a small number of parts for lot sampling. By examining these parts destructively, the service life for the remaining lot could be guaranteed with higher confidence and the proper operation of the coining process could be confirmed.

PART 2
THE USE OF DRILL-UP FOR ON-LINE DETERMINATION OF DRILL WEAR

Drill-up is an instrument designed at the National Bureau of Standards to prevent breakage of small diameter drills used on automatic feed drilling machines with spindle retract ability. This original application and the measurement method utilized are described in the literature.[1,2] Detailed information on the construction and operation of a single drilling speed drill-up is available in the literature.[3]

The discussion below details the potential of drill-up for use as a drill failure sensor. The instrument determines that breakage is imminent and signals the drilling machine to retract the drill. The input sensor is a piezoelectric accelerometer that is mechanically coupled to the workpiece. Potential drill breakage is determined by time domain analysis of the accelerometer signal. A calibration routine, in the single-chip microcomputer that implements the function, automatically adjusts to the amplitude of the signal from normal drilling. Large amplitudes or accelerations, synchronous with the drill rotation, are indicative of improper drilling. The detection method depends on the deflection of a long, slender column with one rigid support on the end opposite an axial load (a drill held in a chuck and loaded by the workpiece).

When the drilling is improper, due to factors such as a worn cutting edge or a hard spot in the workpiece, the material cannot be removed as fast as the drill is being fed. When this occurs, the drill deflects as a column and induces a vibration signal as it scrapes on the side or bottom of the hole. If this is allowed to continue, the buckling deflection will increase, resulting in drill failure. The detection method recognizes that this scraping in the hole is synchronous with the rotation of the spindle.

For many applications, particularly those using drill sizes greater than about 3 mm (0.125 in.), breakage is not a normal failure mode and determination of wear or, more specifically, wear out (end of life) is of interest. Many techniques have been tried for measuring wear including: spindle power consumption;[4] torque;[5] signature analysis of vibrations;[6-9] and spectral analysis of vibrations or sounds.[10-13] Of these, power consumption appears to be the only available cost effective technique for on-line measurements. It is applicable only for large drill sizes where a measurable amount of power is consumed by a worn drill over that used for the normal machining operation. It is not feasible for small drill sizes because most of the power is used to turn the spindle.

The advent of the fully automated factory makes on-line tool wear sensing more important than before. Without an operator to determine when tooling must be changed because it is worn out, drills must be changed based on the number of holes or parts made. This is an inefficient method since the life of drills, even from the same lot, in terms of holes for any given process varies widely (perhaps by a factor of 3 or more).[14]

This means that a tool has to be changed after drilling a number of holes that is *less than* the shortest life expectancy of any tool in a lot. Not only will most of the useful life be wasted for a majority of tools, but excessive time will be lost during tool changing. Worse yet, a machine might not have enough locations in the tool magazine to operate for the required unattended time period. The on-line detection of drill failure that can be used to initiate a tool change will be of great value.

The use of drill-up to determine drill failure is a practical application. A dull drill does not produce as distinct a change in the vibration signal (detected by the accelerometer) as a breaking drill. However, it does produce a signal that can be detected and used as an indication of a worn drill. In many applications, a dull drill is readily detectable by drill-up. It will be shown that drill-up can detect worn drills during the period of rapid wear that has been noted at end of service life.[5,6]

Block Diagram Description of Drill-Up

Since tailoring drill-up for a drill wear application requires changing some parameters from the original drill breakage design, an understanding of the circuit and the computer algorithm is required. An explanation of the diagram of the circuit for a single drilling speed version is given below. Those interested in more details, including flow charts and the microcomputer program, should see reference 3. As shown in Fig. 6, the accelerometer signal is amplified by a high input impedance voltage amplifier with a selectable gain. A remotely selectable, adjustable attenuator has been included to accommodate a second tool (such as a center drill) operating at the same drilling speed, which causes significantly larger normal drilling vibrations.

The attenuator is switched into the circuit by a signal from the machine controller, indicating that the second tool is in use. The accelerometer signal is amplified further by a variable gain amplifier. The signal is AC coupled to one input of an analog comparator. The reference threshold for the comparator is supplied by a digital-to-analog converter. The reference is determined by the microcomputer calibration routine described below.

Signal peaks that exceed the threshold trigger a monostable multivibrator (one-shot) that generates a pulse with a duration equal to 40 percent of the drill rotation period. Figure 7 shows four signal peaks (that are larger than the threshold) and the pulses generated. These pulses are applied to the input of a microcomputer.

The microcomputer continuously monitors this input until a pulse is found. It then delays for one-half pulse width and resamples the input after a time interval equal to the period of rotation. If pulses (caused by successive accelerometer signals greater than the threshold) are found, the input is repetitively sampled at rotation period intervals until a selected number (N) have been found. For the breakage design, N was empirically chosen to be 4. If, at any sample, a pulse is not found, the system resets. After four pulses in a row have been found, the system generates a signal that closes a relay connected to the emergency retract circuit of the machine controller. When the relay closes, the machining center retracts the spindle and interrupts the N/C machine program.

A calibration routine in the microcomputer allows the instrument to automatically adapt to the accelerometer signal during normal operation. This routine is activated by a switch that provides a hardware interrupt to the microcomputer. Prior to this, the operator must determine that drilling is normal (good drill, normal coolant and material). This calibration compensates for variables including the accelerometer sensitivity, coupling to the workpiece, amplifier gain settings, and the factors that make up the normal machining process (tool geometry, workpiece material, drilling speed, feed rated and coolant). The routine reduces the threshold from a starting value (chosen to be less than the full scale value, 10 volts, divided by a factor XL discussed below) in increments at fixed time intervals. For this design, the increments are 0.4 V at 1.5 s intervals.

This continues until the peaks of the signal begin to exceed the threshold. The microcomputer determines that the threshold has been exceeded by counting the pulses from the multivibrator. When pulses of a predetermined number have occurred (chosen to equal the drilling speed in rpm divided by 30, which gives at least two seconds of pulses), the threshold level for normal drilling has been found. The system then sends a signal to the digital-to-analog converter that sets the operating threshold to a value XL times the threshold level for normal drilling. For breakage detection, XL was empirically chosen to be five. The threshold is displayed on a digital voltmeter. The system resumes detecting potential breakage immediately on termination of the calibration cycle.

During normal drilling, the accelerometer signal appears on an oscilloscope as a low level noise signal with occasional larger peaks. There are no significant signals that appear synchronous with the drilling speed. All the factors given in the previous description of the calibration routine affect the operating threshold level when the calibration is run for a specific normal drilling signal. This includes the following parameters: voltage increment; time interval between level changes; XL; and the predetermined number of multivibrator pulses. The operating level for a given normal drilling signal is the important result.

Many other calibration routines with different parameters could be developed for similar operating levels. Manual setting of the operation threshold allows fine tuning and eliminates the need to recalibrate if power to the microcomputer is removed, since nonvolatile memory is not used. The parameters XL and N will be changed later to tailor the circuit for wear detection, but it must be emphasized that the operating threshold resulting from the automatic calibration, and the overall performance, is dependent on all the other factors (this calibration algorithm is not unique).

An alternate design allows the drilling speed to be varied. The operating speed is read from thumbwheel switches by the microcomputer and the necessary timing values are calculated. The pulse with a width that is 40 percent of the drill rotation period is replaced by (1) a short fixed pulse and (2) the 40 percent timing value implemented in the software by sampling the input for this time period in the computer. This eliminates the need for hardware changes when the operating speed is changed. Significant computation is required for the timing and calibration factors, making advantageous an alternate microcomputer better suited for division and multiplication.

Use of Drill-Up for Drill Wear Out Detection

Drill-up has many potential advantages when compared to alternatives for determining drill wear out on an automated drilling machine. It provides on-line monitoring of the drilling process, not of the drill alone. It can provide a signal to the machine tool controller indicating that drilling is no longer normal. This is usually caused by a worn out drill.

For any specific operation, this signal might be used to stop the machine immediately or, since drill wear is gradual in many applications, to stop the machine after the current part or operation is completed. A new drill of the same size could then be called from the tool magazine. If the replacement is the same type drill, no adjustment should be required. A change to a different drill geometry, material or

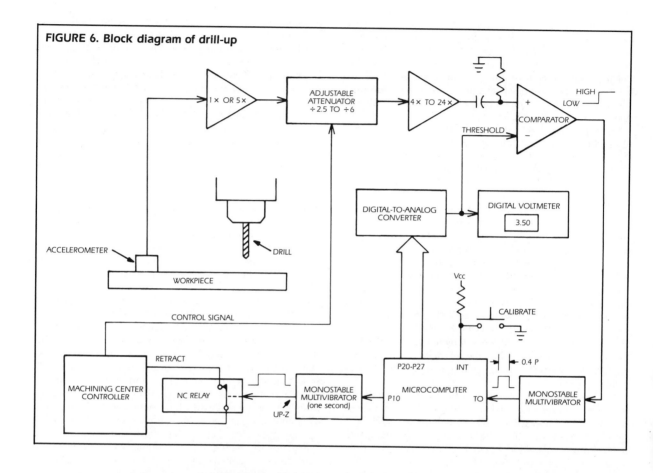

FIGURE 6. Block diagram of drill-up

FIGURE 7. Amplified accelerometer signal (top) and pulses generated

even size, requires recalibration, but no change in N or XL if the drilling speed remains the same. Since drill-up measures the *process*, it detects a similar degree of improper drilling whether a low cost, short life drill or a coated, longer life drill is used.

Relating improper drilling to wear out or failure is not straightforward. The maximum acceptable width of the drill flank wear land will vary widely depending on the application and material. It is important to consider which characteristics (of the finished hole in the part) and which tolerances should be used as criteria to determine the conditions of a drill or drilling operation. Hole diameter, hole surface finish and hole straightness might be the limiting factors in determining allowable drill wear. Measuring the drilling process, not the drill, seems a productive approach — that is what is done by a machinist who judges tool wear by listening to the drilling or by examining the chips.

The potential of drill-up for detecting wear out was initially evaluated by drilling in various materials with drills that were considered worn out (the wear land was wide enough to qualify the drills for removal by most machinists). Many holes were drilled and the parameters in drill-up were altered to achieve detection based on N pulses in a row that exceeded a threshold during a time interval corresponding to the period of drill rotation.

It was found that, depending on the hardness of the workpiece, the appropriate values for these parameters are different from the original values, ($N = 4$, $XL = 5$) chosen to prevent breakage. Specifically, when using the original values, a drill would have to be severely worn (well beyond commonly acceptable limits) before it could be detected in a mild carbon steel, such as AISI 1015 with Brinell hardness number (BHN) of 190. Basically, a worn drill will cause improper drilling at a lower level of wear if the material is hard and will be more readily detectable from the accelerometer signal.

To make drill-up more sensitive, N was reduced to three and XL was reduced to two, allowing lower amplitude signals from the accelerometer to be included in the count and reducing the number of signals that must occur on successive revolutions of the drill to cause detection. With these values for N and XL, drills that were considered to have reasonable amounts of wear could be detected in the AISI 1015 steel. It was also found that in a harder material, AISI 4150 steel with a hardness of 290 BHN, the worn drills were readily detected with the original values.

The more sensitive version, for wear detection in soft steel, was significantly more susceptible to false alarms (detection when the drill was not worn out) than the breakage version, in which premature detection is a rate occurrence. It was decided that a single detection is not a suitable criterion for determining that a drill is worn out. A criterion that a detection occurs in the drilling of each of two consecutive holes was selected to indicate wear out. An experiment was then conducted to determine the feasibility of using drill-up for wear out detection of drills in the AISI 1015 and 4150 steels.

Experimental Procedure for Wear Out Detection

Results from tests using more than one drill size and several samples of each size were desired. Drills of 6.5 mm (0.25 in.), 9.5 mm (0.4 in.), and 13 mm (0.5 in.) diameters were chosen to cover a range of common sizes for which wear out, not breakage, is the common failure mode. Five samples of each size were selected since, in terms of holes, a large variation in the life of drills from the same lot was expected.

The main objective was to demonstrate the detection of wear out in the soft AISI 1015 steel because the initial experiments had shown this to be difficult. Since the expected life (in terms of holes drilled at normal drilling speeds and feed rates[15] in this material) was estimated to be in the thousands, an accelerated wear test was necessary. Wear out detection was also to be tested in the harder AISI 4150 steel and drilling in that material was used to dull the drills before testing them for wear out in the AISI 1015 steel.

Groups of five holes were drilled alternately in each material. Two drill-up circuits were used. The first circuit had the wear parameters $N = 3$ and $XL = 2$ and was used to monitor wear in the soft material. This circuit was connected to an alarm. The second circuit had the original wear parameters ($N = 4$ and $XL = 5$) and was used to control excessive drill wear in the harder material. Normal drilling speeds and feed rates were used in the soft material.

In the harder material, normal drilling speeds were used, but higher than normal feed rates were used to accelerate wear. A feed rate was selected for each different sized drill which would result in wear out of that drill after drilling 50 to 150 holes. This was necessary to complete the experiment in the available machine time. The higher feed rates caused wear to progress very rapidly near failure in the harder material, and with no warning drills wore out within a few seconds of any audible indication of a drilling problem. The wear often became excessive before drilling could be manually terminated. Therefore, the circuit monitoring the drill in the harder material was connected to the machine controller retract circuit to automatically stop the drilling.

The drilling conditions are summarized in Table 1. Blind holes, as deep as practical, were made in the 25 mm (1 in.) thick steel plates. The drills were flood cooled with a cutting oil. The 6.5 mm (0.25 in.) and 9.5 mm (0.4 in.) drills were uncoated high speed steel, each size from a single lot. The 13 mm (0.5 in.) drills were obtained from three different manufacturers. All were high speed steel, one uncoated and two black oxide coated. The different types of drills were used to demonstrate that, since it measures the *process*, drill-up can detect a similar degree of wear even if the tools are different.

The drill-up circuit monitoring the wear was calibrated using a drill that had drilled about ten holes in the harder plate. A drill with a small degree of wear was chosen because the vibration signal produced by a new drill decreases during initial wear which can cause the problems detailed below. After the circuit was calibrated, a test was conducted to evaluate the ability of this circuit to detect wear out in the soft material (AISI 1015 steel).

Drilling was begun in the harder material and, when improper drilling was detected, the drill was automatically retracted. This same drill was then tested for wear out in the soft material to see if the criterion of a detection in two consecutive holes was met. If not, drilling in the harder plate was continued. Operator judgment was used to determine if drilling in the harder plate was to be terminated at each subsequent retract.

A drill was removed whenever improper drilling was detected in two consecutive holes in the soft material. The location of the holes made with each drill was noted for later gaging. This procedure was continued until five worn out

TABLE 1. Cutting conditions used for drill wear

	Drill Diameter millimeters (inches)	Speed meters (feet) per second	Feed millimeters (inches) per revolution
AISI 4150 steel (290 BHN)			
Accelerated cutting conditions	6.5 (0.25)	12.5 (41)	0.18 (0.007)
	9.5 (0.38)	13.5 (45)	0.36 (0.014)
	13 (0.5) uncoated	16 (52)	0.33 (0.013)
	13 (0.5) coated		0.41 (0.016)
Normal cutting conditions	9.5 (0.38)	13.5 (45)	0.13 (0.005)
AISI 1015 steel (190 BHN)			
Normal cutting conditions	6.5 (0.25)	24.5 (80)	0.15 (0.006)
	9.5 (0.38)	23.5 (77)	0.2 (0.008)
	13 (0.5)	25 (82)	0.25 (0.01)

drills of each size were collected. The drills that were used to arrive at suitable accelerated feed rates in the harder material were excluded from the test.

A subsequent test was conducted to evaluate the ability of the drill-up circuit (with the original parameters $N = 4$ and $XL = 5$) to detect wear out in the harder material drill. The evaluation test was conducted using speed and feed rates suitable for the harder material. Each of the five 9.5 mm (0.4 in.) drills that had been previously detected as worn out in the soft material was tested for wear out in the harder material.

Experimental Results and Discussion of the Use of Drill-Up

For the three sizes of drills used, worn out drills were readily detected by the wear version of drill-up in the AISI 1015 steel using the criterion of a detection in two consecutive holes. Detection usually occurred in the bottom part of the hole where poorer coolant penetration and chip buildup make cutting more difficult. In a shallower hole, somewhat more wear could have occurred before wear out was detected.

The important questions are: (1) are the drills worn out? and (2) does the degree of wear constitute failure? These question can be answered only for specific applications. However, the degree of wear is a reasonable concern.

Figure 8 shows four of the 13 mm (0.5 in.) diameter drills, detected and declared worn out by drill-up, surrounding a reference drill with only slight wear. These four worn out drills include samples of the three different types of 13 mm (0.5 in.) drills. Close-ups of the reference drill and a worn drill are in Fig. 9. The degree of wear shown is typical of the 6.5 mm (0.25 in.) and 9.5 mm (0.4 in.) drills.

FIGURE 8. Reference drill (center) and worn out drills detected by drill-up

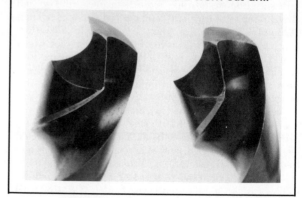

FIGURE 9. Reference drill and a worn out drill

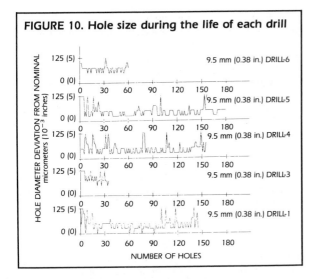

FIGURE 10. Hole size during the life of each drill

For all the drills, the width of each wear land at the outside of the flank is approximately equal to or slightly wider than the margin. Visual inspection of the drills shows that the wear is near the maximum allowed by many machinists. However, depending on the application, a drill that appears to be worn out may or may not still be usable. Therefore, hole diameter was selected, in addition to visual inspection of wear on the drill, as an indicator of drill wear. Detection of worn drills with a specific wear land width was not an objective for this experiment and the wear lands were not measured.

To determine wear on the five 9.5 mm (0.4 in.) drills, the top of each hole in the AISI 1015 steel was measured with a plug gage. The results are shown in Fig. 10, which gives a plot of the deviation from the nominal drill diameter versus the number of holes drilled by each of the five drills. It should be noted that the hole number corresponds to a pair of holes, one in the 1015 steel and one in the 4150 steel. Also, recall that the holes were drilled alternately in groups of five in each material. Drill 2 was damaged by operator error during the test and is excluded from the test.

The plots shown in Fig. 10 suggest several conclusions. The main point is that the diameters of the holes at failure, as determined by drill-up in the last two holes for each drill, were never larger than the maximum hole diameter that occurred during useful life. This indicates that, as far as hole size is concerned, the drills were not worn excessively. Secondly, in terms of holes drilled, the life of these drills from the same lot did vary widely: 145, 35, 155, 180 and 60 holes. Similar variation was observed for the 6.5 mm (0.25 in.) drills, which were also from a single lot.

The data also indicate that sharp new drills make larger diameter holes at the beginning of life than they do during the remaining useful life, a phenomenon that has been observed by others.[5,6] This is most obvious for drills 3, 5 and 6. The effect may have been reduced by the fact that five holes were drilled with each drill in the harder material before drilling the first hole shown in the plots of Fig. 10. A larger vibration signal was observed for new drills during approximately the first 10 percent of life. This larger signal was often detected by the drill-up calibrated on a drill that had been used past the initially larger vibration period. This problem is discussed in detail below. Sample holes made with the 6.5 mm (0.25 in.) drills were gaged and showed similar results.

Since the 9.5 mm (0.4 in.) drills (declared worn out in the soft material) were immediately detected as worn out by the drill-up control circuit ($XL = 5$) while drilling at normal speeds and feed rates in the harder material, gradually wearing new drills should be readily detected as worn out with no more than this degree of wear. With harder workpiece materials, drills should be replaced as soon as wear is detected. Once wear reaches the detectable stage, it begins to progress more rapidly and out-of-tolerance holes may result. Also, the criterion for detection in two consecutive holes may not be necessary, since false alarms are uncommon at the higher operating threshold.

If a more sensitive criterion is needed to detect drills at a lower level of wear, the XL factor could be reduced. However, because of the problems discussed later, a factor lower than $XL = 2$ would be practical only in limited cases. Conversely, if more wear is desired, the circuit can be made less sensitive by increasing XL or N.

Potential Problems and Solutions

A number of potential operating problems were observed. For drill-up to work, the speed of the drill must remain within the allowable tolerance (± 5 percent if $N = 4$ and ± 8 percent if $N = 3$). If the expected variation is not within the tolerance due to poor motor speed regulation during loading, the wear detecting drill-up can be programmed for a nominal drilling speed (between the no-load and full load speeds) to make better use of the available tolerance.

The variance in amplitude of the signal from the sensor must be minimized. A wear circuit that uses a low value for XL is affected much more by amplitude variations because they are indistinguishable from drilling variations. The rigidity of the workpiece and its mounting on the machine table also affect the amplitude of the signal. In addition the amplitude of the sensor signal decreases as the distance to the drilling location increases. This problem is significant when drilling locations are far apart. A solution is to place the sensor as far from the drilling locations as practical to minimize the variation of the coupling path to each hole.

The wear detecting drill-up must be able to distinguish between the amplitudes of the accelerometer signals produced by drilling and those produced by inherent vibrations from the table. Therefore, the amplitude of the signal resulting from the vibrations during a drilling operation must be larger than that resulting from the vibrations of the table during nondrilling operations, but with the spindle turning. For experiments with the 9.5 mm (0.4 in.) and 13 mm (0.5 in.) drills, the sensors were mounted on the table about 300 mm (12 in.) from the drilling locations.

In general, larger drills and harder workpiece materials produce larger amplitude signals that permit the sensor to be placed farther away from the drilling location than is possible when using smaller drills and softer materials. For drilling operations in the soft materials using 6.5 mm (0.25 in.) drills, the sensors had to be placed on the jaw of the vise holding the part in order to pick up a sufficient accelerometer signal. The use of small drills, 3 mm (0.125 in.) and smaller, will usually require mounting of the sensor directly on the part, or on a clamp that holds the part. This procedure, however, introduced other problems.

If the sensor is mounted near the drilling location, long chips may wrap around the body of the drill, hit the accelerometer, or scrape on the surface of the workpiece. This causes large vibrations that are synchronous with the drilling speed and result in false alarms. The primary solution to this problem is to alter the drilling conditions (for example, the feed rate can be increased to prevent long chips). Alternatively, the sensor may be mounted on the underside of a part, such as a plate. A mechanical shield can be placed between the drill and the sensor, provided measures are taken to ensure that the shield vibrations will not be coupled to the sensor or the workpiece. Such a shield may be constructed of lead lined with foam to provide the required isolation.

In addition, the higher vibration signal from a new drill may be a problem. If the calibration can be done on the higher level from a new drill and still result in a suitable degree of wear when detected as worn out, there is no problem. Otherwise, detections during the first 10 percent of expected drill life may have to be ignored by the controller. If breakage is not a concern, this should not cause problems.

Many of the problems appeared only when the lower operating threshold ($XL = 2$) was used. Random false alarms occurred, perhaps due to local hard spots in the material or chips that did not clear the hole. Multiple detections occurred while drilling in the same hole. The criterion of a detection in two consecutive holes distinguishes wear out from these random events. A wear out criterion of this type would best be evaluated by software in the machine controller, but could be implemented by additional hardware circuits if a signal indicating a new hole is available.

In a through-hole, drill wear out detection frequently occurred on breakthrough at the bottom of the part, regardless of the amount of wear on the drill. The two-in-a-row criterion did not help. Similarly, if a center drill was used and did not leave a hole that matched the point angle of the drill, detection frequently occurred at the top surface until the drill cutting edges were in full contact with the material. However, this was not a problem if a drill started its own hole.

The only known solution to these problems is to use Z axis information from the machine controller or external sensors and to ignore detection that occurs at the top or bottom surface of the workpiece. This could readily be done as an integral part of a control program.

Conclusions

The potential of drill-up as a drill wear out sensor has been demonstrated for a limited range of drill diameters and two hardness levels of steel. The results should extend to smaller sizes and down to sizes where breakage, rather than wear out, is the predominant factor that determines end of service life. For larger size drills, no specific limitations have been identified.

The parameters chosen for the wear detecting drill-up are dependent on the workpiece material, primarily its hardness. Worn drills are much easier to detect if the material is hard. In soft, low carbon steel, automated wear out detection is possible if techniques are used to overcome the practical problems. In harder materials, wear out should be readily detectable. In softer materials such as aluminum, wear out detection may not be possible.

PART 3
MATERIAL DEPENDENCY OF CHIP FORM DETECTION USING ACOUSTIC EMISSION

In automated manufacturing, especially in untended manufacturing, metal cutting machines must operate for a long time without operator intervention. Under these conditions, chip control is a major concern. In single point turning, long continuous chips (unbroken chips) can be a serious problem. Long chips may entangle with the tooling or the workpiece, damaging the surface finish of the part or interfering with workpiece or tooling changes. An equally important problem in untended manufacturing is chip removal. Long chips make automated chip disposal impossible.

A direct correlation between the average chip formation rate (chips per second) and the acoustic emission event rate (bursts of signal whose amplitude exceeds a threshold during a time interval) has been demonstrated.[16,17] An instrument was built at the National Bureau of Standards to demonstrate the technique in the machining of various metals. The ability to identify chip form in turning three types of metal (6061-T6 aluminum, 1018 carbon steel and 4340 alloy steel) has been previously reported.[18] The results from the turning of 303 and 304 stainless steels and copper alloy 360 (brass) will be presented here in addition to the original three. Successful chip form detection is shown to be material dependent.

Acoustic Emission from Chip Formation

Acoustic emission is generated by materials undergoing deformation or fracture, both of which occur when chips are formed in machining. Continuous acoustic emission signals are generated in the shear zone, at the chip/tool interface and at the tool flank/workpiece interface. Bursts of acoustic emission are generated, in addition, when a chip breaks. The acoustic emission signal from single point turning, therefore, consists of a continuous component (whenever chips are being formed) with additional burst events (whenever a chip breaks). Counting the events per unit time (average chip breaking frequency) is the means of determining chip form.

Acoustic Emission Cutting Monitor

An acoustic emission cutting monitor was designed to demonstrate the feasibility of monitoring chip form. Several significant features were incorporated into the acoustic emission cutting monitor. First, the root mean square (rms) value of the acoustic emission signal from the transducer is generated before the presence of an event (burst) is detected. This reduces the signals from the order of 100 kHz to the order of 1 kHz, simplifying the circuitry and reducing the cost. Second, a floating threshold is generated that tracks the level of the continuous component of the acoustic emission signal. Bursts that exceed a threshold, which is a selected multiplication factor larger than the continuous signal level, are counted as events.

This technique ensures that events can be detected even if the level of the continuous signal changes significantly. This component may vary due to factors such as cutting conditions (speed, feed, depth of cut, coolant and so on), tool geometry (including chip breakers and grooves), tool wear and workpiece geometry or material variations. A fixed or absolute threshold is not suitable because of the changes in signal level. It is worth noting that these same factors, along with others such as built-up edge, determine the chip form during a cut.[19]

Cutting Monitor Operation

The circuit, shown in Fig. 11, operates as follows. The signal from a broadband piezoelectric acoustic emission transducer, mechanically coupled to the cutting tool, is amplified and band-pass filtered from 20 to 400 kHz. The amplified acoustic emission signal is fed to two root mean square circuits, one with a slow response (100 ms averaging time constant) and one with a fast response (0.5 ms averaging time constant). These time constants are chosen so that the rms slow signal will not respond to the acoustic emission bursts while the rms fast signal will generate the rms value during the burst. Chip form is determined by analyzing these rms signals.

Figure 12 shows the amplified acoustic emission signal (from turning of aluminum) with a burst from a chip breaking and the resulting rms fast signal (the input and output of the rms fast converter circuit). The rms slow signal would have continued with the nearly constant level that is shown before the burst in the oscillogram.

The rms slow voltage is digitized by an eight-bit analog-to-digital converter controlled by the microcomputer. If the measured value is less than about 0.7 V, the condition is *not cutting*. During this condition, acoustic emission bursts are ignored. When the signal increases above the 0.7 V level, the circuit assumes there is cutting and continues normal operation (monitoring the chip form). The *chips not breaking* or

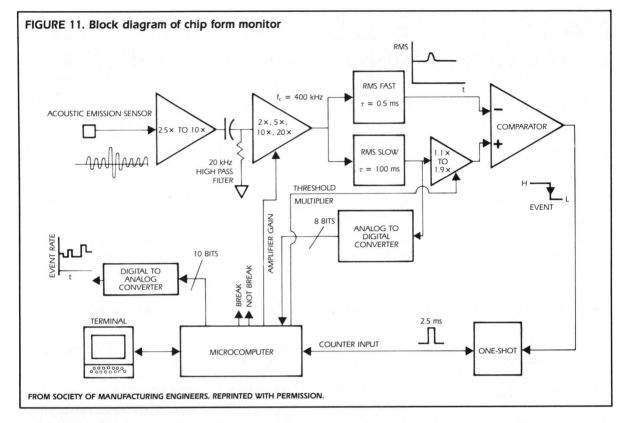

FIGURE 11. Block diagram of chip form monitor

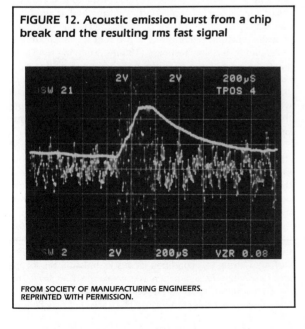

FIGURE 12. Acoustic emission burst from a chip break and the resulting rms fast signal

FROM SOCIETY OF MANUFACTURING ENGINEERS. REPRINTED WITH PERMISSION.

chips breaking condition is determined by monitoring the event rate when cutting is taking place.

The event rate is the number of bursts of acoustic emission signal for which the rms value exceeds a selected threshold level during a selected time interval. The burst causes a pulse on the rms fast signal but not on the rms slow signal. The threshold is set at a value that is the *threshold multiplier* times the present rms slow value. The threshold level tracks changes in the rms slow level. Figure 13 shows an acoustic emission signal with a burst like that in Fig. 12. The threshold for a multiplier of 1.6 is also shown. The rms fast signal exceeds this threshold and is counted as an event.

Figure 14 shows a slow change in the level of rms slow signal due to an increase in the continuous component of the acoustic emission. The threshold tracks the change, always being larger than the rms slow level by the multiplier value. The threshold is applied to one input of an analog comparator. The rms fast signal is applied to the other comparator input. When the rms fast signal exceeds the threshold, the comparator output switches levels and this is sensed by the microcomputer. If an absolute threshold had been used, the rms fast level corresponding to Fig. 14 (even without a burst), would have increased above the threshold level and remained, preventing a burst from being detected. With the tracking threshold, a burst can still be detected in addition to the increased continuous component of acoustic emission.

The computer counts the number of pulses that occur during a *time base* interval, usually 1.0 second. The circuit determines that the *chips not breaking* condition exists if the event rate is less than an event count threshold for three consecutive intervals. Chip form is indicated by an external alarm or status device, or might be transmitted to the machine tool controller for corrective action.

Experimental Procedure for Cutting Monitoring

The acoustic emission sensor was initially mounted on the face of a 25 mm (1 in.) square tool holder, as far as possible from the cutting insert, on an 18 kW (25 horsepower) computer numerical control (CNC) turning center. The turret has eight tool positions but only one was used during these tests to avoid tangling the wire from the sensor to the monitor. The tools were 13 mm (0.5 in.) internal circle, 80 degree diamond inserts that were held in a tool holder with a 5 degree lead angle. Uncoated carbide was used with the 6061-T6 aluminum (hardness of 107 BHN). An aluminum oxide coated carbide was used with the 1018 carbon steel (hardness of 190 BHN) and 4340 alloy steel (hardness 230 BHN).

The insert had a chip breaking groove. The lower half of the tool turret and the face of the tool holder towards the spindle were covered with sheet metal shields that were isolated from the turret with double stick foam tape. The shielding reduces extraneous event counts from chips hitting the turret. The reported data were taken under these ideal conditions. The shields were subsequently removed and limited tests repeated. Cutting was stopped 50 mm (2 in.) from the face of the jaws to prevent chips, particularly continuous ribbons, from becoming entangled.

Cuts were made at various speed and feed rates that produced chips with different forms. Chips ranging from an infinite ribbon or helix to Cs or 6s could be produced in each of the metals. Chips up to about 75 mm (3 in.) long may be considered broken if chip management is the concern. Cuts were 1.3 mm (0.05 in.) deep and made at a constant surface speed appropriate for the material. At heavier depths of cut, chips break more readily and the acoustic emission signal levels should be larger.

At smaller depths of cut, the desired range of chip forms may not have been possible. The feed rate was varied from 0.1 to 0.3 mm (0.004 to 0.012 in.) per revolution. A water soluble oil flood coolant was used, except with the brass, which was discharged through an opening in the tool turret towards the tool tip. Representative chips, formed at the

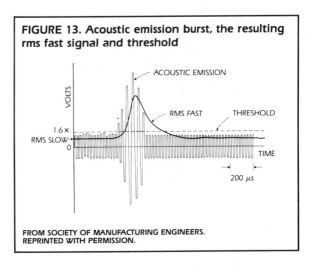

FIGURE 13. Acoustic emission burst, the resulting rms fast signal and threshold

FROM SOCIETY OF MANUFACTURING ENGINEERS. REPRINTED WITH PERMISSION.

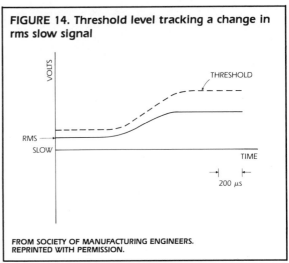

FIGURE 14. Threshold level tracking a change in rms slow signal

FROM SOCIETY OF MANUFACTURING ENGINEERS. REPRINTED WITH PERMISSION.

various feed rates for aluminum are shown in Fig. 15. The chips in the figure formed at 0.1 to 0.15 mm (0.004 to 0.006 in.) per revolution are cut from longer pieces.

Cuts were made at feed rates of 0.1, 0.15, 0.2, 0.25 and 0.3 mm (0.004, 0.006, 0.008, 0.01 and 0.012 in.) per revolution for various threshold multiplier values at the cutting speed previously chosen for the material. The resulting event rates and chip forms were noted. This was repeated for each metal. After a threshold multiplier and event count threshold were chosen, the series of cuts at different feed rates were repeated for each metal, to verify correct chip for classification by the monitor. Some additional cuts were then made without the shields on the turret and with the sensor on the turret rather than the tool holder.

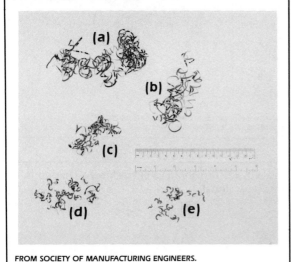

FIGURE 15. Chips formed at various feed rates in 6061 aluminum, 8.75 m (25 ft) per second and 1.3 mm (0.05 in.) deep: (a) 0.1 mm (0.004 in.) per revolution; (b) 1.5 mm (0.006 in.) per revolution; (c) 0.2 mm (0.008 in.) per revolution; (d) 0.25 mm (0.01 in.) per revolution; and (e) 0.3 mm (0.012 in.) per revolution

FROM SOCIETY OF MANUFACTURING ENGINEERS. REPRINTED WITH PERMISSION.

Results and Discussion of Cutting Monitoring

Figures 16 through 20 show the event rate for the various feed rates and the various threshold multipliers for each of the metals, except the copper alloy. The chips change from not broken, ribbon or helix, on the left to Cs or 6s in the broken region on the right. In the center (transition) region, chips of various sizes were observed depending on the material. The effect of the threshold multiplier value is quite different depending on the metals. In Fig. 16 for aluminum, the lower multiplier value results in somewhat higher event rates and the higher multiplier value results in lower event rates, as expected, but any of the three multipliers would be satisfactory.

The best event count threshold, chosen to separate between broken and not broken, is different depending on the multiplier selected. If the middle curve (1.6 multiplier) is chosen and the chips in the middle region at 0.2 mm (0.008 in.) per revolution are classified as broken, then an event count threshold of 20 (indicated by the horizontal dashed line) would be suitable. In Fig. 17 for 1018 low carbon steel, the multiplier value has very little effect on the curves. Any

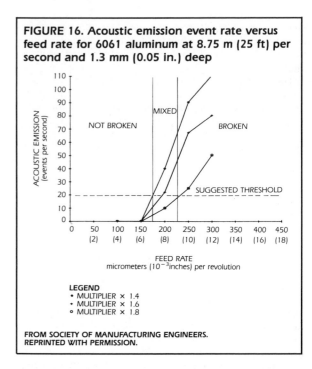

FIGURE 16. Acoustic emission event rate versus feed rate for 6061 aluminum at 8.75 m (25 ft) per second and 1.3 mm (0.05 in.) deep

LEGEND
• MULTIPLIER × 1.4
• MULTIPLIER × 1.6
○ MULTIPLIER × 1.8

FROM SOCIETY OF MANUFACTURING ENGINEERS. REPRINTED WITH PERMISSION.

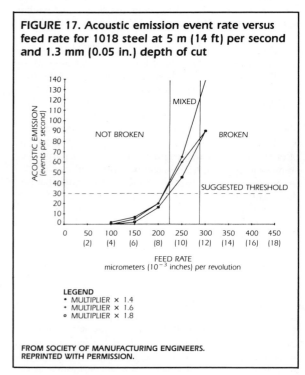

FIGURE 17. Acoustic emission event rate versus feed rate for 1018 steel at 5 m (14 ft) per second and 1.3 mm (0.05 in.) depth of cut

LEGEND
• MULTIPLIER × 1.4
• MULTIPLIER × 1.6
○ MULTIPLIER × 1.8

FROM SOCIETY OF MANUFACTURING ENGINEERS. REPRINTED WITH PERMISSION.

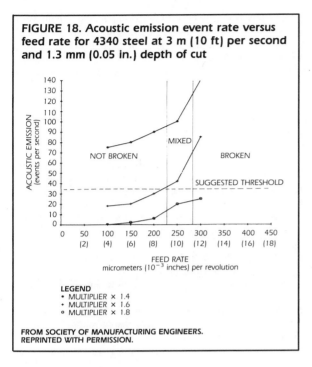

FIGURE 18. Acoustic emission event rate versus feed rate for 4340 steel at 3 m (10 ft) per second and 1.3 mm (0.05 in.) depth of cut

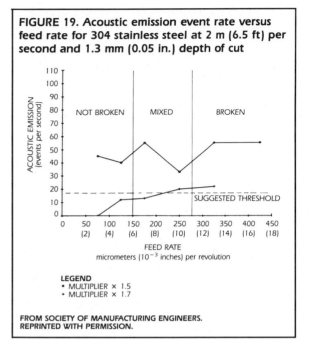

FIGURE 19. Acoustic emission event rate versus feed rate for 304 stainless steel at 2 m (6.5 ft) per second and 1.3 mm (0.05 in.) depth of cut

of the three multipliers would be suitable. If the chips in the middle region at 0.25 (0.01 in.) per revolution are classified as broken, an event count threshold of 30 would be suitable along with any of the multiplier values.

The curves in Fig. 18 for 4340 alloy steel are markedly different. The lowest multiplier value (1.4) is clearly too low since large event rates occur in the not broken region on the left. The highest multiplier (1.8) is too high since the event rate is very low for all regions. The 1.6 multiplier value is certainly the choice. If the chips in the middle region at 0.25 mm (0.01 in.) per revolution are classified as broken, an event count threshold of 35 would be suitable. Note, in addition, that the slope of this curve is not as steep in the middle region as for the other two metals.

Less reliable indication of the correct chip form would be expected for the 4340 than for the 6061 or 1018, since the change in event rate is not as large for a change from not broken to broken. The prospect for successful chip form detection is even worse for the 304 nonmagnetic stainless steel in Fig. 19. The best multiplier (1.7) results in almost no change in event rate as the chips change from broken to not broken. A high multiplier value (1.9) resulted in virtually zero counts at all feed rates. The free-machining version of this material, 303 stainless steel, has much better potential for successful chip form detection. As seen from Fig. 20, the results are dependent on the multiplier but 1.4 would give good results with an even count threshold of 10. A multiplier

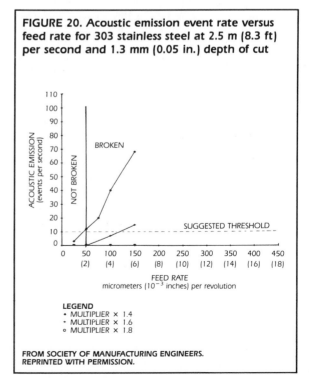

FIGURE 20. Acoustic emission event rate versus feed rate for 303 stainless steel at 2.5 m (8.3 ft) per second and 1.3 mm (0.05 in.) depth of cut

of 1.2 resulted in a large number of counts even when chips were not breaking. Note that the chips break at very low feed rates compared with other metals.

There is no figure showing the results for the copper alloy. The material was cut without coolant at 6 m (20 ft) per second. Feed rates below 0.2 mm (0.007 in.) per revolution produced unbroken chips while feed rates above that value produced broken chips. For the first cut on a new bar stock workpiece (surface hardness 160 BHN), a significant number of events per second were detected (using a multiplier of 1.6) when the chips broke, and near zero when chips were not breaking. However, on subsequent passes at depths of 1.3 mm (0.05 in.), the event rate was always near zero, even while chips were breaking. The hardness below the surface was measured at 130 BHN.

The differences for the various metals and the prospects for successful chip form detection result from the relative amplitude of the acoustic emission bursts (from the chip breaking) and the continuous component of the acoustic emission from the chip formation (cutting process). The most important factor appears to be the energy required to form the chip that determines the amplitude and noisiness of the continuous acoustic emission. The burst amplitude is dependent on the cross-sectional area of the chip and the hardness of the metal.

Figure 12 for aluminum shows a well-defined peak in the rms fast signal that is about five times the amplitude of the continuous component. In Fig. 16, this is one of the better materials for chip form detection. For the 304 stainless steel, the peak to continuous ratio was only about 2 and for the soft copper alloy about 1.3. Another limiting factor is the noisiness (size of the peaks) of the continuous component of acoustic emission. The hard-to-machine metals (304 stainless and 4340 alloy steel) have peaks in the continuous component that are counted, particularly with small multipliers.

These peaks, along with more caused by chips hitting the tool holder and turret, make chip form detection impractical for these two hard-to-machine metals. The soft copper has a different problem. Apparently the acoustic emission burst created by the chip break does not have enough energy to be distinguished from the continuous acoustic emission.

The materials that are good for acoustic emission chip form detection (aluminum and carbon steel), allow successful chip form detection without the shields and with the sensor on the turret rather than the tool holder. The additional interface and the turret geometry attenuate and distort the acoustic emission signals.

Conclusions

An acoustic emission cutting monitor can detect differences in chip form in turning aluminum, low carbon steel, free-machining stainless steel and (in ideal conditions) unhardened medium carbon alloy steel. The chip form monitor works best for easy-to-machine metals (aluminum and carbon steel) where chip breakage is more difficult and tangles from long, unbroken chips are more likely. It does not work in soft materials that do not generate significant acoustic emission due to the chip breaking.

In most metals tested, a single threshold multiplier value (1.6) and a single time base value were satisfactory. This multiplier and an event count threshold must be determined empirically for the particular application. A limited number of test cuts, at feed rates near the transition from broken to not broken chip forms, are required to select a suitable count threshold. The threshold multiplier needs to be changed if the event rates, at all feed rates, are too high or too low. Similarly, the time base needs to be changed only if the resulting event count threshold is too large or too small.

The trend of better performance for easy to machine materials is expected to extend to other similar metals (other aluminum alloys, harder copper alloys, and so on) and the trend for poorer performance for hard-to-machine materials (high strength alloy steels, magnetic stainless steel, titanium alloys) is also expected. Some problems due to thrown chips are expected when cutting close to chuck jaws. Facing cuts in various materials have not been evaluated. The chip form monitor has been demonstrated to sense chip form, independent of tooling and cutting parameters for some metals.

PART 4
THE ROLE OF ACOUSTIC EMISSION IN MANUFACTURING PROCESS MONITORING

The requirements for process monitoring techniques for automated manufacturing operations include: (1) reliability; (2) ease of application; and (3) a close correlation between the output of the technique and the characteristics of the operation. The acoustic emission generated during manufacturing processes has been found to provide insight to the state of the manufacturing process for use in process control.

The goal of this application to detect tool fracture or wear and chip formation (continuous and discontinuous) in metal cutting or wheel contact in grinding, for example. Some of the basic requirements for process monitoring in manufacturing automation are detailed below, as are the application of acoustic emission monitoring of machining operations.

The improvement of manufacturing productivity depends to a great degree on the successful automation of metal removal and forming operations. Successful automation of these processes relies on the availability of data about the state of operation from process monitors. It is necessary that these monitors be reliable, relatively easy to apply, and that they yield an output closely correlated to the characteristics of the operation under surveillance. Acoustic emission generated during manufacturing processing has been analyzed to (1) quantify the relationship between process fundamentals and the emission detected; and (2) to evaluate the sensitivity of the signal to discontinuities or process changes of interest.

Acoustic emission is usually one of two distinct types: continuous (arriving at the transducer in such large numbers that distinct events cannot be detected) and burst signals (distinct emission events can be observed). In the machining process, for example, continuous acoustic emission signals are generated in the shear zone, at the chip/tool interface and at the tool flank/work surface interface while burst signals are generated by chip breakage during or after formation or by insert fracture. More information on metal working and potential acoustic emission sources in other metal working processes is available elsewhere in this volume and in the literature.[20-25]

Detailed below are some of the basic requirements for process monitoring in manufacturing automation (specifically metal cutting) as well as illustrations of acoustic emission monitoring of machining operations for the detection of: (1) tool wear and fracture; (2) chip formation (continuous and discontinuous); and (3) wheel contact and sparkout in grinding.

The Function of Automation

Since the acoustic emission from metal cutting operations is due to metal deformation, friction and rubbing, the analysis of the signal provides insight into the state of the manufacturing process for use in process control schemes. To better understand the eventual use of the information generated it is necessary to understand the function of automation when applied to a manufacturing process.

It is helpful to consider what functions in a process are to be automated and what that implies from the process automation point of view. To be successful, any scheme for process automation must achieve more than merely replacement of the human operator. It has been suggested that that the motivation for automation of a process often includes replacement *and extension* of the human operator's capabilities.[7] With that concept in mind, the functions required for automation can be related to the functions of the operator before any process automation is implemented. This is most seen in Table 2, based on an analysis of the welding automation problem.[26]

In the table, the functions of the human operator are divided into three broad areas: (1) *sense* (or information detection); (2) *experience* in and knowledge of the operation and process (or information processing); and (3) *technique* (operation). For a turning process on a lathe, these functions are easily illustrated. The operator's senses are used as follows: sight to observe chip form and color, surface finish and part form; hearing to detect chatter, machine motor overload, tool failure; and so on. The operator's experience and knowledge determine which physical evidence indicates a deterioration in surface finish and what adjustments are needed to improve surface finish; what chip form and color are desirable; which parameters most influence them. The operator's technique (what must be physically done to adjust the speed, feed, flow of lubricant, and so on) is, in turn, based on an interpretation of the data observed with the senses.

TABLE 2. Considerations in the automation of manufacturing, classified from the point of view of the machine operator

Operator's Function	Related Automation Functions Required
Senses (information detection)	Development of methods and instrumentation for monitoring the manufacturing process: state of process variables; rate of change of variables; characterization of process output
Experience (information processing)	Research into manufacturing process fundamentals: effects of process on materials; process modeling; process mechanics
Technique (operation)	Development of new processes, control of manufacturing processes

AFTER OKADA. SEE REFERENCE 26.

FIGURE 21. Time versus rms acoustic emission signal, mean flank wear and maximum depth of crater

LEGEND
■ MAXIMUM DEPTH OF CRATER
▲ MEAN FLANK WEAR
● RMS ACOUSTIC EMISSION SIGNAL

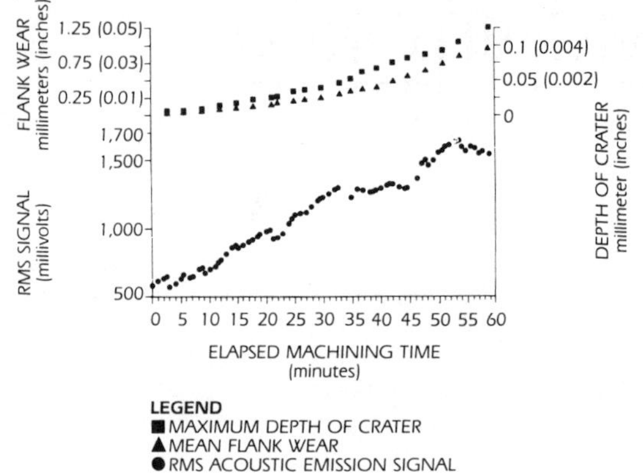

FROM SOCIETY OF MANUFACTURING ENGINEERS. REPRINTED WITH PERMISSION.

As shown in the table, the correlative functions of automation break down into (1) process sensors to replace or extend the operator's senses; (2) qualitative or quantitative relationships describing the characteristics of the manufacturing process to be replaced/extended in the operator's experience; and (3) process control and manipulative techniques to replace/extend the operator's hands. The basic requirement of all attempts to truly automate a process is therefore the availability of process sensors that reliability present information characterizing the process's state and rate of change of state. Details on the feasibility of using acoustic emission techniques in process monitoring of metal cutting are given below.

Tool Wear and Fracture Monitoring with Acoustic Emission

Wear Monitoring

Numerous techniques have been proposed and evaluated for the monitoring of tool wear and fracture but many have limitations that make their application to on-line sensing infeasible.[27,28] Acoustic emission monitoring has proven to have an advantage over many of these techniques because the output acoustic emission signal is directly related to the basic mechanisms of the cutting process.[29]

Generally the term *tool wear* refers to (1) the degradation of the cutting tool with respect to the gradual wear of the cutting or clearance surface of the tool; (2) fracture; and (3) reduction of the tool mechanical properties due to the high temperature. Here, however, only flank (or clearance surface) wear is considered. The effect of flank wear detectable by acoustic emission is the rubbing between the tool flank surface and the machined surface. The energy (rms) of the acoustic emission signal is measured during machining.

Tool wear tests were conducted with a metal grade K68 insert in a tool holder (5 degree rake) machining SAE 4340 steel under the following conditions: 1.5 m (5 ft) per second; 0.08 mm (0.003 in.) per revolution feed rate (producing a continuous chip) and 2.5 mm (0.1 in.) depth of cut. A broadband acoustic emission transducer was mounted at the end of the tool shank. Flank wear was measured with a toolmaker's microscope and crater wear was also measured.

During machining, a significant amount of the acoustic emission signal is due to chip breaking and the impact of the chip on both the workpiece and the cutting tool. Acoustic emission signals due to these effects are typically large amplitude burst signals compared to the acoustic emission generated during normal cutting. If the cutting process is very uniform, the variation of the rms signal energy contributed by the cutting process is small.

Tests were conducted and interrupted at two to three minute intervals to measure flank and crater wear. The tests were run until final insert failure. The results of one test are shown in Figs. 21 and 22, illustrating acoustic emission rms signals versus time and tool wear, respectively. A trend of dramatically increasing rms signal with flank wear is apparent. For flank wear between 0.4 mm (0.015 in) and 0.6 mm (0.022 in.), the rms signal values do not increase, primarily

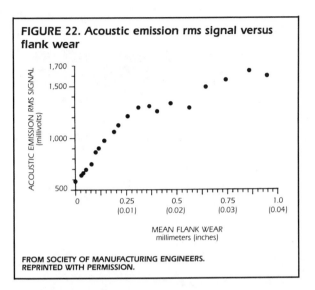

FIGURE 22. Acoustic emission rms signal versus flank wear

FROM SOCIETY OF MANUFACTURING ENGINEERS. REPRINTED WITH PERMISSION.

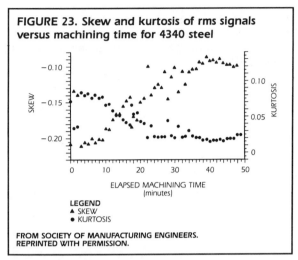

FIGURE 23. Skew and kurtosis of rms signals versus machining time for 4340 steel

LEGEND
▲ SKEW
● KURTOSIS

FROM SOCIETY OF MANUFACTURING ENGINEERS. REPRINTED WITH PERMISSION.

due to the effect of crater wear. For this time period, the maximum crater depth exceeded 0.05 mm (0.002 in.).

Because of the stochastic nature of acoustic emission, analysis has often been approached statistically. It has been demonstrated that the skew and kurtosis of an assumed beta distribution for the rms acoustic emission signal is sensitive to the progressive flank wear of the tool.[22] The same technique was used to analyze the rms data of the tests described here. For the skew and kurtosis analysis, a fast rms response (10 ms time constant) was used. The skew and kurtosis of the rms data from the test described by Figs. 21 and 22 are plotted in Fig. 23 as a function of machining time. The skew illustrates an increasing trend with increasing tool wear while the kurtosis decreases with elapsed time. The skew and kurtosis of the acoustic emission signals have also been sensitive to the change in the signals due to severe crater wear on the rake face.

The spectral characteristics of the acoustic emission generated during machining were obtained using a fast Fourier transform (FFT) program with digitized acoustic emission data. Figure 24a shows the power spectrum of the signal sampled during cutting with a new insert and Fig. 24b the spectrum due to cutting with a flank wear of approximately 1 mm (0.038 in.) under conditions similar to those described.

Fracture Monitoring

Tool fracture may be caused by (1) excessive loading due to metal removal rates; (2) inclusions in the workpiece; or (3) deterioration of the tool geometry due to wear. Fracture generates a significant level of acoustic emission. This suggests the application of acoustic emission monitoring to in-process fracture detection. In contrast to the normal cutting process, burst acoustic emission signals occur during crack formation and propagation in the insert. A significant burst of energy is generated at the moment of tool fracture. The amplitude of the rms signal energy at fracture is dependent on the area of the fracture in the insert.[30]

Two experimental methods were used for accelerating tool failure under realistic cutting conditions: (1) machining a hardened workpiece; and (2) slotting the insert. In both cases, carbide inserts were used. Heat treated SAE 4340 steel (R_C 45) was machined and fracture of the cutting tool was detected under the following conditions: 2.0 m (6.5 ft) per second surface speed; 0.25 mm (0.01 in) per revolution feed; and 1.0 mm (0.04 in.) depth of cut.

The second test method included machining SAE 1018 steel with a slotted insert. The slot, produced by electrodischarge machining, was on the rake face of the insert with a width of about 0.2 mm (0.007 in.) and a depth of about 0.8 mm (0.03 in.). The slotted insert accelerated fracture during machining under the following cutting conditions: 1.5 m (5 ft) per second speed; 0.075 mm (0.003 in.) per revolution feed; and 2.5 mm (1 in.) depth of cut for continuous chip formation and 0.25 mm (0.01 in.) per revolution feed and 1.5 mm (0.06 in.) depth of cut for discontinuous chip formation.

The output of a broadband acoustic emission transducer was amplified and recorded on a videotape recorder for further analysis. During the fracture tests the tangential and feed forces were also measured using a three-component dynamometer. The rms energy of the acoustic emission signals was determined with a dual rms/TMS module (rms is the root of the mean of the squares; TMS is the true mean of the squares or rms^2). The raw acoustic emission signal was digitized using a digital oscilloscope and the spectral characteristics of the signal from the fracture were calculated with an FFT program using the digitized raw acoustic emission data. An insert fractograph was obtained for each test with a

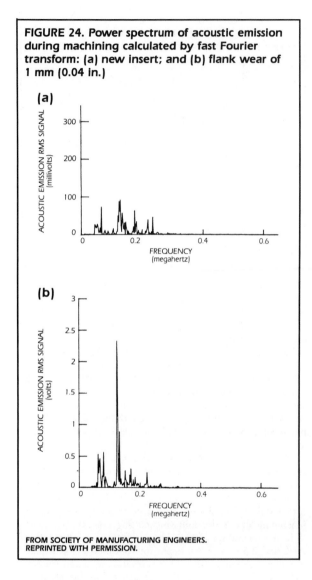

FIGURE 24. Power spectrum of acoustic emission during machining calculated by fast Fourier transform: (a) new insert; and (b) flank wear of 1 mm (0.04 in.)

FROM SOCIETY OF MANUFACTURING ENGINEERS. REPRINTED WITH PERMISSION.

FIGURE 25. Broken surface of slotted insert

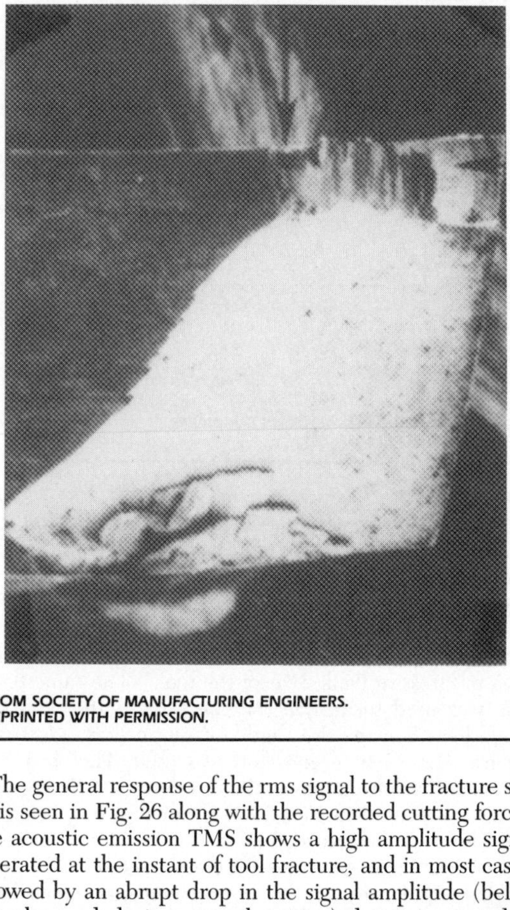

FROM SOCIETY OF MANUFACTURING ENGINEERS. REPRINTED WITH PERMISSION.

scanning electron microscope and the area of the fracture surface was measured using a planimeter.

A typical fracture surface resulting from the breakage of a tool (with a slot) is shown in Fig. 25. A significant burst of acoustic emission is generated at the moment of tool fracture. Although the response time of the raw acoustic emission signal to the fracture is only a few microseconds, rms analysis is used in this investigation because the acoustic emission rms signal is sensitive to tool fracture as well as tool wear. Thus the threshold level for determining the tool fracture can be easily adjusted to offset the effect of progressive wear.

The general response of the rms signal to the fracture signal is seen in Fig. 26 along with the recorded cutting forces. The acoustic emission TMS shows a high amplitude signal generated at the instant of tool fracture, and in most cases, followed by an abrupt drop in the signal amplitude (below that observed during normal cutting) due to a period of noncutting action that immediately follows tool breakage. Following the insert fracture, the rms signal may become irregular or may start increasing because of acoustic emission generated by machining with the rough profile of the broken insert. If the amplitude squared of the rms signal energy is plotted as a function of the cross-sectional area of the fracture for the three test conditions, as shown in Fig. 27, an increase in the rms amplitude squared with increasing fracture surface is seen.

The raw acoustic emission signal (on which the rms energy analysis is based) also responds to the insert fracture and the energy of the event is most clearly seen in the amplitude of the signal at the instant of fracture. Figure 28 shows the power spectrum of acoustic emission signals generated before, at, and after the tool breakage, calculated from the digitized data recorded while machining the 4340 steel. The dominant signal lies in the frequency range of 80 to 150 kHz

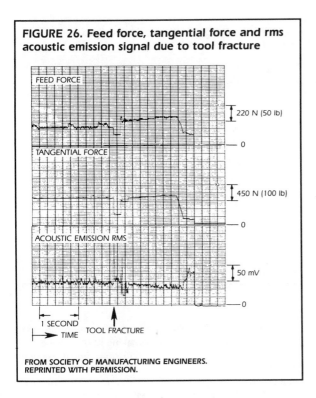

FIGURE 26. Feed force, tangential force and rms acoustic emission signal due to tool fracture

FROM SOCIETY OF MANUFACTURING ENGINEERS. REPRINTED WITH PERMISSION.

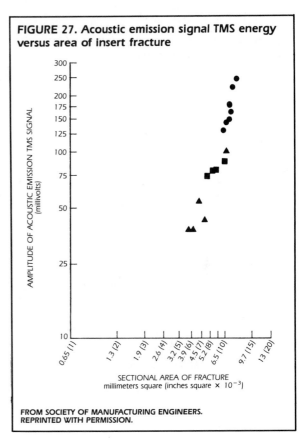

FIGURE 27. Acoustic emission signal TMS energy versus area of insert fracture

FROM SOCIETY OF MANUFACTURING ENGINEERS. REPRINTED WITH PERMISSION.

which is consistent with published results.[29] The high amplitude of the power spectrum at fracture, Fig. 28b is seen in this frequency range. The abrupt change in the geometry of the cutting tool due to fracture will also result in a change in the power spectrum due to machining with the fractured tool.

These data collected under realistic machining conditions indicate the sensitivity of the acoustic emission signal to in-process tool fracture.

Chip Formation Monitoring

During machining, a significant amount of the acoustic emission signal generated is due to fracture of the chip and typically appears as burst emission. Since one of the critical requirements for the operation of unattended machine tools is the production of discontinuous chips (to prevent chip tangling and catastrophic tool failure as well as to ensure chip removal by mechanized chip handling equipment) sensing techniques for chip form are needed. The utilization of acoustic emission signal analysis to sense chip form during machining is discussed earlier in this chapter and in the literature.[31] The correlation of acoustic emission event rate to the chip breaking frequency is discussed below.

Acoustic Emission and Chip Breaking Frequency

Tests were conducted with grade K68 carbide inserts with tool holders at 5 degree rake machining SAE 1018 steel under the following conditions: 1.4 m (4.7 ft) per second speed; 0.08 mm (0.003 in.) per revolution feed for continuous chips; and differing feed rates from 0.13 to 0.5 mm (0.005 to 0.02 in.) per revolution for breaking chips and 1.5 mm (0.06 in.) depth of cut.

An acoustic emission transducer was mounted on the end of the tool holder and a three-component dynamometer was used to measure the machining forces on the cutting tool. Different chip forms resulted from the change in feed rates. The estimated chip breaking frequency was calculated based on the metal removal rate, the number of chips in a sample collected during a period of machining, and the weight of the sample. The event count of the acoustic emission signals were measured with a computer based acoustic emission system with the rms values of the acoustic emission signals determined by an rms module.

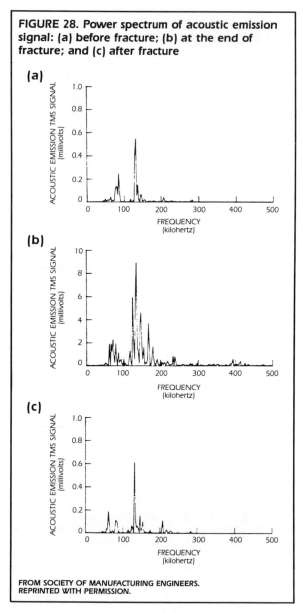

FIGURE 28. Power spectrum of acoustic emission signal: (a) before fracture; (b) at the end of fracture; and (c) after fracture

FROM SOCIETY OF MANUFACTURING ENGINEERS. REPRINTED WITH PERMISSION.

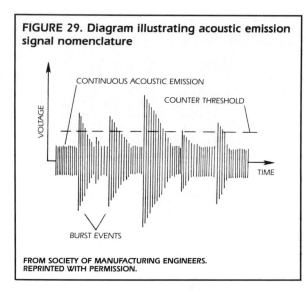

FIGURE 29. Diagram illustrating acoustic emission signal nomenclature

FROM SOCIETY OF MANUFACTURING ENGINEERS. REPRINTED WITH PERMISSION.

During machining, the acoustic emission signals from chip breaking are usually large amplitude burst signals compared with the acoustic emission generated during normal cutting (Fig. 29). Since each significant burst event represents the breaking of a chip, the event count (the number of burst events rising above the threshold level) gives the number of chip breakages during the sampling interval.

The event count per second indicates the frequency of chip breaking. Because of the stochastic nature of the machining process, the chip size cannot be uniformly controlled and some scatter in chip breaking frequency occurs. The sensitivity of the acoustic emission signals to chip breaking behavior is evident when compared with the estimated chip breaking frequency. The average event rate calculated from the data is plotted with estimated chip breaking frequency as a function of feed rate in Fig. 30. The result shows an excellent correlation between the measured event rate and the chip breaking frequency.

Measurement of force variation during machining has been suggested as a means of detecting chip breaking since, when chip breaking takes place, the cutting force fluctuates due to variation of chip load on the tool. Both cutting (tangential) and feed (longitudinal) forces were measured during the machining tests under a variety of chip breaking conditions. The feed force was found to be more sensitive to the chip breaking behavior than the tangential force.

Because of the irregularity of chip formation and breaking during cutting, as well as the low frequency vibrations of the machine tool itself, spectral analysis of the cutting forces for determination of chip breaking frequency in real-time applications does not appear to be as reliable as acoustic emission signal analysis.

Wheel Contact and Sparkout in Grinding

The two major difficulties in implementing adaptive control optimization systems in grinding have been identified

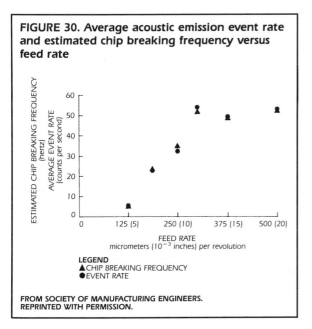

FIGURE 30. Average acoustic emission event rate and estimated chip breaking frequency versus feed rate

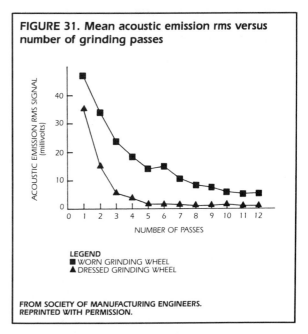

FIGURE 31. Mean acoustic emission rms versus number of grinding passes

as: (1) specifying suitable index of performance to accompany an effective and efficient control scheme; and (2) developing sensors that can measure the necessary process parameters in a manufacturing environment.[32] The index of performance must reflect such operating characteristics as constraints on wheel wear, work surface burn, wheel dressing or sparkout. The potential for using acoustic emission signal analysis as a process monitoring technique in grinding has also been investigated.[33] The investigation described below utilizes acoustic emission from surface grinding to measure sparkout (or loss of contact) between the wheel and the work surface.

Grinding tests were conducted on a surface grinder with a 240 mm (9.6 in.) diameter wheel under the following conditions:

1. width of wheel engagement: 5 mm (0.2 in.);
2. wheel speed (v_s): 22.9 m (75 ft) per second;
3. work feed speed (v_w): 0.29 m (0.95 ft) per second; and
4. work material: AISI 1045 steel.

A three-component crystal dynamometer was used to measure the normal and tangential grinding forces during the tests. A linear variable differential transformer (LVDT) accurately determined the position of the grinding wheel over a range of radial wheel engagements. The acoustic emission monitoring instrumentation included a broadband transducer (100 to 800 kHz) attached to the workpiece via a base mounted in a threaded hole. Although this arrangement gives varying wheel-to-transducer distances during the operation, the results obtained for emission energy during various portions of a single grinding stroke indicated the effect is insignificant. The transducer output was recorded on a modified video recorder for later analysis. The rms values of the acoustic emission signal were measured using an rms/TMS voltmeter.

After an initial radial depth of cut was set at 5 μm (0.0002 in.), a series of tests was conducted for which, a sparkout condition was established with successive strokes of the grinding wheel for dressed and worn wheels. The acoustic emission signal level (rms_{AE}) and grinding forces were measured.

Sparkout is the completion of metal removal to a desired depth as indicated by the absence of sparks generated during motion of the wheel over the workpiece. Sparkout can be determined from observation of the acoustic emission signal. As seen in Fig. 31, after the fifth pass of the dressed wheel and the tenth pass for the worn wheel, no additional reduction in emission level was observed (sparkout had occurred).

Although this can often be detected by observing the process visually, the emission signal is a more consistent indicator. It is also important to determine the first contact of the grinding wheel with the work surface to assist in establishing the location of the surface for measuring depth of engagement, for example.

To determine the sensitivity of the emission signal to work-wheel contact, a series of tests was run with the grinding wheel at various heights above the work surface. As can

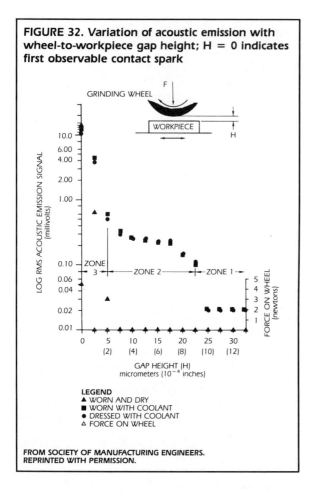

FIGURE 32. Variation of acoustic emission with wheel-to-workpiece gap height; H = 0 indicates first observable contact spark

be seen in Fig. 32, the emission signal (rms_{AE}) is substantially more sensitive to the wheel-work gap than is the measured force, especially for operation with a coolant. The emission level over the range of gap heights H to contact ($H = 0$) as indicated by the first spark can be divided into the three regions listed below.

1. *No significant interaction*: gap height $H > 22.9$ μm. The emission level is essentially the instrumentation noise level.
2. *Fluid-wheel interaction*: gap height from 5.1 to 22.9 μm. This is typified by slightly increasing emission levels as the gap height is reduced and the abrasive grit and fluid, and fluid and work, interact (fluid is air or water in this case).
3. *Wheel-work interaction*: gap height $H \leq 5.1$ μm. Turbulent interaction between wheel, fluid and work surface occurs. Initial work wheel contact (undetectable by force or visible spark) occurs between dominant abrasive grits and the work surface.

The gap heights listed are typical of the recorded test data but vary with grinding conditions such as coolant viscosity, wheel density and composition, wheel speed, and so on.

Conclusion

Three different applications of acoustic emission have been developed in response to the need for advanced sensing techniques for process automation. Those clearly indicate the basic importance of reliable sensors, the output of which can be closely correlated with the characteristics of the process under supervision. In all three of the applications, acoustic emission from the process is first analyzed to investigate the relationship between the process fundamentals and the emission and to evaluate the sensitivity of the signal to the discontinuities or process changes of interest.

Acknowledgments

This research was supported by grants from the National Science Foundation (production research division) and the National Bureau of Standards (automated production technology division) and tooling donated by Kennametal, Inc.

PART 5
ACOUSTIC EMISSION FROM MULTIPLE INSERT MACHINING OPERATIONS

Acoustic Emission from Milling

In milling (rotating multiple toothed cutters), the cutting process is discontinuous with varying chip load and relative tool work velocity. There is a potential for chip congestion and tool rubbing noise. The basic acoustic emission signal due to material deformation inherent in chip formation and chip/tool contact is complicated by these other sources of noise and the periodic interruptions of the cutting process.

This is in contrast to acoustic emission from relatively stationary, single point turning. In both cases, tool work velocity and metal removal rate are significant parameters affecting the energy of the acoustic emission signal. The major sources of acoustic emission identified in previous investigations of metal cutting include:

1. plastic deformation of workpiece material in the shear zone;
2. plastic deformation and sliding friction between chip and tool rake surface;
3. sliding friction between workpiece and tool flank surface; and
4. collision, entangling and breakage of chips.

In face milling, because of the interrupted cutting conditions, additional sources include: (1) shock wave generated in tool entry; and (2) sudden unloading and chip break off at tool exit.

Because of the geometry of tool work interaction, there are additional variations in the acoustic emission generated, mainly resulting from chip thickness and tool velocity changes as the tool rotates through the work material. If tool deterioration (wear) or catastrophic failure (fracture) occurs, the acoustic emission signal during cutting is either modified, or in the case of wear, exhibits a dramatic spike from the energy released at fracture.

Tests have been conducted over a range of conditions on a vertical milling machine using a milling cutter, with grade K420 inserts machining AISI 1018 steel. Figure 33 illustrates the geometry of the test setup and Fig. 34 shows a diagram of the instrumentation. A signal processing technique based on time difference averaging (TDA) was developed to overcome the difficulties of analyzing the acoustic emission signal from milling due to strong periodic components and noise.[34]

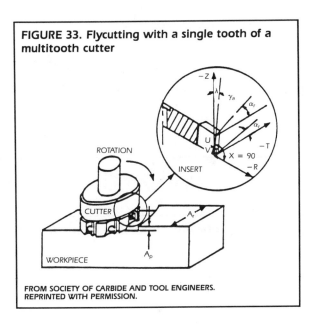

FIGURE 33. Flycutting with a single tooth of a multitooth cutter

FROM SOCIETY OF CARBIDE AND TOOL ENGINEERS. REPRINTED WITH PERMISSION.

Data Analysis of Acoustic Emission from Milling

Time Difference Averaging

A typical acoustic emission root mean square energy (rms) signal from a flycutting operation is shown in Fig. 35. The TDA methodology decomposes a compound repetitive signal into a periodic component that is time locked to the basic signal period and a residual. The output of a TDA analysis of acoustic emission during flycutting, including mean and residual, is shown in Fig. 36. The magnitude of the acoustic emission rms is plotted as a function of rotation of the milling cutter. The initial acoustic emission peak due to tool contact, and the final peak due to toll exit and chip breakoff, are clearly seen.

The acoustic emission related to chip formation is seen at the center of the active region. The residuals from the sampled data indicate the degree of randomness in the signal due to variations in entry and exit as well as noise due to chip conjection.

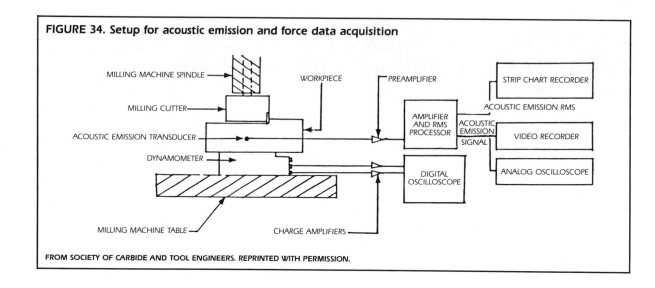

FIGURE 34. Setup for acoustic emission and force data acquisition

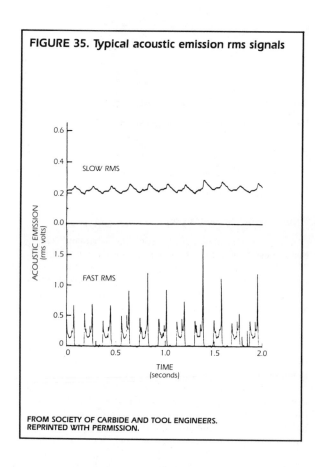

FIGURE 35. Typical acoustic emission rms signals

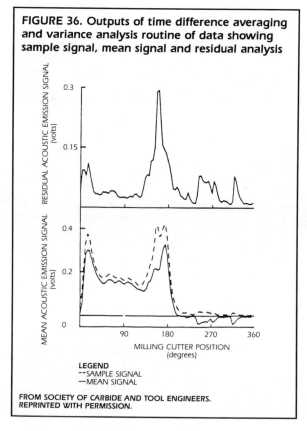

FIGURE 36. Outputs of time difference averaging and variance analysis routine of data showing sample signal, mean signal and residual analysis

Insert Wear

Data on flank wear of inserts are shown in Fig. 37. For the tests AISI 4340 steel in both as-received (25 R_C) and heat treated (35 R_C) conditions was machined. A general trend of increasing mean acoustic emission rms energy and specific cutting force was seen after an initial period of instability corresponding to 0.2 mm (0.008 in.) of measured flank wear.

This instability is most likely related to the breakdown of the sharp insert edge during the initial minutes of machining. Following this instability, however, the acoustic emission rms shows good sensitivity to increasing flank wear. The cumulative change in the acoustic emission signal from initial cutting to a flank wear of 0.4 to 0.5 mm (0.015 to 0.02 in.) is 150 to 200 mV and shows good repeatability.

Insert Fracture

A primary objective was to evaluate the sensitivity of the acoustic emission signal to insert fracture and chipping during single and multitoothed face milling. A theoretical analysis of the acoustic emission generated at fracture had indicated that the acoustic emission was dependent primarily on the area of fractured surface, as well as material geometry constraints.[35]

Tests were conducted to determine the appropriateness of this relationship and to assess the sensitivity of acoustic emission to insert fracture during realistic machining operations.

These tests included both single and multiple insert milling operations on hardened steel (AISI 4340 at 40 R_C) for a natural fracture and on AISI 1018 with notched inserts for artificially induced fracture.[36] The same tooling was used for this set of tests as in the milling insert wear tests described previously. Figure 38 shows the response of feed force and rms acoustic emission signal to the fracture of the insert in flycutting under the following conditions: velocity of 1.3 m (4.4 ft) per second; feed of 2.7 mm (0.11 in.) per revolution; and axial depth of cut of 2 mm (0.08 in.). The dramatic change in the level of both force and acoustic emission rms at fracture is evident.

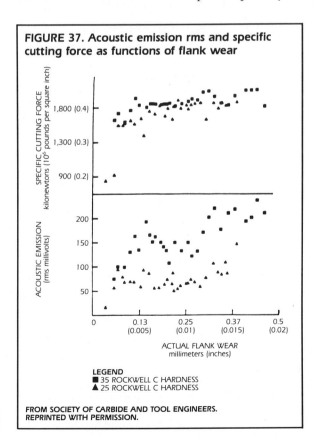

FIGURE 37. Acoustic emission rms and specific cutting force as functions of flank wear

FROM SOCIETY OF CARBIDE AND TOOL ENGINEERS. REPRINTED WITH PERMISSION.

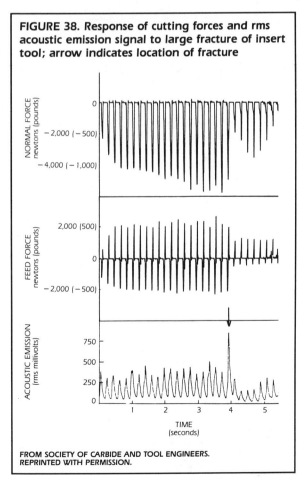

FIGURE 38. Response of cutting forces and rms acoustic emission signal to large fracture of insert tool; arrow indicates location of fracture

FROM SOCIETY OF CARBIDE AND TOOL ENGINEERS. REPRINTED WITH PERMISSION.

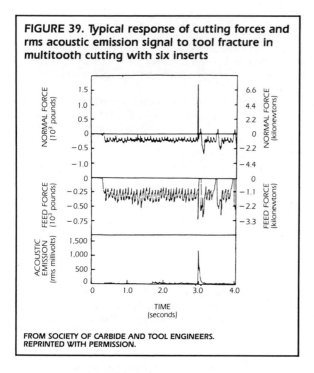

FIGURE 39. Typical response of cutting forces and rms acoustic emission signal to tool fracture in multitooth cutting with six inserts

FROM SOCIETY OF CARBIDE AND TOOL ENGINEERS. REPRINTED WITH PERMISSION.

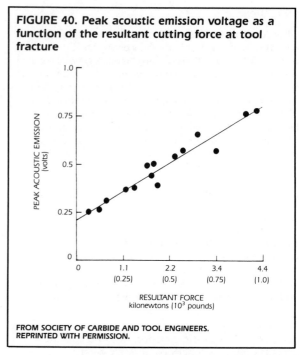

FIGURE 40. Peak acoustic emission voltage as a function of the resultant cutting force at tool fracture

FROM SOCIETY OF CARBIDE AND TOOL ENGINEERS. REPRINTED WITH PERMISSION.

Although the presence of additional cutting inserts complicates the process, as seen in Fig. 39, the fracture event is clearly distinguished above the background level of cutting generated acoustic emission. A similar response is seen in the force data. Conditions for this test were: 0.5 m (1.6 ft) per second velocity; 0.7 mm (0.027 in.) feed per revolution; and axial depth of cut of 1 mm (0.04 in.).

The acoustic emission signal analysis was based on the peak value of the acoustic emission rms voltage produced at fracture. The fractured area was measured from a scanning electron microscope photograph of the insert using a planimeter. Figure 40 is a graph of peak acoustic emission rms voltage versus resultant cutting force at fracture. As expected, the higher the fracture energy, reflected in the force at fracture, the more acoustic emission energy observed. Similar correlation had been seen for testing of previous single tools.[36]

Acoustic Emission from Chip Formation

The effectiveness of acoustic emission signal analysis in the determination of continuous and discontinuous chip forms has been demonstrated.[37] The effectiveness of linear discriminant function techniques to categorize patterns of sensor information for the detection of continuous and discontinuous chip forms has also been studied.[38] The key sensor in this case is a transducer monitoring acoustic emission generated by chip fracture during metal cutting.

The generation of acoustic emission for the fracture of discontinuous chips has been experimentally confirmed in a number of studies. These studies show that there is an excellent correlation between the event rate of the acoustic emission signal and the frequency of chip formation for discontinuous chip machining of steel and aluminum materials. The burst acoustic emission activity due to discontinuous chip formation is superimposed over the continuous acoustic emission from the plastic deformation and rubbing in the machining process.

A methodology for analysis of this acoustic emission has been developed based on the event rate of the rms voltage of the acoustic emission above a predetermined threshold to distinguish between continuous and discontinuous chip forms.[39] However, it is difficult in some cases to identify the exact chip forming condition near the transition from a continuous to discontinuous chip based on any single factor. Further, under differing machining conditions, the definition of a desirable discontinuous chip form may vary. Thus, additional signal analysis capabilities are required.

The use of linear discriminant functions, a pattern recognition technique in the artificial intelligence field, was evaluated for the analysis of acoustic emission signals and related process variables to better determine the chip forming

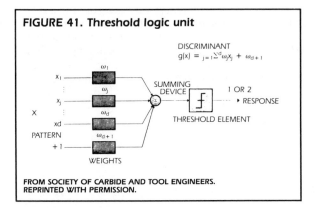

FIGURE 41. Threshold logic unit

FROM SOCIETY OF CARBIDE AND TOOL ENGINEERS. REPRINTED WITH PERMISSION.

state during machining. The function was adopted specifically to help identify the continuous or discontinuous chip forming boundary. Discriminant functions, including the event rate of the rms of acoustic emission, are obtained through a training process using a set of pattern vectors, the classifications of which are known.

The discriminant functions are then applied to the pattern vectors of test cuts. A more efficient discriminator (the threshold logic unit or TLU) was also successfully applied to classification of chip form. Figure 41 shows a schematic of the TLU. Tests were done to distinguish the cutting of states between either continuous or discontinuous chip formation. The following parameters were selected to characterize a cutting pattern:

1. cutting speed: 5.3 m (16.7 ft) per second;
2. feed rate: 0.025 mm (0.001 in.) per revolution
3. depth of cut: 0.25 mm (0.01 in.), and
4. event rate of the rms of acoustic emission using 1.1 times average rms level as threshold (Hz).

The training patterns were picked up such that they represented machining at a variety of cutting speeds, feed rates and depths of cut. Especially, cutting conditions near the boundary of continuous or discontinuous chip formation for this tool geometry and work material were included. Seven training pattern vectors were used to determine the weight vectors. Cutting experiments were carried out under various conditions to check the validity of the recognition system for the cutting state classification. Table 3 shows the results of the discriminant functions in sixteen different cutting tests.

The test cutting conditions are also listed. The classifications are matched with the observed states for all the cases.

Including the event rate of the rms of acoustic emission for the sensing technique with a pattern recognition algorithm, it was seen that the linear discriminant function works to distinguish between continuous and discontinuous chip formation conditions. Further, the components of a successful pattern to characterize the cutting state are cutting speed, feed rate, depth of cut, and event rate of the rms of acoustic emission.

The event rate of the rms is the most important factor in chip form classification. The feed rate ranks second. The depth of cut is not as crucial as the above two, and since it is often not easily obtainable, tests have shown that depth of cut can be excluded as a pattern component without reducing the reliability of the classification. Finally, the single member discriminant function (TLU) works very efficiently in a two-category classification such as studied here.

Conclusion

The work summarized here illustrates the potential for applying acoustic emission techniques to analysis and monitoring of metal cutting processes. As a result of this work, several observations have been made that lead to the concept of an intelligent acoustic emission sensor for the process monitoring described in detail in the next part of this chapter.[40] The intelligent sensor concept is a design that allows the sensor to effectively provide input for a machine control system. It must do more than convert phenomenon related high frequency stress waves into an electrical signal for digitization and transmission to the control system. The sensor must also include provisions for signal conditioning and processing. It should also provide for self-calibration and diagnostics.

The intelligent sensor includes some internal processing capabilities, like the TLU discriminator for chip form detection discussed above. The output of the intelligent sensor is information on one aspect of the state of the process. Action may be taken based on this output but no action component (actuator, for example) is part of the sensor. The action may be commanded by a controller that has utilized the sensor information or could be the action of hardware set to respond to a threshold level crossing. The utilization of a sensing element in a manufacturing system with these capabilities will make the goal of untended manufacturing systems much more feasible.

TABLE 3. Chip forms classified by linear discriminant functions for 6061-T6 aluminum cut by carbide insert tool; the criteria to distinguish chip form are as follows: (1) for the two-member discriminant function, $g_1(X) > g_2(X)$ yields continuous chip cutting and $g_2(X) > g_1(X)$ yields discontinuous chip cutting; (2) for the threshold logic unit, $g(X) < 0$ yields continuous chip cutting and $g(X) > 0$ yields discontinuous chip cutting.

Cutting Parameters/Discriminant Functions					Two-Member Discriminant Function				Threshold Logic Unit	
					With Depth of Cut as a Component		Without Depth of Cut as a Component		With Depth of Cut as a Component	Without Depth of Cut as a Component
Cutting Speed meters per second (1,000 feet per minute)	Feed Rate micrometers (10^{-3} inches) per revolution	Depth of Cut millimeters (10^{-2} inches)	Event Rate of rms of Acoustic Emission (hertz)	Chip Type (observed from cut test)	$g_1(X)$ for continuous chip	$g_2(X)$ for discontinuous chip	$g_1(X)$ for continuous chip	$g_2(X)$ for discontinuous chip	$g(X)$	$g(X)$
7.5 (1.5)	250 (10)	1.25 (5)	44.9	Discontinuous	−36.32	161.12	−7.01	121.81	122.23	74.42
7.5 (1.5)	150 (6)	1.25 (5)	11.7	Continuous	78.65	−28.25	38.37	2.03	−64.05	−18.80
7.5 (1.5)	300 (12)	1.25 (5)	44.4	Discontinuous	−13.36	141.16	8.59	109.21	94.97	58.59
7.5 (1.5)	350 (14)	1.25 (5)	45.9	Discontinuous	0.19	134.61	19.72	105.08	81.80	50.06
7.5 (1.5)	475 (19)	1.25 (5)	51.3	Discontinuous	16.01	137.59	36.60	107.00	72.35	41.77
7.5 (1.5)	400 (16)	1.25 (5)	46.9	Discontinuous	16.10	124.70	31.96	98.84	65.11	39.71
6.5 (1.25)	250 (10)	1.25 (5)	54.7	Discontinuous	−84.30	228.20	−31.46	165.36	192.97	112.65
7.5 (1.5)	350 (14)	1.5 (6)	47.4	Discontinuous	4.31	135.49	16.36	111.44	80.08	55.54
7.5 (1.5)	50 (2)	1.25 (5)	6.3	Continuous	62.84	−31.24	21.49	0.11	−54.60	−10.50
7.5 (1.5)	150 (6)	1.0 (4)	13.7	Continuous	58.07	−5.67	33.89	10.51	−37.68	−11.49
6.6 (1.3)	500 (20)	1.25 (5)	53.2	Discontinuous	26.18	134.82	44.83	106.17	63.39	36.65
7.5 (1.5)	150 (6)	1.5 (6)	12.7	Continuous	85.12	−30.72	36.13	6.27	−69.29	−15.15
6.5 (1.25)	200 (8)	1.25 (5)	23.6	Discontinuous	−24.46	130.16	−7.65	103.35	96.31	64.21
7.5 (1.5)	175 (7)	1.25 (5)	36.1	Discontinuous	−25.83	127.04	−9.03	100.24	95.87	63.30
9.2 (1.8)	200 (8)	1.0 (4)	5.9	Continuous	117.63	−76.23	68.84	−35.44	−118.43	−56.91
7.5 (1.5)	175 (7)	1.0 (4)	16.6	Continuous	54.73	5.47	34.63	17.57	−29.13	−7.91

PART 6
DEVELOPMENT OF AN INTELLIGENT ACOUSTIC EMISSION SENSOR FOR UNTENDED MANUFACTURING

The interest in developing the capability for untended manufacturing systems is growing along with the implementation of advanced flexible manufacturing systems. To remove the need for observation of the manufacturing processes and minimize the time lost due to repair or correction of unexpected failures in the system, new sensing methodologies and sensor based control schemes are being proposed and evaluated. The major roadblock to implementation of true untended manufacturing is the lack of suitable sensors for process monitoring. In spite of many sensing technologies in existence, few totally untended operations exist.

The sensing tasks for untended manufacturing operations are substantial. In addition to being economical, sensors must be nonintrusive, rugged and resistive to the hazardous environment in which they operate, accurate, exhibiting a high sensitivity to the feature or phenomena being monitored, linear with respect to the sensed fracture if possible, and highly repeatable and reliable. Few sensors can meet all of these specifications.

It seems most reasonable that, to be effective in untended manufacturing, combinations of sensors are needed to provide corroborative information on the state of the manufacturing operation. This requires efficient methodologies for integration of sensor information. Available information on the operation of the process (such as speeds or feeds in machining) need to be considered as well. Some efforts have demonstrated the potential of combining information from several sources in process monitoring.[41]

Several reviews of sensing techniques for untended manufacturing have detailed the requirements for sensing, the phenomenon being measured, implementation on types of machines and different sensing methodologies.[42,43] These reviews cover a wide range of sensing methodologies, including acoustic emission techniques.

Intelligent Acoustic Emission Sensors

Much attention has been directed to the characterization of intelligent manufacturing processes. Intelligent machines have been defined as machines that can perform specified functions without detailed input from operators and regardless of how the material characteristics vary.[44] This intelligence can be related to several aspects of the manufacturing process:[45]

1. material related (changes in raw material properties);
2. geometry related (macrogeometric and microgeometric changes in surface, form, position);
3. quantity related (production rate, efficiency, load);
4. place related (translation, rotation, orientation); and
5. time related (sequence).

Information generation and flow play an important role in the performance of manufacturing systems. The interest here is to define more clearly the structure of an intelligent sensor acting as one of the information generators in a manufacturing system.

The ideal sensor should have certain features to ensure that its performance is acceptable as a process monitor in a manufacturing environment. These include:

1. linear output with respect to sensed phenomenon;
2. appropriate resolution;
3. good sensitivity to phenomenon;
4. appropriate accuracy;
5. repeatability and reliability;
6. ruggedness or resistance to environment; and
7. nonintrusive operation.

The concept of an intelligent control system has been defined to include the input from intelligent sensors.[41] Figure 42 illustrates the optimization of the performance of a machine tool by including two stages: first, achieving performance within a broadly defined acceptable range of operation based on *a priori* information about the structure of the process to be controlled and its likely performance, and; second, by modifications of the parameters of the process on-line by the use of some optimization algorithms.

FIGURE 42. Schematic of intelligent control system

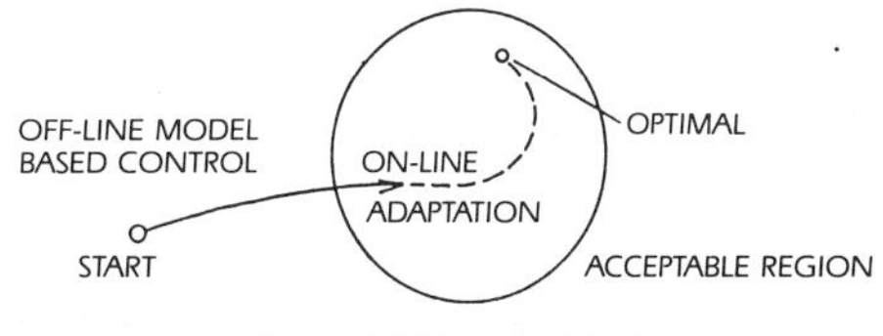

Examples of the implementation of this concept applied to control of machine tools may be found in the literature.[46] The input to the intelligent control system and the basis for determining whether or not the desired machining state has been obtained is from sensors.

To effectively provide input for the machine control system, the sensor must do more than convert phenomenon related stress, strain, vibration or temperature into an electrical signal for digitization and transmission to the control system. The sensor must also include provisions for signal conditioning and processing. The basic elements of an intelligent sensor, one that includes these provisions, are defined in Fig. 43. The intelligent sensor is composed of four major components.

1. *Sensing element*: a piezoelectric crystal in the case of a load cell, accelerometer or acoustic emission transducer.
2. *Electronics*: includes preamplifiers or amplifiers to guarantee acceptable signal-to-noise ratio of the sensing element output for transmission and further processing.
3. *Basic signal conditioning*: includes rms energy measurement; conversion to engineering units of measure (such as charge to force) and processing that visually or digitally displays trends, levels and individual events.
4. *Local qualification and decision making*: based on experience with the process, the output of the sensor (suitably conditioned) can indicate the occurrence of an event or the direction of a trend. This information can provide the basis on which a decision is made regarding the process.

The output of the intelligent sensor is information on one aspect of the state of the process. Action may be taken based on this output but no action is part of the sensor. The action may be commanded by a controller that has utilized the sensor information or could be the action of hardware set to respond to a threshold level crossing, for example. Similarly, an intelligent controller could utilize this information in combination with other information to attempt to optimize the performance of the system.

Acoustic Emission Intelligent Sensing

An example of a typical sensor technique that meets the definition of the intelligent sensor is the application of acoustic emission sensing to manufacturing process monitoring. Acoustic emission analysis has been proven effective as a sensing methodology for machine tool condition monitoring in Japan[47,48] and in the United States.[49-53] The problems of detecting tool wear and breakage in single point turning motivated much of the work but some applications have included chip form detection, forming and forging, grinding and surface finishing and welding process monitoring.

In acoustic emission tests, all aspects of the intelligent sensor can be defined. The sensing element itself is a ceramic piezoelectric crystal, the output of which (a voltage) is sent through a preamplifier and amplifier (the electronics) to get a signal of high enough level for further processing. This voltage output reflects the source of the emission, the characteristics of the sensor (frequency and damping) and the characteristics of the electronics. The basic signal conditioning referred to in Fig. 43 could be display of the raw acoustic emission signal on an oscilloscope, or more likely for some manufacturing applications, the measurement of the root mean square (rms) energy of the signal and display on an oscilloscope.

Previous work has shown a dependency of the rms of the acoustic emission signal on the contact area between the flank face of a cutting tool and the work surface (flank wear)[48,50] and the dependency of the spike height of the rms of acoustic emission on the fracture of the cutting tool and the fracture area.[54,55] Thus, local qualification or decision making in this sensor system is the determination of the occurrence of tool fracture based on spike height of acoustic

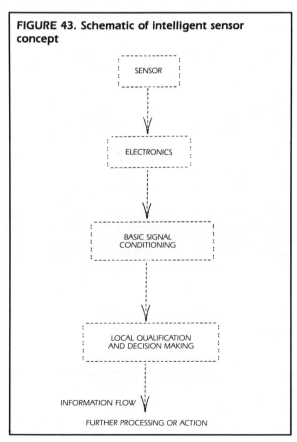

FIGURE 43. Schematic of intelligent sensor concept

emission rms or the existence of excessive tool wear based on the level of mean rms of the acoustic emission.

The spike height threshold as well as the acoustic emission wear limit threshold are determined in a variety of ways, usually based on experience. At this point, the output of the intelligent sensor indicates the state of the cutting tool in this example. Whether or not any action is taken as a result of this information is dependent on the further processing of the intelligent sensor output. This could be retraction of the tool (if fracture), change tool after part is completed (for tool wear), or do nothing if no problem is detected. The point is that the action is not taken by the intelligent sensor but by some other control action at a higher level. This higher level controller may consider the information from other intelligent sensors before a decision on an action is made.

Development of Intelligent Acoustic Emission Sensors

The description of intelligent sensors emphasizes the need for viewing the sensing technique from the basic sensor design, through the output of a decision regarding the phenomena being monitored. Below is a review of research on the development of intelligent acoustic emission sensors for manufacturing process monitoring in three areas: (1) punch stretching and metal forming; (2) chatter detection in end milling; and (3) abrasive disk grinding

Intelligent Sensors for Punch Stretching

Earlier study of the potential of acoustic emission monitoring of manufacturing[56] has identified the feasibility of analyzing acoustic emission signals from punch stretching to determine both the variations in lubrication conditions on the punch as well as the occurrence of localized deformation (necking) preceding failure of the stretched sheet.

To establish the basis for acoustic emission monitoring of punch stretching and metal forming, tests have been run to determine the sensitivity of the acoustic emission signal to stretching process characteristics, notably the change in the signal as a function of penetration (and, hence, the elastic, plastic, final fracture and post-fracture stages of a stretching operation). Characteristics of the acoustic emission generated depend on the properties of the sheet metal (hardness, toughness), the experimental conditions (strain rate, lubrication, punch geometry) and the nature of the acoustic emission source mechanisms (plastic deformation, fracture, rubbing contact) and the instrumentation.

The study utilizes signal processing schemes in both the time and frequency domain to distinguish between the states of stretching and fracture. Earlier studies have demonstrated that the power spectral density of each burst of acoustic emission can be helpful to identify the source mechanism. Thus the frequency domain analysis of the acoustic emission signal from stretching is of interest for analysis and prediction. The shift of continuous emission spectral towards higher frequencies as plastic strain increased has been observed during tensile tests of some specimens.[57] The work detailed here investigates this effect for metal forming to develop a methodology for deconvolution of the raw acoustic emission signal to minimize the influence of transducer characteristics and electronic coloring on the acoustic emission output.

Tests were run on aluminum 1100-0 and 1100-H14 sheet specimens in a typical punch stretching setup (Fig. 44).[58] The punch was mounted on a load cell to record the thrust force, an LVDT measured the displacement of the punch and an acoustic emission transducer was attached to the blankholder. The rms energy of the acoustic emission signal was measured. A typical load displacement curve for this type of experiment is shown in Fig. 45 for an aluminum 1100-0 specimen. Figure 46 shows the variation in the acoustic emission rms voltage as a function of the deformation ratio r (ratio of displacement at a point to the displacement at specimen fracture).

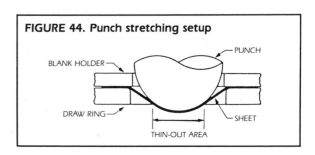

FIGURE 44. Punch stretching setup

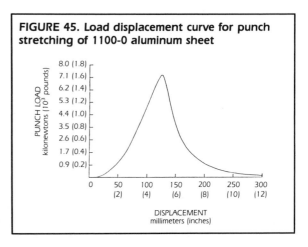

FIGURE 45. Load displacement curve for punch stretching of 1100-0 aluminum sheet

Three distinct regions of acoustic emission rms can be observed from Fig. 46. First, a peak due to punch contact and initial rapid rise of signal level due to the yielding of the specimen out to about $r = 0.32$. Second, strain hardening of the specimen and an accompanying decrease in the level of acoustic emission rms between $r = 0.32$ and $r = 1.0$. And third, after fracture (at $r = 1.0$), a constant level of acoustic emission rms due to sliding contact between the punch and the fractured specimen.

Frequency domain signal analysis was also done. The results showed that over the range of plastic deformation, until fracture, the mean frequency increases and the mean spectral amplitude decreases. The post-fracture rubbing period exhibits a lower mean frequency and amplitude.

Intelligent Sensors for Chatter Detection in End Milling

The sensitivity of acoustic emission to the onset and development of chatter in machining has been observed during turning operations.[59] Described below are details of the results of slot milling tests with two-fluted end mills in 7075-T6 aluminum. Acoustic emission generated during the milling process was recorded using sensors mounted on the spindle of the milling machine as well as on the workpiece. Chatter was induced by variations in the feed rate as well as by changes in the stiffness of the workpiece due to clamping.

Figure 47 shows the variation of the acoustic emission rms with time for the transition from stable machining to chatter as recorded by the sensor on the workpiece. During the portion marked section-3, the feed rate was reduced to inhibit chatter but the amplitude of chatter increased as evidenced by the increase in acoustic emission. Section-4 marks the withdrawal of the tool from the workpiece.

Frequency analysis of the raw acoustic emission signal generated during stable machining and chatter conditions indicated a shift in the signal to lower frequency along with the expected increase in amplitude of the signal. Attenuation at the sensor mounted on the spindle of the machine was significant (approximately 20 dB). Significant attenuation in the frequency domain was observed due to transmission through the spindle bearings. The attenuation in amplitude is consistent with other observations.[55]

Intelligent Sensors for Abrasive Disk Machining

Development continues on feedback techniques for grinding and other surface finishing tasks to ensure removal of material or generation of the desired surface characteristics. This work has concentrated on force feedback, often from load cells mounted on robot end effectors with sanding disks for weld bead removal, lash cleanup on castings and removal of machining burrs on workpieces. Although sensitive to the metal removal process, force feedback is often not sensitive enough to small depths of engagement or fine burrs on workpieces. Research has shown the potential for using acoustic emission feedback in grinding operations for sparkout, wheel wear and loading.[60]

Abrasive disk grinding of mild steel was achieved with a depressed center grinding disk mounted in the spindle of a milling machine (Fig. 48). The disk was inclined at an angle of 15 degrees to the work surface and the acoustic emission

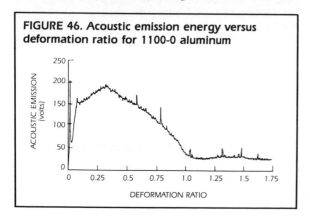

FIGURE 46. Acoustic emission energy versus deformation ratio for 1100-0 aluminum

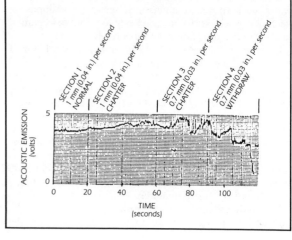

FIGURE 47. Acoustic emission energy during chatter in end milling of 7075-T6 aluminum with the following machining conditions: spindle speed 2,850 rpm at 3.8 m (12.5 ft) circumferential velocity per second; depth of cut 13 mm (0.5 in.); feed rates as labeled; no lubricant

sensor was attached to the workpiece. Data from tests run to determine the sensitivity of the acoustic emission signal to depth of engagement, sparkout and work surface burn are shown in Figs. 49 and 50.

The rms energy of the acoustic emission sensor output clearly indicates sensitivity to: (1) depth of engagement (increasing with increased depth as seen in Fig. 49); (2) sparkout (energy decreasing with subsequent passes until contact between the wheel and the work surface is lost as seen in Fig. 50); and (3) work surface burn (the distinct variation in the energy as a function of travel across the work surface).

Surface burn is quite different from that seen in Fig. 49 during normal grinding. This was due to burning of the surface and no material removal at the center of the workpiece with no burning and continued removal of material at the ends. Hence, the energy level at the ends drops off due to reduced depth of engagement.

The variation in the acoustic emission rms during the removal of burrs on the workpiece is observable as well. In these tests the acoustic emission signal exhibited better sensitivity to the abrasive grinding process than force (normal and feed) measurements recorded simultaneously.

Conclusion

The intelligent sensor includes some additional processing of the sensor output to give a local decision with regard to the state of the process. This concept was illustrated by examples of past and present research on the development of acoustic emission monitoring techniques.

In all three examples, the acoustic emission technique demonstrates good sensitivity to the phenomenon under investigation, good resolution, good repeatability and reliability, and can be packaged to be resistant to the manufacturing environment. The output of the transducer can be clearly related to a major change in a characteristic of the process thus providing the necessary information for a control action or response.

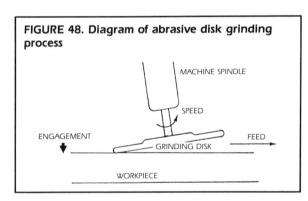

FIGURE 48. Diagram of abrasive disk grinding process

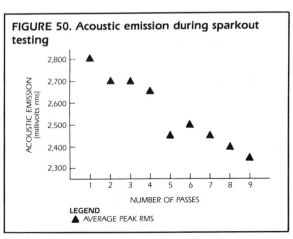

FIGURE 50. Acoustic emission during sparkout testing

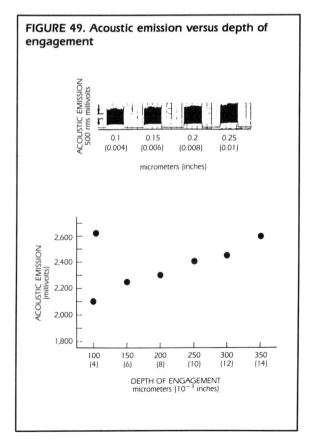

FIGURE 49. Acoustic emission versus depth of engagement

PART 7
CURE MONITORING OF COMPOSITE MATERIALS USING ACOUSTIC TRANSMISSION AND ACOUSTIC EMISSION

Reinforced thermosetting resin composites are fabricated into final form using elevated temperature and pressure. Optimum control of temperature and pressure (cure cycle) must be achieved in order to produce parts with desired chemical and mechanical properties. Historically, the cure cycle has been determined through trial and error. Computer models speed up the trial and error process. A technique that can be used to measure the critical parameters (resin viscosity during cure and residual stress during cool down) is described below. Both techniques can be used to modify computer models to achieve a closed loop control system.

Manufacture of Thermosetting Resin Composites

Fiber reinforced thermosetting resin composites are cured using an autoclave or heated press. The uncured material is placed on a metal tool plate then either covered with

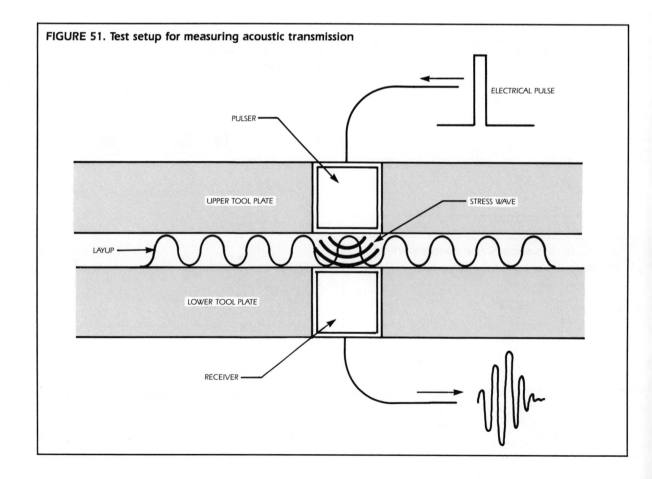

FIGURE 51. Test setup for measuring acoustic transmission

a vacuum bag for autoclave or with an upper metal tool plate for heated press. Elevated temperature induces a chemical reaction in the resin, changing it from a viscous liquid to a viscoelastic solid.

Pressure is used to squeeze out excess resin, compact the plies and minimize void content. The term *cure cycle* refers to the amount of temperature and pressure the layup is exposed to and over what period of time. The appropriate cure cycle must be used in order to ensure chemical and mechanical properties of the final product. The appropriate cure cycle is either chosen from trial and error with the same fiber resin system or through intelligent processing[61] which involves continual adjustment of the cure process model based on sensor output. Two techniques based on acoustic transmission and acoustic emission show practical value as sensors.

Combination Acoustic Emission and Acoustic Transmission Technique

Acoustic transmission is a term that describes the use of a pulser on one side of the layup and a receiver on the opposite side. In principal, the stress wave induced by the pulser is influenced by changes in the medium between it and the receiver. In the case of cure monitoring, the amount of amplitude and energy detected at the receiver is proportional to the resin viscosity.[62-64] The technique is also called *ultrasonic attenuation*.

Figure 51 illustrates a test setup for measuring acoustic transmission. The pulser and receiver are mounted in tool plates above and below the layup. The pulser is excited with a short duration electrical pulse that in turn excites a piezoelectric crystal and launches a broadbanded stress wave. The stress wave, made up of both shear and longitudinal components, propagates through the layup and is detected by the receiver on the opposite side. The receiver also contains a piezoelectric crystal that produces a voltage signal proportional to the amount of energy it receives.

Acoustic Emission Technique

Acoustic emission involves a receiver only. Excitation of the receiver is the result of spontaneously generated stress waves from sources such as matrix cracking and delamination. Stress waves from these and other sources can travel a significant distance within the layup so that a single receiver can monitor a relatively large volume.

During cool-down of the layup, the fiber and resin exhibit significant residual stresses because of their different coefficients of thermal expansion. If the layup is cooled too rapidly, the residual stress is relieved through macrofracture mechanisms such as through-lamina cracks.[63] Conversely, if the layup is cooled too slowly, the residual stress will be

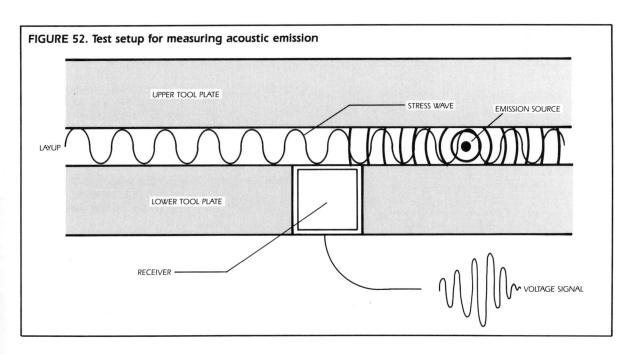

FIGURE 52. Test setup for measuring acoustic emission

locked within the finished part, causing dimensional instability.[65] The optimum cooling rate allows residual stress to be relieved through microfracture mechanisms (such as matrix cracking) distributed throughout the volume of the part. Acoustic emission is also generated during the heat-up portion of the cure cycle. Among the sources that produce emission during heat-up is outgassing.[66]

Figure 52 illustrates a test setup for measuring acoustic emission. Note that it is identical to Fig. 51, with the pulser removed. During heat-up, acoustic emission data are collected in the time period between the acoustic transmission pulses. They are collected continuously during cool-down, when the acoustic transmission pulser is not used.

Experimental Setup, Procedure and Results

Figure 53 illustrates the components for measuring acoustic transmission during heat-up and acoustic emission during cool-down. A list of components and settings is found in Table 4.

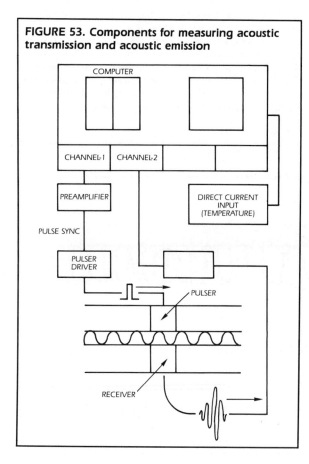

FIGURE 53. Components for measuring acoustic transmission and acoustic emission

TABLE 4. Components for measuring acoustic transmission and acoustic emission during cure and cool-down

Component	Description
Pulser driver	100 volt output 1 pulse per second
Pulser	150 kHz resonance −70 dB peak sensitivity 550 °C (1,000 °F) maximum temperature
Receiver	150 kHz resonance −70 dB peak sensitivity 550 °C (1,000 °F) maximum temperature
Layup	Graphite polyimide 5 ply laminate
Pulser preamplifier	40 dB gain 150 kHz bandpass filter Used to decouple 28 V power and amplify the pulse sync trigger into channel-1 for acoustic transmission measurement
Receiver preamplifier	40 dB gain 150 kHz bandpass filter Used to amplify the received signal and drive a long cable
Computer	Used for data display in real-time and for storage of data on floppy disks; recorded features of each signal include peak amplitude, energy, counts, rise time, duration and velocity timing
DC Input	Used to input a parametric voltage into the data set to represent temperature (manually adjusted)

Signal	Channel	Gain	Frequency	Threshold
Cure	channel-1	20 dB	100 to 300 kHz	1 V
	channel-2	20 dB	100 to 300 kHz	1 V
Cool down	channel-1	not used		
	channel-2	30 dB	100 to 300 kHz	1 V

A typical cure cycle is illustrated in Fig. 54. The pulser is activated during heat-up to measure acoustic transmission. During cool-down the pulser is turned off to measure only acoustic emission.

Typical rheometric analysis of graphite polyimide is illustrated in Fig. 55. Note that two dips in the viscosity curve are shown. The first dip is *imidization* and the second dip is *polymerization and crosslinking*.

Figure 56 illustrates typical acoustic transmission curves, measured as energy counts, during heat-up. Note that the acoustic transmission curve follows the expected viscosity curve for graphite polyimide indicated in Fig. 55. Variation in the acoustic transmission (viscosity) curve in Fig. 56 is expected because of differences in the manually controlled temperature and pressure adjustments. The spurious data (not shown) below the acoustic transmission (viscosity) curve are acoustic emission from sources such as outgassing. No further evaluation of these sources was performed.

Figures 57 and 58 illustrate the acoustic emission detected during cool-down of two samples. The rate of cool-down was controlled using water and air flowing through the heater plates. The rapidly cooled sample produced a significantly higher rate of emission indicative of macrofracture mechanisms in the part. The slowly cooled sample produced a steady rate of emission indicative of controlled stress relief through microfracture mechanisms.

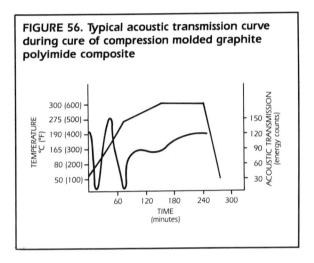

FIGURE 56. Typical acoustic transmission curve during cure of compression molded graphite polyimide composite

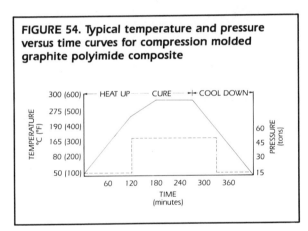

FIGURE 54. Typical temperature and pressure versus time curves for compression molded graphite polyimide composite

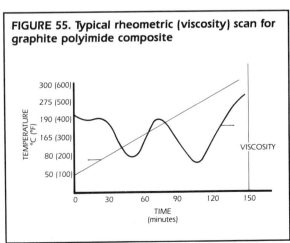

FIGURE 55. Typical rheometric (viscosity) scan for graphite polyimide composite

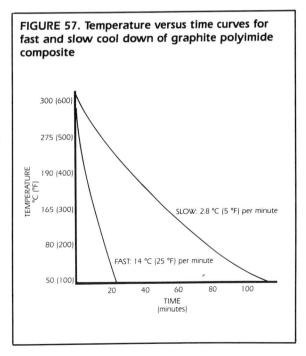

FIGURE 57. Temperature versus time curves for fast and slow cool down of graphite polyimide composite

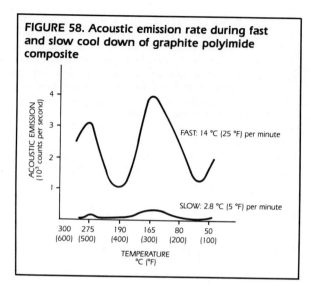

FIGURE 58. Acoustic emission rate during fast and slow cool down of graphite polyimide composite

Conclusion

Acoustic transmission is a sensitive, qualitative measurement of resin viscosity that can be economically added to production cure processes. The acoustic transmission (viscosity) measurement can be used in real-time to modify the cure process model so as to optimize quality and increase production.

During cool-down, acoustic emission is a qualitative measure of fracture mechanisms. Acoustic emission can be used to control residual stress, prevent dimensional instability and improve quality.

REFERENCES

1. Yee, K.W. and D.S. Blomquist. *An On-Line Method of Determining Tool Wear by Time-Domain Analysis.* Technical paper MR82-901. Dearborn, MI: Society of Manufacturing Engineers (1982).
2. Yee, K.W. and D.S. Blomquist. "Checking Tool Wear by Time Domain Analysis." *Manufacturing Engineering.* Vol. 88, No. 5 (1982): pp 74-76.
3. Yee, K.W. *A Guide for the Construction and Operation of Drill-Up.* NBSIR 82-2590. Gaithersburg, MD: National Bureau of Standards (October 1982).
4. Beer, Larry D. *Power Consumption — A Measure of Tool Performance.* Technical paper MR79-398. Dearborn, MI: Society of Manufacturing Engineers (1979).
5. Fabris, N.S. and R.K. Podder. "Optimization of Drill Life: Influence of Cutting Conditions on Tool Wear." *Proceedings of the Tenth North American Manufacturing Research Conference.* Dearborn, MI: Society of Manufacturing Engineers (1982).
6. Braun, S., E. Lenz and C.L. Wu. "Signature Analysis Applied to Drilling." Paper 81-DET-9. *Proceedings of the Design Engineering Technical Conference.* New York, NY: American Society of Mechanical Engineers (September 1981).
7. Micheletti, G.F., S. Rosetto and M. Ponti. "Tool Vibration Pattern and Tool Life on Automatic Screw Machine." *Advances in MTDR.* Vol. A (1970): pp 145-159.
8. Young, F.W. *An Investigation of Available Signals for Adaptive Control Machine Tools.* MS thesis. Boston, MA: Massachusetts Institute of Technology (May 1970).
9. Gaudreau, M. *An On-Line Technique for Tool Wear Measurement.* MS thesis. Boston, MA: Massachusetts Institute of Technology (June 1975).
10. Edwin, A. and T. Vlach. "A New Approach to Tool Wear Monitoring." *Proceedings of the Twenty-Seventh Annual Conference and Exhibitors.* Research Triangle Park, NC: Instrument Society of America (October 1972).
11. Weller, E.J. and B. Welchbrodt. *Listen to Your Tools — They're Talking to You.* Technical paper MR67-444. Dearborn, MI: Society of Manufacturing Engineers (1967).
12. Weller, E.J., H.M. Schrier and B. Welchbrodt. "What Sound Can be Expected from a Worn Tool?" *Journal of Engineering for Industry.* New York, NY: American Society of Mechanical Engineers (August 1969).
13. Lutz, J. *Process Measuring Wear on a Drilling Tool.* US Patent 3,714,822 (1973).
14. Private communication. Augusta, GA: TRW Drill and End Mill Division (October 1981).
15. *Machine Data Handbook.* Cincinnati, OH: Metcut Research Associates (1972).
16. Dornfeld, D.A. and M.S. Lan. "Chip Form Detection Using Acoustic Emission." *Proceedings of the Eleventh North American Manufacturing Research Conference.* Dearborn, MI: Society of Manufacturing Engineers (1983): pp 386-389.
17. Dornfeld, D.A. and C.S. Pan. "Study of Continuous/Discontinuous Chip Formation Using Acoustic Emission Signal Analysis." *Proceedings of the ASM Metals Congress.* Metals Park, OH: American Society for Metals (1984).
18. Yee, K.W., D.S. Blomquist, D.A. Dornfeld and C.S. Pan. "An Acoustic Emission Chip-Form Monitor for Single-Point Turning." *Proceedings of the Twenty-Sixth Machine Tool Design and Research Conference.* Manchester, England: University of Manchester Institute of Science and Technology (1986): pp 305-312.
19. Nakamura, S., et al. *Chip Control in Turning.* Technical paper MR82-235. Dearborn, MI: Society of Manufacturing Engineers (1982).
20. Dornfeld, D. "Acoustic Emission and Metalworking — Survey of Potential and Examples of Applications." *Proceedings of the Eighth North American Manufacturing Research Conference.* Dearborn, MI: Society of Manufacturing Engineers (May 1980).
21. Kannatey-Asibu, E. and D. Dornfeld. "Quantitative Relationships for Acoustic Emission from Orthogonal Metal Cutting." *Journal of Engineering for Industry.* Vol. 103, No. 3. New York, NY: American Society for Manufacturing Engineers (1981): pp 330-340.
22. Kannatey-Asibu, E. and D. Dornfeld. "A Study of Tool Wear Using Statistical Analysis of Metal Cutting Acoustic Emission." *Wear.* Vol. 76, No. 2: pp 247-261.
23. Dornfeld, D. and E. Diei. "Acoustic Emission from Simple Upsetting of Solid Cylinders." *Journal of Engineering Materials Technology.* Vol. 104, No. 2 (1982): pp 145-152.
24. Dornfeld, D. "Investigation of Machining and Cutting Tool Wear and Chatter Using Acoustic Emission." *Proceedings of the AF/DARPA Review in Progress in Quantitative NDE.* New York, NY: Plenum Press (August 1981): pp 475-483.
25. Dornfeld, D. "Investigation of Metal Cutting and Forming Process Fundamentals and Control Using

Acoustic Emission." *Proceedings of the Tenth NSF Conference on Production Research and Technology*. National Science Foundation (March 1983).

26. Okada, A. "Automation in Arc Welding — State of the Art and Problems." *Journal of the Japanese Society of Mechanical Engineers* (March 1979).

27. Micheletti, G., W. Koenig and H. Victor. "In Process Tool Wear Sensors for Cutting Operations." *Annals of the CIRP*. Vol. 25 (1976): pp 483-496.

28. Birla, S. "Sensors for Adaptive Control and Machine Diagnostics." *Machine Tool Task Force Report*. Vol. 4. Livermore, CA: Lawrence Livermore National Laboratory (October 1980): pp 712.

29. Lan, M. and D. Dornfeld. "Experimental Studies of Tool Wear via Acoustic Emission Analysis." *Proceedings of the Tenth North American Manufacturing Research Conference*. Dearborn, MI: Society of Manufacturing Engineers (May 1982).

30. Lan, M. and D. Dornfeld. "In Process Tool Fracture Detection." *Journal of Engineering Materials and Technology*. New York, NY: American Society of Manufacturing Engineers (1983).

31. Dornfeld, D. and M. Lan. "Chip Form Detection Using Acoustic Emission." *Proceedings of the Eleventh North American Manufacturing Research Conference*. Dearborn, MI: Society of Manufacturing Engineers (May 1983).

32. Amitay, G., S. Malkin and Y. Koren. "Adaptive Control Optimization of Grinding." *Journal of Engineering for Industry*. Vol. 103, No. 1. New York, NY: American Society of Manufacturing Engineers (1981): pp 103-108.

33. Dornfeld, D. and H. Cai. "An Investigation of Grinding and Wheel Loading Using Acoustic Emission." *Journal of Engineering for Industry*. Vol. 106, No. 1. New York, NY: American Society of Manufacturing Engineers (1984): pp 28-22.

34. Diei, E. and D. Dornfeld. "Acoustic Emission Sensing of Tool Wear in Face Milling." *Journal of Engineering for Industry*. New York, NY: American Society of Manufacturing Engineers (1986). Vol. 109, No. 3 (1987): pp 234-240.

35. Diei, E.N. *Investigation of the Milling Process Using Acoustic Emission Signal Analysis*. PhD thesis. Berkeley, CA: University of California (February 1985).

36. Lan, M.S. and D.A. Dornfeld. "In-Process Tool Fracture Detection." *Journal of Engineering for Industry*. Vol. 106, No. 2. New York, NY: American Society of Manufacturing Engineers (1984): pp. 111-118.

37. Dornfeld, D. and M.S. Lan. "Chip Form Detection Using Acoustic Emission." *Proceedings of the Eleventh North American Manufacturing Research Conference*. Dearborn, MI: Society of Manufacturing Engineers (May 1983): pp 386-389.

38. Pan, C.S. and D. Dornfeld. "Determination of Chip Forming States Using Linear Discriminant Function Technique with Acoustic Emission." *Proceedings of the Thirteenth North American Manufacturing Research Conference*. Dearborn, MI: Society of Manufacturing Engineers (May 1985): pp 299-303.

39. Yee, K., D. Blomquist, D. Dornfeld and C.S. Pan. "An Acoustic Emission Chip-Form Monitor for Single Point Turning." *Proceedings of the Twenty-Sixth International Machine Tool Design and Research Conference*. Manchester, England: University of Manchester Institute of Science and Technology (September 1986).

40. Dornfeld, D.A. "Acoustic Emission Process Monitoring for Untended Manufacturing." *Proceedings of the Japan-US Symposium on Flexible Automation*. Osaka, Japan: JAACE (July 1986): pp 831-836.

41. Matsushima, K. and T. Sata. "Development of the Intelligent Machine Tool." *Journal of the Faculty of Engineering*. Vol. 35, No. 3. Tokyo, Japan: University of Tokyo (1980).

42. Birla, S. "Sensors for Adaptive Control and Machine Diagnostics." *Machine Tool Task Force Study: Machine Tool Controls*. G. Sutton, ed. Vol. 4. Dearborn, MI: Society of Manufacturing Engineers (October 1980): pp 7.12.1-1.12.70.

43. Tlusty, J. and C. Andrews. "A Critical Review of Sensors for Unmanned Machining." *Annals of CIRP*. Vol. 32, No. 2 (1983): pp 563-572.

44. Suh, N. "The Future of the Factory." *Robotics and Computer-Integrated Manufacturing*. Vol. 1, No. 1 (1984): pp 47.

45. Spur, G. "Growth, Crisis and Future of the Factory." *Robotics and Computer-Integrated Manufacturing*. Vol. 1, No. 1 (1984): pp 21.

46. Tomizuka, M., J. Oh and D. Dornfeld. "Model Reference Adaptive Control of the Milling Process." *Control of Manufacturing Processes and Robotic Systems*. D. Hardt and W. Book, eds. New York, NY: American Society of Mechanical Engineers (1983): pp 55-64.

47. Iwata, K. and T. Moriwaki. "An Application of Acoustic Emission to In-Process Sensing of Tool Wear." *Annals of CIRP*. Vol. 26, No. 1 (1977): pp 21-26.

48. Moriwaki, T. "Detection for Cutting Tool Fracture by Acoustic Emission Measurement." *Annals of CIRP*. Vol. 29, No. 1 (1980): pp 35-40.

49. Kannatey-Asibu, Jr., E. and D. Dornfeld. "Quantitative Relationships for Acoustic Emission from Orthogonal Metal Cutting." *Journal of the Engineering Industry*. Vol. 103, No. 3. New York, NY: American Society of Mechanical Engineers (1981): pp 330-340.

50. Kannatey-Asibu, Jr., E. and D. Dornfeld. "A Study of Tool Wear Using Statistical Analysis of Metal Cutting Acoustic Emission." *Wear*. Vol. 76, No. 2 (1982): pp 247-261.

51. Dornfeld, D. and M. Lan. "Chip Form Detection Using Acoustic Emission." *Proceedings of the Eleventh North*

American Manufacturing Research Conference. Dearborn, MI: Society of Manufacturing Engineers (May 1983): pp 386-389.
52. Diei, E. and D. Dornfeld. "Acoustic Emission Sensing of Tool Wear in Face Milling." *Journal of Engineering for Industry*. New York, NY: American Society of Manufacturing Engineers (1986). Vol. 109, No. 3 (1987): pp 234-240.
53. Pan, C. and D. Dornfeld. "Modeling the Diamond Turning Process with Acoustic Emission for Monitoring Applications." *Proceedings of the Fourteenth North American Manufacturing Research Conference*. Dearborn, MI: Society of Manufacturing Engineers (1986).
54. Dornfeld, D. "Acoustic Emission Monitoring and Analysis of Manufacturing Processes." *Proceedings of the Twelfth NSF Conference on Production Research and Technology*. National Science Foundation (1985).
55. Kakino, Y. "In-Process Detection of Tool Breakage by Monitoring Acoustic Emission." *Proceedings of the International Conference on Cutting Tool Materials*. Fort Mitchell, KY (September 1980): pp 29-43.
56. Dornfeld, D. "Acoustic Emission and Metalworking — Survey of Potential and Examples of Applications." *Proceedings of the Eighth North American Manufacturing Research Conference*. Dearborn, MI: Society of Manufacturing Engineers (May 1980): pp 207-213.
57. Rouby, D., P. Fleischmann and F. Gobin. "An Acoustic Emission Source Model Based on Dislocation Movement." *Internal Friction and Ultrasonic Attenuation in Solids*. Tokyo, Japan (1977).
58. Kalpakjian, S. *Manufacturing Processes for Engineering Materials*. Reading, MA: Addison-Wesley Publishing (1984): pp 429.
59. Dornfeld, D. "Investigation of Machining and Cutting Tool Wear and Chatter Using Acoustic Emission." *Proceedings of the AF/DARPA Review of Progress in Quantitative NDE*. New York, NY: Plenum Press (August 1981): pp 475-483.
60. Dornfeld, D. and H. Cai. "An Investigation of Grinding and Wheel Loading Using Acoustic Emission." *Journal for the Engineering Industry*. Vol. 106, No. 1. New York, NY: American Society of Mechanical Engineers (1984): pp 28-33.
61. Servais, R., C. Lee and C. Browning. "Intelligent Processing of Composite Materials." *SAMPE Journal* (September-October 1986).
62. Houghton, W., R. Shufford and J. Sprouse. "Acoustic Emission as an Aid for Investigating Composite Manufacturing Processes: Materials and Processes for the Eighties." *National SAMPE Technical Conference Series*. Vol. 11 (November 1979).
63. Hinton, Y., R. Shufford and W. Houghton. "Acoustic Emission During Cure of Fiber Reinforced Composites." *Proceedings of the Critical Review: Techniques for the Characterization of Composite Materials*. AMMRC-MS-82-3 (May 1982).
64. Brown, S., P. Jarrett and T. Rose. *In-Situ Cure Monitoring of 2-D Carbon Fiber Reinforced Phenolic Composites*. Sacramento, CA: Aerojet Strategic Propulsion Company.
65. Delacy, T. *Method and Apparatus for Nondestructive Testing of Composite Materials*. US Patent 4,494,904 (January 1985).
66. Wadin, J. "Monitoring of Outgassing During a Graphite Epoxy Cure Cycle." *Proceedings of the Nineteenth Acoustic Emission Working Group Meeting* (December 1978).

SECTION 15

SURVEY OF COMMERCIAL SENSORS AND SYSTEMS FOR ACOUSTIC EMISSION TESTING

Keith Mobley, Technology for Energy Corporation, Knoxville, Tennessee

Carol Bailey, Technology for Energy Corporation, Knoxville, Tennessee
Kevin Bierney, Acoustic Emission Associates, San Clemente, California
Harold Berger, Industrial Quality Incorporated, Gaithersburg, Maryland
Mark Goodman, UE Systems, Elmsford, New York
John Judd, Vibra•Metrics, Hamden, Connecticut
Charles Lewis, Staveley Technologies, East Hartford, Connecticut
John Rogers, Acoustic Emission Technology Corporation, Sacramento, California
Stewart Slykhous, Dunegan Corporation, Irvine, California

INTRODUCTION

For reasons of objectivity, the *Nondestructive Testing Handbook* rarely contains trade names — such information is deleted from manuscripts as a matter of routine. At the inception of this book, however, the Acoustic Emission Method Committee of ASNT decided to include the following text, despite its commercial references. This exception was warranted by: (1) a desire to inform interested readers about all aspects of the acoustic emission testing industry; (2) the comparative newness of acoustic emission techniques; and (3) the correspondingly limited number of American manufacturers.

Twenty manufacturers of acoustic emission and acoustic signature equipment were contacted in order to get a broad base of information for this chapter. The following text comprises all the contributions received on the subject.

This material is intended to provide a basic familiarity with the types of systems commercially available for acoustic analysis. It does not include information for all commercially available systems nor does it address all of the applications for acoustic analysis; it does, though, provide a representative cross section of both.

The vendor information given here includes instrumentation that covers a range of noise analysis methodologies, including acoustic emission, acoustic signature, ultrasonic and vibration analysis techniques. The fundamental difference between these methods is the frequency range of the acoustic signature being analyzed (in some cases, the frequencies overlap from one method to another). For purposes of this chapter, the four techniques can be segregated as follows.

Vibration signature analysis is normally used to identify degradation or the failure mode of rotating or reciprocating machine trains. This technique typically evaluates frequencies between 0.1 Hz and 20 kHz and uses a combination of displacement, velocity and acceleration transducers.

Ultrasonic signature analysis evaluates frequency ranges between 1.0 kHz and 1.0 MHz. Because of these higher frequencies, ultrasonic analysis uses acceleration transducers (accelerometers).

Acoustic signature analysis is the interpretation of sound energy emitted by an object in order to indicate degradation in structures. Often, ultrasonic accelerometers are used to acquire data in the 1.0 kHz to 1.0 MHz range for this kind of analysis.

Acoustic emission testing, unlike the other noise analysis methods, uses the characteristics of acoustic energy emitted by stressed test objects to detect and perhaps locate structural or material discontinuities in the objects.

With the exception of vibration analysis, these techniques use versions of the accelerometer (defined later in the chapter) to convert the noise profile into an equivalent electrical signal. Conditioning and the method of converting and analyzing the acquired signal are the other distinct differences between the noise analysis methods.

The following text provides an overview of the types of sensors (accelerometers) that are commercially available for the four analysis techniques, and an overview of standard data acquisition and analysis systems that can be used for acoustic emission and acoustic signature tests.

PART 1
ACOUSTIC SENSORS, TRANSDUCERS AND ACCELEROMETERS

Piezoceramic Crystals Used in Acoustic Emission Sensors

Piezoceramic Formulations

Three basic piezoceramic types are used as sensing elements in common acoustic emission transducers: barium titanate (BaTiO); lead zirconate titanate [Pb(0.4Zr0.6Ti)O$_3$ to Pb(0.9Zr0.1Ti)O$_3$]; and modified lead metaniobate (PbNb$_2$O$_5$). These piezoceramic elements are made up from high purity chemical powders that are mixed and processed into sintered multicrystalline, isotropic materials having a microstructure of unbalanced, moveable dipoles.[1] As a result, the ceramics possess ferroelectric properties (see Table 1) and tend to change the orientation of their dipoles with the application of a DC potential.

Piezoelectric Properties

With the application of metallic, thick film electrodes and a large DC voltage in a polarizing process, the ferroelectric ceramics also become anisotropic in the polarizing field direction and exhibit piezoelectricity. Derived from the Greek *piezo* ("pressure from force"), *piezoelectricity* is the material property of producing an electrical charge when subjected to a mechanical force or a mechanical force when subjected to an electrical charge.

It is through the direct piezoelectric effect that these special piezoceramics become useful as acoustic emission sensors. Their relatively high conversion efficiencies allow them to receive very small mechanical displacements in a rapid and repetitive sequence, and to convert them to reasonably clear electrical outputs.

Electrical and Electromechanical Properties

In multicrystalline piezoceramics, the direction of the piezoelectric response depends on the direction of the polarizing field. In notation, specific piezoelectric property symbols are often expressed with subscripts, relating response direction to polarizing field direction. Likewise, many piezoelectric properties are elastic, relating stress and corresponding strain, so that two subscripts apply. In addition, a superscript is used to define the electrical boundary conditions on which the property is measured. Mechanical and electromechanical properties are published for each manufacturer's piezoceramic material. Below is a a review of the basic parameters considered in transducer design.[2]

Curie temperature (T_C): the temperature above which there is a transition from ferroelectric to paraelectric material phases, and a corresponding loss of all piezoelectric activity. The magnitude of the Curie temperature varies among materials and can range from 150 °C (300°F) to more than 350 °C (660 °F) for most commonly used piezoceramics. As

TABLE 1. Typical property ranges of piezoceramic materials at room temperature and low drive

	Lead Zirconate-Lead Titanate	Barium Titanate	Lead Metaniobate	Lead Titanate
Free dielectric constant K_{33}^T	600 to 3,500	700 to 1,500	185 to 750	190 to 280
Direct piezoelectric constant d_{33} (m/V · 10^{-12})	160 to 600	80 to 180	60 to 170	50 to 100
Reverse piezoelectric constant g_3 (Vm/N · 10^{-3})	19 to 28	12 to 20	24 to 34	30 to 36
Coupling coefficient squared k_{33}^2	0.47 to 0.49	0.19 to 0.25	0.20 to 0.32	0.33 to 0.40
Frequency constant N_3t (Hz·m)	1,850 to 2,100	2,000 to 2,800	1,580 to 2,570	1,950 to 2,200
Mechanical quality factor Q_m	70 to 1,100	100 to 600	150 to 750	400 to 850
Curie temperature TC (°C)	180 to 360	90 to 135	250 to 450	275 to 330

the Curie temperature is approached, a reduction in piezoelectric effect takes place because of a partial dipole domain disorientation.

Free dielectric constant (K_{33}^T): the ratio of the field applied in the 3-direction (poling direction) to the resulting dielectric displacement in the 3-direction under a boundary condition of no restraint. This constant is useful in selecting the best electrical capacitance match for the piezoceramic configuration when considering the electrical impedance considerations of the associated receiver electronics.

Direct piezoelectric constant (d_{33}): the ratio of the strain developed along the poling axis to the field applied along the poling axis with all other stresses being zero. When considered with the reverse piezoelectric constant, a selection can be made for relative sensitivity of the piezoceramic material.

Reverse piezoelectric constant (g_{33}): the ratio of the field developed along the poling axis to the stress applied along the same axis when all other stresses are zero.

Coupling coefficient squared (k_{33}^2): represents the ability of the ceramic to convert electrical energy to mechanical energy and vice versa. Also represents the ratio of transformed energy in the poling direction relative to electrical energy input to the electrodes of the ceramic.

Frequency constant (N_{3t}): the constant value for a given ceramic formulation in thin plate configuration thickness mode, defined as the resonant frequency multiplied by the controlling dimension (thickness). From the frequency constant, the sensor thickness is established for a given center frequency response.

Mechanical quality factor (Q_m): the ratio of reactance to resistance in the equivalent series circuit representing a mechanical vibrating resonant system. This value is affected by the geometric shape of the piezoceramic element and is also affected by the dielectric lossiness of the dipole domain structure. Generally, high Q_m ceramics tend to produce an adverse ringing phenomenon, giving a long train of electrical response to a single mechanical impulse.

Special Geometry Piezoceramic Element

The National Bureau of Standards has developed a single event, broadband (50 kHz to 1 MHz), high sensitivity, point contact transducer that is free of the resonant condition frequency responses of most commercial, flat, disk-shaped acoustic emission transducers. This transducer was designed to respond only to physical displacements that are normal to the contact surface.[3]

To achieve these qualities, a lead zirconate titanate receiving element was designed as a truncated cone. The piezoceramic element was first formed into a disk of about 6 mm (0.25 in.) diameter with a 2.5 mm (0.1 in.) thickness. The disk was lapped flat and parallel, electroded with fired silver metallization and polarized in the thickness direction. Both electrodes were then polished to a flatness of several lightbands of monochromatic light. Finally the disk was machined into the truncated cone shape, having a 1.0 mm (0.04 in.) tip diameter and 45 degree base angles (see Fig. 1).

Because the acoustic impedance of the ceramic was close to that of brass (about 35 rayls), a slug of about 25 mm (1 in.) diameter was also lapped to several lightbands flatness to serve as the acoustic backing on which to mount the truncated conical element. Both mating surfaces were thoroughly cleaned and a low viscosity, two-part electrically conductive epoxy formed a very thin, uniform bond line between the ceramic cone and brass backing.

Figure 2 shows the transducer's amplitude versus frequency response curve. Note that the curve holds flatly uniform over the frequency range from 100 kHz to 1.0 MHz. The conical transducer has proven comparably sensitive to most flat commercial acoustic emission transducers.

Because of the small tip diameter, this transducer is more delicate and more subject to damage than flat sensors. Recently, a modified design of this truncated cone transducer was produced with a 2.5 mm (0.1 in.) tip diameter, 5 mm (0.2 in.) height and a 6 mm (0.25 in.) base diameter. This modified design has reasonably uniform response qualities as shown in Fig. 3. The larger tip and thickness dimensions afford some protection against failure from rough handling but at the cost of a decrease in the response curve at higher frequencies.

The ceramic components for acoustic emission transducers, including the NBS design, are manufactured by the EBL Division of Staveley NDT Technologies, East Hartford, Connecticut.

Flat Frequency Sensor

A frequency response with an amplitude flat within ±3 dB over the frequency range 50 kHz to 1 MHz is possible in acoustic emission transducers. The transducer design has been successfully used in research work and is now commercially available. One valuable approach has been to use this transducer in a laboratory to determine frequency ranges typically emitted in a practical test and then to employ a conventional sensor with the frequency response peaked at the appropriate frequency.

This transducer is based on the design developed at the National Bureau of Standards (see above)[4] and the device is free of significant resonances in the working range. The contact area is 1 mm (0.04 in.) and this small size eliminated the aperture effect, its attendant nulls and losses of high frequencies. The wear plate and other resonant structures have been eliminated.

The conical piezoceramic element is intimately coupled to an impedance matched brass backing of extended physical size and this effectively eliminates resonances of the element itself. The large size and lossy character of the backing, in the radial and the axial directions, result in improved mechanical impedance match between the element and the backing. The combined effect of these design features is a transducer that faithfully reproduces displacement waveforms on the surface of a structure.

The transducer is furnished with a shield and a matching preamplifier that provides a low impedance (50 ohms) and permits the use of cables up to 15 m (50 ft). The use of the shield and battery power minimizes the possibilities of electrically induced interference. A photograph of the transducer assembly is shown in Fig. 4. Specifications of the sensor system are given in Table 2.

Applications for this novel transducer include: (1) acoustic emission monitoring of structural components; (2) comparison calibration (as a secondary standard for acoustic

FIGURE 1. Components of a conical acoustic emission transducer: (a) active element side view; (b) top view of backing; (c) active element top view; and (d) side view of backing

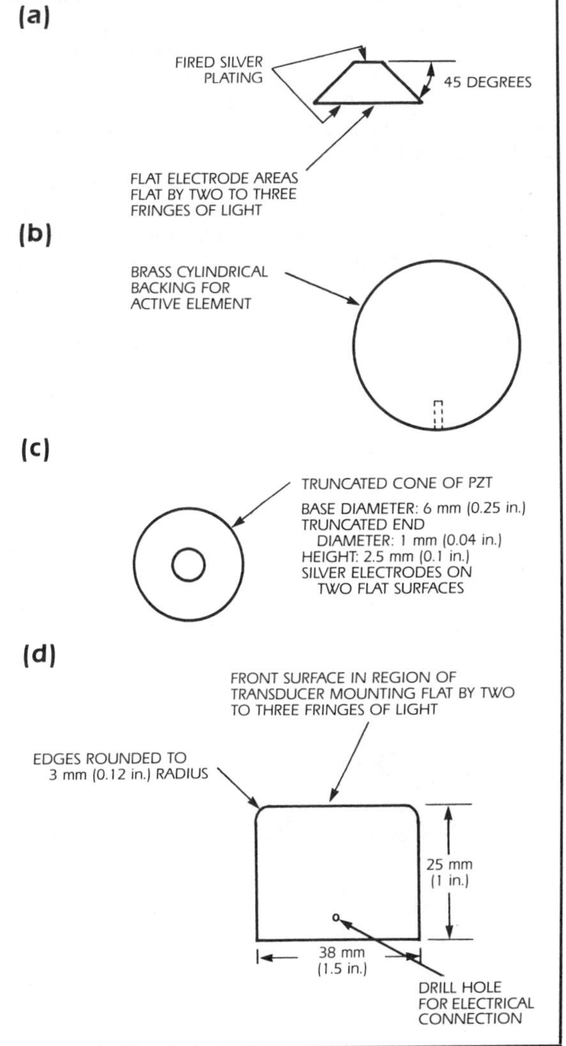

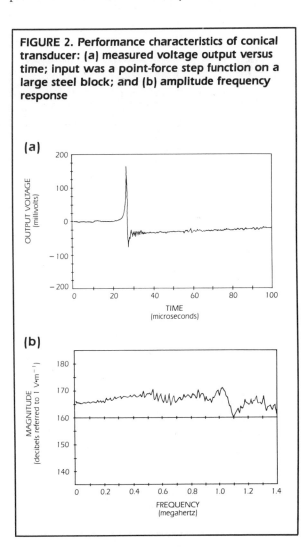

FIGURE 2. Performance characteristics of conical transducer: (a) measured voltage output versus time; input was a point-force step function on a large steel block; and (b) amplitude frequency response

emission or ultrasonic transducers); (3) calibration of complete acoustic emission measurement systems; (4) high fidelity ultrasonic detection in solids; and (5) calibration of complete ultrasonic spectroscopy systems.

The flat frequency conical transducer is available from Industrial Quality Incorporated, Gaithersburg, Maryland.

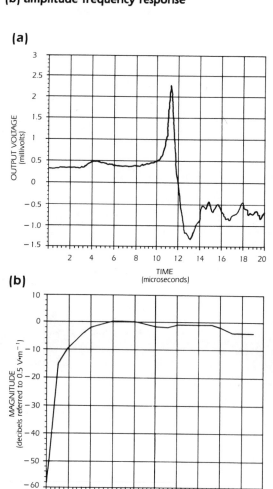

FIGURE 3. Performance characteristics for modified conical transducer: (a) measured voltage output versus time; input was a point-force step function on a steel block; and (b) amplitude frequency response

Vibration Transducers with Integral Electronics

A number of transducers and accelerometers are available in single or triaxial configurations with integral electronic circuitry. These devices are subject to the general characteristics of piezoelectric accelerometers but they differ in performance and use.

An integrated sensor includes impedance matching or conditioning electronic circuitry inside the transducer housing. This eliminates the need for an intermediate cable between the sensor and its impedance matching electronics. The interface between the sensing element and its electronics is controlled during manufacture and is not influenced by the user environment. Outputs are low impedance. Cables are conventional and do not appreciably affect transducer sensitivity or frequency response. Internal circuits are electronically protected against damage due to shock or reverse voltage. In their most basic form, these devices are reliable and simple to use.

Figure 5 illustrates the basic impedance matching, two-wire circuit normally used in the design. The crystal voltage developed across R is impressed on the gate of the mosFET. The mosFET is powered from a constant current DC supply of 1 to 12 mA and 15 to 28 V. The FET circuit will bias off at about 12 V in the quiescent state.

As the system vibrates, voltage develops across the crystal and is impressed on the gate of the mosFET. This voltage causes a linear increase or decrease in the impedance of the mosFET and this in turn causes the line bias voltage to change in a proportional noninverting manner. The voltage change is coupled out of the system through capacitor C. As

FIGURE 4. Conical dynamic surface displacement transducer; frequency response is flat within ±3 dB over the range of 50 kHz to 1 MHz

TABLE 2. Specifications for conical acoustic emission transducer

	Transducer Head	Shield, Cover and Preamplifier
Shape:	Cylindrical	Cylindrical
Dimensions:		
Diameter	44 mm (1.75 in.)	150 (6 in.)
Height	32 mm (1.13 in.)	73 mm (2.88 in.)
Contact size:	1 mm (0.04 in.)	
Weight:	280 g (10 oz)	700 g (25 oz)
Material:	Brass, piezoceramic and plastic	Aluminum with electronic circuit
Electrical:		
Power supply	9 V battery	
Output impedance	50 ohms (BNC connector)	
Maximum output voltage	2 V peak to peak	
Response:	Amplitude response flat within ±3 dB from 50 kHz to 1 MHz	
Displacement sensitivity:	2×10^8 V·m^{-1} (nominal)	

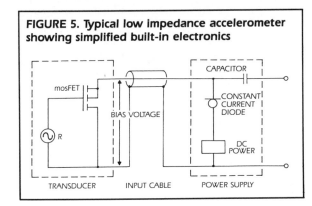

FIGURE 5. Typical low impedance accelerometer showing simplified built-in electronics

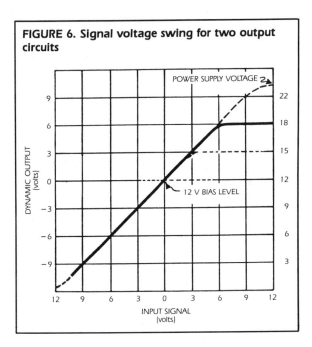

FIGURE 6. Signal voltage swing for two output circuits

in a conventional voltage amplifier, the low frequency corner is controlled by the value of R and the internal capacity of the piezoelectric crystal.

The gain of a typical system is about 0.95 and is linear within ±2 percent over most of its range. A crystal sensitivity of 10.5 mV·G^{-1} produces a 10 mV·G^{-1} output. The sensitivity is virtually independent of power supply voltage and cable length. The internal matching impedance is normally chosen to yield an RC time constant of 0.5 or longer, to avoid influencing low frequency response of the system.

Output Bias

Under normal operation, a bias voltage appears from the output signal lead to ground. There are two basic mosFET configurations in use. One exhibits a 7 to 8 V bias and the second a 9 to 12 V bias. Operation of the two circuits is identical except for the available signal swing. The low voltage version typically exhibits a lower wide band noise floor (10 to 20 µV versus 15 to 30 µV) and higher output impedance (1,000 ohms versus 100 ohms). Available signal voltage swing for the two circuits is shown in Fig. 6.

The bias voltage level changes if the cable to the power supply is broken or shorted and this provides a convenient means for monitoring cable integrity and proper mosFET operation. A coupling capacitor is used in the power supply output circuit to block the DC bias component of the output signal.

Dynamic Range

The dynamic measurement range of these integrated transducers is determined by several factors, including: (1) sensitivity; (2) bias voltage level; (3) power supply voltage; and (4) noise floor. For example, a system has the following parameters: (1) sensitivity of 100 mV·G^{-1}; (2) bias voltage level of 12 V; (3) power supply of 22 V; and (4) a noise floor of 20 μV rms wide band.

If the desired wide band signal-to-noise ratio is 3:1, then the minimum measurable signal is 60 μV (wide band is typically 5 Hz to 25 kHz). The minimum acceleration is the voltage (60 × 10^{-6} V rms) divided by the sensitivity (100 × 10^{-3} V·G^{-1}) or 6 × 10^{-4} G rms. This minimum can be reduced if the bandwidth of the measurement is reduced.

The maximum measured signal is the maximum linear voltage swing (9 V from Fig. 6) divided by the accelerometer sensitivity (100 × 10^{-3} V) or 90 G. Note that if the power supply voltage is reduced, the maximum positive voltage is reduced, making the limits asymmetric.

Low Frequency Response and Time Constant

Low frequency response is determined by the distributed capacity of the sensing crystal and the input resistance of the mosFET amplifier circuit. Low frequency corners of a fraction of a hertz are easily achieved but practical considerations require that input RC time constants be limited to allow bleed off of statically induced signals from pyroelectric or mechanical strain.

High Frequency Response and Cable Effects

High frequency response is usually determined by the manufacturer up to a specific frequency, typically 5 kHz or 10 kHz. Note that high frequency sensitivity is affected by insertion loss (due to cable resistance) and current limiting (due to power supply and cable capacity). Figure 7 illustrates the response of a mosFET circuit with a 50 ohm output impedance and various cable lengths. Curves show response with 2 mA and 12 mA constant current at differing drive levels.

Noise Immunity

The general rules of good instrumentation practice apply when using low impedance integrated electronic sensors. The immunity to airborne electrical noise is improved over conventional open crystal sensors by the inverse ratio of the output impedance.

While ground isolation is required to avoid ground loop errors, care should be taken not to use internally isolated units. It is important that external Faraday shielding be maintained around the high impedance interface between the crystal and its mosFET electronics (including the signal common coaxial return shield). When properly done, the

FIGURE 7. High frequency response of a mosFET circuit with 50 ohm output impedance and various cable lengths: (a) 2 mA constant current source; and (b) 12 mA constant current source

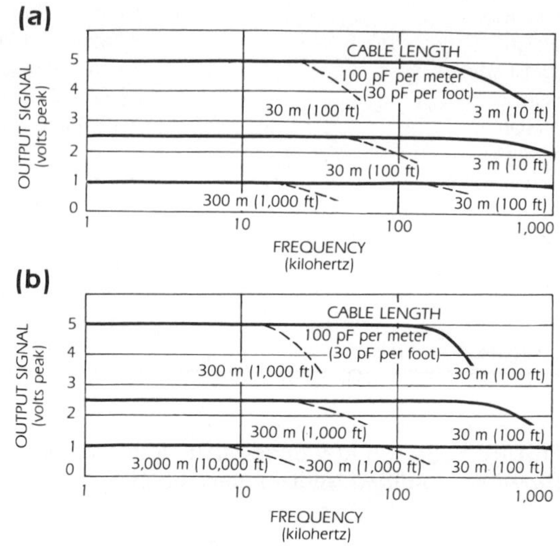

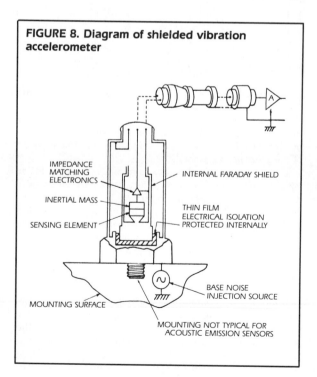

FIGURE 8. Diagram of shielded vibration accelerometer

system's immunity to airborne and base-coupled noise is significantly improved. Figure 8 shows a system with proper shielding.

Steady State Temperature Effects

Integrated electronic transducers are limited to temperature environments of about -50 to $120\ °C$ (-58 to $250\ °F$) for reliable continuous operation. These limits are imposed by the characteristics of the electronic components. Changes in sensitivity over this operating range are typically less than ± 5 percent.

Internal Temperature Compensation

The ideal sensor has a constant conversion sensitivity independent of temperature within its operating range. Most piezoelectric materials exhibit a change in their sensitivity with temperature. Internal electronics may be included that automatically compensate for these changes in sensor temperature. Figure 9 is the equivalent circuit of the piezoelectric accelerometer and its internal crystal capacitance.

For many of the modern ceramic materials, the value of internal crystal capacitance increases as the temperature increases. The temperature compensating capacitive element is made up of modified lead zirconate material and decreases as a function of temperature. With proper choice of the compensation element, the changes in voltage sensitivity may be held to less than 2 percent from ambient to $120\ °C$ ($250\ °F$). Figure 10 illustrates a typical variation in voltage sensitivity from ambient to $120\ °C$ ($250\ °F$), before and after compensation.

Transient Temperature Effects

When using accelerometers with internal electronics, it is important to note that temperature transients can cause shifts in operating bias level due to quasistatic pyroelectric voltages. Bias shifts may produce reduced dynamic range and this may lead to signal clipping.

In severe cases, internal chip saturation may occur, resulting in complete loss of data. It is therefore advisable to avoid exposing accelerometers to direct hot or cold air. A simple covering of lightweight foam or the thermal boots provided by most manufacturers can often eliminate the problems of transient temperatures.

Effects of Moisture, Shock and Cabling

The effect of moisture or other noncorrosive contaminants is minimal because of the low impedance outputs of this sensor's amplifier. For this reason, these units make excellent devices for use outdoors, in underwater environments or in industrial monitoring applications. In cases where prolonged exposure to moisture is expected, sealing of connectors or cables may be required. It is recommended that manufacturers be consulted for specific applications.

No special noise treated cables are required. Signals may be run over any conventional instrumentation grade two-wire, twisted pair shielded or coaxial cable. Modern integral electronic accelerometers include protection circuits that preclude electronic damage due to use beyond specified range or accidental dropping.

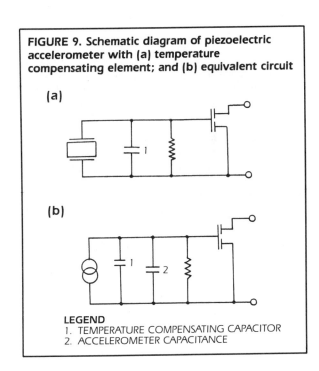

FIGURE 9. Schematic diagram of piezoelectric accelerometer with (a) temperature compensating element; and (b) equivalent circuit

LEGEND
1. TEMPERATURE COMPENSATING CAPACITOR
2. ACCELEROMETER CAPACITANCE

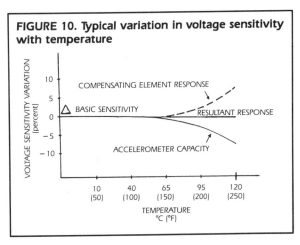

FIGURE 10. Typical variation in voltage sensitivity with temperature

Overload Recovery, Power Supply Effects and Amplifiers

Recovery from clipping due to over-ranging is typically less than one millisecond. Recovery from quasistatic overloads that generate high DC bias shifts are controlled by the accelerometer input RC time constant which is fixed during manufacture. Individual manufacturers normally provide this information, but it may be estimated by using the following:

$$\frac{1}{RC} = 2\pi f \qquad \text{(Eq. 1)}$$

where f is the low frequency 3 dB corner.

The nominal power supply voltage recommended by most manufacturers is 15 to 18 V. Units may be used with voltages up to 28 V. Sensitivity variations caused by voltage change are typically 0.05 percent per volt (see Fig. 11). Power supply ripple should be less than 1 mV rms. Some models may be damaged by reversed power supply polarity.

Almost any charge or voltage amplifier can be modified for use with low impedance accelerometers. Figure 12 illustrates a modification using a simple voltage divider tied to the regulated internal power supply of the charge amplifier. For unregulated supplies, a 15 V zener diode may be used. A zener diode also may be used to eliminate noise and electrical variations normally associated with 28 V aircraft power supplies (see Fig. 13).

Velocity Outputs

Some piezoelectric sensors with internal electronics provide output voltages directly proportional to velocity and acceleration. In addition to simplifying the user's instrumentation requirements, these transducers offer advantages over conventional self-generating velocity transducers in that they are smaller, have wider frequency response, better resolution, no moving parts and are virtually insensitive to pickup.

They do require a source of constant current power for activation. Figure 14 shows a transducer with the required power connections.

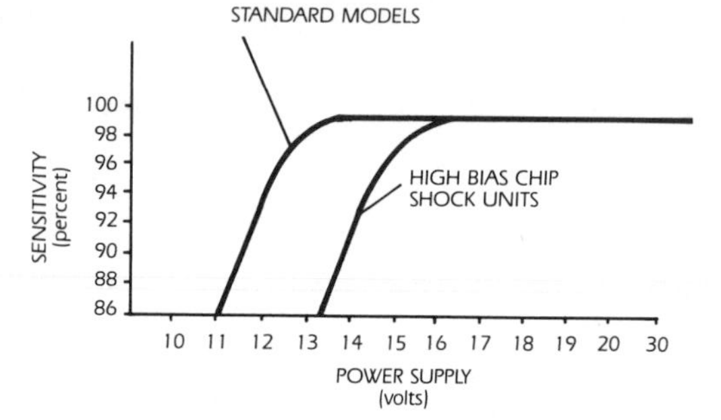

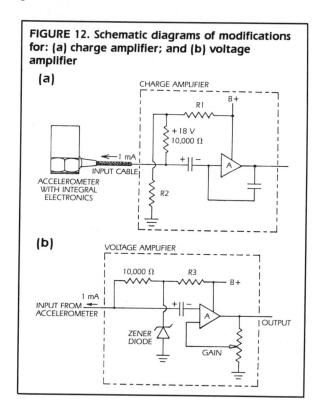

FIGURE 12. Schematic diagrams of modifications for: (a) charge amplifier; and (b) voltage amplifier

FIGURE 11. Typical variation in accelerometer sensitivity versus power supply voltage

FIGURE 13. Schematic diagram showing modification to accommodate aircraft (28 volt) DC power supply

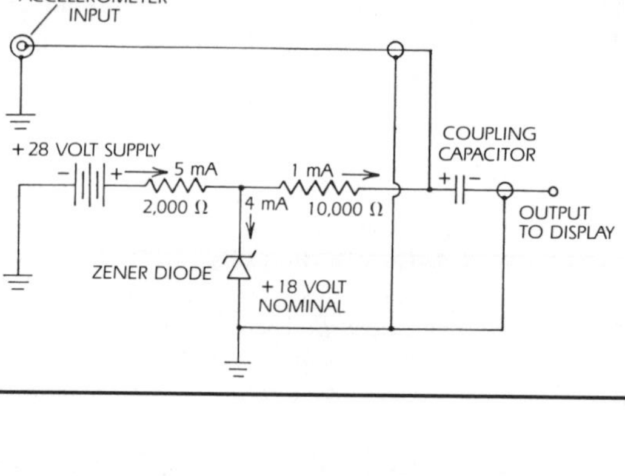

Filtering and Equalization

Figure 15 illustrates a piezoelectric sensor with an internal impedance matching FET circuit. The sensor is shown with both high pass and low pass components. These components are sometimes called *prefilters* because they precede the mosFET electronics.

The low pass filter may be used to equalize the natural increase in accelerometer sensitivity due to its own electromechanical resonance frequency response. This technique may be used to extend the high frequency limit to one-third of the accelerometer's resonance frequency.

The low pass filter may also be used to prevent unwanted high frequency from reaching the impedance matching FET (Fig. 16). This technique is particularly helpful when low level, low frequency vibration must be measured in the presence of high level, high frequency energy. The filter allows maximum sensitivity without exceeding the FET bias clipping level.

Internal Calibration and Testing

When sensors are used in large arrays, it is often desirable to test and calibrate individual sensors in place. Signals may be inserted to calibrate system recorders and to identify proper operation of the sensor's internal electronics. Methods include built-in oscillators operating at frequencies within the passband or at frequencies above the passband. The oscillators are activated as required. When the frequencies are above the passband, the oscillators run continuously and the signal is taken through special high pass filters to the gain calibration and test circuits.

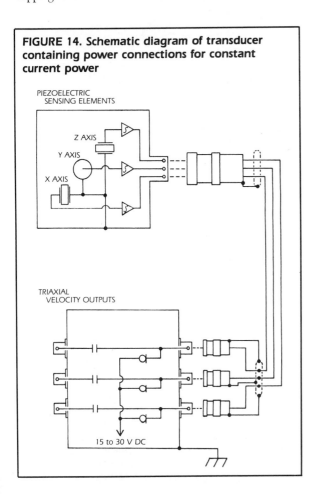

FIGURE 14. Schematic diagram of transducer containing power connections for constant current power

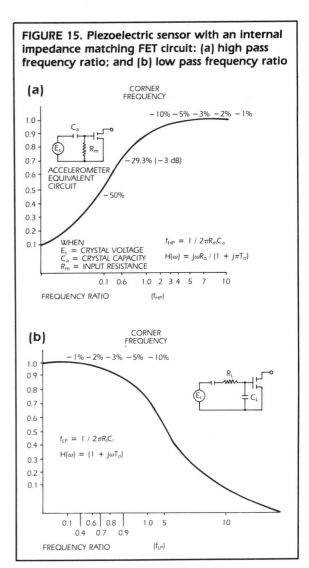

FIGURE 15. Piezoelectric sensor with an internal impedance matching FET circuit: (a) high pass frequency ratio; and (b) low pass frequency ratio

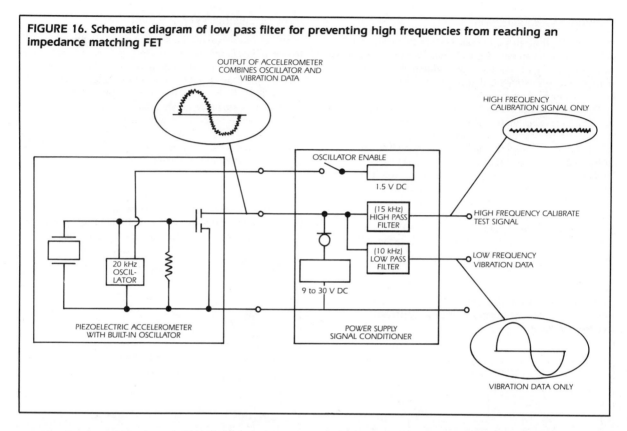

FIGURE 16. Schematic diagram of low pass filter for preventing high frequencies from reaching an impedance matching FET

Other techniques permit the calibration signal to be injected into the electronics from an external source (Fig. 17). These techniques verify that the crystal is connected, has the proper capacity and the correct shunt resistor. They also establish the proper operation of the FET and check the system connections. The techniques may also be used to check the system gain at various frequencies.

These accelerometers are produced by Vibra•Metrics Incorporated, in Hamden, Connecticut.

Sensors for Particular Applications

The sensors described below may be selectively used for either acoustic emission or acousto-ultrasonic applications. The sensor converts an acoustic emission stress wave propagating through the material under examination into a time-varying voltage signal for further processing. This manufacturer's sensors include: resonant anticoincident sensors; broadband sensors; miniature sensors; directional sensors; severe environment and high temperature sensors; integral preamplifier sensors; intrinsically safe sensors; and underwater sensors.

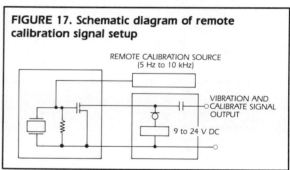

FIGURE 17. Schematic diagram of remote calibration signal setup

Standard AC Sensor Series

The standard anticoincident (AC) or differential sensor is available with frequency (peak) responses of 30, 75, 175, 375, 750, 1,000 and 1,500 kHz. The AC series sensors are highly sensitive and omnidirectional in response. When coupled with preamplifiers, each sensor provides 40 dB of common mode noise rejection, yielding a high signal-to-noise ratio. These AC series sensors are provided with a flush

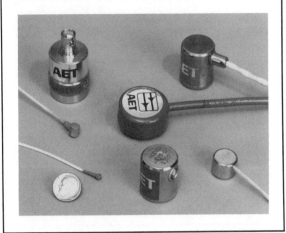

FIGURE 18. Acoustic emission testing sensors: AC series, miniature and broadband

side-mounted differential connector but can be supplied with other connector styles on request.

The operating temperature range is −215 to 160 °C (−355 °F to 325 °F), but can operate at 200 °C (400 °F) for brief periods. The sensors are made from lead zirconate titanate (PZT) material to ensure the highest sensitivity. The AC series sensors are fully shielded, including across the face, and have epoxy wear plates for electrical isolation.

Broadband Sensors

Two models of broadband sensors (single ended and differential) are shown in Fig. 18. These sensors are omnidirectional, highly sensitivity and produce nearly flat response from 100 kHz to 2 MHz, making them ideal for tests where frequency spectral analysis is required. The highly damped signal response makes these sensors ideal as pulsers and as receivers for impact testing of materials. The service temperature range of these sensors is −15 to 105 °C (5 to 225 °F). Both are completely shielded and have a ceramic wear plate for electrical isolation.

Miniature Sensors

The miniature sensors pictured in Fig. 18 are among the smallest, high sensitivity devices available. They are designed for use where a larger sensor might be inappropriate but where high performance is still required. All have an integral cable and some are completely encapsulated in an epoxy resin to provide an hermetic seal. The cable is normally supplied with a differential connector but other connectors are available on request. The sensors have special high flexibility cables.

Directional Sensors

Some sensors have a response that varies depending on the direction of the source from the sensor. One such sensor is 10 dB more sensitive to stress waves that strike the sensor along its long axis. Another is 10 dB more sensitive to stress waves that travel across its long axis.

Proper orientation of the sensor in relation to known noise sources, such as fretting in a complex structure, can ensure the operator that the stress wave originating from the known source will be detected at 10 dB less sensitivity. These sensors are differential, fitted with highly flexible integral cable and an epoxy wear isolation shoe. These sensors are hermetically sealed. Both sensors are resonant at 500 kHz and, because of their high sensitivity, are ideal for fatigue testing of materials.

Severe Environment and High Temperature Sensors

Special acoustic emission sensors have been widely and successfully used in long term monitoring of rotating machinery and have survived extended periods of use in petrochemical plants. Fully encapsulated, the sensor has an integral signal cable covered by a complete stainless steel braid and polyurethane outer covering. The sensor is single ended and resonant at 100 kHz.

High temperature sensors are manufactured to operate up to 260 °C (500 °F). These sensors are generally available in two frequencies: 175 kHz with a sensitivity of −46 dB referred to 1 $V \cdot Pa^{-1}$ (−66 dB referred to 1 $V \cdot ubar^{-1}$); and 375 kHz with a sensitivity of −48 dB referred to 1 $V \cdot Pa^{-1}$ (−68 dB referred to 1 $V \cdot ubar^{-1}$). Sensitivities are at room temperature.

Integral Preamplifier Sensors

These sensors are designed for field applications and have special use in applications where the placement and restraint of a preamplifier is difficult (see Fig. 19). Sensor/preamplifier combinations are available in 40, 60, 110 and 150 kHz versions. They can be used with 300 m (1,000 ft) of coaxial cable.

Remote pulsing for calibration is made possible through unique bypass circuitry contained in the integral preamplifier. The 40 dB preamplifier makes these sensors small and compact. The stainless steel housing protects the sensor and allows for easy bonding and removal. The case also ensures the user that the sensor is highly immune to electromagnetic interference. These sensors are repairable.

Intrinsically Safe Sensor

This sensor is certified to *EEx-ia-IIB-T4* (see Fig. 20). The *EEx* means that the device has passed both United

FIGURE 19. Integral preamplifier sensors (left to right): 40 kHz, 60 kHz and 150 kHz

FIGURE 20. Intrinsically safe front-end acoustic emission sensor

Kingdom and CENELEC (European) standards. The *ia* indicates that the sensor can be used in category *a* zone 0 locations where explosive air/gas mixture is either continually present or present for long periods. The *IIB* refers to the gas group most usually encountered and includes: acetylene, hydrogen, blue water gas, coke oven gas (depending on proportions) and carbon disulphide. The *T4* indicates that the sensor can be used in areas where the maximum surface temperature, either in normal operation or in fault condition, does not exceed 135 °C (275 °F).

This sensor was designed for use in areas where there is the possible presence of gas with an associated risk of explosion. The intrinsically safe front end is available in two housings: for magnetic attachment to ferromagnetic material and for a waveguide. The front end is also supplied with a zener barrier for voltage protection.

Underwater Sensor

An underwater sensor was first developed for use in the North Sea. This sensor is encased and potted to withstand 5.5 MPa (800 psi) maximum pressure and 700 kPa (100 psi) operating pressure. The preamplifier provides 40 dB of amplification and requires 28 V direct current for operation. Resonance of this sensor is at 325 kHz and sensitivity is -52 dB referred to 1 V·Pa^{-1} (-72 dB referred to 1 V·ubar^{-1}). The sensor has an integral cable coated with neoprene for protection. An underwater connector is available on the cable or a standard connector can be installed if the cable is run to the system.

Wheeled Sensors

Dry coupling wheeled sensors have been developed for use with acousto-ultrasonic systems and these additionally have found application in detection of acoustic emission from parts that are in motion. The wheeled sensors are available in 175 and 375 kHz versions. The 175 kHz sensor has a sensitivity of -66 dB referred to 1 V·Pa^{-1} (-86 dB referred to 1 V·ubar^{-1}). The 375 kHz sensor has a sensitivity of -66 dB referred to 1 V·Pa^{-1} (-88 dB referred to 1 V·ubar^{-1}).

The sensor is designed so that the piezoelectric element is kept at the same orientation to the part under inspection, thus minimizing variations in response.

A wet coupled wheel with a larger timer footprint is also available. Any standard frequency (from 30 kHz to 1.5 MHz) AC series sensor can be mounted within the internal axle.

These sensors for special applications are all manufactured by Acoustic Emission Technology Corporation, Sacramento, California. Table 3 is a compilation of the manufacturer's sensors and their performance characteristics.

Variety of Sensors from a Single Manufacturer

Resonant Sensors

The resonant sensor is the most popular and most sensitive available. Some models have response at a nominal frequency of 150 kHz in the quiet zone of the acoustic spectra and similar frequency response to both longitudinal and surface excitation.

TABLE 3. Specifications for various acoustic emission sensors

RESONANCE (peak frequency response to kHz)	SENSITIVITY decibels referred to 1 V/Pa (1 V/ubar)	DIAMETER millimeter (inch)	HEIGHT millimeter (inch)	WEIGHT gram (ounce)	CONNECTOR
AC Series Sensors					
30	−50 (−70)	22 (0.875)	29 (1.14)	35 (1.24)	LEMO
75	−60 (−80)	22 (0.875)	29 (1.14)	26 (0.92)	LEMO
175	−46 (−66)	22 (0.875)	29 (1.14)	28 (1.00)	LEMO
375	−48 (−68)	22 (0.875)	29 (1.14)	26 (0.92)	LEMO
375	−48 (−68)	22 (0.875)	19 (0.74)	26 (0.92)	LEMO
750	−55 (−75)	22 (0.875)	25 (0.98)	26 (0.92)	LEMO
1,000	−52 (−72)	22 (0.875)	27 (1.06)	24 (0.85)	LEMO
1,500	−60 (−80)	22 (0.875)	26 (1.02)	20 (0.71)	LEMO
Broadband Sensors					
Flat 100 to 2,000	−70 (−90)	25 (1.0)	45 (1.75)	170 (6.0)	BNC
Flat 100 to 1,000	−55 (−75)	25 (1.0)	45 (1.75)	170 (6.0)	BNC
Miniature Sensors					
175	−60 (−80)	15 (0.60)	12.4 (0.49)	14 (0.5)	LEMO
300	−68 (−88)	6 (0.26)	6.6 (0.26)	7 (0.25)	LEMO
425	−60 (−80)	8 (0.32)	9.9 (0.39)	9 (0.3)	LEMO
500	−58 (−78)	4 (0.15)	2.5 (0.10)	3 (0.1)	LEMO
Directional and Severe Environment Sensors					
500	*	32 (1.25) to 13 (0.5)	8 (0.31)	14 (0.5)	LEMO
500	*	32 (1.25) to 13 (0.5)	8 (0.31)	14 (0.5)	LEMO
100	−52 (−72)	30 (1.2)**	16 (0.63)	35 (1.25)	TNC
175	−46 (−66)	28 (1.1)	64 (2.5)	57 (2.0)	LEMO
375	−48 (−68)	28 (1.1)	64 (2.5)	57 (2.0)	LEMO
Integral Preamplifier Sensors					
40	−15 (−35)	21 (0.81)†	75 (2.96)	96 (3.38)	BNC
60	−15 (−35)	21 (0.81)†	75 (2.96)	96 (3.38)	BNC
150	−5 (−25)	21 (0.81)†	40 (1.6)	53 (1.87)	BNC

* DIRECTIONAL
** DIAGONAL
† HEXAGONAL

In one popular design, a single pill piezoelectric crystal is housed in a small multilayer stainless steel case that is 17 mm (0.7 in.) high by 16 mm (0.6 in.) diameter. Electrical connection is provided by a standard coaxial microdot connector. To electrically isolate the sensor from the test specimen, a nonconductive ceramic wear plate is used. The high temperature epoxies used in the construction of the sensor allow this unit to operate continuously in environments ranging from about −100 to 200 °C (−150 to 390 °F) without degradation.

Calibration curves for both surface wave excitation (spark impulse calibration) and longitudinal wave excitation (ultrasonic calibration) are given in Fig. 21. Frequency responses of 60, 500 and 800 kHz are available in this sensor when a crystal of different geometry is placed in the same housing. In these other resonant frequency sensors, the frequency response for the two modes of excitation will not match as closely. In general, the longitudinal frequency is a function of the diameter-to-thickness ratio of the crystal, while the surface wave excitation is only a function of the diameter.

The sensor has been widely used in testing of composite materials; the 150 kHz frequency response provides a good balance between attenuation and mechanical background noise. The sensitivity makes this sensor ideal for most laboratory conditions, material testing or corrosion monitoring.

Miniature Sensors

In many applications, the 16 mm (0.6 in.) diameter of a general purpose sensor is too large. To test smaller specimens or to perform very accurate source location studies, a microminiature sensor is required. Combining small size and light weight with good sensitivity, such a sensor is useful for applications such as: disk media testing; short beam shear tests; complex multidimensional tests; and triangulation studies. The small size of the transducer results in slightly reduced sensitivity but yields broad frequency response to both surface and longitudinal waves.

The miniature sensor contains a tiny single PZT crystal in an aluminum housing. Its integral 0.6 m (2 ft) coaxial cable terminates at a standard BNC coaxial connector. The housing is 2.4 mm (0.09 in.) high by 3.6 mm (0.14 in.) in diameter. The housing is grounded and isolated from an anodized aluminum wear plate. The recommended operating temperature ranges from −54 to 100 °C (−65 to 212 °F). Recommended shock limits should not exceed 10,000 peak G (gravity) in any direction. The frequency response curve of this sensor is given in Fig. 22.

Slightly larger designs are also available, occupying 7 mm (0.28 in.) diameter by 8 mm (0.31 in.) high. These units are supplied with an integral coaxial cable, 1 m (3 ft) in length terminating in a standard BNC. The small size of the single piezoelectric crystal yields moderately wide response with a slight peak resonance near 300 kHz. The larger crystal provides a sensitivity increase of nearly 10 dB (see Fig. 23).

A still larger package occupies 8.8 mm (0.35 in.) in diameter and 8.4 mm (0.33 in.) in height. The single pill crystal is slightly larger and results in a slightly lower frequency response (see Fig. 24). The larger crystal also provides better sensitivity with narrower frequency response. Electrical connection is provided through a standard coaxial microdot

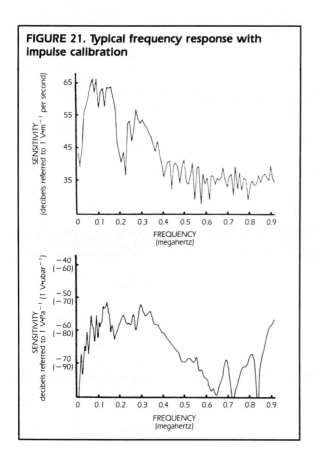

FIGURE 21. Typical frequency response with impulse calibration

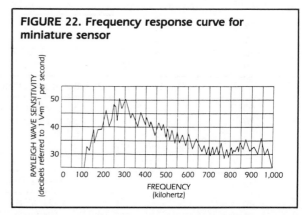

FIGURE 22. Frequency response curve for miniature sensor

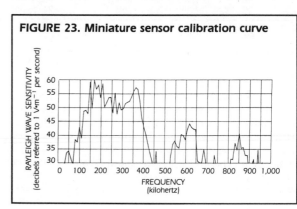

FIGURE 23. Miniature sensor calibration curve

connector. To electrically isolate this sensor from the test specimen, a small nonconductive ceramic wear plate is used. High temperature epoxies yield temperature limits of −100 and 200 °C (−148 and 390 °F). This sensor is also available with a higher frequency package or differential crystal designs.

High Temperature Sensors

In many applications, the upper temperature range of 200 °C (390 °F) offered by sensors with certain PZT crystals is not sufficient. Other crystal materials such as lead metaniobate, cadmium titanate or cadmium bismuth have higher Curie temperatures and provide wider temperature performance. The sensor described below provides excellent sensitivity up to temperatures of 540 °C (1,000 °F).

These high temperature sensors contain a crystal of bismuth titanate in a hermetically sealed stainless steel housing. They occupy a footprint of 20 mm (0.8 in.) in diameter with a 20 mm (0.8 in.) height. Hard line integral cables are included in the package.

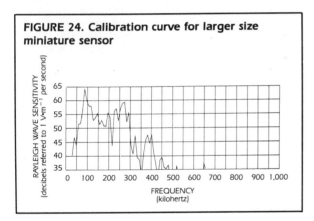

FIGURE 24. Calibration curve for larger size miniature sensor

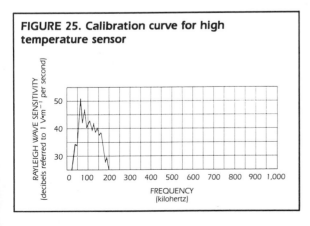

FIGURE 25. Calibration curve for high temperature sensor

Electrical connection is provided through the microdot termination of the coaxial cable. Connection to a standard preamplifier requires an additional microdot-to-BNC coaxial cable. Electrical isolation from the test object is achieved by grounding the housing and isolating it from the stainless steel wear plate.

The frequency response to surface wave spark calibration for this high temperature sensor is shown in Fig. 25. The package is also available in a higher frequency version and optional differential crystal construction.

Wideband and Flat Frequency Sensors

An ideal sensor would have maximum sensitivity, maximum bandwidth and perfect displacement performance, but wideband acoustic emission sensors tend to fall short of these goals. The simplest way to increase the bandwidth is to mass load a crystal, suppressing all resonances, as is done with ultrasonic sensors. Unfortunately, sensitivity is reduced and bandwidth is only extended for longitudinal wave excitation. For surface wave excitation, mass loading the crystal lowers the frequency response and narrows the frequency performance.

However, a sensor has been manufactured with a mass loaded crystal that offers broadband performance to longitudinal waves and excellent performance as a pulser. Broadband frequency response and high sensitivity can be achieved using multiple crystal technology.

The simplest sensor of this type is a differential wideband sensor. Two PZT crystals are used to extend the response of each crystal. The housing is stainless steel and measures 16 mm (0.6 in.) in diameter with a 18 mm (0.7 in.) height. The sensor is electrically isolated from the test object by a ceramic composite wear plate. Electrical connection is provided through integral triaxial cable terminating in a differential BNC connector. High temperature epoxy construction allows the sensor to operate over the temperature range of −100 to 150 °C (−148 to 300 °F). Pill crystals provide excellent longitudinal performance to match the broad frequency response of the surface wave calibration (see Fig. 26).

A more complicated design uses three crystals, each tuned with an inductor. The crystal material is again PZT but each crystal is made by layering PZT wafers. This geometry accentuates the sensor performance for surface wave sensitivity. Figure 27 is the spark impulse calibration of this sensor and shows its combination of excellent sensitivity and broad frequency response. The sensor measures 25 mm (1 in.) in diameter with a 25 mm (1 in.) height and a weight of 62 g (2.2 oz). A ceramic wear plate is sealed in epoxy to electrically isolate the sensor from the test object. The housing is made of stainless steel and electrical connection is provided through a side mounted microdot twin axial connector. The shock rating is 10,000 peak G in any direction.

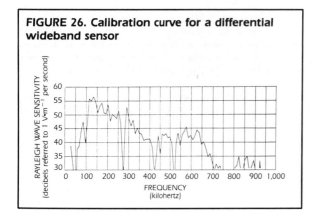

FIGURE 26. Calibration curve for a differential wideband sensor

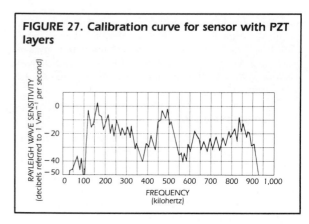

FIGURE 27. Calibration curve for sensor with PZT layers

Integral Electronic Sensors

The use of a sensor and preamplifier combination is sometimes cumbersome and time consuming in the field. Putting the preamplifier in the same housing as the sensor not only simplifies mounting but also minimizes radiofrequency noise. The key elements of a good integral electric sensor for field use are: (1) narrow frequency band response to eliminate unwanted noise; and (2) the ability to drive long cables with minimum signal loss.

One such sensor combines a low noise FET input preamplifier with a standard high sensitivity resonant sensor. A narrow 100 kHz frequency preamplifier is used to minimize environmental noise on the low end and radiofrequency interference on the high end. A 50 ohm cable drive circuit allows a 300 m (1,000 ft) cable drive within 2 dB signal loss. Longer cables can be used with minimal loss of sensitivity.

The housing is made of multilayered copper, nickel and steel coatings in a stainless steel cavity. Electrical isolation from the test object is provided by a ceramic shoe. The electrical connection is provided by a coaxial BNC connector. The sensor is rated for temperatures from -45 to $80\ °C$ (-50 to $175\ °F$). The amplifier has 40 dB gain with a noise level of less than $1.0\ \mu V$ rms. The amplifier produces output voltage greater than 15 V into 50 ohm termination. Power requirements are 28 V at 15 mA for background activity.

Two other narrow frequency bands are available at 60 kHz for low frequency detection and 500 kHz for higher frequency response. The integral electronic sensor uses the low frequency sensor combined with a band-pass filter from 20 to 100 kHz. The integral electronic sensor uses the higher frequency sensor with a band-pass filter from 300 to 600 kHz. All of these sensors from a single manufacturer are produced by Physical Acoustics Corporation, Lawrenceville, New Jersey.

PART 2
ACOUSTIC EMISSION AND ACOUSTIC ANALYSIS SYSTEMS

Acoustic Emission Simulator

Natural acoustic emission signals are variable and unpredictable in time, amplitude and waveform. An acoustic emission simulator (see Fig. 28) provides a user defined, repeatable signal to allow functional tests and performance analyses of acoustic emission systems. Variable waveform controls include: carrier frequency, amplitude, rise rate and decay rate (dB•m•s^{-1}).

Two independent channels may be used with a variable trigger time delay to simulate precise time differential (Δt) data for location functions. Mixer circuitry may be used to create composite waveforms. The addition of an external trigger controller allows a controlled number of signals to be generated at a defined rate for evaluating system data rate handling capability and accuracy.

The acoustic emission simulator produces the sort of transient waveforms that an acoustic emission system is designed to measure. The basic waveform is an exponentially rising, exponentially decaying oscillation with variable rise rate, decay rate, amplitude and carrier frequency.

The waveform characteristics are independently variable on each of two channels over the range that a typical acoustic emission system is likely to encounter in actual operation. The simulator's mixer allows complex waveforms to be built by combining the signals from two or more channels. In this way, wave precursors and reflections can be produced, providing a simulation closer to the true character of natural acoustic emission signals.

Linking the channels on one or more simulators provides a precise Δt for testing the source location performance of multichannel acoustic emission systems. The delay on each channel is independently variable from microseconds to seconds.

Waveform Controls

The waveform controls are independent for each channel and include the following characteristics. The carrier frequency is variable from 1 kHz to 1.4 MHz and is continuously adjustable in six overlapping ranges. An LED display is switchable to give direct readout of carrier frequency on either channel to the nearest 100 hertz.

Amplitude ranges from 40 to 99 dB with preamplifier output of 0.1 to 89 mV and system output from 10 to 8,900 mV in 1 dB steps. Accuracy is from 1 dB up to 500 kHz and uncalibrated from 500 kHz to 1.4 MHz. System output accommodates the preamplifier drive current produced by typical acoustic emission mainframes.

The simulator's rise rate is adjustable from 1 to 999 dB•ms^{-1} and is accurate to within 5 percent. On fast settings, the rise rate reaches peak in less than 5 μs.

Decay rates are from 1 to 999 dB•ms^{-1} and accurate to within 5 percent.

Triggering Controls

Signal production on each channel may be triggered (1) by hand; (2) by the channel's own repetition rate control; or (3) by another channel with controlled delay.

The time between signals is continuously variable from 0.1 ms to 10 s in two overlapping ranges. Maximum repetition rate exceeds 5,000 signals per second. Automatic protection is provided against signal overlapping. The LED display is switchable to give direct readout of repetition rates on either channel to the nearest hertz.

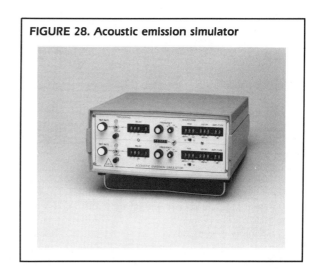

FIGURE 28. Acoustic emission simulator

Digital switch settings give three-digit precision over five decade ranges of delay time. The lowest range is 0 to 999 μs, the highest is 0 to 9.99 s. Accuracy depends on setting but is typically less than 1 percent.

A sync/async switch, in the sync position, produces envelope triggering phase locked to the carrier. In the async position, the envelope is triggered independently of carrier phase.

Simulator's Input and Output

A three-channel mixer gives the linear sum of three cable-connected inputs. The system output is 10 ohms and the preamplifier output is 50 ohms from each channel. Triggering signals provided for multichannel operation are TTL trigger input and output for each channel.

The oscillator out is an unmodulated carrier signal, 1 V peak nominal, for each channel. The DC output voltage for each channel is an approximately linear function of frequency, furnished for swept frequency applications. The simulator's mixer has three inputs of 50 ohms and one output at 50 ohms.

The acoustic emission simulator is manufactured by Acoustic Emission Associates in San Clemente, California.

Signal Acquisition and Analysis System

If a consistent and known mechanical transduction has been made by an acoustic emission sensor, then the two basic requirements are: (1) accurate signal measurements; and (2) accurate analysis of the measured signals. Figure 29 illustrates a four-channel system for achieving these requirements.

Accurate Signal Measurement

This first requirement is met by a signal acquisition module (SAM): a device that changes preamplified sensor signals into a digital format for subsequent analysis by a main processor unit (MPU). All SAM controls and settings may be varied by the MPU and the operating system in use. Any number of transducer signals may be measured by adding SAMs to the system. In a multi-SAM application, global timing provides accurate time delay measurements for signals handled by separate SAMs.

In a specific application, the maximum number of SAMs or channels is determined by the MPU limitations, the operating system and the chosen digital input/output link between SAMs and MPU.

Repeatability and reliability are designed into such a module and enhanced by vibration isolation of the operational circuitry.

Data acquisition speed with sustained accuracy should be considered. Depending on the signal parameters to be measured, a SAM can measure up to 10,000 signals per second continuously until its buffer is full. Additional buffer capacity may be added to store up to 500,000 signals before any are transferred to the MPU. The module also provides measurement of two back-to-back signals (if a second signal occurs one microsecond after termination of the first, both are accurately measured and stored).

An acoustic emission system user needs a clear understanding of the transient data rates and average data rates being generated by the specimen under test. The common sporadic nature of acoustic emission signals may produce thousands of signals per second, with an *average* rate several orders of magnitude lower. If a system cannot handle transient high data rates, potentially critical signals may be lost while the system is processing spurious data.

This signal acquisition unit is equipped with independent front-end high speed counters to provide an indication of lost data due to excessive rates. A real-time display is

FIGURE 29. Signal acquisition module

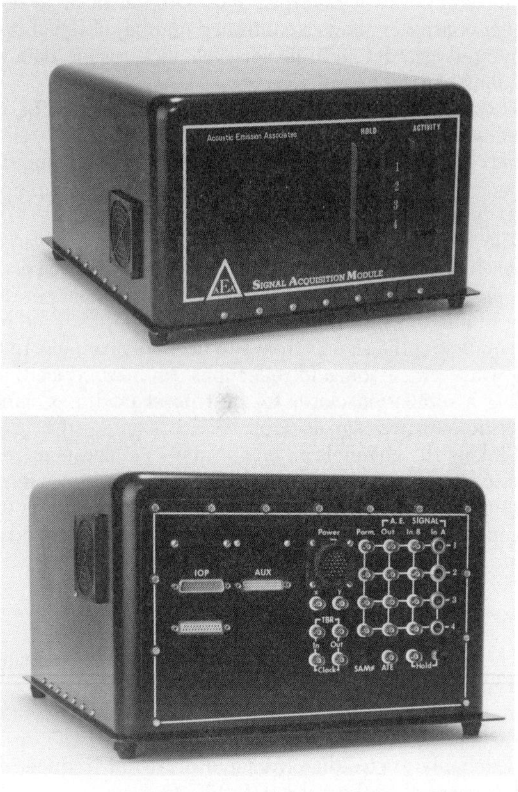

provided to indicate signals occurring at high speed before more comprehensive analysis by the MPU has been completed.

In multichannel applications, high input data rates may exist and systems can produce a biasing effect when certain channels measure more signals than others. Careful design can eliminate this effect, except for the case of signals occurring within one clock cycle of the signal timing circuitry. In the case of a 5 MHz clock, this SAM shows no bias effect until simultaneous signals occur within 0.2 μs.

The unit provides other features, including: true polarity of peak amplitude; positive and/or negative threshold selection; and fifth channel for nonacoustic signals. The potential user of such a signal acquisition system is advised to consult published technical data and to discuss specific applications with the manufacturer.

Analysis of Measured Signals

Analysis of the received signals is the second basic requirement of a system and is performed by the main processing unit. The unit provides an interface for setting controls and observing the status of the signal acquisition module. It also supplies permanent storage of raw and analyzed data.

The analysis software is designed to be used both by the novice and the more experienced operator without unnecessary menu steps. The test starts with a main menu providing six major functions: signal acquisition setup; parametric calibration; graphic display setup; utility functions; post-test analysis; and run test. Published data are again recommended for specific details.

The acoustic emission measurement and analysis system described above is manufactured by Acoustic Emission Associates in San Clemente, California.

Loose Part Monitoring System

In addition to vibration analysis of machinery, loose parts monitoring is receiving considerable attention by both industry and regulatory agencies. Unfortunately, the consequences of implementing an *inadequate* loose parts detection (LPD) system in a nuclear plant, for example, can be worse than having no system at all. If alarm warnings are given without enough information to evaluate the nature of the event, the result can be detrimental to plant availability, personnel exposure and safety.

The US Nuclear Regulatory Commission (NRC) Regulatory Guide 1.133 *Loose Part Detection Program for the Primary System of Light-Water-Cooled Reactors* emphasizes this need, stating that a well developed system should enable discrimination of the signals induced by the impact of a loose part from those signals caused by normal plant maneuvers, and that there should be diagnostic procedures to determine the significance of a loose part. The NRC guide describes a methodology for detecting and evaluating a potential safety related loose part during "preoperational testing and the start-up and power operation modes."

In addition to recommended system characteristics (including sensor requirements, sensitivity and data acquisition modes), the guide discusses the formulation of a loose parts detection program for submittal to the NRC. The program description should include system characteristics; definition of alert logic; methods of data acquisition; a summary of the available diagnostic procedures; a description of surveillance requirements ensuring channel operability; and guidelines for the report to be submitted to the NRC within two weeks of the initial notification of the presence of a loose part. These requirements reflect the need to establish loose parts surveillance and diagnostic programs built and supported with the most current technology.

The LPD system described below has been seismically qualified by type testing to IEEE Standard 344. The systems are manufactured under a nuclear quality assurance program that complies to ANSI N45.2.

The following sections contain a description of a loose parts detection system and its specifications, as well as some information on the theory of operation, installation procedures, operating procedures and methods of evaluating loose parts data.

Description of the Loose Parts Detection System

The system is a modular acoustic detection, analog processing and annunciation system, providing real-time detection of impact noises in primary coolant systems of nuclear power reactors (see Fig. 30). Frequency range, signal-to-noise ratios and rate of occurrence are subjected to manually programmable conditioning for optimization of system sensitivity and alarm logic.

The configuration of this LPD system comprises up to six monitoring channels per rack. Each channel consists of a field mounted sensor, a signal preamplifier and the signal processor.

The field mounted equipment consists of an accelerometer at each of the monitoring sites. A charge converter services each of the sensor sites in-line between the accelerometer and the control room equipment.

The signal is then routed to control room signal processing equipment mounted in a standard rack. From there, the signal travels directly to a differential transducer amplifier that transmits to an impact detector for signal processing and bussing.

Also included in this configuration are: a loose parts monitoring control module; annunciator module; audio monitor

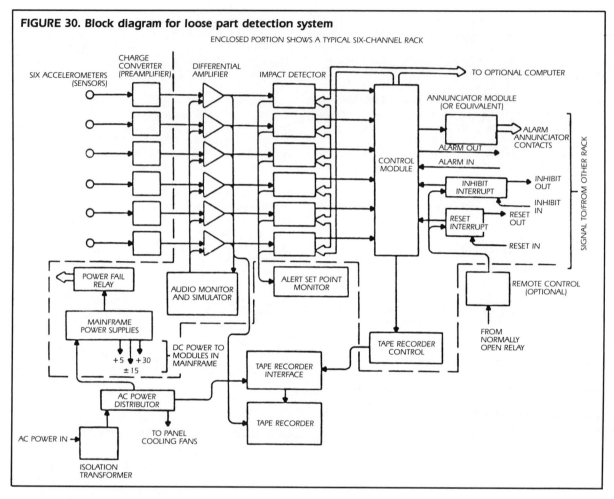

FIGURE 30. Block diagram for loose part detection system

and simulator; power monitor; alert set point monitor; tape recorder interface; and tape recorder.

The first four of these modules are housed in the rack. The rack is provided with every LPD system and contains all necessary power supplies as well as the tape recorder controller, the inhibit interface, the reset interface, and the power fail relay. The field mounted accelerometers and charge converters are connected to their corresponding filter amplifiers through the rack rear input/output panel which in turn provides terminations for all field mounted charge-to-voltage converters.

The system has been seismically qualified with the portion of the system that is outside containment being mounted in a seismic cabinet. Whether or not the purchaser exercises the option for the seismic cabinet, the manufacturer supplies the seismic test report as evidence of seismic qualification. The seismic test report documents the test response spectra for which the equipment is qualified.

Acoustical signals excite the piezoelectric transducer and are converted to an electrical charge proportional to detected acceleration (G). Charge-to-voltage converters change the charge to a proportional voltage variation driving the signal transmission lines connecting to the filter amplifiers.

Each amplifier controls several key functions in the signal detection process, as noted below.

1. The sensor and charge converter bias current is supplied by a regulated current source in the amplifier.
2. Test (or simulator) signal insertion is made through remotely controlled relays at the front end of the amplifier.
3. High pass filtering is supplied for low frequency noise rejection.
4. Low pass filtering is supplied.

The amplifier outputs are bussed to several locations within the system (Fig. 30). Each amplifier output is routed directly to a selection switch within the audio monitor. An audio amplifier provides a means of listening to the output signals of a selected channel through a small, built-in speaker. In addition to driving the audio monitor, each amplifier output is routed to the input of an impact detector. The function of the impact detector is rms measurement of signals present at the amplifier outputs.

The impact detector also separates impact signals from background noise. This component addresses the problem of reliably detecting loose part impacts regardless of variations in background noise. This feature makes possible reliable monitoring during periods of high background noise such as start-up, while providing greater sensitivity in the relatively quiet periods of operation. Impact signals of sufficient form and strength to be detected as impacts cause the generation of a digital pulse (called an *alert*) from the unit.

The background and threshold signals for the sensor channels are bussed to the alert set point monitor. These voltage signals can be displayed on the front panel's numerical LED for any channel in the active mode.

The alert output of the impact detector is also sent to the control module. The control module continuously monitors the alert pulse outputs of up to six impact detectors and includes a first alert display on the front panel. This display identifies by number the first impact detector module to transmit an alert signal. The alert data are stored in a temporary latched memory within the control module which drives the display for identification of the first alert channel.

The first alert data are latched by the impact detector's alert pulse output. Once stored, first alert data can be cleared only by a system reset or an automatic reset (if enabled). The system reset condition is an initial response by the operator at power-up or when starting a new monitoring period. The reset clears all previous data from the system logic. The automatic reset (if enabled) occurs if a preselected alert rate is not met. If a qualified alarm has been detected, the automatic reset will be inhibited and the first alert information will be displayed.

In conjunction with the inhibit interface, the control module also contains circuitry to inhibit the alert and alarm features of the LPD system. This is particularly useful in the event of a deliberate plant maneuver (such as control rod stepping) that could otherwise be detected as a loose parts impact event.

In addition, the control module provides communication and data control to other system modules as well as other LPD racks. It has alarm outputs used to initiate the tape recorder controller auto-start function; to activate the annunciator module; and to notify other LPD racks of an alarm condition.

Sounds from loose parts, flow or background noise received by the LPD system sensors can be heard via an audio monitor and simulator. Individual channels can be monitored using the selector switch. The other major function of the audio monitor is its short term and long term signal simulation feature. When enabled, the component injects a constant or long term signal comprised of band-limited white noise into all channels. A front panel momentary contact provides a 10 kHz burst that simulates the short term signals common to loose parts impact signals from accelerometers. Thus, the monitor is an effective diagnostic testing instrument for the LPD system as well as an on-line audio monitor.

The AC power for the system is first sent through an isolation transformer before it is distributed to the system components. The isolation transformer minimizes the effects that power line noise and transients can have on the operation of the system.

Sensors for Loose Parts Detection

The sensors used for this system are piezoelectric accelerometers. When the accelerometers move upward, the mass of the internal piezoelectric element and its backing is slow to follow the motion, thus compressing the element. Conversely, downward motion tends to expand the thickness of the element. An electric charge is generated by the piezoelectric element as a result of the dynamic stress.

The sensors in the typical LPD configuration are high temperature, nuclear rated, self-generating piezoelectric transducers requiring no external power for operation. These accelerometers are specially designed for use in reactor vibration and loose parts monitoring systems. Measurements may be made in high radiation environments at temperatures up to 370 °C (700 °F).[5]

The loose part monitoring system described above is manufactured by Technology for Energy Corporation, Knoxville, Tennessee.

Valve Flow Monitoring System

Acoustic monitoring is used to determine the open and closed position of valves, monitoring the status of pressurized safety valves and power operated relief valves, in both pressurized water reactors (PWR) and boiling water reactors (BWR). This monitoring function is interfaced with an alarm function that is activated when the valves are open. Steam flow across valves can also be detected by acoustic monitoring.

The monitoring of flow through a valve in the frequency range of 15 to 50 kHz provides a sensitive detection method while limiting the spurious indications from mechanical noise. Studies on valve flow demonstrate that acoustic excitation increases significantly above background when flow is present (about 60 G maximum).

Flow Monitoring System Description

For each monitoring point, a high temperature accelerometer is connected to a charge converter. The charge converter amplifies the small picocoulomb signal from the accelerometer. The output of the charge converter is routed by shielded, twisted pair cable to a signal processor. Among a number of responsibilities, the signal processor: (1) provides power to the remote charge converter from a built-in current source; (2) performs signal processing; (3) displays ten levels of flow indication; and (4) provides digital outputs corresponding to a lower trip level, an alarm level and an upper trip level. These levels are sent to alarm modules for actuation.

Such systems should be environmentally and seismically qualified by type testing to IEEE Standards 323 and 344. The system in this discussion is manufactured under the Nuclear Quality Assurance Program in accordance with ANSI N45.2.

The configuration for a typical three-channel valve flow monitoring system comprises up to three monitoring channels per rack. Each channel consists of a field mounted sensor and preamplifier. Signal processing electronics are located in the control room cabinet (see Fig. 31).

The field mounted equipment consists of an accelerometer as the sensor. Charge converters serve as sensor preamplifiers for each of the sensor sites and are located in-line between the sensors and the control room equipment.

The acoustic signal is routed to the control room signal processing equipment, then travels directly to a monitoring module that conditions and processes the signal. This module displays and distributes any signal causing an alarm indication.

Figure 32 shows an instrumentation rack containing (1) the modules for monitoring six valves; and (2) a ganged alarm system.

Field Mounted Equipment Sensors

The sensors used in the valve flow monitoring system are piezoelectric accelerometers. These are strapped to a pipe near the valve being tested (see Fig. 33).

The recommended accelerometer must be designed and qualified to function in a nuclear plant environment. The illustrated sensor has a mounted frequency resonance of

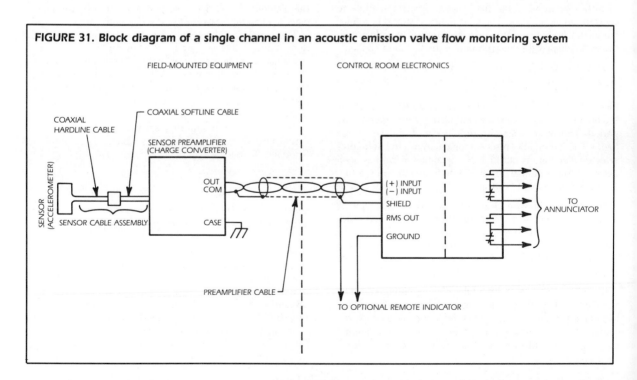

FIGURE 31. Block diagram of a single channel in an acoustic emission valve flow monitoring system

about 27 kHz ±1 kHz, and a charge sensitivity of 10 pC·G^{-1} (±10 percent). The charge sensitivity increases above the nominal value for frequencies approaching the resonance frequency. The sensor frequency response and subsequent electronics determine the optimum frequency range of the vibrations analyzed in the system.

The accelerometers are connected to the charge converters with hardline and softline cabling. The hardline cable is connected to the accelerometer on one end and terminates at the other end in a male microdot. A microdot-to-BNC adapter connects the hardline cable to a BNC-terminated softline cable that in turn runs to a transient shield containing the charge converters (see Fig. 34)

Charge Converter, Shield and Cabling

The charge converter provides signal amplification for the accelerometers and is a charge sensitive feedback amplifier that serves as a sensor preamplifier. The output of the converter is a DC bias voltage with an AC signal voltage superimposed. The AC signal voltage is directly related to the charge input to the amplifier.

The charge converter in this system has a gain of 2 mV·pC^{-1} (±5 percent). In addition to supplying signal gain, the sensor preamplifier also provides the very low impedance necessary for an accelerometer to operate in the charge mode. This low input impedance prevents signal loss between the accelerometer and the charge converter due to cable noise. This converter is powered by a 9 mA constant current superimposed on its output signal terminal.

Charge amps do have an effective low impedance, but it should be noted that it is a very high input resistance and large input capacitance that lead to a low impedance. The large input capacitance effectively swamps the effects of transducer and cable capacitance. Charge amps do not minimize cable noise. Increased length of cable (and hence its capacitance) into a charge amplifier increases its noise floor.

The transient shield (see Fig. 35) is a passive shield assembly composed of a stainless steel shell filled with loose vermiculite insulation. The shield houses the charge converters, protecting them from chemical spray, high pressure and high temperature environments. Inputs are BNC receptacles and outputs are ring tongue terminals. The input and output connections are sealed with nuclear grade heat shrink to ensure their integrity in harsh environments.

Because of the charge converter's low output impedance and the presence of large signal levels, it is recommended that a shielded twisted pair cable be used for each sensor channel. Two signal wires are used in the cable because of the differential input of the field mounted electronics. Connections to the charge converter and other field mounted electronics are made using ring tongue terminals.

The valve flow monitoring system described above (see Fig. 36) is manufactured by Technology for Energy Corporation, Knoxville, Tennessee.

System for Detection of Air-Borne Acoustic Signals

Operating equipment, including fluid and gas systems, all emit acoustic signatures that may be defined as recognizable

FIGURE 32. Instrumentation rack containing a ganged alarm system and the modules for monitoring six valves

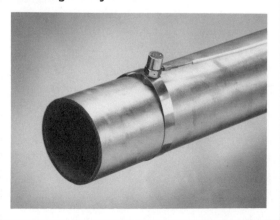

FIGURE 33. Accelerometer mounted on pipe for monitoring nearby valve

FIGURE 34. Diagram of mounting configuration for valve flow monitoring system sensor and preamplifier

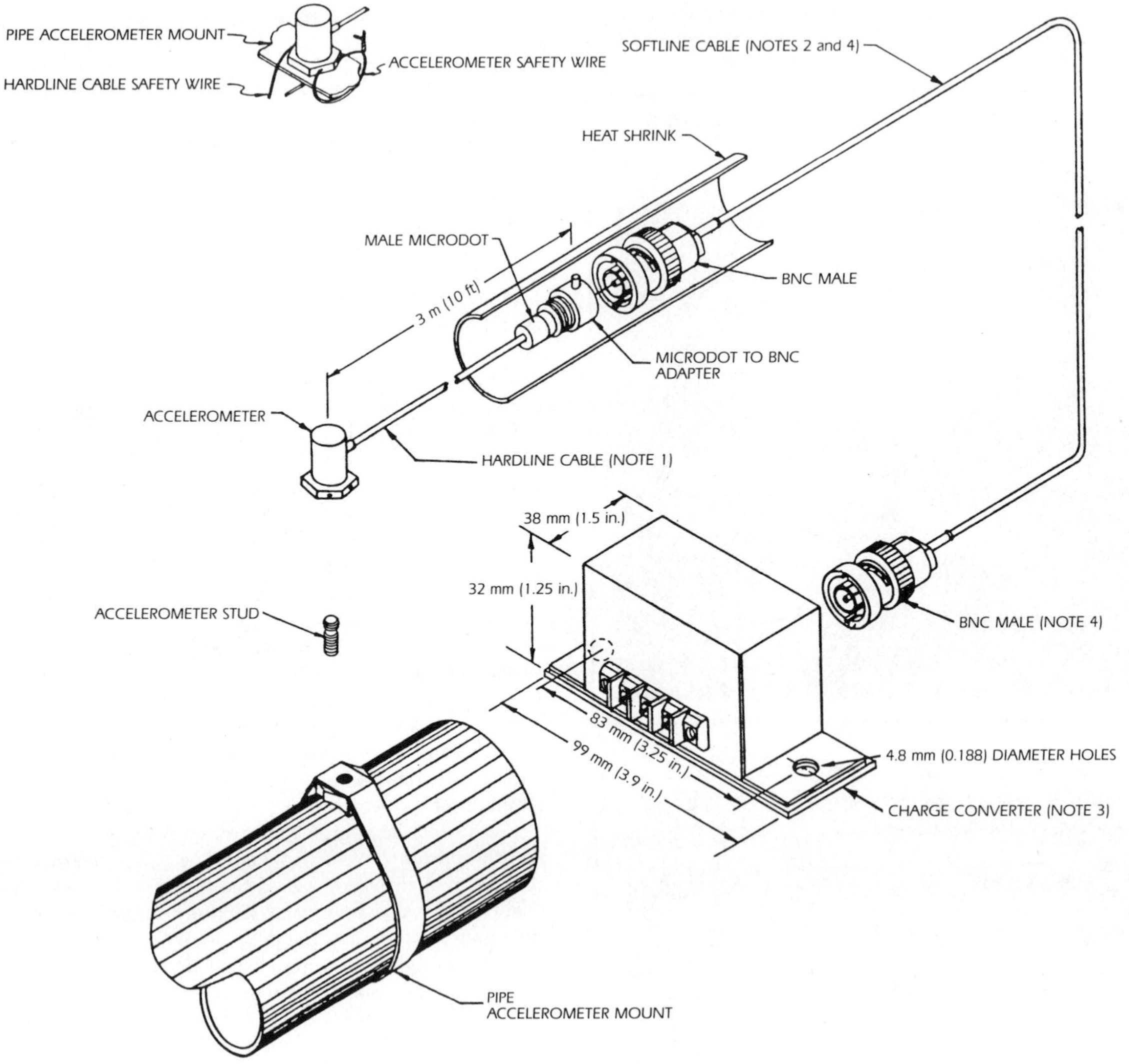

LEGEND
1. HARDLINE CABLE MUST BE SUPPORTED BY STRAPPING TO PIPE OR SOME OTHER METHOD.
2. SOFTLINE CABLE SHOULD BE SUPPORTED.
3. CHARGE CONVERTER SHOULD BE PLACED INSIDE TRANSIENT SHIELD OR ENCLOSURE
4. SOFTLINE CABLE LENGTH DEPENDS ON APPLICATION.

FIGURE 35. Stainless steel passive shield assembly filled with loose vermiculite

and consistent sound patterns generated under normal working conditions. When an apparatus or component enters the initial stages of failure, there are changes in the acoustic signature and systems are available for detecting these changes.

This instrumentation is almost exclusively receptive in nature and is sensitive to a range of sound in the low kilohertz. Such systems are accurate in spotting potential bearing failure as well as for locating all types of pressure leakage and vacuum leakage. In addition, they are effective for detecting faulty steam traps, valves and electric discharge.

Instrumentation for Signal Detection

Portable instruments use electronic circuitry to convert a narrow band of sound (between 20 and 100 kilohertz) into the audible range. The operator hears the qualitative sounds of operating equipment through headphones. In order to maintain accurate reproduction, most of these instruments heterodyne the signal. Signal strength is usually displayed on an analog meter.

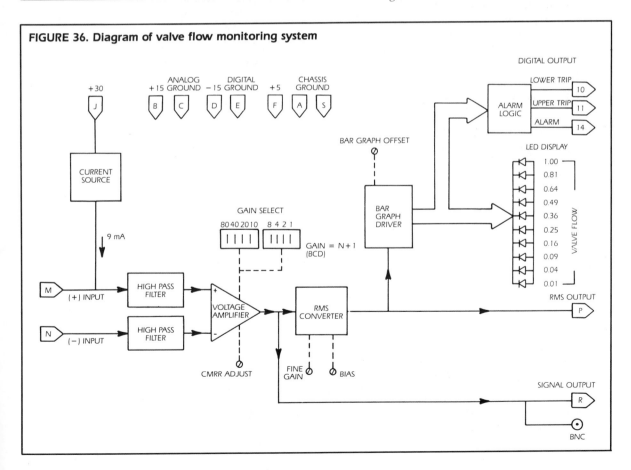

FIGURE 36. Diagram of valve flow monitoring system

Certain models provide additional flexibility with features such as variable frequency (to filter out spurious acoustic signals) and meter mode selection (to regulate the meter response from real-time to averaging). Some instruments contain volume controls; others have a sensitivity selection dial that provides adjustment for detectability of both weak and strong signals. In some instruments, the sensitivity selection is numerically calibrated for repeatability.

These acoustic detectors may be used in either a scanning, noncontact mode or in a typical contact mode.

Use as Air-Borne or Structure-Borne Detectors

As scanners, these instruments are often used to detect leaks. Because they are sensitive only to certain frequencies, their applications are not limited (as are most leak detectors). For this reason, they can detect all types of pressure leaks. In addition, such systems are often used to locate vacuum leaks (sometimes called *air in-leakage*) and electrical discharge problems (such as corona).

These instruments may include ultrasonic transmitters or tone generators. In this application, the transmitter is usually placed within a container while the scanner is used to detect areas of acoustic penetration along the container's surface. This process will often detect leakage in systems before they are put into service.

In the contact mode, a metal rod acts as a waveguide. Applications include the detection of impending bearing failure and the location of faulty steam traps or leaking valves.

Bearing Failure

In most plants, the condition of bearings is an important consideration. Fortunately, in most kinds of equipment, bearings can indicate their condition with structure-borne acoustic signals, long before problems are detectable through heat or vibration monitoring.

By the time a bearing becomes noisy enough to detect audibly, it has usually reached failure. Bearing wear is a progressive process and even good bearings emit characteristic sounds. This acoustic signal changes and its intensity increases perceptibly, long before a bearing actually fails.

For bearing analysis, an operator uses the contact mode, touching the bearing housing with the waveguide. A characteristic rushing sound is heard in the headphones. As the raceway, balls or rollers begin to fatigue, they deform slightly. This increases the acoustic signal by a factor of 12 to 50, long before any heat or vibration can be detected.

Brinelling of bearing surfaces also produces an increase in acoustic signal amplitude. As the bearings become more out of round, a repetitive ringing develops as flat spots hit the raceway. Lack of lubrication is indicated by a uniform increase in the acoustic signal and is heard as a rushing sound,

somewhat louder and rougher than that of a properly lubricated bearing. Scarred raceways are often perceived as a grinding or crackling noise.

These characteristic sounds are used in the two fundamental methods for testing bearings: comparative and historical techniques.

Because a good bearing also generates sound, the operator must compare each signal with one from a bearing known to be operating normally. By comparing bearings of the same type under the same operating stresses, any deviation from normal indicates malfunction.

A better procedure for structure-borne bearing inspection is the development of an historical record. As regular inspections are made, a record is kept containing the following information: (1) the bearing number; (2) location of the probe; (3) amplitude control setting; (4) meter reading; and (5) a short description of the sound. Since settings are fixed for a large number of similar bearings in the same service, compiling this report is not time consuming.

The importance of the equipment as well as past replacement and service records dictate how often routine acoustic inspections should be made.

Pressure Leak Detection

Acoustic leak detection depends on the fact that forcing a fluid through a small opening creates turbulence on the downstream side. That turbulence generates strong acoustic characteristics. While most of the audible sounds of a leak might be masked, high frequency signals may be detected even in the presence of loud ambient plant noises.

When nearby equipment emits competing noise, it is usually simple to minimize its effect. Because these are short-wave signals, noise can often be blocked by something as simple as a clip board. In addition, air-borne detection instruments often use a rubber probe to shield the scanner against spurious signals.

To locate leaks, the scanning mode is used on the side of a pressure vessel or tubing. A definite jump in volume occurs in the earphones and is indicated on the meter when a leak is present. It is most accurate when the probe is close to the surface being inspected. By increasing the sensitivity setting, however, an operator can observe gross leaks from a considerable distance. This is particularly useful when the pressurized gas is dangerous, as in the case of superheated steam or when an operator must inspect a piping run along a ceiling or in a deep pit.

Locating Vacuum Leaks

Vacuum leakage produces turbulence in the same manner as pressure leakage, except that it is generated within the vessel. Some sound escapes through the leak itself, but the intensity is much lower than a pressure leak. In many instances, this problem may be overcome by positioning the instrument close to the test area or by increasing the sensitivity setting.

Heat Exchange Tube Leaks

There are three approaches to heat exchanger testing: on-line, tone generation and pressurization. The latter two are performed while the unit is off-line.

On-line testing is performed in the contact mode. The surface area of the heat exchanger is divided into three or four equal zones along the length of the shell and these sections are monitored for a normal flow. Should a leak occur in the tubes, the change in turbulence is immediate and is usually characterized by popping sounds. This technique is an effective way of detecting the beginning of leakage conditions. Small leaks are more readily detected when they occur in the peripheral tubes. In the central tubes, gross leaks are detected.

Tone generation is an ultrasonic technique performed with the heat exchanger off-line. Not only leaks may be found, but tube-to-tube sheet integrity may also be tested in this manner.

Pressure tests are another off-line test. The heat exchanger is pressurized with air and standard leak detection methods are incorporated. A scan of the tube sheet for a tell-tale rushing sound indicates leakage.

Leaky Valves and Steam Traps

As gas leaks, a poorly seated valve reveals itself through the sound created by fluid flow. With the valve truly seated, there is no such flow. The noise of a leaky valve is more evident on the downstream side of a valve. This is because of the turbulence created when the gas or liquid flows from the high pressure side, through the leak, to the low pressure side. Therefore, sensors should be located downstream from the valves.

Tests may also be performed on a wide variety of steam traps operating under different conditions. Before testing, the type of trap and its operating conditions should be researched. Generally speaking, steam traps may be loosely categorized as either intermittent flow or continuous flow.

Intermittent traps normally operate in an open/close cycle and often fail in the open position. The traps in this category have particular methods of operation and patterns of opening and closing. Intermittent traps include: inverted bucket; bucket; thermodynamic (disk); bimetallic; and thermostatic traps.

Continuous flow traps allow condensate to continually flow. They usually modulate according to condensate load. This type of trap usually fails in the closed position; the absence of sound indicates failure of the valve. Continuous flow traps include: float; float and thermostatic; and bellows traps.

Inverted bucket traps normally fail in the open position because the trap loses its prime. This condition means complete blow-through, not a partial loss. The trap will no longer operate intermittently.

Float and thermostatic traps normally fail in the closed position. A pinhole leak produced in the ball float will cause the float to be filled with water and weighted down. Since the trap is totally closed, no sound is heard. An additional check of the thermostatic element is important for verification. If the trap is operating correctly, this element is usually quiet. If a rushing sound is heard, either steam or gas is moving through the air vent.

Thermodynamic or *disk traps* should cycle (hold, discharge, hold) four to ten times per minute. When a disk trap fails, it usually fails in the open position, allowing continuous blow-through. Rapid switching between the rushing sound and silence usually indicates a cycling problem.

Thermodynamic traps operate on the difference in temperature between condensate and steam. The trap accumulates condensate and the temperature of condensate drops to a level below saturation temperature in order for the trap to open. By backing up condensate, the trap will tend to modulate open or closed. An acoustic test indicates failure by changes in a constant condition: either failed open, in which case a constant sound is heard, or failed closed when no sound is heard.

Bimetallic traps are usually intermittent and fail open with a characteristic constant rushing sound.

The descriptions above are simplifications of the test procedures. There are many factors to consider because of the unique ways that different traps operate.

Electrical Discharge

The integrity of an electrical system may be tested using acoustic instruments. In most instances, the scanning mode is utilized to sweep over a test area. An electric arc or corona produces a buzzing or crackling sound. Various acoustic signals indicate loose connections and other openings that force the current to jump a gap. The sound of arcing at switches and relays is easily detected. Insulation on high voltage lines may be tested for corona discharge.

Circuit breakers, junction boxes, electric panels and transformers are among the components tested with this technique.

Summation

Air-borne and structure-borne acoustic instruments are easy to use, portable, accurate and versatile. The techniques are effective for reducing maintenance costs and improving operational efficiencies when incorporated into a preventive maintenance program. The acoustic test systems described above are manufactured by UE Systems Incorporated, Elmsford, New York.

Assorted Acoustic Testing Systems

Acousto-Ultrasonic Inspection Unit

An acousto-ultrasonic testing instrument designed to NASA specifications provides a measure of the integrated effect of distributed discontinuities in composite and adhesive bonded structures (Figs. 37 and 38). The instrument operates as a pulser/receiver for acousto-ultrasonic tests and as a single channel receiver for acoustic emission tests.

The pulser/receiver features two pulsing modes; adjustable pulse rate and amplitude; signal gate; and oscilloscope sweep rate. The acoustic signal processor is broadband from 40 kHz to 1 MHz. Features include counts; events (cumulative or rate); signal level measurements; selectable gain; and threshold.

Outputs include LED display, oscilloscope and DC voltage output for counts, event and signal level A variety of probe types are available, including dry coupled wheeled probes, wet coupled wheeled probes, a motor-driven dry coupled probe fixture for plastic pipe joint inspection, and point probes for narrow bondlines. The system operates on battery or AC power.

Single-Channel Acoustic Emission Units

Single-channel, portable acoustic testing instruments have been designed for a variety of applications ranging from laboratory experiments to field monitoring (Figs. 39 and 40). These units use a patented floating/fixed threshold for maximum versatility in eliminating unwanted noise.

System measurements include total counts, total events, count rate, event rate, rms and threshold. Outputs include three half-point digital LED displays and DC voltage outputs for events, counts and rms.

One variation of this unit has an extended bandwidth (10 kHz to 1.5 MHz). Another variation is designed for low frequency geotechnical applications (30 Hz to 200 kHz). Options include a waterproof case and a multiplexer for simultaneous or sequential monitoring of eight channels.

Data Acquisition Systems

A multichannel computerized data acquisition system has been designed with a sixteen bit or thirty-two bit computer

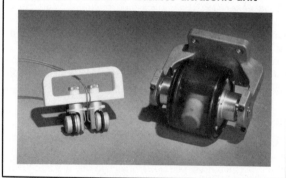

FIGURE 38. Probes for acousto-ultrasonic unit

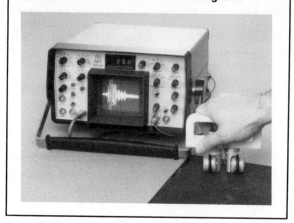

FIGURE 37. Acousto-ultrasonic testing unit

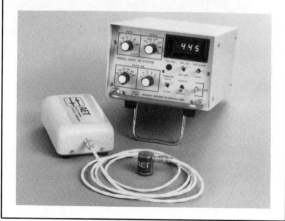

FIGURE 39. Single channel acoustic emission testing unit with extended bandwidth

architecture and distributed processing (see Fig. 41). The multibus (IEEE 796) front end processor incorporates up to twenty-four single channel signal characterization boards with a sixteen bit microcomputer to acquire and buffer over two thousand events per second.

FIGURE 40. Single channel acoustic emission testing unit with eight-channel multiplexer

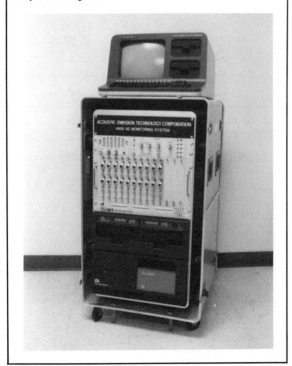

FIGURE 41. Multichannel acoustic emission data acquisition system

Multiple front end processor subsystems (up to sixty-two channels) can be linked to the host computer, a thirty-two bit microcomputer with seventy-one megabyte hard disk storage and ninety-five megabyte cartridge tape drive. The system terminal is a 350 mm (14 in.) green phosphor display terminal with ASCII keyboard. The unit provides full signal characterization, including: events; counts; amplitude; rise time; duration; energy; signal level; and seven analog DC inputs. Real-time source location programs include: linear; cylindrical (planar); and zone (adaptive location for complex structures).

The system processes, displays and stores independent channel data as well as computed source location events in real-time. Real-time graphic displays can be configured from hundreds available in system software. Source location plots show location of sensors and computed source location events in real-time. This system meets and exceeds all requirements for testing of metal pressure vessels in accordance with the proposed ASME Section V Code requirement (BC 85-033).

Another data acquisition system has modular expandability by function, and two to eight channels in an integral package (see Fig. 42). Full signal characterizations include: events; counts; rise time; signal duration; peak amplitude; energy; rms (average signal level); peak signal level; slope (amplitude divided by rise time); and three analog DC inputs (expandable). The system features real-time graphics display; real-time signal waveform discrimination; and real-time source location.

Location programs include linear; zone (adaptive learning program); spherical; and three-dimensional. Advanced post-processing and user programming software options are available for personal computers. Recording options include: dual 350 kilobyte quad density floppy drives (5.5 in.); or fifteen megabyte hard disk with a 350 kilobyte floppy back-up. The sixteen bit, 128 kilobyte main processor can handle more than three thousand events per second with a four thousand event dynamic buffer.

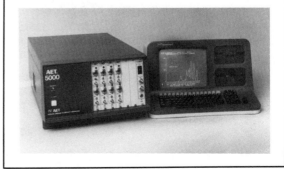

FIGURE 42. Computerized acoustic emission data acquisition system

Particle Impact Noise Detector

This system tests electronic devices to detect the energy generated when a loose particle strikes the wall of a component cavity (see Fig. 43). The test specimen is attached to an acoustic emission sensor mounted on top of an electromechanical vibrator. During vibration, impacts between particles and the cavity wall generate stress waves that are transferred through a viscous couplant into the acoustic emission sensor. The acoustic signals are then electronically processed. Mechanical shocks are imposed to free particles that may be held captive either electrostatically or mechanically inside the test specimen.

Procedures that establish the PIND methodology to be used are incorporated into military standards such as MIL-STD 883 Test Method 2020 (semiconductors), MIL-STD 750 Test Method 2052 (discrete active components) and MIL-STD 202 Test Method 217 (relays).

Composite Tester

This system is an eight-channel or sixteen-channel microcomputer acoustic emission system designed for field or laboratory testing of composites (see Fig. 44). Its features include a 175 mm (7 in.) CRT; ASCII keyboard; dual microfloppy (3.5 in.) disk recording; integral preamplifier sensors (40, 60, 110 and 150 kHz) in stainless steel housings; four analog parameter inputs; channel selectable audio and scope outputs; and independent channel analysis. It processes events, counts, peak amplitude, and event duration by channel.

The user can program high and low amplitude thresholds and long duration event thresholds. The software maintains statistics by channels for up to ten selectable time phases during a test. The system also produces cumulative statistics on grouped channels. Over 400 graphic displays are preprogrammed. Software is available for general testing of fiberglass structures; aerial lift testing (ASTM F914-85); and for data transfer to personal computer systems in ASCII format.

Multichannel Acoustic Emission System

This is a unit designed for research or on-line industrial applications (see Fig. 45). Up to seven channels can be configured on a single bus. Signal measurements include rms; counts; events; energy; and peak amplitude. Programmable alarm modules with relay closure outputs can be configured for each channel, based on counts, events or events above a selected amplitude.

A dual channel plotter module provides strip chart or XY outputs for counts, events, energy and amplitude distribution. Other modules include audio, pulser, Δt (location) and RS232C computer interface.

FIGURE 44. Acoustic emission system designed for field testing of composites

FIGURE 45. Multichannel acoustic emission testing system

FIGURE 43. Particle impact noise detector

Systems and components manufactured by Acoustic Emission Technology Corporation of Sacramento, California are discussed above, beginning with the acousto-ultrasonic inspection unit.

Signal Analysis Software

A complete acoustic emission signal analysis program has been designed for personal computer systems. A plug-in transient recorder board with 25 MHz, eight bit analog-to-digital conversion is used to provide digital capture and storage of signals, which are then analyzed by the intelligent classifier engineering package (ICEPAK) program.

This comprehensive analysis program provides signal processing, feature extraction and statistical classification capabilities. Feature extractions include 108 features in five domains: time; power (spectral); phase; cepstral; and autocorrelation. Four statistical classifiers are available: linear discriminant; K-nearest neighbor; empirical Bayesian; and minimum distance.

Signal groups can be split into training and testing sets to optimize the classifier approach in signal recognition. The software performs automatic feature ranking, as well as manual selection, and additionally provides both feature histogram analysis and cluster analysis. ICEPAK was produced by Tektrend International, Montreal, Canada and is marketed by Acoustic Emission Technology Corporation, Sacramento, California.

Variety of Acoustic Emission Systems by a Single Manufacturer

Multiple Microprocessor System

A valuable acoustic emission system offers distributive and parallel data processing to avoid slow acquisition and to provide expandability in channels, computing power and software (see Fig. 46). The system is a multibus (IEEE 796) multiple microprocessor system that can be expanded to 128 channels through the addition of modular boards containing two channels of analog interface and a digital preprocessor bus organizer. Each of the front end processors for two channels communicates through the multibus to a sixteen bit central processing unit that organizes the data flow to the high end multiprocessor computer.

Each channel operates completely independently to provide a high quality thirty-two byte digital representation of each acoustic emission signal on each channel. This entire data set, or any portion of it, is then available for data acquisition and display. The unique independent architecture allows each channel to time signal arrivals with global clock accuracy down to 0.1 microsecond. The analog processing circuitry employs both linear and logarithmic amplifiers to simultaneously provide wide dynamic range and accurate signal parameter measurements.

Data are recorded and displayed by a modular computer system employing multiple eight bit and sixteen bit processors for both high speed data acquisition and high resolution full color displays. Parallel processing allows the high end processor to dedicate an eight bit processor to disk handling, a sixteen bit graphics engine for high resolution display, and a main sixteen bit central processing unit for data handling prior to display or storage. Modularity and flexibility allow the user to upgrade the system as instrumentation technology advances.

Locator Analyzer

A locator analyzer combines sophisticated electronics and compact packaging to meet the requirements of a laboratory where six channels (with expansion to fourteen) are adequate (see Fig. 47). Four channels of lockout (master and guard) capability are available in addition to the standard six channels. Up to four channels of external parametrics can be

FIGURE 46. Multichannel microprocessor acoustic emission unit

FIGURE 47. Acoustic emission location analyzer

used to combine acoustic emission data with other test measurements. Parallel processing allows simultaneous data acquisition and data storage using separate processors.

Two-channel analog processing boards provide high quality thirty-two byte digital representation of the analog acoustic emission signal. These acoustic emission data are then processed, displayed and recorded by a high speed sixteen bit computer to generate high quality graphics and high speed performance.

The analyzer can achieve transient data storage of more than two thousand hits (complete thirty-two byte description) at data rates exceeding three thousand hits per second. The steady state throughput is limited by graphics but can exceed one thousand hits per second for twenty thousand hits. Higher data rates can be achieved by reducing the data set to include less information about each acoustic emission occurrence below the thirty-two byte standard.

Standard features of this system include: real-time clock; alarm; two 720 kilobyte floppy disks; high resolution, eight gray scale graphics (640 × 250 point resolution); software; scope monitor; modem output; pulser output; audio monitor; full thirty-two byte data description; color printer port; and choice of operating system. Options include expansion to fourteen channels; twenty megabyte hard disk; color video display; digitizer; IEEE 488 interface; and many software options.

Portable Acoustic Emission Monitor

Acoustic emission signal processing and storage can be provided by a portable, battery operated, multifunction, single channel, real-time signal system shown in Fig. 48. The low power consumption of this system allows twelve hours of normal use without recharging, while retaining collected acoustic emission in RAM memory for more than one month. The system can be programmed to monitor acoustic emission test data on a per test basis for sample data at a preset test interval. Information is output to a small battery operated printer.

FIGURE 48. Portable signal processor and storage device

The acoustic emission signal is input through a rear panel BNC connector. The system includes small 20 dB integral electronic sensors to allow cable drive up to 30 m (100 ft). Test setups and control are handled through front panel keypad entry. Test setups, collected data and alarm conditions are viewed on a front panel liquid crystal display. The system can process data for acoustic emission events, acoustic emission counts, acoustic emission energy and acoustic emission rms signal level. Data can be accumulated and measured on a time-base rate mode.

The combination of light weight and battery power make this design practical for geotechnical field applications, leak detection, bearing monitoring or corona discharge monitoring.

Transient Recorder and Analyzer

A transient recorder and frequency analyzer module can provide a complete digital record of the acoustic emission waveform tagged to the event number, as recorded by specially designed acoustic emission computer systems. The sample rate is variable from 1 kHz to 25 MHz with nine bit analog-to-digital converter resolution for both positive and negative signals. The event size can be preset from eight kilobyte to two megabyte samples. Up to three two-channel modules can be used in any of three trigger modes: first hit channel; scan mode (scans both channels and locks on the channel with conversion above threshold); or all-hit mode. A pretrigger can be set between 0 to 100 percent of the selected sample size.

The software package for waveform analysis allows the user to display the waveform itself; display an enhanced waveform; filter the recorded data with high pass, low pass or band-pass filter; print all data or store to disk; and set up sample parameters (see Fig. 49). In addition to displaying the waveform, the software package can perform a digital fast Fourier transform on the waveform and display the frequency spectrum.

These various systems from a single manufacturer are produced by Physical Acoustics Corporation, Lawrenceville, New Jersey.

Particle Impact Noise Detector

Particle impact noise detection (PIND) systems provide users with a method of identifying loose particles in electronic components prior to service (see Fig. 50). PIND techniques have been incorporated into reliability programs for critical components in applications ranging from cardiac pacemakers to manned spacecraft. In the pictured system, microprocessor control of an integrated feedback loop (for shock and vibration) is employed simultaneously with acoustic emission detection of loose particles. Programming and display of test conditions allow the user to tailor the system to meet rigorous internal procedures to qualify parts for MIL-STD 883C, Method 2020.6 and MIL-SPEC 202F, Method 217.

In this system, a shaker automatically steps through programmed vibration sequences. The feedback control circuitry maintains programmed shock and vibration values while an integral accelerometer and charge amplifier display the actual measured values. Vibration values up to 20 G and shock values from 200 to 2,000 G can be programmed. The integral acceleration feedback loop generates the correct shock and vibration drive values and measures the actual environmental conditions to ensure proper performance of the instrument.

Both visual and audio outputs of the acoustic emission signal are provided. Integral to the system is an audio monitor that allows the user to listen to the loose particles as they vibrate. Automatic threshold detectors provide indications as required by military standards. A monitor provides visual display of detected signals for further confirmation of detections. This unit is manufactured by Dunegan Corporation in Irvine, California.

Acoustic Emission Software

Simultaneous Data Recording

A practical form of acoustic emission software is designed to display acoustic emission data and simultaneously record them on disk or computer memory. The user can preset a limited number of graphs (generally ten or twenty from over two thousand possible plots) to display data in real-time while they are being collected and recorded on disk.

Any of the acoustic emission parameters (the four external parametrics, time and events) can be plotted against each other with no limitation on the available plots. To further enhance the graphics capability, the program gives the user the ability to alter the axes of graphs in many ways. The choice includes linear or logarithmic displays; summation or rate plots; correlation; histogram or line plots; user selection of minimum and maximum axes values; or automatic scaling of the axes. Furthermore, all these graph features are available in either real-time or replay.

The software includes source location capabilities based both on hit sequence and time of arrival algorithms. The user can filter the data at the front end, based on acoustic emission parameter value. Graphical filtering is based on spatial filtering of any parameter that can be displayed. For fatigue studies, a cycle counter is included to display acoustic emission versus cycle number.

The software package makes complete use of a computer's memory access capability, giving the user high data storage by bypassing the normally slow computer operating system. The user can choose to automatically and continuously transfer data to the disks or to manually transfer data from computer memory for tests expected to generate a small number of events.

In addition, this software package can (1) retain many test setups for ease of retesting; (2) take forced samples of the parametrics on a time-driven basis; (3) set special channel selection and discrimination; (4) set guard sensors; (5) change time of test resolution; (6) perform internal system diagnostics; and (6) set the dead time. The basic idea of these programs is to give the user complete but simple control of a sophisticated computer system.

Post-Test Plotting

A post-test plotting package is used for the display of acoustic emission test data and to aid in interpreting the results. A practical package runs under either CP/M or

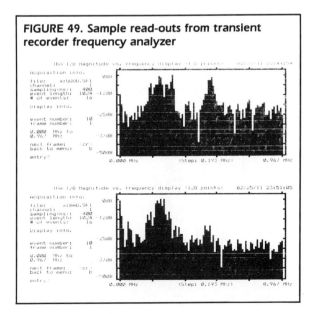

FIGURE 49. Sample read-outs from transient recorder frequency analyzer

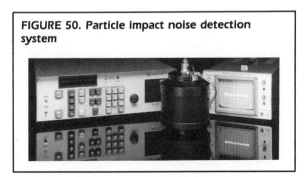

FIGURE 50. Particle impact noise detection system

MS/DOS operating systems. The program takes advantage of the high resolution graphics capabilities of dot matrix printers so that the acoustic emission data may be plotted with high resolution and in three dimensions. This plotting package has all of the software's standard features, plus the following enhancements.

1. High resolution (400 × 600 dots) on 215 × 280 mm (8.5 × 11 in.) paper.
2. Three-dimensional plots for location, planar, grid and all other parameters.
3. The axes variables may be relationships between data parameters: any parameter may be plotted versus (or operated by) any parameter and the operators can be any of the four arithmetic operators (addition, subtraction, multiplication and division).
4. User defined axes labels.
5. Five-digit floating point arithmetic allows greater flexibility in specifying axes limits and boundaries (some programs restrict the lower and upper limits of the axes to multiples of 10.)
6. Time intervals are normalized to one second. Thus, time plots accurately reflect the time resolution of the acquired data.
7. Ability to generate data plots through a submit or command file. This feature allows the user to create a file that contains the information necessary to print a number of data plots and submit the file. The system then generates the requested plots (overnight, for example).

Sample plots are shown in Figs. 51 and 52. Both of these software packages are available from Physical Acoustics Corporation.

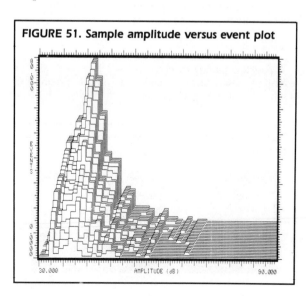

FIGURE 51. Sample amplitude versus event plot

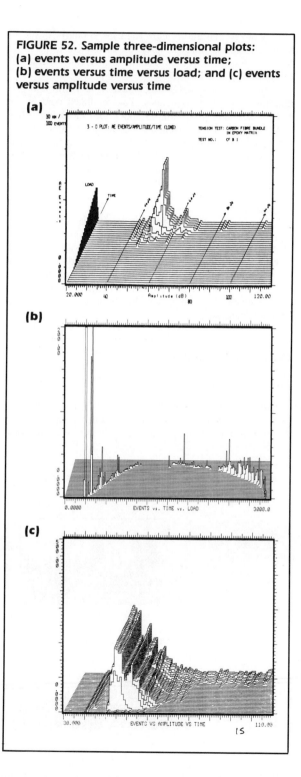

FIGURE 52. Sample three-dimensional plots:
(a) events versus amplitude versus time;
(b) events versus time versus load; and (c) events versus amplitude versus time

REFERENCES

1. Jaffe, B. and W.R. Cook. "Piezoelectric Ceramics." *Nonmetallic Solids*, J.P. Roberts, ed. New York, NY: Academic Press (1971).
2. *Piezoelectric Technology Data for Designers*. No. S-30. Bedford, OH: Vernitron Corporation (1985): pp 1-11.
3. Proctor, T.M., Jr. "Some Details on the NBS Conical Transducer." *Journal of Acoustic Emission*. Vol. 1, No. 3 (1982): pp 173-178.
4. Proctor, T.M. "An Improved Piezoelectric Acoustic Emission Transducer." *Journal of the Acoustical Society of America*. Vol. 71, No. 5 (May 1982): pp 1,163-1,168.
5. *Instruction Manual for Piezoelectric Accelerometers*. No. 101. Irvine, CA: Dunegan/Endevco Corporation.

SECTION 16

SPECIAL ACOUSTIC EMISSION APPLICATIONS

Gilbert Dewart, ESD Geophysics, Pasadena, California (Part 1)
Marvin Hamstad, University of Denver, Denver, Colorado (Parts 3 and 6)
Edmund Henneke, Virginia Polytechnic Institute, Blacksburg, Virginia (Part 4)
Hisashi Kishidaka, Kawasaki Steel Corporation, Chiba, Japan (Part 5)
Masato Kumagai, Kawasaki Steel Corporation, Chiba, Japan (Part 5)
Emmanuel Papadakis, Ford Motor Company, Redford, Michigan (Part 2)
Adrian Pollock, Dunegan Corporation, Irvine, California (Part 5)
Richard Stiffler, Virginia Polytechnic Institute, Blacksburg, Virginia (Part 4)
Ryoji Uchimura, Kawasaki Steel Corporation, Chiba, Japan (Part 5)

PART 1
ACOUSTIC EMISSION STUDIES OF GLACIERS

The test methods described in this *Nondestructive Testing Handbook* are based on acoustic emission that is structural, occurring mainly between 50 kHz and 3 MHz frequencies. The text below details geophysical phenomena and the frequencies of interest are between 0.1 and 50 Hz. In addition, the vocabulary of seismology is used here and no where else in the volume. Despite these differences, this study of glaciers is presented as a unique and special form of acoustic emission testing.

The Microseismicity of Glaciers

Sources of Acoustic Emission

The surface of a glacier is commonly a very noisy place. Some of that noise contains valuable information that, once decoded, can tell a great deal about the state of the glacier and perhaps about its immediate or even distant future. Many of the sounds emanating from the glacial environment reveal little or nothing about ice deformation processes. While sometimes of interest for other reasons, these sounds include the noise of wind, running surface water, rock falls on adjacent valley slopes and superficial thermal effects. Such sounds are generally considered interference, degrading the desired acoustic signals.

Some audible and subaudible glacial acoustic phenomena can be observed and measured and are intimately associated with the mechanics of ice movement. These phenomena provide a promising investigative technique for augmenting our understanding of glacial rheology and structure. The study of acoustic emission from glaciers, ice sheets, ice shelves and related snow, firn and ice masses, is analogous to the use of earthquakes in the study of the tectonic evolution of the earth's crust and upper mantle.

The term acoustic emission is defined as transient elastic oscillations produced by the rapid release of strain energy internally stored in a natural or manufactured structure that is undergoing a deformational process. A number of other terms have been used to describe this low level radiation: *stress wave emission*, *seismoacoustic activity* (favored by European and Soviet geophysicists), *elastic impulses*, *elastic shocks*, *subaudible rock noise* (SARN in geotechnical terminology), *sonic pulses* and *microseisms*. The last term is used in this discussion because of its currency in regard to geological materials and its seismological connotation. Other nomenclature specifically referring to glacier noise has entered the literature, including: *cryoseisms*, *firnquakes* and *icequakes*. The latter appears to be increasing in acceptance.

The Nature of Microseisms

The microseismic phenomena discussed here fall toward the lower end of a broad range of ground motion amplitudes and periods of oscillation of natural vibratory phenomena. This range extends from tiny pulses (released by slip dislocations on the scale of single mineral crystals) up to the surface waves, mantle waves and normal modes of free oscillation of the Earth (set up by great earthquakes). Seismologically, the magnitude of a microearthquake (M on the Richter scale) is by definition less than 3. Most of the events discussed here are several orders of magnitude smaller and come under the category of ultramicroearthquakes or *nanoseisms* ($M < 1$).

Because acoustic emission is closely associated with the instability of structures and is intrinsically related to processes of material deformation and fracture, its applications have been directed toward the detection of increasing instability and the prediction of structural failure.

The growing glaciological interest in instability phenomena has given considerable impetus to the study of the microseismicity of glaciers. Investigations of kinematic waves and glacier surges[1,2] and speculation about large scale unstable behavior in major ice sheets, with potentially profound geographical and social consequences[3,4] have heightened the demand for new methods of monitoring and forecasting the dynamic behavior of moving ice masses. Avalanches of glacier ice involving block, slab and ice bedrock failure can also have catastrophic results. They represent another, more localized and concentrated form of progressive instability.[5,6] Microseismic monitoring provides an effective approach to both the practical and theoretical aspects of research in this field.

History of Acoustic Emission's Use in Glacier Studies

Origins of the Technique

Geological investigations of acoustic emission originated in hard rock mines threatened by structural failures such as subsidence and rockbursts. These studies were initiated in the late 1930s in the South African gold field, in Canada by the seismologist E.A. Hodgson, and by L. Obert and W.I. Duvall of the US Bureau of Mines. Obert and Duvall discovered that stressed rock structures produced acoustical radiation.[7] They found that by passive observation of the rate and the amplitudes of the small elastic pulses (which they called microseisms), domains of high stress could be defined before rupture took place.

Count Rate and Wave Modes

The count rate of the shocks was found to be proportional to the time rate of change of measured pressure in the rock mass, whether the pressure was increasing or decreasing. The factor of proportionality was found to be greater for increasing stress. The rise in count rate began before the onset of visible cracking in the rock and the rate varied with the type of rock material under observation. These general principles, established in the early stages of acoustic emission research, have retained their validity and are the foundation of later work in the field. The primary forms of measurement, though now enhanced by more detailed descriptions of waveforms, also retain their value. Counts and count rates of events with amplitudes above a threshold (usually determined by signal-to-noise ratios) are still among the most valuable measurements.

Important discoveries were made during the 1950s about the response of various substances to different directions of applied load. New knowledge was also gained about the accompanying strain-related modes of elastic wave emissions. In particular, a kind of irreversibility in plastic deformation was found to be exhibited by a large number of common engineering materials.[8] Known as the *Kaiser effect*, this phenomenon expresses the retention of a stress memory by a restressed body.

In experiments on the degree to which cracking contributes to the different stages of creep, pulses of acoustic emission were observed coming from blocks of ice undergoing mechanical and thermal stress in the laboratory.[9,10] The ice was subjected to unconfined compressive load and at the stage of tertiary creep microfractures were found developing in planes perpendicular to the direction of principal tensile stress. The microfractures were accompanied by a sharp rise in the count rate of tiny pulses of elastic waves. This was probably the first demonstration of the formation of microcrevasses and their associated icequakes.

Detection of Acoustic Emission from Ice Fracturing and Movement

In connection with the International Geophysical Year, a network of standard short period seismograph stations was installed on the Antarctic continent. This soon lead to the observation of microseismic activity apparently originating in the Antarctic glaciers and ice sheets. Local tremors with equivalent Richter magnitudes of 3 or below were recorded almost every day at Scott Base.[11]

Some of these tremors were evidently associated with the mild volcanic activity of nearby Mount Erebus but most of these shallow focus events were unexplained by either volcanism or tectonism. Certain of the prominent low velocity sinusoidal waves with frequencies of 0.3 to 0.6 Hz and estimated M values of 3 or 4 were interpreted as being air coupled waves associated with flexural wave propagation through the Ross Ice Shelf to Scott Base.[12] It was suggested that these waves were caused by fracturing near the edge of the shelf prior to the calving of icebergs.

Microseismic vibrations recorded by the Soviet Antarctic Expedition exhibited emissive onsets and weak transverse phases with frequencies of 1 to 2 Hz.[13] Combining seismic observations with measurements of sea ice fluctuations, the Soviet researchers concluded that the tremors were ocean/sea ice coupled waves, which, like the similar but lower frequency waves observed at Scott Base, emanated from the ocean ice sheet boundary where processes of glacial disintegration were taking place.

Similar small shocks have also been described as originating from McMurdo Sound and the Dry Valley, and the Japanese Syowa base.[14,15] These signals were readily distinguishable from ordinary earthquakes by their monochromatic waveforms. They too were ascribed to sources associated with movements of the Antarctic ice.

Field investigations of microseismic phenomena in high latitude temperate and subpolar valley glaciers began in the mid 1960s. Short duration microseismic signals of 60 Hz were recorded on the terminus and near the firn line of the Lovenbreen Glacier in West Spitsbergen using an exploration geophone with relatively high frequency response characteristics.[16] These disturbances were at first called *eigen impulses* (similar signals are now called *type-I emission*) and the ones not caused by atmospheric effects were attributed to spasmodic motion of the glacier, modulated by the lubricating action of a subglacial water layer, as proposed in 1964.[17]

Detection of Crevasse Propagation

A three-component standard short period seismograph system, with peak response of about 1 Hz, and a strong motion seismometer for violent local shocks were installed on the medial moraine in the upper part of the ablation zone at a major confluence of a large glacier in the St. Elias Mountains.[18]

Four types of microseismic noise were originally recorded: continuous, earthquake, long duration and impulsive noise. A reexamination of the records by the author indicates a fifth pulse shape: a front loaded envelope with a high initial burst of energy, followed by a fairly rapid decay. This noise resulted from the shallow bore hole explosions of a seismic reflection/refraction survey being carried out on the glacier at the time of the seismograph operation.

The continuous oscillations were largely associated with various wind, pressure and temperature effects at the glacier surface. The more distant shocks, that had definite compressional phases and shear phases, were assumed to be local tectonic earthquakes (a few were identified with aftershocks of the Great Alaska Earthquake of March 27, 1964). Some of the extended disturbances were definitely correlated with large ice, snow and rock avalanches that were frequently visible and audible in the surrounding mountains.

The short impulses with frequencies of 2 to 3 Hz were thought to be chiefly due to elastic wave radiation from the initiation and propagation of cracks in the ice.[19] Strain rate measurements on the glacier indicated that the seismograph station was located within a zone of tensile stress and there was active crevasse formation and enlargement occurring in the region.[20]

Acoustic Emission of Surface Rupture

Vertical component geophones, in surface arrays and bore holes, were positioned on the temperate Athabasca Glacier in Alberta.[21] A large number of small, discrete shocks were observed, sometimes occurring in swarms. The array geometry enabled the location of many emission sources with good accuracy. The major activity appeared to occur up-glacier from visible crevasse fields and the microseisms were associated with extensional rupture near the surface of the glacier. No evidence of emission was observed from deep within the glacier, such as might be caused by active shear faulting.

Numerous microseismic events were observed in the College Fiord/Harvard Glacier region of Alaska.[22] These events displayed weak compressional phases or emissive onsets; prominent monochromatic (nondispersive) sinusoidal shear wave trains; frequencies of about 1.5 Hz; and equivalent magnitudes of 2.0 to 2.5 (representing release of mechanical energy in the form of elastic waves of around $10^{7.5}$ joules). On the basis of epicentral locations, energy considerations and frequency of occurrence, these events were believed to result from sudden, discontinuous ice motion that constituted a significant component of the movement of the nearby glaciers. This tended to confirm earlier conclusions.

In the course of a microearthquake survey with portable seismographs, directed toward tectonic relationships in northwest British Columbia and southeast Alaska, a large number of low frequency events were recorded.[23] The events were of unusual form, with epicenters restricted to regions containing large glaciers. A nontectonic and probable glacial origin was hypothesized for these clusters of microseisms.

Comparison of Magmatic and Glacial Emissions

In the 1970s a series of highly productive investigations of microseismic activity was carried out on the glaciated strato-volcanoes of the Cascade Mountains. Arrays of high and low frequency response seismometers were set up and telemetered to a central station recording with magnetic tape on Mount St. Helens.[24] A persistent type of shallow focus event characteristic of the glacier covered side of the volcano was apparently caused by discrete glacier movement rather than volcanic activity.

The typical signal, similar to the type-B events registered from Japanese volcanoes, consisted of a train of low frequency (1 to 5 Hz) waves. The first motion was strong at short epicentral distances but became emersive farther away. Poorly developed compressional and shear phases with a long, dispersive coda were also observed. The dispersion was apparently path related, due to the highly stratified velocity structure of the propagating media.

Further studies extended to other glaciated volcanoes in the same range reinforced the theory that these consistently recognizable events resulted from stick/slip motion involving localized increases in basal shear stress around obstacles at the bed of the glaciers.[25] It was suggested that microseismic monitoring of this process might serve as a forecasting aid for the start of a glacial surge. The investigators also confirmed that the peculiarities in waveform were indeed due to the properties of the propagation path: the coda became monochromatic if the waves were trapped in an essentially homogeneous channel of glacier ice decoupled from the layered bedrock.

An interesting adjunct to this investigation was the recording of an increasing count rate and amplitude level of emissions for two days prior to a major ice avalanche on Mount Baker. This crescendo-like time sequence is a typical feature of acoustic emission behavior in materials stressed to the point of rupture.

An intriguing alternate explanation for these type-II glacial events attempts to link them to the indistinguishable type-B volcanic tremors that are presumably associated with magmatic effects.[26,27] A hydraulic mechanism is believed to be involved in the abrupt stoppage of the water flow in an intraglacial or subglacial conduit and a consequent autoresonating water hammer vibration occurs on the sides of the cavity.

Expanded Instrument Capabilities

The sensing instruments used in all of these glacier studies were various models of intermediate frequency geophones or seismometers, which operate as inductive velocity gages. Little has been done in the range up to a kilohertz or beyond, where acoustical attenuation becomes a serious limiting factor.

A highly portable miniaturized magnetic tape recording system has been designed and assembled for these applications. Its sensor is a piezoelectric transducer acting as an accelerometer (the output increases as the square of the wave frequency). This device is intended to detect high frequency emissions under favorable circumstances of small attenuation rate and low noise level in the desired band. It has been field tested on the Valdez Glacier in the Chugach Mountains of southern Alaska. Its advantages are its lightness, ease of coupling and its ability to operate effectively at any orientation.[28]

Renewed Focus on Glaciers and Ice Sheets

Recently, attention has shifted back to the microseismic activity of polar glaciers and ice sheets. A portable seismograph is being used to detect many shocks with type-II waveforms and frequencies of 1 to 1.7 Hz on the Greenland ice sheet.[29] These events are readily distinguishable from earthquakes and teleseisms. They exhibit the familiar emergent onsets and monochromatic codas of emissions generated by smaller temperate glaciers, but with much greater durations and amplitudes (equivalent Richter magnitudes of 1.5 to 3.1 were estimated). Most of these signals seemed to originate near the terminus of the Jakobshavn outlet glacier, a large high speed ice stream that calves vigorously into Disko Bay. Hence there seems to be a suggestive reprise of the polar ice shelf calving conditions postulated as a source of microseisms in Antarctica.[12]

In the course of these glacier studies, two distinct types of signals have been the focus of attention: impulsive signals and extended signals. Two kinds of proposed source mechanisms have also been advanced: (1) jerky motion at the bed of the glacier; and (2) the opening of fractures and their development into crevasses near the glacier's surface. No discriminatory correlation of signal to source mechanism has emerged, and wave path effects have been found to greatly influence the form of the observed signal. The question of possible glaciotectonic movements within the body of the glacier and any emission that might be associated with it is still very open.

Processes and Problems in the Use of Acoustic Emission in Glacier Studies

Sources of Acoustic Emission

The radiation of elastic waves from localized domains of a stressed polycrystalline solid may be due to one or more possible microscopic or macroscopic mechanisms. None of the mechanisms are fully understood or described, theoretically or empirically, at the present time. Among the physical processes commonly related to acoustic emission generation are phase transformations, dislocation coalescence, cavity collapse, friction impact and the opening and propagation of fractures. Crack formation, impact and friction appear to predominate among the likely causes of microseismic radiation from geologic structures.

However, in a body consisting of a stressed aggregate of ice crystals, it is conceivable that a combination of microstructural dislocations associated with the process of plastic flow could result in an avalanche effect that releases a significant amount of elastic wave energy.

Acoustic emission observed in both geotechnical and mechanical engineering studies tends to fall into the two general categories of equivalent phenomena observed on glaciers: (1) type-I emission is impulsive or burst emission, in which each individual pulse is a discrete envelope of more or less sinusoidal waves that increase sharply in amplitude then decay more gradually; and (2) type-II emission is continuous noise, consisting of long wave trains of variable coherence that slowly change amplitude level.

There is empirical evidence from laboratory tests that impulse radiation emanates from the formation of microfractures and their subsequent enlargement, while continuous emission at relatively low energy levels is associated with dislocation sources.[30]

Regardless of initial form, all acoustic emission necessarily undergoes alteration as it propagates through transmitting media that are inhomogeneous, anelastic or anisotropic. The geometry of the environment and variations in impedance and attenuation constants greatly complicate attempts to examine source mechanisms. Reflection and mode conversion at boundaries, refraction and waveguide concentration, dispersion and scattering all tend to blur original waveforms that could provide clues to the nature of the generative process. The potency of some of these effects has been demonstrated in the microseismic studies on glaciers. However, the relative homogeneity of glacier ice affords some encouragement that judiciously positioned sensing devices and sophisticated signal analysis may be able to overcome some of the inherent difficulties of interpretation.

The application of array seismometry for obtaining fault plane solutions of source movements (which require the delineation of compression and dilatation sectors) may be feasible. This is done in microearthquake studies[31] and may also prove effective for large ice sheets where the question of glacier tectonics arises, such as with movement along well defined shear zones.[32,33]

Environmental Concerns

In this environment there are serious problems of acoustic coupling and high attenuation in snow (observed in investigations of acoustic emission capabilities for predicting snow avalanches).[34] Sensors on polar ice sheets or the accumulation zones of glaciers may have to be placed in a way that will permit penetration of the snow firn layers.

Instruments may be lowered into bore holes or attached to high conductivity waveguides, such as those used to acoustically monitor instability in other high attenuation materials.[35]

Interpretation of Acoustic Emission Data

It is important to pick up severely damped, high frequency radiation and it is increasingly recognized that a broad frequency range should be observed, from the seismic through the sonic to the ultrasonic band. There is little information available on acoustic attenuation in firn and ice in different states of cohesiveness and heterogeneity, and values for these materials would be a useful subject for *in situ* measurements.

Acoustic emission data gathering can take many forms. Simple counting of recorded events per unit time gives a rough indication of the intensity of the deformational process. The frequency distribution of all events magnitudes can be obtained by preselecting a discrete set of amplitude thresholds and counting the number of events whose maximum amplitudes exceed each of the preset levels. This is standard procedure for most industrial acoustic emission tests and is equally effective in this unique application.

Alternatively, an arbitrary amplitude threshold can be established and the number of individual waves that exceed it can be counted. This ringdown count is a function of instrument sensitivity, threshold level and signal amplitude. A spectral analysis can be made by calculating the Fourier transform of the signal as a function of time and then using the frequency spectrum of the transform magnitude and/or phase to help interpret the event.

Even though signal information may be inadequate for determining parameters directly relevant to source mechanism (first motion polarity, frequency spectrum, amplitude distribution, radiation pattern), the general character of the events and their count rates, maximum equivalent magnitudes and source locations can help identify local regions of high strain rate, tensile fracturing, internal creep or frictional sliding, giving a qualitative indication of the nature and dimensions of the process.

Suggested Developments

Microseismic investigations are concerned with inverse geophysical problems: looking for characteristic signatures in received signals in an effort to determine the nature of the mechanical process that generated them. There are a number of problems in glacier dynamics whose solutions might be significantly advanced by the application of acoustic emission techniques of measurement and analysis.

It would be very helpful to detect and identify emission coming from stick/slip movement at the bed and sides of a glacier and to differentiate their signatures from those of signals emanating from other sources. If this were practical, it would be possible to map the existence and intensity of this contribution to glacier movement throughout the ice rock contact area. Monitoring this activity would provide a temporal dimension related to changes in flow and could eventually have predictive value.

Similarly, it would be helpful to identify intraglacial noise sources, especially in concentrated configurations, that might represent glaciotectonic processes such as active shearing zones at depth. Passive bore hole logging of acoustic emission activity could provide correlation with structural and petrofabric evidence of these conditions, especially where great depths (as in the interiors of continental ice sheets) and high frequency, strongly attenuated waves are concerned.

There is still a great deal to learn about the relationship between the character and intensity of microseismic activity and the strength and type of stress field in which it originates. Studies of the comparative microseismicity in regions of compressive and extensional stress would be valuable in developing analysis criteria. The acoustical phenomena associated with vortical and turbulent ice flow are practically unexplored.

To illuminate the transition from creep to brittle fracture, there must first be a development of the empirical relationship between (1) the count rate, amplitude distribution and wavelength of acoustic emission; (2) the strain rate of the ice; and (3) the size distribution of microcracks and macrocracks in a region where intact ice is subjected to active crevasse initiation. There may be a critical value of strain rate or crack size at which significant changes take place in certain key parameters and relations, including count rate, spectral distribution and the magnitude/frequency relationship of microseisms. This might also be a suitable setting for investigation of the presumed association of type-I emission with fracture and type-II with internal creep.

The applicability of the Kaiser effect to a mass of ice subjected to various and repeated stress loading and unloading as it is transported along the glacier could be tested with microseismic arrays and strain networks at appropriate sites. Modification of the Kaiser effect by the healing of cracks and recrystallization could also be investigated.

Tensile and shearing fracture mechanisms raise some interesting questions with implications for microseismic activity. What is the pattern of precursor microfractures at a bergschrund, ice fall, incipient ice slab avalanche or internal shear zone? Is there an evolution from a random spatial distribution of microcracks to an increasing concentration of coalescing cracks along the ultimate fracture surface? How early can a critical configuration of cracks or emission sources be discerned? Perhaps fractal analysis involving scale invariant similarities of pattern complexity could be applied to these situations, as has been attempted in the case of earthquake fault zone asperities.[36]

The effects of internal structure and heterogeneity on microseismic activity in glacier ice should be examined, since these factors are significant in controlling acoustic emission in most materials.[37] Pronounced foliation, petrofabric and preexisting fields of preferentially oriented microfractures could strongly affect activity thresholds and rates. They might also influence the relationship between the size of the fractures and emissive events, and the frequency of both. This would be expressed in changes in the constants a and b of the Richter relation:

$$\log N = a + bM \qquad \text{(Eq. 1)}$$

where N is the number of events and M is the instrumental magnitude.

The very process of cracking and crack extension alters the effective elastic moduli of the media through which the resulting acoustic emission propagates. The opening of fracture patterns in the ice, whether they are filled by air or water, creates a new, oriented two-phase material that is elastodynamically anisotropic for emission wavelengths much longer than the dimensions of the cracks.[38] Thus, a stress-induced system of microcracks can cause azimuthal variations of attenuation, variations of the velocity ratio of compressional to shear waves, and polarization anomalies in both kinds of waves. This phenomenon, if not taken into account, can have misleading effects on hypocentral locations, but it can be detected by an array of three component sensors, in which the extent of a fractured zone and its development through time might be gaged.

The elements involved in microseismic production may vary with the temperature of the ice mass undergoing deformation. It is especially desirable that all of these recommended investigations be extended to high altitude and polar environments. The large, fast outlet glaciers of Greenland and the Transantarctic Mountains and the well-defined ice streams of continental ice sheets, such as those of the interior Ross Sea embayment region, may exhibit species of acoustic emission that are unknown on warmer glaciers, and they may be diagnostic of unfamiliar modes of ice movement.

Polar ice shelves are also of great interest, particularly where there are conditions of structural singularity like tidal hinge zones, actively calving ice fronts and bedrock islands that impede and divert the flow of ice. The roles of the ice shelf as a waveguide, in air-coupled and water-coupled wave propagation of microseismic signals also warrants further investigation.

Conclusions

This discussion outlines acoustic emission techniques used in studies of glaciers and ice structures and indicates a commonality between these studies and industrial acoustic emission methods. Although the test objects are very large and exotically located, glaciers do exhibit the expected acoustic emission response to applied load. The goal of the testing is also the same: locating characteristic signals in order to determine the nature of the mechanical processes that generated them.

This unique application additionally shares the general goals of typical acoustic emission test procedures: establishing appropriate data gathering criteria; determining accurate measurement and analysis techniques; establishing productive signal-to-source relationships; and devising successful interpretation methodologies.

PART 2
ACOUSTIC EMISSION STATISTICS AND THE CHOICE OF REJECT LIMITS

Statistical analysis has been applied to acoustic emission data taken from batches of ceramic parts under stress and it was discovered that the acoustic emission data in each batch fell along two distributions. It is believed that the distribution skewed toward large acoustic emission activity is representative of nonconforming parts.

This discussion proposes a method for assessing the proper acoustic emission activity for use as a go/no go test limit. The presence of the two distributions also leads to the hypothesis that the lower distribution represents controlled conditions while the upper distribution represents simultaneous out-of-control occurrences.

The acoustic emission technique is used on an empirical basis to accept or reject materials or indicate the need for repairs. This discussion focuses on the difference in acoustic emission from acceptable and from nonconforming parts within a batch of ceramic parts. The usual practice is to set an upper limit on some quantitative measure of acoustic emission activity such as the ringdown count or the number of acoustic emission events during a proof testing cycle.

Stressed discontinuities which propagate generally produce acoustic emission and the detection of acoustic emission is used by investigators as an indicator of the discontinuities' presence. Some discontinuities are small or structurally benign and some parts register acoustic emission activity even though they may fulfill acceptable performance criteria. It is commonly presumed that nonconforming parts produce more acoustic emission activity in the same proof test and, accordingly, the upper limit for acceptable parts can be established.

Presented below is a statistical method for investigating the acoustic emission output of various parts in a batch and for selecting the upper limit of acoustic emission activity to assign to acceptable parts. Very little research has been published on acoustic emission statistics but the existing work points to the possibility of advantageously using nonparametric statistics.[39,40] It is probable that many acoustic emission applications have yielded data that could be reevaluated in the manner proposed below.

Method for Assignment of Upper Acceptance Limit

Consider the acquisition, production or disbursement of nominally identical parts grouped into batches. Examples could include a tub of castings, an hour's production from a screw machine or a load from a heat treating furnace. Permit the batch to be inspected by any method *other than acoustic emission* which may detect and reject some of the structural discrepancies that would produce acoustic emission in a future proof test. This inspection step is not critical to the argument that follows.

Next, apply a stress test identically to every part while recording its acoustic emission over a definite portion of the stress cycle. The acoustic emission activity is the information to be analyzed statistically.

Finally, arrange the acoustic emission data from the entire batch in ascending order by acoustic emission activity. Assign the data median ranks from tables in the literature,[41] and plot the data on probability paper. Try both normal and log normal graph scales. It is probable that the log normal configuration will give better results.

Data on acoustic emission from four batches of ceramic parts are shown in Figs. 1 through 4. The parts in each batch were nominally identical while the batches represented different designs or different steps in chemical processing. The statistical properties of the data are evident from the graphs.

First, the data fall on one distribution (solid line) from the lowest acoustic emission activity up to a breakpoint. Second, the data above the breakpoint fall on a different distribution

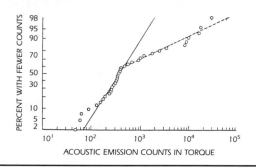

FIGURE 1. Graph of the percentage of thirty-seven ceramic specimens with fewer acoustic emission counts versus the total acoustic emission counts recorded; log normal distribution in first sixty percent of the specimens indicates a random process

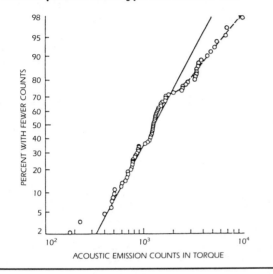

FIGURE 2. Graph of the percentage of seventy-seven ceramic specimens with fewer acoustic emission counts versus the total acoustic emission counts recorded; analogous to Figure 1 with ceramic parts of same type but another size

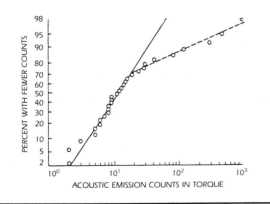

FIGURE 3. Graph of the percentage of twenty-six ceramic specimens with fewer acoustic emission counts versus the total acoustic emission counts recorded; specimens of the type and size of Figure 1, but coated and plated; note that the wash coat and plating have the effect of decreasing the emissions in the lower portion of the distribution by a factor of ten, but still allow large count rates in the other specimens

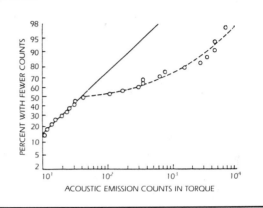

FIGURE 4. Graph of the percentage of thirty ceramic specimens with fewer acoustic emission counts versus the total acoustic emission counts recorded; uncoated ceramic parts of a different structure but the same cross-sectional area as in Figure 1; the log normal distribution is again present

(dashed curve) skewed to much larger acoustic emission activities. It is hypothesized that the presence of the second distribution at much higher acoustic emission activity represents nonconforming characteristics of the parts in the second distribution. However, the change in acoustic emission activity is not discontinuous: there is no gap between low emitters and high emitters. A breakpoint exists but not a step function. Unfortunately, durability data are not available from the specimens in these illustrations.

Choosing the level of acoustic emission for a practical go/no go test is the next step. The breakpoint is not the proper location. It can be seen that some points above the breakpoint might lie on the first distribution (solid line) if they were not pulled to the right by other points at very high acoustic emission values. After all, if all the parts were uniform, it is likely that all the points should lie along the first distribution. Log normal is a typical distribution for quantities not smaller than zero and not limited for very large magnitudes.[42,43]

With this consideration in mind, a statistical approach could be followed, setting the go/no go limit at a point equivalent to the upper 2σ or 3σ limit of the initial (lower activity) distribution. For clarity, consider the graph in Fig. 5, where the points are omitted and the line segments are labeled, and perform the following steps.

1. Draw the initial distribution AB.
2. Draw the upper distribution BC (may be straight or curved).
3. Extend the initial distribution along BD.

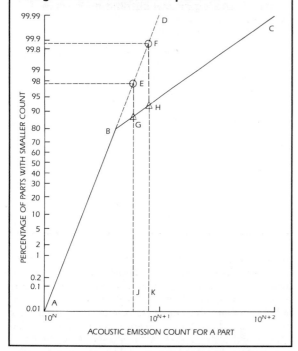

FIGURE 5. Graph of cumulative distribution on log scale; the geometric interpretation given in the text shows how to choose the acoustic emission level or rejection of parts based on the change in the distribution at point-B

4. Find the points E at 97.72 percent smaller for 2σ or F at 99.86 percent smaller for 3σ along the segment BD.
5. Drop the verticals EGJ and FHK: these define the last acceptable point along the upper distribution (G or H, depending on the chosen confidence limits) and the corresponding go/no go limits of acoustic emission activity (J or K).
6. Consider the points above G or H along BC to represent nonconforming material.

The above steps are taken to analyze acoustic emission data from batches of parts, to screen out nonconforming parts when there is no step function difference between conforming and nonconforming parts but only a change in slope (a breakpoint). While the methodology may not work in all cases, it provides a framework for logically examining the data for trends and significance. It is suggested that acoustic emission data be analyzed in this manner before decisions on limits are finalized for practical tests.

Batch Analysis Extended to Continuous Production

It is always desirable to establish standards and calibration levels so that a test can be used repetitively without being set up with new values for every batch.

The batch analysis given above can be extended to continuous production if the following assumption holds true: the distribution AB in Fig. 5 represents the controlled process and is repeatable from batch to batch (continuously). *Repeatable* is taken to mean a duplication of slope from batch to batch and a duplication of intercept at some moderately low exceedance probability such that the point B never reaches a value that low.

The point B need not remain where it is drawn in Fig. 5 (at 80 percent), but may wander up and down the line AD. The curve BC represents an out-of-control condition that can engulf a varying number of parts (actually above points G or H) as time goes on. While it is unusual to consider simultaneous controlled and out-of-control conditions, this anomaly is suggested by the nonparametric statistics used here.

If the slope and the low exceedence intercept remain constant, then the line AD is defined and the points J and K are constants. Thus, J or K could be used as the go/no go calibration for the test system as long as the line AD represents the controlled condition.

If the process goes so grossly out-of-control that point B is forced down to point A in Fig. 5, then only about 2 percent of the production would be accepted by a set point at K. This can be seen by moving line BC downward, parallel to itself in Fig. 5. On the other hand, set point K would reflect less that 0.2 percent of production if control became so perfect as to move point B above point F.

If slope AD changes or if line AD translates up or down, the system must be recalibrated using the six steps given above.

Conclusion

The statistics of this method presuppose a double distribution, partially controlled and partially out-of-control, existing simultaneously in batches of parts production.

The concept of simultaneous controlled and out-of-control conditions is novel in statistical process control but may lead to a deeper understanding of subtle nonconforming conditions influencing abstruse measurements such as acoustic emission activity.

PART 3
ACOUSTIC EMISSION IN DEVELOPMENT OF COMPOSITE MATERIALS

A guide to the use of acoustic emission testing in the research and development of composite materials is presented below. Topics covered include:

1. unique information to be gained from acoustic emission data;
2. sources of acoustic emission in composites;
3. procedures for achieving successful composite acoustic emission testing;
4. measurement and interpretation of acoustic emission from composites;
5. specific applications of previous acoustic emission data from composites; and
6. future directions of acoustic emission research in the field of composites.

Characteristics of Composite Materials

Composite materials can be divided into three broad categories: dispersion strengthened, particle reinforced and fiber reinforced.[44] In each category, the composite is made of a matrix material and of a second phase material that is distributed throughout the matrix.

Dispersion strengthened composites are distinguished by a small diameter (10^{-2} to 10^{-1} μm or 0.4 to 4 μin.) hard second phase (volume concentration less than 15 percent) dispersed throughout the matrix material. These materials are distinct from precipitation alloy systems in that (1) they are normally made by techniques such as powder metallurgy and (2) the second phase does not go into solution when the material is heated near the melting temperature of the matrix.

Particle reinforced composites are distinguished from dispersion strengthened composites by larger dispersoid size (greater than 1 μm or 40 μin.) and increased dispersoid volume concentration (greater than 25 percent). An additional difference is that particle reinforced composites are strengthened by the inherent relative hardness (compared to the matrix) of the dispersoid and by dispersoid constraint of matrix deformation. This strengthening mechanism is distinct from that in dispersion strengthened composites. In this latter category, restriction of the motion of dislocations by the second phase provides the strength enhancement.

The distinct microstructural difference of *fiber reinforced composites* is that the second phase (the fiber) has one dimension that is larger than the other two dimensions. This characteristic leads to anisotropic properties rather than the isotropic properties of the first two composite categories.

Acoustic Emission in Composite Tests

As in other acoustic emission examinations, data are obtained from a composite structure when the specimen is significantly loaded to cause a stress field. Because research and development of composite materials nearly always includes the application of mechanical tests or thermal loads, acoustic emission tests may often simply be added on to an existing test procedure.

Acoustic emission data provide information that is presently unobtainable by any other technique. There are several aspects of acoustic emission data that give it this unique character. First, acoustic emission data give a real-time record of the progressive damage that occurs. This record is revealed in the form of rapid energy releases as the composite structure moves toward failure. Such information is valuable for developing damage, hardening and failure models for composite materials.

Secondly, acoustic emission signals are very sensitive. In metals for example, discontinuities as small as about 10^{-11} mm (10^{-12} in.) have been observed by acoustic emission.[45] The fractures of single filaments of organic fibers with diameters of 10^{-2} mm (4×10^{-4} in.) are also easily observed. Because acoustic emission responds to microstructural variations, it can provide very sensitive checks on process control.

Thirdly, the acoustic emission technique is often sensitive to energy releases within the entire test object. This characteristic is an advantage over real-time measurement techniques that make measurements on the surface or over a limited portion of the volume of a composite.

Finally, acoustic emission methods have great potential to nondestructively determine the strength or integrity of composite parts and to determine the location of growing discontinuities. This potential exists because observations indicate that many fiber reinforced composites fail as a result of discontinuities generated by progressive damage that occurs during loading.[46]

Because of the unique information obtainable from acoustic emission tests, researchers use the technique in many standard mechanical tests of fiber reinforced composite specimens. The method's use in field testing of composites is now well-documented. For the purpose of study, the list below itemizes some of the original papers detailing the research capabilities of acoustic emission techniques in the development of composites materials and in the study of their characteristics.

1. *Tensile tests*: (strands) references 47 and 48; (NOL rings) references 49 and 50; and (tensile bars) references 51 through 60.
2. *Bending tests*: references 51 and 61.
3. *Failure mechanisms*: references 62, 63 and 64.
4. *Fatigue tests*: (strands) reference 48; (NOL rings and pressure vessels) reference 49; and (tensile bars) reference 65.
5. *Pressure vessels*: references 47, 49, 50, 66 and 67.
6. *Stress rupture tests*: (strands) reference 48; and (NOL rings and pressure vessels) reference 49.
7. *Fracture studies*: references 68 and 69.
8. *Structures*: references 60, 71 and 72.

Sources of Acoustic Emission in Composite Materials

The application of mechanical or thermal stress to a composite material results in elastic energy being stored in the material. The sudden local release of such energy is the basic cause of acoustic emission. To observe the acoustic emission from this energy release, it must occur over a short enough period of time. Secondly, the release must be of sufficient size to generate a stress wave that propagates through the material. The stress wave must in turn reach the acoustic emission transducer with sufficient magnitude to give an output above the electronic noise level in the acoustic emission system. The local energy release is called an *acoustic emission event* and the signal that comes out of the transducer is the electrical equivalent.

The specific mechanisms that produce acoustic emission events in composite materials are numerous and include the following: (1) second phase or fiber cracking and failure; (2) second phase or fiber interfacial debonding; (3) second phase or fiber plastic deformation; (4) matrix plastic deformation and cracking; (5) interlaminar debonding; and (6) rubbing of the second phase or fiber against the matrix.

Typically, the acoustic emission from a fiber reinforced composite material is of significantly larger amplitude than that from a metal. In fact, it is not unusual to audibly detect acoustic emission from such a specimen. Acoustic emission signals from a transducer are often 2 to 50 mV for fiber reinforced composites. Signals from metals are typically 500 μV or less.

The amplitude of acoustic emission from dispersion strengthened and particle strengthened composites is often similar to that from typical metals. Hence, the signals from these composites will probably be continuous acoustic emission[72] related to dislocation events in the metal matrices, except when high energy release mechanisms (such as shearing of the dispersed second phase particles) are present. Under these latter conditions, burst emission is expected.

Typical Factors Affecting Acoustic Emission Tests of Composites

Eliminating Noise

The single most important factor in a successful acoustic emission test of composites is the elimination of noise generated outside of the material under study. Since acoustic emission is such a sensitive tool, it is not always easy to successfully eliminate such noise. Too often acoustic emission field tests are run without adequate checks on noise sources. The net result is that the data are incorrectly interpreted.

Typical sources of noise in composite and other materials are: (1) specimen grips (wedge grips); (2) tabs on composite tensile specimens; (3) the test system load train; (4) the loading mechanism; (5) hydraulic pumps, leaks or gas expansion (during gas pressure vessel testing); (6) noisy nearby equipment; (7) spurious electronic signals; (8) ground loops; (9) overhead cranes; and (10) friction or rubbing during fatigue tests.

Two methods may be used to check for extraneous noise sources: (1) reload a composite test sample that exhibits the Kaiser effect;[72] and (2) replace the composite specimen with a high strength metal specimen (this substitute specimen should remain well in the elastic region when loaded and should not produce detectable acoustic emission). If significant extraneous noise is found by these methods and cannot be eliminated, electronic filtering is recommended.

Transducer Placement for Composites

Acoustic emission transducers should be placed directly on the composite specimen rather than on specimen grips, metal bosses or other attachments to the specimen. This procedure provides the most consistent results since coupling and interface losses are decreased. Filtering and attenuation of the stress wave during propagation to the transducer is also decreased. Rubber bands are a convenient way of securing the transducer in place when permitted by the environment and when it is inconvenient to bond the transducer to the test piece.

When epoxy bonding is not used, a viscous couplant is important to achieve consistently good transmission. Use of a waveguide should be limited to specimen environments in which the acoustic emission transducer cannot function properly (temperature cycling of some commercial transducers may cause extraneous acoustic emission) or when failure of the part leads to excessive costs for replacement of broken transducers.

Establishing Bandwidth and Gain for Composite Tests

The frequency band-pass used in acoustic emission tests of many fiber reinforced composites and metals is 100 to 300 kHz. Most experimenters who have examined acoustic emission frequency distributions in fiber reinforced composites report the majority of high amplitude signals are at frequencies less than 400 kHz. Significant attenuation has been found above 400 kHz for S-glass fiber epoxy.[47] This result encourages the use of lower frequencies. Except for special tests, the lower frequency of the bandwidth should generally be set as low as possible while still eliminating significant extraneous noise.

For the two other categories of composites, bandwidths typically used with metals are appropriate. Again decreasing the low frequency cutoff is encouraged because of an expected increase in attenuation in composites over that present in metals.

The electronic gain and amplifiers must be compatible with the amplitude of the acoustic emission signal out of the transducer. To obtain acoustic emission signals that are not distorted, the input and output limiting voltages of the amplifiers must not be exceeded. This can be a problem in acoustic emission systems that have their threshold level at 1 V. Since typical acoustic emission amplifiers saturate at 20 V peak-to-peak, the result is that only a 14 dB range of undistorted signal exists above the threshold level. Often the range of signal amplitudes in fiber reinforced composites exceeds this 14 dB range.

Measurement of Acoustic Emission Signals from Composites

Characteristics of Signals from Composites

The most commonly reported acoustic emission from fiber reinforced composites is burst emission. This might be because of the fact that the electronic gains convenient for processing burst emission are too small to allow the observation and measurement of continuous acoustic emission. Since burst emission dominates, it is not expected that the summation of acoustic emission in composites depends on strain rate. Also, the count rate for burst emission depends on strain rate. The root mean square (rms) voltage also has a complicated dependence on the strain rate.

For dispersion and particle strengthened composites where continuous acoustic emission is expected, the summation of counts and count rate techniques is highly dependent on strain rate. The rms measurement depends on the square root of the plastic strain rate, provided the total number and amplitude of acoustic emission events are independent of strain rate.[73]

Measurement Techniques

Techniques useful for measuring acoustic emission signals from composites are: (1) visual observation on an oscilloscope; (2) listening to an audio loudspeaker; (3) ringdown counting both for count rate and summation of counts above a fixed or variable threshold; (4) event counting both for count rate and summation of counts; (5) the DC output of a true rms voltmeter; (6) amplitude distributions; (7) energy measurements; and (8) frequency distributions.

The particular technique used in a given test depends on the proposed use of the acoustic emission data. The technique most used for fiber reinforced composites is summation of ringdown counts. For composites that exhibit continuous acoustic emission, the use of rms measurements has the same advantages that it has for metals.[73]

Interpretation of Acoustic Signals from Composites

Attenuation in Composites

Several key factors must be kept in mind when interpreting acoustic emission signals from composites. First, the attenuation of the acoustic emission signal as it propagates in the composite is normally much higher than in most metals. Amplitudes and amplitude distributions of the signals are not as meaningful as they are in tests with metals unless the acoustic emission stress waves travel a small distance or unless all the stress waves have traveled about the same distance under the same composite structural conditions.

To analytically assess this effect, it is necessary to have data on the attenuation properties of the composite in the various directions where composite properties differ.

Frequency Composition

As with metals, the frequency composition of acoustic emission in composites undergoes changes as the wave propagates from the location of the event to the transducer. For the most part, these changes are little known and are in general more difficult to determine in composites than in metals. In addition, resonances of the acoustic emission transducer also appear in the signal, as they do in metal tests. This presents an additional complicating factor in attempts at frequency analysis.

Electronic Gain

The test system electronic gain determines which acoustic emission mechanism the system will respond to. Since by their nature composites can be expected to emit stress waves over a large amplitude range, it is quite possible to set the gain too low and thus miss signals from certain low amplitude acoustic emission mechanisms.

When acoustic emission signals vary in amplitude up to 60 dB, the choice of gain depends on the intended use of the acoustic emission data. For studies of deformation and microfailures, a relatively high gain should be used. For macroscopic failure warnings, a relatively low gain should be used.

Data Plotting

Acoustic emission test data should be plotted against some other macroscopic or microscopic engineering parameter. These parameters should in turn be related to the possible mechanisms that generate the acoustic events. Plots of acoustic emission versus stress or strain are very useful for interpreting tensile tests of composites. Plots of acoustic emission versus time are convenient for studying composite stress rupture properties.

Precision of Interpretation

The degree of precision required in the interpretation of acoustic emission data depends on the intended use of the results. For example, if the acoustic emission data are used as an indicator of macroscopic failure, then it is not necessary to know the source and mechanism that generated the signal. The only requirement for this kind of interpretation is the ability to recognize the acoustic emission signals that indicate the approach to failure. Similarly, when acoustic emission is used for process control, only significant changes in the acoustic emission pattern need to be recognized.

At the other extreme, when acoustic emission is used to study the deformation of a composite material, it is often required that sources and mechanisms be identified. In such cases, it is best to first characterize and identify the sources from each of the separate materials that make up the composite.

This requires generating stress fields similar to those in the actual composite. Next, the acoustic emission from each possible composite mechanism is characterized. Simple composite specimens should be designed to study each different mechanism (see references 53 to 64 and 66). After this characterization, it is much more likely that the acoustic emission data from the real composite can be reasonably interpreted.

Applications in Research and Development of Fiber Reinforced Composites

Design Applications of Acoustic Emission Tests of Composites

Acoustic emission tests can be used in composite structure design and for comparisons of different structures designed to exhibit the same measured properties. Figure 6 is an example of this application, showing the count rate and count summation for an organic fiber epoxy strand and for an elongated NOL ring of the same material.[50] The early damage in the ring (Fig. 6b) is indicated by the early peak in count rate data and does not occur in the strand (Fig. 6a). It is unlikely that the tensile strengths determined from these two tests will be directly comparable; strand results will not be directly translatable to elongated NOL rings.

An application of acoustic emission to filament wound pressure vessel design also provides unique design data. Figure 7 shows this information for organic fiber epoxy pressure vessels.[67] The rapid rise in count rate and count summation just before failure in Fig. 7a is due to the fact that the uniformly stressed hoop wraps were highly loaded with a substantial number of breaks before macroscopic vessel failure (vessel failure does not always occur in the hoop wraps in such cases).

The absence of this rapid rise in the acoustic emission data shown in Fig. 7b indicates that the hoop wraps were not highly loaded before vessel failure and were not used to their full potential strength. Using this acoustic emission information, the number of hoop wraps can be reduced until they are used to their full potential. This results in an increase in vessel performance (failure pressure times volume divided by mass).

Acoustic emission data have also proved useful in describing the design changes in a filament wound pressure vessel. A reduction in burst strength occurred when the hoop and longitudinal wraps of an S-glass epoxy vessel were interspersed.[46] The acoustic emission signals indicated an extreme amount of early fiber damage resulting in a pronounced loss of strength. Figure 8 shows this graphically by comparing summation of acoustic emission counts for a typical, interspersed vessel versus a typical, noninterspersed vessel.

The location of acoustic emission sources in composite structures[47] is also used to find stress concentrations in composite prototypes. Stress concentrations are then corrected by redesigning the structure. Acoustic emission technology

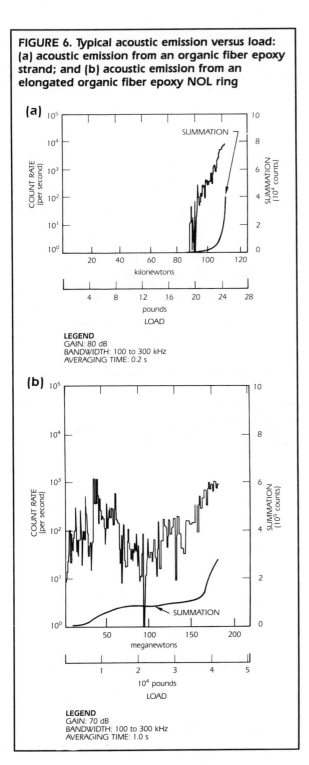

FIGURE 6. Typical acoustic emission versus load: (a) acoustic emission from an organic fiber epoxy strand; and (b) acoustic emission from an elongated organic fiber epoxy NOL ring

is also used to set design or operating load levels below those where acoustic emission data indicate significant damage.

Applications of Acoustic Emission in the Manufacture of Composites

Because acoustic emission tests provide microinformation on damage accumulation in composites, acoustic emission signals have significantly different patterns when damaging changes occur in the manufacturing process. Improper

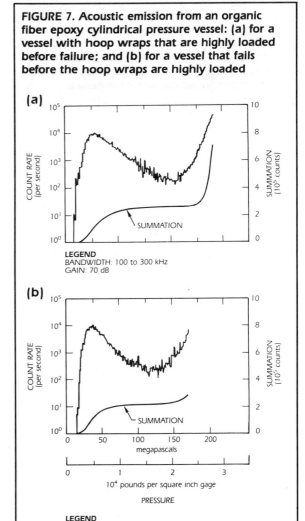

FIGURE 7. Acoustic emission from an organic fiber epoxy cylindrical pressure vessel: (a) for a vessel with hoop wraps that are highly loaded before failure; and (b) for a vessel that fails before the hoop wraps are highly loaded

epoxy cures, wrong resin mixtures and many other changes in process variables can be detected by obtaining acoustic emission data during cure cycles or subsequent mechanical testing.

As an example of this, Fig. 9 shows the changes in acoustic emission data when a pressure vessel is wound with graphite fiber that tends to fray during the winding.[67] The early acoustic emission, which is caused by the breaking of the frayed graphite, allows identification of poorly constructed vessels, during a proof test to about 20 percent of design strength.

Certain acoustic emission procedures have been designed based on tests of specimens containing artificially induced discontinuities. These procedures are then used to sort composite pressure vessels containing anomalies induced by

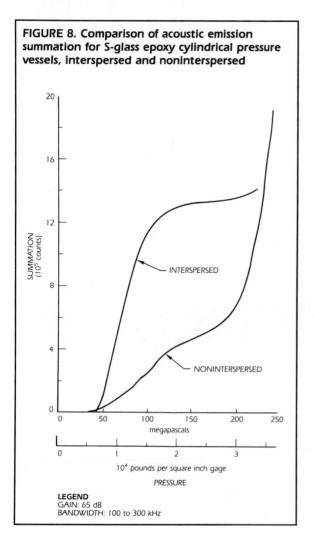

FIGURE 8. Comparison of acoustic emission summation for S-glass epoxy cylindrical pressure vessels, interspersed and noninterspersed

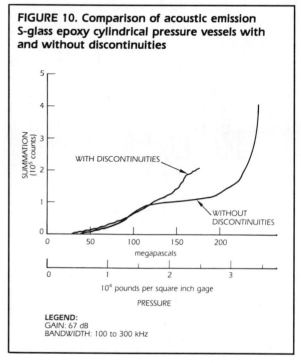

FIGURE 9. Acoustic emission from frayed and unfrayed graphite epoxy cylindrical pressure vessels

FIGURE 10. Comparison of acoustic emission S-glass epoxy cylindrical pressure vessels with and without discontinuities

improper handling during manufacturing. This technique is demonstrated in Fig. 10. The summation of acoustic emission for an S-glass epoxy pressure vessel containing significant discontinuities is compared to a typical vessel containing no discontinuities.[46] If acoustic emission techniques were used during a proof test to about 30 percent of the normal failure pressure of such vessels, the rejectable vessel could be sorted out.

Development of Models for Composite Behavior

Acoustic emission has a unique application in the development of models for composite damage mechanisms. The ability to progressively monitor the entire composite structure on a real-time basis is unique. Several examples serve to illustrate the type of information that can be applied to model development.

A model to characterize the behavior of composite, elongated NOL rings under tensile loads would be unlikely to correlate with real data unless it included a mechanism to account for the early damage that is so clearly demonstrated in Fig. 6b. Similarly, a model for the deformation of an organic fiber epoxy pressure vessel that exhibits the acoustic emission characteristics shown in Fig. 7 must include a mechanism to explain the early peak in count rate data (reference 50 presents a mechanism that seems to explain this early peak in the acoustic emission data).

Models that predict the accumulation of broken fibers during stress rupture of organic fiber epoxy strands must correspond to the damage accumulation shown in Fig. 11. Figure 12 shows the correlation of large amounts of acoustic emission with the knee in a stress strain curve.[56] This example also indicates the unique characteristics of acoustic emission which only begins beyond the elastic region.

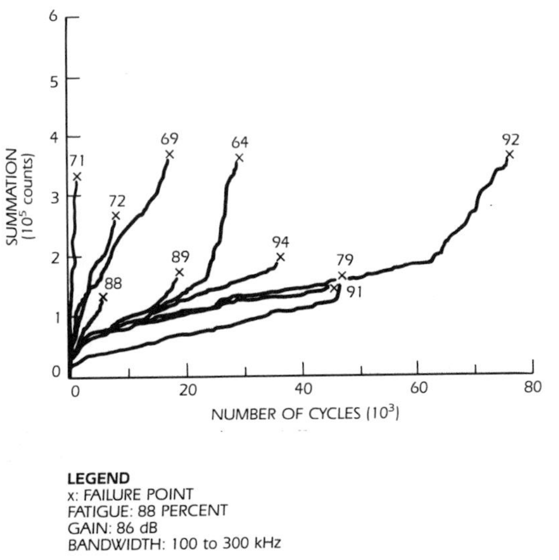

Determination of Composite Properties

Acoustic emission can be used to determine composite properties that would be difficult or impossible to determine by other techniques. For example, acoustic emission has been used to stop loading just prior to macroscopic failure of composite specimens. After rapidly unloading such specimens, the characteristics of test generated discontinuities in various advanced stages can be studied by other techniques. Without acoustic emission, it would be virtually impossible to find damaging discontinuities at such early

FIGURE 11. Number of cycles versus summation of acoustic emission above 75 percent of tensile failure load during fatigue tests; organic fiber epoxy strand specimen numbers shown

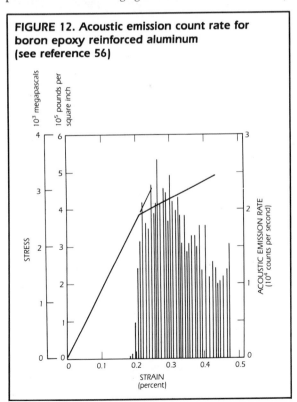

FIGURE 12. Acoustic emission count rate for boron epoxy reinforced aluminum (see reference 56)

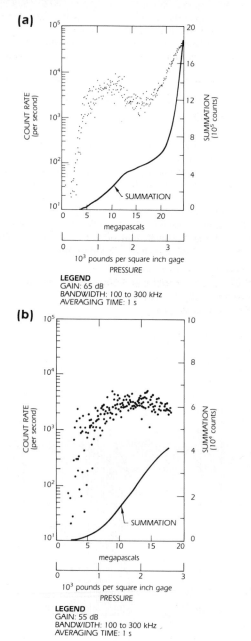

FIGURE 13. Typical summations of acoustic emission: (a) acoustic emission from an S-glass epoxy cylindrical pressure vessel; and (b) acoustic emission from a graphite fiber epoxy cylindrical pressure vessel

times. This technique greatly enhances the success of failure mechanism studies in composites.

Another composite property determined from the results of acoustic emission tests is the relative density of fiber failures in filament wound pressure vessels made with different fibers.[67] Figure 13 shows the typical summation of acoustic emission from a graphite fiber epoxy cylindrical pressure vessel and a similar S-glass fiber epoxy pressure vessel. The researchers were unable to obtain the rapid rise in summation just prior to failure in the graphite vessels even when they substantially reduced the number of hoop wraps. They concluded that the more brittle graphite composite failed macroscopically at a much lower average density of fiber breaks.

Conclusion

Further research would open many new applications of acoustic emission on composites. The most obvious additional research involves the extension of acoustic emission testing to particle and dispersion strengthened composites. Immediate applications would be in the areas of advanced process control, detailed model development and composite properties.

The present state of acoustic emission testing presents a technique uniquely suited to the study of composites, in the design phase, in the laboratory and in the field (see the next part of this chapter and the volume index for details of acoustic emission tests of composites in critical industries).

PART 4
APPLICATION OF POLYVINYLIDENE FLUORIDE AS AN ACOUSTIC EMISSION TRANSDUCER FOR FIBROUS COMPOSITE MATERIALS

The application of polyvinylidene fluoride (PVDF or PVF_2), a polymer exhibiting strong piezoelectricity, shows promise as an acoustic emission transducer for fibrous composite materials. Although not as sensitive as lead zirconate titanate (PZT), there is no disadvantage in using PVF_2 for composite materials because of the relatively energetic acoustic emission events caused by fiber fracture, delamination and matrix cracking.

Polyvinylidene fluoride has a number of features that give it a distinct advantage over conventional piezoelectric transducers, including: low cost, low mass, flexibility, wide band frequency response and high internal damping. In addition, PVF_2 is sensitive to in-plane displacements and responds to flexural modes and in-plane displacements of the composite structure when bonded to the surface of a structure.

A technique for making PVF_2 transducers and the results of tests comparing PVF_2 transducers with commercial transducers are discussed below.

Characteristics of Polyvinylidene Fluoride

In fibrous composites, there are four primary sources of acoustic emission: fiber breakage, matrix cracking, fiber matrix disbonds and delamination. These events are manifested as elastic waves propagating to the surface of the structure and can be detected by piezoelectric transducers. A material that has been used with success as an acoustic transducer for composite materials is the semicrystalline polymer PVF_2.

This polymer has a molecular conformation of two fluorine atoms opposite two hydrogen atoms along the carbon backbone of the PVF_2 molecule.[74] Because of the strong electronegativity of fluorine, the PVF_2 molecule possesses a large dipole moment that gives it its piezoelectric properties after processing.

The processing of PVF_2 includes electrical poling, unidirectional mechanical stretching and thermal treatment. This results in a piezofilm with maximum in-plane displacement sensitivity in the stretched direction as well as some out-of-plane displacement sensitivity. Polyvinylidene fluoride is commercially available in preprocessed sheets, with nickel electrodes, in thicknesses from 9 to 110 μm. Some typical properties of PVF_2 are given in Table 1.[75] Much research has been done on the electrical and mechanical properties[76-83] of PVF_2 and on its applications.[84-85] One review of the properties and applications of PVF_2 cites more than one hundred references.[76]

Polyvinylidene fluoride has several advantages over conventional transducers, as outlined below.

1. Its flexibility allows it to conform easily to surveyed surfaces.
2. The size and shape of the transducer can be tailored for a specific application.
3. Because of its low cost, a disposable transducer can be economically made.
4. Because of its low mass, a PVF_2 transducer will not drastically affect the mechanical response of the specimen.
5. Its high internal damping reduces transducer ringing and flattens frequency response.

The in-plane sensitivity of PVF_2 can be used to detect in-plane displacements of a structure surface when bonded to the surface with an adhesive. Acoustic emission in a thin layered material such as a fibrous composite can manifest itself as plate waves (having the largest displacement in the plane of the specimen) or as flexural modes. Experimental observations have shown that PVF_2 is as sensitive as a standard commercial ceramic acoustic emission transducer to acoustic emission in composite materials. This may be due to the fact that acoustic emission in composites apparently excites major in-plane displacement components.

Transducer Construction

Because leads cannot be soldered directly to the electrodes of the PVF_2 transducer without damaging the polymer and electrode, an alternate method must be used. A method is outlined below that has proven satisfactory as far

TABLE 1. Typical properties of polyvinylidene fluoride (PVF$_2$) and lead zirconate titanate (PZT)

Characteristic	Symbol	PVF$_2$	PZT	Units
Static piezoelectric strain constant	d_{13}	20 to 25	110	pC·N^{-1}
	d_{33}	20 to 22		
Static voltage output coefficient	g_{31}	0.19 to 0.23	0.01	Vm·N^{-1}
	g_{33}	0.19 to 0.21		
Electromechanical coupling factor	k_{31}	9 to 15	30	percent
Relative dielectric permittivity	ϵ/ϵ_O	12.0	1,200	percent at 1 kHz
Dielectric loss factor	tan δ	0.015 to 0.02		percent at 1 kHz
Young's modulus of elasticity	E			
stretched direction		1.5 to 3	83.3	10^9 N·m^{-2}
transverse direction		1.1 to 2.4		
Density	ρ	1.8	7.5	g·cm^{-3}
Velocity of sound	c	1.5 to 2.2	5.0	km·s^{-1}

as durability, cost and ease of fabrication. The required materials consist of PVF$_2$, flat wire, bondable strain gage terminals and silver filled epoxy. A small soldering iron, some dental tools, binder clips and pressure pads consisting of aluminum plates and rubber pads are the required tools.

First, the strain gage terminal strip (Fig. 14a) is cut (Fig. 14b) to make the contacts and terminals for the wiring harness. The excess backing is trimmed away from the contacts (Fig. 14c). Next, two pieces of flat wire are soldered to the terminal leads (Fig. 14d). The contacts are soldered to the ends of the flat wire with each of the copper tabs facing the opposite direction (Fig. 14e).

Next, the transducer element is cut to the desired size from the sheet of PVF$_2$ (Fig. 14f). One edge of both sides is slightly buffed using 320 grit emery paper (Fig. 14g); this helps provide a good bond between the contacts of the wire harness and the transducer element. Care must be taken to minimize the amount of the electrode abraded away.

A small amount of silver filled epoxy is then placed on the contacts. The PVF$_2$ element is placed between the two contacts and the assembly is sandwiched (Fig. 15a) between the rubber pads and the aluminum plates and is then clamped together with binder clips (Fig. 15b).

After the epoxy has cured, the PVF$_2$ transducer is carefully removed from between the rubber pads and excess epoxy is removed to prevent a short across the electrodes. The transducer is then ready to be bonded to a specimen and to have leads soldered to the terminal strip (Fig. 15c).

FIGURE 14. Construction of a PVF$_2$ transducer's wire harness and transducer element: (a) terminal strip; (b) terminal; (c) contacts; (d) flat wire on terminal leads; (e) wire harness; (f) transducer element; and (g) buffed edge

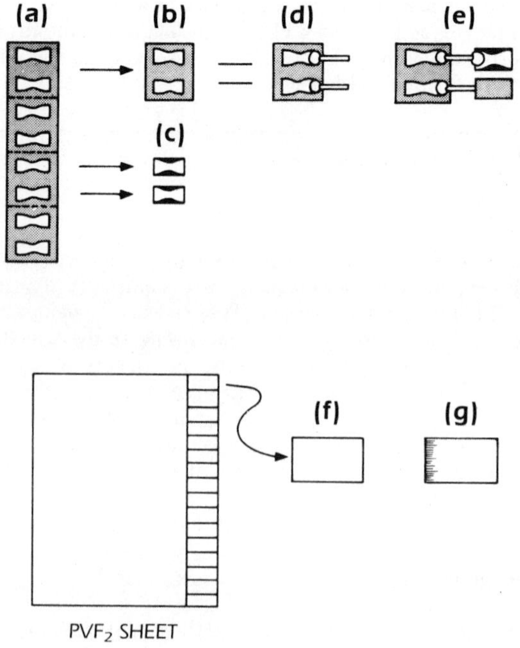

Tests of Sensitivity

Several sensitivity tests were performed using PVF$_2$, PVF$_2$ with brass backing bonded to the transducer, and a commercially available acoustic emission transducer. The sensing elements of the PVF$_2$ transducers were 6.5 by

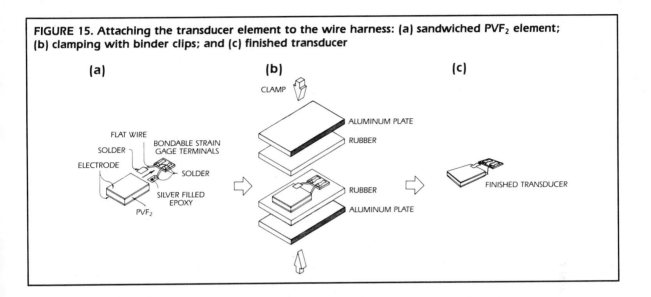

FIGURE 15. Attaching the transducer element to the wire harness: (a) sandwiched PVF$_2$ element; (b) clamping with binder clips; and (c) finished transducer

6.5 mm (0.25 by 0.25 in.) and 110 μm (4.4 mil) thick. The PVF$_2$ transducers were mounted using a double backed transfer adhesive. The commercial transducer was mounted using a water soluble couplant.

Because of the coupling between the in-plane displacement and the electric field perpendicular to the plane, and because of polyvinylidene fluoride's relatively large compliance, the PVF$_2$ transducer without the brass backing (mounted with the double backed adhesive) has the greatest sensitivity to in-plane displacements when compared to the other transducers. Brass backing was used for inertial loading for out-of-plane displacements so this transducer is more sensitive to out-of-plane displacements than the unbacked one. However, with the brass backing constraining the in-plane displacements, this transducer is not as sensitive to in-plane displacements as the unbacked one. The commercial transducer is sensitive only to out-of-plane displacements and does not respond to in-plane displacements. Knowing how the transducers respond, they can be ranked according to their sensitivity as shown in Table 2.

Two types of tests were conducted to evaluate the transducers. In the first type, the transducers were mounted on an aluminum plate 76 mm (3 in.) thick. A piece of capillary tubing was slowly loaded until it fractured, producing a step unloading. This establishes a Rayleigh wave on the surface of the plate with the largest displacements out-of-plane.

The two transducers were positioned 38 mm (1.5 in.) from the epicenter of the wave. The signals were captured with a two-channel transient recorder and plotted on an XY plotter (see Fig. 16). Several tests were performed to ensure that there was consistency in the data. The amplifications used were: PVF$_2$ at 200×; PVF$_2$ with brass backing at 50×;

TABLE 2. Relative sensitivity of transducers

	In-Plane Displacements	Out-of-Plane Displacements
Most sensitive	PVF$_2$	Commercial transducer
	PVF$_2$ with brass	PVF$_2$ with brass
Least sensitive	Commercial transducer	PVF$_2$

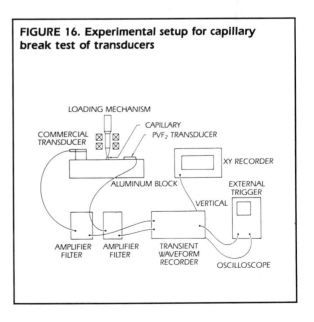

FIGURE 16. Experimental setup for capillary break test of transducers

and the commercial transducer at 20×. The signals were band-pass filtered from 1 kHz to 0.3 MHz.

Figure 17 shows a typical test result. The relative sensitivities normalized with respect to the PVF_2 are given in Table 3.

The second type of test was performed by mounting the transducers on a $[0, \pm 45]_s$ graphite/epoxy composite specimen. The specimen was placed in a closed loop hydraulic tensile testing machine and pulled in quasistatic tension. The amplifications used were: PVF_2 at 20×; PVF_2 with brass backing at 20×; and the commercial transducer at 100×. The signals were band-pass filtered from 1 kHz to 0.1 MHz.

Figure 18 shows a typical acoustic emission event detected using a PVF_2 transducer with a brass backing and one

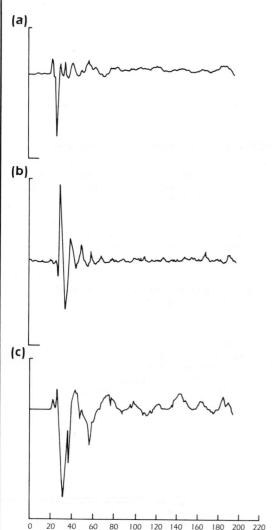

FIGURE 17. Typical acoustic emission waveforms from the capillary break test: (a) PVF_2 with gain of 200; (b) PVF_2 with brass backing and gain of 50; and (c) commercial transducer with gain of 20

TABLE 3. Observed sensitivities of transducers in capillary break test

Transducer	Sensitivity
PVF_2	1
PVF_2 with brass	4
Commercial transducer	10

TABLE 4. Observed sensitivities in acoustic emission tests of composite specimen

Transducer	Sensitivity
PVF_2	1
PVF_2 with brass	1/2
Commercial transducer	1/10

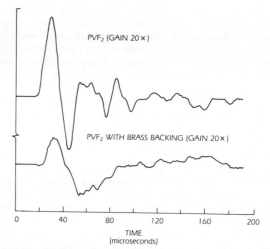

FIGURE 18. Acoustic emission event occurring at a load of 14,700 N (3,296 lb) detected using an unbacked PVF_2 transducer and one with a brass backing, both at a gain of 20

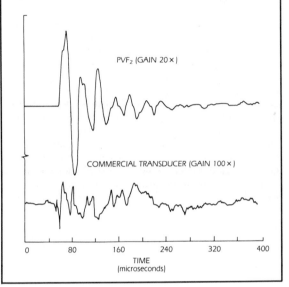

FIGURE 19. Acoustic emission event occurring at a load of 17,283 N (3,880 lb) detected using an unbacked PVF$_2$ transducer and a commercial transducer

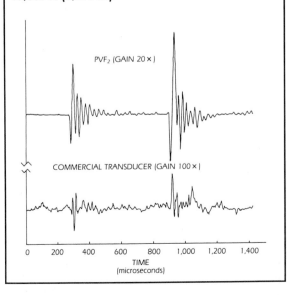

FIGURE 20. Two acoustic emission events, detected with the commercial transducer and an unbacked PVF$_2$ transducer, occurring at 17,660 N (3,970 lb)

without the backing. Figure 19 shows a typical event detected with an unbacked transducer and a commercial transducer. The relative sensitivities are shown in Table 4.

Figure 20 shows two acoustic emission events detected with a PVF$_2$ transducer and a commercial transducer.

Fatigue Tests with the Polyvinylidene Fluoride

Polyvinylidene fluoride has also been used to monitor acoustic emission during fatigue. The fatigue test was performed using an S-2 glass epoxy composite specimen with a stacking sequence of $[0,90]_2$. It was cycled in tension-tension at 4 Hz with an R of 0.1 and a maximum stress of 150 MPa (2.2×10^4 psi). Strain was measured using a 50 mm (2 in.) clip gage and load was measured with a load cell. The stress and strain data were input to a data acquisition unit that calculated the elastic modulus and specific damping as a function of the number of cycles (Fig. 21).

Acoustic emission was monitored using an unbacked PVF$_2$ transducer bonded to the specimen. The signal was band-pass filtered from 10 to 30 kHz and amplified 60 dB. The conditioned signal was the input to a digital counter to obtain the count rate (number of excursions per second made by the signal above a given threshold level). This was output to a strip chart recorder giving the acoustic emission count rate versus time.

Figure 22 shows the specific damping, dynamic secant modulus and the envelope of the acoustic emission count rate toward the end of the specimen life. The specimen

FIGURE 21. Experimental setup for monitoring stiffness, damping and acoustic emission during fatigue

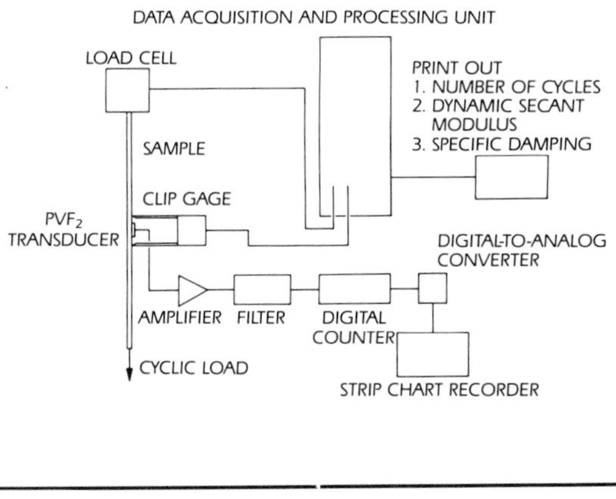

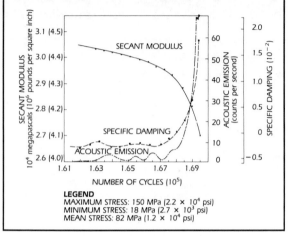

FIGURE 22. Secant modulus, specific damping and acoustic emission toward the end of the life of an S-2 glass specimen undergoing fatigue loading

failed at 169,000 cycles. The increase in acoustic emission and specific damping and the decrease in stiffness occur at the same time.

Conclusions

While PVF_2 showed the smallest response for the glass capillary test on an aluminum plate, it showed the best sensitivity when used with a composite material. By knowing how the transducers respond to in-plane and out-of-plane displacements, it may be inferred that, with composite materials, the greatest components of displacement during an acoustic emission event are in the plane of the composite. Further work will be conducted to see if this is the case.

Polyvinylidene fluoride was also used with success to monitor acoustic emission during fatigue loading. Because of these results and the advantages noted earlier, it appears that this material offers distinctive benefits for monitoring acoustic emission from advanced filamentary composite materials.

PART 5
STUDIES OF FRACTURE BEHAVIOR IN REFRACTORIES USING ACOUSTIC EMISSION TECHNIQUES

Fracture behaviors of various refractories have been studied using acoustic emission techniques. The effects of thermal shock conditions, refractory materials, specimen dimensions and restraining conditions on acoustic emission characteristics were investigated. Measurements of total ringdown counts on various types of refractories demonstrate that each brick material exhibits different acoustic emission characteristics corresponding to the difference of thermal shock damage behavior.

Fracture behaviors of brittle inorganic materials such as ceramics and refractories have also been investigated using acoustic emission techniques. Fractures by thermal shock have been studied by a number of investigators[86-89] but their use of acoustic emission techniques were limited in most cases to crack monitoring. Using the acoustic emission technique for this kind of study required that two basic problems be addressed: (1) the uncertainty about the emission mechanism for cracking; and (2) the lack of quantitative treatments of acoustic emission signals.

The experiment detailed below evaluates the thermal shock resistance of refractories through the study of acoustic emission characteristics under several thermal shock conditions. There were many problems in earlier evaluations of thermal shock resistance.[90,91] These problems resulted from a lack of good correspondence between experimental thermal shock conditions and actual service conditions of refractories.

Experimental Procedure for Study of Refractories

Generating Thermal Shock

Thermal shock was produced in two ways. Method-1 was panel heating, in which rectangular specimen bricks and insulating bricks were assembled into a panel. One face of the panel was heated by a programmable electric furnace. The testing apparatus for method-1 is shown in Fig. 23.

The testing brick was placed at the center of the furnace door and bricks having similar thermal properties were arranged around the specimen brick to ensure one-dimensional heat flow. The acoustic emission sensor was mounted on a stainless steel waveguide attached to the specimen brick face opposite the heated face. Contacts between brick, waveguide and sensor were kept secure by high temperature silicone lubricant. A thermocouple for temperature control of the furnace and for measurement of brick surface temperature was set on the hot face of the testing brick.

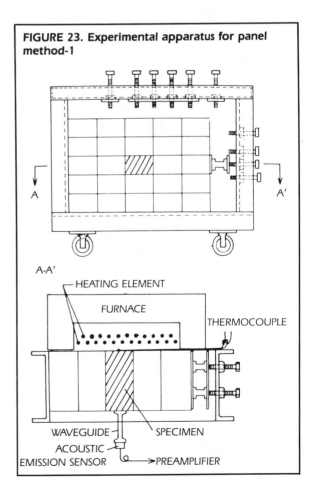

FIGURE 23. Experimental apparatus for panel method-1

In thermal shock method-2, a hollow cylinder was cut from a testing brick and its inner surface was heated with a silicon carbide heating element. The testing apparatus for method-2 is shown in Fig. 24. In this case, a thermocouple for temperature control and measurement was located on the inner surface. The acoustic emission sensor was mounted on the outer surface of the specimen with a stainless steel waveguide.

Test System and Parameters

A block diagram of the acoustic emission system is shown in Fig. 25. Measurements of acoustic emission were carried out under the following conditions: a single-end transducer with center frequency at 140 kHz; preamplifier gain of 40 dB; filter bandwidth at 0.1 to 1 MHz; and main amplifier gain of 25 dB.

The acoustic emission ringdown counts were above 1 V threshold. In some of the method-2 experiments, amplitude distributions were measured. Thermal shock experiments together with acoustic emission measurements were carried out during heating at a constant rate.

Characteristics of Refractories

The refractories used in this study were commercial bricks used in various processes of iron and steelmaking. Standard specimen size was 114 by 65 by 230 mm (4.5 by 2.5 by 9 in.) for method-1. The hot face was 114 by 65 mm (4.5 by 2.5 in.). In method-2, the inner diameter of the cylinder was 30 mm (1.2 in.), its height was 100 mm (4 in.) and the outer diameter was between 50 mm (2 in.) and 150 mm (6 in.) with a wall thickness between 10 mm (4 in.) and 60 mm (2.4 in.).

Thermal shock experiments by method-1 were run for four types of bricks. Properties of the bricks are listed in Table 5. Figure 26 shows the relationships between the bricks' surface temperature and total ringdown counts during heating at the rate of 20 °C (36 °F) per minute. In the illustration, the acoustic emission characteristics can be seen to differ from brick to brick. Roughly speaking, there are two patterns of acoustic emission that increase as temperature increases. One is a pattern showing gradual increase, such as magnesia-dolomite brick-D and magnesia-chromite brick-MC. The other is the pattern showing one or more sharp rises in count at certain temperature ranges, such as high-alumina brick-A and magnesia brick-M. These patterns are unique to each type of brick and indicate the difference of fracture behaviors during thermal shock.

It was necessary to confirm that the acoustic emission characteristics given in Fig. 26 corresponded to thermal shock fracture. For this purpose, the following experiments were performed on magnesia brick-M and magnesia-dolomite brick-D, where they were subjected to thermal shock by method-1 at heating rates of 13.3 °C (24 °F) per minute and 10 °C (18 °F) per minute, respectively.

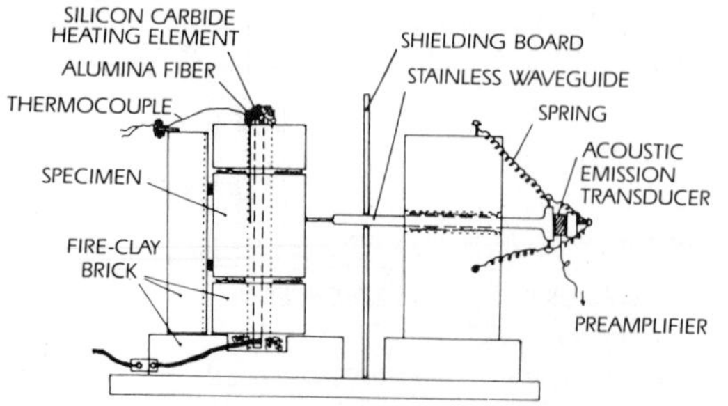

Results of the Procedure

Results of total counts measured through the period of heating are shown in Fig. 27. For brick-M, the first brick was heated to 800 °C (1,470 °F), the temperature just below the first sharp rise of acoustic emission counts. The second brick was heated to 850 °C (1,560 °F), just above the first sharp rise of acoustic emission counts. The third and fourth bricks were heated to 1,000 °C (1,830 °F) and 1,650 °C (3,000 °F), the temperatures just below and above the second sharp rise of acoustic emission counts, respectively.

All bricks were cooled slowly from each temperature and cut to bar specimens of 20 by 20 by 120 mm (0.8 by 0.8 by 4.7 in.), as shown in Fig. 28. For each bar specimen, three-point bending tests were carried out. As shown in Fig. 29, the second brick (heated to the temperature above the first sharp rise of acoustic emission counts) produced bar specimens having no strength around a center axis vertical to the hot face (Fig. 29a). A low strength area extended to the circumference of the brick heated to the temperature above the second sharp rise of acoustic emission counts.

For the specimens cut in parallel to the hot face, strength decreased markedly between 40 and 60 mm (1.5 and 2.4 in.) behind the hot face (Fig. 29b). By comparison, brick-D showed gradual increases of acoustic emission count as the temperature increased. Brick-D was heated to 900, 1,000 and 1,100 °C (1,650, 1,830 and 2,010 °F), then cooled slowly from each temperature. The bending strength for brick-D decreased gradually, as shown in Fig. 30.

From these two experiments, it was verified that acoustic emission ringdown counts correspond to fracture and that crack propagation behaviors during thermal shock are different in each brick. Furthermore, it was shown that crack propagation is unstable if accompanied by a sharp rise in total counts and that crack propagation is stable if accompanied by a decrease in total counts.

FIGURE 24. Experimental apparatus for hollow cylinder test method-2

TABLE 5. Properties of refractories tested with acoustic emission techniques

Property	Brick-A	Brick-M	Brick-MC	Brick-D
Porosity (percentage)	16.4	13.8	14.9	15.0
Specific gravity (g/cm^3)	3.24	3.01	3.11	2.96
Compressive strength (MN/m^2)	158	54.5	56.2	26.3
Modulus of rupture (MN/m^2)				
room temperature	31	21.5	6.5	11.9
800 °C (1,470 °F)		16	6.9	8.6
1,100 °C (2,010 °F)		5.7	7.8	5.8
Young's modulus (GN/m^2)	130	121	44	75
Chemical composition (percent by weight)				
MgO		94	78.6	90.5
CaO		0.8	0.7	8.1
SiO$_2$	4.21	2.6	0.7	1.3*
Al$_2$O$_3$	92.3	0.6	4.9	
Cr$_2$O$_3$	2.81		9.3	
Fe$_2$O$_3$		0.5	3.3	

*SiO$_2$ + Al$_2$O$_3$ + Fe$_2$O$_3$

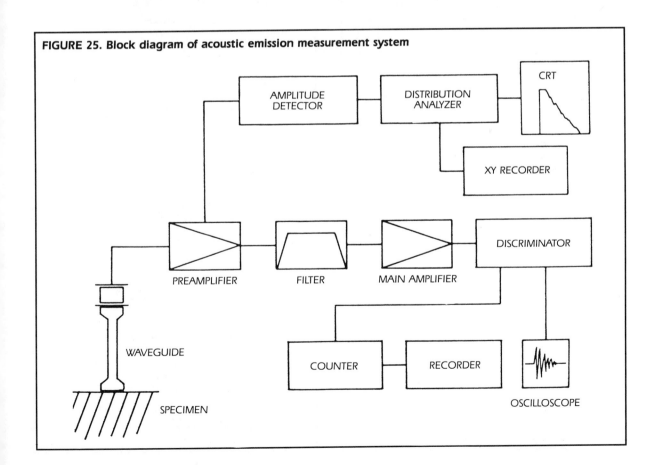

FIGURE 25. Block diagram of acoustic emission measurement system

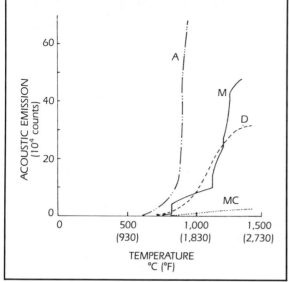

FIGURE 26. Total ringdown counts during heating for: high-alumina brick-A, magnesia brick-M, magnesia-chrome brick-MC and magnesia-dolomite brick-D; heating rate of 20 °C (36 °F) per minute by method-1

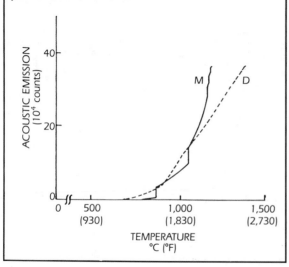

FIGURE 27. Total ringdown counts during heating for magnesia brick-M and magnesia-dolomite brick-D brick; heating rate 13.3 °C (24 °F) per minute for brick-M and 10 °C (18 °F) per minute for brick-D

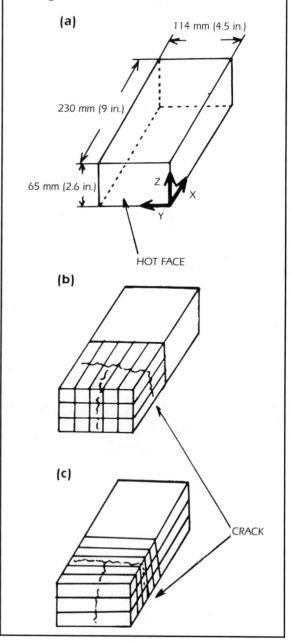

FIGURE 28. Original specimen and specimen configuration for bending strength determination: (a) original specimen; (b) preparation of specimen for bending in X direction; and (c) preparation of specimen for bending in Y direction

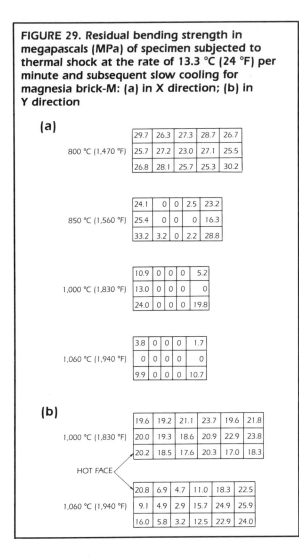

FIGURE 29. Residual bending strength in megapascals (MPa) of specimen subjected to thermal shock at the rate of 13.3 °C (24 °F) per minute and subsequent slow cooling for magnesia brick-M: (a) in X direction; (b) in Y direction

(a)

	29.7	26.3	27.3	28.7	26.7
800 °C (1,470 °F)	25.7	27.2	23.0	27.1	25.5
	26.8	28.1	25.7	25.3	30.2

	24.1	0	0	2.5	23.2
850 °C (1,560 °F)	25.4	0	0	0	16.3
	33.2	3.2	0	2.2	28.8

	10.9	0	0	0	5.2
1,000 °C (1,830 °F)	13.0	0	0	0	0
	24.0	0	0	0	19.8

	3.8	0	0	0	1.7
1,060 °C (1,940 °F)	0	0	0	0	0
	9.9	0	0	0	10.7

(b)

	19.6	19.2	21.1	23.7	19.6	21.8
1,000 °C (1,830 °F)	20.0	19.3	18.6	20.9	22.9	23.8
	20.2	18.5	17.6	20.3	17.0	18.3

HOT FACE

	20.8	6.9	4.7	11.0	18.3	22.5
1,060 °C (1,940 °F)	9.1	4.9	2.9	15.7	24.9	25.9
	16.0	5.8	3.2	12.5	22.9	24.0

FIGURE 30. Residual bending strength (MPa) in X direction of brick-D subjected to thermal shock at the rate of 10 °C (18 °F) per minute and subsequent slow cooling

	6.4	1.5	7.7	0	6.1
900 °C (1,650 °F)	2.2	8.3	0	0	0
	1.1	2.6	0	0	0

	0	0	0	0	0
1,000 °C (1,830 °F)	0	0	0	0	0
	11.6	3.3	4.0	3.0	0

	0	0	0	0	0
1,100 °C (2,010 °F)	0	0	0	0	0
	0	0	0	0	0

Effect of Heating Rate

In service, refractory bricks are subjected to various kinds of thermal shock and therefore the effect of heating rate on fracture behavior must be investigated. Thermal shock experiments under different heating rates were run using method-1 for brick-*M* and brick-*D*. The results showed different crack propagation behaviors for the different bricks and these are illustrated in Figs. 31 and 32.

Curves of acoustic emission counts for brick-*M* change into more of a step pattern with an increase in heating rate, but total ringdown counts at the end of heating become smaller with higher heating rate. Cracks at the surface of the brick after thermal shock were more clearly visible with higher heating rate. Finally, bricks heated at 26.7 °C (15 °F) per minute were broken into two pieces at 40 to 50 mm (1.6 to 2.0 in.) behind the hot face. The degree of damage caused by thermal shock becomes greater with higher heating rate and this corresponds the step-like acoustic emission pattern from increased heating rates.

For brick-*D* with stable crack propagation, a continuous rise of acoustic emission counts is observed even at higher heating rate. The number of acoustic emission total counts at the end of heating is almost unchanged for all heating rates (Fig. 32). But the number of counts per unit time (count rate) necessarily increased with increased heating rate and this might correspond to the degree of damage.

Effect of Specimen Dimensions

In the experiments with method-1, the effect of specimen dimensions on thermal shock damage was examined using brick-*M*. The hot face dimensions were 65 by 55 mm (2.6 by 2.2 in.), 30 by 114 mm (1.2 by 4.5 in.) and 65 by 114 mm (2.6 by 4.5 in.); the length for all specimens was 230 mm (9 in.). As shown in Fig. 33, the number of total counts decreases with the decrease of the hot face area and with the change of its shape from a rectangular to a square.

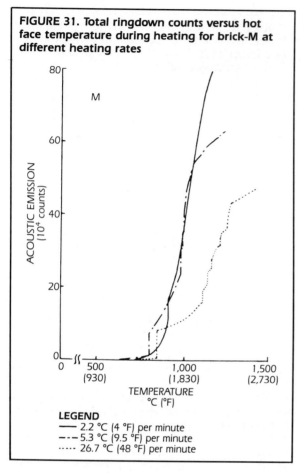

FIGURE 31. Total ringdown counts versus hot face temperature during heating for brick-M at different heating rates

LEGEND
— 2.2 °C (4 °F) per minute
--- 5.3 °C (9.5 °F) per minute
····· 26.7 °C (48 °F) per minute

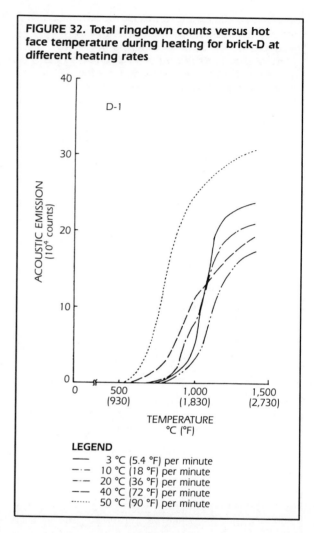

FIGURE 32. Total ringdown counts versus hot face temperature during heating for brick-D at different heating rates

LEGEND
— 3 °C (5.4 °F) per minute
-·- 10 °C (18 °F) per minute
--- 20 °C (36 °F) per minute
-- 40 °C (72 °F) per minute
····· 50 °C (90 °F) per minute

It is more important to realize that the curve becomes more gradual (that crack propagation becomes more stable) with a decrease in hot face area or with a change in its shape. These facts are useful in making improvements to new furnace designs.

In the experiments with method-2, size effects were more noticeable. These studies were performed with various high-alumina bricks. Wall thickness was changed from 10 mm (0.4 in.) to 60 mm (2.4 in.). The heating rate of the specimen's inner surface was constant in the series at 40 °C (72 °F) per minute. Typical results for four kinds of bricks are shown in Fig. 34.

Effect of Restraining Conditions

When refractories are in service, many bricks are used as construction materials for furnaces. Bricks are restrained from thermal expansion by neighboring bricks. Effects of the restraining condition on thermal shock damage was tested, using method-1. The tested brick was restrained and compressed by bolts installed on the panel frame in horizontal (Y) and vertical (Z) direction. Results obtained for brick-M and and brick-D are shown in Figs. 35 and 36.

In brick-M, the restraining force was increased from the original Y direction to the Y and Z directions and was further strengthened by compression in the Y and Z directions. Curves of total ringdown counts show pronounced step-like patterns. In brick-D, patterns of acoustic emission counts do not change with the increase of restraining force.

Note that crack propagation behaviors in brick-M became more unstable with strengthening of the restraint, and that both brick-M and brick-D were broken about 50 mm (2.0 in.) behind the hot face after the test. These facts have

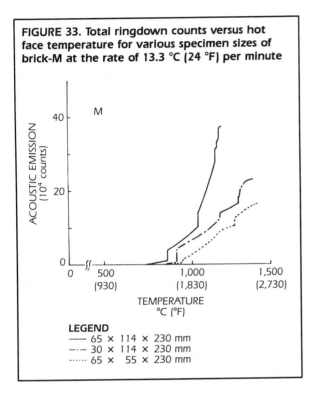

FIGURE 33. Total ringdown counts versus hot face temperature for various specimen sizes of brick-M at the rate of 13.3 °C (24 °F) per minute

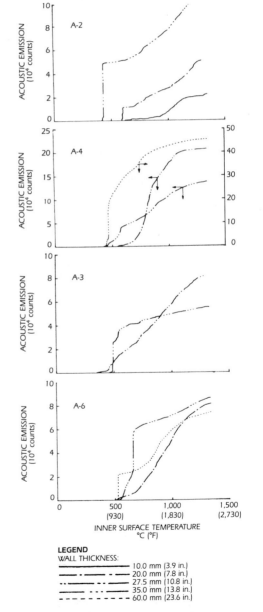

FIGURE 34. Total ringdown counts versus inner surface temperature for four high-alumina bricks of different wall thicknesses (by method-2) with heating rate of 40 °C (22 °F) per minute

also contributed to the improvement of furnace brickwork design in the steelmaking process.[93]

Conclusions

Advantages and Disadvantages of Acoustic Emission Techniques

Crack propagation in refractory bricks can be monitored with high sensitivity by measuring total acoustic emission counts. In early investigations, crack propagation behaviors were studied using the relations between temperature differences and bending strength losses of bar specimens subjected to sudden temperature changes.[94,95] By comparison, studies with acoustic emission methods have many advantages. It was determined, for example, that thermal shock conditions can simulate actual service conditions and crack propagation behaviors can be studied after only one thermal shock.

The degree of damage during thermal shock cannot be determined by the number of acoustic emission ringdown counts, because the attenuation characteristics differ in each refractory material and the mechanism of acoustic emission generation have not been clarified.

Theoretical Verification of Results

For five kinds of magnesia-dolomite bricks made by different manufacturers, tests were run under the following conditions: the heating rate was 20 °C (36 °F) per minute; no restraint; specimen dimensions were 65 by 114 by 230 mm (2.5 by 4.5 by 9 in.); and a total gain of 65 dB was used. Properties of the bricks and results of thermal shock tests are shown in Table 6 and Fig. 37. From Fig. 37, the tendency of thermal shock degradation is evaluated in the order: D-7 >> D-8 > D-5 > D-6 > D-9.

A proposed theoretical thermal shock resistance parameter for ceramics and bricks with stable crack propagation is as follows:[96]

$$R_{st} = \sqrt{\frac{\gamma}{\alpha^2 E}} \qquad (Eq.\ 2)$$

Where:

R_{st} = thermal shock resistance;
γ = the fracture energy;
E = Young's modulus; and
α = the thermal expansion coefficient.

Bricks having large R_{st} values have good thermal shock resistance. The thermal expansion coefficient of the five tested bricks were equal and the relation between $(\gamma/E)^{1/2}$ and the total acoustic emission count are plotted in Fig. 38. In the illustration, bricks having large $(\gamma/E)^{1/2}$ values show small acoustic emission total counts. This verifies that the evaluation by acoustic emission counts is theoretically acceptable.

In method-2, the effect of wall thickness on crack propagation behaviors is clear. It can be supposed that bricks showing stable crack propagation at heavy wall thicknesses should be superior to bricks showing unstable crack propagation at thin wall thicknesses. To prove this assumption, ten kinds of high-alumina bricks were tested. The relations between wall thickness and crack propagation behaviors are summarized in Table 7. The critical wall thickness at which crack propagation changes from stable to unstable may be a parameter of thermal shock damage resistance.

A proposed theoretical parameter of thermal shock damage resistance for bricks with unstable crack propagation is as follows.[96]

$$R'''' = \frac{\gamma E}{\sigma^2}(1 - v) \qquad (Eq.\ 3)$$

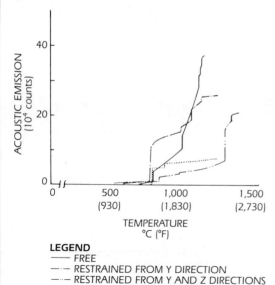

FIGURE 35. Total ringdown counts versus hot face temperature for brick-M under different restraining conditions at the heating rate of 13.3 °C (24 °F) per minute

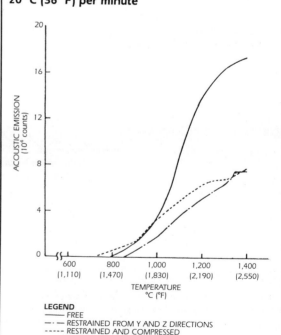

FIGURE 36. Total ringdown counts versus hot face temperature for brick-D under different restraining conditions at the heating rate of 20 °C (36 °F) per minute

where σ is fracture strength and v is Poisson's ratio. The relationship between the R'''' value and the critical wall thickness for high-alumina bricks is plotted in Fig. 39. Correlations between theoretical and experimental values verify the advantages of this acoustic emission testing methodology.

Testing for Spalling Resistance of Refractories

The testing apparatus and method discussed below were designed to make a quantitative measurement of spalling resistance of refractories with relative ease. The test results correlate well with the spalling behavior under actual operating conditions. The usefulness of the method is enhanced by the fact that it is also applicable to brickwork for: (1) determining optimum brick shape; (2) setting up optimum joint width; and (3) establishing drying conditions suitable for monolithic linings.

Theoretical Background of the Test

Generally, when refractories are used for the lining of furnaces or vessels of molten metal, the working surface is subject to high temperature, while its opposite surface is cooled by the safety lining and shell. Under these conditions, there arises a certain temperature gradient within the refractory. When the temperature at the working surface changes rapidly, various thermal stresses are generated because of the inhomogeneous distribution of temperature.

A typical example of the calculated results of thermal stresses produced within a refractory by the finite element method is shown in Fig. 40. One side of the refractory was

FIGURE 37. Total ringdown counts versus hot face temperature for five magnesia-dolomite bricks at the heating rate of 20 °C (11 °F) per minute

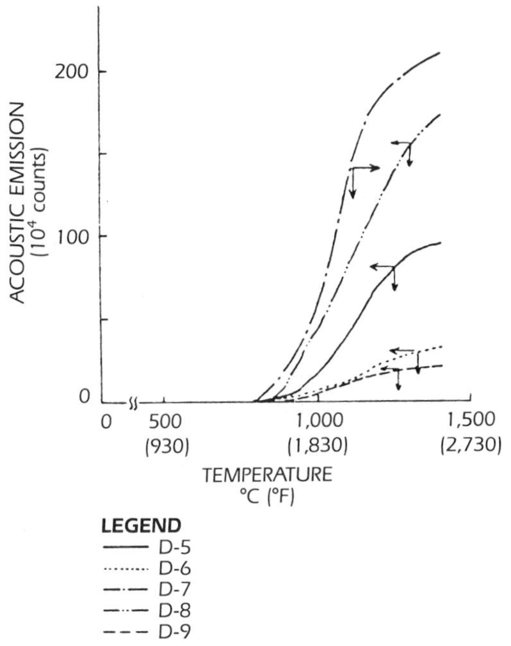

TABLE 6. Typical properties of magnesia-dolomite bricks thermal shock tested with acoustic emission techniques

Property	D-5	D-6	D-7	D-8	D-9
Porosity (percent)	15.1	13.6	12.0	14.4	15.0
Specific gravity	2.95	2.96	3.08	2.96	2.96
Modulus of rupture (MN/m^2)					
room temperature	11.3	12.3	12	15.3	15.8
800 °C (1,470 °F)	13.9	18.2	14.2	15.4	18.6
1,100 °C (2,010 °F)	8.3	9.8	6.2	8.6	8.2
1,400 °C (2,550 °F)	3.6	6.7	2.5	4.6	4.8
Young's modulus (GN/m^2)	75	83	102	97	85
Fracture energy (J/m^2)		108	72	115	114
$[\gamma/E]^{1/2}$ (10^{-3} m$^{1/2}$)		3.64	2.68	3.48	3.70
Chemical composition (weight percent)					
MgO	91.5	90.5	90.5	87.0	87.0
CaO	6.5	8.1	8.1	11.0	11.0
SiO$_2$ + Al$_2$O$_3$ + Fe$_2$O$_3$	1.0	1.3	1.3	2.0	2.0

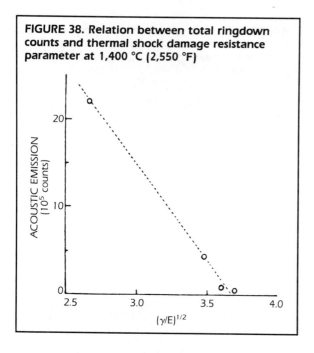

FIGURE 38. Relation between total ringdown counts and thermal shock damage resistance parameter at 1,400 °C (2,550 °F)

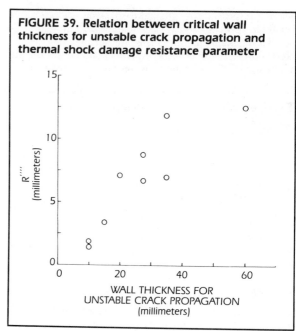

FIGURE 39. Relation between critical wall thickness for unstable crack propagation and thermal shock damage resistance parameter

TABLE 7. Effect of specimen dimension on the crack propagation behavior

Sample Dimensions ($r_i \times r_o \times l$)	A-1	A-2	A-3	A-4	A-5	A-6	A-7	A-8	A-9	AD
30 × 50 × 100 mm	I	S		S	S	S	S	U	S	
30 × 70 × 100 mm	U	U		S	S	S	S	U	U	I
30 × 85 × 100 mm			S		S	U	U			U
30 × 100 × 100 mm		U	U	S	U	U	U	U	U	U
30 × 150 × 100 mm				U		U				

S: stable
I: intermediate
U: unstable

rapidly heated (the length of the line denotes the magnitude of stress and the arrow denotes direction).

In the illustration, a large tensile stress occurs at a distance from the working surface. When this stress exceeds the tensile strength of the refractory, cracks begin. If the thermal stress is slightly larger than the tensile strength, the cracks will alleviate the stress. If the stress is great enough, cracks will propagate. In the case illustrated, cracks are initiated at the center of the refractory, parallel to the working surface, and are extended to its periphery.

In various conventional spalling tests, the only observable phenomena are: (1) cracks that extend to the surface and (2) the peeling off of the working side of the refractory. Conventional methods cannot determine cracks formed inside the refractory.

The advantage of the acoustic emission testing method lies in its ability to detect the initiation and propagation of cracks within the refractory. This ensures a much higher sensitivity and accuracy than conventional visual testing methods.

Testing Procedure

The testing apparatus consists of a furnace, its temperature control unit, a cooling device, the acoustic emission measuring system and a steel frame in which a refractory panel is constructed. One of the faces of the refractory panel, composed of testing brick, is put in contact with the open face of the furnace. Ordinarily, the test during heating is performed by rapidly heating the furnace, and the test

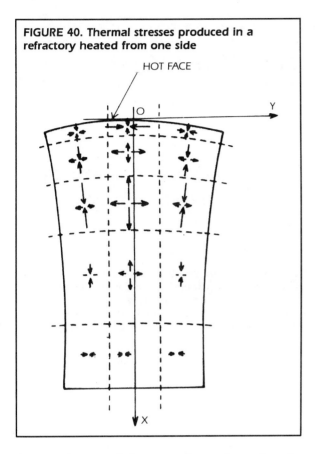

FIGURE 40. Thermal stresses produced in a refractory heated from one side

during cooling is made by moving the panel away from the furnace to the front of the cooling device (usually a fan).

For acoustic emission measurement, the sensor is put in contact, through a waveguide, with the face opposite the heating and cooling face of the brick. Measurement of brick with high thermal conductivity is achieved by rapidly exposing one face of the brick to the preheated open face of the furnace. Measurement of brick with less thermal conductivity begins simultaneously with the heating of the furnace at a constant rate.

When spalling during cooling is a matter of concern, heating at the lowest possible rate, up to a predetermined temperature, is recommended. After being held at a temperature that gives a stationary temperature gradient within the brick, the panel is moved to the cooling device and acoustic emission is measured during rapid cooling. Usually, only one measurement is required for evaluating the spalling tendency of a refractory.

Advantages of the Technique

The panel acoustic emission spalling test has an advantage in that bricks are tested in the form of a wall structure. Behavior of the structure can be studied, with emphasis on the effect of temperature, joint thickness, number of joints and surrounding restraints. This aids in the determination of the effect of joints on the absorption of thermal expansion and the effect of brick shape on differences in spalling resistance. This information is very useful for actual furnace building techniques.

Spalling resistance of unshaped refractories, such as castable and plastic refractories, can be tested in the same way as brick using a molded block. Tests for cracking and explosive spalling during drying can be performed on these undried refractories.

Spalling resistance of cylindrical refractories, such as nozzles and runners, can be evaluated by different heating methods depending on operating conditions. For example, the nozzle interior is heated rapidly and the resultant acoustic emission signals are counted in the same way as mentioned above.

PART 6
ENERGY MEASUREMENT OF CONTINUOUS ACOUSTIC EMISSION

The measurement of energy in a continuous acoustic emission signal by a true rms (root mean square) voltmeter or a tms (true mean square) voltmeter is reviewed below. Included in this discussion are: (1) a comparison of typical rms and tms voltmeters; (2) a discussion of how these voltmeters measure energy; (3) a discussion of the subtraction of background noise from the true acoustic emission; (4) a presentation of the advantages of the rms or tms measurement techniques; and (5) a discussion of some practical factors in such energy measurements.

Definition of Continuous Emission

Measuring the energy contained in acoustic emission signals is one of the primary means of presenting acoustic emission data quantitatively. For so-called continuous acoustic emission, the DC output from either a true rms (root mean square) voltmeter or a tms (true mean square) voltmeter can be used to make an energy measurement. The term continuous emission is used qualitatively to describe the appearance of a sustained signal level resulting when the average time interval between the beginning of emission signals of comparable amplitude is less than or comparable to the duration of the emission signals.[97]

This definition infers that when a continuous acoustic emission signal is observed on an oscilloscope, it cannot be determined at what time one acoustic emission event began or ended (the term *event* refers here to the actual physical process that generates a stress wave originating at a certain location in the specimen).

Figure 41 is an oscilloscope indication of a continuous acoustic emission signal generated during a tensile test of 7075-T6 aluminum. It represents a typical continuous acoustic emission signal at a certain stress level for a sufficiently long period that an rms or tms measurement can be made.[98] It is the indistinguishable events and sustained signal level that make it possible to measure continuous acoustic emission with an rms or a tms voltmeter.

The circuits designed to measure the energy of burst type acoustic emission do not give a true measure of the energy in continuous acoustic emission. Reference 99 describes one use of special circuits for measuring the energy in continuous acoustic emission.

Theory and Operation of Voltmeters

The rms voltage of a signal $v(t)$ is defined as:

$$\mathrm{rms} \equiv \sqrt{\frac{1}{\Delta t}\int_0^{\Delta t} v^2(t)dt} \qquad \text{(Eq. 4)}$$

where Δt is the averaging time and $v(t)$ is the function of voltage with respect to time. Since $v(t)$ is a random signal for continuous acoustic emission, a true rms voltmeter rather than an AC voltmeter must be used for its measurement. In a similar way, the tms voltage of $v(t)$ is defined as:

$$\mathrm{tms} \equiv \frac{1}{\Delta t}\int_0^{\Delta t} v^2(t)dt \qquad \text{(Eq. 5)}$$

or

$$(\mathrm{rms})^2 = \mathrm{tms} \qquad \text{(Eq. 6)}$$

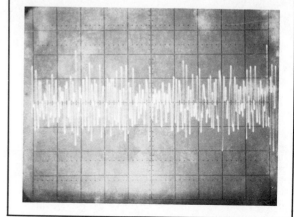

FIGURE 41. Photograph of an oscilloscope showing continuous acoustic emission at about four percent strain from a tensile test of 7075-T6 aluminum; the X axis is 0.1 ms·cm^{-1} and the Y axis is 1 V·cm^{-1}

An alternate definition for the rms voltage is used in electronics. This definition takes the rms voltage of an arbitrary signal $v(t)$ over some time interval Δt to be equal to the value of the DC voltage that would produce the same average heating value over the time interval Δt when both signals are put into a given resistance. From this definition, we can show that the rms of $v(t)$ over time intervals Δt is such that:

$$\Delta E \propto (\text{rms})^2 \Delta t = (\text{tms}) \Delta t \qquad (\text{Eq. 7})$$

where ΔE is the energy contained in the signal over the time interval Δt. Applying the techniques of calculus to Eq. 7 gives:

$$\frac{dE}{dt} \propto (\text{rms})^2 = \text{tms} \qquad (\text{Eq. 8})$$

The tms as a function of time gives the rate of energy contained in the continuous acoustic emission signal. Integrating Eq. 8 gives:

$$E \propto \int_{t1}^{t2} (\text{rms})^2 dt = \int_{t1}^{t2} (\text{tms}) dt \qquad (\text{Eq. 9})$$

where t_2 is greater than t_1 and E is the total energy contained in the acoustic emission signal between time t_1 and time t_2.

Typical Meter Characteristics

A typical true rms meter with a DC output has a frequency range of 10 Hz to 10 MHz and uses a heat comparison bridge. The signal $v(t)$ to be measured is applied to a measuring thermocouple and a feedback control system is used to match the heating power of a DC voltage to the heating power of $v(t)$. The meter deflection is proportional to this DC voltage.

The transient response and averaging time of the meter are therefore controlled by the characteristics of the DC controller and the heat sensors. By using 150 kHz tone bursts, it has been determined that the averaging time of some meters is on the order of 200 to 400 ms. For a strain rate of 10^{-2} per second, this time interval is equivalent to a strain increment of about 0.02 percent. Since in most cases continuous acoustic emission does not change rapidly with strain, this response time is usually not a limiting factor in the accuracy of rms measurements. Figure 42 shows the results of the tone burst test.

A typical tms voltmeter with a DC output proportional to the true mean square voltage has a frequency range of 10 Hz to 20 MHz and uses diodes as a square law detector. These detector diodes, when operated near the origin of their characteristic voltage versus current curve, result in a nearly perfect square law relationship between the input signal and the output DC voltage.

The DC voltage is applied to a resistance capacitance filter section, which averages the DC voltage over a fixed averaging time constant (usually either 250 ms or 500 ms). This average DC voltage, after proportional modulation and demodulation, results in a DC current that can be passed through a resistor to give the final DC output voltage or the tms voltage over the selected time constant. Response times are often similar to the rms meters so that these meters are not limited by response time in the usual measurements of continuous acoustic emission.

Most voltmeters will not correctly measure the rms or tms when the input signal has a DC component. Continuous acoustic emission is assumed here to have zero mean values.

Signal Amplification

For many cases of continuous acoustic emission, the signal from the acoustic transducer is quite small. High electronic gains (90 to 105 dB or more) are often required. The result is that the continuous acoustic emission, as seen on an oscilloscope or as measured by an rms meter, is really the superposition of the real acoustic emission and electronic noise from the preamplifier and transducer system. In the presence of such preamplifier noise, the continuous acoustic emission signal $v(t)$ can be taken to be:

$$v(t) = v_1(t) + n(t) \qquad (\text{Eq. 10})$$

where $n(t)$ is the noise function and $v_1(t)$ is the real continuous acoustic emission. Substituting Eq. 10 into Eq. 4 gives:

$$\text{rms} = \sqrt{\frac{1}{\Delta t}\int_0^{\Delta t} v_1^2(t)dt + \frac{2}{\Delta t}\int_0^{\Delta t} v_1(t)n(t)dt + \frac{1}{\Delta t}\int_0^{\Delta t} n^2(t)dt} \qquad (\text{Eq. 11})$$

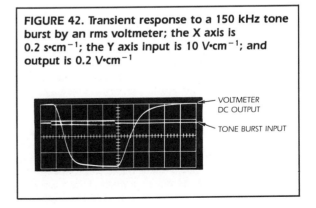

FIGURE 42. Transient response to a 150 kHz tone burst by an rms voltmeter; the X axis is 0.2 s·cm^{-1}; the Y axis input is 10 V·cm^{-1}; and output is 0.2 V·cm^{-1}

Using the definition of rms and squaring Eq. 11 gives:

$$\text{rms}^2 = \overline{\text{rms}}^2 + \frac{2}{\Delta t}\int_0^{\Delta t} v_1(t)n(t)dt + \text{rms}_n^2 \quad \text{(Eq. 12)}$$

where $\overline{\text{rms}}$ and rms_n are the root mean squares of the real continuous acoustic emission and the preamplifier noise, respectively, over time Δt. If $v_1(t)$ and $n(t)$ are taken to be statistically independent (see reference 100) and if $n(t)$ has a zero mean value, then the cross correlation function of $v_1(t)$ and $n(t)$ is zero and the second term of the right side of Eq. 12 is zero. Therefore:

$$\overline{\text{rms}} = \sqrt{\text{rms}^2 - \text{rms}_n^2} \quad \text{(Eq. 13)}$$

and

$$\overline{\text{tms}} = \text{tms} - \text{tms}_n \quad \text{(Eq. 14)}$$

where $\overline{\text{tms}}$ and tms_n are true mean squares for the real continuous acoustic emission and the preamplifier noise, respectively. Equation 14 indicates the advantage that tms meters have over rms meters: with the tms meter, the DC output corresponding to tms_n can be set at zero on the plotter. This procedure results in an immediate plot of $\overline{\text{tms}}$, which is proportional to the energy rate of the acoustic emission signal. As Eq. 13 indicates, such a simple procedure cannot be used to subtract the background level from the DC output of an rms meter.

Advantages of the Techniques

Figures 43 and 44 compare rms and tms measurements with the traditional counting techniques. For a tensile test of 7075-T6 aluminum, Fig. 43 compares (1) rms versus strain and (2) both the classical count rate and the summation of ringdown counts above a 1 V threshold versus strain. Figure 44 shows rms versus strain and $\overline{\text{tms}}$ versus strain, both taken during a tensile test of 7075-T6 aluminum. The dots on the $\overline{\text{tms}}$ curve represent values of $\overline{\text{rms}}^2$ that were calculated from the rms values using Eq. 13.

These figures demonstrate one of the advantages of rms and tms measurements: rms measurement smooths the acoustic emission data. This means that it is easier to pick out distinguishing features such as the strain level at the peak of the acoustic emission. Smoothing also increases the ease of fitting experimental acoustic emission data to analytical functions.

These mean square measurements have further advantages. First, the rms or tms techniques of acoustic emission data presentation are much less sensitive to small changes in the system electronic gain or in the transducer coupling efficiency.[100] Moreover, the extreme sensitivity of the count rate technique to small changes in the threshold level does not appear at all. Experimental work has shown that the counting techniques are exponentially dependent on the threshold level for continuous acoustic emission.[101]

A second advantage is that the shape of curves for summation of counts versus strain or the count rate versus strain

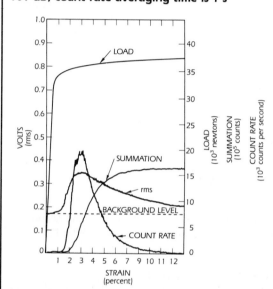

FIGURE 43. Typical curve of acoustic emission and load versus strain for 7075-T6 aluminum; gain is 101 dB; count rate averaging time is 1 s

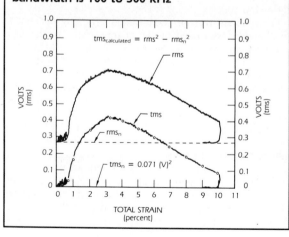

FIGURE 44. Strain versus rms and tms for tensile test of 7075-T6 aluminum; gain is 100 dB; bandwidth is 100 to 300 kHz

does not depend on the threshold level. Such dependence would be unacceptable for work that attempts to relate the shape of these curves to basic dislocation theory.[101]

Thirdly, the rms or tms acoustic emission data can be directly related to the energy contained in the continuous acoustic emission signal.[98,102] Such a simple relationship does not exist at present for count rate or summation of acoustic data. Finally, as detailed below, with the rms or tms techniques corrections can be made for strain rate differences and specimen volume differences.

Practical Factors in Energy Measurements

Volume and Plastic Strain Rate

Continuous acoustic emission is dependent on the plastic strain rate as well as the volume of the specimen undergoing deformation. During tensile testing of 7075-T6 aluminum, a relatively simple relationship was discovered between $\overline{\text{rms}}$ and these two factors:

$$\overline{\text{rms}} \propto \dot{\epsilon}_p^{1/2} V^{1/2} \quad \text{(Eq. 15)}$$

or

$$\overline{\text{tms}} \propto \dot{\epsilon}_p V \quad \text{(Eq. 16)}$$

where $\dot{\epsilon}_p$ is the plastic strain rate and V is the volume of the reduced section of the specimen. Figures 45 and 46 are examples of experimental data from 7075-T6 aluminum tensile tests that exhibit these two dependencies. The data in Fig. 45 came from work described in references 98 and 99. Before the $\overline{\text{rms}}$ values were plotted in Fig. 46, they were corrected according to $\overline{\text{rms}} \propto \dot{\epsilon}_p^{1/2}$. After this correction, the 7075-T6 aluminum data followed the predicted half-power dependence on specimen volume.

The analytical derivation[98] of the results in Eq. 15 implies that these values will hold for all materials that exhibit continuous acoustic emission, as long as (1) for strain rate differences, the total number and amplitudes of the acoustic emission events do not change; and (2) for volume differences, the density of acoustic emission events does not change as the volume of the specimen reduced section does. Other experimental work has shown that $\overline{\text{rms}}$ depends on the square root of the test rate, for three different steels and single crystals of aluminum and copper.[104,105]

Count Rate and Summation of Counts

No simple relationships such as those in Eqs. 15 and 16 exist for count rate or summation of counts data. In fact, if it is assumed that the continuous acoustic emission signal is a relatively narrow band random signal (such that the probability density distribution of amplitude peaks resembles a Rayleigh distribution), then it can be shown that the strain rate or volume dependency expressed by Eqs. 15 and 16 leads to exponential dependence for the count rate or summation of counts techniques.

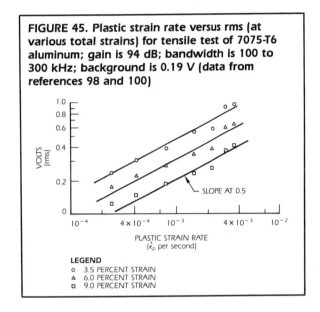

FIGURE 45. Plastic strain rate versus rms (at various total strains) for tensile test of 7075-T6 aluminum; gain is 94 dB; bandwidth is 100 to 300 kHz; background is 0.19 V (data from references 98 and 100)

FIGURE 46. Specimen reduced section volume versus rms for tensile tests of 7075-T6 aluminum; rms values corrected to the same strain rate used in the smallest volume specimens; gain is 10 dB; bandwidth is 100 to 300 kHz (data from reference 103)

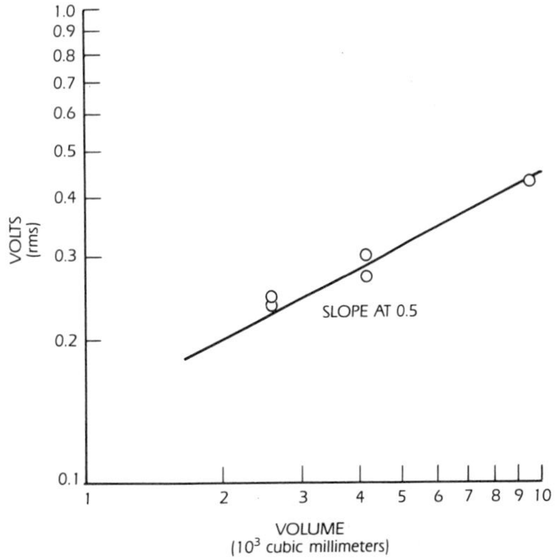

To show this, consider an arbitrary small strain increment with corresponding time increment Δt such that rms is independent of strain. Now the Rayleigh distribution function[106] for the amplitude peaks is:

$$p(v_2) = v_2 \, \text{rms}^2 \exp[-1/2(v_2/\text{rms})^2] \quad \text{(Eq. 17)}$$

where v_2 is the amplitude of a given signal peak and the rms value is over the small strain increment. Let v_t be the discriminator threshold level. Then the expected number of peaks exceeding v_t is:

$$\sum N = (\Delta t)(f) \int_{v_t}^{\infty} p(v_2) dv_2 \quad \text{(Eq. 18)}$$

where f is the apparent frequency of the signal. Substituting from Eq. 17 and carrying out the integration:

$$\sum N = (\Delta t)(f) \exp[-v_t^2/2(\text{rms})^2] \quad \text{(Eq. 19)}$$

Now take Δt to be the averaging time for the count rate system. The count rate $\dot{N}$ over this strain interval is:

$$\dot{N} = \sum \frac{N}{\Delta t} \quad \text{(Eq. 20)}$$

Substituting Eq. 19 in Eq. 20 gives:

$$\dot{N} = f \exp[-v_t^2/2(\text{rms})^2] \quad \text{(Eq. 21)}$$

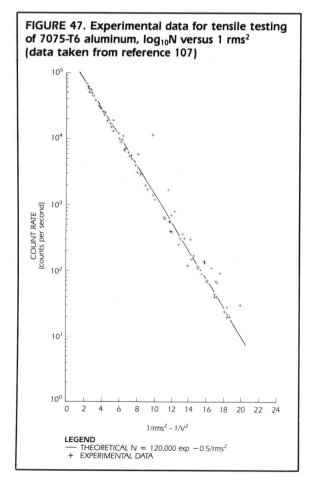

FIGURE 47. Experimental data for tensile testing of 7075-T6 aluminum, $\log_{10} N$ versus $1/\text{rms}^2$ (data taken from reference 107)

LEGEND
— THEORETICAL $N = 120{,}000 \exp -0.5/\text{rms}^2$
+ EXPERIMENTAL DATA

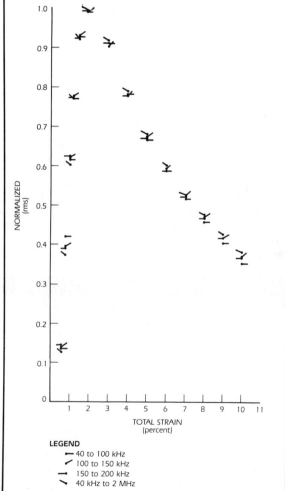

FIGURE 48. Plot of normalized rms versus strain for various band-pass frequencies for tensile testing of 7075-T6 aluminum; gain is 100 dB (data taken from reference 103)

LEGEND
— 40 to 100 kHz
— 100 to 150 kHz
— 150 to 200 kHz
— 40 kHz to 2 MHz

Now, from Eq. 15

$$\overline{rms}^2 = k \dot{\epsilon}_p V \quad \text{(Eq. 22)}$$

where k is an arbitrary constant. Substituting Eq. 13 and then Eq. 22 into Eq. 21 gives the final result:

$$\dot{N} = f \exp[-v_t^2/2(k\dot{\epsilon}_p V + rms_n^2)] \quad \text{(Eq. 23)}$$

The adequacy of the functional relationship between $\dot{N}$ and rms expressed in Eq. 21 can be judged from Fig. 47. The experimental data in this figure were taken from reference 107. The theoretical fit is based on the experimental constant $v_t = 1$ V and a best fit choice of $f = 210{,}000$ Hz.

Results obtained during tensile testing of 7075-T6 aluminum over relatively wide frequency ranges (between 40 kHz and 2 MHz) indicate that the shape of the rms versus strain curve does not change with frequency band-pass (see Fig. 48). All the data in the figure were normalized so that $\overline{rms}$ equals 1 V at the peak of the curve. Except for very narrow frequency windows, these data indicate that the rate of energy in the continuous acoustic emission is proportional to $\overline{rms}^2$, regardless of the band-pass in the material.

Other Factors

Four other practical factors should be mentioned. First, the $\overline{rms}$ or $\overline{tms}$ values should be plotted against the parameters associated with the mechanism expected to generate the continuous acoustic emission. Second, for continuous acoustic emission where $\overline{rms}$ or $\overline{tms}$ measurements are made, a lightly damped ringing transducer will give the best signal-to-noise ratio.[97] Third, it is not possible to accurately estimate the total number of separate acoustic emission events that combine to produce the continuous acoustic emission measured by the rms or tms meters over a given time period. Finally, when an rms or tms voltmeter is used in the presence of burst emission, the acoustic emission data are not related (in a simple fashion) to energy release rates.

Conclusion

To extend the usefulness of the rms or tms techniques (for example, in dislocation theory studies), research should be advanced in two areas. First, investigation is needed to determine if measurements of tms or rms over very limited frequency ranges can be used to isolate individual acoustic emission resulting from different simultaneous acoustic emission generation mechanisms.

Second, research should be directed toward developing a technique to determine how many acoustic emission events have combined to result in the continuous acoustic emission measured over a given time or strain interval. This information is vital to the development of quantitative theories for predicting the actual rms or tms values of continuous acoustic emission as a function of stress or strain.

REFERENCES

1. Bindschadler, R., W.D. Harrison, C.F. Raymond and R. Crosson. "Geometry and Dynamics of a Surge-Type Glacier." *Journal of Glaciology*. Vol. 18. (1977): pp 181-194.
2. Kamb, B., C.F. Raymond, W.D. Harrison, H. Engelhardt, K.A. Echelmeyer, N. Humphrey, N.M. Brugman and T. Pfeffer. "Glacier Surge Mechanism: 1982-1983 Surge of Variegated Glacier, Alaska." *Science*. Vol. 227 (1985): pp 469-479.
3. Hughes, T.J. "Is the West Antarctic Ice Sheet Disintegrating?" *Journal of Geophysical Research*. Vol. 78. (1973): pp 7,884-7,910.
4. Schubert, G. and D.A. Yuen. "Initiation of Ice Ages by Creep Instability and Surging of the East Antarctic Ice Sheet." *Nature*. Vol. 296, No. 5,853 (1982): pp 127-130.
5. Scheidegger, A.E. *Physical Aspects of Natural Catastrophes*. Amsterdam: Elsevier Publishing (1975): pp 289.
6. Perla, R.I. "Avalanche Release, Motion and Impact." *Dynamics of Snow and Ice Masses*. S.C. Colbeck, ed. New York, NY: Academic Press (1980): pp 397-462.
7. Obert, L. and W.I. Duvall. "Microseismic Method of Determining the Stability of Underground Openings." *US Bureau of Mines Bulletin 573*. (1957): pp 18.
8. Lord, A.E., Jr. "Acoustic Emission." *Physical Acoustics: Principles and Methods*." W.P. Mason and R.N. Thurston, eds. Vol. 11. New York, NY: Academic Press (1975): pp 289-253.
9. Gold, L.W. "The Cracking Activity in Ice during Creep." *Canadian Journal of Physics*. Vol. 38, No. 9 (1960): pp 1,137-1,148.
10. Gold, L.W. "Crack Formation in Ice Plates by Thermal Shock." *Canadian Journal of Physics*. Vol. 41, No. 10 (1963): pp 1,712-1,728.
11. Hatherton, T. "A Note on the Seismicity of the Ross Sea Region." *Geophysical Journal of the Royal Astronomical Society*. Vol. 5 (1961): pp 252-253.
12. Hatherton, T. and F.F. Evison. "A Special Mechanism for Some Antarctic Earthquakes." *New Zealand Journal of Geology and Geophysics*. Vol. 5 (1962): pp 864-873.
13. Smirnov, V.N. and E.M. Lin'kov. "On a Source of High-Frequency Seismic Disturbances and Flexural-Gravitational Waves in the Antarctic." *Izvestia, Earth Physics*. Vol. 8 (1967): pp 88-92.
14. Adams, R.D. "New Zealand National Report (1967-70)." *Proceedings of the International Association of Seismology and Physics of the Earth's Interior*. No. 17, Part 2 (1971): pp 20.
15. Kaminuma, K. "Seismicity in Antarctica." *Journal of Physics of the Earth*. Vol. 24 (1976): pp 381-395.
16. Oelsner, C. "Seismoakustik, eine neue Messmethode für die Gletchermechank." *Polarforschung*. Bd. 6, J. 35, Ht 1/2 (1967): pp 19-26.
17. Weertman, J. "The Theory of Glacier Sliding." *Journal of Glaciology*. Vol. 5, No. 39 (1964): pp 287-303.
18. Dewart, G. *Seismic Investigation of Ice Properties and Bedrock Topography at the Confluence of Two Glaciers, Kaskawulsh Glacier, Yukon Territory, Canada*. Report No. 27. Columbus, Ohio: The Ohio State University Institute of Polar Studies (1968): pp 207.
19. Knopoff, L. and F. Gilbert. "First Motions from Seismic Sources." *Bulletin of the Seismological Society of America*. Vol. 50, No. 1 (1960): pp 117-134.
20. Anderton, P.W. *Structural Glaciology of a Glacier Confluence, Kaskawulsh Glacier, Yukon Territory, Canada*. Report No. 26. Columbus, Ohio: The Ohio State University Institute of Polar Studies (1973): pp 109.
21. Neave, K.G. and J.C. Savage. "Icequakes on the Athabasca Glacier." *Journal of Geophysical Research*. Vol. 75, No. 8 (1970): pp 1,351-1,362.
22. Van Wormer, D. and E. Berg. "Seismic Evidence for Glacier Motion." *Journal of Glaciology*. Vol. 12 (1973): pp 259-265.
23. Rogers, G.C. "A Microearthquake Survey in Northwest British Columbia and Southeast Alaska." *Bulletin of the Seismological Society of America*. Vol. 66 (1976): pp 1,843-1,655.
24. Weaver, C.S. and S.D. Malone. "Mount Saint Helens Seismic Events: Volcanic Earthquakes or Glacial Noises?" *Geophysical Research Letters*. Vol. 3 (1976): pp 197-200.
25. Weaver, C.S. and S.D. Malone. "Seismic Evidence for Discrete Glacier Motion at the Rock-Ice Interface." *Journal of Glaciology*. Vol. 23 (1979): pp 171-184.
26. St. Lawrence, W. and A. Qamar. "Hydraulic Transients: A Seismic Source in Volcanoes and Glaciers." *Science*. Vol. 203 (1979): pp 654-656.
27. St. Lawrence, W. and A. Qamar. "Transient Water Flow in Sub-Glacial Channels." *Journal of Glaciology*.

Vol. 23 (1979): pp 432-433.
28. Dewart, G. "Measurements of Glacier Noise." *Proceedings of the Twenty-Second Meeting of the Acoustic Emission Working Group*. Boulder, CO (1981): pp 9.
29. Qamar, A. and W. St. Lawrence. *An Investigation of Icequakes on the Greenland Icesheet near Jakobshavn Icestream, June-July, 1980*. Boulder, CO: World Data Center-A for Glaciology (1983): pp 38.
30. "Proceedings of the Second Conference on Acoustic Emission/Microseismic Activity in Geologic Structures and Materials." *Transcripts of Technical Publications*. H.R. Hardy and F.W. Leighton, eds. Clausthal, Germany (1980).
31. Lee, W.H.K. and S.W. Stewart. *Principles and Applications of Microearthquake Networks*. New York, NY: Academic Press (1981): pp 293.
32. Gow, A.J. and T. Williamson. "Rheological Implications of the Internal Structure and Crystal Fabrics of the West Antarctic Ice Sheet as Revealed by Deep Core Drilling at Byrd Station." *Geological Society of America Bulletin*. Vol. 87 (1976): pp 1,665-1,677.
33. Colbeck, S.C., W. St. Lawrence and A.J. Gow. "Creep Rupture at Depth in a Cold Ice Sheet." *Nature*. Vol. 275 (1978): pp 733.
34. Sommerfeld, R.A. and H. Gubler. "Snow Avalanches and Acoustic Emissions." *Annals of Glaciology*. Vol. 4 (1982): pp 271-276.
35. Koerner, R.M., A.E. Lord, Jr. and W.M. McCabe. "Acoustic Emission Monitoring of Soil Stability." *Journal of the Geotechnical Engineering Division*. Vol. 104, No. GT5. New York, NY: American Society of Civil Engineers (1978): pp 571-582.
36. Smalley, R.F., Jr., D.L. Turcotte and S.A. Solla. "A Renormalization Group Approach to the Stick-Slip Behavior of Faults." *Journal of Geophysical Research*. Vol. 90, No. B2 (1985): pp 1,894-1,900.
37. Mogi, K. "Rock Fracture." *Annual Review of Earth and Planetary Sciences*. F.A. Donath, ed. Palo Alto, CA: Annual Reviews, Inc. (1973): pp 63-84.
38. Crampin, S. "A Review of Wave Motion in Anisotropic and Cracked Elastic Media." *Wave Motion*. Vol. 3, No. 4 (1981): pp 343-391.
39. Papadakis, E.P. "Sonic and Ultrasonic Methods for Nondestructive Testing of Catalytic Converter Substrates." *Materials Evaluation*. Vol. 34, No. 2. Columbus, OH: The American Society for Nondestructive Testing (1976): pp 25-31.
40. Papadakis, E.P. "Empirical Study of Acoustic Emission Statistics from Ceramic Substrates for Catalytic Converters." *Acoustica*. Vol. 48 (1981): pp 335-338.
41. Lipson, C. and N.J. Sheth. *Statistical Design and Analysis of Engineering Experiments*. New York, NY: McGraw Hill Publishing Company (1973): pp 456-461.
42. *High-Flow-Volume Frequency Curves for Selected Durations: Lower Missouri River Tributaries*. Omaha, NE: US Corps of Engineers, Missouri River Division (September 1966).
43. Seeman, H.J. and W. Bentz. "Untersuchungen uber den Einfluss des Gefuges aus der Extinktion von Ultraschallwellen in Metallischen Stoffen." *Z. Metallkde*. Vol. 45 (1954): pp 663-669.
44. Krock, R.H. and L.J. Broutman. "Principles of Composites and Composite Reinforcement." *Modern Composites Materials*. Reading, MA: Addison-Wesley Publishing Company (1967): pp 3-26.
45. Tatro, C.A. "Design Criteria for Acoustic Emission Experimentation." *Acoustic Emission*. R.G. Liptai, D.O. Harris, and C.A. Tatro, eds. STP 505. Philadelphia, PA: American Society for Testing Materials (1972).
46. Zweben, C. "A Bounding Approach to the Strength of Composite Materials." *Engineering and Fracture Mechanics*. Vol. 4, No. 1 (1972).
47. Rathbun, D.K., A.G. Beattie and L.A. Hiles. *Filament Wound Materials Evaluation with Acoustic Emission*. Report SCL-DC-70-260. Albuquerque, NM: Sandia Laboratories (1971).
48. Hamstad, M.A. and T.T. Chiao. *Acoustic Emission from Stress Rupture and Fatigue of an Organic Fiber Composite*. Report UCRL-74745. Livermore, CA: Lawrence Livermore Laboratory (1974).
49. Liptai, R.G. "Acoustic Emission from Composite Materials." *Proceedings of the Second ASTM Conference on Composite Materials*. STP 497. Philadelphia, PA: American Society for Testing Materials (1972): pp 285-298.
50. Hamstad, M.A. and T.T. Chiao. "A Physical Mechanism for the Early Acoustic Emission in an Organic-Fiber/Epoxy Pressure Vessel." *SAMPE Quarterly*. Vol. 5, No. 22. Azusa, CA: Society of Aerospace Material and Process Engineers (1974).
51. Tutans, M.Y. and Y.S. Urzhumtsev. "Seismoacoustic Prediction of the Fracture Process in Glass-Reinforced Plastics (1). Load Dependence of the Seismoacoustic Emission." *Polymer Mechanics*. Vol. 3, No. 377 (1972). Translated from *Mekhanika Polimerov*. Vol. 3, No. 421 (May-June 1972).
52. Swanson, G.D. and J.R. Hancock. *Effects of Interfaces on the Off-Axis and Transverse Tensile Properties of Boron-Reinforced Aluminum Alloys*. Technical Report 2 to ONR. Kansas City, MO: Midwest Research Institute (1971).
53. Takehana, M. and I. Kimpara. "Internal Fracture and Acoustic Emission of Fiberglass Reinforced Plastics." *Mechanical Behavior of Materials: Proceedings of International Conference on Mechanical Behavior of Metals*. Vol. 5. Japanese Society of Metals Science (1971).

54. Harris, D.O., A.S. Tetelman and F.A.I. Darwish. "Detection of Fiber Cracking by Acoustic Emission." *Acoustic Emission*. STP 505. Philadelphia, PA: American Society for Testing Materials (1972): pp 238-249.
55. Pattnaik, A. and A. Lawley. *Role of Microstructure on Acoustic Emission in the Deformation of Al-CuAl$_2$ Composites*. Technical Report No. 10. Philadelphia, PA: Drexel University (1973).
56. Henneke, E.G., II, C.T. Herakovich, G.L. Jones and M.P. Renieri. *Acoustic Emission from Composite Reinforced Metals*. Blackburg, VA: Virginia Polytechnic Institute (October 1973).
57. Stone, D.E.W. and P.F. Dingwall. "The Kaiser Effect in Stress Wave Emission Testing of Carbon Fibre Composites." *Nature Physical Science*. Vol. 241, No. 68 (1973).
58. Grenis, A.F. and F.J. Rudy. *Fracture Characteristics of Boron Fiber-Reinforced Aluminum 2024 as Detected by Acoustic Emission*. Report 73-57. Watertown, MA: Army Materials and Mechanics Research Center (1973).
59. Lloyd, D.J. and K. Tangri. "Acoustic Emission from Al$_2$O$_3$-Mo Fiber Composites." *Journal of Materials Science*. Vol. 9, No. 482 (1974).
60. Lingenfelder, P.G. *Acoustic Emission Monitoring of Graphite/Epoxy Composite Structures*. Report NDT-185. Sunnyvale, CA: Lockheed Missiles and Space Company (1974).
61. Owston, C.N. "Carbon Fibre Reinforced Polymers and Nondestructive Testing." *British Journal of Nondestructive Testing*. Vol. 15, No. 2 (1973).
62. Mehan, R.L. and J.V. Mullin. "Analysis of Composite Failure Mechanisms Using Acoustic Emissions." *Journal of Composite Materials*. Vol. 5, No. 266 (1971).
63. Rothwell, R. and M. Arrington. "Acoustic Emission and Micromechanical Debond Testing." *Nature Physical Science*. Vol. 233, No. 163 (1971).
64. Mullin, J.V. and R.L. Mehan. "Evaluation of Composite Failures through Fracture Signal Analysis." *Journal of Testing and Evaluation*. Vol. 1, No. 3 (1973): pp 215.
65. Kim, H.C., A.P. Ripper Neto and R.W.B. Stephens. "Some Observations on Acoustic Emission During Continuous Tensile Cycling of a Carbon Fibre/Epoxy Composite." *Nature Physical Science*. Vol. 237, No. 78 (1972).
66. Hamstad, M.A. "Acoustic Emission from Filament Wound Pressure Bottles." *Nonmetallic Materials*. Vol. 4. Azusa, CA: Society of Aerospace Material and Process Engineers (1972): pp 321.
67. Hamstad, M.A. and T.T. Chiao. "Acoustic Emission Produced During Burst Tests of Filament Wound Bottles." *Journal of Composite Materials*. Vol. 7, No. 3. Livermore, CA: Lawrence Livermore Laboratory (1973): pp 320.
68. Fitz-Randolph, J., D.C. Phillips, P.W. Beaumont and A.S. Tetelman. "The Fracture Energy and Acoustic Emission of a Boron-Epoxy Composite." *Journal of Materials Science*. Vol. 7, No. 289 (1972).
69. Ahlborn, H., J. Becht and J. Eisenblatter. *Application of Sound-Emission Analysis in Investigations of the Fracture Toughness of Fibrous Composite Materials*. Livermore, CA: Lawrence Livermore Laboratory. UCRL-10785 (1974).
70. Green, A.T., C.S. Lockman and R.K. Steele. "Acoustic Verification of Structural Integrity of Polaris Chambers." *Modern Plastics* (1964): pp 137.
71. Robinson, E.Y. "Acoustic Emission Studies of Large Advanced Composite Rocket Motor Case." *Proceedings of the Twenty-Eighth Annual Technical Conference*. Sec. 12-D. New York, NY: Society of the Plastic Industry (1973): pp 1-6.
72. "Recommended Acoustic Technology." *Acoustic Emission*. STP 505 Philadelphia, PA: American Society for Testing Materials (1972): pp 336.
73. Hamstad, M.A. and A.K. Mukherjee. *Acoustic Emission vs. Strain and Strain Rate: Measurement and a Dislocation Theory for a Dispersion-Strengthened Aluminum Alloy*. Report UCRL-76077. Livermore, CA: Lawrence Livermore Laboratory (1974).
74. Sussner, H. "The Peizoelectric Polymer PVF$_2$ and Its Applications." *Ultrasonics Symposium Proceedings*. New York, NY: Institute of Electrical and Electronic Engineers (1979): pp 491-498.
75. *Electrets*. TDS 7181. King of Prussia, PA: Pennwalt Corporation.
76. Sessler, G.M. "Piezoelectricity in Polyvinylidene Fluoride." *Journal of the Acoustic Society of America*. Vol. 70, No. 6 (December 1981): pp 1,596-1,608.
77. Murayama, N., K. Nakamura, H. Omara and M. Segaw. "The Strong Piezoelectricity in Polyvinylidene Fluoride (PVDF)." *Ultrasonics*. Vol. 14, No. 1 (January 1976): pp 97-105.
78. Bui, L.N., H.J. Shaw and L.T. Zitelli. "Study of Acoustic Wave Resonance in Piezoelectric PVF$_2$ Film." *Transactions on Sonics and Ultrasonics*. Vol. Su-24, No. 5. New York, NY: Institute of Electrical and Electronics Engineers (September 1977): pp 331-336.
79. Leung, W.P. and K.K. Yung. "Internal Losses in Polyvinylidene Fluoride (PVF$_2$) Ultrasonic Transducers." *Journal of Applied Physics*. Vol. 50, No. 12 (December 1979): pp 8,031-8,033.
80. Linvill, J.G. "PVF$_2$ Models, Measurements and Devices." *Ferroelectrics*. Vol. 28, No. 1 (January 1980): pp 291-296.
81. Das-Gupta, D.K. and K. Doughty. "Dielectric Relaxation Spectra of Poled Polyvinylidene Fluoride." *Ferroelectrics*. Vol. 28, No.1 (January 1980): pp 307-310.

82. Wagers, R.S. "PVF$_2$ Elastic Constants Evaluation." *Ultrasonics Symposium Proceedings*. New York, NY: Institute of Electrical and Electronic Engineers. (1981): pp 464-469.
83. Ravinet, P., J. Hue, G. Vollvet, P. Hartmann, D. Broussoux and F. Micheron. "Acoustic and Dielectric Loss Processes in PVF$_2$." *Ultrasonics Symposium Proceedings*. New York, NY: Institute of Electrical and Electronics Engineers (1981): pp 1,017-1,022.
84. Evans, A.G. *Proceedings of the British Ceramic Society*. Vol. 25, No. 217 (1975).
85. Shiraiwas, T., et al. *Proceedings of the Fourth Acoustic Emission Symposium*. Tokyo, Japan (1978).
86. Leers, K.L. and O. Schmidt. *Tonind. Z.* Vol. 100, No. 9 (1976): pp 325.
87. Williams, R.S. and E.M. Anderson. *Proceedings of the ASNT Fall Conference*. Columbus, Ohio: The American Society for Nondestructive Testing (1977): pp 3.
88. Padgett, G.C. *Ref. J.* Vol. 4 (September 1978).
89. Esper, F.J. and H.M. Wiedenmann. *Ber. D.K.G.* Vol. 55, No. 12 (1978): pp 507.
90. DIN 51068.
91. *Method of Panel Spalling Testing Refractory Brick*. ASTM C38. Philadelphia, PA: American Society for Testing Materials (1984).
92. Uchimura, R. M. Kumagai, T. Morimoto and A. Harita. *Taikabutsu*. Vol. 31, No. 257 (1979): pp 11.
93. Larson, D.R. and D.P.H. Hasselman. *Journal of the British Ceramic Society*. Vol. 74, No. 59 (1975).
94. Nakayama, J. *Fracture Mechanics of Ceramics*. R.C. Bradt, et al, eds. New York, NY: Plenum Press (1973): pp 759.
95. Hasselman, D.P.H. *Journal of the American Ceramic Society*. Vol. 46, No. 11. Columbus, OH: American Ceramic Society (1963): pp 535.
96. Hasselman, D.P.H. *Journal of the American Ceramic Society*. Vol. 52, No. 11. Columbus, OH: American Ceramic Society (1969): pp 600.
97. *US Acoustic Emission Working Group Subcommittee Report*. J.R. Frederick, ed. Ann Arbor, MI: University of Michigan, (August 1974).
98. Hamstad, M.A. and A.K. Mukherjee. *Acoustic Emission versus Strain and Strain Rate: Measurement and a Dislocation Theory Model for a Dispersion Strengthened Aluminum Alloy*. Report UCRL-76077. Livermore, CA: Lawrence Livermore Laboratory (1974).
99. Harris, D.O. and R.L. Bell. *The Measurement and Significance of Energy in Acoustic Emission Testing*. Technical Report DE-74-3A. San Juan Capistrano, CA: Dunegan Endevco (1974).
100. Hamstad, M.A. and A.K. Mukherjee. "The Dependence of Acoustic Emission on Strain Rate in 7075-T6 Aluminum." *Experimental Mechanics*. Vol 14, No. 33 (1974).
101. Jax, P. and J. Eisenblatter. "Acoustic Emission Measurements During Plastic Deformation of Metals." *Battelle Information*. Vol. 15, No. 2 (1972).
102. Imanaka, T. K. Sano, and M. Shimizu. "Acoustic Emission, a Critical Review." *Bulletin for the Japanese Institute for Metals*. Vol. 12, No. 12 (1973): pp 871. Lawrence Livermore Laboratory translation UCRL-10747.
103. Hamstad, M.A. Personal communication. Livermore, CA: Lawrence Livermore Laboratory (November 1974).
104. Sano, K., T. Imanaka, T. Funakoshi and K. Fujimoto. "Acoustic Emission During Deformation of Tempered Martensitic Steels." *Proceedings of the Second Acoustic Emission Symposium*. Tokyo, Japan: Japan Industrial Planning Association (1974): pp 23.
105. Shimizu, M., K. Sano, T. Imanaka and K. Fujimoto. "Acoustic Emission During Plastic Deformation in Aluminum, Copper and 3% Silicon-Iron." *Proceedings of the Second Acoustic Emission Symposium*. Tokyo, Japan: Japan Industrial Planning Association (1974): pp 1.
106. *Vibration and Acoustic Measurement Handbook*. M.P. Blake and W.S. Mitchell, eds. New York, NY: Spartan Books (1972).
107. Harris, D.O. Personal communication. San Juan Capistrano, CA: Dunegan Endevco (December 1974).

INDEX

A

AC weld monitoring 282
   nugget size control 283-284
   nugget size detection 283
Acoustic emission
   aluminum and 48
   characterization 27-33
   continuous, energy measurement 586-591
   comparison with other techniques 12, 187, 201, 220-221, 287
   definition 12
   early observations 2-3
   frequency range 313-314
   geotechnical applications 312-325
   nuclear industry use 6, 226-232
   relative amplitude 13
   research 4-6
   sources, deformation 47
   sources in trailer tubes 164
   steel and 48
   waveform characteristics 428
Acoustic emission data
   batch analysis 560
   statistical analysis 558-560
   upper acceptance limit 558-560
Acoustic emission detection 64
   criteria 69-70
   modeling 66-70
   time dependence 64
Acoustic emission simulator 531-532
Acoustic emission testing
   advantages 13, 136
   applications 13-14
   characteristics of 137
   continuous 131-144
   equipment 14-16, 46-47, 513-549
   macroscopic 46-59
   other techniques and 13, 136
   personnel 36
   petrochemical industries 220-221
   pressure vessels and 21-22
   procedures 36, 39-42
   response amplitude factors 13
   standardizing 430
   standards 36-38
   structures 333
   terminology 34-36
   test sensitivity 16-17
   zone location method 137-138
Acoustic emission testing systems 14-16, 46-47, 513-549
   calibration 21
   choice of 20
   computers and 21, 22
   diagram 371
   electron beam weld 289
   electronics of 371
   expansion 20-21
   feasibility study stages 244
   geotechnical 315

   instrumentation 22, 46
   loose part monitoring 533
   metal matrix composites 433-434
   multichannel 544
   organic composites 430-431
   particle impact noise detector 544, 546
   requirements 443
   signal acquisition and analysis 532
   simulator 531
   software 545, 547-548
   valve flow monitoring 535
Acoustic Emission Working Group 7
Acoustic impedance 95
Acoustic transmission 505
Acoustic-ultrasonic inspection unit 542
Airborne acoustic signal detection system 537-542
Aircraft alloy testing 418-420
   crack growth in 419-420
   emission sources 418
   heat treatment effect 420
   signal amplitudes 418-419
   temperature effect 420
Aircraft structure monitoring 421-424
   crack presence and noise 422
   findings 421
Aircraft structure monitoring, in-flight 421-424
   component noise 422
   crack advance 422-423
   observations 421-424
   signal characteristics 423-424
Airframe structure tests 434-443
   acoustic emission setup 436-437
   auxiliary support beam 438
   damage detection 434
   graphite curved frame 438-439
   graphite roof structure 429-441
   limit load techniques 437-438
   structure complexity and 434
   transition section 441-442
Aluminum alloy inclusions 74, 81
   aluminum alloy testing 418-420
   electron beam melting and 88
American Petroleum Institute standards 168, 172-173
American Society of Mechanical Engineers
   pressure vessel and boiler code 36-37, 167-168, 170-173
   standards 36-37, 167, 170, 173, 203, 209-210
American Society for Nondestructive Testing 36
American Society for Testing and Materials 34, 37, 170, 203, 209, 268-271, 291
Ammonia spheres, testing 170-172
Ammonia storage tanks 187
Amplitude analysis 329
Antarctic seismic studies 553
Attenuation 313-314
   organic composites and 428
   wave 103-107
Automation and monitoring 485

B

Background noise. See *Noise*
Bearing failure 216-217
   detection 540
Blowout preventer monitoring 168
Bonding, tape-automated, testing 361-364
Bonding, thermocompression 358
   crack detection 358-360
Bone fracture 452-454
   animal 451-453
   human 453-454
Bone studies 449
Bridges, monitoring of 333-339
   concrete 341-342
   welds 336-339
Brittle fracture source 41
Bucket truck inspection 267-271
   ASTM standard 271
   instrumentation 268
   test procedure 268-269
   typical test data 270-271
Butane storage spheres 188-193
   acoustic testing instrumentation 190-191
   sensor locations 190-192

C

CANDU reactors
   hydride cracks 253, 255-256
   leak detection 253-259
   seal leakage 256-259
CARP 37
Cable monitoring 341
Capacitor, ceramic 349
   crack detection 353-355
   discontinuity detection 349-350
   failure modes 349
   test methods 349-351
Capacitor, high voltage 352-353
Catalytic cracking unit testing 182
Center of compression equation 110-111
Ceramic substrate cracking 358-360
   acoustic emission data detection 359
   bond head transducer 359
Ceramics. See *Capacitor, ceramic; Piezoceramic materials*
Chip capacitor. See *Capacitor, ceramic*
Chip formation. See *Metal chip formation*
Cleavage microfracture 79
Coherence techniques 142-144
Coining 469
   emission sources 469
   monitoring benefits 470-471
   monitoring research 469-470
Composite manufacture
   emission data and 565-566
   properties determination 567-568
Composite materials. See also *Pipe, composite; Plastic vessels, fiber reinforced*
   bandwidth used 543
   characteristics of 561
   cure monitoring 504-508
   data plotting 564
   emission sources 562
   emission tests 561-562
   failure mechanisms 429
   metal matrix 432-433
   measurement techniques 563

   noise elimination 562
   polyvinylidene fluoride sensor 569
   properties determination 567-568
   signal attenuation 563
   signal characteristics 563, 564
   thermosetting resin manufacture 504-505
   transducer placement 562-563
   vessel testing 203
Composite tester 544
Composites, behavior models 567
Composites, fiber reinforced
   Emission data and design of 564-565
Composites, metal matrix 432
   acoustic emission testing 432-433
   aerospace applications 432
   failure mechanisms 432-433
   proof load testing 433
   test instrumentation 433
Composites, organic 425
   acoustic emission profiles 429
   acoustic emission propagation 428
   aerospace applications 425
   attenuation in 428
   characterization testing 427
   discontinuities in 429
   failure mechanisms 427-429
   in-flight 426-429
   in-service 426
   monitoring 426
   test instrumentation 430-431
Composites, thermosetting resin
   curing 504-508
   emission techniques 505-506
   manufacture 504-505
Concrete monitoring 341-342
   corrosion activity 341-342
Continuous emission, definition 586
   of pressure vessels 160
Conductor motion signal projection 393-399
Continuous emission energy measurements 586-591
   count rate 589-590
   emission technique advantages 588-589
   volume and plastic strain rate 589
Corrosion, stress cracking and 58-59
Count rate
   corrosion and 58
   crack growth and 56
Coupling coefficient squared 516
Coupling transducers 39
Crack growth 48-54
   aircraft alloys 419-420
   airframe noise and 422
   ammonia spheres 171
   count rate and 56
   die shear testing and 382-385
   fatigue 55-56
   in-flight 422-423
   pressure vessels 157, 159
Cracking
   delayed hydride 263
   fatigue 55, 261
   stress corrosion 58-59, 228, 261-263
   surface oxide 261
   thermocompression bonds 358-360
Crystal deformation, single 73-74

Curie temperature 124, 515-516
Cutting monitoring 479-484
   experimental procedure 481-482
   metal type differences 484-484
   operation 479-481
Cyclic noise 24

D

Data acquisition systems 542-543
Data analysis 27-33
   airframe composite testing 434-435
   characteristics of discrete emission 27
   waveform parameters 28
   characterization methods 29-31
   reject limits 558-560
   severity estimates 230
   source characterization 31-32
   structural steel 329
Die shear testing, crack growth and 382-385
Diffuse wave fields 117-118
Dilatational waves. See *Longitudinal waves*
Direct piezoelectric constant 515
Discontinuity location 329-331
   linear 330
   planar source isolation 330-331
Dislocation sources, microscopic 71-76
   detectability 71-73
   grain size effects 74-75
   precipitation effects 75-76
Discrete source location 145, 148-152
Disturbance energy quantification 399
Drill wear detection 473-478
   drill-up use in 473-474, 476-477
   experimental procedure 475-476
   problems 477-478
   test results 476-477
Drill-up 472
   block diagram 472-473
   drill wear detection use 474-477
Ductile materials 52-54
Dunegan corollary 17-18

E

Earthen dam monitoring 324
Electric waves 64-65
Electrical discharge testing 541
Electromagnetic interference 25-26
Electron beam welding 285
   acoustic emission testing 286-287
   burst emission from 288
   continuous emission from 287-288
   discontinuities in 286-286
   emission signal analysis 289-290
   instrumentation for 288-290
   sensor for 288
Electronic component testing problems 348
Electronic repeater, undersea 353-357
   crack detection 353-355
   housing 355-356
Electroslag weld emission testing 291-301
   data collection 294
   discontinuities in 293-294
   equipment for 292-294
   procedure for 298-299
   process control and 297-300
   results of 294-297
   source location 300
   specimens for 291
   weld changes and 299-300
   weld flux and 301
Electroslag welding process 291-292
Embankment monitoring 324
Energy analysis 329
European Working Group on Acoustic Emission 7, 37
Extruder shaft rubbing 217, 218

F

Fatigue 55-57
   low-cycle monitoring 56
Feature analysis 30
Felicity effect 18, 208-209
Fluid catalytic cracking unit 183-186
   testing sequence 184
   weld monitoring 183, 185
Foundation monitoring 323
Fracture 49-54
   criteria for 52-54
   ductile materials and 52-54
   linear elastic model 51-52
Fracture sources, microscopic 77-83
   amplification factors 82-83
   detectability 77-78
   environmental factors 82
Free dielectric constant 516
Frequency analysis 329
Frequency constant 516
Fretting 23-24

G

Gas jet source 41
Gas pipeline tests 166
Gas trailer tubing 161-165
   discontinuities in 164-165
   physical properties 162
   recertification 161-162
   test advantages 163-164
   test procedure 162-163
Gear discontinuities 217, 219
Geophones 317, 320
Geotechnical materials testing 312-322
   equipment for 317-322
   laboratory behavior 314-316
   monitoring devices 319-322
   sensors for 317
   signal acceleration formula 313
Glacier studies 553-557
   count rate of shocks 553
   crevasse propagation detection 554
   data interpretation 556
   emission sources 552, 555-556
   microseismicity 552-553
   surface rupture emissions 554
Green's function approach 66
Griffith cracking 409-410
Grinding control systems 490-492
   sparkout 491-492
   wheel contact 490-492

H

HEP quadrupole signals 398-399
Half-space sources 112-114
Harmonic waves 93-94

INDEX/599

Heat exchange tube leak detection 541
Hot plate shock test 365-369
Hybrid microcircuit package 365
    testing 365-369
Hydroformer reactor vessels 173, 177, 183
    sensor layout 179-182
Hydrogen embrittlement monitoring 59
Hydrophones 317, 318
Hydrostatic testing 194-197

I

Ice sheets, microseismic activity 555
Impulsive sources, equation 118
Incipient failure detection 213, 217
Integrated circuit cracking 370-388
    manufacturing process and 370
Integrated circuits, tape-bonded, testing 362-364
Intelligent sensors 499-503
    abrasive disk machining 502-503
    applications 499, 500-501
    components 500
    development 501
    end milling and 502
    features of 499
    punch stretching 501-502
Intergranular microfracture 79

K

Kaiser effect 4, 14, 16-18
    deformation and 48
Kaiser principle 18
Knee replacement 461-463
Knoop indentation 375
Knowledge representation method 30

L

Lamb's problem 112-114
Laser spot welding 302
    acoustic emission 305-307
    experimental correlation for 307-308
    geometry of 302-303
    insulation effect 303
    materials for 303
    mechanics of 303-304
    real-time evaluation theory 305-308
Leak detection 194-202, 256
Limb equalization research 457-460
Linear elastic fracture model 51-52
Linear source location technique 145
Liquid-solid transformations, emissions and 87-88
Locator analyzer 545-546
Longitudinal waves 91-94
    propagation 93
    reflection 99
Loose part monitoring system 533-535
    sensors for 535

M

Machinery monitoring 213-219
    case histories 215-219
    cavitation and 216
    incipient failure detection 213-215, 219
    signal above threshold 213-214
    spectrum analysis 214

Machining, abrasive disk, monitoring 502
Magnets, adiabatic 390-391
Magnets, cryostable 390-391
Magnets, superconducting
    dipole 393-398
    emission sources 391
    HEP quadrupole 398-399
    normal zone detection 398-399
    quench identification 391-392
    uses of 390
Manufacturing process monitoring
    automation role 485-486
    requirements for 485
    tool fracture 487-499
    tool wear 486-487
    untended 499
Martensitic transformation detectability 84-87
    magnetic effects 87
    steel coldworking and 86-87
Material characteristics, count rate and 28
Mechanical quality factor 516
Metal chip formation 479
    acoustic emission 479, 496-497
    breaking frequency 489-490
    cutting monitor 479-484
    monitoring 489-490
Metal storage tank, testing 172-182, 187-193
Microcrack detectability 77-78
Microelectric welds 409
    Griffith cracking 409-410
Micromechanical modeling of acoustic emission 65-70
Microscopic sources of acoustic emission 64, 71-85
Microseismic studies 552-557
    environmental problems 556
    noise sources 556
Microvoid coalescence 81-82
Milling
    acoustic emission in 493-498
    chip formation and 496-498
    end 502
    insert fracture 495-496
    insert wear 495
    time difference averaging 493-494
Modeling, micromechanical 66-70
Monitoring
    aircraft structures 421
    blowout preventer 168
    chip form 479-484
    coining 469
    cure of composites 504-508
    cutting 482
    drill wear 472-478
    electron beam weld 285-290
    fossil fuel power plants 244
    foundation 323
    in-flight 421-423
    laser spot welding 302-309
    mine monitoring 324
    machinery condition 213-219
    machining 493-498
    nuclear power plants 226-232
    organic composites 426
    process 485-492
    spot weld 276-285
    steel structures, in-service 331
    structures 326-339
    superconducting magnets 389-401

Monitoring devices
  in-service 227
  multichannel 319-320, 544-545
  portable 546
  single-channel 319, 320, 542
  structures 332-335
  weld 336-339
Motion equations 92-93
Multiple processor system 545

N

Noise 16-17, 23-26
  acoustic emission and 158, 229
  airframe loading and 422
  background 17
  cyclic 24
  compensation for 17
  control of 24
  electric 332
  hydraulic 23
  mechanical 23-24
  rise time and 24
  sources in pressure vessels 158
  suppression 429-430
  welding and 24-25
Nuclear power plant monitoring 226-232
  applications 233-236
  discontinuity detection 228-229
  instrumentation 227-228
  noise interference 229-230
  severity estimate 230-231
  signal identification 230

P

Particle impact noise detection ... 402, 544, 546-547
  amplitude versus frequency 405-406
  effectivity 406-407
  failure analysis and 407
  history of 402-403
  instrumentation for 403
  loose particle detection 407-408
  military standards and 403
  particle characteristics 403-406
  particle elimination 406
Particle microfracture 79-81
Passivation/metal layer structure testing . 381-382
Pattern recognition use 30, 31
Peening. See *Coining*
Pekeris solution 112-114
Petrochemical industries, acoustic emission testing in 220-221
Phase transformation, microscopic 84-88
  martensitic 84-87
  solid state 84
Piezoceramic element
  flat frequency sensor 516-518
  material properties 515
  National Bureau of Standards 516
  special geometry 516
Piezoelectric pulses 41
Pinchweld testing 355-357
PIND testing. See *Particle impact noise detection*
Pipe, composite, testing 205
Pipeline, gas, testing 166
Pipeline, leak location 194-202
Pipeline, metal, testing 166
  hydrostatic tests 194-197

Pipeline, off-shore crude oil 198-200
Pipeline, plastic
  pneumatic testing 197-198
  transducer locations 199
Pipeline, steam condensate 200
Piping, stainless steel
  monitoring research 236-243
Planar source isolation 330-331
Plane waves 93
Plastic deformation 97
  model 49, 52
Plastic vessels, fiber reinforced
  acoustic emission techniques 209-210
  chemical industry applications 203, 205
  evaluation criteria 212
  felicity effect 208-209
  testing 203-212
  zone location 205-207
Plate sources 115-116
Platform, off-shore, testing 339
Polyvinylidene fluoride
  characteristics 569
  fatigue tests and 573-574
Polyvinylidene fluoride transducer 569-570
  sensitivity tests using 570-573
Power plant, fossil fuel
  acoustic emission monitoring in 244-245
Power plant, hydroelectric
  acoustic emission monitoring in 245-252
Power plant, nuclear
  acoustic emission monitoring in 226-244
Process monitoring. See *Machinery monitoring; Manufacturing process monitoring; Monitoring*
Pressure leak detection 540
Pressure vessels 21-22, 188
  acoustic emission tests 157-160
  codes 167-168, 170-173
  continuous monitoring 160
  evaluation standards and 170, 173
  failure causes 156-157
  in-service testing 159-160
  metal, testing 167-186
  new, testing 157-158
  sensor locations 176-179
  testing procedures 203
Proof testing 331-332
  metal matrix composites 433
Punch stretching 501-502

Q

Quality control card 42
Quenching, magnets and 390-393

R

Radiographic testing for particles 408
Raleigh calibration 128
Raleigh waves 95-96
Reactor pressure vessel testing 233-244
Refractories
  acoustic emission techniques and 581
  experimental procedure 575-576
  spalling resistance 583-585
  theory of test results 582

Refractories fracture behavior
   heating rate and 579
   restraint and 580-581
   specimen dimension and 579-580
   thermal shock and 575-576
Repeater, undersea 352-357
Resistance spot weld monitoring. See *Spot weld monitoring*
Reverse piezoelectric constant 516
Rock, acoustic emission and stress 315

S

Seal failure 217
Sensor calibration
   aperture effect and 128
   P-wave 128
   primary 126-130
   reciprocity 130
   secondary 131-132
   step-force 128-130
   surface 128
   terminology 126
   through- 128
Sensor mounting 125, 255
   mines and 318
   techniques for 318-319
Sensors 16-17, 435-436
   airframe composite testing 435-436
   AC standard 524-525
   bonds 124
   broadband 525
   ceramic 124
   couplants 124
   cryogenic 399-400
   definition 122
   design 123
   differential 124, 524-525
   dimensions 123-124
   directional 525
   flat frequency 516-518, 529
   frequency range 122-123
   high temperature 529
   integral electronic 530
   integral preamplifier 525
   integrated 518-524
   intelligent 499-503
   intrinsically safe 525-526
   miniature 529-529
   motion sensitivity 122
   piezoceramic material properties 515
   piezoelectric 122, 518-524
   polyvinylidene fluoride 569
   preamplifier 125
   remote 136
   resonant 526-527
   specifications 527
   spectral sensitivity 68-69
   temperature effects 124, 125
   types 122-123
   underwater 526
   wheeled 526
   wideband 529
Service monitoring 331-332
Signal acquisition module 532-533
Signal amplitude measurement method 138-140

Signal characterization and processing
   airframe composite testing 434-435
   characteristics of discrete emission 27
   characterization methods 29-31
   choice of reject limits 558-560
   severity estimates 230
   source characterization 31-32
   structural steel 329
   waveform parameters 28
Signals
   acquisition 532-533
   airborne 537-542
   aircraft in-flight 423-424
   amplitude of 29
   analysis of 434-435, 533
   characterization 27-33
   classification 31-33
   data interpretation 27
   energy release and 29
   guidelines 32-33
   material properties and 28
Slip 73-74
SNT-TC-1A Recommended Practice 36
Software, data classifying 545
   acoustic emission 547-548
   post-test plotting 547-548
   simultaneous data recording 547
Soils, acoustic emission tests 315-317
Solid state bonding, noise production 25
Source activation 331-332
Source location 136-153
   attenuation and 137
   characterization 31-32
   coherence techniques 142-144
   computers and 21
   cross-correlation approach 141-142
   discrete, error in 148-152
   linear 145-146
   mapping distortions 149
   nuclear plant 229
   remote sensors for 136
   sensor positions 150
   signal amplitude measurement 138-139
   special array configurations 148
   structures 331
   three-dimensional 152-153
   time difference measurement 151-152
   timing techniques 140-141
   two-dimensional 140-147
   weak 149-150
   zone restriction 137-138, 145-147
Source mechanisms 12
   plastic deformation 47, 49, 52
Sources 12, 46-59
   continuous 137-144
   crack growth 49-54
   directivity 70
   discrete 145-153
   dislocation 71-76
   fatigue 55-57
   fracture 51-54, 77-83
   half-space 112-114
   martensitic transformation 84-87
   microscopic 64-88
   plastic deformation 47, 49, 52
   wave 108-111

Spalling resistance tests 583-585
    emission technique 585
    procedure for 584-585
Spherical indentation testing 381-382
Spot weld monitoring 22
    acoustic emission applications 280-282
    emission monitoring 276-277
    expulsion emission 277-278, 282-282
    instrumentation for 279-280
    nugget formation emission 277, 281
    phase transformation and 278
    post-weld cracking and 278-279
Standards
    API 168, 172-173
    ASME 36-37, 167, 170, 173, 203, 209-210
    ASNT 36
    ASTM 34, 37, 170, 203, 209, 268-271, 291
    European 37
    Japanese 38
    military 402
    pressure vessels and 167-168, 170-173
    SPI (CARP) 37, 170, 203
Steam trap leak detection 541
Steel, galvanized, spot welding 282-283
Steel, martensitic transformation 84-87
    particle microfracture 79-81
Steel, structural
    crack propagation 327
    emission detectability 327
    emission sources 326-328
    failure 326
    source types 332
    weld monitoring 340
Storage tank, testing 172-182, 187-193, 247-249
Storage area, underground, monitoring 324
Structure monitoring 326-340
    discontinuity location 329-331
    early use 333-334
    instrumentation 332-333
    nature of emission 328
    noise and 332
    signal processing 329
    transducer use 328-329
    welded 340
Superconducting dipole signals 393
    conductor motion and 393-393
    estimation of 397
    frictional model 394-397
    frictional power dissipation 397-398
    test results 393-394
Superconductivity 389
Superconductors, applications for 389-390
Surface waves 95-96

T

Tank, flat bottom 301-302
Tank, horizontal 188-189
Tank, oil 249
Tank, storage
    ammonia 187
    butane 188-193
    metal, testing 187-193
    sensor locations 174-176, 187
Test data interpretation 17
Thermal shock test 365-369
    characterizing 365-366
    monitoring 366-369

Thermocompression bonding 358-360
Timing techniques 140-141
Tin cry 2-3
Tool wear monitoring 486-487
Toutle River Bridge 334-335
Transducers 64, 318-319
    bandwidth 69
    broadbanded 336
    bond head 359
    characteristics required 39
    characterization 40
    conical 519
    coupling 39-40
    guard 338-339
    high fidelity 131
    ideal 435
    installation 321-322
    piezoceramic element 40, 516
    propagation medium 41
    reference sources 41
    standard 435-436
    structure testing and 328-329
    verification 40-42
    vibration 518-524
Transient recorder and analyzer 546
Transmission plus acoustic emission technique 505
Tunnel monitoring 323-324
Turbine rotor monitoring 248, 250

V

Vacuum leak detection 540-541
Valve flow monitoring system 535-537
    charge converter 537
    sensors for 536-537
    valve leak detection 541
Valves, cast steel gate 168-169
Vibration testing for particles 407-408
Vibration transducers 518-524
    calibration, internal 523-524
    dynamic range 520
    filtering 523
    frequency response 520
    moisture effects 521
    power supply 522
    temperature effects 521
    velocity outputs 522
Vickers indentation 375-381
Voltmeters
    characteristics 587
    operation 586-587
    signal amplification 587-588

W

Wafers
    backdamage process 373
    biaxial bending tests 372-374
    diamond indentation experiment 374-375
    fracture occurrence 373-373
    Knoop indentation 375
    stepped passivation layers and 373-374, 377
    Vickers indentation 375-381
Water dam gate monitoring 245
Water seepage tests 316, 318

Wave attenuation
   diffraction and 104-105
   dispersion and 103-104
   energy loss and 105
   geometric 103
   real structures and 106-107
   scattering and 104-105
   source location and 107, 137
Wave fields, diffuse 117-118
Wave reflection, interfacial 97-102
   energy partition 99-101
   longitudinal 99
   solid interface 101-102
   transverse 97
Wave sources, idealized 108-111
   concentrated force 108-109
   double force 110-111
Waveform parameters 27-30
   energy release and 29
   limitations 29

Welding. See also specific types
   acoustic emission bridge monitor 336-339
   microelectronic 409
   pinchweld analyzer 356
   structure monitoring 336-340

Welding noise 24-25
   grounding and 24

Wire rope testing 340-341

Z

Zirconium alloy 260-266
   delayed hydride cracking 263-265
   fatigue and 261
   plastic deformation and 259-260
   stress corrosion cracking 263
   surface oxide cracking 261-262

LIBRARY
ST. LOUIS COMMUNITY COLLEGE
AT FLORISSANT VALLEY